SZYCHER'S HANDBOOK OF
POLYURETHANES

SZYCHER'S HANDBOOK OF
POLYURETHANES

Michael Szycher, Ph.D.

CRC Press
Boca Raton London New York Washington, D.C.

Library of Congress Cataloging-in-Publication Data

Szycher, M. (Michael)
    [Handbook of polyurethanes]
    Szycher's handbook of polyurethanes / Michael Szycher.
        p. cm.
    Includes bibliographical references and index.
    ISBN 0-8493-0602-7
    1. Polyurethanes--Handbooks, manuals, etc. I. Title. II. Title:
Handbook of polyurethanes.
TP1180.P8S98 1999
668.4′239—dc21   98-52786
                                                   CIP

No claim to original U.S. Government works
International Standard Book Number 0-8493-0602-7
Library of Congress Card Number 98-52786
Printed in the United States of America   2 3 4 5 6 7 8 9 0
Printed on acid-free paper

Preface

A handbook is a compilation of data from many sources, presented in a logical and easy-to-follow sequence. The data must be compiled from disparate sources; the sources may be strictly academic, commercial, or promotional. It may come from technical publications, seminars, or the patent literature.

The U.S. patent literature is one of the most comprehensive sources of technical information. The technical information provided in patents is exhaustive, current, and represents the most valuable technology discovered by corporations, universities, and independent inventors. Because of the wealth of information contained in patents, this handbook features many full-text patents. These patents have been carefully selected by the author to best illustrate the complex principles involved in polyurethane chemistry and technology.

This handbook is not composed of original articles; instead, it is based on hundreds of published references. The author has tried to credit original sources by providing an extensive bibliography. The reader is encouraged to refer to the original sources for more complete information and insight.

Polyurethanes are arguably the most complex family of polymers. Polyurethanes range from soft elastomeric polymers to hard elastoplastics that rival metals. Polyurethanes are used as structural materials, coatings, adhesives, and sealants. Polyurethanes can be synthesized as thermoplastics, thermosets, and curable compositions by either heat or UV energy—and all this by molecular design, as opposed to compounding by the addition of plasticizers or other modifiers.

Dr. Michael Szycher

About the Author

Michael Szycher, Ph.D., is Chairman and Chief Executive Officer of CardioTech International, Inc., a manufacturer of medical products based on specialized polyurethanes. He holds a Ph.D. from Boston University School of Medicine and an MBA from Suffolk University.

Dr. Szycher is a recognized international authority on polyurethanes and blood-compatible polymers. Author of more than 100 research articles and a pivotal force in the creation of the Medical Plastics Division of the Society of Plastics Engineers (SPE), he is the editor of several other books,

- *Biocompatible Polymers, Metals and Composites*
- *Synthetic Biomedical Polymers*
- *Blood Compatible Materials and Devices: Perspectives Towards the 21st Century*
- *High Performance Biomaterials: A Comprehensive Guide to Medical/Pharmaceutical Applications*
- *Szycher's Dictionary of Biomaterials and Medical Devices;* and
- *Szycher's Dictionary of Medical Devices.*

He is also editor-in-chief of the quarterly *Journal of Biomaterials Applications.*

Acknowledgment

This handbook summarizes the published work of many polyurethane chemists, lecturers, researchers, and technologists. The contributions of these outstanding authors grace the pages of this handbook.

Dedication

This handbook is dedicated to my wife, Laurie, whose unwavering support and dedication made this work possible.

Contents

1

Introduction

1.1 Historical

The year 1987 marked the 50th anniversary of the introduction of polyurethanes. Prof. Otto Bayer was synthesizing polymer fibers to compete with nylon when he developed the first fiber-forming polyurethane in 1937. His invention ranks among the major breakthroughs in polymer chemistry, but the polymer was dismissed as impractical by his superiors at I.G. Farbenindustrie. For more than 20 years, Germany had been at the forefront of synthetic fibers technology, beginning with the introduction of polyvinyl-chloride fibers in 1913.

Germany remained preeminent in the fibers field until 1935, when Carothers in the U.S. discovered the nylons; E. I. DuPont in America introduced and began marketing nylon fibers, protected by a barrage of patents that proved impossible to overcome. Nothing as versatile and practical as the polyamides was available, prompting Bayer to investigate similar polymers not covered by the impenetrable DuPont patents.

At the end of January 1938, Rinke and collaborators were successful in reacting an aliphatic 1,8 octane diisocyanate with 1,4 butanediol to form a low-viscosity melt from which they were able to draw fibers. These early efforts resulted in what are now known as polyurethanes: the esters of carbamic acid. These polyurethanes could be spun from the melt; yarns and monofilaments that could be made from their new polymer were of high quality. Rinke and Associates were awarded the first U.S. patent on polyurethanes in 1938.[1]

Like many developments, polymer chemistry, which began as a small specialized branch of organic chemistry, began to grow rapidly and adopted a new nomenclature, much as biochemistry had done before. Table 1.1 presents the recognized names of two important linkages found in polymers, comparing classical organic chemistry nomenclature to that used in polymer chemistry and biochemistry.

TABLE 1.1   Names of Some Important Nitrogen-Containing Polymer Linkages

Linkage	Organic Chemistry	Polymer Chemistry	Biochemistry
-NHCO-	Amide	Nylon	Peptide
-NHCOO-	Carbamate	Urethane	Not applicable

The first I.G. Farbenindustrie polyurethane had a melting point of 185° C and became available under the trade names Igamid U for synthetic fabrics, and Perlon U for producing artificial silk or bristles. A softer version was also available under the trade name Igamid UL. Foams were also produced by adding water to isocyanates in the presence of hydroxyl-terminated polyesters to form carbonamides and release carbon dioxide as the blowing agent. These foams, named Troporit M, were used to produce aircraft propeller blades and rigid, foam-filled landing flaps and skis.

DuPont[2] and ICI[3] recognized the elastomeric properties of the polyurethanes which led to production on an industrial scale in the 1940s.[4] Water was used as the chain extender, and the diisocyanate was naphthalene-1,5-diisocyanate (NDI).

DuPont surged to the forefront of polyurethane technology in the U.S., receiving patents in 1942 covering the far-reaching reactions of diisocyanates with glycol, diamines, polyesters, and certain other active hydrogen-containing chemicals. From these humble beginnings emerged the polyurethanes, one of the most versatile polymers in the modern plastics armamentarium.

1.2 Polyurethanes

Polyurethanes are among the most important class of specialty polymers. But, ironically, the term *polyurethane* leads to a great deal of confusion. The term is more one of convenience than of accuracy, because polyurethanes are not derived from polymerizing a *urethane* monomer, nor are they polymers containing primarily urethane groups. The polyurethanes include those polymers containing a plurality of urethane groups in the molecular backbone, regardless of the chemical composition of the rest of the chain. Thus, a typical polyurethane may contain, in addition to the urethane linkages, aliphatic and aromatic hydrocarbons, esters, ethers, amides, urea, and isocyanurate groups.

Polyurethanes are used in a surprising array of commercial applications. Figure 1.1 presents the universe of polyurethane applications. For convenience, we have divided the applications into seven major groups: flexible slab, flexible molded foams, rigid foams, solid elastomers, RIM, carpet backing, and two-component formulations.

The chemistry of urethanes makes use of the reactions of organic isocyanates with compounds containing active hydrogens. When polyfunctional isocyanates and intermediates containing at least two active hydrogens per mole are reacted at proper ratios, a polymer results that can produce rigid or flexible foams, elastomers, coatings, adhesives, and sealants. An isocyanate group reacts with the hydroxyl groups of a polyol to form the repeating urethane linkage, as shown in Reaction 1.1.

REACTION 1.1 Classical Urethane Linkage Reaction

N=C=O	+	H–O	$\rightarrow$ NHCOO
isocyanate group		hydroxyl group	urethane linkage

The isocyanates also react with amines to form substituted urea linkages; they will react with water to form carbamic acid, which is an unstable intermediate, and it decomposes readily to evolve carbon

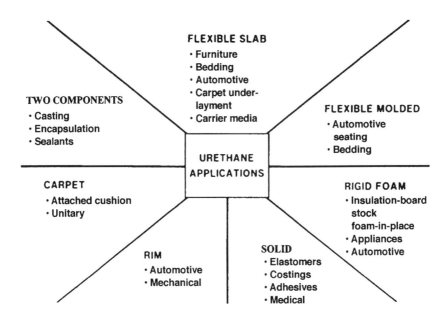

FIGURE 1.1 The polyurethane universe.

dioxide and an amine. This amine, in turn, reacts with additional isocyanate to form disubstituted urea. In addition, a number of cross-linking reactions may take place, depending on the reaction conditions such as temperature, the presence of catalysts, the structure of the isocyanate, alcohols, and amines involved. These reactions form linkages of allophanate (reaction between urethane-isocyanate), biuret (reaction between substituted urea and isocyanate), and isocyanurate (trimerization of isocyanate groups). Isocyanates can also be polymerized to form dimers (uretidine diones), carbodiimide, and 1-nylon.

The repeating urethane linkage is the basis for the generic name: *polyurethane*. However, the use of the generic term *polyurethane* is deceiving in that all useful polyurethane polymers contain a minority of urethane functional groups. Thus, polyurethane is more a term of convenience rather than accuracy, since these polymers are not derived by polymerizing a monomeric urethane reactant, nor are they polymers containing primarily urethane linkages. In fact, other groups such as ethers, amides, biurets, and allophanates are the majority linkages in the molecular chain. Urethane linkages represent the minority of functional groups as long as the polymers contain a significant number of urethane linkages. The name *polyurethane* may be correctly ascribed to these polymers.

Polyurethanes and the closely related polyureas are the products of the reaction of isocyanates (–N=C=O) with the active hydrogen compounds (R-OH) or (R-NH$_2$). Alternative chemistry to the isocyanate reactions was explored by Hoff and Wicker[5] as they developed chemistry to prepare polyurea from bis chloroformate and diamines.

The polyurethanes are a heterogeneous family of polymers unlike PVC, polyethylene, or polystyrene. The polyurethanes comprise an array of different products, ranging from rigid foams to soft, millable gums.

Figure 1.2 presents a summary of the structure-property relationship for polyurethanes; branching/cross-linking is plotted on the ordinate, and intermolecular forces on the abscissa. Under these

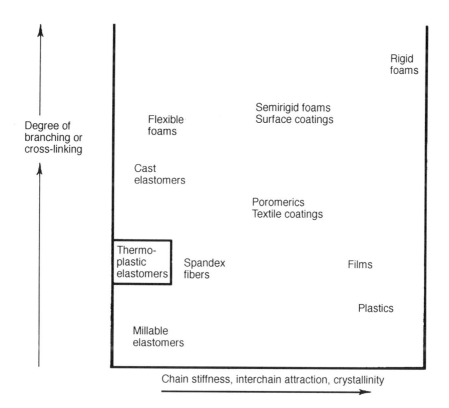

FIGURE 1.2 Structure-property relationships in polyurethanes.

conditions, we can encompass all the commercially available polyurethanes and define which elastomeric products are of the greatest importance in current commercial applications. At the right corner, representing extreme branching and chain stiffness, are the rigid urethane foams. Therefore, at this extreme, the thermoset rigid foams occupy the highest rank of crosslinking and chain stiffness.

Table 1.2 summarizes some of the most important events in the historical development of the polyurethanes.

TABLE 1.2 Summary of Events in the Historical Development of the Polyurethanes

Year	Event	Reference
1849	Isocyanate reaction with an alcohol	Wurtz, A. Ann. 71,326 (1849)
1937	I.G. Farbenindustrie applies for first polyurethane patent	German Patent 728,981
1938	First U.S. patent awarded to Rinke, et al.	U.S. Patent 2,511,544
1942	DuPont receives patents for reaction of polyisocyanates with glycols, diamines, and polyesters	
1942	Introduction of Igamid U, Perlon U, and Igamid UL in Germany	
1943	Vulkollan polyester-based elastomers introduced in Europe	
1945	Allies reorganize German industry and create Farbenfabriken Bayer A.G.	
1954	Patent on Lycra® spandex elastomeric fiber awarded	U.S. Patent 2,692,893
1954	Bayer and Monsanto Co. form Mobay Chemical Co.	
1955	Patent on Estane® thermoplastic elastomers awarded	U.S. Patent 2,871,218
1955	Union Carbide develops first one-shot foam; Dow Chemical introduces polyether polyols; Wyandotte introduces polyfunctional polyether polyols for rigid foams	
1956	Teracol 30-PTMEG introduced	Bulletin HR-11 DuPont
1959	First use of chlorofluorocarbons as blowing agents	
1971	Patent on medical-grade polyurethane-silicone elastomer	U.S. Patent 3,562,352
1993	Patent on aliphatic biostable polyurethane elastomer	U.S. Patent 5,254,662

1.3   Overview of Polyurethane Markets

The worldwide demand for polyurethanes is estimated to approach 16 billion pounds in 1995, or about 5% of total world consumption of plastics. Polyurethanes, although considered to be a specialty, are behind only such commodity plastics as polyethylene, polyvinyl chloride, polypropylene, and polystyrene in overall volume.

The U.S. market for polyurethanes in 1995 is estimated at about 3.6 billion pounds and increasing at a rate of close to 6% per year. Figure 1.3 presents the breakdown of polyurethane usage; as the figure indicates, the majority of polyurethanes are used in the production of flexible foams, followed by rigid foams and elastomers. A surprising portion of polyurethanes (14.1%) is used for specialty applications, such as protective coatings, adhesives, caulks, sealants, etc.

Table 1.3 presents a breakdown of polyurethane pattern of consumption in the U.S. The polyurethane market may also be considered in light of a world market, since many products—both raw materials and end-use—are traded around the world. While statistics are not available, we can safely assume that the North American market represents about 32% of world polyurethane consumption.

The flexible foam market is composed primarily of the following applications:

· Automotive seating and crash pads
· Carpet underlay and cushions
· Bedding
· Furniture

One of the large outlets for flexible polyurethane foam is for furniture cushions. Lighter weight, greater strength, and ease of fabrication as compared to latex foam are some of the deciding factors in its success.

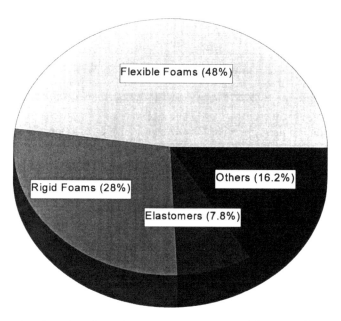

FIGURE 1.3 U.S. markets for polyurethanes (1995 consumption, est. 3.6 billion pounds).

TABLE 1.3 Polyurethane Pattern of Consumption (U.S.)

Markets	Millions of pounds
Flexible foam:	
Bedding	240
Furniture	742
Carpet underlay	194
Transportation	401
Other	182
Total	**1759**
Rigid foam:	
Building insulation	519
Refrigeration	195
Industrial insulation	92
Packaging	88
Transportation	57
Other	77
Total	**1028**
RIM elastomers:	
Transportation	134
Other	76
Elastomers	71
Others (sealants, adhesives, coatings)	553
Total	**834**
Grand total	3621

In addition, the fabrication and application of slab stock foam is easier and faster than the use of animal hair, bird feathers, or other filling materials. Improved molding techniques of flexible foam are responsible for its acceptance in furniture with unusual shapes. Molded rigid foam has made great inroads into the furniture industry. Shortages of select hardwood and a scarcity of skilled wood workers has encouraged the furniture industry to look for replacement materials. Among other candidate materials, rigid polyurethane foam has found acceptance in decorative parts, mirror frames, chair shells, and the like.

Millions of polyurethane foam mattresses are being produced in the United States yearly. They have found acceptance because of their superior durability, freedom from odor, nonallergenic properties, ease of cleaning, resistance to dry cleaning solvents, oils, perspiration, and the fact that they are only one-fourth the weight of a comparable innerspring, and one-half to one-third the weight of latex foam.

Another important use of flexible foam is in the automotive industry for seat cushioning, instrument-panel trim, safety pads, arm rests, floor mats, sunvisors, underlays, roof insulation, weather stripping, air filters, etc. It has been estimated that the 1995 model cars will use an average of 30 lb of foam per automobile. One of the first uses of flexible polyurethane foam was in seating for aircraft where its light weight and flame retardance were of special importance.

The use of flexible foam as a bonding material for fabric primarily started in 1961, when the apparel industry began to employ it. Polyester foam is used for this application, wherein the foam is bonded to the fabric by the flame-lamination technique. In this process, the surface of foam is heated until the surface layer fuses and becomes soft and tacky. It is then bonded under pressure to the fabric. Adhesives can also be used for bonding the foam to the fabric. A foam lining for the garments makes the fabric dimensionally stable and provides high insulating qualities. Other advantages include excellent hand drape and outstanding crease and wrinkle resistance.

Another important area for flexible polyurethane foams is in carpet underlay, where they provide cushioning, are nonskidding, and do not mat down. They also impart a luxurious feel, even to low-cost carpeting.

One of the major uses of rigid polyurethane foam is in home refrigerators. Most major manufacturers are currently using rigid urethane foam as insulation in their lines. Because of the superior insulating characteristics of the fluorocarbon-blown foams, manufacturers build refrigerators with thinner insulation and, therefore, larger inside capacity. All types of refrigerated trucks such as milk trucks, ice cream trucks, and trailers are insulated with rigid polyurethane foams. Besides having good insulating properties, rigid foams contribute to the structural strength of the bodies of the trucks, have low moisture pickup, and can withstand gasoline and temperatures up to 100° C.

Potentially, the biggest market for rigid polyurethane foam is in the building industry. The areas of utilization in this field encompass curtain-wall construction, preformed rigid panels, spray-applied wall construction, and roofing insulation, either sprayed or in preformed panels. This market includes residential homes and commercial and industrial buildings such as large refrigerated warehouses.

The use of rigid polyurethane foam in marine applications such as flotation equipment is growing. Many modern boats utilize rigid foam to help support the boat in the water. Larger ships have used rigid polyurethane foam as void fillers, and it is also used in lifeboats and refrigerator ships.

The major applications for rigid foams are

- Building insulation
- Appliance insulation
- Packaging

1.4   Elastomers

The dictionary defines *elastomer* as a "material which at room temperature can be stretched repeatedly, and upon immediate release will return to its approximate original length." Since natural rubber was the original elastomer, in polymeric nomenclature, synthetic materials that approximate or exceed cured natural rubber in physical properties are called *elastomeric*. Because natural rubber exhibits such an excellent combination of physical properties (i.e., tensile strength or 4000 psi, 400 to 600% ultimate elongation), the term *elastomeric grade* is frequently used to characterize and describe those synthetic materials with the highest physical performance.

Elastomers exhibit initial elastic moduli in the range of 0.1 to 0.4 lb/in², and instantaneous and nearly complete reversible extensibility. As the temperature is lowered, the extensibility decreases significantly, and below its glass transition temperature (Tg) the elastomers become brittle. Other properties of polyurethane elastomers, along with some trade names, are shown in Fig. 1.4.

The U.S. polyurethane elastomers market is divided principally between the castable resins and the thermoplastic polyurethane elastomers (TPUs). Figure 1.5 presents the estimated 1995 usage for elastomers in the U.S. Elastomers represent a heterogeneous class of polyurethanes. Among the most typical applications, we can cite:

1. Solid and microcellular elastomers
   Footwear
   Automotive and transportation
   Material handling
   Industrial
   Medical devices

Trade Names	Manufacturer
* ChronoFlex	CT Biomaterials
* ChronoThane	CT Biomaterials
• Bayflex	Miles Inc.
• Estane	B. F. Goodrich Chemical
• Isoplast	Dow Chemical
• Mor-Thane	Morton
• Neuthane	New England Urethane
• Pellethane	Dow Chemical
• Tecoflex	Thermedics Inc.
• Tecothane	Thermedics Inc.
• Texin	Miles Inc.

* Medical-grade

General Description

Polyurethanes are produced by the condensation reaction of an isocyanate and a material with a hydroxyl functionality, such as a polyol. PU can have the chemical structure of either a thermoplastic or thermoset and can have the physical structure of a rigid solid, a soft elastomer, or a foam. The chemical composition of PU can also vary widely, depending on the specific polyol and isocyanate bearing species which are reacted to form the PU. The many different chemical structures and physical forms possible for PU make it a versatile, widely used polymer. Specialty grades available include flame retardant, clay, silica, and glass filled. In 1994, the price of PU ranged approximately from $2.50 to $6.50 per pound at truckload quantities.

General Properties

The major benefits offered by PU are that it retains its high impact strength at low temperatures, it is readily foamable, and it is resistant to abrasion, tear propagation, ozone, oxidation, fungus, and humidity. Although thermoplastic PU is attacked by steam, fuels, ketones, esters, and strong acids and bases, it is resistant to aliphatic hydrocarbons and dilute acids and bases. The highest recommended use temperature of thermoplastic PU is approximately 220°F (104°C), rendering it inappropriate for most high temperature applications. **Aromatic** thermoplastic PU has poor weatherability stemming from its poor resistance to UV degradation. Since PU can be painted with flexible PU paints without pretreatment, it has found use in many automotive exterior parts.

Typical Properties of Polyurethane		
	American Engineering	**SI**
Processing temperature	385-450°F	196-232°C
Linear mold shrinkage	0.004-0.014 in/in	0.004-0.014 cm/cm
Melting point	400-450°F	204-232°C
Density	69.9-77.4 lb/ft^3	1.12-1.24 g/cm^3
Tensile strength, yield	4.9-35.0 lb/in^2 x 10^3	3.4-24.6 kg/cm^2 x 10^2
Tensile strength, break	4.9-35.0 lb/in^2 x 10^3	3.4-24.6 kg/cm^2 x 10^2
Elongation, break	100.0-500.0%	100.0-500.0%
Tensile modulus	0.6-45.0 lb/in^2 x 10^5	0.4-31.6 kg/cm^2 x 10^4
Flexural strength, yield	6.0-60.0 lb/in^2 x 10^3	4.2-42.2 kg/cm^2 x 10^2
Flexural modulus	0.1-0.4 lb/in^2 x 10^5	0.0-0.2 kg/cm^2 x 10^4
Compressive strength	1.2-29.5 lb/in^2 x 10^3	0.8-20.7 kg/cm^2 x 10^2
Izod notched, R.T.	1.5 ft-lb/in-no break	8.1-0.0 kg cm/cm
Hardness	A55-A95 Rockwell	A55-A95 Rockwell
Thermal conductivity	1.7-2.3 BTU-in/hr-ft^2-°F	0.25-0.33 W/m°-K
Linear thermal expansion	1.8-8.4 in/in-°F x 10^{-5}	3.2-15.1 cm/cm-°C x 10^{-5}
Deflection temp. @ 264 psi	100-330°F	38-166°C
Deflection temp. @ 66 psi	115-370°F	46-188°C
Continuous service temp.	180-220°F	82-104°C
Dielectric strength	430-730 V/10^{-3} in	1.7-2.9 V/mm x 10^4
Dielectric constant @ 1MHz	4.4-5.1	4.4-5.1
Dissipation factor @ 1MHz	0.060-0.100	0.060-0.100
Water absorption, 24 hr	0.10-0.60%	0.10-0.60%

Typical Applications

• Automotive   facias, padding, seats, gaskets, body panels, bumpers

• Medical   implantable devices, tubing, blood bags, dialysis membrane

• Machinery   bearings, nuts, wheels, seals, tubing

• Consumer   furniture padding, mattress goods, roller skate wheels, athletic shoes

FIGURE 1.4   Thermoplastic polyurethane (TPU) manufacturers and trade names.

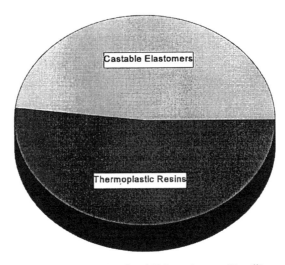

FIGURE 1.5   Polyurethane elastomer consumption (total U.S. market, est. 71 million pounds, 1995).

2. Cast elastomers
   Pump and pipe linings
   Chute liners
   Industrial tires
   Seals and o-rings
   Casters and wheels
   Sonar windows
   Bearing and ski pads
   Vibration and shock mounts
   Bowling balls
   Ski boots
   Buoys and dock fenders

Among the "others" classification, we include adhesives and sealants as well as protective, decorative coatings and finishes, as shown below:

1. Adhesives and sealants
   Foundry sand
   Rebonded carpet underlay
   Construction
   Transportation
   Footwear
   Textile
   Automotive caulks and glazing compounds
2. Protective/decorative coatings and finishes
   Textile
   Automotive and transportation finishes
   Architectural
   Industrial
   Anticorrosive metal finishes

In terms of production capacity, the isocyanates and the polyols are the most important because of their volume. The isocyanate usage is dominated by the two "workhorses," namely diphenylmethane-4,4'-diisocyanate (MDI) and toluene diisocyanate (TDI). The worldwide production of MDI is estimated at about 3.5 million pounds in 1995, while the worldwide production of TDI is estimated at approximately 2.5 million pounds for the same year.

Close to 45% of the production capacity of MDI is located in Western Europe, 38% in North America, 13% in the Far East, and the remaining 4% in the rest of the world. Close to 40% of the production capacity of TDI is located in Western Europe, 27% in North America, 12% in the Far East, and the remaining 21% fairly evenly distributed between Eastern Europe, Latin America, and the rest of the world.

Polyether polyols constitute the lion's share of polyurethane raw materials production. The 1995 production of polyether polyols is estimated to be close to 8 million pounds, with 36% of production facilities found in Western Europe, followed by 35% in North America and 16% in the Far East.

The production capacity of polyester polyols is estimated at about 900,000 pounds in 1995, with more than half produced in Western Europe, and another 25% produced in North America.

Many other raw materials are not included in this market survey, such as catalysts, chain extenders, blowing agents, surfactants, flame retardants, stabilizers, lubricants, etc., because their production capacity is small compared to the isocyanates and polyols already mentioned.

References

1. Rinke, H., Schild H., and Siefken, W. (I.G.Farben). U.S. Patent 2,511,544 (1938).
2. Christ, A.E. (DuPont). U.S. Patent 2,333,639.
3. British Patents (ICI.) 580,524 (1941) and 574,134 (1942).
4. Pinten, P. (Dynamit A.G.). German Patent 932,633 (1943).
5. Hoff, G.P., Wicker, D.B., I.G. and Farben P.B., Report 1122, Sept. 12, 1945.

2

Basic Concepts in Polyurethane Chemistry and Technology

2.1   Overview

Polyurethanes are extremely large and complex molecules produced by combining a large number of simpler molecules called monomers. Monomers are compounds whose properties (molecular weight, boiling point, melting point, crystallinity, etc.,) are discrete. Polyurethanes, on the other hand, typically do not have discrete properties but have average properties that represent a range of molecules with differing molecular weight and often slightly differing structure. Table 2.1 shows the comparison and contrast of the properties of typical small molecules with those of a polyurethane polymer.

Table 2.1 Properties of Small Molecules vs. Polyurethane Polymers

Small Molecules	Polyurethane Polymers
Three discreet states: solid, liquid, gas	Three states: liquid, quasisolid, solid
Discreet formula	Average formula
Discreet molecular weight	Average molecular weight
Discreet properties: melting point, density, crystalline/glassy	Range of properties: melting range, density range, microcrystalline

The molecular weight of polyurethanes can greatly affect the physical properties of a polymer.[1] This is shown graphically in Fig. 2.1. Molecular weight distribution can also have a significant effect upon polyurethane characteristics, especially processing and rheological characteristics.

The polymer chains have a spacial architecture. They may be linear, branched, or networked. Polyurethanes display stereo microstructure. Polyurethanes exist as homopolymers and copolymers. Copolymers may be random, alternating, segmented, block, or graft types.

Polyurethanes can be crystalline solids, segmented solids, amorphous glasses, or viscoelastic solids. With respect to mechanical properties, polyurethanes are nonideal solids. The mechanical properties of polyurethanes are time dependent. For every excitation, there are two responses: a viscous response and an elastic response, i.e., a time-dependent and a nontime-dependent response. The properties of the linear polyurethanes are also very temperature and moisture dependent. Figure 2.2 shows the inverse relationship between modulus of elasticity and frequency in a typical polyurethane. Figure 2.3 shows a similar inverse relationship between modulus of elasticity and increasing temperature. Note that in both instances there is a sharp decrease in modulus at a critical frequency or temperature.

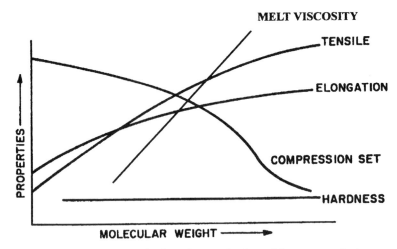

FIGURE 2.1   Effect of polyurethane molecular weight on properties.[1]

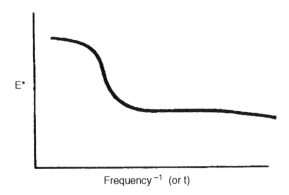

FIGURE 2.2   Modulus vs. frequency (T = constant).

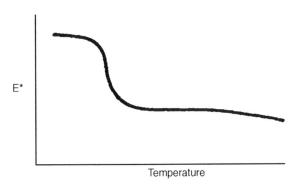

FIGURE 2.3   Modulus vs. temperature (t = constant).

2.1.1   Traditional Polymer Classification

Polymers were traditionally classified according to their polymerization method as either addition-type or condensation-type polymer. Recently, the distinction between polymer types has been drawn along the lines of the major differences in the kinetics of polymerization. In this handbook, the terms *chain polymerization* and *step-growth polymerization* replace the terms *addition polymerization* and *condensation polymerization,* as shown in Table 2.2.

Table 2.2 Modern Classification System

By mode of polymerization	By kinetics of polymerization
Addition polymers	Chain addition polymers
Condensation polymers	Step addition polymers

There are major differences between the kinetics of chain and step polymerizations. The chain polymerization and the high polymer is formed very early in the polymerization. The locus of the polymerization is only on those few chains containing an active propagating center. In step polymerization, all species present can react with all other species present. Polymerization proceeds as dimer, trimer, tetramer, pentamer, etc. High molecular weight polymer does not result until late in the reaction. In contrast, ionic polymerizations proceed in a nearly linear fashion. Most polyurethane reactions proceed to polymerization via step-addition reactions, as depicted in Fig. 2.4.

Step-Addition Polymers

In step polymerization, monomers are typically difunctional or trifunctional. In a typical step polymerization, a single reaction and the propagation reaction is responsible for the formation of the polymer.

Step polymerization typically occurs in or more difunctional or trifunctional monomers that bear two distinctly different functionalities. A single chemical mechanism is responsible for the formation of the polymer. Chain transfer and termination reactions are inherently absent in step polymerizations, although side reactions and reactions with contaminants can have similar effects as transfer or termination.

In a step system, all monomer units and growing polymer chain ends participate in the propagation reaction throughout the polymerization process. Initially, monomers reacts with monomers to form dimers (dimers with monomers or dimers with dimers to form timers or tetramers, etc.). Monomers are depleted very early in the polymerization; however, high polymers are not formed until late in the reaction. Table 2.3 provides examples of some typical step addition polymerizations.

Condensation polymerizations are step polymerizations where the repeat unit of the finished polymer does not contain the same structures as the monomers from which it was prepared. Certain polymers such as polycaprolactones, polycaprolactams, polyurethanes, and polyureas are also classified as condensation polymers, even though the definition does not fit the routes commonly used to prepare them.

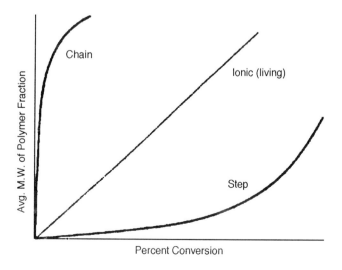

FIGURE 2.4 Chain addition vs. step addition.

Table 2.3 Step Addition Polymers

Polymer Type	Group A	Group B	Example
Condensation	R-OH	R'-COO	Polyester
	R-OH	R'-COCl	Polyester
	R-OH	R'COOR	Polyester
	R-OH	COCl$_2$	Polyester
	R-OH	R'-NCO	Polyurethane
	R-NH	R'-NCO	Polyurea
	R-NH	R'COH	Polyamide
	R-NH	R'-COCl	Polyamide
	Si-R$_2$-Cl$_2$	H$_2$O	Silicone
	Ar-OH	COH$_2$	Phenolic resin
	Ar-OH	R-CH-O-CH$_2$	Epoxy resin

Condensation polymerization may well be thought of as a functional group polymerization. Thus, functional groups on each end of monomer molecules (e.g., -NCO) react with functional groups on the ends of other monomers (e.g., -OH) to form functional group linkages (e.g., urethanes). Monomers for the condensation polymerization may consist of the structure AB where molecules of molecules of A-B react with other A-B molecules to form the polymer.

There are no inherent initiation or chain transfer reactions (as in chain addition reactions)—only propagation. Any monomer of functionality A may react with any monomer of functionality B from the outset. Polymerization thus proceeds by successive monomer, dimer, trimer, and tetramer formations until it reaches a high conversion, a molecular weight polyurethane polymer is produced. Table 2.4 summarizes the characteristics of polyurethane reactions.

Table 2.4 Characteristics of Polyurethane Polymerizations

1. A single reaction (propagation) is responsible for polymer formation.
2. Monomer system contains A, B, or C functional groups, any one of which may react at random.
3. Monomers disappear early in the polymerization.
4. Molecular weight rises slowly. High molecular weights achieved only at high conversions.
5. Molecular weight distribution is very broad throughout the polymerization until the very end, when all oligomers have reacted.
6. Precise stoichiometric ratios are required.
7. High-purity monomers are required.
8. Reactions are often reversible; equilibrium driven to high polymer by oligomer depletion.

2.2 Kinetics of Polyurethane Polymerization

The extent of reaction of a given polyurethane system (p) is given by:

$$p = \frac{No - N}{No} \tag{2.1}$$

where No = initial number of monomer molecules, N = remaining number of monomer molecules, or N = No $(1 - p)$ so that, at 100% conversion (i.e., N = 0), p = 1.

The degree of polymerization, DP, is defined as the ratio of the initial number of molecules of monomer to the remaining number of molecules, or

$$DP = \frac{No}{N} \quad \text{or} \quad DP = \frac{1}{1 - p} \tag{2.2}$$

Now consider a reaction pushed to 90% conversion, or 90% yield.

$$DP = \frac{1}{1 - 0.9} = 10$$

Therefore, it requires a 98% conversion to reach a DP = 50, which is approximately the threshold value for minimal mechanical properties in polyurethanes. The effects of a stoichiometric imbalance (whether by impurity, weighing error, or design) is given by

$$DP = \frac{1 = r}{1 + r - 2rp} \tag{2.3}$$

where r = NA/NB for A-A/B-B system, or r = NA/NB = 2NB' for A-A/B-B system, with monofunctional monomer B' or A-B system with B"-B' added. From the foregoing discussion, we can calculate that, at 98% conversion, a 2% excess of monomer BB will reduce DP from 50 to 33. Or with a 2% excess of B-B, it would require a conversion of just over 99% to restore DP to 50. One should readily recognize that, while most organic chemists may consider a 95% conversion as a very good yield, the polyurethane chemist at the same conversion ratio would end up with only a sticky mess.

2.2.1   E. Polyurethane Depolymerization Mechanisms

All polymers can be depolymerized; polyurethanes are no exception. There are seven ways polyurethanes can be depolymerized chemically.[2] See Table 2.5.

Table 2.5 Depolymerization Reactions of Polyurethanes

Hydrolysis	Thermolysis	Oxidation
Photolysis	Pyrolysis	Microbial
Solvolysis	Biologically-induced environmental stress cracking	

Hydrolysis is defined as *a chemical reaction in which water reacts with another molecule to form two or more substances.* Thermolysis reactions are those that occur due to heat. Oxidation is a reaction in which oxygen combines chemically with another substance. Oxidation can be initiated with heat (thermooxidation) or by light (photooxidation). Photolysis is the decomposition of a chemical compound into smaller molecular weight units caused by the interaction with light. Pyrolysis is the transformation of a substance into other substances by heat alone, i.e., without oxidation. Chemical depolymerizations caused by the attack of microorganisms are called *microbial degradations.* Attack on polyurethanes by solvents, e.g., alcohols, can cause a surface degradation referred to as *solvolysis.* Biologically induced environmental stress cracking is a special surface degradation caused by exposing polyurethanes to enzymes secreted by certain inflammatory cells when the polyurethane is implanted within a living system for prolonged periods of time.

Polyurethanes, like all major plastics, are the objects of considerable recycling efforts. Several industry groups have been formed to develop recycling technology based on depolymerization reactions. These groups include the Polyurethane Recycling and Recovery Council (PURRC) and the American Plastics Council (APC). Current recycling technology generally falls into three basic categories: reuse of scrap, chemical depolymerization, and incineration.

At present, the most impressive story in recycling of polyurethane scrap lies with flexible foams, since much of the flexible-foam scrap in the U.S. is reused in carpet underlay. Polyurethane scrap is compounded into "rebond" foam (the type of foam used in carpet underlay) by chopping the scrap foam with a TDI binder and polyol and then compressing the mass in a mold.

Another recycling approach is to hydrolyze (or glycolize) the polyurethane back into the precursor raw materials. Developmental methods use an alcohol under high-temperature conditions to depolymerize the polyurethane and obtain a somewhat modified version of the original polyol. The reconstituted polyol usually has a lower molecular weight distribution and cannot be used at 100% concentrations in the polyurethane reaction. However, it is possible to use reconstituted polyols as a 10 to 15% blend with virgin polyols and produce good parts.[3]

First, let us examine the bonds involved in some depolymerization reactions (Fig. 2.5). The three bonds most susceptible to hydrolytic degradation are the ester, urea, and urethane (Fig. 2.6). The ester reverts to the precursor acid and alcohol; this precursor acid further catalyzes ester hydrolysis, and thus the reaction becomes autocatalytic. Because of the autocatalytic nature of ester hydrolysis, this is the most prevalent hydrolytic degradation reaction. The urea bond can hydrolyze to form a carbamic acid and an amine. The carbamic acid normally is unstable and typically undergoes further reaction. The urethane linkage, although somewhat less susceptible, may undergo hydrolysis to yield a carbamic acid and the precursor alcohol.

Comparing various polyurethane systems, Fig. 2.7 shows that polyethylene adipate glycol/MDI/BD systems hydrolyze quite rapidly, two times faster than polybutylene adipate glycol/MDI/BD systems. Conversely, using the more stable polycaprolactone glycol, there is a significant increase hydrolytic stability. The greatest hydrolytic resistance is obtained by the use of polytetra methylene ether glycol (PTMEG). Although not shown in this date, the choice of isocyanate also influences hydrolytic stability. Thus, a polyester/MDI-BD system will display greater hydrolysis resistance than that of a polyester/TDI BD system.

Another important environmental influence on hydrolysis is temperature. At 50° C, the tensile half life of a polyester/TDI/diamine system may be four or five months, while that of a polyether/TDI/amine appears to be almost two years. At 70° C, however, these half lives fall to two weeks and five weeks, respectively. And at 100° C, the ester-based polymers can be expected to degrade in a matter of days.

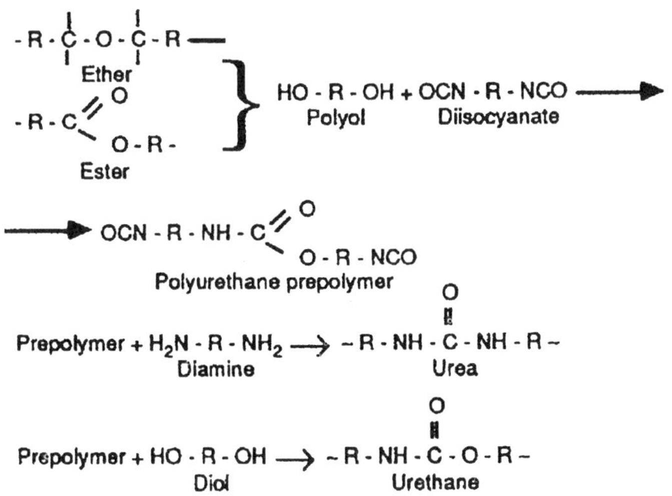

FIGURE 2.5   Chemical reactions involved in polyurethane preparation.

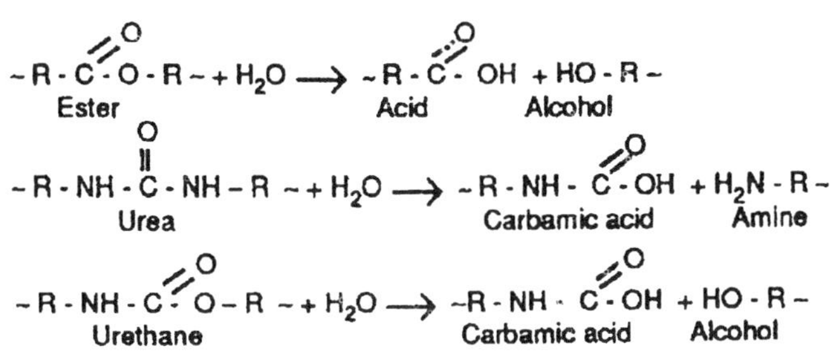

FIGURE 2.6   Bonds susceptible to hydrolytic attack.

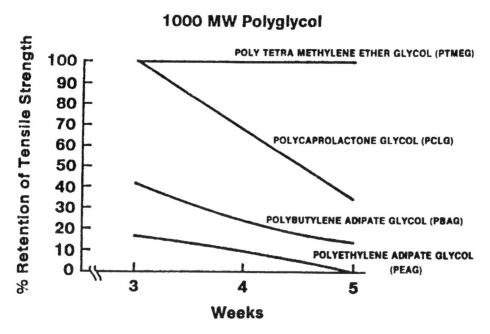

FIGURE 2.7 Polyurethane hydrolytic performance, humidity aging at 80° C/95% RH, tested one week after removal.

In general, polyesters do not fare well in hydrolysis resistance. If a polyester must be used in a wet environment, carbodiimides help in prolonging their longevity. Carbodiimides act as acid scavengers. As seen in Fig. 2.8, the acid and carbodiimide reacts to form an intermediate, which rearranges to give an N-acyl urea. This consumes the acid thereby preventing further hydrolytic autocatalysis.

Figure 2.9 shows the marked increase in life span of a polyester with a 2% addition of polycarbodiimide. It should be remembered, however, that the carbodiimide is being consumed and eventually will be totally depleted resulting in the onset of hydrolysis.

Polyurethane Thermostability

Heat can cause the degradation of polyurethanes. The onset of allophanate disassociation is around 100 to 120° C. The dissociation temperature of the biuret linkage is around 115 to 125° C. These reactions are dissociations and somewhat reversible. They revert to the urethane or urea from which they were formed. The aromatic-based urethane bond begins its thermal dissociation around 180° C, which is prior to the urea linkage which is about 160 to 200° C. The urethane can dissociate into the isocyanate and

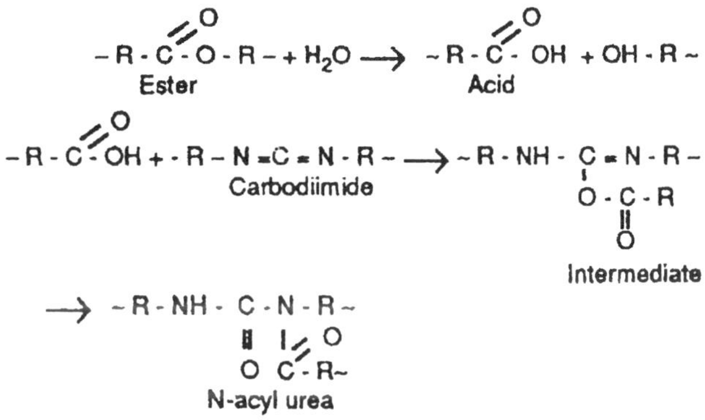

$$-R-C(=O)-O-R-+H_2O \longrightarrow \sim R-C(=O)-OH + OH-R\sim$$
Ester   Acid

$$-R-C(=O)-OH + \cdot R-N=C=N-R- \longrightarrow \sim R-NH-C=N-R-$$
Carbodiimide   O-C-R, O

Intermediate

$$\longrightarrow \sim R-NH-C-N-R\sim$$
O O, C-R~

N-acyl urea

FIGURE 2.8 Carbodiimide stabilization of polyester-based polyurethanes.

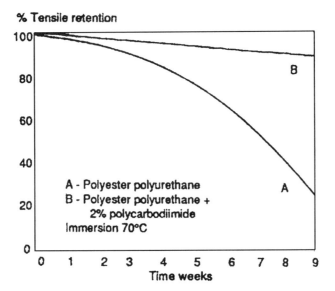

FIGURE 2.9   Effect of carbodiimide on a polyester-based polyurethane.

polyol from which it was formed. This reaction is reversible as long as the isocyanate is not lost to a side reaction. The second reaction produces a primary amine and an olefin. The third reaction produces a secondary amine. Since these latter reactions generate CO_2, which is lost as a gas, they are irreversible. These thermal dissociation relationships are summarized in Table 2.6.

Table 2.6 Thermal Dissociation Temperatures of Linkages Found in Polyurethanes

	Onset of dissociation	
Linkage	°C	°F
Aliphatic allophanate	85–105	185–220
Aromatic allphanate	100–120	212–250
Aliphatic biuret	100–110	212–230
Aromatic biuret	115–125	240–260
Aliphatic urea	140–180	285–355
Aromatic urea	160–200	320–355
Aliphatic urethane	160–180	320–355
Aromatic urethan	180–200	355–395
Disubstituted urea	235–250	455–480

Referring to the urethane bond depicted in Fig. 2.10, it is clear that the urethane linkage may undergo three separate types of thermal degradation: (1) the formation of the precursor isocyanate and the precursor alcohol; (2) cleavage of the oxygen of the alpha CH2 group and association of one hydrogen on the second CH2 group would lead to the carbamic acid and an olefin with subsequent carbamic acid

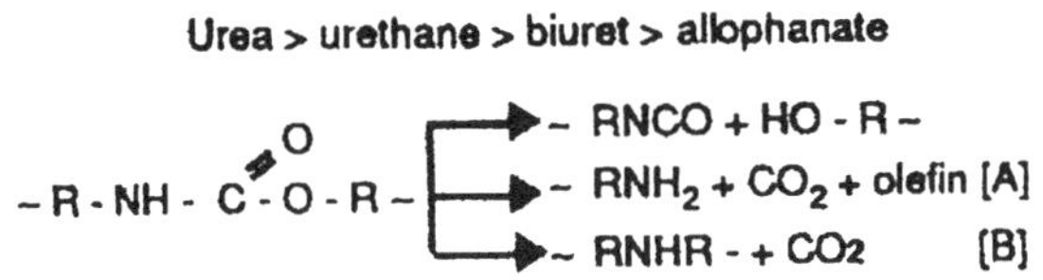

Urea > urethane > biuret > allophanate

$$-R\cdot NH - \overset{O}{\overset{\displaystyle\|}{C}} - O - R - \begin{cases} \sim RNCO + HO - R - \\ \sim RNH_2 + CO_2 + \text{olefin [A]} \\ \sim RNHR - + CO_2 \quad\quad \text{[B]} \end{cases}$$

FIGURE 2.10 Thermal degradation of urethane linkages.

decomposition to give a primary amine and CO_2 depicted in Fig. 2.11, and (3) the formation of a urethane and a secondary amine, as seen in Fig. 2.12.

Figure 2.12, or mechanism B, refers to writing the urethane structure to include the first CH2 group of the polyol, thus visualizing how the secondary amine would be formed. Cleavage of the oxygen CH2 bond with association of the CH2 hydrogen to the NH group would force cleavage of the nitrogen carbonyl carbon bond splitting out CO_2. This would result in a secondary amine. Which of these thermodegradation reactions takes place, and to what extent, depends on the structure of the urethane, the reacting conditions, and the environment.

Polyurethane Oxidation

Oxidation is simply degradation that occurs due to the reaction with oxygen. Oxidation reactions may be heat initiated or light initiated. Heat-initiated oxidation is called *thermooxidation,* and light-initiated oxidation is called *photooxidation.* Thermooxidation and photooxidation will be discussed separately. The discussion of photooxidation will include photolysis reactions since they are closely related.

Thermooxidation

Previously, we demonstrated the ester to be the weak link in hydrolysis. Now it is the ether that is the weak link in thermooxidation. (See Fig. 2.13.) Thermooxidation proceeds via a free radical mechanism. Heat causes a hydrogen extraction at a carbon alpha to the ether linkage. This radical is subject to oxygen addition and forms a peroxide radical. The peroxide radical then extracts another hydrogen from the backbone to form a hydroperoxide. The hydroperoxide radical then decomposes to form an oxide radical and the hydroxyl free radical. The order of thermooxidation stability is ester > urea > urethane >> ether.

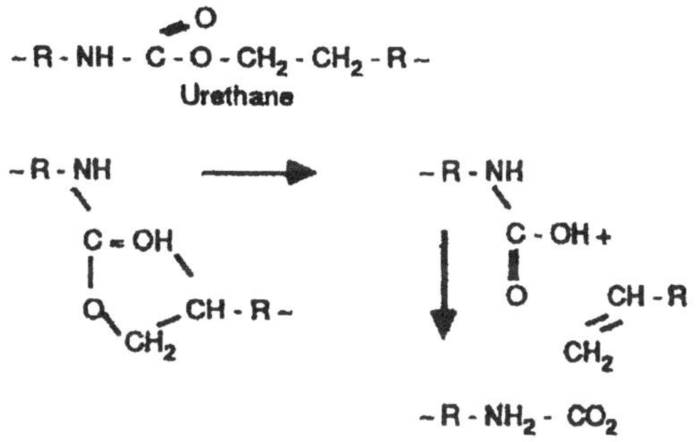

FIGURE 2.11 Thermal degradation of urethane linkages leading to an olefin and CO_2.

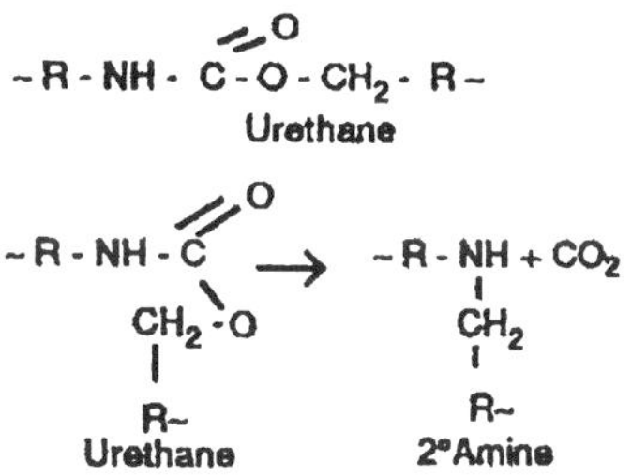

FIGURE 2.12 Thermal degradation of urethane linkages leading to a urethane and a secondary amine.

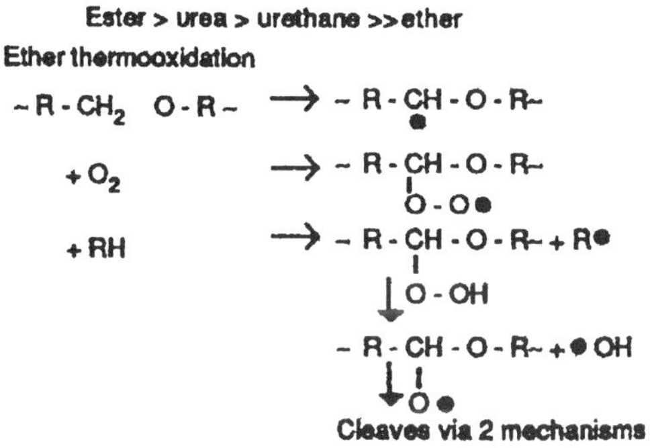

FIGURE 2.13 Thermooxidation.

The oxide radical will cleave at either of two places. (See Fig. 2.14.) One, it may cleave at the carbon bond adjacent to the oxide radical. If so, formates are formed. Two, if the cleavage is at the carbon-oxygen bond, aldehydes are formed. The order of stability of polyethers to thermooxidation is PTMEG is more stable than poly(ethylene oxide) glycols, that are, in turn, more stable than poly(propylene oxide) glycols.

Photooxidation

The exact mechanism of photolytic degradations is controversial. Photooxidation is believed to take place in MDI and TDI aromatic urethane via a quinoid route. The urethane bridge oxidizes to the quinone-imide structure as seen for MDI in Fig. 2.15. This structure is a strong chromophore resulting in the yellowing of urethanes. Further oxidation produces the diquinone-imide structure that is amber in color and is responsible in part for the browning of urethanes. To prevent the discoloration, non-quinoid structures should be used.

A second scheme in the photolysis of polyurethane is the scission of the urethane bond. There are two possible bonds susceptible to scission, as seen in Fig. 2.16. When the nitrogen to carbon bond breaks,

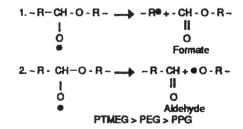

FIGURE 2.14   Thermooxidation-oxide cleavage.

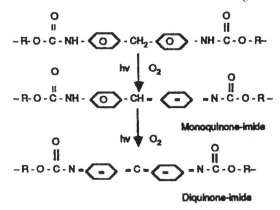

FIGURE 2.15   Photolysis scheme 1: photooxidation.

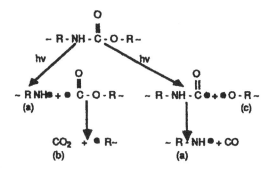

FIGURE 2.16   Photolysis scheme 2: urethane scission.

it results in an amino radical and a formate radical. The formate radical will liberate CO_2, and an alkyl radical will result. Should the carbon to oxygen bond cleave, a carbamyl radical and an alkoxy radical will be formed. The carbamyl radical decomposes to generate the amino radical and CO_2. The net result of these urethane scission reactions is three radicals (amino, alkyl, and alkoxy) that may undergo further reactions.

Those further reactions are seen in Fig. 2.17. In reaction 1, two amino radicals react to form an intermediate that, in turn, reacts with the alkoxy radical to form diazo products. These are again chromophoric materials that are responsible for polyurethanes turning brown in sunlight. The second reaction demonstrates how olefins are formed in these processes. The third reaction is an oxidation process wherein aldehydes are produced. In reaction 4, the alkoxy radical undergoes scission to produce formaldehyde and another alkyl radical which may then be used in reaction 2 or 3.

A third photo-degradative process is known as a *rearrangement*. A sequence of this reaction for a TDI based urethane can be seen in Fig. 2.18. This rearranged product undergoes additional degradation to colored azo-containing products.

These radical-producing processes can be prevented or ameliorated by the use of stabilizers. Certain families of compounds known as antioxidants and UV stabilizers have been effective in inhibiting degradation in polyurethanes. Hindered phenols and aromatic amine compounds act as radical chain terminators. Thioethers and phosphites are peroxide decomposers. Any of these compounds will disrupt the degradation process. Benzotriazoles, along with certain hindered amines and benzophenones, will absorb the UV light and use its energy in a nondestructive sequence.

Certain combinations of these compounds act synergistically. Typically a UV stabilizer such as benzotriazole, used in conjunction with an antioxidant such as a hindered phenol, will inhibit discoloring and property loss for longer periods of time than either one used alone.

Molecular Forces and Chemical Bonding

Polyurethanes are characterized by the forces at work within and between molecules. Of these, covalent bonds are the strongest and most significant. The energy needed to break covalent bonds—energy of dissociation—helps predict how polymers will react to heat degradation. A list of covalent bonds commonly found in polymers, along with their dissociation energies (kilocalories per mole), is shown in Table 2.7 in order of decreasing strength. The carbon-carbon bond in polyethylene, for example, will break first upon overheating, whereas it takes an additional 16 kilocalories per mole to undo carbon-

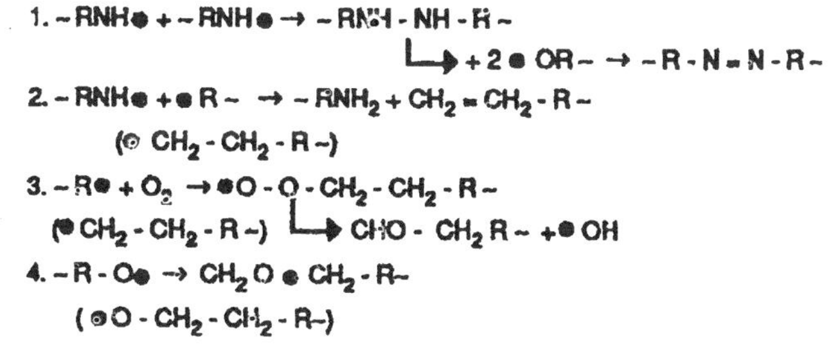

FIGURE 2.17   Photolysis scheme 2B: reactions of radicals.

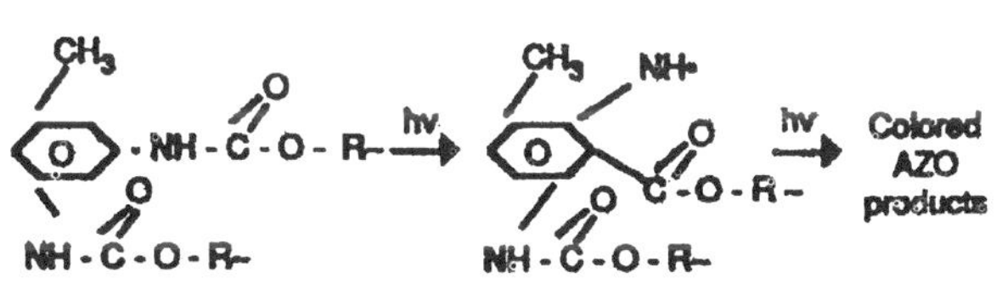

FIGURE 2.18   Photolysis scheme 3: rearrangement.

hydrogen bonds. The first indication of degradation is a reduction in molecular weight, i.e., smaller polymer chains through cleavage. In PVC, the carbon-chlorine bond will succumb first to dissociation, which, in turn, affects the nearby hydrogen atom bonded to the same carbon atom. Evidence of PVC degradation is the generation and evolution of chlorine or chlorine gas.

Table 2.7 Strength of Common Covalent Bonds

Covalent Bond	Dissociation Energy (kilocalories per mole)
C≡N	213
C=O	174
C=C	146
C–F[2]	103–123
O-H	111
C–H	99
N–H	93
C–O	86
C–C	83
C–Cl	81
S–H	81
C–N	73
C–S	62
O–O	35

*Dissociation energy increases as additional fluorine atoms are substituted on the same carbon atom.

To gain a fuller understanding of the nature of polyurethanes, we must account for secondary bonding forces that act between individual polymer molecules. Although much weaker than covalent bonds, they nonetheless directly affect a material's physical properties, such as viscosity, surface tension, frictional forces, miscibility, volatility, and solubility. In order of increasing strength, these secondary forces are classified as van der Waals forces, dipole interaction, hydrogen bonding, and ionic bonding, as shown in Fig. 2.19.

Van der Waals forces are responsible for the short-range natural attraction of similar molecules. When they are overcome by heating, softening and melting follow. Slightly stronger dipole forces are generated by polar groups in the backbone (–C–O–C–) or in side chains (–C–N,–C–C–C–NH$_2$). Polar groups arise from the development of slightly positive and negative charges due to unequal sharing of electrons in covalent bonds. The slightly charged atoms attract oppositely charged atoms on adjacent molecules.

Hydrogen bonding, often considered a strong form of dipole interaction, is a third category of secondary bonding forces. Hydrogen bonding is associated with the group in backbones and the –OH or NH$_2$ groups in the side chains found in nylons, polyurethanes, polyvinyl alcohols, and butadiene-acrylonitrile copolymers. As in dipole interaction, oxygen and nitrogen atoms attract positively charged hydrogens of other molecules. Polymers with hydrogen bonding usually are compatible with small molecules such as those that constitute plasticizers, solvents, and water.

Ionic bonds, the strongest of all, are attractive forces between positively and negatively charged ions. Because of their atomic arrangement, certain atoms tend to accept or release electrons to reach a more stable electron configuration—much like inert gases, and become positively or negatively charged in the process. Ionic bonds between polymers actually are half-polar in nature (ion-dipole bonds). A positive ion (Zn^{++}) is positioned between negatively charged polar groups (O$^-$) of two molecules, linking them together.

Weaker than covalent bonds, secondary bonds will yield before covalent bonds under heat. Furthermore, there is evidence that secondary forces dissipate in groups and not in sequence like a zipper. Thus, polymers with strong secondary forces exhibit high viscosities and are more difficult to process. Polytetrafluoroethylene (Teflon®) is a good example. One of the most chemically inert of all plastics, it is a linear polymer that is melt-processible with a ram extruder; however, complex parts can be made only

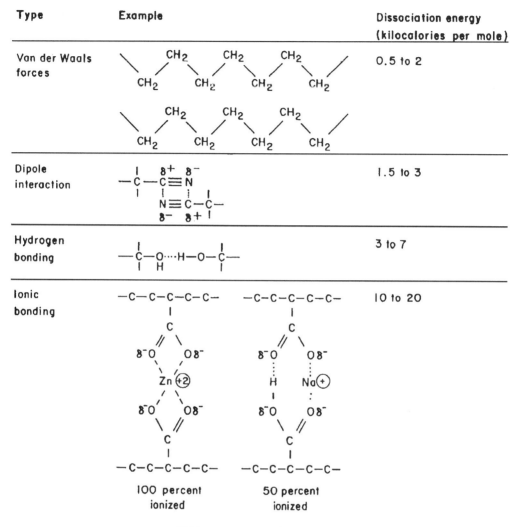

Type	Example	Dissociation energy (kilocalories per mole)
Van der Waals forces		0.5 to 2
Dipole interaction		1.5 to 3
Hydrogen bonding		3 to 7
Ionic bonding		10 to 20

FIGURE 2.19 Secondary bonding.

by machining or sintering. When covalent bonding occurs between polymer molecules, a network (thermoset) molecule is produced. Once these cross-linking bonds have formed, they cannot be resoftened by heating.

Until now, we have assumed that all synthetic polymers are plastics, but that is not true. Polymers do not exhibit the strength associated with plastics until a certain number of repeating units have polymerized—in effect, the degree of polymerization (DP). Such polymers are sometimes referred as *high polymers* to distinguish them from polymers below the critical chain length, but in the plastics industry they are simply *polymers*. Polymers with network bonding or significant secondary forces have much lower DP values than simple linear polymers.

Typically, polyamides (nylon) start gaining strength at a DP of 40, cellulosics at 60, and vinyls at 100. Polyamides have very strong hydrogen bonds, whereas vinyl molecules are secondary bonded by the much-weaker van der Waals forces. High polymers (plastics) usually have DP values around 600. Above the DP value, mechanical properties increase rapidly and eventually plateau. Polymer chain length greatly influences viscosity. Flow resistance rises greatly at a DP of 1000, presumably because of increased molecular entanglements and secondary forces. The point at which viscosity behavior changes rapidly is also used to define the minimum DP for a high polymer.

Amorphous Polymers

Completely amorphous polymers such as atactic polystyrene exist as long, randomly coiled, interpenetrating chains that are capable of forming stable, flow-restricting entanglements at sufficiently high molecular weight. In the melt state, thermal energy is sufficient for long segments of each polymer chain to move in random micro-Brownian motions. As the melt is cooled, a temperature eventually is reached at which all long-range segmental motions cease. This characteristic temperature is called the *glass-transition temperature,* or Tg, which varies widely with polymer structure (shown for several typically amorphous polymers in Table 2.8.

Table 2.8 Glass-Transition Temperatures of Amorphous Polymers

Polymer	Repeat Structure	Glass-Transition Temperature (Tg), °C
Polydimethylsiloxane	$-\!\left[\!\mathrm{Si(CH_3)_2\!-\!O}\!\right]_n\!-$	−123
Cis–1,4–polyisoprene	$-\!\left[\!\mathrm{CH_2\!-\!C(CH_3)\!=\!CH\!-\!CH_2}\!\right]_n\!-$	− 73
Poly(vinyl acetate)	$-\!\left[\!\mathrm{CH_2\!-\!CH(O\!-\!C(CH_3)\!=\!O)}\!\right]_n\!-$	− 72
Polystyrene	$-\!\left[\!\mathrm{CH_2\!-\!CH(C_6H_5)}\!\right]_n\!-$	100
Poly(methyl methacrylate)	$-\!\left[\!\mathrm{CH_2\!-\!CH_2(C(CH_3)\!=\!O\!-\!O\!-\!CH_3)}\!\right]_n\!-$	105
Polycarbonate	$-\!\left[\!\mathrm{O\!-\!C_6H_4\!-\!C(CH_3)_2\!-\!C_6H_4\!-\!O\!-\!C(=O)}\!\right]_n\!-$	150
Polysulfone	$-\!\left[\!\mathrm{O\!-\!C_6H_4\!-\!C(CH_3)_2\!-\!C_6H_4\!-\!O\!-\!C_6H_4\!-\!S(=O)_2\!-\!C_6H_4}\!\right]_n\!-$	190
Poly(2,6–dimethyl–1,4–phenylene oxide)	$-\!\left[\!\mathrm{C_6H_2(CH_3)_2\!-\!O}\!\right]_n\!-$	220

In the glassy state (below Tg), the only molecular motions that can occur are short-range motions of several contiguous chain segments or motions of substituent groups. These processes are called *secondary transitions*. Examples of main-chain secondary relaxation processes are two proposed versions of a crankshaft-rotation model where by several contiguous bonds are rotated around the main-chain axis, as illustrated in Fig. 2.20.

Crystalline Polymers

No polymer is completely crystalline. Even the most crystalline polymers, such as high-density polyethylene, have lattice-defect regions that contain unordered, amorphous material. Crystalline polymers may therefore exhibit both a Tg corresponding to long-range segmental motions in the amorphous regions

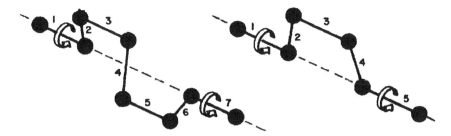

FIGURE 2.20 Proposed crankshaft secondary-relaxation motions of the polymer chain.

and a crystalline melt temperature, Tm, at which lamellae are destroyed and an amorphous, disordered melt is formed. For many polymers, Tg is approximately one-half to two-thirds of Tm (in K). Typical values of Tg and Tm for crystalline polymers are given in Table 2.9.

Table 2.9 Thermal Transitions of Crystalline Polymers

Polymer	Repeat Structure	Glass-Transition Temperature (Tg), °C	Melting Temperature (Tm), °C
Polycoprolactone	$\{O-(CH_2)_5-\overset{\overset{O}{\|}}{C}\}_n$	−60	61
Polyethylene (high density)	$\{CH_2-CH_2\}_n$	−125	135
Poly(vinylidene fluoride)	$\{CH_2-\overset{\overset{F}{\|}}{\underset{\underset{F}{\|}}{C}}\}_n$	−45	172
Poly(oxymethylene)	$\{CH_2-O\}_n$	−85	195
Poly(vinyl alcohol)	$\{CH_2-\overset{\overset{OH}{\|}}{CH}\}_n$	85	258
Poly(hexamethylene adipamide)	$\{NH-(CH_2)_6-NH-\overset{\overset{O}{\|}}{C}-(CH_2)_4-\overset{\overset{O}{\|}}{C}\}_n$	49	265
Poly(ethylene terephthalate)	$\{O-CH_2-CH_2-O-\overset{\overset{O}{\|}}{C}-\langle O \rangle-\overset{\overset{O}{\|}}{C}-O\}_n$	69	265
Polytetrafluorpethylene	$\{CF_2-CF_2\}_n$	none	327

For a given polymer, the extent of crystallization attained during melt processing depends on the rate of crystallization and the time during which melt temperatures are maintained. Above Tm, some polymers such as poly(ethylene terephthalate) and polycaprolactone exhibiting slow rates of crystallization can be quenched rapidly to an amorphous state. Other polymers having much more rapid rates of crystallization, such as polyethylene, cannot be quenched quickly enough to prevent crystallization.

Polymers differ in their rate of crystallization. In polyurethanes, the rate of crystallization depends on the crystallization temperature, as illustrated in Fig. 2.21. At Tm, crystalline lamellae are destroyed as fast as they are formed from the melt, and therefore the rate of crystallization is zero. Since, at Tg, all large-scale segmental mobility required for chain folding cease, the crystallization rate is again zero. At intermediate temperature, Tmax reaches an optimum balance between chain mobility and lamellar growth.

The measurement of Tm and Tg can be performed in a differential scanning calorimeter (DSC). This method uses individual heaters to maintain identical temperatures for a sample and a reference. The differential power needed to maintain both sample and reference is recorded as a function of temperature. Figure 2.22 shows an idealized DSC thermogram.

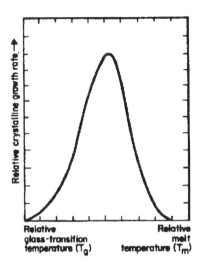

FIGURE 2.21 Crystalline growth rate depends on crystallization temperature.

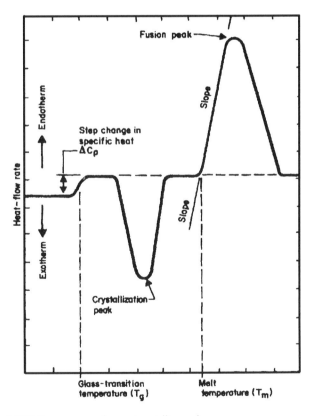

FIGURE 2.22 Idealized DSC thermogram of a semicrystalline polymer.

Segmented Polymers

One reason for the excellent physical properties displayed by polyurethane elastomers is their tendency to pack themselves into tight, stereoregular molecular chains—a phenomenon referred to as *crystallinity*. A crystalline polymer is usually the opposite of an amorphous polymer; however, it is now known that polyurethane elastomers consist of a mixture of crystalline and amorphous domains, a state described as *segmentation.*

Polymer science characterizes intermolecular order as the geometrical relationship between adjacent polymer molecules and their arrangement in the total mass of polymeric material. At present, we recognize three distinctly different states of intermolecular order: crystalline, amorphous, and segments, as shown in Fig. 2.23.

When polymer molecules are arranged in completely random, intertwined coils, the unordered structure is known as the *amorphous* state. When polymer molecules are so neatly arranged that each atom falls into precise position in a tightly packed repeating regular structure, this highly oriented structure is described as the *crystalline* state. Polyurethane elastomers are a two-phase structure, where the hard segments separate to form discrete domains in a matrix of soft segments. This arrangement is termed *segmented.* Polyurethane elastomers are segmented polymers. The rigid segments act as bridges, and as filler particles, reinforcing the soft segment matrix.

An example of this type of structure is in the segmented polyurethane elastomers obtained from polyether [poly(oxytetramethylene)] and polyester (a copolyester) based elastomers incorporating MDI and extended with ethylene diamine. In the relaxed state, spatially separated hard and soft segments can be shown (by X-ray diffraction) to exist in the material. The hard segments are considered held together in discrete domains through the action of van der Waals forces and hydrogen bonding interactions. This concept is demonstrated in Fig. 2.24, where the crystallization of soft segments is accomplished by elongating the polymer by 200%. Even greater crystallization is observed at 500% elongation as shown in Fig. 2.25.

Once crystallinity has been achieved, an additional phenomenon occurs within the polyurethane chains: hydrogen bonding. Polyurethanes contain basic electronegative ions with semiavailable unshared pairs of valence electrons, such as nitrogen and oxygen ions. Nitrogen and oxygen donate these valence electrons to the hydrogen atoms of adjacent molecules to produce hydrogen bonding between the two molecules. Hydrogen bonding between adjacent polymer chains significantly increases the physical properties of polyurethane elastomers. This gives rise to a three-dimensional "virtually cross-linked" molecular domain structure.

Interchain attractive forces between rigid segments are far greater than those present in the soft segments, due to the high concentration of polar groups and the possibility of extensive hydrogen bonding. Hard segments significantly affect mechanical properties, particularly modulus, hardness, and tear strength. The performance of elastomers at elevated temperatures is very dependent on the structure of the rigid segments and their ability to remain associated at these temperatures. Rigid segments are considered to result from contributions of the diisocyanate and chain extender components.

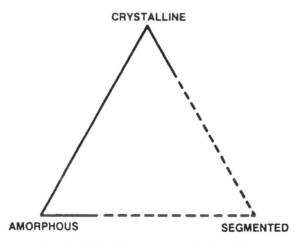

FIGURE 2.23   States of intermolecular order.

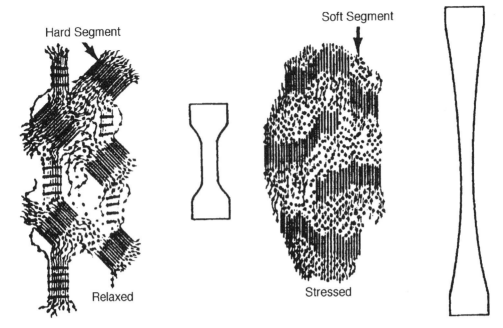

FIGURE 2.24 Strain-induced elongation of poly-ether soft segments in a segmented polyurethane elastomer by elongating to 200% elongation.

FIGURE 2.25 Segmented polyurethane elastomer at 500% extension and placed in warm water at 80° C.

The lateral effect of all the foregoing states and forces, particularly crystallinity and hydrogen bonding, is to tie together or "virtually cross-link" the linear primary polyurethane chains. That is, the primary polyurethane chains are cross-linked in effect, but not in fact. Concurrently, of course, the virtual linkages also lengthen the primary polyurethane chains. The overall consequence is a labile infinite network of polymer chains that displays the superficial properties of a strong rubbery vulcanizate over a practical range of use temperatures.

Virtual cross-linking is a phenomenon that is reversible with heat and, depending on polymer composition, with solvation, that offers many attractive processing alternatives for thermoplastic polyure-thanes. Thermal energy great enough to (reversibly) break virtual cross-links, but too low to appreciably disrupt the stronger covalent chemical bonds that link the atoms in the primary polymer chains, can be used to extrude or mold the polymers. The solvent that solvates the polymer chains, reversibly insulating the virtual cross-links, carries the primary polymer chains into solution separate and intact for such application as coatings or adhesives.

Present views on the morphology and structure of segmented polyurethane polymers are as follows:

- Because the hard and soft blocks are partly incompatible with each other, the elastomers show a two phase morphology, although there significant level of mixing of the hard and soft blocks.

- The soft segments containing the macroglycol form an amorphous matrix in which the hard segments are dispersed.

- The hard domains containing the chain extender act as multifunctional cross-link sites or "virtual crosslinks," resulting in elastomeric behavior. Hydrogen bonding can occur between hard and soft blocks although the extent to which this is responsible for physical properties is not certain.

- Hydrogen bonding occurs between individual hard blocks giving rise to a three-dimensional molecular domain structure.

- These domains may themselves be in a larger, ordered arrangement including both soft and hard blocks, the hard blocks being built up in a transverse orientation to their molecular axis leading, in cases, to the appearance of spherulites in the polymer.

- The morphology is unstable with respect to temperature and is dependent on both the chemical constitution and thermal history of the polymer.

In summary, crystallinity refers to any highly ordered arrangement of atoms. Polymers can be visualized as long meandering chains of atoms that occasionally find themselves in patterns of highly oriented, close proximity.

These small regions of high order are referred to as *spherulites* or *crystallites*. Regions containing randomly oriented and widely spaced chains are described as being *amorphous*. The degree of crystallinity is a measure of the frequency of these highly ordered crystallites. It is important to emphasize that no polymer is totally crystalline, and that an amorphous material can exhibit crystalline characteristics. In practice, polymers are classified as being amorphous, segmented, or crystalline according to their various degrees of crystallinity and their tendencies to form crystallites, as shown in Tables 2.10 and 2.11.

Table 2.10 Common Polymers Classified as Either Crystalline or Amorphous

Crystalline	Amorphous
Polyacetal	ABS
Polyamide	Polyamide
Polybutylene terephthalate	Polyacrylate
Polyethylene terephthalate	Polycarbonate
Polyetherether ketone	Polyetherimide
Polyphenylene sulfide	Polyphenylene oxide
Polyethylene	Polysulfone
Polybutylene terephthalate	
Polypropylene	
Nylon 6/6	

Table 2.11 Selected Characteristics of Crystalline and Amorphous Polymers

Crystalline	Amorphous
High strength	More pronounced glass-transition temperature
Increased stiffness	Transparency
Increased density	Reduced mold shrinkage (0.8 ± 0.4 vs. 2.0 ± 1.0)
Resistance to organic solvents	More uniform mold shrinkage
Opacity	Decreased dimensional response to temperature gradients
Resistance to dynamic fatigue	Low density
Increased temperature range with reinforcement	Good impact strength
Pronounced melting point	Melting range
Low viscosity melt	
Chemical resistance	

The probability that a given polymer will exhibit crystalline structure is determined primarily by the chemical nature of the polymer chains. Polymer chains of low molecular weight, or that possess high flexibility, favor crystallinity; e.g., polyphenylene sulfide is composed of many flexible sulfide linkages adding to its tendency toward crystallinity. Other flexible units include ether and ester linkages. Polyurethane elastomers are mixtures of crystalline and amorphous regions.

If the polymer is capable of forming intermolecular bonds, and if these bonds are advantageously distributed along the polymer chain, crystallinity is more likely. These forces include hydrogen bonding, as in polyurethane. Homopolymers present more ideal conditions for crystalline structure than random copolymers, whose chemistry will result in an uneven distribution of intramolecular forces.

Because crystallites consist of closely packed chains, it is correct to assume that polymer chains containing bulky side groups (as in polystyrene), or branching (as in low-density polyethylene), would inhibit close packing and interfere with the formation of crystallites. This is also known as *steric hindrance*.

References

1. Szycher, M. "Opportunities in Polyurethane Elastomers: Industrial and Medical Applications," Technomic Seminar, Boston, MA, June 13, 1989.
2. Gajewski, V. Chemical degradation of polyurethane, *Rubber World*, pp.15–18, September, 1990.
3. Biermann, T.F., and Markovs, R.A. *Modern Plastics Encyclopedia*, 1995, B-46, 1995.

3

Structure-Property Relations in Polyurethanes

The most specific definition of a *polyurethane* is any polymer with a plurality of carbamate (urethane) linkages. The term *polyurethane* or *polyisocyanates* are general names for the segment of the plastics industry that manufactures use. The name polyurethane is almost universally applied to the final (polymeric) products of that segment of the industry.[1]

Polyurethanes, also called *urethanes* are used to form a broad range of products including flexible and rigid foams, elastomers, coatings and adhesives. For our purposes, we will define a *polyurethane* as any of a large family of polymers, based on the reaction products of an organic isocyanate with compounds containing a hydroxyl group.

Polyurethane characteristics are controlled by the molecular structure and include degrees of flexibility/rigidity, density (foamed or solid), cellular structure, hydrophilicity or hydrophobicity, processing characteristics, and end-use properties. The processing characteristic is controlled by their basic plastic nature, i.e., whether the material is thermoplastic (linear molecular structure) or thermoset (cross-linked molecular structure). Figure 3.1 presents the universe of structure-property relationships in polyurethane polymers.

The general principles of the structure-property relationship can be summarized as follows:

1. *Molecular weight.* As the molecular weight increases some properties such as tensile strength, melting point, elongation, elasticity, glass transition temperature, etc. increase up to a limiting value and the remain constant.
2. *Intermolecular forces.* The weaker bonds such as hydrogen bonding, polarizability, dipole moments, and van der Waals forces may form, in addition to the primary chemical bonds, and these weaker bonds are affected by temperature and stress. If there is repulsion between like charges or bulky chains, or if there is high cross-link density, the effect of intermolecular forces will be reduced. Figure 3.2 shows the effect of hydrogen bonding in the hard segments of polyurethanes. Figure 3.1 is an idealized three-dimensional representation of hydrogen bonds in an aromatic polyurethane.
3. *Stiffness of chain.* Presence of aromatic rings stiffens the polymer chains and causes a high melting point, hardness and a decrease in elasticity. On the other hand, the presence of flexible bonds (such as ether bonds) favors softness, low melting point, low glass transition temperature, and elasticity.
4. *Crystallization.* Linearity and close fit of polymer chains favor crystallinity, which leads to reductions in solubility, elasticity, elongation, and flexibility, and increases in tensile strength, melting point, and hardness. Figure 3.3 presents an idealized view of crystallization in polymers, including polyurethanes (bundle crystallization). Crystallization depends on the level of stretch induced in the polymers. Figure 3.2 shows the transient morphologies of segmented polyurethanes on stretching.

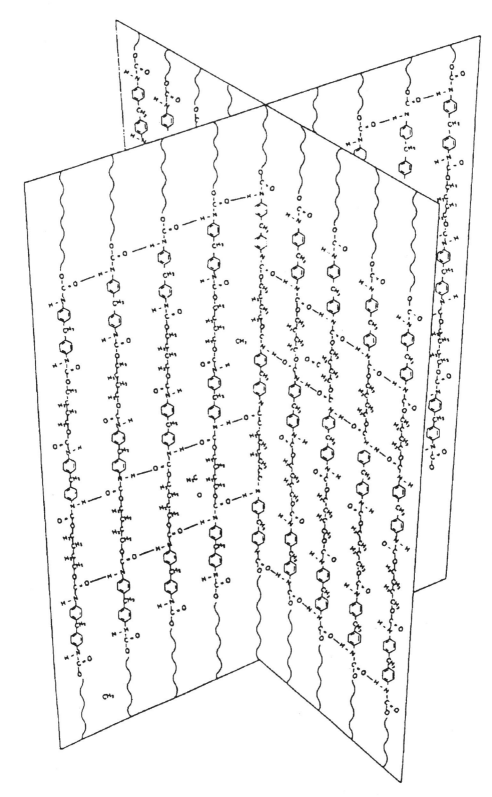

FIGURE 3.1   Three-dimensional representation of hydrogen bonds. *Source:* from R. Bonart, L. Morbitzer, and E.H. Muller, *J. Macromol. Sci. Phys.* B9(3), 447 (1974).

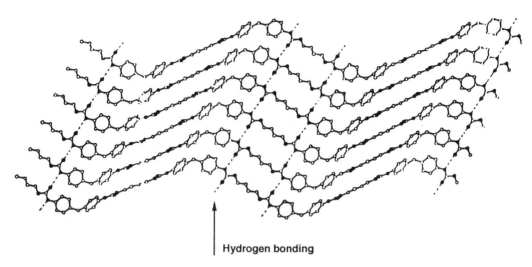

Hydrogen bonding

FIGURE 3.2   Polyurethane elastomer structure, crystallization, and hydrogen bonding.

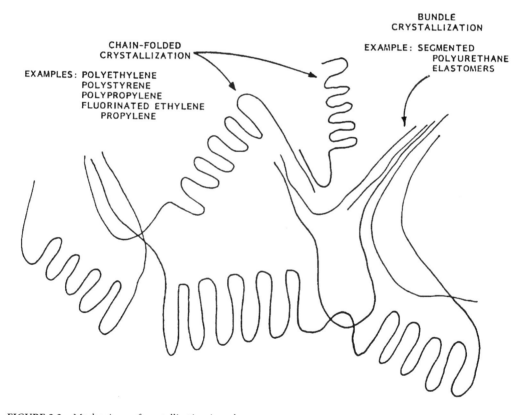

FIGURE 3.3   Mechanisms of crystallization in polymers.

5. *Cross-linking.* Increases in the degree of cross-linking cause an increase in rigidity, softening point, and modulus of elasticity for amorphous polymers, and reduces elongation and swelling by solvents.

Table 3.1 summarizes key structure-property interactions in polyurethane polymers, and the key effects related to polymer structure.

TABLE 3.1 Key Structure vs. Property Interactions

Key Effect	
Isocyanate type	Environment stability
	Softening point
	Flexibility
	Dynamic properties
	Solubility
Short chain diol type	Thermal stability
	Flexibility
	Dynamic properties
	Softening point
	Solubility
Macroglycol	Environmental stability
	Chemical stability
	Fluid swell
	Flexibility
	Tg
Weight percent hard segment	Solubility
	Flexibility
	Softening point
	Fluid swell
	Dynamic properties
Macroglycol mw	Tg
	Softening point
	Solubility
	Fluid swell
	Dynamic properties
Macroglycol mwd	Dynamic properties
Polymer molecular weight	Dynamic properties
	Softening point
NCO/OH	mw

3.1 Monomers, Reactions, and Structures

Commercially, the most attractive feature of polyurethanes is the unparalleled versatility offered by isocyanate chemistry for tailoring properties to specific end-use requirements. At room temperature (and up to 50° C), isocyanates react with hydroxyl groups to produce polyurethanes, while with amines and/or water they will produce urea linkages. At higher temperatures (up to 150° C), further reactions produce allophanate, biuret, and isocyanate linkages. Polyurethanes can be considered as mixed amide esters of carbamic acid, and thus their properties are between polyesters and polyamides.[2] This versatility has allowed polyurethanes to be utilized in a broad range of applications including automotive, architectural, aerospace, office machinery, agriculture, mining, marine, electrical, communication, footwear, and medical.

The primary reactions of isocyanates with active hydrogen compounds occur with remarkable ease at ordinary temperatures with the evolution of heat. If a linear polyurethane is desired, the application of the lowest temperature is required. If high cross-linking and branching through secondary reactions is desired, higher temperatures (100 to 150° C) are needed. At about 180° C, aromatic polyurethanes start to revert. For aliphatic polyurethanes, this reversion takes place at around 160° C. At temperatures above 250 C, all polyurethanes start to decompose. Decomposition yields free isocyanates, alcohols, free amines, olefins, and carbon dioxide.[3]

Polyisocyanates are very reactive substances, even reacting with water. As a result, the reaction between polyisocyanates and polyhydroxy compounds is complicated by the presence of moisture. The presence of water first causes formation of unstable carbamic acid, which then disintegrates into amine and carbon dioxide. The formation of gas gives rise to foam production. Further reaction between amine and isocyanate lead to the formation of urea groups.

Polyurethane elastomers can be prepared by two basic processes. The simplest and most obvious method is to mix a liquid diol, a polyol, and diisocyanate, and to cast the mixture in a mold while still liquid. Curing of the cast mixture yields an elastomeric product. To obtain a thermoplastic elastomer, the reactants should be chosen such that they produce a linear structure. This is called the *one-shot* process. Figure 3.4 presents a schematic representation of the differences between the prepolymer method and the one-shot method of polyurethane preparation. Figure 3.5 compares thermal proprieties of polyurethanes prepared by the one-shot versus prepolymer methods.

The second method involves the reaction of a linear hydroxy-terminated polymer with an excess of diisocyanate to form an isocyanate-terminated polymer called a *prepolymer*, as shown in Fig. 3.6. A prepolymer is either a viscous liquid or a low-melting solid. Figure 3.7 is a schematic representation of the equipment used for the full-scale preparation of prepolymers.

The next step is chain extension and network formation with a small molecular weight polyol or amine called a *chain extender*. If the NCO/OH ratio is greater than1, then this step is usually accompanied by

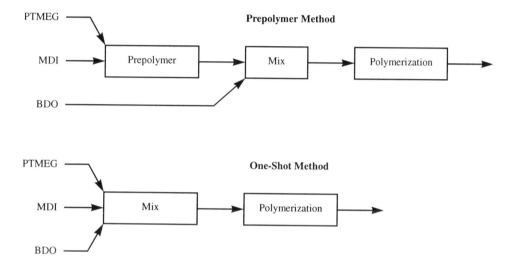

FIGURE 3.4 Prepolymer and one-shot methods of polymerization.

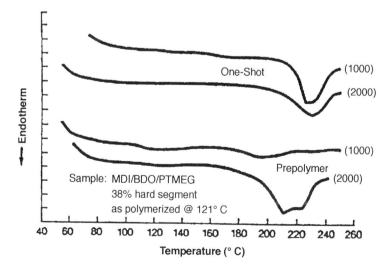

FIGURE 3.5 DSC comparing one-shot to prepolymer-based elastomers using PTMEG 1000 and 2000 MW.

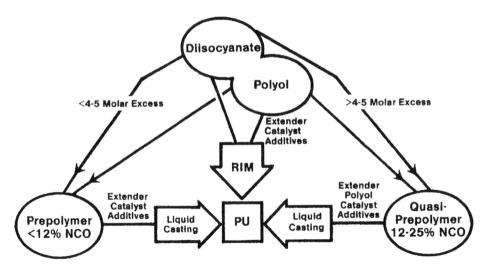

FIGURE 3.6   Preparation of polyurethane elastomers.

some allophanate and/or biuret branch point formation. Figure 3.8 shows the theoretical relationship between molecular weight and NCO/OH ratio.

The development of cross-linking in elastomers also depends on the reaction of some isocyanate groups with atmospheric moisture to form urea groups, which then react with other isocyanate groups to form biuret cross-links. To obtain the required amount of cross-linking, a more useful procedure is to use formulations with at least one component having more than two reactive end groups.

Relatively simple linear polymers may be obtained by using compounds with two active groups such as diisocyanate and diols. In each polymerization reaction, however, secondary reactions take place to a certain extent. Polyurethanes contain urea, allophanate, biuret groups, and aromatic rings in addition to urethane groups. These groups affect the properties of the resultant polymer.

The reactions between a diisocyanate, a linear long chain dial, and a low molecular weight chain extender lead to the production of elastomers. The properties of the elastomers are determined mainly by the chain structure, the degree of branching of the polymeric intermediate, and the stoichiometric balance of the components. The ratio of NCO to OH for optimum mechanical strength is usually 1.0 to 1.1. As the ratio falls below 1.0 the mechanical strength, hardness and resilience decrease and elongation and compression increase very sharply. Resultant linear polyurethanes behave as cross-linked polymers, and are termed *virtually cross-linked*, as shown in Fig. 3.9.

Useful polyurethanes are produced over a Durometer range of 55A to 75D. They can be solubilized in a variety of solvents or made resistant to solvents and other fluids. Polyurethanes can be environmentally stable or biodegradable, hydrophobic or hydrophilic. Useful service temperature ranges from below -100 to over $+125°$ C.

Polymer chemists can synthesize polyurethanes in many interesting ways. However, commercial manufacture is based primarily on the addition reactions of polyols with polyisocyanates. Diols and diisocyanates produce linear polymers. Triols, triisocyanates, and/or monomers of still higher functionality produce branched and cross-linked polymers. These reactions generally proceed readily at room temperature, particularly with the help of amine catalysts and/or organotin catalysts. This is the most general reaction for the production of polyurethanes, as shown in the following reaction:

$$\langle\!\bigcirc\!\rangle\text{--N=C=O} + \text{R-OH} \longrightarrow \langle\!\bigcirc\!\rangle\overset{\displaystyle \overset{O}{\|}}{\underset{\displaystyle \underset{H}{|}}{\text{--N--C--O--R}}}$$

Urethane formation (simple addition reaction)

isocyanate   alcohol   **urethane**

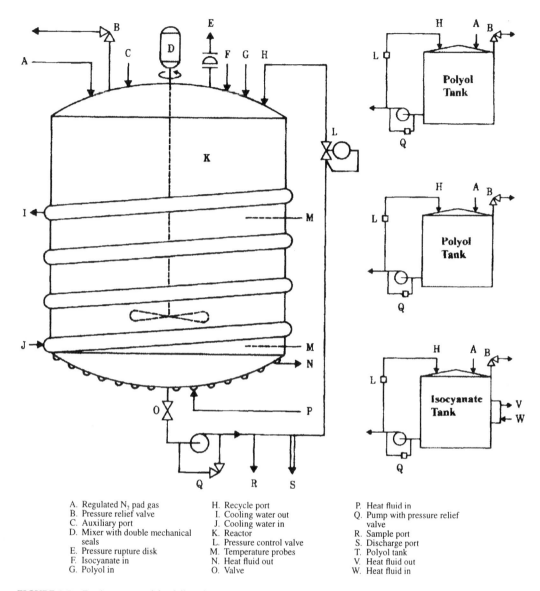

A. Regulated N$_2$ pad gas
B. Pressure relief valve
C. Auxiliary port
D. Mixer with double mechanical seals
E. Pressure rupture disk
F. Isocyanate in
G. Polyol in
H. Recycle port
I. Cooling water out
J. Cooling water in
K. Reactor
L. Pressure control valve
M. Temperature probes
N. Heat fluid out
O. Valve
P. Heat fluid in
Q. Pump with pressure relief valve
R. Sample port
S. Discharge port
T. Polyol tank
V. Heat fluid out
W. Heat fluid in

FIGURE 3.7 Equipment used for full-scale preparation of prepolymers. *Source:* Dow Plastics, Form No. 109-00895-290-SAI (1990).

The properties associated with the polyurethanes are due to the presence of urethane linkages in the molecular backbone. The urethane linkage is shown in Fig. 3.10. The type of diisocyanate, glycol, and solvent used may affect the rate and the type of the reaction as well as the properties of the product. Polyols give the high flexibility to the backbone of the network chains. Therefore they are called *soft domains* or *soft segments.* Conversely, isocyanate and chain-extender components give rigidity to the chains and are called *hard domains* or *rigid segments.* Physical and mechanical properties of polyurethane that contain these types of segments can be explained in terms of morphological structure, i.e, rigid domains dispersed in a flexible segment matrix. The two isocyanates most frequently used in the synthesis of polyurethanes are MDI and TDI. The isomeric structures of TDI are shown in Fig. 3.11. These diisocyanates are exhaustively covered in Chapter 4.

Because the hard and soft blocks are partly incompatible with each other, the elastomer show a two-phase morphology, although there is a significant level of mixing of the hard and soft blocks. At low

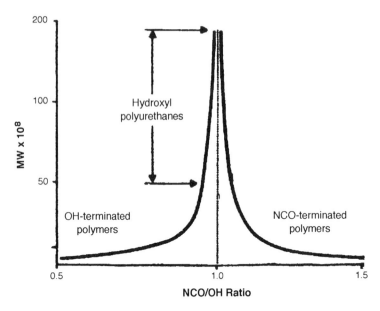

FIGURE 3.8   The theoretical relationship between molecular weight and NCO/OH ratio.

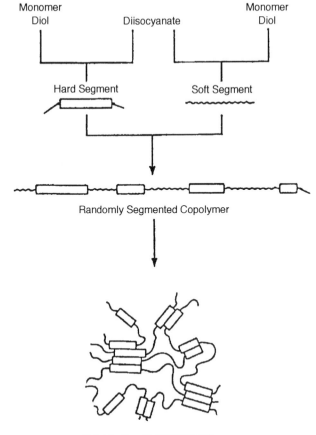

FIGURE 3.9   Virtually cross-linked network.

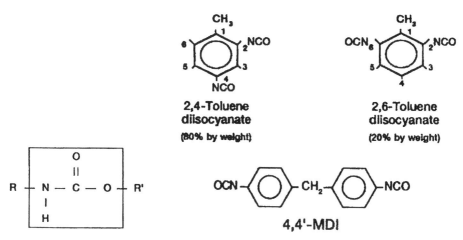

FIGURE 3.10 Urethane linkage.

FIGURE 3.11 Difunctional aromatic isocyanates.

temperatures, the soft matrix having a low Tg influences the properties. Hard segments in the domains act as cross-link points as well as reinforcing filler entities, and these govern the mechanical performance of the material at elevated temperatures.

3.1.1 Polyols

Polyol compounds used in polyurethane production are generally compounds with molecular weights in the range of 400 to 5000 Daltons. Depending on the chain length of these diols or glycol, the properties of the polyurethane changes. If the polyol has a low molecular weight, it creates hard plastics, and if it has a high molecular weight, it creates flexible elastomers. The reactivities are not the same for all hydroxyl groups. Primary alcohols react readily at 25 to 50° C, while the secondary and tertiary alcohols are about 0.3 and 0.005 times less reactive than the primary ones. Figure 3.12 presents some commercially important polyols.

The reaction between a hydroxyl group and an isocyanate is catalyzed by mild and strong bases, by many metals, and by acids. Bases such as sodium hydroxide, sodium acetate, tertiary amine, and certain metal compounds (especially tin compounds such as dibutyltin dilaurate and stannous octoate) are the most commonly used catalysts.

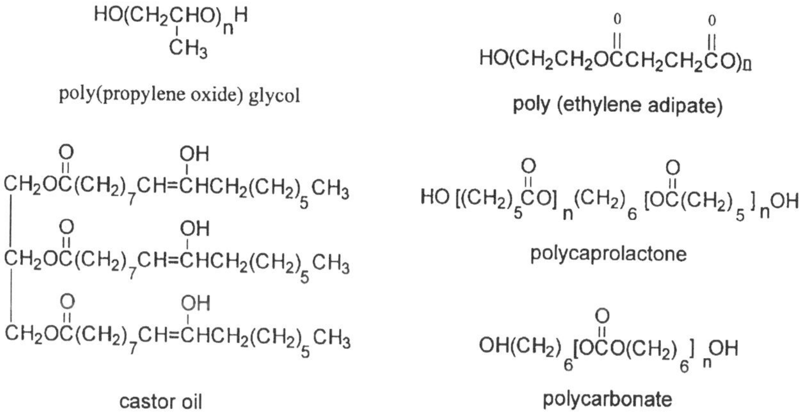

$HO(CH_2CHO)_nH$
CH_3

poly(propylene oxide) glycol

$HO(CH_2CH_2O\overset{O}{\overset{\|}{C}}CH_2CH_2\overset{O}{\overset{\|}{C}}O)_n$

poly (ethylene adipate)

$CH_2O\overset{O}{\overset{\|}{C}}(CH_2)_7CH=\overset{OH}{\overset{|}{C}}HCH_2(CH_2)_5CH_3$
$CH_2O\overset{O}{\overset{\|}{C}}(CH_2)_7CH=\overset{OH}{\overset{|}{C}}HCH_2(CH_2)_5CH_3$
$CH_2O\overset{O}{\overset{\|}{C}}(CH_2)_7CH=\overset{OH}{\overset{|}{C}}HCH_2(CH_2)_5CH_3$

castor oil

$HO[(CH_2)_5\overset{O}{\overset{\|}{C}}O]_n(CH_2)_6[O\overset{O}{\overset{\|}{C}}(CH_2)_5]_nOH$

polycaprolactone

$OH(CH_2)_6[O\overset{O}{\overset{\|}{C}}O(CH_2)_6]_nOH$

polycarbonate

FIGURE 3.12 Some important polyols.

Polyfunctional alcohols (those containing more than one hydroxyl group) are known as *polyols*. Polyols are in the production of polyurethanes. The following three families of structures are commonly used to provide the polyols for manufacture of polyurethane: polyethers, polyesters, and hydrocarbon backbones.

Polyethers

The most common polyether polyol is poly(oxypropylene) glycol made by alkaline polymerization of propylene oxide; this provides good flexibility and low cost. the molecular architecture of poly(oxypropylene glycol) is shown in Fig. 3.13.

$$HO(CH_2CHO)_nH$$
$$|$$
$$CH_3$$

FIGURE 3.13 Poly(oxypropylene glycol).

For higher strength, poly(oxytetramethylene) glycol, better known as polytetra methylene ether glycol (PTMEG) is frequently used, as shown in Fig. 3.14.

$$HO(CH_2CH_2CH_2CH_2O)_nH$$

FIGURE 3.14 Poly(oxytetramethylene glycol).

PTMEG and is made by acid polymerization of tetrahydrofuran. Polyurethane products made from PTMEG can be characterized as high-performance polymers exhibiting high engineering toughness, resiliency, high abrasion resistance, inherent hydrolysis resistance, and superior low-temperature properties, when compared to other polyether-based polyols.[4]

Polyesters

Linear aliphatic polyesters are commonly structures such as poly(ethylene adipate) shown in Fig. 3.15.

$$\overset{O}{\overset{||}{}} \qquad \overset{O}{\overset{||}{}}$$
$$HO(CH_2CH_2OCCH_2CH_2CH_2CH_2CO)\underline{n}CH_2CH_2OH$$

FIGURE 3.15 Poly(ethylene adipate).

These polyols are used to provide high strength to PU elastomers. Polyester-based urethanes thermal behavior is dependent upon the concentration of ester groups on the polyester. An increase in ester group concentration leads to reduced flexibility at low temperatures, higher hardness, higher modulus, and a marked increase in permanent elongation. Conversely, reducing the ester group concentration improves the low temperature flexibility and reduces the tear strength. Glass-transition temperatures of polymers prepared with poly(1,4-butylene adipate) and poly (1,5-pentylene adipate) are significantly lower than that of the poly(ethylene adipate) elastomer.

Table 3.2 shows the effect of isocyanate structure on the properties of polyester urethanes that are prepared by using 0.1 equivalents polyethylene adipate mol. wt 2000, 0.32 equivalents diisocyanate, and 0.2 equivalents, 1,4 butanediol chain extender.[5]

Polyester structure also affects the properties of the product. For a special polyurethane series prepared from diphenylmethane diisocyanate, 1,4-butanediol (as chain extender) and different polyesters, the properties can be summarized as shown in Table 3.3.

For improved resistance to hydrolysis, polyester polyols such as polycaprolactone or polycarbonate are use; the structures of these polyol is shown in Fig. 3.16.

A naturally-occurring polyester polyol often used in polyurethane is castor oil, shown in Fig. 3.17.

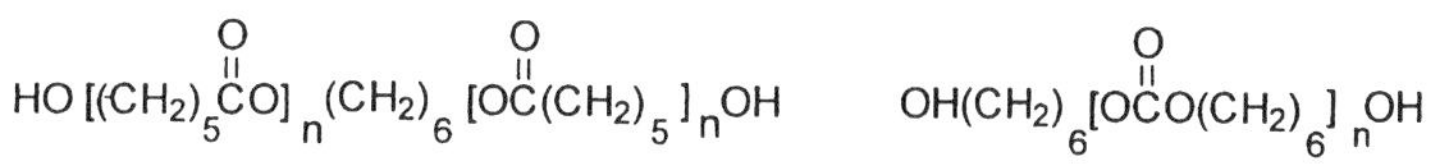

$$HO\,[(CH_2)_5\overset{O}{\overset{||}{C}}O]_n(CH_2)_6\,[O\overset{O}{\overset{||}{C}}(CH_2)_5\,]_n OH \qquad OH(CH_2)_6[O\overset{O}{\overset{||}{C}}O(CH_2)_6\,]_n OH$$

FIGURE 3.16 Polycaprolactone (left) and polycarbonate (right).

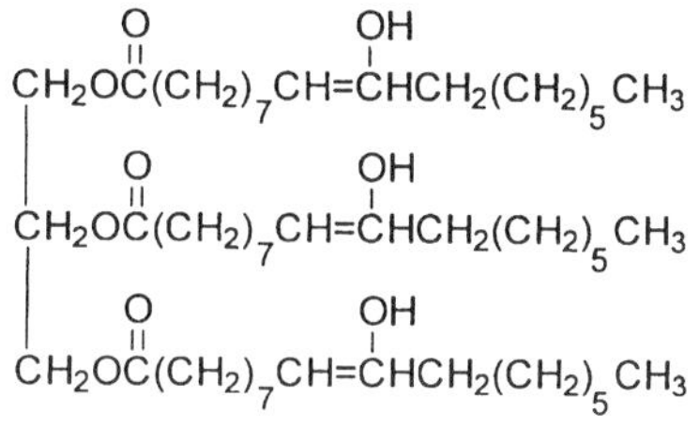

$$CH_2O\overset{O}{\overset{||}{C}}(CH_2)_7CH=\overset{OH}{\overset{|}{C}}HCH_2(CH_2)_5CH_3$$
$$|\quad O \qquad\quad OH$$
$$CH_2O\overset{O}{\overset{||}{C}}(CH_2)_7CH=\overset{OH}{\overset{|}{C}}HCH_2(CH_2)_5CH_3$$
$$|\quad O \qquad\quad OH$$
$$CH_2O\overset{O}{\overset{||}{C}}(CH_2)_7CH=\overset{OH}{\overset{|}{C}}HCH_2(CH_2)_5CH_3$$

FIGURE 3.17 Structure of castor oil.

TABLE 3.2 Effect of Isocyanate Structure on Polyester-based Polyurethanes

Diisocyanate	Tensile Strength (psi)	Elongation (%)	Modulus at 300% (psi)	Tear Strength (lb/in)	Hardness (Shore B)
NDI	4300	500	3000	200	80
p-PDI	6400	600	2300	300	72
TDI	4600	600	350	150	40
MDI	7900	600	1600	270	61
TODI	4000	400	2300	180	70
DMDI	5300	500	600	40	47
DPDI	3500	700	300	90	56

TABLE 3.3 Influence of Polyester Structure on Polyurethane Properties

Polyester	Tensile Strength (psi)	Elongation (%)	Tear Strength (lb/in)	Hardness Shore B
Poly (ethylene adipate)	6900	590	240	60
Poly (1,4 butylene adipate)	6000	510	280	70
Poly (1,3 butylene adipate)	3200	520	100	58
Poly (1,5 pentylene adipate)	6300	450	60	60
Poly (ethylene succinate)	6800	420	200	75
Poly (2,3 butylene succinate)	3500	380	ND	85

Hydrocarbons

For lower polarity, better electrical insulation, and higher resistance to hydrolysis, aliphatic hydrocarbon polyols are sometimes synthesized in structures such as hydroxy-terminated polybutadiene glycols,

$$HO(CH_2CH=CHCH_2)OH$$

and their hydrogenated derivatives,

$$HO(CH2CH2CH2CH2)n$$

Triols and Higher Functionality Polyols

Polyols with three hydroxyl groups, such as glycerine, are known as *trihydroxyl polyols* or *triols*. When triols are reacted with an isocyanate, the resulting polyurethane is cross-linked. The amount of cross-linking affects the stiffness of the polymer. If a rigid foam is required, the polymer structure must be highly cross-linked; for flexible foams, less cross-linking is needed, as shown in Fig. 3.18a.

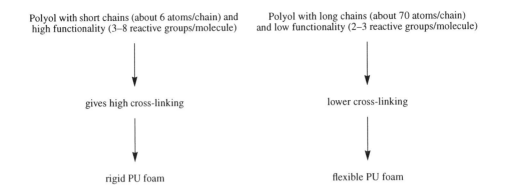

FIGURE 3.18a   Cross-linking requirements

Branching and cross-linking are most commonly accomplished by use of higher-functionality polyols. These are most conveniently based upon a triol or hexol backbone as shown in Fig. 3.18b.

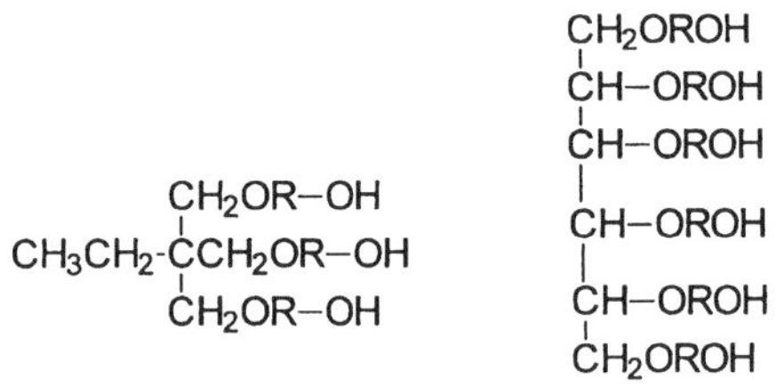

$$
\begin{array}{cc}
\text{CH}_2\text{OR-OH} & \text{CH}_2\text{OROH} \\
\text{CH}_3\text{CH}_2\text{-CCH}_2\text{OR-OH} & \text{CH-OROH} \\
\text{CH}_2\text{OR-OH} & \text{CH-OROH} \\
 & \text{CH-OROH} \\
 & \text{CH-OROH} \\
 & \text{CH}_2\text{OROH}
\end{array}
$$

FIGURE 3.18b   Triol (left) and hexol (right).

3.1.2   Isocyanates

The most common diisocyanates are toluene diisocyanate (TDI) and methylene bis diphenylisocyanate (MDI), plus higher oligomers for increased functionality and cross-linking. For resistance to ultraviolet light and outdoor weathering resistance, aliphatic polyisocyanates such as hexamethylene diisocyanate (HDI) and hydrogenated MDI (HMDI) are useful, even though these aliphatic diisocyanates involve lower polymerization reactivity and much higher cost.

The reactions between a diisocyanate, a linear long chain polyol, and a low-molecular-weight chain extender lead to the production of elastomers. The properties of the elastomers arc determine mainly by the chain structure, the degree of branching of the polymeric intermediate, and the stoichiometric balance of the components. The ratio of NCO to OH for optimum mechanical strength is usually 1.0 to 1.1. As the ratio falls below 1.0 the mechanical strength, hardness and resilience decrease, and elongation and compression set increase very sharply.

Reactivity of isocyanates depends on their chemical structures. Aromatic isocyanates are generally more reactive than aliphatic ones. The presence of electron withdrawing substituents on the isocyanate molecule increases the partial positive charge on the isocyanate carbon and moves the negative charge farther away from the site of reaction. This makes the transfer of the electron from the donor substance to the carbon easier, thus causing a faster reaction. On the other hand, the presence of electron-donating substituents on isocyanate compounds can cause slower reactions. Bulky groups in the ortho positions of aromatic isocyanates, and bulky and branched groups in aliphatic isocyanates, retard the reaction because of strike hindrance. In the reactions of diisocyanate, the reactivity of the second isocyanate decreases significantly after the first has been reacted. The difference in the reactivities is less if the two isocyanates belong to different aromatic rings or separated with an aliphatic chain.

Prepolymers

Mixing all the reactants at once—a one-shot process—provides the fastest, simplest, and most economical manufacturing technique, and is most often favored in highly competitive commodity fields such as flexible foam. Conversely, two-step or three-step processes give the manufacturer much greater control over toxicity, reactivity, structure, properties, processibility, and finished product quality. In a typical two-step process, the first step is production of a prepolymer, as shown in Fig. 3.19.

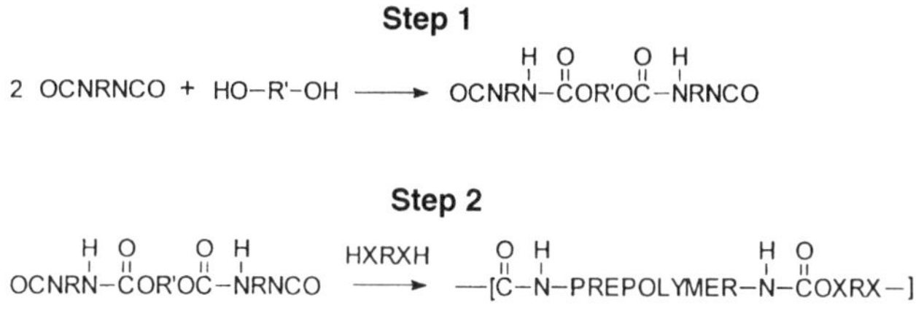

Step 1

$$2 \ OCNRNCO + HO\text{–}R'\text{–}OH \longrightarrow OCNRN\overset{H}{\underset{}{}}\overset{O}{\underset{}{}}COR'OC\overset{O}{\underset{}{}}\overset{H}{\underset{}{}}NRNCO$$

Step 2

$$OCNRN\overset{H\ O}{\underset{}{}}COR'OC\overset{O\ H}{\underset{}{}}NRNCO \ \xrightarrow{HXRXH} \ \text{—}[\overset{O\ H}{\underset{}{}}C\text{–}N\text{–}PREPOLYMER\text{–}N\overset{H\ O}{\underset{}{}}COXRX\text{—}]$$

FIGURE 3.19 Typical prepolymer production and subsequent reaction. Step 2 is the reaction of the prepolymer with a chain extender.

Prepolymers (isocyanate intermediates between monomers and the final polymer) are the reaction product of a polyol or blend of polyols with *excess* isocyanate. A *prepolymer* describes a polyol-isocyanate adduct with a free isocyanate content of 1 to 15 percent by weight. Prepolymers are made by the slow addition of the polyol or blends of polyols to the isocyanate at a controlled temperature, usually 60 to 70° F, followed by reaction to a constant free isocyanate content. Catalysts may be used to accelerate the reaction.

Prepolymers are useful in the production of urethane elastomers, coatings, adhesives, sealants, reaction injection molding, binders, etc., where the lower reaction exotherm produced when using a prepolymer is a crucial advantage. Since the reaction between a hydroxyl group (-OH) and an isocyanate (-NCO) to form a urethane linkage develops approximately 25 kcals/mole.[6] The use of a prepolymer is distinctly advantageous to lower the overall reaction exotherm.

The last step is a reaction of the prepolymer with a *chain extender* such as a diol or a diamine to produce high molecular weight polymers. In many cases, controlled cross-linking is introduced as the second or third step.

Quasiprepolymers

A quasiprepolymer is the reaction product of a polyol or blend of polyols with a large excess of isocyanate. The term quasi-prepolymer is used to describe a polyol-isocyanate adduct with free isocyanate contents between 16 and 32 percent by weight. Quasiprepolymers are extensively utilized in the production of foams. It is important to distinguish a prepolymer from a quasiprepolymer; this is done primarily by observing the nomenclature regarding free isocyanate contents as follows:

	Prepolymer	Quasiprepolymer
Percent free isocyanate	1 to 14	16 to 32
	Elastomers, coatings Sealants, adhesives	Foams

3.1.3 Chain Extenders and Cross-Linking Agents

For stepwise extension (cure) reactions, the three most common additives are 1,4-butanediol, diamines, and water. Thus, these additives function as chain-extender and/or cross-linking agents, depending on the functionality of the prepolymer and the overall stoichiometry of the reaction.

In *vulcanizing* polyurethane elastomers, rubber compounders also use conventional sulfur cure, producing sulfide and polysulfide links between polyurethane chains, and sometimes peroxide cure, producing direct C–C links between the polyurethane chains.

Indigenous Cross-Linking Reactions

Aside from higher-functionality polyols, polyisocyanates, and/or polyamines, there are several reactions that can contribute significant cross-linking during polymerization and cure of polyurethane. There are three indigenous cross-linking reactions: allophanate, biuret, and isocyanurates.

Allophanate
When a polyurethane is formed in the presence of excess polyisocyanate, the urethane group itself can supply an active hydrogen to react with the isocyanate, thus forming a branch point. Obviously, a diisocyanate could similarly form a cross-link between two vicinal polyurethane chains. These cross-links are not as stable as the conventional cross-links formed from polyfunctional polyols and polyisocyanates; allophanate cross-links are thermally labile and open quite easily at higher temperature.

Biuret
When polyurea is formed in the presence of excess polyisocyanate, the urea group can supply active hydrogens to react with the isocyanate, thus forming a branch point. Obviously, a diisocyanate could similarly form a cross-link between two polymer chains. These cross-links form more readily than allophanates and are somewhat stabler than allophanates, but they are still thermally labile and open fairly easily at higher temperatures.

Isocyanurate
Under proper conditions, excess isocyanate forms cyclic timers that are called isocyanurates. When the isocyanate is a diisocyanate or higher polyisocyanate, these isocyanurate rings act as extremely stable cross-links in the polyurethane network. Isocyanaurates are frequently used to produce polyurethane of high heat stability and flame retardance.

Block Copolymers
When small comonomer units are assembled randomly into a polyurethane molecule:

$$\sim\sim\sim\sim\sim ABAABABBBAAABBAB\sim\sim\sim\sim\sim$$

the resulting random copolymer has an overall average structure that is fairly uniform and forms a single homogeneous phase containing this average composition and structure. When the growth of a copolymer molecule produces fairly large areas (blocks) of one monomer structure alternating with fairly large areas (blocks) or another monomer structure:

$$\sim AAAABBBB\sim$$

these blocks will tend to separate into microphases or *domains*, and each type of domain will contribute independently to the properties of the block copolymer. In polyurethane, the polyol generally forms fairly large blocks even before they are reacted with the polyisocyanate. Furthermore, in steps synthesis of polyurethane, the first-stage prepolymer forms one type of block (often called the *soft* block), while the reaction of short chain-extender with isocyanate forms another type of block (often called the *hard* block). Thus, polyurethane are block copolymers. In many cases, the separation of these blocks into domains has a major synergistic effect on the properties of the resultant polymer.

3.1.4   Hetero-Block Copolymers

The stepwise synthesis of polyurethane, with active hydroxyl, amine, and/or isocyanate end-groups remaining after each intermediate step, provides the polyurethane chemist with the additional possibility of combining these polyurethane blocks with blocks of other polymer structures, to unite the best properties of polyurethane with the best properties of the other polymers as well. Some of the most common may be described briefly as follows:

- *Acrylic esters.* Acrylic esters that contain hydroxyl groups are readily combined with polyurethane that contain isocyanate groups to produce acrylic-urethane block copolymers.

- *Epoxy resins.* Some epoxy resins contain hydroxyl groups, and most of them form hydroxyl groups during cure. These react readily with excess isocyanate groups of polyurethane to form block copolymers. Conversely, polyurethane containing terminal hydroxyl or amine groups can act as curing agents for epoxy resins, again forming block copolymers.
- *Drying oils.* Drying oils are often transesterified with glycerol to form mono- and di-glycerides, containing both drying oil groups and hydroxyl groups. These copolymerize readily in the presence of catalysts with excess isocyanate groups in polyurethane to form drying-oil urethanes, the most common form of polyurethane coatings.
- *Silicones.* Cure of silicone elastomers and resins proceeds through a transient intermediate silanol stage. At this point, the silanol can react with the hydroxyl and isocyanate end-groups of polyurethane to form silicone-urethane hetero-block copolymers, which combine some of the best properties of each polymer.

Thermal analysis comprises a group of techniques in which a physical property of a substance is measured as a function of temperature while the substance is subjected to a controlled-temperature program. A complete modern thermal analysis instrument measures temperature and energy of transitions, weight loss, dimensional changes, and viscoelastic properties.

In differential thermal analysis (DTA), the temperatures of the sample and a thermally inert reference material are measured as a function of temperature. Any transition that the sample undergoes will result in liberation or absorption of energy by the sample, with a corresponding deviation of its temperature from that of the reference. This tells us whether the transition is endothermic or exothermic.

In differential scanning calorimetry (DSC), the sample and a reference material are subjected to a closely controlled, programmed temperature variation. In case of a transition in the sample, thermal energy is added or subtracted from the sample or reference container to keep both at the same temperature. Recording of this balancing energy yields a direct calorimetric measurement of the transition energy.

Thermogravimetric analysis (TGA) provides a quantitative measurement of any weight change associated with a transition.

Thermomechanical analysis (TMA) provides measurements of penetration, expansion, contraction, and extension of materials as a function of temperature.

Dynamic mechanical analysis (DMA) detects transitions associated with movement of polymer chains. The technique involves measuring tic resonant frequency and mechanical damping of a material forced to flex at a selected amplitude. Mechanical damping is the amount of energy dissipated by the sample as it oscillates, while the resonant frequency defines Young's modulus of stiffness. Loss modulus and the ratio of loss modulus to elastic modulus can be calculated from the raw frequency and damping data. In general, modulus and frequency, as well as damping, change more dramatically than heat capacity of thermal expansion during secondary transitions For example, dynamic mechanical analysis is helpful in determining the effectiveness of reinforcing agents and fillers used in thermoses resins.

Since polyurethanes are made up of three types of ingredients (isocyanates, polyhydroxy compounds, and chain extenders) that can be selected from a wide variety of chemicals, generalization of their thermal properties is not a simple task. Still, three endothermic peaks, one for the glass transition temperature of the soft segments and two for the dissociation of the short- and long-range order of the hard segments, can be observed in most polyurethane. For a two-phase block copolymer, tic width of the soft-phase glass transition zone provides a qualitative measure of soft-phase homogeneity. At higher temperatures, melting of the hard-segment microcrystallites may also be observed. In the DSC of a MDI, 1,4 butanediol and polytetramethyleneoxide (mol. wt 1000) polyurethane these four peaks were observed at 35, 71, 176, and 193° C, respectively. However, the lack of high-temperature endotherms related to hard segment ordering are commonly reported for MDI-based polyurethane block copolymers, especially when asymmetric chain-extender diols are used.

The chemical structure and the linkages present in polyurethane have a significant effect on thermal behavior. If diamines are used as chain extender, urea linkages will form rather than urethane linkages, which leads to an increase in hydrogen bonding and results in stronger resistance to thermal disruption. The hydrogen bonding promotes greater cohesiveness of the hard-segment domains. In DMA analysis of polyurethaneimides, an increase in the storage modulus *E* was observed with an increase in the fraction of the hard phase.

All of this polyurethane chemistry produces a broad spectrum of polymer structures that contribute many interesting and useful properties for many fields of technology. On the one hand, study of polyurethane structure and properties has contributed greatly to our understanding of polymer structure-property relationships in general. On the other hand, systematic organization of our understanding of polymer structure-property relationships in general contributes to a sounder understanding of the specific practical relationships in polyurethane polymers and end-products. These are best reviewed from smallest to largest structural feature, starting with submolecular structure and continuing on to larger and larger structural features such as molecular weight, molecular flexibility, intermolecular order, intermolecular bonding, and finally supermolecular structures.

3.1.5 Submolecular Structure: Atoms and Functional Groups

Polyurethane may contain aliphatic, aromatic, ether, ester, carbonate, urethane, hydroxyl, amine, urea, biuret, isocyanate, and isocyanurate groups during processing and/or use of the finished product. They may also contain plasticizers and other additives along with variable amounts of water. All of these contribute individually to many properties. A brief review of some of the more important effects is instructive.

Processibility

Polyurethanes are the leading prototype of reactive processing, including all their foams and reaction injection molding (RIM), along with most of their elastomer, castings, coatings, and adhesives. These are high-speed processes based on the high reactivity of aromatic isocyanates. Reactions are fastest with aliphatic amine, slower with aromatic amine. Reactions with polyols and water are fairly competitive, depending on the choice of catalyst (amine favor water, organotins favor hydroxyl groups). Primary hydroxyl groups are more reactive than secondary, whereas aromatic hydroxyls (phenols) are slower and more likely to revert on heating. Reactions with urea and urethane groups are more difficult, but they can form biuret and allophanate cross-links at higher temperature. Use of aliphatic isocyanates in place of aromatic isocyanates, for UV and weather resistance, generally requires considerably more time, temperature, and a higher concentration of catalysts to obtain full polymerization.

Moisture is a ubiquitous impurity that causes hydrolytic waste of isocyanates, unbalanced stoichiometry, and poor properties in end products. Thus, all ingredients and process conditions must be scrupulously dry, and most processors routinely add several percent excess isocyanate to account for all wasteful side reactions.

Thermal reversion is most noticeable when urethane groups are formed from phenols, somewhat less with allophanate, and still less with biuret, but significant in all these structures. Even normal urethane groups can open at higher process temperature. Such reversion can actually be useful when the rubber processor wants to recycle lightly thermoses scrap.

The N–H groups in urethane and urea structures are still reactive enough for derivatization, grafting, and cross-linking reactions. These are used in isocyanate and biuret cross-linking, epoxy copolymerization, heparinization for antithrombotic *in vivo* plastics, and controlled release of drugs.

Mechanical Properties

Mechanical properties are most open controlled by molecular flexibility, crystallinity, and cross-linking. It is possible to soften polyurethane elastomers by addition of compatible plasticizers such as isodecyl pelargonate (Table 3.4), similarly to conventional PVC technology; but such techniques have not been of commercial importance. Water absorption provides plasticization here as it does in nylons but, since most polyurethane are already flexible, the added plasticization is not usually significant, except in medical applications, where the softening effect of water can provide advantageous properties.

TABLE 3.4 Isodecyl Pelargonate Plasticizer in Polyether PU elastomers[7]

IDP content (%)	Glass Transition Temperature (°C)
0	−52.5
5	−57.5
10	−61.5
15	−63.5
20	−66.5

The mechanical properties of a material are concerned with the effects of stress on the material. Materials can react in a number of various ways to an applied stress. Laboratory tests for measuring mechanical properties generally are

1. *Static tests.* Loads are applied slowly enough that quasistatic equilibrium of forces is maintained.
2. *Cyclic tests.* Loads are applied and then (a) partly or wholly removed or (b) reversed a sufficient number of times to cause the material to behave differently than under static loading.
3. *Impact tests.* Loading is applied so rapidly that the material must absorb energy rather than resist a force.

The simplest mechanical test is the tensile test, generally carried out with an Instron tensile tester, in which the specimen is gradually elongated under an applied stress, and the resulting changes in length are recorded. The stress is the force applied per unit cross-sectional area of specimen. The strain is the fractional change in the length of the specimen. It can be measured as linear strain, ε, which is the change of length per unit length, or as shearing strain, λ, which measures the change in angle from the original right angle. In a tensile test, a specimen is subjected to a progressively increasing tensile force until it fractures. At the beginning, the test material extends elastically. The strain is directly proportional to the stress, and the specimen returns to its original length immediately on the removal of the stress. In this region, loads are not great enough to cause permanent shifting between the atoms. Beyond the elastic limit, the applied stress produces plastic deformation, so a permanent extension remains after the removal of the applied load. The material either fractures or undergoes some change in shape due to flow of the material. In the elastic region the stress and strain are linearly proportional to each other. Stress is called *normal* stress or, when it acts perpendicular to a given area and the slope, E, of the line defined as $E = \delta/\varepsilon$ where E represents Young's modulus. When a shear stress that acts in a plane other than normal to the plane is applied, the relation between the shear stress, τ, and shear strain, γ, is given as $G = \tau/\gamma$, where G is the shear modulus.

The amount of deformation in any direction depends on the magnitude and direction of the loading, and on the condition of the material. When the stress reaches a critical value, the atomic bonds fail across an atomic plane. The relationship between stress and strain departs from linearity, and plastic deformation begins accompanied by a reduction in the cross section known as *necking*. As deformation continues, the applied load increases until the tensile strength or ultimate strength is reached. This strength is the maximum point shown on the stress-strain curve. Polymers have a wide spectrum of mechanical properties ranging from hard brittle materials to gel-like structures, from flexible elastomers to tough materials, from porous foam structures to nonporous rigid materials. The applied force may cause fractures on the material. The fractures form either under constant-stress (*creep*) or fluctuating-stress (*fatigue)* conditions.

Creep tests are carried under certain combinations of stress and temperature. All materials, when subjected to a constant stress, will exhibit an increase of strain with time. This phenomenon is called *creep*. Most materials creep to a certain extent at all temperatures. High temperatures lead to a rapid creep, which is often accompanied by microstructural changes.

Fatigue test is the measurement of the failure of a structure under the repeated application of a constant stress smaller than that required to cause failure in one application. The material initially suffers some microstructural damage. Eventually, the cyclic application of load leads to the formation of cracks, which grow larger with every application of load. A series of specimens of the material are tested to failure by application of different values of stress. Properties such as fatigue lifetime, number of cycles to failure after crack initiation, permanent set, and total elongation are also measured. Generally, logarithms of the stress are plotted versus the logarithms of the number of cycles to failure.

Impact test measures the brittleness of the material. In this test, a standard notch is cut in a standard test specimen, which is then struck under impact conditions by a heavy weight forming the end of a pendulum. The notch serves to introduce triaxial tensile stresses into the specimen, encouraging brittle failure to occur. The weight is released from a known height so as to strike the specimen on the side opposite the notch and to induce tensile stresses in it. After breaking the specimen, the pendulum swings on, and the height to which it rises on the other side is measured. Thus, the energy absorbed in breaking the material under high-speed loading can be determined and, if it is low, the specimen is called *brittle*.

The techniques of continuous and intermittent stress relaxation measurement have been used successfully for studying the thermal behavior of elastomers. In these tests, with continuous stress relaxation, a strip specimen of an elastomer is held at a fixed extension while it is being exposed to a constant elevated temperature. The tensile force or stress at constant extension and temperature is monitored continuously. If the temperature is high enough to induce random thermal session of the elastomers' network chains, the measured force of stress will decrease with time at a rate that is proportional to the rate of chain session. Such a stress decay process is often referred to as a *chemorheological* or simply a *chemical stress relaxation*.[14]

Continuous stress relaxation data are often fitted to a simple exponential Maxwell model, expressed as

$$f_t = f_0 e^{(9 - t/\tau)}, \qquad E_\tau = E_0 e^{(-t/\tau)}$$

where f_0 is the initial tensile force measured at constant extension and constant temperature, and τ is the tensile force measured after some relaxation time t. In the second expression, E_0 and E_τ are relaxation modulus values initially and at time t. τ is a characteristic relaxation time constant. Stress relaxation technique measures the rate of network chain session but does not give information about the rate of reversible reactions. Recombination of thermally cleaved chains might occur in a relaxed condition. Therefore, continuous relaxation measurements are often supplemented by intermittent relaxation measurements. The specimen is maintained in a stretched position at widely spaced intervals of time, and the resulting stress is rapidly measured, after which the strip is immediately returned to its unstretched state.

The mechanical properties of polyurethanes can be controlled by using components of different chemical structure at appropriate molar ratios. The final reformational properties of segmented polyurethane rubber result from the combination of segment flexibility, cross-linking, chain entanglement, orientation of segments, rigidity of aromatic units, hydrogen bonding, and other van der Waal's forces. These parameters affect the applied force and lead to different types of deformations. There is a correlation between morphological structure and stress-strain behavior. The modulus decreases with decreasing hard segment content in poly(ethylene oxide)-based elastomers extended with either ethylenediamine or phenylenediamine. Such trends are explained in terms of greater degree of domain-formation with increasing hard segment content. Decrease in hard segment content causes a decrease in modulus but, in some cases where the hard content value is less than 10%, crystallization of soft segments may occur and create an unexpectedly high modulus.

Temperature affects the tensile properties and demonstrates a decrease of stress at a given strain with increasing temperature. This negative temperature coefficient of stress can be explained in terms of the viscoelastic softening of hard segment domains, resulting in a decrease in effective physical cross-links.

Due to dissimilarity in chemical structure of the hard and soft segments, there is a thermodynamic incompatibility between these segments. There is a driving force that causes them to form separate phases. This effect is, however, limited due to the presence of covalent links. It forms domains or leads to microphase separation.

Experiments carried out with small-angle X-ray scattering showed that microphase separation for segmented poly(urethane urea) is improved with an increase of average molecular weight of poly(tetramethylene glycol). Some processing parameters of the casting method also affect the morphology. The temperature of the prepolymer synthesis as well as the temperature of the mold, strongly influence the phase segregation. Generally, the spherulites are bigger if the temperature of the mold is higher.[8]

Polyurethane, especially the linear segmented polyurethanes, show significant time-dependent changes in their physical properties. Time dependency is observed in stress-strain behavior. For the samples annealed at certain temperature and then rapidly cooled, it was observed that the Young's modulus value was much lower than before annealing. With increase in time after annealing, Young's modulus increases, but complete recovery is not observed.[9]

Stress-relaxation measurement is a practical method for the examination of viscoelasticity of the polymer. Processes of breaking and rebuilding polymer network as well as changes occurring in molecular structures can be examined by this method. In stress relaxation processes, many reactions may happen. Some of these reactions are

1. Breaking of urethane or allophanate-type linkages
2. Disruption of hydrogen bonds or other secondary bonds
3. A decrease in the number of free entanglements

During stress-relaxation, it has been observed that cleavage of polymer chain occurs most readily in groups having nitrogen atoms. When the length of the diols increase, it is observed that relaxation speed increases. When cross-linking density and hard-segment content increase, relative relaxation speed decreases with a parallel influence on the disappearance of viscoelastic properties.[10]

The molecular weight of the macroglycol is another parameter that affects the properties of the polyurethane. For example, for a sample prepared with polyoxyethyleneglycol, MDI and 1,5-pentanediol, the molecular weight of polyether was changed between 600 and 2000. With longer polyether chains, the system was more flexible and more hydrophilic, and had lower elastic modulus. Percent water absorption was increased from 9.9 to 62.4, with an increase in the molecular weight of polyether from 600 to 2000. For the same polyurethane, experiments to measure shear modulus were carried out in saline solution at 37° C, where relaxed and unrelaxed modulus showed an increase from 0.33 to 0.92 GPa and 0.57 to 2.58 GPa, respectively. An increase in shear modulus values with a decrease in the molecular weight of the polymers was observed.[11]

Like the isocyanates, polyols have an important effect on the properties of the polyurethane. The mechanical properties of polyurethane prepared from different polyols as the backbone for polyurethane are given below. The preparation mixture consisted of two moles of DMI, 1 mol. backbone diol, and 1.02 moles of either 1,3-propane diol or 1,butane diol as chain extender[12] as shown in Table 3.5.

TABLE 3.5 Effect of Molecular Weight and Diol Structure on MDI-MOCA Polymers

Polyol	Polyethylene polypropylene adipate (MW 1040)	Polycaprolactone (MW 1050)	Poly (tretramethylene etherglycol) (MW 808)	Poly (trimethylene glycol) (MW 658)
Tensile strength (psi)	5650	6410	5840	3875
100% modulus (psi)	705	655	765	780
Ultimate Elongation (%)	1035	1000	930	830
Hardness, Shore A	81	81	82	83
Melt temp (° C)	200	197	172	140

Changing the backbone from polyester to polyether glycol leads to different properties.[13] Such differences are shown in Table 3.6 for polyethylene adipate and polypropylene glycol urethane. For the given set, it is observed that polyester urethane is far superior to its polyether counterpart of the same molecular weight. However, reducing the molecular weight of the polypropylene glycol improved the properties, although the tensile strength was still greatly inferior to the polyester compound.

A more detailed study on the effect of the polyether polyurethane molecular weight on the mechanical properties involved polyurethane prepared from a range of polypropylene glycol of different molecular

TABLE 3.6 Effect of Molecular Weight and Diol Structure on MDI-MOC Polymers

Diol	Tensile Strength (psi)	100% Modulus (psi)	Elongation (%)	Hardness Shore A
Polypropylene glycol				
MW2000	148	650	560	83
MW1500	2400	1060	690	88
Polyethylene adipate				
MW2000	5050	880	930	88

weights, 2,4-toluene diisocyanate and 4,4'-methylene bis(o-chloroaniline) (MOCA).[14] As observed in the previous example, a decrease in the molecular weight leads to increases in the tensile and tear strengths, and in the modulus. The results are tabulated in Table 3.7.

TABLE 3.7 Effect of Molecular Weight on Properties of MDI-MOCA Polyurethanes

Properties	1000 MW	1250 MW	1500 MW	2000 MW
Tensile strength (psi)	5050	4500	3500	1200
Elongation (%)	N/D	860	N/D	N/D
300% modulus (psi)	2100	1000	600	400
Tear strength, Graves, (lb/in)	310	240	225	125
Hardness, Shore A	88	77	67	60

In polyurethanes, cross-linking can occur by use of trifunctional chain extenders, by allophanate formation or by biuret formation. It is assumed that cross-linking by isocyanurate formation is of little importance. Increasing the amount of cross-linking by decreasing the molar ratio of chain extender results in hysteresis loss, lower heat build-up, and lower flex-fatigue resistance. The molecular weights between branch points (Mc) and modulus has been found to vary in opposite directions for some polyurethanes[15] as shown in Table 3.8.

TABLE 3.8 Effect of Cross-Linking on Properties of Polyurethane Elastomers

	Hexanetriol	Polypropylene triol MW 700	Polypropelene triol plus pentane diol	Polypropylene triol plus polyprolylene diol (MW 425)
Equivalents	1.0	1.0	0.6 plus 0.4	0.6 plus 0.4
Mc	2090	3700	5800	6150
Tensile strength (psi)	475	555	1340	945
Elongation (%)	235	380	645	625

Chemical cross-linking can be controlled by varying the proportion of diol to triol. The data in Table 3.9 was obtained by substituting trimethylol propane for butanediol, and it shows the effect of increasing average molecular weight per cross-link unit.

TABLE 3.9 Effect of Mc on the Properties of Polyurethane Elastomers

Mc	Tensile Strength (psi)	Elongation (%)	100% Modulus (psi)	Tear Strength, Graves (lb/in)	Hardness, Shore B	Compression Set (%)
2100	1800	170	540	30	57	1.5
3100	1750	200	420	25	53	16
4300	1450	280	300	30	49	10
5300	2800	350	270	30	46	5
7100	4500	410	330	40	51	25
10900	5600	490	460	60	55	40
21000	5500	510	500	140	56	45
∞	6750	640	630	300	61	55

The number of repetitions and the length of the chains of the hydrocarbon, ether, and ester groups appear to control the properties. Table 3.10 summarizes some known effects of the various groups. To control properties, the technologist must not only have the means to correctly and accurately identify qualitatively and quantitatively these groups in the end polymer, but must also be able to evaluate them as they form in the polymerizing mass.

TABLE 3.10 Effects of Different Groups on Properties

	Hardness	Elongation	Tensile Strength	Tear Strength	Abrasion Resistance	Chemical Resistance	Heat Resistance
Hydrocarbon CH^2	G	NK	NK	NK	NK	G	G
Aromatic C_6H_4	E	F	E	F-G	E	G	G
Urea (carbamide)	E	P	G	G	F	G	G
Disubstituted urea	E	P	G	G	F	G	G
Alphanate	F	G	P	P	G	F	P
Biuret	F	F	F	F	G	F	P
Substituted biuret	F	F	F	F	G	F	G
Substituted urea	E	P	G	G	F	G	G
Acyl urea	G	P	G	G	G	G	G
Ester	P	E	E	E	NE	F	P
Ether	P	G	F	F	NE	NK	P
Amide	G	G	G	G	G	F	F

Note: NE = no effect, G = good, NK = no known effect, F = fair, E = excellent, P = poor

Thermal Properties

Thermal stability of polyurethane depends primarily on the polymerization ↔ depolymerization equilibria of the functional groups in the polymer molecule. Urethane groups made from phenols revert quite readily at higher temperatures. Allophanate and biuret cross-links also reopen quite readily on heating. Conventional urethane and urea links decompose at considerably higher temperatures, and isocyanurate rings are the most stable.

Flammability of polyurethane is a frequent concern, particularly in flexible foam bedding and upholstery and in rigid foam insulation. Burning is effectively retarded either by reactive flame retardants (such as polyols containing phosphorus, chlorine, or bromine) that are built right into the polyurethane molecule during polymerization, or by additive flame retardants such as phosphate esters, halogenated phosphate, and halogenated hydrocarbons, either liquid (plasticizers) or solid (fillers). When a halogen is used, either as an reactive or additive flame retardant, addition of antimony oxide produces excellent synergism.[16]

Electrical Properties

Polyurethanes contain many polar groups that tend to orient in an electrical field, and most polyurethane molecules have enough flexibility to permit their polar groups to orient in this way, producing high dielectric constants. On the other hand, molecular flexibility, and the resulting polar group mobility, are very sensitive to frequency and temperature, so dielectric *constants* are far from constant, and electrical-mechanical hysteresis produces considerable and variable dielectric loss. Thus, polyurethanes are not generally used as high-performance electrical insulation. However, they are often used as outer sheathing to protect electrical insulation from abrasion and attack by fuel and oil.

Optical Stability

Aromatic polyurethanes absorb ultraviolet light from the sun but are unable to cope with the excess energy, and they degrade to quinoid structures that discolor and often suffer loss of mechanical properties. Where weather resistance is important, particularly in coatings, most manufacturers replace aromatic isocyanates by aliphatic isocyanates to solve this problem.

Chemical Properties

Strong polarity and hydrogen bonding make polyurethanes highly resistant to hydrocarbon fuel and oil, a major advantage for polyurethane elastomer over conventional hydrocarbon rubbers. They also enjoy another major advantage because their saturated structure is resistant to ozone and exudative aging, whereas most diene-based rubbers have serious problems in this respect.

Linear polyurethanes dissolve, making these elastomers useful in solution processing of fibers, coatings, and adhesives. Cross-linked polyurethane only swell, in polar organic solvents. But in general, polyurethanes are not widely used in the production of chemically resistant products.

Water absorption and hydrolysis, especially at higher temperature, cause aging problems in polyurethane, particularly polyester urethane. In polyester-based PUs, hydrolytic resistance can be improved by use of polycaprolactone, polyethers, or hexamethylene polycarbonate based polyols (see Table 3.11). More often, this problem is minimized by changing from polyester to polyether polyols, and occasionally even to hydrocarbon-based polyols.

TABLE 3.11 Hydrolytic Stability of PU Elastomers[17]

Polyol	Hydrolytic Stability
Esters	
Poly(diethylene glycol adipate)	Poor
Poly(ethylene adipate)	Fair
Poly(butylene 1,4 adipate)	Good
Polycaprolactone	Good
Ether	
PTMEG	Excellent
Polycarbonate	
Poly(hexanediol 1,6 carbonate)	Excellent

Biodegradation can attack the aliphatic polyester segments of polyurethane, degrading them to much smaller molecular units via hydrolytic attack. This is a problem in long-term implantable products, but it is useful in controlled release of drugs and a possible solution to solid waste disposal problems. Biostability is greatly increased by changing to polyether polyols and by addition of biocides and biostats. Biodegradation can be promoted by use of short straight-chain aliphatic polyester units, and by inoculation, humidity, and temperature control in solid waste disposal units.

3.1.6 Molecular Weight

As the molecular weight increases, some properties (e.g., tensile strength, melting point, elongation, elasticity, and glass transition temperature) increase up to a limiting value and the remain constant.

A high molecular weight is the most distinctive structural feature that differentiates between polymer molecules and all other types of materials. The effects of molecular weight on polymer properties are commonly generalized by the rule that *lower molecular weight is easier to process, but higher molecular weight gives better end-use properties.* Many properties change from low to medium molecular weight but then approach an asymptote and become fairly constant from high to very high molecular weight. In the polyurethane field, it is generally assumed that final molecular weights are in the high to very high range, and therefore molecular weight is not a significant variable affecting polyurethane properties.

While this may be true in respect to many properties, there are some properties that depend significantly on molecular weight. These deserve careful consideration and are discussed in the sections below.

Processibility

Most polyurethane processing is based on the use of low-molecular-weight liquid monomers or prepolymers, which are easily handled at atmospheric or low pressure in lightweight equipment, before rapid reactions convert them into solid end products. Liquid viscosities are a direct function of prepolymer molecular weight, so specific process techniques dictate the optimum molecular weight that can be used.

Solution processing of fibers, coatings, and adhesives depends on optimum solution viscosity, which is controlled by polyurethane molecular weight and solution concentration. Generally, it is desirable to work at maximum concentration to minimize solvent problems such as flammability, toxicity, recovery, and cost. Thus, lower molecular weights are generally desirable. Often, these are achieved by use of reactive monomers.

Melt processing viscosities depend on polymer molecular weight. This is important in extrusion of fibers and in extrusion and injection molding of thermoplastic elastomers. Often, the lower molecular weight required for easy melt processing conflicts with the higher molecular weight required for end-use properties. Some *thermoplastic* elastomers are designed with some residual reactivity, which permits them to be melt processed at lower molecular weight and then polymerized or cross-linked up to higher molecular weight to improve end-use properties. Reaction injection molding takes the extreme position, mixing and injecting low-molecular-weight prepolymers and then polymerizing and cross-linking them rapidly in the mold to produce the best end-use properties.

The largest usage of polyurethane is in the production of flexible and rigid foams by reactive processing. Here, low-molecular-weight reactive prepolymers are mixed and poured, with polymerization, foaming, and curing happening in rapid overlapping succession. The dynamic balance between increasing viscosity and gas bubble formation is very critical for optimum foam formation. If gas bubbles form at too low a viscosity, they grow irregularly, burst, and collapse. If they form at too high a viscosity, the gas pressure is insufficient to produce full expansion, and the foam is too dense. The optimum dynamic balance is achieved in flexible foam by balancing catalysis of the isocyanate-polyol action vs. the isocyanate-water reaction. In rigid foams, this delicate equilibrium it is achieved by balancing polymerization rate and exotherm against volatility of the physical blowing agent.

Mechanical Properties

Modulus, strength, extensibility, creep resistance, lubricity, and abrasion resistance generally increase with increasing molecular weight. This is particularly true in thermoplastic polymers, and these effects are particularly applicable to thermoplastic polyurethane elastomers. In thermoset polymers of infinite molecular weight, they would tend to approach high-level value, which would then become a function of cross-linking rather than of molecular weight itself.

Thermal Stability

Thermal stability is generally greater at higher molecular weight, because higher molecular weights have less mobility and reach asymptotic properties that are less sensitive to scission. Such effects might be observed in thermoplastic polyurethane but, in cross-linked polymers, stability depends primarily on cross-link concentration rather than on the molecular weight itself.

Solubility

In many linear polymers, low-molecular-weight products are soluble in a fairly broad range of solvents; as molecular weigh increases, the choice of solvents becomes more restricted (as predicted from thermodynamic theory), and solution viscosity rises rapidly. This is undoubtedly true in solution processing of thermoplastic polyurethane. For cross-linked polyurethanes, solubility becomes irrelevant; if they are exposed to solvents of similar polarity and hydrogen bonding, the degree of swelling is proportional to the degree of cross-linking. Nevertheless, as a rule, the greater the cross-link density, the lower the degree of swelling.

3.1.7 Molecular Flexibility

- *Intermolecular Forces.* The weaker bonds, such as hydrogen bonding, polarizability, dipole moments, and van der Waals forces, may form, in addition to the primary chemical bonds. These weaker bonds are affected by temperature and stress. If there is repulsion between like charges or bulky chains, or if there is high cross-link density, the effect of intermolecular forces will be reduced.

- *Stiffness of Chain.* Presence of aromatic rings stiffens the polymer chains and causes a high melting point, hardness, and a decrease in elasticity. On the other hand, the presence of flexible bonds (such as ether bonds) favors softness, low melting point, low glass transition temperature, and elasticity.

The inherent flexibility of the individual polymer molecule is an important theoretical concept, and it has many consequences in practical properties. The subject is best divided into these two separate aspects.

Effect of Structure on Molecular Flexibility

The structure of the individual polymer molecule determines its inherent flexibility (Table 3.12). In the absence of crystallinity, intermolecular attractions, or cross-linking, the polymer molecule is free to exhibit all its inherent flexibility, and this unrestricted freedom has major effects on polymer properties. Even in the presence of such conflicting factors, the inherent flexibility of the polymer molecule is still very important, but the evidence for it is more obscure, and the practical effects are more complex. To better understand this phenomenon, let us consider the simple concept of the inherent flexibility of the individual polymer molecule.

TABLE 3.12 Molecular Flexibility of Ether Groups

Glycol Reacted with Hexamethylene Diisocyanate	Polyurethane Melting Point
$HO(CH_2)5(OH)$	151° C
$HO(CH_2)_{20}(CH_2)_2OH$	120° C
$HO(CH_2)_{25}(CH_2)_2OH$	132° C

Molecular flexibility depends on freedom of rotation about the single bonds in the main chain of the polymer molecule; restriction of rotation reduces molecular flexibility. A linear aliphatic chain, as shown in Fig. 3.20, is fairly free to rotate about its C–C bonds but is restricted by (a) the 109° angle between C–C–C bonds and (b) the electropositive repulsion between adjacent H atoms requiring some energy to rotate them past each other. When a –CH$_2$- is replaced by an oxygen (—CH$_2$-O-CH$_2$—), rotation around the C–O bond does not bring H atoms into conflict with each other, so rotation is easier, and the molecule is more flexible; this occurs in polyethers, polyesters, and even polyurethane linkages.

$$—[CH_2CH_2CH_2CH_2CH_2CH_2CH_2—]—$$

FIGURE 3.20 Linear aliphatic chain.

Sulfide links (—CH$_2$-S-CH$_2$—), although less common, have a similar effect. When methyl side groups are attached to the aliphatic main chain in an amorphous polymer, as shown in Fig. 3.21, steric hindrance restricts rotation around the main chain, and the molecule becomes stiffer. It should be noted that, in crystalline polymer, random methyl side groups reduce regularity and crystallinity and thus make the solid *mass* of polymer molecules more flexible. When aromatic groups are present, particularly in the main chain, they introduce large flat rigid units that greatly reduce molecular flexibility (see Table 3.13).

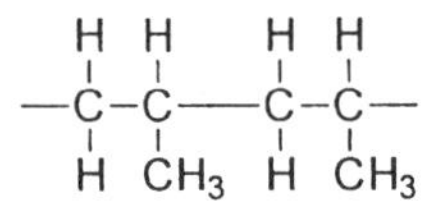

$$\begin{array}{cccc} H & H & H & H \\ | & | & | & | \\ —C & –C & —C & –C— \\ | & | & | & | \\ H & CH_3 & H & CH_3 \end{array}$$

FIGURE 3.21 Amorphous polymer.

TABLE 3.13 Stiffening of Polyester Urethane Foams Due to Aromatic Rings

Aromatic Structure (%)	Tensile Strength (Psi)	Ultimate Elongation (%)
21.8	55	10
16.6	35	60
15.9	27	160
14.7	26	270
12.6	25	3500

When the aromatic rings are conjugated with adjacent unsaturated groups in the main chain, the entire conjugated resonating unit becomes much larger, and the stiffening effect is much greater, as shown in Fig. 3.22.

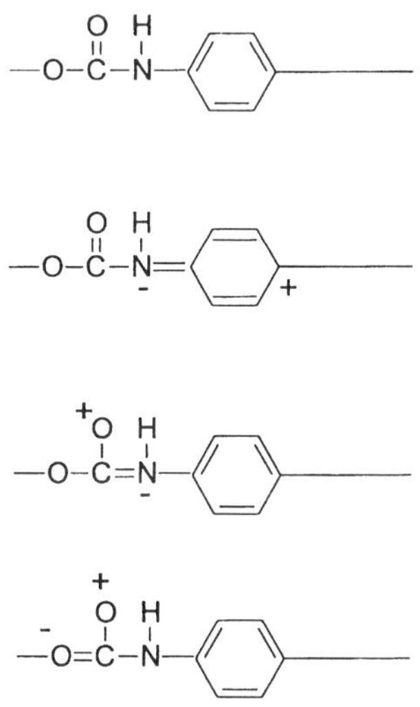

FIGURE 3.22 Stiffening due to aromatic rings in main chain.

Likewise, in polyaromatic systems such as naphthalene units, the stiffening effect extends further and becomes much greater Table 3.14. These effects are most clearly seen in amorphous polymer of low intermolecular attraction and no cross-linking; when these complications occur, they interact with molecular flexibility and often overpower it so that its effects are no longer clearly seen.

TABLE 3.14 Polyaromatic Stiffening of Polyester Urethane Elastomer

Diisocyanate Used in Making Polyurethane	Tensile Strength (psi)
2,4 Toluene Diisocyanate	3200
1,5 Naphthalene Diisocyanate	4400
2,7 Fluorine Diisocyanate	6200

Processibility

In liquid and melt processing, flexible polymer molecules coil, uncoil, and disentangle more easily and give lower viscosity and easier processing. Similarly, in solution processing, flexible polymer molecules form a random coil of small diameter, low viscosity, and easy processing; whereas stiff polymer molecules remain extended in solution, presenting much larger end-to-end dimensions, higher viscosity, and more difficult processing. Thus, molecular flexibility generally brings easier processing.

Mechanical Properties

Without the complications of crystallinity, intermolecular attraction, and cross-linking, molecular flexibility generally permits the polymer molecules to disentangle and flow more easily when mechanical stress is applied. This generally produces lower hardness, modulus, strength, creep resistance, and lubricity, along with higher extensibility and friction. See Tables 3.15 through 3.17.

TABLE 3.15 Poly(oxypropylene) Diol Molecular Flexibility in Polyurethane Elastomers

Poly(oxypropylene) diol MW	1000	1250	1500	2000
Hardness, shore A	88	77	67	60
300% Secant modulus, psi	2100	1000	600	400
Tensile strength, psi	5050	4500	3500	1200
Graves tear strength, lb/in	310	240	225	125
Bashore resilience,%	24	20	22	38
Taub abrasion loss, mg/1000 cycles	85	60	25	8
Tinius Olsen brittle point, °C	−40	−50	−55	−56

TABLE 3.16 Molecular Flexibility of Poly(oxtetramethylene) Glycol-Based Polyurethane Elastomers[18]

	650 (MW)	1000 (MW)	2000 (MW)
Hardness, Shore A	78	76	75
100% secant modulus, psi	605	521	446
Tensile strength, psi	3500	2980	2530
Ultimate elongation, %	488	640	710
Tear strength, pli	385	361	344
Rebound, %	33	51	55
Abrasion loss, mg/5000 cycles	13	9	8
Coeff. of kinetic friction	0.57	0.63	0.59

TABLE 3.17 Molecular Flexibility of Poly(ethylene adipate) Glycol-Based Polyurethane Elastomers

	650 (MW)	1000 (MW)	2000 (MW)
Hardness, Shore A	74	73	71
100% Secant modulus, psi	462	376	341
Tensile strength, psi	2320	1620	1350
Ultimate elongation, %	643	735	856
Tear strength, pli	330	315	306
Rebound, %	21	39	45
Abrasion loss, mg/5000 cycles	9	9	6
Coeff. of kinetic friction	0.47	0.58	0.52

Thermal Properties

For any given molecular structure, increasing temperature means increasing atomic and molecular motion, greater free volume, higher energy to overcome electronic repulsion and steric hindrance, and thus greater molecular flexibility. Thus, practical molecular flexibility is the summation of (a) inherent molecular flexibility plus (b) thermal mobility. Conversely, polymers with inherent molecular flexibility retain this flexibility down to lower temperatures, giving lower glass transition temperature, retaining flexibility or toughness down to lower temperature, but suffering lower heat deflection temperature, and greater loss of strength and creep deformation at high temperatures.

Thermal stability tends to correlate with molecular rigidity, because chemical reaction and degradation depend on molecules moving and meeting each other, and this becomes more difficult when the molecules have less flexibility.

Electrical Properties

As noted earlier, molecular flexibility permits polar groups to orient in an electrical field, producing high dielectric constant; molecular rigidity prevents orientation, producing low dielectric constant. Since practical molecular flexibility and rigidity depend not only on molecular structure but also on temperature and frequency, polyurethane do not have constant dielectric *constant*. In the transition region, they have considerable dielectric loss, so they are not generally used in highly demanding electrical insulation applications.

Infrared Spectroscopy

Bending, stretching, and rotation about bonds in the polymer molecule occur at frequencies and energies primarily found in the infrared region of the electromagnetic spectrum. Since each bond absorbs energy at very specific frequencies, this provides sensitive analytical techniques to characterize the structure of polyurethanes. Conversely, for rapid preheating during processing, infrared radiation with these specific frequencies provides a very convenient technique for bringing polyurethane up to processing temperatures.

Some of these properties depend only on the inherent flexibility of the individual polymer molecule. Other structural features, however (particularly crystallinity, intermolecular attraction, and cross-linking) can immobilize polymer molecules and thus exert controlling effects on many practical properties. These will now be considered in detail.

3.1.8   Intermolecular Order: Crystallinity

The linearity and close fit of polymer chains favor crystallinity, which leads to reductions in solubility, elasticity, elongation, and flexibility, and to increases in tensile strength, melting point, and hardness.

Factors Affecting Crystallinity of Polyurethanes

In the melt and in solution, polymer molecules tend to form random coils more or less entangled with each other but relatively free to disentangle, uncoil, recoil, and exhibit more or less liquid flow. On cooling from the melt, evaporation, or coagulation of the solution, irregular molecular structures [such as poly(oxypropylene) as shown in Fig. 3.23 and urethanes from 2,4 toluene diisocyanate] remain in the form of random entangled amorphous coils and come closer together, with less free volume and less mobility, gradually becoming rubbery and then glassy in properties.

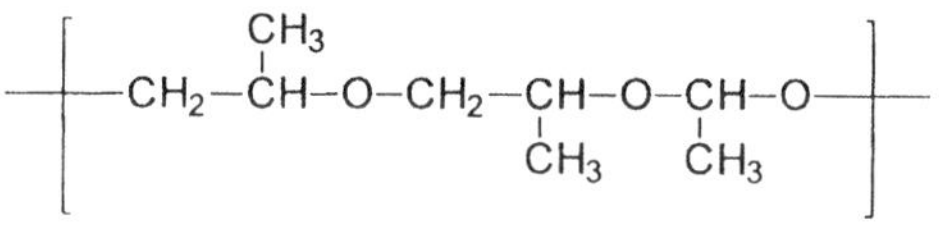

FIGURE 3.23 Polyoxypropylene.

In contrast, regular molecular structures of reasonable flexibility, such as poly(oxytetramethylene), polyesters, and urethanes from diphenyl methane 4'4-diisocyanate (MDI), on cooling from the melt or coming out of solution, tend to organize and pack into regular dense crystalline lattice structures, which immobilize the polymer molecules and greatly restrict their *inherent* flexibility, producing much harder, stronger, chemically resistant products. Even in elastomeric polyurethane, such crystallization may become very apparent in low-temperature stiffening and embrittlement. Thus, crystallinity plays a major roll in practical properties, often overpowering the significance of inherent molecular flexibility.

Several types of structural features affect the ability of regular polymer structures to fit and pack into the tight regular lattice required for crystallization. On the positive side,

1. A reasonable degree of molecular flexibility makes it possible for polymer molecules to disentangle from random coils and conform and fit into the precise positions required for lattice formation.
2. Intermolecular attractions help to bind the polymer molecules into the crystal lattice, making it stronger and more resistant to mechanical, thermal, and chemical stresses.
3. Monomer units with an even number of atoms in the main chain tend to pack and fit more neatly and densely than those with an odd number, giving stronger crystalline structures.

Several other types of structural features have negative effects on the ability of polyurethane to fit and pack into the crystal lattice:

1. Short side groups such as CH$_2$ make it more difficult for polymer backbones to fit neatly and tightly into the crystal lattice and thus reduce or prevent crystallinity.
2. When side groups become much longer, they may tend to form little crystallites among themselves, generally referred to as *side-chain crystallization*, which restrict molecular mobility and usually give stiffer, more waxy type of properties.
3. Introduction of cross-linking into a regular linear polymer produces branch points that reduce its regularity, decreasing and even preventing crystallinity. The result is a *softening* of mechanical, thermal, and chemical properties at first; however, as the degree of cross-linking is increased further and further, it eventually immobilizes the polymer molecules so effectively that properties revert and become *harder* again.

Effects of Crystallinity on Properties

Processibility

Crystalline polymers are held together by a high accumulation of intermolecular attractions, which requires more energy to separate the polymer molecules and permit them to flow during processing. In melt processing, this means higher temperatures. In solution processing, it means much more difficulty in dissolving the polymer molecules into the solvent and keeping them there. In highly crystalline polyurethane, this can make processing difficult to impossible, thus placing limits on the enthusiasm of the synthetic organic polymer chemist. On the other hand, fast crystallization from the melt can solidify molded and extruded products much faster than the gradual stiffening of amorphous polymers and thus helps to shorten processing cycles handsomely.

Mechanical Properties

Crystallization packs and immobilizes even highly flexible polymer molecules, making them much more resistant to mechanical stress. This produces higher hardness, modulus, strength, lubricity, and resistance to creep. All these are desirable in rigid products, but loss of softness and flexibility can be a major concern in flexible foams and elastomers.

Thermal Properties

The attractive forces between polymer molecules tend to hold them into the crystal lattice. Temperature represent the vibrations of atoms and molecules that tend to free them from the crystal lattice and into the mobility of the random coil. Thus, the balance of these factors determines the melting point of the crystal structure, $T = \Delta H/\Delta S$, and the contrast between properties above and below this point.

Molecules with high inherent flexibility tend to remain amorphous, soft, and flexible at ambient temperatures. If they are regular in structure, cooling reduces vibrations and encourages crystallization, producing serious stiffening and embrittlement of flexible elastomeric products. In fact, random uncrystallizable structures can retain their soft, rubbery properties down to much lower temperature.

Molecules with high intermolecular attractions tend to form more stable crystals and resist thermal vibrations up to higher temperatures, retaining higher hardness, modulus, strength, and creep resistance, before they eventually melt. However, melting point is often accepted as direct evidence of intermolecular attraction.

Chemical Resistance

Polar solvents are attracted to polar polyurethane molecules. In random amorphous coils, these solvents can easily penetrate the free volume in the random coils, separate them from each other, and dissolve them. In a tightly packed crystalline lattice, on the other hand, there is no room for solvent molecules to penetrate, and the accumulation of attractive forces between polymer chains in the crystal lattice is too great for solvent molecules to compete with it. Thus, crystalline polymers have much greater solvent resistance.

3.1.9   Intermolecular Attraction

In a solid product made of linear polymer molecules, the ability to resist mechanical, thermal, electrical, and chemical stress depends on the attractive forces between the polymer molecules, which transfer these stresses from one molecule to another throughout the solid mass. These secondary attractive forces cover a spectrum from weak long-distance attractions up to fairly strong close-range forces (see Table 3.18).

TABLE 3.18 Molar Cohesive Energies of Functional Groups in Polyurethanes

Group		Cohesive Energy, kcal/mole
$-CH_2-$	(hydrocarbon)	0.68
$-O-$	(ether)	1.00
$-COO-$	(ester)	2.90
$-C_6H_4-$	(aromatic)	3.80
$-CONH-$	(amide)	8.50
$-OCONH-$	(urethane)	8.74

The order of intermolecular attractions may be arranged as follows:

1. *London Dispersion Forces.* London dispersion forces are due to the interaction between nonpolar electron cloud bonds, typically the C–C and C–H bonds in the hydrocarbon portions of polymer molecules. They are generally 1 to 2 kcal/mol. in strength and 3 to 5 Å in length.

2. *Permanent Dipoles.* Permanent dipoles exist in polyurethane bonds such as C–O, C=O, C–N, O–H, and N–H. Attractions between such dipoles are typically about 3 kcal/mol. in strength and 3 Å in length.

3. *Hydrogen Bonds and Ionic Bonds.* Hydrogen bonds form when an electronegative atom in one molecule pulls electrons away from a hydrogen, leaving the hydrogen relatively electropositive and electron deficient, and an electronegative atom in another molecule shares its extra electrons with the electropositive hydrogen atom. Such hydrogen bonds are very important in polyurethane, and are typically 1.5 to 6 kcal/mol. in strength and about 3 Å in length (See Fig. 3.24).

FIGURE 3.24 Hydrogen bonds.

Ionic bonds may occur occasionally in polyurethanes and are typically 10 to 20 kcal/mol. in strength and 2 to 3 Å in length.

Total Intermolecular Attraction

Total intermolecular attraction is a critical concept, and may be expressed in the form

$$I = S \times C \times M \times F$$

where I is the total intermolecular attraction per unit volume, S is the strength of a single intermolecular attraction as given above, C is the concentration of such groups in the polymer molecule (the size of the repeat unit), M is the molecular weight of the polymer (which determines the number of such groups in the molecule), and F is the frequency with which such potential intermolecular attractions come close enough to each other to actually form. Thus, for example, London dispersion forces individually have low-strength S, but their concentration C is so ubiquitous that they add up to a major portion of the total intermolecular attraction in all polymer molecules; in fact, in hydrocarbon polymers, they alone account for the useful properties of such widely used polymers as polyethylene, polypropylene, and polystyrene.

For another example, in polyurethane the strength S of polar and hydrogen bonding attractions is great enough to accumulate and give useful properties even at rather modest molecular weight M, whereas in polyolefins, the lower strength S of London dispersion forces requires much higher molecular weights M before they can accumulate to produce useful properties. Additionally, in the random coils of amorphous polyurethane, the polar and hydrogen-bonding groups (S) of adjacent polymer molecules only rarely come close enough to function (low F), leaving such polymers soft and flexible, whereas, in the neatly packed area of crystalline polyurethane, the same groups (S) occur close to each other with high frequency (F), producing much greater total attraction and much greater strength and heat resistance. In general, it is the total intermolecular attraction that best explains effects on practical properties.

General Effects on Properties

In general, increasing intermolecular attraction makes procession more difficult, requiring higher melt temperatures and more polar, higher-boiling solvents to overcome these attractions and produce good flow. As intermolecular attractions increase, they bring increasing hardness, modulus, strength, and creep resistance. They decrease low-temperature flexibility but increase hot strength and melting point. They also increase resistance to nonpolar fuels and oils, a major advantage over conventional hydrocarbon elastomer and foams (see Tables 3.19 and 3.20).

TABLE 3.19 Intermolecular Attraction of Urethane Groups in Elastomers

Urethane wt., %	8.4	9.3	10.9	13.6	17.0
Tensile strength, psi	110	110	140	250	2000
Ultimate elongation, %	190	180	220	340	720
Glass transition, °C	−51	−48	−43	−34	−24
Swelling in benzene, %	524	492	450	368	300

TABLE 3.20 Intermolecular Attraction of Urethane Groups in Flexible Foams

Urethane wt., %	10.5	8.3	5.3	3.1	2.7
100% Secant modulus, psi	30	16	12	7	6
Tensile strength, psi	30	21	18	22	15
Ultimate elongation, %	100	130	155	295	340
Compression load at 75% psi	8.3	4.3	2.1	1.6	0.9
Rebound,%	16	15	44	42	—
% swelling in dimethylacetamide (DMAc)	145	170	237	350	—

Typical Effects

A few examples from polyurethane structure-property relations can serve to illustrate some of these principles.

- Polyesters alone have oxygen to serve as electron donors, but only meager amounts of electropositive hydrogen to serve as electron acceptors; for this reason, aliphatic polyesters tend to be weak and low melting. When aliphatic polyesters are used in polyols to make polyurethane, however, the N–H groups provide electropositive hydrogen to act as electron acceptors; the resulting hydrogen bonding in polyester urethanes makes them the optimum choice for high-strength elastomers.

- Hydrogen bonding in polyurethane is sufficient to provide good properties, but hydrogen bonding in polyureas is much stronger and gives much higher strength and melting points. Thus, in making high-performance polyurethane elastomers, the use of diamines to create urea group contributes even more to performance than the urethane groups, which take total credit in conventional nomenclature. Even in flexible polyurethane foams, use of water plus diisocyanate to produce CO_2

for foaming also incidentally introduces diamine, which then reacts to introduce polyurea units into the foam and contributes greatly to its strength.

• Introducing ionic bonding between polyurethane chains has been observed to increase hardness, modulus, strength, and glass transition temperature, along with decrease in elongation (see Table 3.21). All of these effects are evidence that ionic bonding serves to immobilize the polymer molecules and make them more resistant to mechanical and thermal stress.

TABLE 3.21 Intermolecular Attraction of Quaternary Ammonium Ions

Degree of quaternization (%)	10	30	60	100
Shore A hardness	48	65	73	82
100% secant modulus	110	320	730	940
Tensile strength (psi)	670	2800	3170	4250
Ultimate elongation (%)	1280	1000	790	710

3.1.10 Cross-Linking

The ultimate intermolecular force is a primary covalent cross-link, and these are generally use in the vast majority of polyurethane products.

Cross-Linking Reactions

Cross-linking increases in the degree of cross-linking ease an increase in rigidity, softening point, and modulus of elasticity for amorphous polymers, and reduces elongation and swelling by solvents.

It is possible to use more than one poly component to vary the formulation, and hence the processing characteristics and the properties, of the product. This affects the degree of branching and also the order of interaction of the polyols with the diisocyanate. It is also possible to vary processing factors such as temperature, viscosity, rate of setup, curing of the casting mix, or incorporation of fillers. Addition of low-molecular-weight polyols such as glycerol or trimethylol propane may introduce branching.

Generally, in the preparation of polyurethane, the excess diisocyanate may first react with polyol to form isocyanate-terminated prepolymer, which is stable in the absence of moisture and catalyst. This prepolymer reacts with more polyol at the casting stage. Therefore, by varying the polyol component added in this stage, it is possible to change the properties of the resulting elastomer. Depending on the chemical composition and the amount of cross-linking and branching, it is possible to obtain products ranging from soft elastomers to hard resinous materials.

The chain structure of the polymerized intermediate determines the level of mechanical properties. Fillers, if used, may increase hardness and modulus of elasticity and reduce elongation, but they do not have a very substantial effect on strength.

For a given set of diols and triols and a given NCO/OH ratio, the molecular weight per branch point may be altered by varying the proportions of diols and triols. The molecular weight per branch point may be varied while retaining the same proportion of polyisocyanate, either by using polyols that are the same equivalent weight or by simultaneously varying the equivalent weights of the polyols to give a combination of the same average equivalent weight. An increase in the proportion of polyisocyanates results in a decrease in the molecular weight per branch point and an increase to tensile strength, modulus of elasticity, and hardness, while decreasing elongation.

Besides the triols, cross-linkage and branching will also take place at the urethane and the urea groups. When the NCO/OH ratio is low, the branching will occur at the urethane linkage. At higher NCO/OH ratios the probability for the formation of urea linkages will be high and, thus, branching will occur at the urea linkages. The presence of higher atmospheric pressure and relative humidity assures the presence of ETO-linkage and branching at the urea groups, and the percentage of urea to urethane will be higher than at lower NCO/OH ratios.

Most often, cross-links are introduced by use of trifunctional or higher-functional polyols as shown in Fig. 3.25.

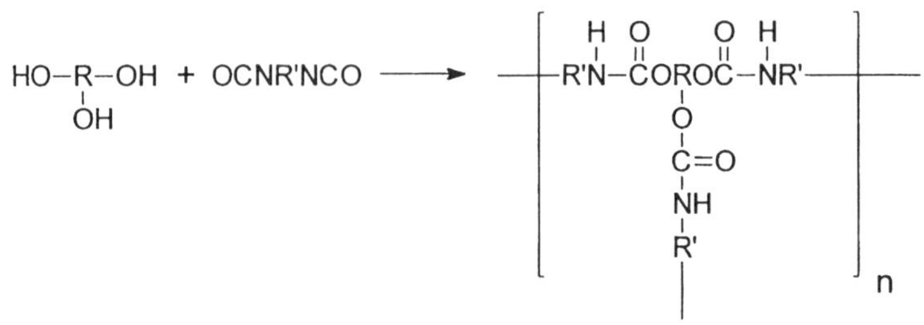

FIGURE 3.25 Use of trifunctional or higher-functional polyols.

In flexible urethane foams and cast elastomers, the polyol is a long, linear Y-shaped triol. In rigid urethane foams, it may be a short-chain polysaccharide hexol. Here, the cross-links have the same chemical composition and stability as the main chains.

Second in importance is the use of low-molecular-weight polyisocyanates made from aniline-formaldehyde oligomers. These are used primarily in production of rigid foams. Here again, the cross-links have the same chemical composition and stability as the main chains.

Often, the cross-links are produced by using excess allophanate in the formulation and allowing it to react with urethane groups already formed to produce allophanate cross-links, or with urea groups already formed to produce biuret cross-links. In this case, the cross-links form somewhat reluctantly and are less heat stable than the main chains, tending to open and revert upon heating.

The stablest cross-links are formed by using excess isocyanate in the formulation and causing it to form cyclic timers called isocyanurates. These are by far the stablest types of cross-links, and they are used to produce products with maximum heat and flame resistance.

Cross-linking may be induced by peroxides, either by abstraction of hydrogen (which is quite difficult) or by introduction of vinyl groups into the polyurethane (which requires extra synthetic effort). Either way, the cross-links formed are C–C bonds, which are as stable as the main chain itself.

Rubber processors sometimes buy thermoplastic polyurethane elastomer gums and cross-link them as they would conventional rubber, using sulfur and accelerators. Such sulfur cross-linking is difficult using saturated elastomers; it can be improved by synthesizing the polyurethane with some vinyl groups to permit conventional sulfur vulcanization reactions. However, such sulfur cross-links appear to be less heat stable than the more common conventional polyurethane cross-linking systems.

Mechanical Properties

The introduction of cross-links between polyurethane molecules produces a definite quantitative restriction of molecular mobility. At low degrees of cross-linking, where long segments of the backbone molecule are still free to move, there is little loss of soft flexible rubbery behavior; and there are distinct improvements in strength and creep resistance (see Tables 3.22 and 3.23). In crystalline polymers, low cross-linking destroys regularity and reduces crystallinity and thus actually produces softer properties. With increasing degree of cross-linking, modulus increases and extensibility decreases, giving tough, flexible products. At still higher degrees of cross-linking, the polymer molecules are thoroughly immobilized and become rigid thermoses plastic (see Table 3.24). Thus, cross-linking alone can develop a broad spectrum of polyurethane properties and products. Figure 3.26 presents the mechanism of cross-linking via dicumyl peroxide.

TABLE 3.22 Low Cross-Linking of Polyether Urethane Elastomers

MW between cross-links	2500	4500	8500	12500
Tensile strength (psi)	200	220	130	160
Ultimate elongation (%)	110	190	260	310
Glass transition (°C)	−59	−59	−59	−59
Swelling in benzene (%)	294	342	384	467

a) **Thermal decomposition**

b) **Free radical transfer to polymer chain**

c) **Polymer crosslinking**

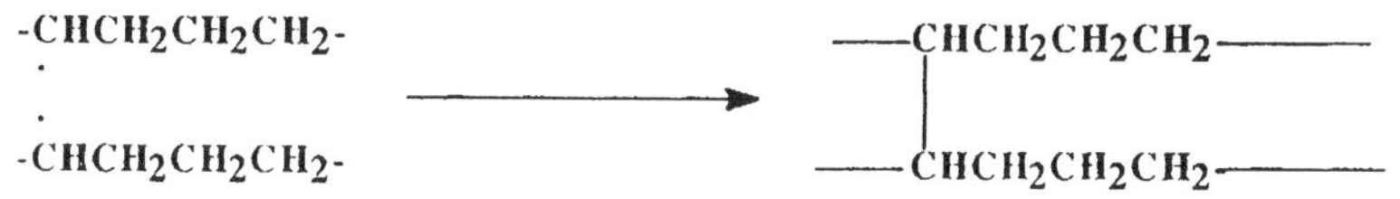

Peroxide	Recommended Cure Temperature, °F
Di-t-butyl peroxide	355
t-butyl cumyl peroxide	338
2,5 dimethyl-2,5-bis-(t-butylperoxy)-hexane	325
dicumyl peroxide	320
n-butyl-4-4-bis (t-butylperoxy)-valerate	315
t-butyl perbenzoate	305
1,1-di-(t-butylperoxy)-cyclohexane	280
Benzoyl peroxide	237

Peroxide cross-linking of polyurethanes is accomplished via free radical dehydrogenation of the polymer followed by covalent bonding of the adjacent chains at the site. The first step is the decomposition of the peroxide into cumyloxy free radicals; while the cumyloxy can decompose into acetophenone, the cumyloxy radical pedominates. The free radicals abstract hydrogens, creating two adjacent polymeric free radicals, which then cross-link.

FIGURE 3.26 Peroxide cross-linking. Example: decomposition of dicumyl peroxide and polymer cross-linking.

TABLE 3.23 Low Cross-Linking of Polyester Urethane Elastomers

MW between cross-links	2100	4300	7100	21000	∞
Shore B hardness	57	49	51	56	61
Tensile strength (psi)	1800	1450	4500	5500	6750
Ultimate elongation (%)	170	280	410	490	640
100% secant modulus	570	300	330	500	630
Tear strength (pli)	30	30	40	140	300
Compression set (%)	2	10	25	45	55

Thermal Properties

Increasing cross-linking draws polymer backbones closer together, reduces molecular mobility, and raises glass transition temperature. Thus, flexible elastomers begin to stiffen at low temperatures. However, where secondary attractions permit weakening and creep at high temperatures, these are both controlled by the introduction of permanent cross-links.

TABLE 3.24 High Cross-Linking Produces Rigid Thermoset
Polyurethane Plastics

Hexol/diol ratio	'80/20	'90/10	'100/0
Shore D hardness	77	85	91
Tensile strength (psi)	4300	8800	13,250
Ultimate elongation (%)	4	2	2
Tensile modulus (psi)	170,000	410,000	600,000
Flexural strength (psi)	7,000	16,700	18,000
Heat deflection temperature (°C)	52	78	87
Swelling in acetone (%)	27.4	17.9	0.3

At higher temperatures, the weaker, less stable types of cross-links tend to reopen and revert back to linear structures. This is noticed first in allophanates, then in biurettes. Sulfur vulcanization cross-links are less stable than conventional poly and polyisocyanate cross-links. When direct C–C cross-links can be formed, these have very good stability. Generally, the stablest are those formed by isocyanate cyclic trimerization into isocyanurate rings. Isocyanurates are most commonly used for heat- and flame-resistant products.

Chemical Resistance

Even the lightest degree of cross-linking will prevent individual polymer molecules from dissolving in a solvent. Since there is still strong attraction between polyurethane and polar solvents, lightly cross-linked polymers will absorb large amounts of solvent and swell to soft gels. Increasing degree of cross-linking produces less free lump for absorption of solvent molecules, and less freedom for polymer chains to move apart to accept them. Thus, there is a precise quantitative relationship between swelling and cross-linking, (Table 3.25), which can actually be used to calculate the concentration of cross-links in the polymer. A high degree of cross-linking can even produce considerable improvement in resistance to polar solvents.

TABLE 3.25 Acetone Swelling of Cross-Linked Polyether Urethane Foam

MW between cross-links	1650	070	690
Swelling in acetone (%)	116	90	83

Water absorption should similarly be reduced by cross-linking. Furthermore, the degrading effects of hydrolysis on molecular weight and properties should be retarded by the presence of cross-links to maintain the molecular weight above the critical level needed for useful properties. The extent of such improvement in individual products must be determined experimentally.

3.1.11 Supermolecular Structure

A number of larger structural features are of major practical importance in the properties and use of polyurethane, and some of them are also fairly well understood at the theoretical level as well.

Latex

Polyurethane coatings, adhesives, and fibers are frequently processed by conventional organic solution techniques. Such use of solvents introduces growing problems of cost, toxicity, flammability, disposal, and recycling, problems that create growing pressure to reduce or eliminate the use of organic solvents. One attractive alternative is to disperse the polyurethane in aqueous medium, apply in place, and remove the water to deposit the solid polyurethane product. A variety of techniques have been studied to accomplish this.

Finished thermoplastic polyurethane can be emulsified and used in latex form. Fatty isocyanates of low polarity can be emulsified in water without suffering hydrolysis and then formulated into reactive systems. Isocyanates can be temporarily blocked to make them stable in water; then, the blocking reaction is reversed to liberate the isocyanate for the curing reaction.

Copolymers

The versatile reactivity of polyols and polyisocyanates invite the preparation of copolymers with various other polymer systems to combine the best properties of each. Thus, there are occasional reports of polyurethane copolymers with other common polymer systems. Most important are probably the polyurethane drying oils, which combine conventional drying oil processibility with the high performance of polyurethane. Also popular are acrylic polyurethanes, which combine the easy radiation cure of the acrylics with the end-use performance of the polyurethane. More specialized are copolymers with epoxy resins (Table 3.26), polyester (Table 3.27), and silicones. Most of these systems are fairly homogeneous one-phase copolymers and thus not truly supermolecular structures.

TABLE 3.26 Copolymers of urethane elastomer with an epoxy resin

Urethane/epoxy ratio	'100/0	'80/20	'60/40	'40/60	'20/80	'0/100
Shore D hardness	39	50	72	84	87	90
Flexural modulus (psi)	4230	2850	51200	238000	338000	380000
Flexural strength (si)	393	270	2270	10300	13300	15600
Volume resistivity	4×10^{11}	7×10^{11}	6×10^{13}	2×10^{14}	4×10^{14}	2×10^{15}
Dielectric constant	7.2	5.7	4.4	3.9	4.4	4.6
Dissipation factor	0.048	0.036	0.018	0.012	0.013	0.015

TABLE 3.27 Block Copolymers of Polyester Urethane Elastomers with Poly(ethylene terephthalate)

	15 mole % PET			35 mole % PET		
	Tensile str. (psi)	Elong. (%)	T_g (°C)	Tensile str. (psi)	Elong. (%)	T_g (°C)
Poly(ethylene succinate)	1560	6.7	5	—	—	—
Poly(ethylene adipate)	3400	700	−60	4060	500	−41
Poly(diethylene adipate)	1700	200	—	1900	113	—
Poly(ethylene azelate)	1650	400	−34	1280	13	−30
Poly(ethylene sebacate)	—	—	−24	2400	27	−15

Block Copolymers

Polyurethanes are generally made up of polyol segments of lower polarity, plus polyurethane and/or polyurea segments of higher polarity. When the segments are fairly short, the polymer may have a fairly random homogeneous structure. When the segments are longer, the resulting block copolymer will tend to separate into microphases. The phase present in larger amount will tend to form the continuous matrix and control most of the properties, while the phase present in smaller amount will tend to segregate as discrete domains and contribute specific properties to the composite structure.

Generally, the poly blocks are lower polarity, long and flexible, and frequently referred to as *soft segments*; they form the continuous matrix and make the majority of polyurethane soft, flexible, and rubbery. In some cases, very regular polyols such as poly(oxytetramethylene) and poly(ethylene adipate) will tend to crystallize, giving higher-strength elastomers, but stiffen when cooled to lower temperatures.

The polyurethane and polyurea blocks are generally higher in polarity and hydrogen-bonding, short and bulky, and are frequently referred to as *hard segments*. They form the dispersed domains, contribute strength and creep resistance, and retain these to higher temperatures. Irregular structures such as 2,4-toluene diisocyanate and polymethylene polyphenyl isocyanates generally give glassy hard segments, held together by steric hindrance, polar attraction, and hydrogen bonding. Regular structures such as diphenylmethane-4,4'-diisocyanate tend to give crystalline hard segments, also held together by steric hindrance, polar attraction, and hydrogen bonding.

Thus, block copolymer structure and microphase separation permit polyurethane to combine the best properties of both the *soft* polyol continuous matrix and the *hard* urethane and/or urea dispersed domains. This supermolecular structure is a major reason for their high performance and versatility.

Polyblends and Interpenetrating Polymer Networks (IPN)

When two polymers are mixed, the requirement for thermodynamic miscibility at equilibrium is a decrease in free energy ΔG:

$$\Delta G = \Delta H - T\Delta S$$

where ΔH is the enthalpy of mixing (attraction between the two polymers) and ΔS is the entropy of mixing (gain in statistical randomness). Two unlike polymers will generally repel each other, making ΔH unfavorable, unless there is some specific group attraction between them such as hydrogen-bonding. Mixing of large polymer molecules does not produce much gain in randomness ΔS, because all the atoms in a polymer molecule remain attached to each other during the process. Thus, most polymer blends are not miscible at the molecular level and tend to separate into microphases. If the interface between these microphases is strongly bonded, each phase may be free to contribute some of its best properties to the blend, and such two-phase blends can benefit from synergistic balance of properties. This has turned polyblending into one of the fastest-growing segments of the polyurethane industry.

Polyurethanes contain aliphatic, aromatic, ether, ester, urethane, and urea groups, offering a wide range of polarities and hydrogen-bonding possibilities, which should promote miscibility, or at least strong interfacial bonding, with a wide variety of other polymers. Blends with acrylonitrile styrene-butadiene (ABS) combine the melt processibility/rigidity and heat defection temperature of the ABS with the classic recovery and abrasion resistance of the polyurethane. Blends with polyvinyl chloride (PVC) combine the rigidity and flame retardance of the PVC with the impact resistance of the polyurethane (Table 3.28); in more miscible blends, the polyurethane acts as a permanent polymeric plasticizer. Blends with epoxy resins combine the rigidity and heat and chemical resistance of the epoxy with the ductility and impact resistance of the polyurethane (Table 3.29).

TABLE 3.28 Flame Retardation of Polyester PU by PVC and Antimony Oxide

PU	PVC	Sb_2O_2	Oxygen Index (%)
'100	'0	'0	'22.0
100	10	0	22.0
100	5	5	29.1
100	0	10	23.5
100	20	0	21.4
100	15	5	31.0
100	10	10	31.3
100	5	15	29.4
100	0	20	23.5
100	40	0	22.1
100	35	5	27.9
100	30	10	32.6
100	25	15	32.2
100	20	20	29.8
100	15	25	31.5
100	10	30	27.5
100	5	35	26.8
100	0	40	22.3

When a polymer is lightly cross-linked and swollen with a second monomer, and the second monomer is then polymerized in the swollen network, the resulting interpenetrating polymer network (IPN) contains the two polymer phases in a controlled degree of dispersion.[18] IPN synthesis benefits from the use of two distinctly different polymerization mechanisms. Since polyurethane are formed by a unique

TABLE 3.29 Polyblends of PU and Phenoxy Resin

PU/Phenoxy Ratio	'100/0	'80/20	'60/40	'40/80	'20/80	'0/100
Melt index (gm/10 sec)	10.4	0.7	0.2	0.5	0.9	1.7
Shore A hardness	79	79	90	92	93	93
Tensile strength (psi)	2300	1800	3680	5080	8230	8440
Elongation (%)	754	526	354	25	8	10
Flex temperature (°C)	−48	−27	0	—	—	—
Heat deflection temperature (°C)	—	—	—	36	54	67

chemical reaction, they can conveniently form one of the two polymer phases in IPNs and have proved to be one of the most popular polymer systems in IPN research. Surprisingly, many polyurethane IPNs, particularly with acrylics and polyesters, have shown remarkable synergistic improvement of tensile strength, as compared with the individual polymers involved. Much IPN research has been directed toward broad spectrum damping of noise and mechanical vibration.

Reinforcing Fillers

The use of fibers to produce high-performance reinforced composites has been developed primarily in epoxy resins and thermoset polyesters. In reaction injection molding (RIM), addition of short glass fibers produces reinforced RIM (RRIM), with much improved rigidity and strength (Table 3.30). Similarly, addition of short glass fibers to rigid urethane foam produces major improvements in modulus and strength (Table 3.31). The most extreme reinforcement is, of course, observed when flexible polyurethane foam is coated onto cloth, for upholstery and clothing, combining the high planar modulus, strength, and dimensional stability of the cloth with the high transverse flexibility, softness, and thermal insulation of the foam.

TABLE 3.30 Reinforced Reaction Injection Molding RRIM

Type of RIM	Glass Fiber (%)	Flexural Modulus (psi)	Coeff. of Thermal Expansion (in/in/°F × 10^{10})
Semi-high modulus	none	60	78
Semi-high modulus	17	120	24
High modulus	none	275	70
High modulus	17	400	19
Rigid foam	none	50	60
Rigid foam	17	170	32

TABLE 3.31 Reinforcement of Rigid Low-Density PU Foam

1/4-in Glass Fiber	Compression Modulus (psi)	Compressive Strength (psi)	Flexural Modulus (psi)	Flexural Strength (psi)
0	537	31	1327	55
10	713	33	2376	76
20	761	35	2947	90
30	856	39	3296	94
40	1349	54	3837	111
50	1697	72	6136	148

Foams

When gas bubbles are dispersed in a solid polymer to form a foam, it is possible to combine some of the best properties of each phase and to produce some synergistic benefits as well. In polyurethane, the gas contributes light weight, thermal, and electrical insulation. The thermal insulation is used in rigid foams for refrigeration and freezer applications in flexible forms for winter outerwear clothing. In closed-cell foams, the gas contributes flotation, rigidity and strength, and impact absorption; while in open-cell foam, the fluidity of the gas contributes softness, elastic recovery, impact absorption, noise damping, filtration, and sponge performance. Thus, the gas phase, negligible in weight but most prominent in volume, contributes many useful properties and uses that the polyurethane alone did not have.

Coatings

Thin polyurethane surface films contribute useful properties far beyond their thin dimensions. They are easy to apply by a variety of techniques and can have a wide range of modulus as desired. They have strong adhesion to many substrates. They are strong, impact resistant, extremely abrasion resistant, and resistant to hot aging, aqueous solutions, fuels, oils, and many other chemicals. When aliphatic isocyanates are used, they are also very resistant to weathering. Aliphatic-based polyurethane coatings find increasing use in corrosion-preventive maintenance coatings.

Adhesives

Polyurethane combine the fluidity needed to wet irregular substrates, the cure reactions needed to convert them into high-molecular-weight materials of strong cohesion, and the range of polarities and hydrogen-bonding needed to form strong adhesive bonds with many types of surfaces. For these reasons, they have proved useful in a wide range of specialty adhesives.

3.2   Surface Analysis

Theoretically, a surface is an infinitely thin layer separating two phases. Surfaces are generally examined with spectroscopy, which involves probing a sample target with a flux of energetic particles and detecting characteristic particles emitted from the surface after interaction.

Activation may be achieved by electromagnetic radiation, as in electron spectroscopy for chemical analysis (ESCA); by a beam of incident ions, as in secondary ion mass spectrometer (SIMS) and ion scattering spectroscopy (IS); or by an incident electron beam, as in Auger electron spectroscopy (AES) or by electron microprobe (X-ray fluorescence) analysis.

Infrared spectrophotometry (IR) measures the energies involved in the twisting, bending, rotating, and vibrating motions of asymmetrical chemical bonds in a molecule. Upon interaction with infrared radiation, portions of the incident radiation are absorbed at particular wavelengths. These wavelengths correspond to particular vibrations of the bonds and thus yield information about the chemical structure of the absorbing species. A Fourier transform infrared spectrophotometer (FT-IR) makes use of computer technology and yields much higher resolution through processing of very large amounts of data obtained through numerous runs. IR and FT-IR give information about the chemical bonds present in the bulk structure. On the other hand, an attenuated total reflectance infrared spectrophotometer (AT-IR or ATR-IR) gives information about tic chemical bonds present on tic surface.

Adsorption and segregation depend on surface chemistry as well as on surface free energy. Determination of contact angles (advancing, receding, or critical) by the use of a goniometer is a convenient and simple method of measuring the surface free energies of the polymeric samples.

Polyurethanes have very widely varying structures, depending on the components used in the formulation. The aromatic isocyanates will give more urea groups in the finished structure than will the aliphatic diisocyanate. Excess NCO in the formulation will react with atmospheric moisture to form urea and barrette groups in the structure. Aromatic diamines will produce a great amount of urea groups in the structure of the polymer, with some amide groups and aromatic rings. Polyester components will give far more urethane groups, along with a few ester groups in some cases, and the urethane linkages will be predominant. Polyether components give a very large number of elephant groups in the polymer, with a few ether groups, and, in some cases, a few ester groups. The amount of these groups in comparison with the urethane linkages can be varied easily by choice of the individual polyether. Other groups such as urea and barrette can be decreased by reacting these hydroxyl components with an aliphatic diisocyanate at very near to the stoichiometric 1:1 ratio by avoiding any further reaction with atmospheric moisture and by using completely anhydrous components.

The structure of the diisocyanate and the structure of the hydrogen donors have a great influence on the final properties. NCO/OH ratios, the manner in which they are reacted, the presence of catalyst, and the cure conditions all have large degree of influence on the end product.

Although the types of ingredients are much more varied in polyurethane than other polymers, some analytical techniques, such as IR, are quite useful in the chemical characterization of polyurethanes. In IR analysis (or AT-IR or FT-IR), the wavelengths of the absorption bands for some groups can be summarized as in Table 3.32.

TABLE 3.32 Wavelengths of Absorption Bands

$-N=C=O$ (diisocyanate)	2275 cm^{-1}	2250 cm^{-1}	1350 cm^{-1}	
$-NH-COO-$ (urethane)	1640 cm^{-1}	1610 cm^{-1}	1650 cm^{-1}	1680 cm^{-1}
$C=C$ (in benzene)	800 cm^{-1}	1610 cm^{-1}	1500 cm^{-1}	
Aromatics	3030 cm^{-1}	1600 cm^{-1}	1500 cm^{-1}	
$-OH$ (hydroxyl)	3200 cm^{-1}	3400 cm^{-1}		
$C-O-C$ (ether)	1150 cm^{-1}	1070 cm^{-1}		
$ArNH_2$ (aromatic amine)	1350 cm^{-1}	1250 cm^{-1}		
$AlNH_2$ (aliphatic amine)	1280 cm^{-1}	1180 cm^{-1}		

However, interferences and shifts caused by the neighboring groups may be encountered. In large polymers such as polyurethane, the characteristic spectra become diffuse, and identification of small amounts of any given functionality becomes increasingly difficult.

The effect of secondary valence internal bonding through polar internees and hydrogen bonding is also manifested in the modulus of elasticity and hardness of the cured product, which increase markedly with increases in the proportion of diisocyanate to glycol. Although this increase is accompanied by some loss in resilience it is possible to obtain very hard products that still retain elastomeric character. In the production of high-grade, high-modulus elastomers, symmetrical diisocyanate such as naphthalene 1,5-diisocyanate, phenylene 1,4-diisocyanate, and diphenylmethane 4,4'-diisocyanate, and symmetrical glycol such as 1,4-butanediol, 1,6-hexanediol, p-bis(β-hydroxy ethoxy)-benzene, 1,5-bis-(β-hydroxy ethoxy)-naphthalene, are useful.

Unsymmetrical diisocyanate and glycol result in very different processing characteristics and physical properties. Low proportions of the diisocyanate-glycol urethane segment yield elastomers that are substantially softer and of lower modulus than analogous compositions based on symmetrical diisocyanate. Interruption attractive forces between rigid segments are very high, because the urethane group is highly polar and exhibits a strong tendency to hydrogen bonding between the $-NH-$ and the $-COO-$ parts of the group. Consequently, it has a strong tendency toward association and a high energy of cohesion. Regular polyurethane are generally similar in character to other polymers containing regularly occurring groups with high energy of cohesion.

References

1. Gillis, H.R. 1994. Polyurethanes. *Modern Plastics Encyclopedia Handbook*, 82. New York: McGraw Hill.
2. Buist, J.M., and Gudgeon, H.A. 1968. *Polyurethane Technology*, London: McClaren and Sons, Ltd.
3. Lyman, D.J. 1960. *J. Pol. Sc.,* 45:49.
4. Pentz, W.J., and Krawiec, R.G. *Rubber Age*, 43. December 1975.
5. Pigott, K.A. Polyurethanes, in *Encyclopedia of Polymer Science and Technology*, 11:506–563.
6. Lovering, E.J., and Laedler, K.J. *Canadian J. of Chem.*, 40, 26 (1972).
7. Saunders, J.H., and Frisch, K.C. *Polyurethanes: Chemistry and Technology.* Ch. 6, 9, 12. New York: Interscience Publishers, 1962.
8. Foks, J., et al. Morphology and Thermal Properties of Polyurethanes Prepared under Different Conditions, *Eur. Polym. J.*, 25:31–37, 1989.
9. Wilkes, G.L., Dziemianowicz, T.S., Ophir, Z.H., and Wildnauer, R., Thermally Induced Time Dependence of Mechanical Properties in Biomedical Grade Polyurethanes, *J. Biomed. Mater. Res.*, 13:189–206, 1979.

10. Dzierza, W., Stress-Relaxation Properties of Segmented Polyurethane Rubbers, *J.Appl. Polym. Sci.,* 27:1487–1499, 1982.

11. Wong, E.W.C. Development of a Biomedical Polyurethane, in *Urethane Chemistry and Applications,* ACS Series, 489–504, 1981.

12. Rausch, K.W., and Sayigh, A.A.R. *Ind. Eng. Chem, Prd. Res. Dev.,* 4(2):92, 1965.

13. Cooper, S.L., and West, J.C. Polyurethane Block Polymers, *Encyclopedia of Polymer Science and Technology,"*16:521–543.

14. Axelrod, S.L., and Frisch, K.C. *Rubber Age,* 88(3):465, 1960.

15. Saunders, J.H. and Frisch, K.C. *Polyurethanes Chemistry and Technology,* Parts I and II, New York: Interscience Pub. Inc., 1962 and 1964.

16. Lyons, J.W. *The Chemistry and Uses of Fire Retardants.* Ch. 8. New York: Wiley-Interscience, 1970.

17. Meckel, W.W., Goyert W., and Wieder, W. *Thermoplastic Elastomers.* Ch. 2. Legge, Holden and Schroeder, eds. Munich: Hanser Publishers, 1987.

18. Deanin, R.D., Murarka, M.R., and Kapasi, V.C. *SPE ANTEC,* 31:1297, 1985.

4

Isocyanate Chemistry

4.1 World Capacity

The modern polyurethane industry is based on isocyanate chemistry. This organic functional group is capable of a surprisingly diverse range of chemical reactions. As the foundation of the polyurethane industry, isocyanates are among the most heavily produced specialty organic chemicals. Table 4.1 presents the 1993 isocyanate world capacity. Table 4.2 presents some of the global trademarks utilized in isocyanates.

TABLE 4.1 1993 Isocyanate Production of Major Manufacturers (1000 tons)

Manufacturer	TDI	MDI	PMDI	Aliphatics
NORTH AMERICA				
BASF	95	12.5	100	—
Dow	70	40.5	163.5	—
ICI (Rubicon)	32.5	25	150	—
Bayer	132.5	35	167.5	16
Olin	110	—	—	—
Total North America	**485**	**113**	**592.5**	**24.5**
EUROPE				
ICI	—	11.5	142.5	—
BASF	17.5	29.5	73	pilot plant
Bayer	206	67.5	209	10
Montedison	121	3.5	88	—
Rhone-Poulenc	126.5	—	—	5
Dow	—	39.5	181.5	—
Total Europe	**530**	**176.7**	**905**	**27.5**
PACIFIC				
Korea Fine Chemical	27.5	—	—	—
Mitsui Toatsu	61	60.5	0.2	—
Takeda	82.5	—	—	0.6
BASF	—	12.5	30	—
Dow Mitsubishi	—	14	15.5	—
Nippon Polyurethanes	—	20	110	—
Sumitomo Bayer	—	12.5	41	—
Sunghua	—	11.5	27	—
Total Pacific	**308**	**775**	**306**	**3.3**
LATIN AMERICA				
Pronor	70.5	—	22	—
Bayer	13	—	22	—
Total Latin America	**102.6**	**7.5**	**44**	**—**

Commercially available organic isocyanates include aliphatic, cycloaliphatic, araliphatic, aromatic, and heterocyclic polyisocyanates. Examples of isocyanates are ethylene diisocyanate; 1,4-tetramethylene diisocyanate; 1,6-hexamethylene diisocyanate; 1,12-dodecane diisocyanate; cyclobutane-1,3-diisocyanate;

TABLE 4.2 Trademarks Utilized by Global Isocyanate Manufacturers

Manufacturer	Trademark	Region
Akzo	Elate	North America
Bayer	Desmodur	Europe
Bayer	Multrathane	North America
Bayer	Mondur	North America
BASF	Ekanate	North America
BASF	Lupranat	Europe
BASF-Schwarzheide	Systanat	Europe
Cariosa	Cortume	Europe
Dainippon	Sothanate	Pacific
Dainichi Saika	Resamine	Pacific
Dow	Isonate	North America
Dow	Voranate	North America
ICI	Suprasec	Europe
Interchem	Prepol	Europe
Lancro	Quasilan	Europe
Lancro	Isocon	Europe
Montedison	Tedimon	Europe
Olin	Olin-TDI	North America
Polymer Chemicals	Polidur	Europe
Reichold	Polylite	North America
Rheinchemie	Rhenodur	Europe
Rhone-Poulenc	Scurane	Europe
Rhone-Poulenc	Tolonate	Europe
Rubicon	Rubinate	North America
Shell	Caradate	Europe
Takeda	Takenate	Europe
Thanex	Poronat	Europe
Thiokol	Solithane	North America

cyclohexane-1,3- and -1,4-diisocyanate, and mixtures of these isomers; 1-isocyanato-3,3,5-trimethyl-5-isocyanatomethylcyclohexane (see, e.g., U.S. Pat. No. 3,401,190); 2,4- and 2,6-hexahydrotoluene diisocyanate and mixtures of these isomers; hexahydro-1,3- and/or -1,4-phenylene diisocyanate; perhydro-2,4'- and/or -4,4'-diphenylmethane diisocyanate (HMDI); 1,3- and 1,4-phenylene diisocyanate; 2,4- and 2,6-toluene diisocyanate and mixtures of these isomers ("TDI"); diphenylmethane-2,4'- and/or -4,4'-diisocyanate (MDI); naphthylene-1,5-diisocyanate; triphenylmethane 4,4',4"-triisocyanate; polyphenyl-polymethylene-polyisocyanates of the type that may be obtained by condensing aniline with formaldehyde, followed by phosgenation (*crude* MDI), which are described, for example, in British Pats. 878,430 and 848,671.

We also have norbornane diisocyanates, such as described in U.S. Pat. No. 3,492,330; m- and p-isocyanato phenyl sulfonylisocyanates of the type described in U.S. Pat. No. 3,454,606; perchlorinated aryl polyisocyanates of the type described, e.g., in U.S. Pat. No. 3,227,138; modified polyisocyanates containing carbodiimide groups of the type described in U.S. Pat. No. 3,152,162; modified polyisocyanates containing allophanate groups of the type described, e.g., in British Pat. 994,890, Belgian Pat. 761,616, and published Dutch Pat. Application 7,102,524; modified polyisocyanates containing isocyanurate groups of the type described, e.g., in U.S. Pat. No. 3,002,973, in German Patentschriften 1,022,789, 1,222,067 and 1,027,394, and in German Offenlegungsschriften 1,919,034 and 2,004,048; modified polyisocyanates containing urea groups of the type described in German Patentschrift 1,230,778; polyisocyanates containing biuret groups of the type described, e.g., in German Patentschrift 1,101,394, U.S. Pat. Nos. 3,124,605 and 3,201,372, and in British Pat. 889,050; polyisocyanates obtained by telomerization reactions of the type described, e.g., in U.S. Pat. No. 3,654,106; polyisocyanates containing ester groups of the type described, e.g., in British Pats. 965,474 and 1,072,956, in U.S. Pat. No. 3,567,763; reaction products of the above-mentioned isocyanates with acetals as described in German Patentschrift 1,072,385; and polyisocyanates containing polymeric fatty acid groups of the type described in U.S. Pat. No. 3,455,883. Structures of important diisocynates are shown in Fig. 4.1.

4.2 Historical

The first synthetic route for isocyanates was reported by Wurtz in 1848.[1] The work of Wurtz clearly demonstrated the preparation of monoisocyanates and the subsequent formation of urethanes and substituted urea linkages by reaction of the isocyanate group with a primary alcohol and a secondary amine respectively.

Subsequent efforts by Hofmann, Curtius, and Hentschel pioneered alternative synthetic approaches.[2] These efforts highlighted the phosgene-amine approach. Staudinger elucidated the structural similarities between isocyanates and ketenes and stimulated interest in this class of compounds.[3] However, it was not until 1945, when the world was pressed for an alternative to natural rubber, that synthetic routes to isocyanates became an area of commercial importance. Since then, scores of different reactions of organic isocyanates have been described in the chemical literature. Most of these have been used experimentally for the synthesis of polymeric materials; however, only a handful have reached industrial significance. The reactions of isocyanates fall into two broad categories:

1. active hydrogen donors
2. nonactive hydrogen reactions

The most important category comprises reactions involving active hydrogens. This category of reactions requires at least one coreagent containing one or more hydrogens that are potentially exchangeable (or labile) under the conditions of reaction. The familiar reaction of isocyanates with polyols, to form polycarbamates, is of the *active hydrogen* type. The active hydrogen groups are, in this case, the hydroxyl

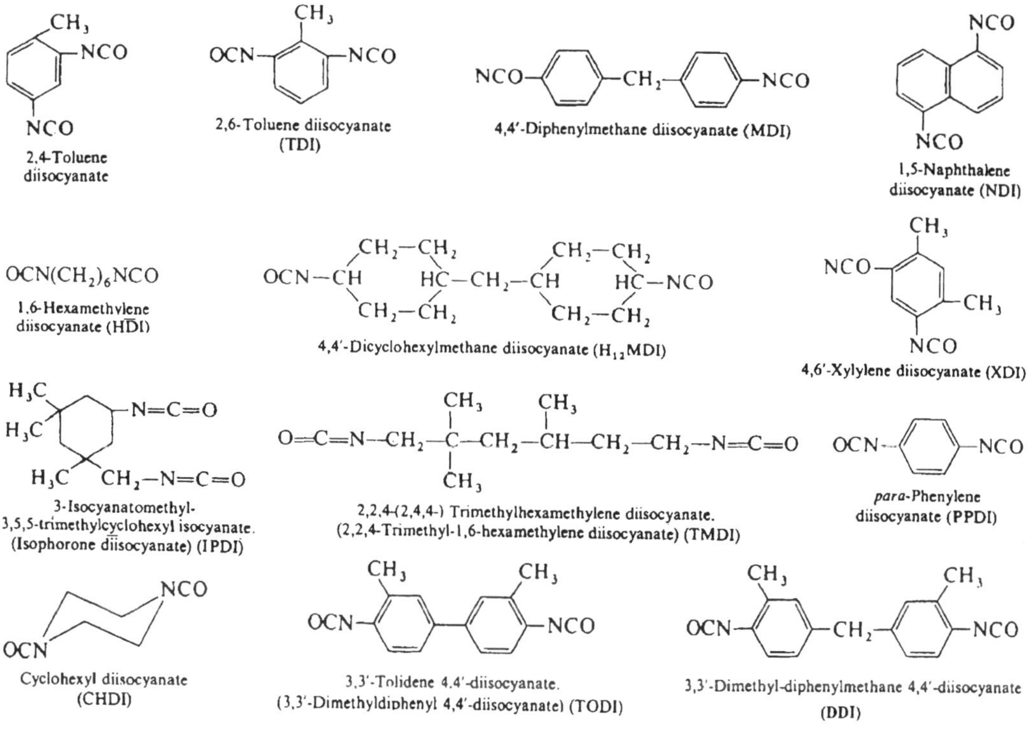

FIGURE 4.1 Structures of important diisocyanates

groups on the polyol. Nonactive hydrogen reactions constitute the second broad category. These include cycloaddition reactions and linear polymerizations (which may or may not involve coreagents). Table 4.3 lists some of the known polymer-forming reactions of isocyanates.

TABLE 4.3 Examples of Polymer-Forming Reactions of Isocyanates

Co-Reactant	Product(s)	Catalysts	Category
Alcohols	Carbamates	3° amines	Active-H
		Tin soaps	
		Alkali soaps	
Carbamates	Allphanates	3° amines	Active-H
		Tin soaps	
		Alkali soaps	
Primary amines	Ureas	Carboxylic acids	Active-H
Secondary amines	Ureas	Carboxylic acids	Active-H
Ureas	Biurates	Carboxylic acids	Active-H
Imines	Ureas	Carboxylic acids	Either
	Amides		
	Triazines		
Enamines	Amides	Carboxylic acids	Either
	Heterocycles		
Carboxylic acids	Amides	Phospholene oxides	Active-H
Amides	Acyl ureas	Acids bases	Active-H
H_2O	Ureas	3° Amines	Active-H
	CO_3	Alkali soaps	
Anhydrides (cyclic)	Imides	Phospholene oxides	Nonactive hydrogen
	CO_2		
Ketones	Imines	Alkali soaps	Either
	Heterocycles	Alkali soaps	Either
	CO_2		
Aldehydes	Imines	Alkali soaps	Either
	Heterocycles	Alkali soaps	Either
	CO_2		
Active methylene	Amides	Bases	Either
(Active methine) compounds	Heterocycles		
Isocyanates	Carbodiimides	Phosphorous-oxygen compounds	Nonactive
	Dimers	Pyridines	Nonactive
	Trimers	Alkali soaps	Nonactive
	Polymers	Strong bases	Nonactive
Carbodiimides	Uretonimines	—	Nonactive
Epoxides	Oxazolidones	Organoantimony iodides	Either
2 or B Hydroxy acids (esters)	Heterocycles	Acids/bases	Active-H
	ROH or H_2O		
2 or B Amino acids (esters)	Heterocycles	Acids/bases	Active-H
	ROH or H_2O		
Cyanohydrins	Heterocycles	Acids/bases	Active-H
2-Cyano amines	Heterocycles	Acids/bases	Active-H
2-Amino esters	Heterocycles	Acids/bases	Active-H
	ROH		
Orthoformates	Heterocycles	Acids/bases	Active-H
	ROH		
Oxazolines; imidazolines	Various heterocycles	Acids/base	Either
Cyclic carbonates	Heterocycles	Bases	Either
	CO_2		
Acetylenes	Various heterocycles	Various	Either
Pyrroles	Amides	Acids/bases	Active-H
Carbamic acid	Ureas	—	Active-H
Amine salts	CO_2		

Source: H.R. Gillis, Polyurethanes, *Modern Plastics Eng. Handbook,* p. 83, 1994.

The distinctions between active-hydrogen and nonactive hydrogen categories are sometimes blurred, The pathway often depends on the reaction conditions and the precise structure of the chemical groups involved. Several types of reactions are frequently involved in any given chemical system. Active hydrogen compounds may catalyze reactions that do not belong to the active hydrogen category.

The diversity of isocyanate chemistry, combined with the availability of selective catalysts, has made it possible to select reactions which "fit" desired modes and rates of processing. The physical form and properties of the polymer can also be tailored by selecting from among a broad range of monomers. Polyurethane polymer parameters and their influence on properties are presented in Table 4.4. Many of the chemical reactions in Table 4.4 are addition reactions that produce no by-products. Others are condensation reactions that produce gases, such as CO_2, in addition to polymer.

TABLE 4.4 Polymer Parameters and their Influences on Properties

	Density		Molecular Weight		Molecular Weight Distribution	
	Increases	Decreases	Increases	Decreases	Broadens	Narrows
Environmental Stress	∇	Δ	Δ	∇	Δ	∇
Impact strength	∇	Δ	Δ	∇	∇	Δ
Stiffness	Δ	∇	—	—	—	—
Hardness	Δ	∇	—	—	—	—
Tensile strength	Δ	∇	Δ	∇	—	—
Permeation	∇	Δ	—	—	—	—
Warpage	Δ	∇	—	—	Δ	∇
Abrasion resistance	—	—	Δ	∇	—	—
Flow processibility	—	—	∇	Δ	Δ	∇
Melt strength	—	—	Δ	∇	Δ	∇
Melt viscosity	—	—	Δ	∇	∇	Δ
Tear strength	—	—	Δ	∇	—	—

Isocyanate-based polymers cover a range of densities extending from about 70 lb/ft^3 to well below 1 lb/ft^3. The range of bulk flexural moduli (77° F) extends below 4000 psi (soft rubber) to well over 400,000 psi (elastoplastic). Much higher flex moduli can be achieved through the use of reinforcing fibers or fillers (i.e., 2,000,000 psi and up). Fiber reinforced structural composites, based on MDI, can attain notched Izod impact strength in excess of 25 ft-lb/inch. With these composites, heat distortion temperatures (264 psi) of over 570° F. can be achieved. Ultimate elongations of PU polymers range from a low of 1 to >1000% in the softest grades.

Combinations of properties that compare favorably with engineering resins (nylons, polyesters, etc.) can be obtained with isocyanate-based polymers. These polymers include polyurea elastoplastics (which contain little or no urethane linkages) and mat-reinforced polyisocyanurate ("trimer") composites, made by the reaction injection molding process (RIM, a fast two-component, mixing-activated, molding process). The RIM process can mold large, complex parts from such materials at rates that exceed one cycle per minute in some systems.

Factors that determine properties include chemical composition (the *soft block* vs. *hard block* ratio), the types of chemical linkages formed (the type of isocyanate reaction involved), density, degree of cross-linking, and a molecular-segregation phenomenon known as *phase separation* or *segmentation*.

The nature of the linkages formed from the reaction between isocyanate and isocyanate-reactive monomers influences the properties of the resulting polymer. Linkages incorporated into polymers to improve various thermal properties (relative to straight urethanes) include ureas, amides, imides, triazines, carbodiimides, oxazolidones, and other heterocyclic species.

The chemical and structural diversity of the many materials called *polyurethanes* is so broad that it is impossible to describe them in terms of nominal or average properties. As the technology has evolved, the ranges of properties that can be achieved with isocyanates are unrivaled by any other polymer.

4.3   The Isocyanate Functionality

Isocyanates are derivatives of isocyanic acid, H-N=C=O, in which alkyl or aryl groups, as well as a host of other substrates, are directly linked to the N=C=O moiety via the nitrogen atom. Structurally, isocyanates (imides of carbonic acid) are isomeric to cyanates, ROCN (nitriles of carbonic acid), and nitrile oxides, RCN-O (derivatives of carboxylic acid).

The isocyanate functionality [-N=C=O] is highly reactive toward proton-bearing nucleophiles. This reaction occurs by nucleophilic addition across the carbon nitrogen double bond. Aromatic isocyanates are generally more reactive than aliphatic isocyanates. The presence of electron withdrawing substituents on the ring of aromatic isocyanates tends to increase reactivity whereas electron donors decrease reactivity. Bulky substituents adjacent to isocyanate groups tend to decrease reactivity due to steric hindrance in addition to inductive effects. The reactions of isocyanates are catalyzed by acids and bases of both Bronsted and Lewis character.

The isocyanate reaction normally proceeds by addition to the carbon-nitrogen double bond. The classical "active hydrogen" reactions involves compounds containing a replaceable active hydrogen, i.e., a hydrogen replaceable by sodium. In the active hydrogen reaction, the hydrogen becomes attached to the isocyanate nitrogen, and the remainder of the active hydrogen radical (R') becomes covalently attached to the carbonyl carbon, according to the reaction below.

REACTION 4.1 Reaction of isocyanates with active hydrogen compounds.

Other reactions not involving active hydrogens usually involve breaking the susceptible carbon-nitrogen bond as well. A summary of commercially important isocyanate reactions is presented in Figs. 4.2 and 4.3, respectively. These reactions are discussed in greater detail in the following sections.

4.3.1   Reaction of Isocyanate with [-OH] Compounds

Alcohols and Phenols

Primary alcohols react vigorously with most isocyanates to form urethanes in quantitative yield. Secondary alcohols are only slightly less reactive than 1^0 alcohols and also react to form urethanes in quantitative yield. Tertiary alcohols, on the other hand, are up to several orders of magnitude less reactive.

The reaction of phenols with isocyanates to form urethanes is significantly slower than with aliphatic alcohols due to the relatively high acidity of the phenols. This reaction proceeds only with the assistance of catalysts and due to the more vigorous reaction conditions employed may be accompanied by side reactions. The urethane formed by this reaction are less stable than those from aliphatic alcohols and may be reversible at temperatures as low as 100° C. This characteristic has been commercially used in the preparation of blocked isocyanates.

4.3.2   Reactions of Isocyanates with [-NH] Compounds

There is a broad range of compounds containing an N-H bond. Such compounds are nearly always reactive with isocyanates. Primary and secondary amines react vigorously with isocyanates to yield urea and substituted ureas.

Due to their strong nucleophilic and autocatalytic nature, the reactions of 1^0 aliphatic amines can be extremely vigorous. In such cases, the second proton on the H can react to form a biuret, reducing the

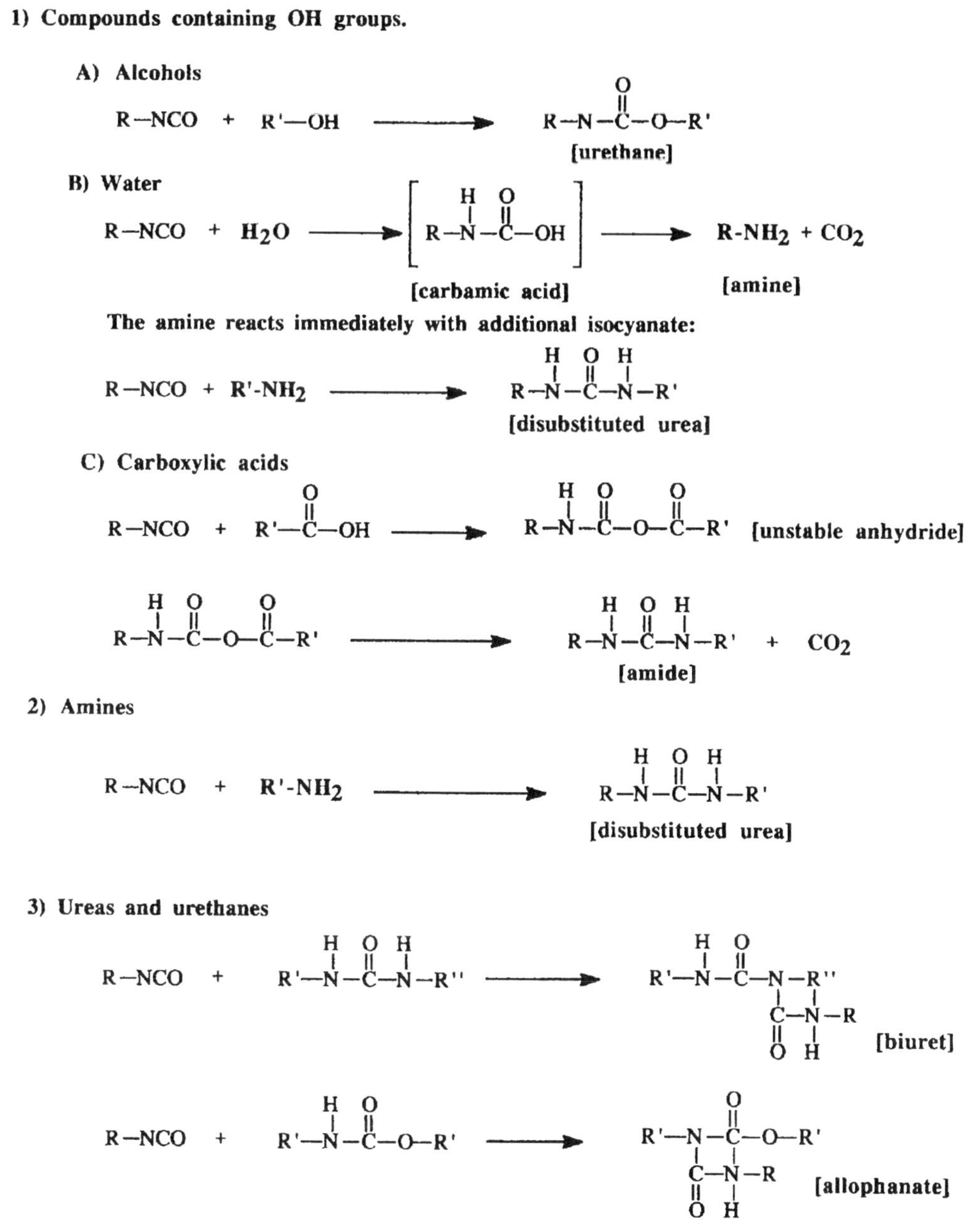

1) Compounds containing OH groups.

A) Alcohols

$$R-NCO \ + \ R'-OH \ \longrightarrow \ R-N-\overset{\overset{\textstyle O}{\|}}{C}-O-R'$$

[urethane]

B) Water

$$R-NCO \ + \ H_2O \ \longrightarrow \ \left[R-\overset{\overset{\textstyle H}{|}}{N}-\overset{\overset{\textstyle O}{\|}}{C}-OH \right] \ \longrightarrow \ R\text{-}NH_2 + CO_2$$

[carbamic acid]   [amine]

The amine reacts immediately with additional isocyanate:

$$R-NCO \ + \ R'\text{-}NH_2 \ \longrightarrow \ R-\overset{\overset{\textstyle H}{|}}{N}-\overset{\overset{\textstyle O}{\|}}{C}-\overset{\overset{\textstyle H}{|}}{N}-R'$$

[disubstituted urea]

C) Carboxylic acids

$$R-NCO \ + \ R'-\overset{\overset{\textstyle O}{\|}}{C}-OH \ \longrightarrow \ R-\overset{\overset{\textstyle H}{|}}{N}-\overset{\overset{\textstyle O}{\|}}{C}-O-\overset{\overset{\textstyle O}{\|}}{C}-R'$$ [unstable anhydride]

$$R-\overset{\overset{\textstyle H}{|}}{N}-\overset{\overset{\textstyle O}{\|}}{C}-O-\overset{\overset{\textstyle O}{\|}}{C}-R' \ \longrightarrow \ R-\overset{\overset{\textstyle H}{|}}{N}-\overset{\overset{\textstyle O}{\|}}{C}-\overset{\overset{\textstyle H}{|}}{N}-R' \ + \ CO_2$$

[amide]

2) Amines

$$R-NCO \ + \ R'\text{-}NH_2 \ \longrightarrow \ R-\overset{\overset{\textstyle H}{|}}{N}-\overset{\overset{\textstyle O}{\|}}{C}-\overset{\overset{\textstyle H}{|}}{N}-R'$$

[disubstituted urea]

3) Ureas and urethanes

$$R-NCO \ + \ R'-\overset{\overset{\textstyle H}{|}}{N}-\overset{\overset{\textstyle O}{\|}}{C}-\overset{\overset{\textstyle H}{|}}{N}-R'' \ \longrightarrow \ R'-\overset{\overset{\textstyle H}{|}}{N}-\overset{\overset{\textstyle O}{\|}}{C}-\overset{\overset{\textstyle |}{N}}{\underset{}{}}-R''$$

[biuret]

$$R-NCO \ + \ R'-\overset{\overset{\textstyle H}{|}}{N}-\overset{\overset{\textstyle O}{\|}}{C}-O-R' \ \longrightarrow \ R'-\overset{\overset{\textstyle |}{N}}{}-\overset{\overset{\textstyle O}{\|}}{C}-O-R'$$

[allophanate]

FIGURE 4.2   Isocyanate chemistry, summary reactions I.

yield of urea. This biuret reaction is discussed further below and in subsequent sections. In the case of aromatic amines the reaction with isocyanates is less vigorous overall and secondary reactions are reduced due to inductive and for steric mechanisms.

4.3.3   Reactions of Isocyanates with Amides

Due to the electron withdrawing influence of the adjacent carbonyl group, the N-H protons of amide and substituted amides are less reactive than aliphatic or aromatic amines. The primary reaction product here are acyl ureas or substituted acyl ureas.

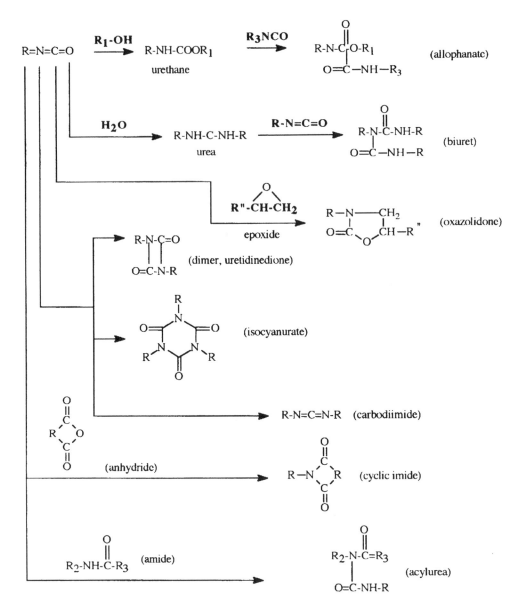

FIGURE 4.3   Isocyanate reactions, summary reactions II.

4.3.4   Reactions of Isocyanates with Urea

The ureas formed by the reaction of 1° amines and isocyanates contain an active proton capable of reacting with isocyanates to form stable biuret cross-links. In contrast with the amide, in this case the inductive influence of the adjacent carbonyl is mitigated by the opposite nitrogen making the urea more reactive than the substituted urea.

4.3.5   Reactions of Isocyanates with Urethane Linkages

Like urea, urethane groups contain an N-H adjacent to a carbonyl group and opposite to an electron donor (-O-). The urethane proton is sufficiently reactive that under moderate to vigorous reaction conditions significant amounts of allophanate are formed.

In polymerizing systems, this reaction like the foregoing biuret formation, results in a branch in the growing chain and results in a stoichiometric imbalance due to the consumption of two moles of isocyanate by one mole of R-OH or R-NH. The allophanate linkage is thermally labile and may dissociate to the precursor isocyanate at temperatures of 150° C or higher.

4.3.6 Reactions of Isocyanates with Water

The reaction of isocyanate with water is vigorous and somewhat complex—vigorous because isocyanates react readily and quickly with water, particularly under the influence of catalysts. Complex because an unstable intermediate is first formed (carbamic acid), with the concomitant evolution of CO_2. The carbamic acid subsequently decomposes to an amine. Recalling that amines react with isocyanates and are autocatalytic, the reaction proceeds to the formation of a biuret cross-link.

In a typical polymerization system, the result of these events are the evolution of CO_2, chain extension followed by branching, and, ultimately, the consumption of three moles of isocyanate for each mole of H_2O. Considering the low molar mass of water compared to the other molecules in a urethane polymerization system, clearly even a small amount of H_2O can have a profound effect on the reaction sequence and ultimate molecular structure.

The most important reaction used in the foam industry is that of polyisocyanates with water to produce CO_2 and a polymeric urea. Other convenient CO_2-producing polymerizations include the self-condensation of isocyanates to produce carbodiimides, and reaction of isocyanates with activated (i.e. hydroxy functional) ketones/aldehydes. Foams are also manufactured by incorporating a volatile liquid (e.g., chlorofluorocarbons), which is evaporated by the reaction exotherm.

4.3.7 Reactions of Isocyanates with Carboxylic Acids

Carboxylic acids like amines are a class of materials that are reactive with isocyanates and are also catalytic toward isocyanate reactions. The reactivity of carboxylic acids is relatively mild compared to amines, alcohols, and water. The pathway of the reaction varies, depending on the nature of the isocyanate and to a less extent the acid.

4.3.8 Reactions of Isocyanates with [-CH-OCH-]; 2-oxazolidone Formation

In the presence of catalysts, isocyanates react with 1,2 epoxides to form 2-oxazolidones.

4.3.9 Dimerization

Aromatic isocyanates undergo dimerization to form uretidine dione. Dimerization may occur spontaneously or may be catalyzed. Certain materials such as trialkyl phosphines strongly catalyze dimer formation. Alternatively, under the influence of selective catalysis, the dimerization of isocyanate yield a carbodiimide group.

One special feature of isocyanates is their propensity to dimerize and trimerize (i.e., undergo oligomerization). The aromatic isocyanates in particular are known to undergo these reactions, even in the absence of a catalyst. The dimerization product bears a strong dependency on both the reactivity and structure of the starting isocyanate. For example, aryl isocyanates dimerize, in the presence of phosphorus-based catalysts, by a crosswise addition to the C=N bond of the NCO group to yield a symmetrical dimer (see Reaction 4.2).

REACTION 4.2 Dimerization reaction in aromatic isocyanates.

Slow dimerization is generally noted to occur in some isocyanates during prolonged storage. The tendency for 4,4'-methylene diphenyl diisocyanate to undergo uncatalyzed dimerization is related to its crystal structure. The molecules of the 4, 4'-MDI align in the solid state, with the NCO groups in close proximity, which leads to slow formation of the dimer at room temperature. The structure of the symmetrical MDI dimer has been verified by X-ray analysis. For this reason, MDI is usually kept refrigerated at 0° C. It has been reported that substituted benzyl isocyanates form mixtures of both dimers and trimers in high yield when 1,2-dimethylimidazole is used as a catalyst. Conversely, acyl isocyanates yield dimers that include the C=X moiety (where X = O, S, NR) in the product ring structure (see Reaction 4.3).

REACTION 4.3 Formation of dimers containing O, S, and NR groups.

Reportedly, simple alkyl isocyanates do not dimerize upon standing. They trimerize to isocyanurates under comparable conditions.[4] Dimerization is catalyzed by pyridine and phosphines. Trialkylphosphines also catalyze the conversion of dimer into trimer upon prolonged standing. Pyridines and other basic catalysts are less selective because the required increase in temperature causes trimerization to compete with dimerization. The gradual conversion of dimer to trimer in the catalyzed dimerization reaction can be explained by the assumption of equilibria between dimer and polar catalyst-dimer intermediates. The polar intermediates react with excess isocyanate to yield trimer. Factors such as charge stabilization in the polar intermediate and its lifetime or steric requirement, are reported to be important. For these reasons, it is not currently feasible to predict the efficiency of dimer formation given a particular catalyst.

4.3.10 Trimerization

Trimerization occurs with both aliphatic and aromatic isocyanates. Dimerization catalysts such as trialkylphosphine also induce trimerization, as do a number of acids, amines, and carboxylates. The product of trimerization is a trisubstituted isocyanurate (Reaction 4.4).

REACTION 4.4 Formation of isocyanurates (1,3,5 trisubstituted hexadydro-s-triazinetrione).

Unlike dimers, trimers are often quite stable. Much work has been done in the area of isocyanate trimerization as the result of the commercial interest in this reaction to produce rigid isocyanurate foam.

Both alkyl and aryl isocyanates trimerize upon heating or in the presence of catalysts to 1,3,5, trisubstituted hexahydro-s-triazinetriones (isocyanurates). Only highly substituted isocyanates, such as tert-butyl isocyanate and tert-octyl isocyanate, fail to trimerize under these conditions.

Commercially, polymeric MDI is trimerized during the manufacture of rigid foam to provide improved thermal stability and flammability performance. Numerous catalysts are known to promote the reaction. Tertiary amines and alkali salts of carboxylic acids are among the most effective. The common thread in all catalyzed trimerizations is the activation of the C=N double bond of the isocyanate group. The example highlights the alkoxide assisted formation of the cyclic dimer and the importance of the subsequent intermediates. Similar oligomerization steps have been described previously for other catalysts.[5]

Interestingly, methyl isocyanate is noted to form unusual trimer products in the presence of trialkylphosphine catalysts. Both the expected triazine and 3,5-dimethyl-2-methylimino-4,6-dioxohexahydro-1,3,5-oxadiazine products are formed.

Diisocyanates undergo anionic homopolymerization at subambient temperatures in polar solvents to yield high-molecular-weight cross-linked isocyanates. This type of polymerization generally has been observed for short-chain, aliphatic diisocyanates, which are structurally conducive to an alternating intermolecular and intramolecular propagation mechanism.

4.3.11 Insertion Reactions

Isocyanates also may undergo insertion reactions with C-H bonds. Acidic compounds, such as 1,3-dicarbonyl compounds, react readily at room temperature to form carboxyamides. At higher temperatures carboxyamides frequently undergo secondary reactions leading to cyclized products,[6] as shown in Reaction 4.5.

REACTION 4.5 Formation of cyclized compounds.

Another common process involves reaction with C=C or C=N species having adjacent CH_2 or CH_3 groups. Initial attack of the isocyanate is on the electron-rich center of the double bond with subsequent migration and insertion of the CONR group into the CH bond. Richter found that suitable reagents include N-alkylated acetamidines, I-methyl dihydroisoguinoline, and 2-methyl-2-oxazoline (Reaction 4.6).

REACTION 4.6 Insertion reactions with C=C and C=N species.

A variety of olefins or aromatic compounds having electron-donating substituents are known to undergo C-H insertion reactions with isocyanates to form amides. Many of these reactions are known to involve cyclic intermediates.

Isocyanates insert into RO and RN bonds. Cyclic ethers, such as oxiranes, are known to undergo reactions with isocyanates to form 2-oxazolidinones in high yield. Similarly, dimethoxymethane or cyclic acetals react to form carbamate in the presence of catalysts (Reaction 4.7).

REACTION 4.7 Formation of carbamates from dimethoxymethane.

Tertiary amines have been shown to react with isocyanates in an analogous fashion to form ureas.[7] Similarly, aziridines (three-membered rings containing nitrogen) are found to react with isocyanates to yield cyclic ureas. Tertiary amines also form labile dipolar I-1 adducts with isocyanates reminiscent of salt formation. In contrast, formaldehyde N,N-acetal aminals form insertion products with sulfonyl isocyanates.

4.3.12 Cycloaddition Reactions

Isocyanates undergo cycloadditions across the carbon-nitrogen double bond with a variety of unsaturated substrates. Addition across the C=O bond is less common. The propensity of isocyanates to undergo cyclization reactions has been widely explored for the synthesis of heterocyclic systems. Substrates with C=O, C=N, C=S, and C =C bonds have been found to yield either 2 + 2, 2 + 2 + 2, or 2 + 4 cycloadducts or a variety of secondary reaction products.

Most reactions of this type were found to involve acyclic 1,4-dipolar intermediates which cyclize to four-membered heterocycles or are intercepted by isocyanate or C=X components, such as C=N, C=S, and C=C, to form a six-membered ring. Variations in the component ratio or judicious choice of reagents are noted to have pronounced control of product type. Additional reaction details, as well as a description of the multiple transformations involving adjacent substituents, have been summarized by Gurgiolo in U.S. Pat. 4,268,683, assigned to Dow Chemical (Reaction 4.8).

REACTION 4.8 Cycloaddition reactions of isocyanates.

The dimerization and trimerization of isocyanates are special cases of the cycloaddition reaction in that they involve reagents of the same type. The uncatalyzed carbodiimidization of isocyanates likely involves a labile 2 + 2 cycloadduct which liberates carbon dioxide.

Acyl isocyanates (13, X = 0, S) have been shown to react as heterodienes in most cycloadduct formations. Notable examples include autodimerization and the addition to imines. Unlike aromatic isocyanates, it is not possible to predict the reaction pathway nor the structure of the products that may arise from a given approach or set or reaction conditions.

A large number of Diels-Alder-type reactions, involving both aromatic and sulfonyl isocyanates, have been reported. Heterodienes having high electron density are found to add to isocyanates to form six-membered heterocycles.

A comprehensive review of reactions of isocyanates with 1,3 dipolar compounds has been published by Van Look.[8] Reaction 4.9 illustrates the reaction of azides and isocyanates to yield tetrazoles.

4.3.13 Polymerization Reactions

The most commonly referenced reaction of isocyanates involves their addition to polyhydroxyl, polyamine, or polycarboxylic acid compounds to yield addition polymers. Due to the wide diversity of

REACTION 4.9 Formation of tetrazoles from 1,3 dipolar compounds.

raw material characteristics and the broad range of functionality, polyurethane polymers having a wide-range of processing and performance characteristics are available.

The reaction of isocyanates with alcohols to form carbamates is catalyzed by amines and a variety of organometallic compounds (Reaction 4.10).

REACTION 4.10 Polymerization reaction.

This important reaction is the bedrock of the polyurethane industry. Certain carbamates are known to reversibly yield the isocyanate and polyol upon heating. This fact has been commercially used to synthesize a number of blocked isocyanates for elastomer and coating applications.

Similarly, thioalcohols and thiophenols react with isocyanates to form thiocarbamates. Although these reactions are generally found to be much slower than that of the corresponding alcohol, alkoxide catalysts have successfully been used to provide moderate levels of rate enhancement.

Conversely, the rate of reaction of isocyanates with amines to yield ureas is both rapid and quantitative. Much has been written concerning the reaction kinetics, solvent effects, and catalysis of this reaction. The rate of reaction is a strong function of the basicity of the amine. Commercially, this relationship has been used to develop a wide variety of sterically hindered or electronically deactivated aromatic diamine chain extenders for reaction injection molding (RIM) and elastomer applications.

Industrially, polyurethane flexible foam manufacturers combine a version of the carbamate-forming reaction and the amine-isocyanate reaction to provide both density reduction and elastic modulus increases. The overall scheme involves the reaction of one mole of water with one mole of isocyanate to produce a carbamic acid intermediate. The carbamic acid intermediate spontaneously loses carbon dioxide to yield a primary amine which reacts with a second mole of isocyanate to yield a substituted urea.

Carboxylic acids react with aryl isocyanates, at elevated temperatures to yield anhydrides. The anhydrides subsequently evolve carbon dioxide to yield amines at elevated temperatures. The aromatic amines are further converted into amides by reaction with excess anhydride. Ortho diacids, such as phthalic acid, react with aryl isocyanates to yield the corresponding N-aryl phthalimides. Reactions with carboxylic acids are irreversible and commercially used to prepare polyamides and polyimides, two classes of high-performance polymers for high-temperature applications where chemical resistance is important. Base catalysis is recommended to reduce the formation of substituted urea by-products

4.4 Isocyanate Adducts and Their Reactions

In many cases, the products of these isocyanate reactions are themselves reactive. These reactions may be viewed as cross-linking reactions, as in the case of allophanate or biuret formation.

The greater reactivity of certain isocyanate adducts has been employed in the preparation of useful intermediates. Among these are carbodiimides, and blocked isocyanate. These reactions will be discussed in the following sections.

4.4.1   Carbodiimides

Carbodiimides are a special class of isocyanate dimer formed through selective catalysis. Carbodiimides are used extensively in organic synthesis as dehydrating agents. Carbodiimides and polycarbodiimides are used as acid scavengers to retard the acid-catalyzed autocatalytic hydrolysis of polyesters. A carbodiimide-containing diisocyanate is shown in Fig. 4.4.

The inherent difficulties encountered when using pure MDI (normally a solid at room temperature, with a tendency to dimerize) led a number of investigators to seek a method of circumventing these problems. Carbodiimide formation finds commercial application in the manufacture of *liquid MDI*. Heating of MDI in the presence of catalytic amounts of phosphine oxides or alkyl phosphates leads to the partial conversion of isocyanate into carbodiimide. This application is more fully explained in the section that follows. Carbodiimides linkages have also been used in producing stable adducted diisocyanates.

Liquid MDI via Carbodiimide Formation

The classical method of producing a storage-stable polyisocyanate composition was disclosed in British Pat. Specification 1,069,858 (May 24, 1967), Novel Storage-Stable Isocyanate Compositions and Processes for their Preparation, assigned to Upjohn Company. The process entails heating the MDI with about 2% by weight of trihydrocarbyl phosphate at a temperature of about 225° C to produce a carbodiimide linkage, thus converting the solid MDI into a liquid MDI.

Methylene bis (phenyl isocyanate), MDI, is one of the diisocyanates commonly used in the preparation of both cellular and noncellular polyurethanes. This material is available commercially either in pure form or in admixtures with related polyisocyanates having higher functionalities. The latter mixtures are generally produced by phosgenation of a mixture of polyamines produced by acid condensation of formaldehyde and aniline (see, for example, U.S. Pats. 2,683,730, 2,950,263, and 3,012,008). Such mixed products containing methylenebis (phenyl isocyanate) are normally liquids at temperatures of 20° C or higher. Therefore, they present no difficulties in handling or dispensing through conventional foam and elastomer formulation mixing machines.

Pure methylene bis(phenyl isocyanate) (MDI), on the other hand, presents a problem in that it is normally a solid at room temperature (≈25° C) having a melting point of the order of 35 to 42° C. This material has to be melted and maintained in a molten state so that it can be transferred by piping arrangements normally employed in the preparation of polyurethanes. Pure methylenebis(phenyl isocyanate), is MDI containing less than 5% by weight of polyisocyanates of higher functionality that are normally produced in the preparation of the crude material.

The objective is to convert MDI which is normally a solid at room temperature to a storage stable liquid product that is suitable for transfer as a liquid using conventional procedures and apparatus for the preparation of polyurethanes.

In general, the physical properties of the polyurethane compositions prepared from the liquid MDI compositions are comparable to the properties of the corresponding compositions obtained using pure MDI.

Liquid MDI via Uretoneimine Formation

Another method of producing liquid MDI is by the formation of uretoneimine linkages. Interestingly, the uretoneimine linkages are obtained by heating carbodiimide-containing MDI with a phosphine catalyst.

Liquid MDI containing carbodiimide species can react with excess isocyanate to form a 2 + 2 cycloaddition product (uretoneimine linkage). The presence of this product in MDI leads to a melting point

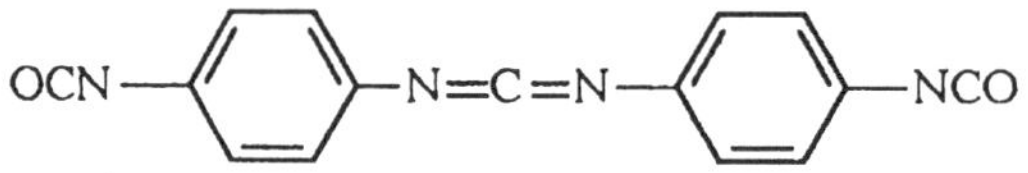

FIGURE 4.4   A carbodiimide-containing diisocyanate.

depression and thus a mixture that is liquid at room temperature. Figure 4.5 shows the ingredients used in the preparation of a uretoneimine-containing diisocyanate by first producing a carbodiimide and subsequently treating it with a phosphine catalyst to produce the uretoneimine structure, according to the disclosure of Findeisen, et al.[9]

Findeisen found that it is possible to subject mono-, di-, and/or polyisocyanates to a partial carbodiimidization reaction to form storage stable isocyanate groups and carbodiimide groups or mixtures having uretoneimine groups formed from these groups by reversible addition, when the phospholine oxide or phospholine sulphide catalysts are used in quantities of from 0.1 to 100 ppm, based on the isocyanate or polyisocyanate to be partially carbodiimidized, and when deactivation takes place using the deactivating agents containing labile chloride or bromide bonds.

To carry out the process, 1 to 60 ppm of the catalyst based on the isocyanate is introduced with stirring at a temperature of from 90 to 110° C, under pressure, into the liquid or dissolved isocyanate. After reaching the desired degree of carbodiimidization, the reaction is terminated by the addition of the deactivating agent.

4.4.2 Blocked Isocyanates

The urethane linkage is thermally reversible to R-OH and R'-N=C=O. Depending on the nature of [R], the presence of catalyst, and the presence of active hydrogen compounds, this reaction can be utilized to block or mask isocyanates such that the blocked isocyanate is stable in the presence of active hydrogens at room temperature but unblocks to liberate free isocyanate at elevated temperature, as shown in Table 4.5.

TABLE 4.5 Typical Blocked Isocyanates

Isocyanate	Blocking Agent	Unblocking Temperature (°C)
TDI	Nonyl phenol	150
TDI	Phenol	150
TDI	Butanone oxime	125
MDI	Butanone oxime	140
HDI	Butanone oxime	150
IPDI	Butanone oxime	150
HDI	Caprolactam	150

Figure 4.6 presents the structure of some commonly used blocking agents. Many of these blocked polyurethanes are used in specialized coatings and adhesives.

Containing 0.1 to 100 ppm of the following phosphorous catalysts:

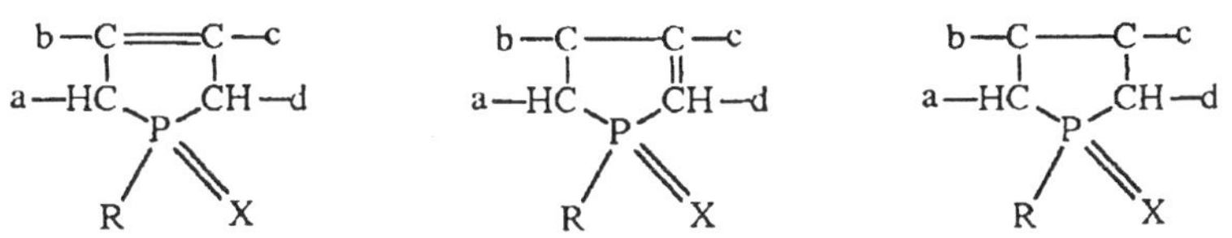

where: R= unsubstituted alkyl radical from 1-4 carbon atoms

a,b,c, and d= hydrogen, halogen, lower alkyl, phenyl, cyclohecxyl

X= oxygen or sulfur

FIGURE 4.5   Production of storage-stable, carbodiimide-containing diisocyanates. *Source:* U.S. Pat. 4,088,665 (May 9, 1978), K. Findeisen, K. Wagner, W. Schafer, and H.J. Hennig, Process for the Partial Carbodiimidization of Organic Isocyanates (to Bayer).

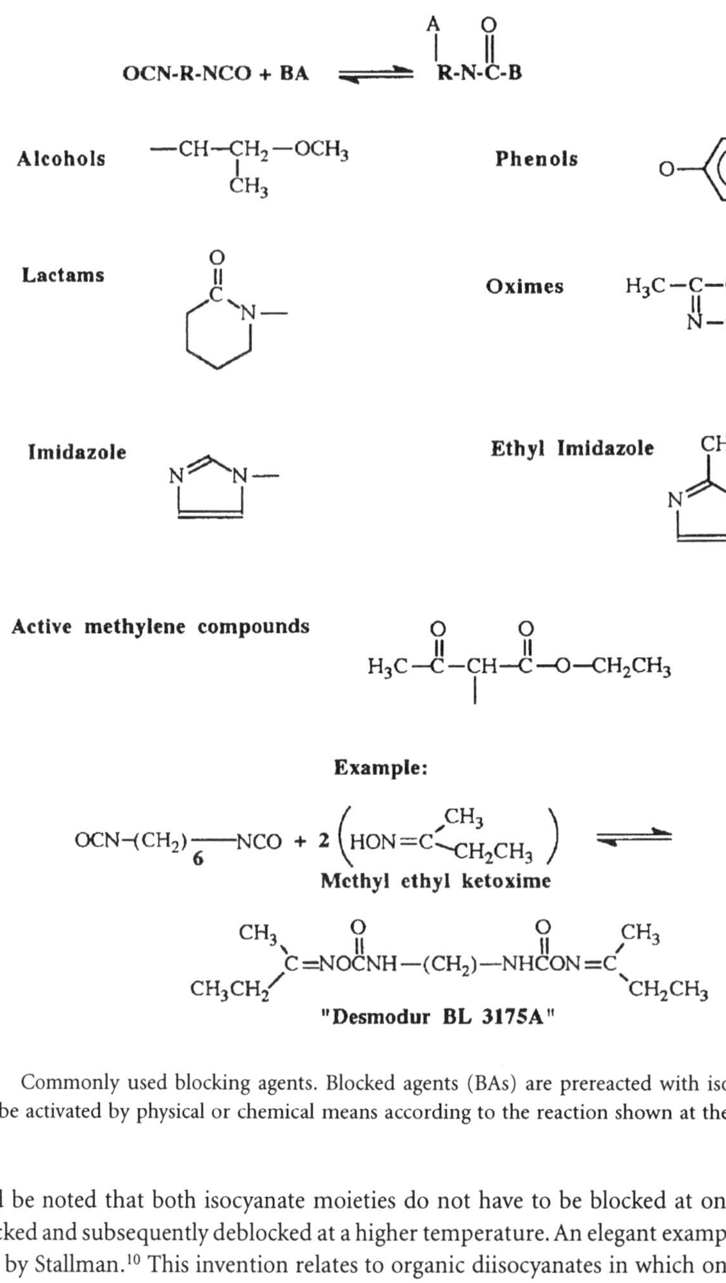

FIGURE 4.6 Commonly used blocking agents. Blocked agents (BAs) are prereacted with isocyanates and subsequently can be activated by physical or chemical means according to the reaction shown at the top of the figure.

It should be noted that both isocyanate moieties do not have to be blocked at once. One isocyanate may be blocked and subsequently deblocked at a higher temperature. An elegant example of this technique is provided by Stallman.[10] This invention relates to organic diisocyanates in which one of the isocyanate groups is hindered or blocked to render it unreactive with most compounds having active hydrogen-containing functional groups below temperatures of about 85° C. It is therefore an object of the present invention to reduce the activity of one of the –NCO groups in organic diisocyanates at temperatures below about 85° C.

Due to their chemical nature, isocyanates are very reactive with groups containing an active hydrogen such as -OH, -COOH, -NH$_2$, etc. In carrying out the reactions with the isocyanates, extreme care must be exercised to prevent undesirable reactions by carefully controlling the various steps such as order of addition, temperature, pressure, presence of moisture, and so on to avoid undesirable side reactions.

Many attempts have been made to reduce the activity of the diisocyanates by various means including blocking or hindering one of the -NCO groups. One such suggested method involves reacting the isocyanate group with a phenol or a compound containing methylene hydrogen, such as a malonic ester, to form adducts that regenerate the -NCO group on heating to about 150 to 180° C. Among the compounds that may be used to form mono adducts of diisocyanates are aceto-acetic ester; diethyl malonate; mercaptans such as 2-mercapto benzothiazole; lactams; imides such as succinimide, phthalimide, and the like; tertiary amyl alcohol; dimethyl phenyl carbinol; and secondary amines such as diphenylamine. These adducts regenerate -NCO groups on heating to 100 to 150° C. It is also known that dimeric aromatic isocyanates such as the dimer of phenyl isocyanate regenerate the original isocyanate on heating to 150 to 180° C. However, it is frequently impossible, due to various reasons (such as shape and dimensions of the object containing them) that preclude placing the object in an oven, or due to the adverse effect on the objects by high temperatures. These requirements for the liberation of the -NCO group obviously place restrictions and inconvenience on their use.

4.5 Industrially Important Isocyanates

Isocyanates are liquids or solids that are highly reactive and undergo addition reactions across the C=N double bond of the NCO group. Reactions with alcohols, carboxylic acids, and amines have been widely exploited in developing a variety of commercial products. Cycloaddition reactions involving both the C=N and the C=O double bond of the NCO group have been extensively studied and used for product development.[11–13]

The basis for the high reactivity of the isocyanates is the low electron density of the central carbon, as indicated by the resonance structures of Fig. 4.7.

Electron-withdrawing or -donating substituents alter the electrophilic nature of the isocyanate. Thus, whereas p-N,N-dimethylaminophenyl isocyanate is a rather slow-reacting material, sulfonyl or acyl isocyanates are noted to be extremely reactive. The reactivity of isocyanates is also manifested in their tendency to react with themselves to form dimers, trimers, or higher oligomers and polymers. Analytically, isocyanates are readily identifiable through derivatization (urea formation) or via spectroscopy using the strong absorbance between 2300 and 2200 cm^{-1}. Many isocyanates are strong lachrymators (tear-inducing agents).

Industrially, isocyanates have become large-volume raw materials for addition polymers such as polyurethanes, polyureas, and polyisocyanurates. By varying the reactants (isocyanates, polyols, polyamines, and others) for polymer formation, a myriad of products have been developed, ranging from flexible and rigid insulation foams, to the high-modulus automotive exterior parts, to high-quality coatings and abrasion-resistant elastomers, unmatched by any other polymeric material. Generally, the name isocyanate is used to mean diisocyanate. However, monoisocyanates are commercially useful, as presented in Table 4.6. For convenience, the Chemical Abstract Service (CAS) Registry Number is also provided.

TABLE 4.6 Industrially Important Specialty Monoisocyanates

Name	CAS Registry Number
methyl isocyanate (MIC)	624-83-9
n-butyl isocyanate (BIC)	111-36-4
phenyl isocyanate (PIC)	103-71-9
3-chlorophenyl isocyanate	2909-38-8
3,4-dichlorophenyl isocyanate	102-36-3
p-toluenesulfonyl isocyanate	4083-64-1

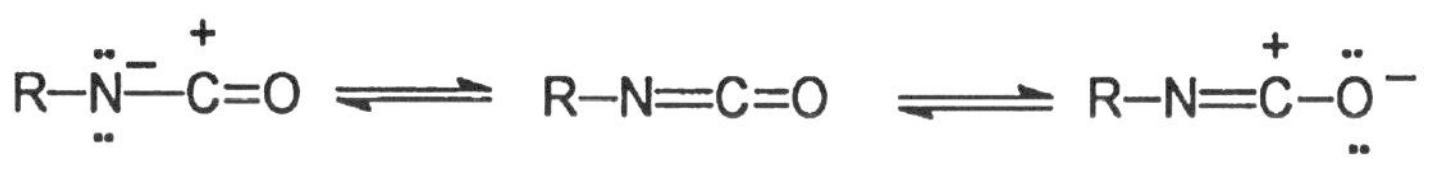

FIGURE 4.7 Resonant structures of isocyanates.

The most important aromatic and aliphatic isocyanates are shown in Figs. 4.8 and 4.9. Figure 4.8 presents some of the most important aromatic isocyanates available in the U.S. Figure 4.9, in contrast, presents some of the most important aliphatic (non-yellowing) isocyanates available in the U.S.

Isocyanates can be produced by four general methods:

1. from amine precursors via phosgenation
2. from nitrene precursors
3. from aryl azides
4. via nonphosgene multisequences, as shown in Fig. 4.10

Note that the phosgenation process is one of the most important commercial processes for the manufacture of diisocyanates.

Isocyanates display varying reactivities toward active hydrogens. This is shown in Table 4.7, where the reactivity of a very simple isocyanate (phenyl isocyanate) is shown against several reactants. Of greater practical importance is the relative reactivities of diisocyanates toward active hydrogens, as shown in Figs. 4.11 and 4.12 and Table 4.8.

TABLE 4.7 Relative Reactivities of Diisocyanates toward Active Hydrogens

MDI	1.00
NDI	0.37
CHDI	0.28
IPDI	0.15
HDI	0.14
HMDI	0.13

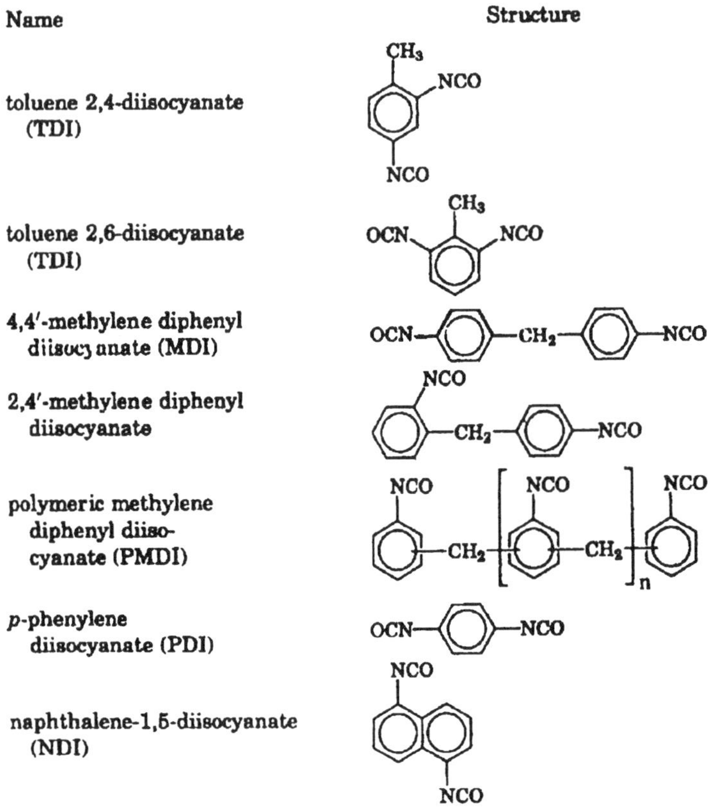

FIGURE 4.8   Aromatic isocyanates.

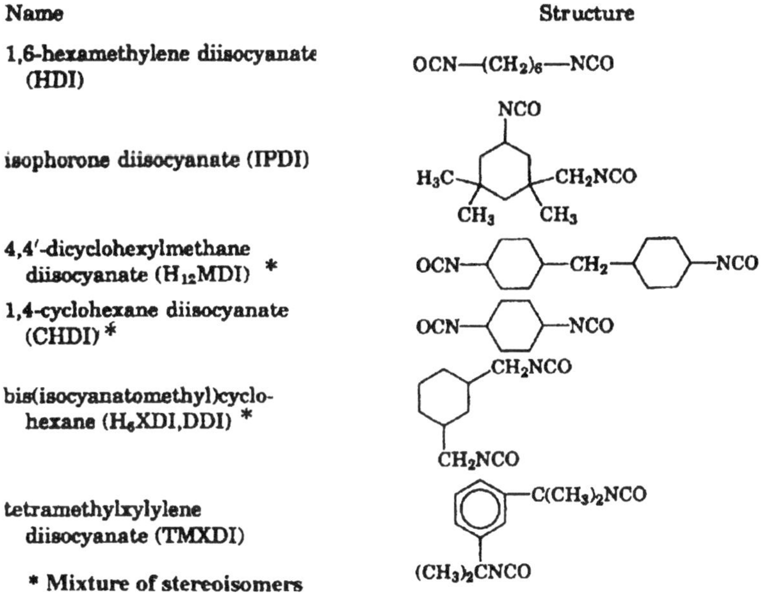
Name
Structure
1,6-hexamethylene diisocyanate (HDI)
OCN—(CH2)6—NCO
isophorone diisocyanate (IPDI)
NCO
H3C
CH2NCO
CH3
CH3
4,4'-dicyclohexylmethane diisocyanate (H12MDI) *
OCN
CH2
NCO
1,4-cyclohexane diisocyanate (CHDI) *
OCN
NCO
bis(isocyanatomethyl)cyclohexane (H6XDI,DDI) *
CH2NCO
CH2NCO
tetramethylxylylene diisocyanate (TMXDI)
C(CH3)2NCO
(CH3)2CNCO
* Mixture of stereoisomers

FIGURE 4.9 Aliphatic isocyanates.

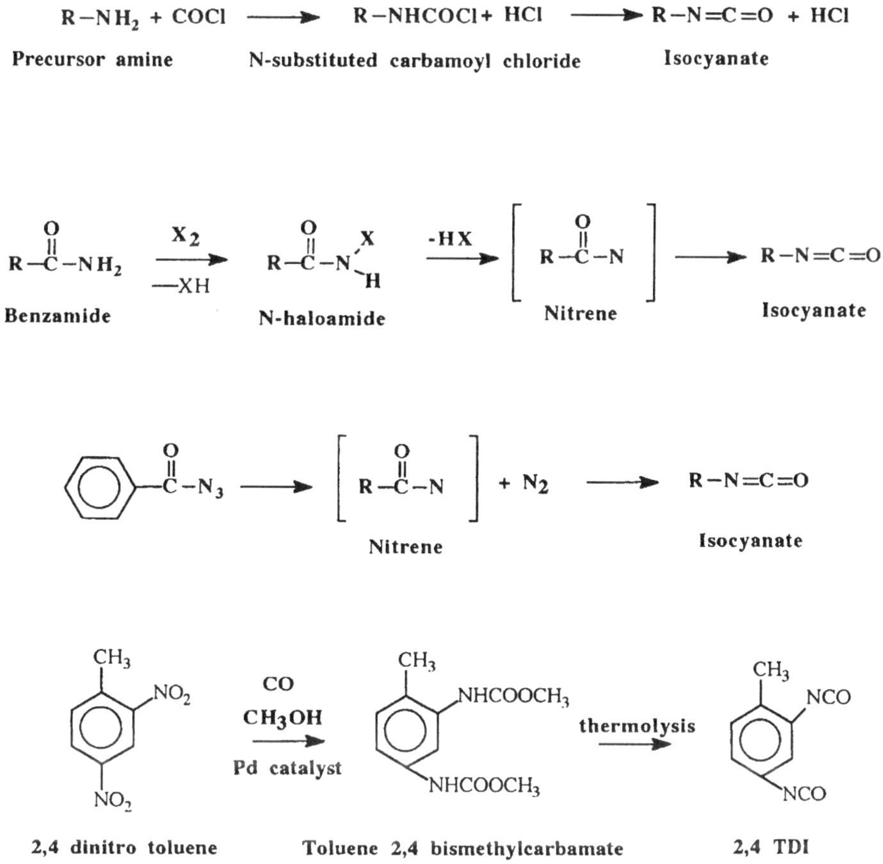
R–NH2 + COCl
R–NHCOCl + HCl
R–N=C=O + HCl
Precursor amine
N-substituted carbamoyl chloride
Isocyanate
X2
—XH
-HX
Benzamide
N-haloamide
Nitrene
Isocyanate
+ N2
Nitrene
Isocyanate
CH3
NO2
NO2
CO
CH3OH
Pd catalyst
NHCOOCH3
NHCOOCH3
thermolysis
NCO
NCO
2,4 dinitro toluene
Toluene 2,4 bismethylcarbamate
2,4 TDI

FIGURE 4.10 Preparation of isocyanates.

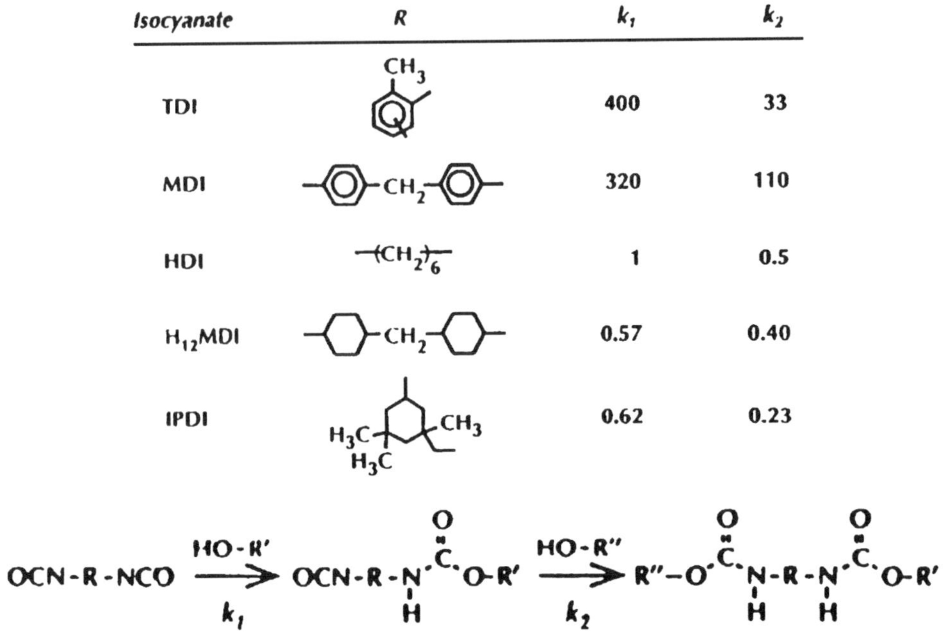

Isocyanate	R	k_1	k_2
TDI	(CH₃-substituted benzene ring structure)	400	33
MDI	—⬡—CH₂—⬡—	320	110
HDI	—(CH₂)₆—	1	0.5
H₁₂MDI	—⬡—CH₂—⬡—	0.57	0.40
IPDI	(H₃C, H₃C, CH₃ substituted cyclohexane structure)	0.62	0.23

$$OCN\text{-}R\text{-}NCO \xrightarrow[k_1]{HO\text{-}R'} OCN\text{-}R\text{-}\underset{H}{N}\overset{O}{\overset{\|}{C}}O\text{-}R' \xrightarrow[k_2]{HO\text{-}R''} R''\text{-}O\overset{O}{\overset{\|}{C}}\underset{H}{N}\text{-}R\text{-}\underset{H}{N}\overset{O}{\overset{\|}{C}}O\text{-}R'$$

FIGURE 4.11 Isocyanate reactivities.

TABLE 4.8 Isocyanate Reactivity (Phenyl Isocyanate in Toluene @ 99% Stoichiometry)

Hydrogen Donor	Rate Constant*		Act. Energy, kcal/mole
	25° C	80° C	
Aromatic amine	10–20	—	—
Primary OH	2–4	30	8–9
Secondary OH	1	15	10
Tertiary OH	0.01	—	—
Water	0.4	6	11
Primary thiol	0.005	—	—
Phenol	0.01	—	—
Urea	—	2	—
Carboxylic acid	—	2	—
Anilide	—	0.3	—
Carbanilate	—	0.02	16.5

*k × 104, 1/mole sec.

4.5.1   Isocyanate Synthetic Routes

Phosgenation Process

The most common method of preparing isocyanates involves the reaction of phosgene and aromatic or aliphatic amine precursors. The initial reaction step, the formation of N-substituted carbamoyl chloride (1), is highly exothermic and is followed by hydrogen chloride elimination, which takes place at elevated temperatures. The phosgenation sequence is shown in Fig. 4.13. Figure 4.14 presents typical phosgenation reactions for the manufacture of TDI, MDI, and NDI.

To suppress the formation of side products, primarily ureas and isocyanurates, excess phosgene is employed. In most small-scale batch processes, diluted amine is added gradually to chilled solutions of phosgene, followed by venting of excess phosgene. Dehydrohalogenation of the resultant carbamoyl chloride is performed at approximately 80 to 100° C. Isolation of the desired isocyanate products is achieved by distillation at reduced pressure. The carbamoyl chlorides of certain alkyl isocyanates are

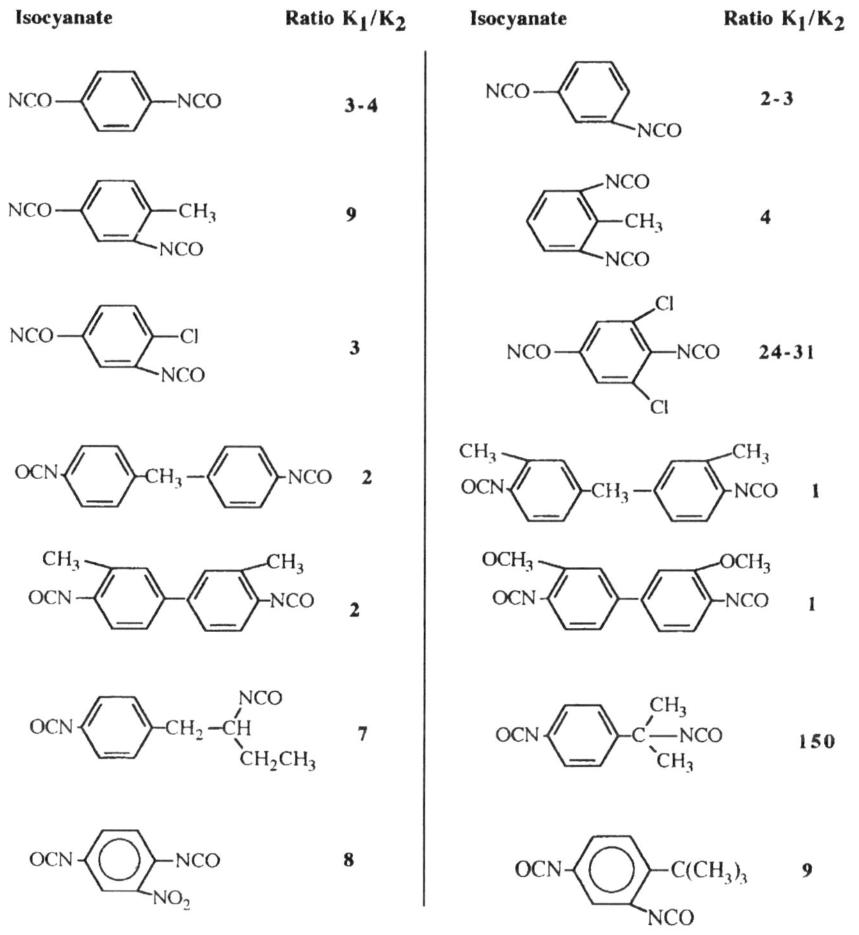

Isocyanate	Ratio K$_1$/K$_2$	Isocyanate	Ratio K$_1$/K$_2$

FIGURE 4.12 Isocyanate reactivity ratios. Difference in reactivity between the isocyanate groups was measured by taking the ratio of the apparent first rate constants with a tenfold excess of 2-ethyl hexanol at 20–25° C. The rate constants were measured at 10--20% (K1) and 70–90% (K2). *Source:* U.S. Pat. 3,180,883 (Apr. 27, 1965), L.C. Case, 4,4-diisocyanato-3-5-di-lower alkyl-diphenylmethane.

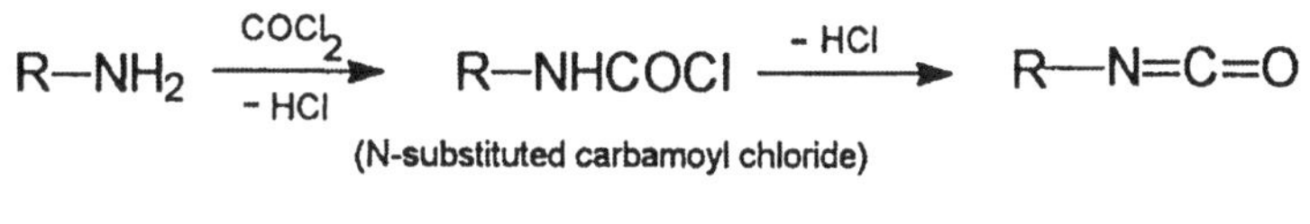

FIGURE 4.13 Phosgenation reaction.

stable (or recombine from isocyanate and hydrogen chloride) and distill without decomposing. Because distillation of the crude alkyl isocyanates does not necessarily yield a purified product, alternative synthetic routes are often preferred.

Grant and Irwin[14] disclosed a continuous process for phosgenating organic amines in inert solvents to form isocyanates, where hydrochloride persistence and hold times were minimized. A schematic diagram of this process is shown in Figure 4.15. In this process, a second stage reactor is provided for the continuous removal of phosgene and hydrogen chloride gases, resulting in an isocyanate solution. The patent claims that a particularly useful isocyanate is produced when the starting amine is 4,4'-methylene-bis(cyclohexylamine).

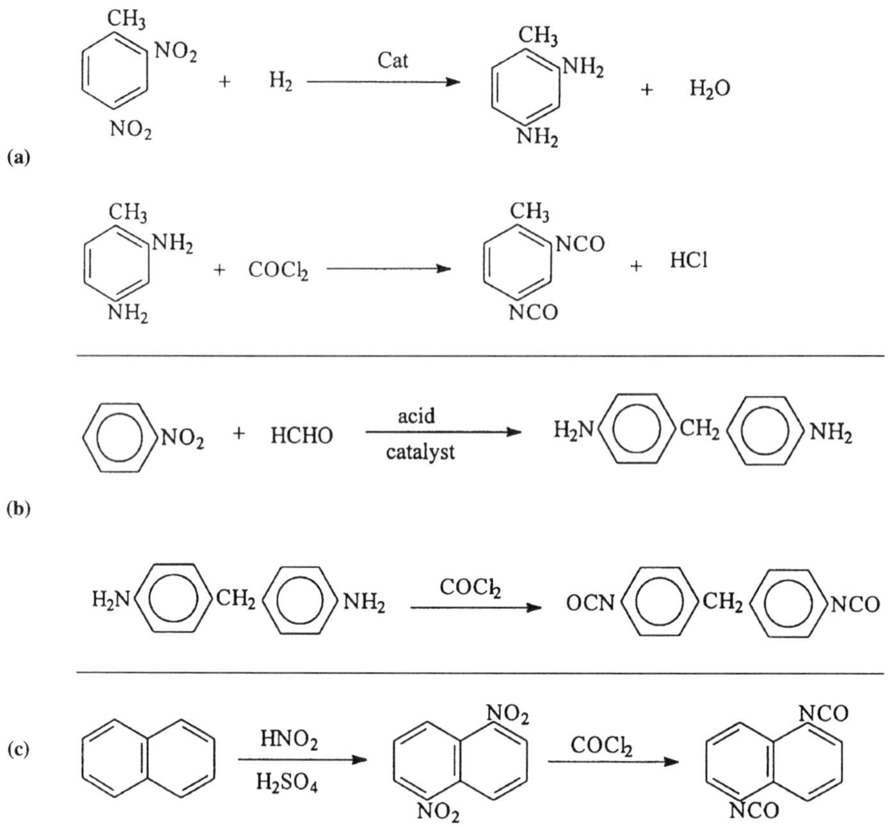

FIGURE 4.14 Manufacturing processes for some important diisocyanates: (a) TDI, (b) MDI, and (c) NDI.

A variation of this method involves the conversion of the amine into the amine hydrochloride prior to treatment with phosgene. This method has the advantage of producing generally cleaner products by retarding the secondary reaction of the free aniine with carbamoyl chloride.

The reaction of isocyanate precursors with reactive groups and phosgene also results in isocyanates. For example, o-aminobenzoic acid and phosgene react to form 2-isocyanatobenzoyl chloride. Interestingly, isocyanatophenols have been synthesized from aminophenols, under controlled conditions, without formation of the corresponding chloroformates.[15]

Instead of amines, sulfonamides have also been used as starting materials, producing the highly reactive sulfonyl isocyanates, which have found applications in the manufacture of drugs for diabetics and as drying agents.

Oligomers of phosgene, such as diphosgene $(COCl_2)_2$, have found use in the laboratory preparation of isocyanates. Carbamoyl chlorides, N,N disubstituted ureas, dimethyl- and diphenylcarbonates, and arylsulfonyl isocyanates have also been used to convert amines into urea intermediates, which are subsequently pyrolyzed to yield isocyanates. These methods have found applications in the preparation of low boiling point aliphatic isocyanates.[16]

Preparation from Nitrene Intermediates

A convenient, small-scale method for the conversion of carboxylic acid derivatives into isocyanates involves electron sextet rearrangements. For example, treatment of benzamide with halogens leads to an N-haloamide which, in the presence of base, forms a nitrene intermediate. The nitrene intermediate undergoes rapid rearrangement to yield an isocyanate. Ureas can also be formed in the process if water is present[17] (Reaction 4.11).

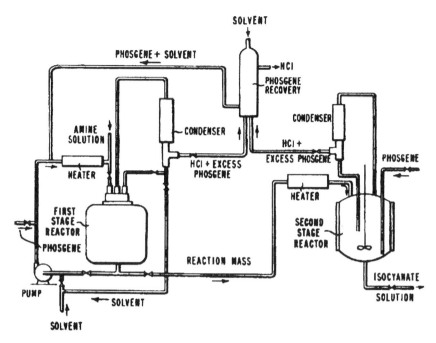

FIGURE 4.15 Two-stage phosgenation for producing isocyanates. *Source:* Irwin Grant, U.S. Pat. 3,574,695.

REACTION 4.11 Isocyanate formation.

More convenient is the use of aryl azides that are readily converted into isocyanates upon heating in nonreactive solvents via the loss of nitrogen. The latter method is useful for the synthesis of isocyanates with additional substituents that cannot be prepared with phosgene (Reaction 4.12).

REACTION 4.12 Isocyanate formation via thermoconversion of aryl azides.

Nonphosgene Preparation

The term *nonphosgene route* is primarily used in conjunction with the conversion of amines (or the corresponding nitro precursor) to isocyanates via the use of carboxylation agents. These multistep approaches are becoming more attractive to the chemical industry as environmental or toxicological restrictions involving chlorine or phosgene are increasingly enforced. For example, 2,4-dinitrotoluene undergoes reductive carbonylation to form 2,4-toluene diisocyanate (TDI) in the presence of palladium catalysts, as disclosed in U.S. Pat. 4,369,141 to Zajacek, McCoy, and Fuger of Arco Chemical. A variation of this process consists in capturing the isocyanate formed with methanol, followed by pyrolysis of the biscarbamate, in German Pat. 52,635,490 to Tsumura, Takaki, and Takeshi of Mitsui-Toatsu, as depicted in Reaction 4.13.

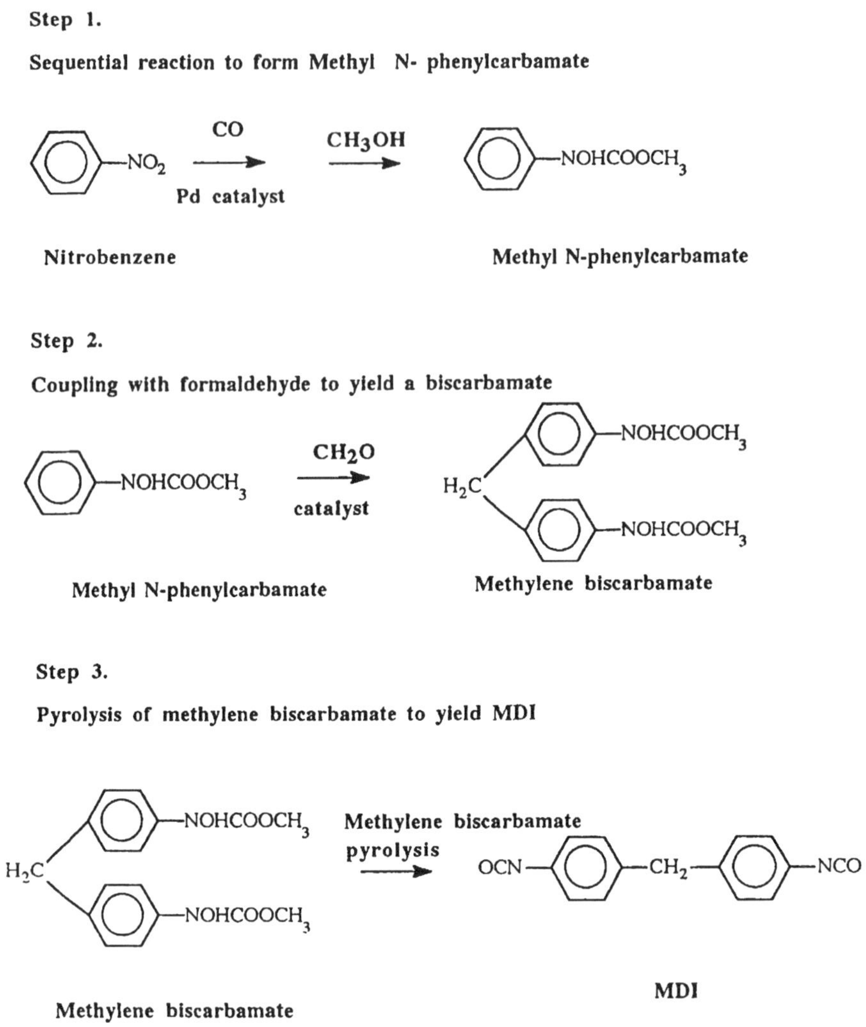
Step 1.
Sequential reaction to form Methyl N- phenylcarbamate
NO2
CO
Pd catalyst
CH3OH
NOHCOOCH3
Nitrobenzene
Methyl N-phenylcarbamate
Step 2.
Coupling with formaldehyde to yield a biscarbamate
NOHCOOCH3
CH2O
catalyst
H2C
NOHCOOCH3
NOHCOOCH3
Methyl N-phenylcarbamate
Methylene biscarbamate
Step 3.
Pyrolysis of methylene biscarbamate to yield MDI
H2C
NOHCOOCH3
NOHCOOCH3
Methylene biscarbamate
pyrolysis
OCN
CH2
NCO
Methylene biscarbamate
MDI

REACTION 4.13 Preparation of MDI via pyrolysis of biscarbamate precursor.

4.6 Commercial Manufacturing Processes

We will divide our discussion of commercial manufacturing processes into two sections: those processes directed at the synthesis of aromatic, and those processes directed at the synthesis of aliphatic isocyanates.

4.6.1 Aromatic Isocyanates

TDI

The most important mixture of TDI is the 80:20 mixture of 2,4-TDI and 2,6-TDI (80-20 TDI). A 65:35 ratio mixture of 2,4-TDI and 2,6-TDI (65:35TDI) is also commercially available. Physical properties of different TDI isomers are shown in Table 4.8.

Preparation of TDI

Reaction 4.14 depicts the classical manufacturing process for the production of TDI. The process depends on the nitration of toluene in the presence of sulfuric acid. The nitrotoluene(s) are then hydrogenated

TABLE 4.8 Physical Properties of Different TDI Isomers

Property TDI isomers	100/0 (2,4/2,6 ratio)	80/20 (2,4/2,6 ratio)	65/35 (2,6/2,6 ratio)
Physical form	Liquid	Liquid	Liquid
Molecular weight	174.16	174.16	174.16
Equivalent weight	87.08	87.08	87.08
Boiling Point C @ 760 mm Hg	251	251	251
Freezing Point C	21.4	14.0	8.5
Fire Point C (Cleveland open cup)	143	143	143

to the corresponding amines, and finally the diamines are phosgenated to the corresponding diisocyanates.

REACTION 4.14 TDI manufacturing process. A classical manufacturing process depends on the nitration of toluene in the presence of sulfuric acid. The nitrotolulene(s) are then hydrogenated to the corresponding diamines and, finally, the diamines are phosgenated to the corresponding diisocynates.

In 1959, Detlef and Muntz[18] were awarded a patent for the production of TDI. The patent relates to improvements in the production of aromatic isocyanates. It is well known to produce aromatic isocyanates by causing phosgene to react with the hydrochlorides of the corresponding amines. In carrying out the conventional process, the dry hydrochloride of the aromatic amine serving as the starting material for the phosgenation is suspended in an inert diluent (such as xylene, chlorobenzene, or o-dichlorobenzene) and treated with phosgene at a temperature ranging from about 130 to 180° C. This method has the disadvantage that it must be effected at temperatures that assist the polymerization of the isocyanate formed, particularly in the case of isocyanates containing more than one isocyanato group. This polymerization accounts for the formation of resinous residues, remaining in the still upon distillation of the diluent and the isocyanate formed. In the case of nondistillable isocyanates, it may be very difficult to separate the same from the resinous polymerization products. On the other hand, the formation of polymerization products may not be remedied by using lower temperatures, since at lower temperatures, longer reaction times (which also act beneficially on the polymerization) are required.

The invention is illustrated by the following example, with all parts being by weight.

Example 1. We mix 122 parts of a mixture of 80% of 2,4 and 20% of 2,6' diamino-toluene prepared by nitrating toluene to form dinitro-toluene prepared by reducing the latter with 600 parts of phenyl-n-butylsulfone. Then, dry hydrogen chloride is passed in at a temperature of about 70° C until the mixture is saturated. Now, a strong stream of phosgene is introduced while raising the temperature to about 90° C. The hydrochloride is thereby dissolved within 1.5 hr. The phosgenation is completed by raising the temperature to 120° C and maintaining the mixture at this temperature for 15 min. By passing in dry carbon dioxide, hydrogen chloride and excess phosgene are removed from the mixture. By distillation in vacuo 171 parts of toluylene-diisocyanate are obtained, which corresponds to a yield of 89%.

MDI

Perhaps no other isocyanate is more responsible for the spectacular commercial success of polyurethane elastomers than MDI. Pure MDI is a solid with a melting point of 38° C, an important property compared to the volatility of TDI. MDI has a functionality of 2.0, making it well suited for the manufacture of high-performance elastomers. The manufacturing process of MDI is considerably more complex than that of TDI, as shown in Fig. 4.16.

Currently, MDI is produced in large quantities, based on nitrobenzene as a starting reactant. Nitrobenzene, carbon monoxide, and methanol can react sequentially in the presence of noble metal catalysts, to produce methyl N-phenylcarbamate (4). The phenylcarbamate is subsequently coupled with formaldehyde to yield the methylene bis(carbamate), which is pyrolyzed to yield methylene diphenyl diisocyanate MDI. This is shown in Figure 4.17.

pMDI

The polymeric MDIs (pMDIs) encompass a broad range of isocyanates with low functionality and low viscosity. These properties enable pMDI to encompass the entire urethane spectrum of rigid foams, high-resiliency foams, integral foams, adhesives, and coatings.

The most important properties of MDI, compared to pMDI, are shown in the Table 4.9.

TABLE 4.9 Most Important Properties of MDI, Compared to pMDI

Property	MDI	pMDI
Physical form	Solid	Liquid
Molecular weight	250	≈450
Equivalent weight	125	≈225
Functionality	2	2+
Boiling point, °C @ 760 mm Hg	170	N/A
Flash point, °C (Cleveland open cup)	213	210–230
Fire point, °C (Cleveland open cup)	N/A	220–250

Polymeric MDI has been modified to meet several applications. The different pMDIs are formulated to provide a variety of functionalites. Table 4.10 presents the properties associated with each type.

Both dimethyl carbonate and diphenyl carbonate have been used in place of carbon monoxide as reagents for the conversion of amines into isocyanates via this route, as described by Gurgiolo in U.S.

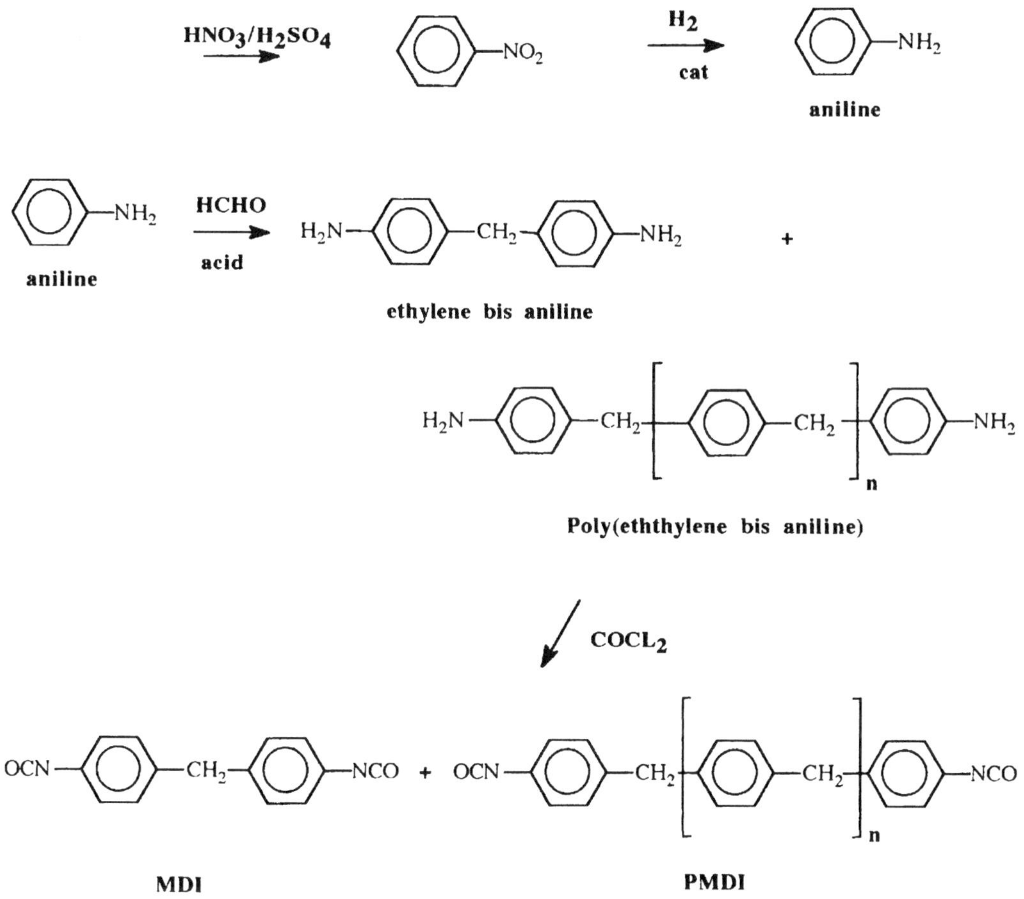

FIGURE 4.16 MDI and PMDI manufacturing process. Following phosgenation, pure MDI is separated by distillation. Polymeric MDI (PMDI) is sold as a crude liquid useful in foams.

TABLE 4.10 Properties Associated with Each Type of pMDI

Functionality	pMDI Description	Application
2.01–2.1	Modified, liquid MDI	Microcellular elastomers, adhesives, coatings
2.1–2.3	Low-functionality polyisocyanate	Rigid and semirigid foams, adhesives
2.5	Low-viscosity polyisocyanate	High-density foams
2.7	Low viscosity polyisocyanate	Low-density rigid foams
2.7–3.1	High-functionality polyisocyanate	Rigid and isocyanurate foams

Pat. 4,268,683, and U.S. Pat. 3,366,662, issued to Kober and Smith of Olin Chemical. Alternatively, aniline, toluene diamines (TDAs), and methylene dianilines (TDAs) have also been used as starting materials in the carbonylations to provide a wide variety of isocyanate monomers.

A simpler nonphosgene process for the manufacture of isocyanates consists of the reaction of amines with carbon dioxide in the presence of an aprotic organic solvent and a nitrogenous base. The corresponding ammonium carbamate is treated with a dehydrating agent. This concept has been applied to the synthesis of aromatic and aliphatic isocyanates. The process relies on the facile formation of amine-carbon dioxide salts using acid halides such as phosphoryl chloride and thionyl chloride, as disclosed by McGhee and Waldman in U.S. Pat. 5,189,205 and issued to Monsanto.

The McGhee patent covers a process for preparing isocyanates consists of (a) contacting carbon dioxide and a primary amine in the presence of an aprotic organic solvent and an organic, nitrogenous base to

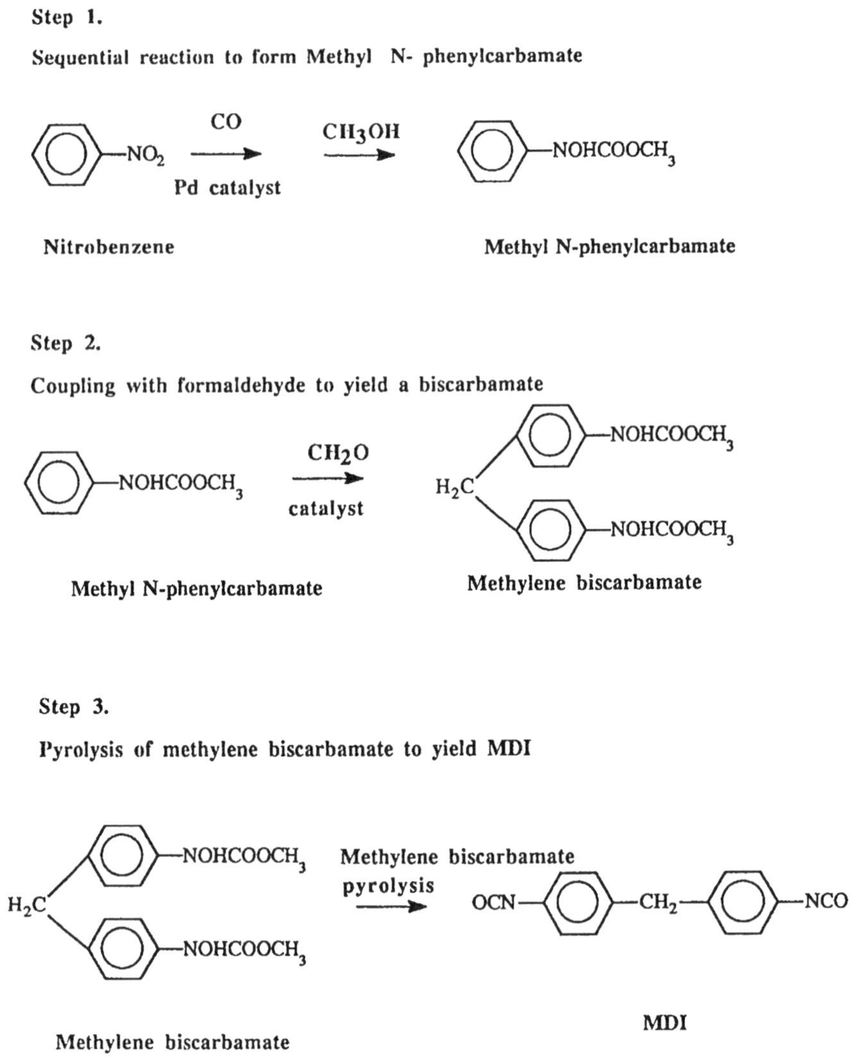

Step 1.

Sequential reaction to form Methyl N- phenylcarbamate

Nitrobenzene   Methyl N-phenylcarbamate

Step 2.

Coupling with formaldehyde to yield a biscarbamate

Methyl N-phenylcarbamate   Methylene biscarbamate

Step 3.

Pyrolysis of methylene biscarbamate to yield MDI

Methylene biscarbamate   MDI

FIGURE 4.17 Formation of MDI.

produce the corresponding ammonium carbamate salt, and (b) reacting the ammonium carbamate salt with an electrophilic or oxophilic dehydrating agent to produce the corresponding isocyanate.

A second embodiment consists of recovering the ammonium carbamate salt of step (a) prior to reacting the ammonium carbamate salt with an electrophilic or oxophilic dehydrating agent in the presence of an aprotic organic solvent and an organic, nitrogenous base.

Commercially, the phosgenation of primary amines is by far the most widely used method for producing isocyanates. The use of phosgene has several disadvantages. The phosgenation route is long, energy intensive, and requires handling highly corrosive materials (e.g., hydrogen chloride, chlorine, and sulfuric acid).

For methylene diphenyl diisocyanate (MDI), the initial reaction involves the condensation of aniline with formaldehyde to yield a mixture of oligomeric amines where n = 1, 2, 3.... The term MDI encompasses a large family of isocyanates based on the methylene diphenylene structure. The basic product, called polymeric MDI (or crude MDI), is a complex mixture of isomers and oligomers with isocyanate equivalent weight of 133 and a number-average isocyanate functionality of 2.7. This liquid material is used heavily in rigid insulation foams and binder applications.

Other MDI isocyanates are derived from polymeric MDI. These include pure 4,4'-MDI, which is used extensively to manufacture elastomeric and thermoplastic polymer articles. "Pure MDI," which has an isocyanate functionality of 2.0, is available as mixtures with its 2,4'-isomer ranging from about 99–1 to about 50–.50. Pure MDI is a solid which melts at 108° F or slightly below. Liquid derivatives of pure MDI are used widely in mixing-activated systems, for the manufacture of elastomeric, elastoplastic, and specialty items. These liquid products are used heavily in molded flexible foam applications and in high-density elastomers for auto exterior parts. Several liquid derivatives of pure MDI are known. Most of these consist of urethane prepolymers (see the section on liquid MDI).

For toluene diisocyanate, amine monomers are prepared by the nitration of toluene and subsequent hydrogenation. These materials are converted to the isocyanate, in the majority of the commercial aromatic isocyanate phosgenation processes, using a two-step approach. TDI is generally offered as a mixture of the 2,4 (predominant) and 2,6 (minor) isomers. Mixtures of several different isomer ratios are commercially available.

In the first step, a solution of amine is mixed with a solution of phosgene. An excess of phosgene is needed to retard by-product formation. The solvents most commonly used in the phosgenation reaction include toluene, xylene, halobenzenes, and decahydronapthalene. The halobenzenes are preferred because of their polarity. In the second step, the resulting carbamoyl chloride-amine hydrochloride slurry reacts with excess phosgene at temperatures in excess of 100° C to yield the isocyanate. The appearance of a clear solution signal the end of the reaction. Distillation of the solvent followed by fractional crystallization, fractional distillation, or sublimation affords pure isocyanate.

An excess of phosgene is used during the initial reaction of amine and phosgene to retard the formation, of substituted ureas. Ureas are undesirable, because they serve as a source for secondary product formation which adversely affects isocyanate stability and performance. By-products, such as biurets and triurets, are formed via the reaction of the labile hydrogens of the urea with excess isocyanate. Isocyanurates, where R = phenyl, toluyl, may subsequently be formed from the urea oligomers via ring closure. These oligomerization steps result in a continuous increase in viscosity of the desired isocyanate and ultimately cause solidification.

The *in situ* generated disubstituted ureas also react with phosgene to yield thermally unstable allophanoyl chlorides and chloroformamides. The allophanoyl chlorides eliminate hydrogen chloride to form the isocyanate. The chloroformamides, however, yield chloroformamidine-N-carbonyl chloride, which decomposes to yield both carbodiimides and isocyanide dichlorides. The carbodiimides simply contribute to yield loss. The isocyanide dichlorides, although present in small amounts, contribute to chlorine-containing impurities that detrimentally affect product performance.

Alternatively, the aromatic amine can first be treated with hydrogen chloride to form a slurry of amine salts, which is subsequently phosgenated. The slurry is processed using a temperature-staged reaction sequence. Excess hydrogen chloride and phosgene are vented to retard the formation of isocyanate recombination products. The isocyanate is purified via solvent separation and fractionation. This method has the disadvantage that gaseous phosgene reacts very slowly with the suspended amine salt; thus, high temperatures and pressures are generally needed.

Some of these isocyanates are commercially available in derivatized form, such as biurets and carbodiimides, to provide materials having improved handling or processing characteristics.

Attempts have been made to develop methods for the production of aromatic isocyanates without the use of phosgene. None of these processes is currently in commercial use. Processes based on the reaction of carbon monoxide with aromatic nitro compounds have been examined extensively. The reductive carbonylation of 2,4-dinitrotoluene to toluene 2,4-dialkylcarbamates is reported to occur in high yield at reaction temperatures of 140 to 180° C under 1000 psi of carbon monoxide. The resultant carbamate product distribution is noted to be a strong function of the alcohol used. Mitsui-Toatsu and Arco have disclosed a two-step reductive carbonylation process based on a cost-effective selenium catalyst.

Two-step approaches based on cocatalysts or alternate catalysts, and one-step approaches that circumvent the formation of the biscarbamate intermediates, have also been disclosed.

Other approaches have explored the reaction of amines with dimethyl carbonate or its precursors. This approach is complicated by the fact that the isocyanate is produced via the thermolytic cleavage of the methyl carbamate. Reactions with the unconverted carbamate cannot be prevented. Much effort has been focused on improving the selectivity of the latter step. MDI is commercially available as three products: pure MD, polymeric MDI, and liquid MDI. These products are shown in Fig. 4.18.

Polymeric MDIs are important monomers in the preparation of foams, RIM elastomers, adhesives, etc. The preparation of pMDI in continuous fashion is an important goal of many suppliers. In 1986, Ohlinger[19] of BASF was awarded a patent for the continuous preparation of organic mono- and/or polyisocyanate. The Ohlinger patent relates to a process for the continuous preparation of organic isocyanates, preferably polyisocyanates, through the reaction of organic amines, preferably polyamines, with phosgene in the presence of organic solvents under pressure—for example, from 5 to 100 bar, at elevated temperatures, for example, from 100 to 220° C. In this manner, the reaction mixture is partially recycled, preferably using natural circulation, and the hydrogen chloride content in the reaction mixture prior to the addition of the amine is less than 0.5% by weight, based on the total weight of the reaction mixture. The molar ratio of phosgene to NH_2 group in the organic amines is from 12:1 to 200:1.

The conventional processes suffer from numerous disadvantages. If low temperatures are used for the initial phosgenation (cold phosgenation), the liberation of large amounts of phosgene while heating the reaction mixture to the final phosgenation temperature (hot phosgenation temperature) presents a problem that is difficult to deal with and that is further aggravated by the high toxicity of phosgene.

1) Pure MDI

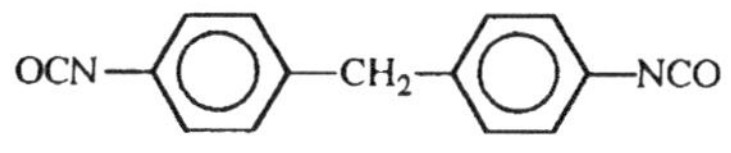

Functionality= 2, with equal reactivities for both isocyanates

2) Polymeric MDI (PMDI)

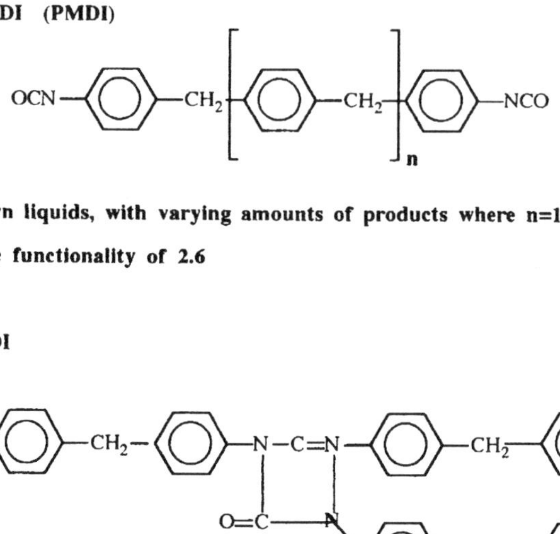

Dark brown liquids, with varying amounts of products where n=1 to 4, with

an average functionality of 2.6

3) "Liquid" MDI

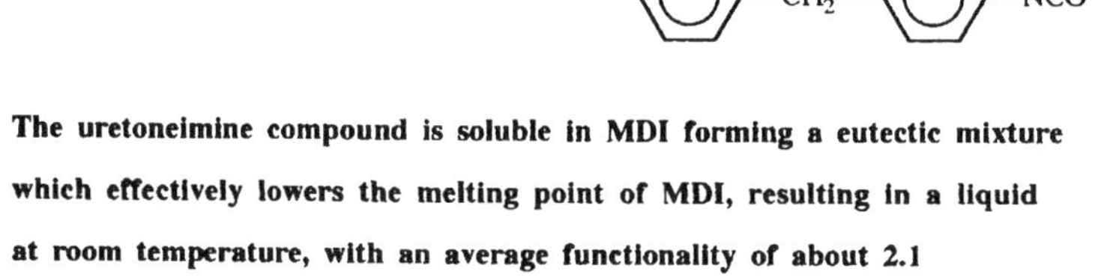

The uretoneimine compound is soluble in MDI forming a eutectic mixture

which effectively lowers the melting point of MDI, resulting in a liquid

at room temperature, with an average functionality of about 2.1

FIGURE 4.18 MDI commercial products.

Low-temperature processes also suffer from a further disadvantage, namely that the rate of reaction is relatively low, so that large reaction volumes have to be handled. In the two-stage processes, the end product and the carbamoyl chloride formed as an intermediate in the first stage can react with some of the starting amine to form substituted ureas and polyureas or other undesirable products.

4.6.2   Aliphatic Isocyanates

Historically, conventional aliphatic isocyanates have been manufactured using the hydrogen chloride salt slurry approach. Exceptions to this are the longer chain aliphatics that, due to the increased solubility, have reaction rates conducive to the free amine process. In the hydrogen chloride salt approach, a fine slurry of salt reacts with phosgene in an agitated autoclave. The reaction must be carried out at temperatures below 150° C to avoid the formation of chlorinated monoisocyanates, as anhydrous hydrogen chloride has been found to displace isocyanate groups. Similar to the aromatic isocyanate processes described, the carbamate salt suspension is treated with phosgene using a series of reactors. Typically, the amine salt-phosgene reaction is carried out at a temperature of 30° C for 12 to 24 hr, then finished using a series of 100° C digestion steps. The resultant isocyanate solution is purified by solvent stripping followed by fractional vacuum distillation.

An alternative approach, generally referred to as a two-phase phosgenation, has gained wide-scale acceptance for the production of aliphatic isocyanates. Typically, a cooled solution of phosgene and amine are mixed in the presence of concentrated sodium hydroxide. Reaction rates are very fast, and overall product yields exceed 90%. Refinements in the reaction conditions have reduced the by-product formation arising from the reaction of the hydroxide with phosgene to less than 10%. This approach allows for the preparation of isocyanates containing labile groups, such as alkoxy (RO), that would be lost in a traditional high-temperature, amine-hydrogen chloride salt phosgenation.

Low-boiling isocyanates, such as methyl isocyanate, are difficult to prepare via conventional phosgenation due to the fact that the N-alkyl carbamoyl chlorides are volatile below their decomposition point. Interestingly, N-ethyl carbamoyl chloride decomposes at its boiling point, whereas the N-propyl carbamoyl chloride is thermolyzed cleanly into isocyanate and hydrogen chloride.

A convenient method for the synthesis of these low-boiling materials consists of the reaction of N,N'dimethylurea with toluene diisocyanate to yield an aliphatic-aromatic urea. Alternatively, an appropriate aliphatic-aromatic urea can be prepared by the reaction of diphenylcarbamoyl chloride with methylamine.

Thermolysis of either of the mixed ureas produces methyl isocyanate in high yield.

H$_{12}$MDI

One of the most commercially successful aliphatic polyurethane is 4,4'-methylenebis(cyclohexyl isocyanate), or 1,1' methylenebis(4, 4' isocyanatocyclohexane), popularly known as *hydrogenated HMDI* or *H$_{12}$MDI*. This diisocyanate is commercially available as Desmodur W, a trademark of Bayer, Inc.

In addition to the Chemical Abstracts name [1,1' methylenebis(4, 4'isocyanatocyclohexane)], the compound is known by many variations of this name, as well as by many abbreviations. Some of the more common abbreviations contained in the literature include H$_{12}$MDI (H$_{12}$-methylene diphenyldiisocyanate), HMDI (hydrogenated methylene diphenyldiisocyanate), PICM (para-isocyanatocyclohexyl methane) and RMDI (reduced methylene diphenyldiisocyanate). Unfortunately, HMDI is occasionally used as an abbreviation for hexamethylene diisocyanate. Hence, care should be taken in using these abbreviations.

Desmodur W is an aliphatic diisocyanate composed of a specific mixture of the three isomers of 1,1' methylenebis(4'isocyanatocyclohexane).[20] As with other isocyanates, Desmodur W reacts with active hydrogen containing compounds such as alcohols, amines, and water. However, because of its cycloaliphatic structure, it reacts more slowly than most other isocyanates. Hence, Desmodur W is generally catalyzed to achieve reasonable reaction rates.

The structure of Desmodur W gives rise to a unique set of property characteristics. First, because of its aliphatic nature, Desmodur W is light stable and produces optically clear parts or coatings. This diisocyanate also yields polyurethane systems with an unusual degree of hydrolytic stability. These systems generally outperform systems based on other diisocyanates, such as diphenylmethane diisocyanate (MDI) or isophorone diisocyanate (IPDI).

The polymer morphology of $H_{12}MDI$ systems is also different from those of other diisocyanates, particularly MDI-based materials. While MDI often forms well defined crystalline hard segments, $H_{12}MDI$ yields smaller amorphous or semicrystalline domains. The consequences of this difference are evident in the tensile properties of these systems. $H_{12}MDI$ elastomers are typically harder and stronger than those based on MDI.

The characteristics that make $H_{12}MDI$ especially suited to certain applications are (1) low reactivity, (2) light stability, (3) hydrolytic stability, and (4) polymer morphology. For instance, $H_{12}MDI$ is used in preparing aqueous polyurethane dispersions (latexes) for coatings. These properties have also enabled it to be used in safety glass and biomedical applications. Desmodur W is an cycloaliphatic diisocyanate composed of a specific mixture of the three geometric isomers of 1,1'methylenebis(4-isocyanato cyclohexane), namely 30% trans-trans, 65% trans-cis, and 5% cis-cis,12 as shown in Figure 4.19.

Reactions of $H_{12}MDI$

As with other isocyanates, $H_{12}MDI$ reacts with active hydrogen-containing compounds, such as alcohols, amines, and water, as shown in Fig. 4.20. The typical reaction conditions used for preparing $H_{12}MDI$ /polyol prepolymers are 100 to 110° C for 3 to 4 hr. In contrast to the conditions used to prepare prepolymers based on other isocyanates, such as 4,4' methylene bis(phenylisocyanate), these conditions require significantly longer reaction times. The lower reactivity of $H_{12}MDI$ is a result of its cycloaliphatic structure, which reduces the reactivity of the isocyanate groups. The relative rates and second isocyanate groups of $H_{12}MDI$ and a variety of other isocyanates are shown in Fig. 4.21.

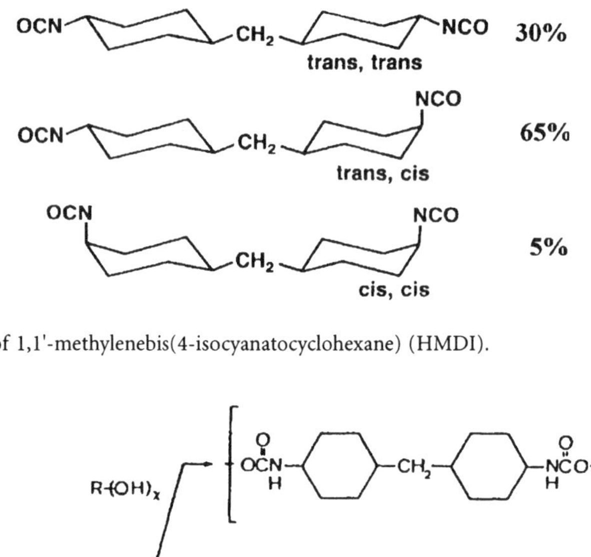

FIGURE 4.19 Isomers of 1,1'-methylenebis(4-isocyanatocyclohexane) (HMDI).

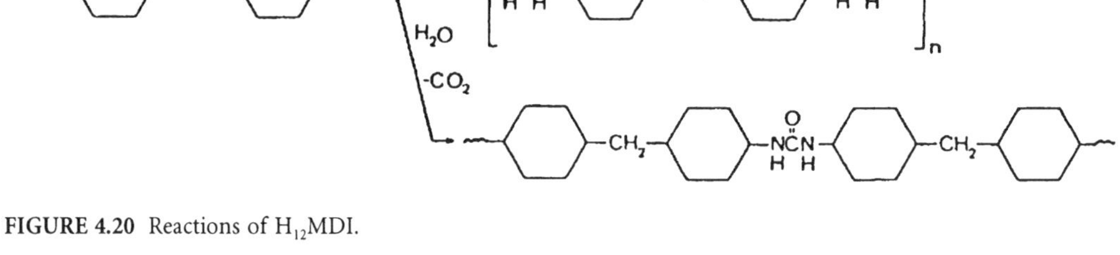

FIGURE 4.20 Reactions of $H_{12}MDI$.

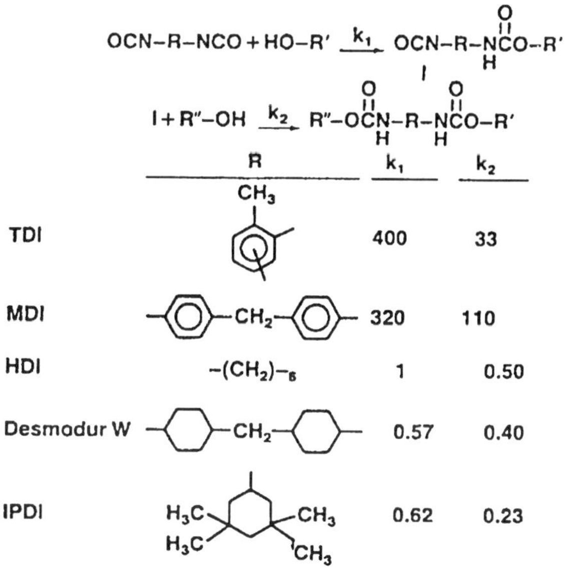

$$OCN-R-NCO + HO-R' \xrightarrow{k_1} OCN-R-\underset{H}{N}\overset{O}{\overset{\|}{C}}O-R'$$
$$\text{I}$$

$$\text{I} + R''-OH \xrightarrow{k_2} R''-O\overset{O}{\overset{\|}{C}}\underset{H}{N}-R-\underset{H}{N}\overset{O}{\overset{\|}{C}}O-R'$$

	R	k_1	k_2
TDI	(methyl-substituted benzene)	400	33
MDI	(phenyl)–CH$_2$–(phenyl)	320	110
HDI	$-(CH_2)-_6$	1	0.50
Desmodur W	(cyclohexyl)–CH$_2$–(cyclohexyl)	0.57	0.40
IPDI	(isophorone structure)	0.62	0.23

FIGURE 4.21 Reactions of HMDI relative to TDI, MDI, HDI, and IPDI.

The relative reactivities of the two isocyanate groups of H$_{12}$MDI are essentially equal, which is in contrast to the other monomers. Although the first isocyanate group of isophorone diisocyanate (IPDI) reacts more rapidly than those of H$_{12}$MDI, the second group reacts more slowly. Overall, IPDI may not react any faster than H$_{12}$MDI

To allow the reaction time for H$_{12}$MDI to be shortened or the reaction temperature to be lowered, catalysts are often added to reactions of H$_{12}$MDI to enhance its reactivity. As with other diisocyanates, tin compounds are generally very effective catalysts (see Table 4.11).

TABLE 4.11 -H$_{12}$MDI+1.4 Butane Diol + Catalyst (1% by Weight)

Catalyst	Relative Reactivity
None	1
Dimethyltin dichloride	103
Dibutyltin maleate	215

Preparation of H$_{12}$MDI

A revealing preparation technique of H$_{12}$MDI was disclosed by Irwin.[21] The Irwin invention relates to aliphatic isocyanate compositions that contain a high percentage of the trans-trans isomer of 4,4'-methylene bis (cyclohexyl isocyanate).

4,4'-methylene bis cyclohexyl isocyanate, also referred to as PICM, is a cycloaliphatic diisocyanate of low volatility. PICM and other aliphatic isocyantes such as tetramethylene diisocyanate, hexamethylene diisocyanate, and 1,3- and 1,4-cyclohexylene diisocyanate are of value in the preparation of non-discoloring polyurethanes in which they are reacted with various non-discoloring glycos or polyols followed by chain extension and/or cross-linking with a glycol, polyol, diamine, polyamine, or water. These isocyanates are useful in the preparation of moisture-cure and two-part polyurethane coatings and fibers.

PICM and its parent diamine bis(4,aminocyclohexyl) methane (PACM) exist in three stereoisomeric forms (trans-trans, cis-trans, and cis-cis) as described for PACM in U.S. Pat. 2,606,925, to Whitman. Commercial grades of PACM normally contain at least a minor proportion of all three of these stereoisomers and, in addition, small amounts of 2,4'-di(aminocyclohexyl) methane normally originating from the 2 to 4% of 2,4' diaminodiphenylmethane normally found in methylene dianiline from which the PACM is prepared.

This example illustrates the preparation of high trans-trans 4,4'-methylenebis(cyclohexyl isocyanate.) About 45 parts of a mixture of stereoisomers of bis(4-aminocyclohexyl) methane having an approximate isomer composition of 80% trans-trans, 16-17% cis-trans, and 3-4% cis-cis is dissolved in 425 parts of o-dichlorobenzene contained in a well agitated reaction vessel. Hydrogen chloride gas is introduced below the surface of the liquid at a rate of 7 parts/hr for 2 hr and then at a rate of 3 parts/hr for 1 hr. During the hydrogen chloride addition, the temperature is maintained within the range of 65to 80° C. Phosgene is then introduced into the resulting amine hydrochloride slurry at a rate of 8.5 parts/hr, and the reaction mass is heated to 150° C as rapidly as possible. While maintaining the phosgene feed rate at 8.5 parts/hr, the temperature is maintained at 150° C for 2 hr, then raised to 160° C and maintained for 2.5 hr, and finally raised to 170° C, where it is maintained for 5 hr. The phosgenation mass is then sparged with a slow stream of nitrogen and is cooled to 50° C or below.

The phosgenation mass is distilled at reduced pressure. After the o-dichlorobenzene has been removed, 46.6 parts of 4,4'-methylenebis-cyclohexyl isocyanate, boiling at about 165° C at 0.7 mm Hg is collected. The material has a freezing point of 72.6° C, corresponding to a melting point about 73° C and a trans-trans isomer content of 80%. A residue of nonvolatile phosgenation by-products amounting to 2.58 parts remains after the distillation. Based on the quantities of distilled product and residue, the yield of PICM is about 95%.

CHDI Preparation

Hentschel, Zengel, and Bergfeld[22] of Akzo developed a process for the stereospecific synthesis of trans-1,4-cyclohexane diisocyanate.

The invention provides an improved method for the preparation of organic mono- and polyisocyanates. The products have the general formula R(NCO)[n], in which R is an optionally substituted aliphatic, cycloaliphatic, aromatic, or heterocyclic residue, and n is 1, 2, or 3. The trisubstituted urea, which undergoes thermal decomposition, has the general formula R(NHCONR'R'') [n], in which R' and R'' are the same or different optionally substituted aliphatic hydrocarbon residues with 1 to 4 carbon atoms.

The process may be characterized in that the reaction is carried out with a hydrogen chloride adduct of the trisubstituted urea, which minimally contains the stoichiometric amount of hydrogen chloride and at most a 10 mole-% excess. The thermal decomposition is carried out in a melt or in the presence of an inert organic solvent in a closed system at temperatures between about 80 and 180° C.

The trisubstituted urea starting materials are easily prepared. They can be synthesized from a primary amine and a disubstituted carbamide acid chloride, or from a secondary amine and a monosubstituted carbamide acid chloride.

The hydrogen chloride adducts employed in the inventive process are the hydrochlorides of the trisubstituted ureas. These adducts do not necessarily exhibit exact stoichiometric compositions; they can contain excess hydrogen chloride. In this process, only those adducts that have at least one hydrogen chloride equivalent for each urea group and, at most, a 10 mole-% excess of hydrogen chloride based on the urea are contemplated. Preferably, the stoichiometric hydrochloride is used. The adduct is easily prepared by reaction of the trisubstituted urea with the corresponding amount of hydrogen chloride. For example, dry hydrogen chloride may be introduced into a solution or suspension of the urea, or alternatively, dry pulverulent urea may be brought into contact in a suitable manner with hydrogen chloride.

The process is predicated upon the following reaction:

$$R(NHCONR'R'')[n] \cdot nHCl \rightarrow R(NCO)[n] + n\ R'R''NH \cdot HCl$$

A wide variety of mono- and polyisocyanates may be prepared according to the inventive process. In general, these include primary, secondary, or tertiaryallphatic, cycloaliphatic, araliphatic, alky laromatic, aromatic, and heterocycllcmonoand polyisocyanates that may, if desired, be substituted by halogen, nitro, alkyl, alkoxy, and/or heterocyclic residues.

This process, based on the conversion of poly(ethylene terephthalate) circumvents the elaborate fractional crystallization procedures required for the existing p-phenylenediamine approaches. The synthesis starts with poly(ethylene terephthalate) (PET) or phthalic acid that is converted to the dimethyl ester and hydrogenated to yield the cyclohexane-based diester. Subsequent reaction of the ester with ammonia provides the desired bisamide. The synthesis of the amide is the key to the selectivity of this route. Typically, dimethyl 1,4-cyclohexane dicarboxylate is dissolved in a solvent. Ammonia is introduced, and methyl alcohol is removed by cracking and distillation. The desirable trans-amide is noted to precipitate upon prolonged reflux under an ammonia atmosphere.

Subsequent chlorination of the amide takes place in a two-phase reaction mixture (a dispersion of diamide in hydrochloric acid) through which a chlorine stream is passed. The temperature of this step must be maintained below 10° C to retard the formation of the product resulting from the Hofmann degradation of amides. Reaction of the N,N dichloroamide with diethylamine in the presence of base yields trans-1,4-cyclohexane-bis-1,3-diethylurea, which is transformed to the urea hydrochloride and pyrolyzed to yield the diisocyanate (Reaction 4.15).

trans 1,4 cyclohexane-bis-1,3 diethylurea

trans CHDI

REACTION 4.15 Preparation of CHDI.

Another method of preparing CHDI was disclosed by Zengel[23] for selectively making trans-cyclohexane-1,4-diisocyanate, trans-cyclohexane-1,4-diamine, a trans-cyclohexane-1,4-diurethane, a trans-cyclohexane-1,4-diurea, and trans-cyclohexane-1,4-disulphonyl urea by reacting ammonia with a mixture of cis and trans cyclohexane-1,4-dicarboxylic acid, a lower alkyl ester, a glycol ester, an oligomeric ester or a polyester to make a solid trans-dicarboxylic acid diamide in a first step.

The object of this inventionis to provide a process for selectively making trans-cyclohexane-1,4-diisocyanate, trans-cyclohexane-1,4-diamine, trans-cyclohexane-1,4-diurethane, trans-cyclohexane-1,4-diurea, and trans-cyclohexane-1,4-sulphonyl urea that does not require separation of the trans- and cis-isomers.

Production of TMXDI

Production of novel cast polyurethane elastomers based on TMXDI, an aliphatic isocyanate, was published by Arendt et al.[24] of American Cyanamid. The isocyanates, known as TMXDI(META) aliphatic isocyanate (tetramethyl xylene diisocyanate) and TMIR (META) unsaturated aliphatic isocyanate (isopropenyl dimethylbenzyl isocyanate) have a variety of uses in polymer applications. These products can be used in performance coatings, roofing compositions, flooring, sealants, adhesives, specialized dispersions, and elastomeric compositions for casting and injection molding.

TMXDI is a monomer containing a reactive double bond and a reactive isocyanate group. These groups can be selectively reacted to produce specialty polymers with unique properties. TMXDI can be copolymerized with acrylic, styrenic, and other monomers to yield polymers with isocyanate functionality. It can also be coreacted with diols, triols, or polyols to yield polymers with vinyl functionality. The physical properties of these isocyanates are shown in Table 4.12.

TABLE 4.12 Physical Properties of Isocyanates

Physical properties	m-TMXDI	P-TMXDI
Empirical formula	$C_{14}H_{16}N_2O_2$	$C_{14}H_{16}N_2O_2$
Molecular weight	244.3	244.3
NCO contents, % by weight	34.4	34.4
Physical form	Colorless liquid	White crystals
Melting point, °C	−10	+72
Boiling point, °C	150° C/3 mm	150° C/3 mm
Vapor pressure, mm Hg at 100° C	0.5	0.4
Pounds per gallon	8.8	9.1 (melt)
Flash point (closed cup)	450° C	450° C

The chemical structures of m and p-TMXDI are shown in Fig. 4.22. They are structural isomers of one another, the principal difference in physical properties being that the *meta* isomer is liquid at room temperature and the *para* isomer is a solid, crystalline material melting at 77° C. While interesting castable elastomers can be made from prepolymers using either the *meta* or the *para* isomer, the best balance of properties is achieved with the *para*, probably because of its highly symmetrical structure. In this section, we confine our discussion to the *para* isomer. p-TMXDI is an aliphatic diisocyanate; that is, the NCO groups are not attached directly to a benzenoid ring. A general characteristic of aliphatic isocyanates is that they react with hydroxyl and amine compounds more slowly than do their aromatic counterparts such as TDI and MDI. This behavior requires the use of catalysts where fast reaction is required; however, it permits a choice of curing agents that would be objectionably fast with the aromatics. This flexibility in choice of curing agents can be a key factor in the achievement of desired performance in the final cured elastomers.

Preparation of TMXDI

In 1982, Singh et al.[25] of American Cyanamid disclosed a process for production of tertiary benzyl isocyanates by reaction at low temperatures of the corresponding halides with an excess of isocyanic acid. This invention relates to the manufacture of tertiary benzylic isocyanates, particularly isocyanates such as the tetramethylxylyenediisocyanates (TMXDI) and chlorolsopropyldlmethylbenzyl isocyanates.

Figure 4.23 presents the synthetic method for the preparation of TMXDI via nonphosgenation chemistry. This preparation leads to the formation of both the m and p isomeric forms of the diisocyanate.

Preparation of IPDI

A popular non-discoloring isocyanate is 3-isocyanatomethyl-3,5,5-trimethylcyclohexyl isocyanate, commonly known as isophorone diisocyanate (IPDI). IPDI is extensively used in the preparation of polyurethane-based, light-stable coatings. Surprisingly, it is also used in military applications, where it is a primary ingredient in the manufacture of binders in many cast-cured solid propellants and plastic-bonded explosives. It excels in these applications. Its desirable pot life and cure rate are due to two isocyanate groups with different reactivities—one hindered and the other unhindered. Generally, isocyanates are toxic, but IPDI's lower volatility makes it easier and safer to handle than other candidates.

The preparation of IPDI via a nonphosgene route was disclosed by Geigel[26] where an isocyanate compound of the general formula R(NCO) n, wherein R is either an aliphatic or in aromatic radical or a combination of both, and n is 1 or a larger integer is prepared by the process comprising dissolving in a high-temperature boiling point solvent contained in an agitated, jacketed reactor vessel, an amine of the formula R(NH$_2$) n, completing a hydrochlorination and precipitation step to form a colloidal suspension of microcrystalline salt particles of amine hydrochloride salt R(NH$_2$) n·nHCl, introducing oxalyl chloride with agitation and by subsurface injection into the amine hydrochloride salt solution to form an intermediate of amine oxamyl chloride in situ, which is terminally decomposed to yield an socyanate of the general formula R(NCO) n, and separating and purifying the isocyanate by distillation. The

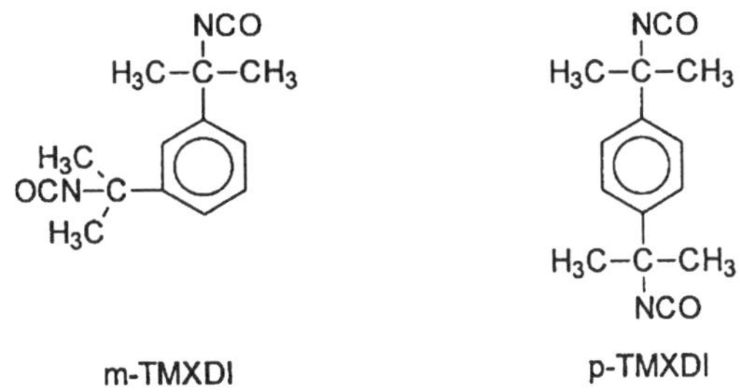

FIGURE 4.22 Chemical structures of m- and p-TMXDI.

TMXDI isomers

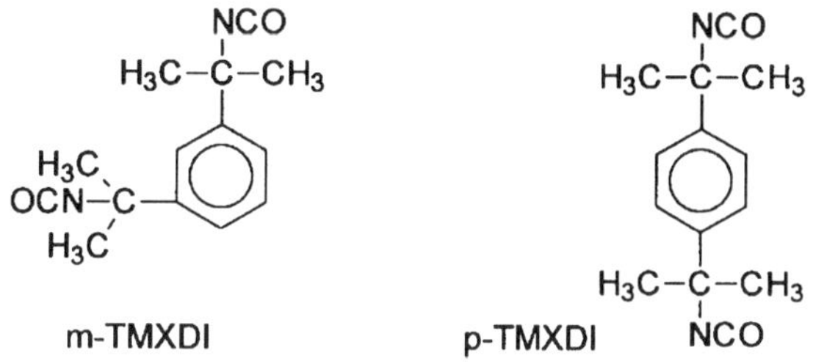

m-TMXDI   p-TMXDI

Non-phosgenation chemistry

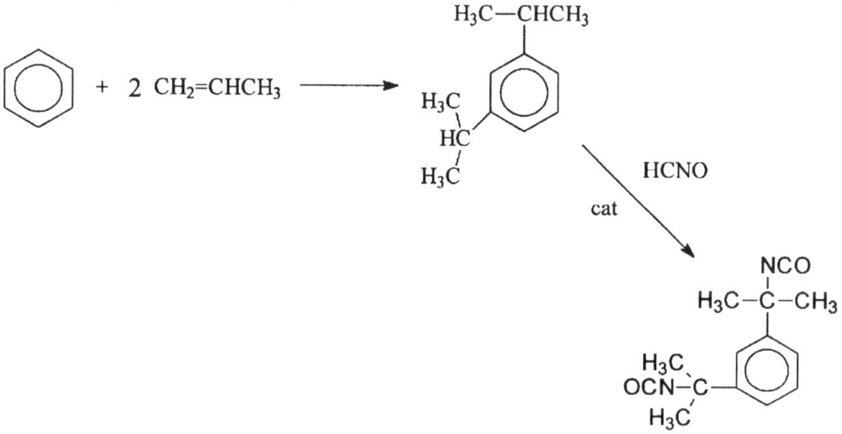

FIGURE 4.23 Synthetic method for the preparation of TMXDI via nonphosgenation chemistry.

unreacted oxalyl chloride and high-temperature reaction solvent are recovered by distillation and by absorption from the HCl by-product and are recycled. The by-product HCl is also partially recycled and used for hydrochlorination.

The Geigel route is presented in Reaction 4.16.

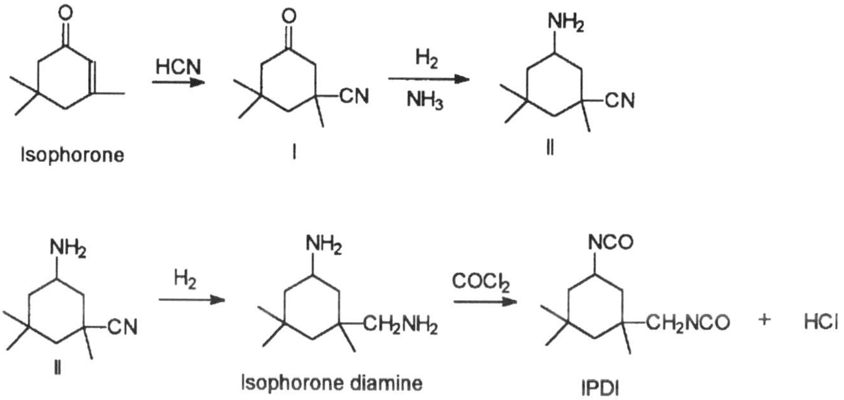

Isophorone   I   II

II

Isophorone diamine   IPDI

REACTION 4.16 Present commercial route to IPDI, U.S. Pat 4,663,473.

4.7 Specialty Isocyanates

Specialty isocyanates are premium-priced isocyanates that can be used to produce polyurethane polymers with unusual properties. Several of these isocyanates, such as Desmodur TH, N, and HH, are really adducts. Others are monomeric isocyanates; one isocyanate, trade named T-1190 by Veba Chemie, is an isocyanurate-containing triisocyanate. The molecular structure of these complex products is shown in Fig. 4.24.

Acyl isocyanates, extensively used in synthetic applications, cannot be directly synthesized from amides and phosgene. Reactions of acid halides with cyanates have been suggested. However, the dominant commercial process utilizes the reaction of carboxamides with oxalyl chloride. Cyclic intermediates have been observed in these reactions that generally give a high yield of the desired products.

Commercially important arenesulfonyl isocyanates are not directly accessible from the corresponding sulfonamides via phosgenation due to lack of reactivity or by-product formation at elevated temperatures.

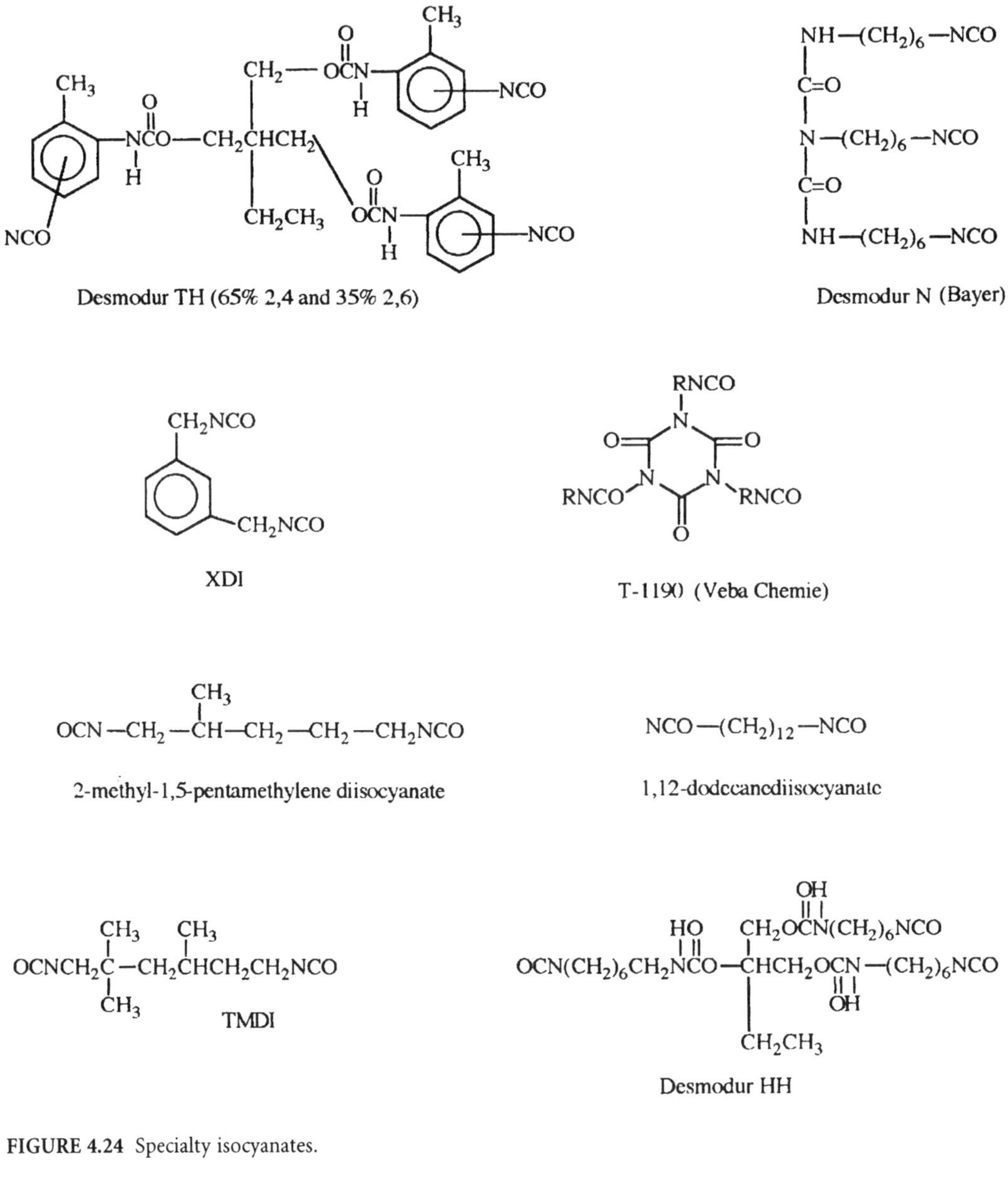

Desmodur TH (65% 2,4 and 35% 2,6)

Desmodur N (Bayer)

XDI

T-1190) (Veba Chemie)

2-methyl-1,5-pentamethylene diisocyanate

1,12-dodecanediisocyanate

TMDI

Desmodur HH

FIGURE 4.24 Specialty isocyanates.

A convenient method for their preparation consists of the reaction of alkyl isocyanates with sulfonamides to produce mixed ureas that, upon phosgenation, yield a mixture of alkyl and arenesulfonyl isocyanates. The desired product can be obtained by simple distillation. Optionally, the oxalyl chloride route has been employed for the synthesis of arenesulfonyl isocyanate.[27]

Of the many other methods leading to isocyanates, only a few are practical enough in regard to availability of starting materials to be of general applicability. One of the more promising approaches utilizes olefinic substrates that add isocyanic acid in Markovnikov fashion to form alkyl isocyanates). This reaction is used to produce 1,4-bis(2-isocyanatoisopropyl)benzene from cumene in commercially useful yields.[28]

One approach uses the slow addition of the olefin to an excess of solvent and isocyanic acid in the presence of a catalytic amount of inorganic acid. Reaction temperatures are preferably maintained between 25 to 80° C. In the case of a diolefin, such as diisopropenylbenzene, the reaction can be controlled to favor the production of either the mono- or the diisocyanate by controlling the stoichiometry of the isocyanic acid in the reaction mixture. Another approach involves the formation of the dichloro intermediate. The dichloro compound reacts at low temperatures with an excess of isocyanic acid in the presence of a Lewis acid.

An alternative approach involves the reaction of an alky carbamate with a tertiary olefin.[29] The resultant carbamates are thermally cracked at temperatures of 150 to 350° C to yield the isocyanate. The isocyanate is generally purified via distillation.

The exchange of halogen by isocyanato groups has also been suggested as a method of preparing isocyanates from chloro- or bromoalkanes. Metal cyanates are the reagents of choice for these exchange reactions, which often entail the formation of oligomers of the desired isocyanates. For example, ethyl isocyanate can be prepared in 90% yield by the reaction of ethyl bromide with potassium isocyanate.[30,31]

The use of polar solvents, such as N,N-dimethylformamide, is noted to result in extensive trimer formation. However, if the isocyanate is trapped using compounds such as alcohols, carboxylic acids, and amines that contain active hydrogens, high yields are obtained.[32]

Pyrolysis approaches can also be used to prepare substituted isocyanates that cannot be prepared using other methods. For example, N, N', N"trichlorocyanuric acid thermally dissociates to yield chloroisocyanate and carbonyl diisocyanate. The carbonyl isocyanate is unstable and polymerizes.[33]

Preparation of α, α-Dimethylbenzyl Isocyanates

Nagato[34] of Showa Denko disclosed the preparation of α, α-dimethylbenzyl isocyanates or its derivatives that comprises reacting the corresponding organic halide in an anhydrous condition in the presence of a specified catalyst with an alkali metal cyanate in an aprotic solvent that forms no salt, with a hydrogen halide. The aprotic solvents that do not form salts with hydrogen halides include chlorinated aliphatic or aromatic hydrocarbons, esters, nitriles, nitro compounds, ketones, and aliphatic or cyclic ethers.

References

1. Wurtz, A., *Compt. Rend.*, 89,242 (1848).
2. Hentschel, W., *Chem. Ber.*, 17,1284 (1884).
3. Staudinger H., and Hauser, E., *Helv. Chim. Acta*, 4, 861 (1921).
4. Trenbeath, S.L., et al., U.S. Pat. 4,377,530 to American Cyanamide.
5. Ulrich, H., in Gum, W.F., Riese, W. and Ulrich, H., Reaction Polymers, Hanser Publishers, New York, 358 (1992).
6. Brederek, H., Simchem, G., and Goknel, E., *Chem Ber.*, 103, 236 (1970).
7. Biener, H., *Liebigs Ann. Chem.*, 686, 102 (1965).
8. Van Look, E., *Ind. Chem. Belg.*, 1974, 661 (1974).
9. Findeisen, K., Wagner, K., and Henning, H.J., U.S. Pat. 4,088,665 (May 9, 1978), "Process for the Partial Carbodiimidization of Organic Isocyanates" (to Bayer Aktiengesellschaft).
10. Stallmann, O. U.S. Pat. 2,729,666 (Jan. 3, 1956) "Alkyl-aryl Diisocynates with Reduced Activity."

11. Rasmussen, J.K., and Hassner, A., *Chem. Rev.,* 76,389 (1976).
12. Gorbartenko, V.I., and Samarai, L.I., *Synthesis,* 1980, 85 (1980).
13. Twitchett, H.J., *Chem. Soc. Rev.,* 3, 209 (1974).
14. Grant, B.R., and Irwin, C.F., U.S. Pat. 3,574,695 (April 13, 1971), A Two-Stage Phosgenation Process for Producing Isocyanates, assigned to E.I. DuPont.
15. Dhar, D.N., and Murthy, S.K., *Synthesis,* 1986, 437 (1986)
16. Ulrich, H.G., *Chem. Rev.,* 65,869 (1965).
17. Smith, P.A.S., *Organic Reactions,* 3,337 (1946).
18. Detlef, D., and Muntz, F., U.S. Pat. 2,901,497 (Aug. 25, 1959), "Production of Aromatic Isocyanates," assigned to Mobay Chemical Company.
19. Olinger, R., Schnez, H., Blumenberg, B., and Raabe, H.J., U.S. Pat. 4,581,174 (April 8, 1986), A Process for the Continuous Preparation of Organic mono-and/or Polyisocyanate, assigned to BASF Aktiengesellschaft.
20. Zimmerman, E.K., and Williams J.L., "Desmodur W: Chemistry and Applications," Industrial Coatings Group, Mobay Corporation Technical Publication, October 29, 1985.
21. Irwin, C.F., Canadian Pat. 971184, "Aliphatic Isocyanate Compositions," assigned to DuPont, July 15, 1975.
22. Hentschel, P., Zengel, H., and Bergfeld, U.S. Pat. 4,223,145, A Process for the Preparation of Organic Mono-and polyisocyanates, assigned to Akzo, N.V., September 16, 1980.
23. Zengel, H., and Bergfeld, M., U.S. Pat. 4,457,871, "Preparation of Transcyclohexane, 1,4-Diisocyanate and Related Compounds," to Akzona Incorporated, July 3, 1984.
24. Arendt, V.D., Logan, R.E., and Saxon, R., J. *Cellular Plastics,* 8,376 (1982).
25. Singh, B., Henderson, W.A., U.S. Pat. 4,361,518, "Manufacture of Isocyanates," assigned to American Cyanamide Company.
26. Geigel, M.A., and Marynowski, C.W., U.S. Pat. 4,663,473, "Isocyanates from Oxalyl Chloride and Amines," assigned to the United States of America.
27. Hoover, F.W., Rothrock, H.S., and Olson, K.E., J. *Org. Chem.,* 29, 143 (1964).
28. Singh, B., and Henderson, W.A., U.S. Pat. 4,361,518 (to American Cyanamid Company).
29. Singh, B., Chang, L.W., and Forgione, P.S., U.S. Pat. 4,439,616 (to American Cyanamid Company).
30. Schaeffer, W.D., U.S. Pat. 3,017,420 to Union Oil of California.
31. Nagato, N., and Naito, T., U.S. Pat. 4,130,577 to Showa Denko.
32. Fukui, K., and Kitano, H., British Pat. 858,810 (1961).
33. Hagemannn, H. German Pat. 2,408,069 to Bayer A.G.
34. Nagato, N., and Naito, T., U.S. Pat. 4,130,577 (December 19, 1978) "Process for Preparing α, α–Dimethylbenzyl Isocyanates" (to Showa Denko K.K.).

5

Polyols

5.1 Introduction

Polyols are reactive substances, usually liquids, containing at least two isocyanate-reacting groups attached to a single molecule. A large variety of polyols are offered, but most of the polyols used fall into two major categories: hydroxyl-terminated or amino-terminated polyols.

Polyols, or macroglycols, have a profound effect on the properties of the finished polyurethane. While associating the properties of the polymers with the urethane linkage, the structure of the polyol has a direct bearing on both processing and finished properties of the polyurethane polymer. In fact, the majority of the linkages found in polyurethanes are derived from the linkages found in the polyol.

Thus, if polyether-based or polyester-based polyols are used, we obtain polyether-based or polyester-based polyurethanes. A similar statement can be made for caprolactone, or polycarbonate-based polyols, which produce caprolactone-based or polycarbonate-based polyurethanes.

Most of the confusion regarding polyurethanes stems from the fact that many technologists tend to associate polyurethanes as a monolithic family of polymers, as opposed to the reality that polyurethanes are poorly described until we modify the name *polyurethane* by the prefix that denotes the type of polyol used. Therefore, there is a world of difference between polyether-based and polyester-based polyurethane foams, or polyether-based and polycarbonate-based polyurethane elastomers.

5.2 Classification

The isocyanate-reactive polyols used to prepare polyurethanes or polyurethane ureas contain at least two active hydrogen atoms per molecule, as determined by the Zerewittinoff method. There are four classes of polyols.

1. *Polyether polyols.* These materials are the products of reaction between a simple starter molecule (also called initiators) such as ethylene glycol, propylene glycol, glycerin, pentaerythritol, trimethylolpropane, sucrose, or sorbitol, and a cyclic ether such as ethylene oxide, propylene oxide, mixtures of ethylene oxide and propylene oxide, or tetrahydrofuran. The reaction between an initiator and an alkylene oxide to produce hydroxy-terminated polyether polyols is shown in Fig. 5.1. The functionality of the resultant polyether depends on the functionality of the chosen initiator; thus, if the initiator is a diol, the resultant polyol will have a functionality of two; conversely, if the initiator is a triol, the resultant polyol will have a functionality of three. This is shown in Fig. 5.2.

   Polyether-based polyols produce very high quality polyurethane foams and elastomers. Among the most important polyether polyols are the polyBD, polytetramethylene etherglycol (PTMEG), polypropylene oxide glycol (PPO) and polybutylene oxide glycol (PBO); the chemical structures of these polyols are shown in Fig. 5.3.

   Polytetramethylene ether glycol (PTMEG) deserves special comment as a reactant in the production of hydrolysis-resistant polyurethane elastomers. The synthetic route to PTMEG from

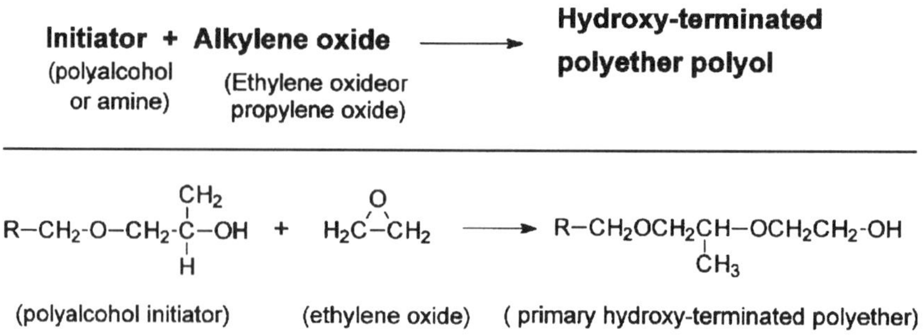

Initiator + Alkylene oxide ⟶ **Hydroxy-terminated polyether polyol**

(polyalcohol or amine) (Ethylene oxideor propylene oxide)

$$R-CH_2-O-CH_2-\underset{\underset{H}{|}}{\overset{\overset{CH_2}{|}}{C}}-OH \ + \ H_2\overset{O}{\overset{\diagup \diagdown}{C}}-CH_2 \ \longrightarrow \ R-CH_2OCH_2\underset{\underset{CH_3}{|}}{CH}-OCH_2CH_2-OH$$

(polyalcohol initiator)   (ethylene oxide)   (primary hydroxy-terminated polyether)

FIGURE 5.1   Reaction between an initiator and an alkylene oxide to produce hydroxy-terminated polyether polyols.

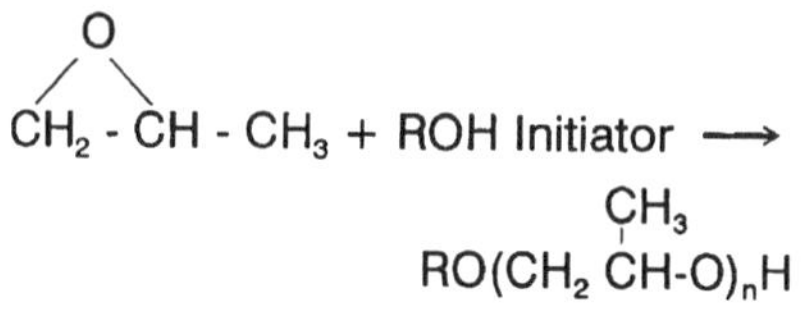

$$\overset{O}{\overset{\diagup \diagdown}{CH_2 - CH - CH_3}} + ROH \ Initiator \longrightarrow$$

$$RO(CH_2 \ \underset{\underset{CH_3}{|}}{CH}-O)_nH$$

FIGURE 5.2   Chemical structures of polyether polyols, least expensive backbone. Functionality of polyether will depend on functionality of the initiator, e.g., diol, triol.

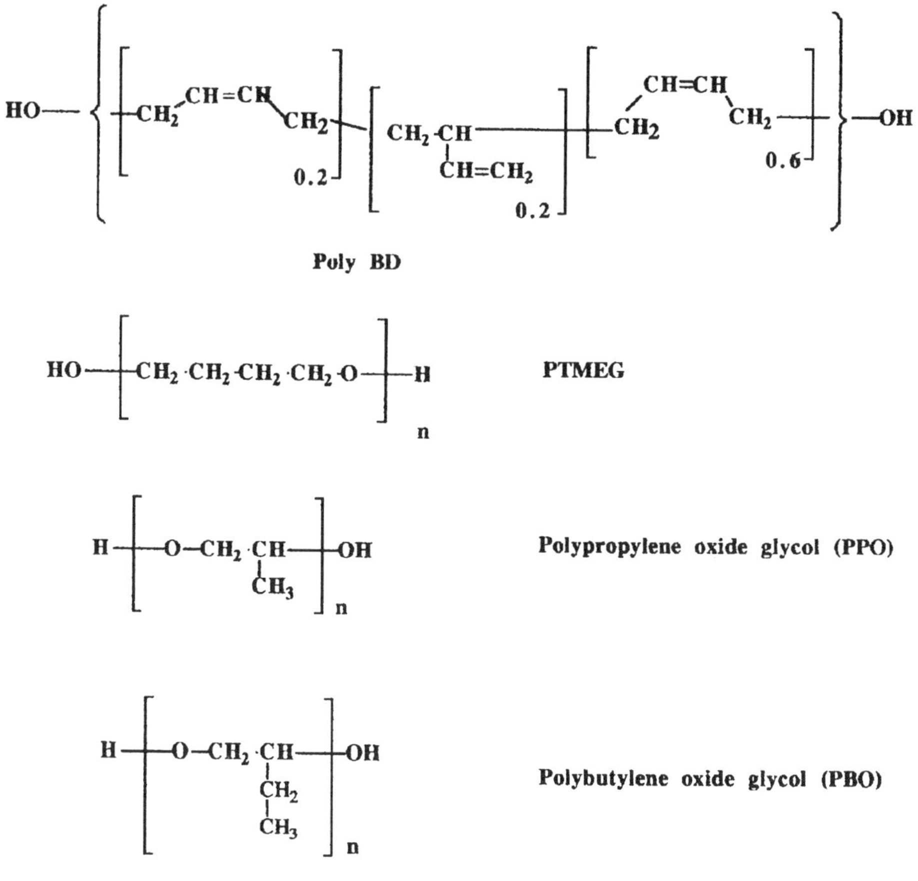

Poly BD

PTMEG

Polypropylene oxide glycol (PPO)

Polybutylene oxide glycol (PBO)

FIGURE 5.3   Some polyether polyol chemical structures.

furfural is shown in Fig. 5.4, along with the expected properties of polyurethane elastomers produced from this polyol.

Also included within this class of active hydrogen-containing materials are polymer polyols, which are compositions in which a polymer has been formed in a polyether polyol medium. The polymers employed in these systems may be polymers of unsaturated monomers such as acrylonitrile or styrene, and also may be copolymers or terpolymers of such monomers. They may, additionally, be polyureas prepared from reaction of polyisocyanates and diamines, or polyurethanes prepared by reaction of polyisocyanates and polyalkanol amines such as triethanol amine.

2. *Amine-terminated polyethers.* These are based on polyether polyols, in which the terminal hydroxyl groups have been replaced by primary or secondary amino functionalities.

3. *Polyester polyols.* These are polyalkylene glycol esters such as polybutylene terephthalate or adipate, or caprolactone polyesters. Polyalkylene glycol adipates are prepared by condensation polymerization of the alkylene glycol and the corresponding diester or diacid. An example would be 1,4-butanediol plus adipic acid to form the "polybutanediol adipate." The most important glycols and polyols used in the manufacture of polyesters are shown in Fig. 5.5. The most important acids are shown in Fig. 5.6.

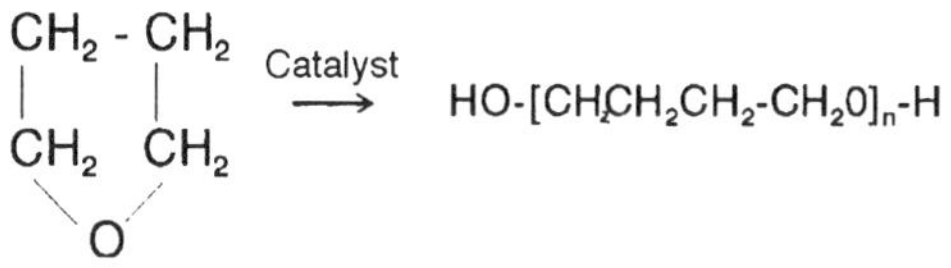

$$\underset{\underset{\displaystyle\diagdown\!\diagup}{O}}{\overset{\displaystyle CH_2 - CH_2}{\underset{\displaystyle CH_2 \quad CH_2}{|\qquad|}}} \quad \xrightarrow{\text{Catalyst}} \quad HO\text{-}[CH_2CH_2CH_2\text{-}CH_2O]_n\text{-}H$$

FIGURE 5.4 Polytetramethylene ether glycol, which offers excellent hydrolytic stability, excellent flexibility and impact resistance, molecular weights ranging from 650 to 3000, and premium pricing.

- Glycols

    ▸ 1,6 Hexanediol
    ▸ Neopentyl Glycol
    ▸ Butanediol (1,4 and 1,3)
    ▸ Ethylene and Propylene Glycols
    ▸ Cyclohexanedimethanol

- Polyols

    ▸ Trimethylolpropane
    ▸ Glycerine
    ▸ Tris (Hydroxyethyl) Isocyanurate

FIGURE 5.5 Important glycols and polyols used in the manufacture of polyesters.

- HOOC-R-COOH + X'S HO-R'-OH $\longrightarrow$
  $$HO\text{-}R'\text{-}O[\overset{\overset{\displaystyle O}{\|}}{C}\text{-}R\text{-}\overset{\overset{\displaystyle O}{\|}}{C}\text{-}O\text{-}R'\text{-}O]_nH$$

- Acids

    ▸ Adipic Acid $HOOC(CH_2)_4COOH$

    ▸ Azelaic Acid $HOOC(CH_2)_7COOH$

    ▸ Phthalic Anhydride 1,3 benzodicarboxilic acid

    ▸ Isophthalic Acid   1,3 benzofurandione

    ▸ Dimethylolpropionic Acid $\underset{\displaystyle CH_3}{OHCH_2\overset{\displaystyle COOH}{\underset{|}{(C)}}CH_2OH}$

FIGURE 5.6 Important acids.

The condensation chemistry is carefully controlled to ensure that these polyols contain terminal hydroxyl groups. The most common acids used in the production of polyesters are adipic, glutaric, and azeloic acids. Figure 5.7 presents the chemical structures of some common polyesters.

Caprolactone polyesters are prepared by ring opening polymerization of caprolactone monomer with a glycol such as diethylene glycol or ethylene glycol, as depicted in Fig. 5.8. Caprolactone polyols are used in those applications where better hydrolytic stability is important.

Castor oil is a polyester-based polyol with good hydrolytic stability. Castor oil-based polyurethanes are frequently used in the production of casting compounds. The structure of castor oil is shown in Fig. 5.9.

4. *Polycarbonates,* such as poly(1,6-hexanediol)carbonate. Such polycarbonates are prepared by condensation of phosgene or alkylene glycol carbonates, e.g., dimethyl carbonate, with alkylene glycols such as 1,6-hexanediol. As in ester chemistry, this chemistry is controlled to ensure that these polycarbonates contain terminal hydroxyl groups. The chemical structure of a hexanediol/ethyl carbonate-based polycarbonate diol is shown in Fig. 5.10.

Hydroxyl-terminated polymer	Repeat unit	Polyester m.p. (°C)
Poly(ethylene glutarate)	$-[-CH_2CH_2OOC(CH_2)_3COO-]-$	5
Poly(ethylene adipate)	$-[-CH_2CH_2OOC(CH_2)_4COO-]-$	50
Poly(ethylene azelate)	$-[-CH_2CH_2OOC(CH_2)_7COO-]-$	46
Poly(trimethylene glutarate)	$-[-CH_2CH_2CH_2OOC(CH_2)_3COO-]-$	39
Poly(tetramethylene glutarate)	$-[-CH_2CH_2CH_2CH_2OOC(CH_2)_3COO-]-$	37
Poly(pentamethylene glutarate)	$-[-CH_2CH_2CH_2CH_2CH_2OOC(CH_2)_3COO-]-$	30
Poly(diethylene glutarate)	$-[-CH_2CH_2OCH_2CH_2OOC(CH_2)_3COO-]-$	29
Poly(diethylene adipate)	$-[-CH_2CH_2OCH_2CH_2OOC(CH_2)_4COO-]-$	4
Poly(triethylene adipate)	$-[-(CH_2CH_2O)_2CH_2CH_2OOC(CH_2)_4COO-]-$	<0
Poly(1,2-propylene adipate)	$-[-CH_2CH\ OOC(CH_2)_4COO-]-$ $\overset{\mid}{CH_3}$	—

FIGURE 5.7   Chemical structures of some common polyesters. *Source:* G. Buist, *Polyurethane Technology,* John Wiley & Sons, 1968.

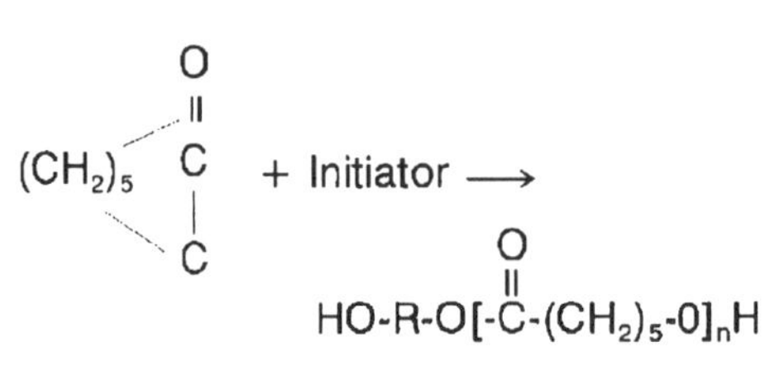

$$(CH_2)_5 \overset{\overset{O}{\parallel}}{\underset{\mid}{\underset{C}{C}}} \quad + \text{ Initiator } \longrightarrow$$

$$\text{HO-R-O}[\text{-}\overset{\overset{O}{\parallel}}{C}\text{-}(CH_2)_5\text{-O}]_n\text{H}$$

Features:
  • Improved hydrolytic stability over conventional polyesters
  • Narrow molecular weight distribution

FIGURE 5.8   Polycaprolactones. Note: if initiator is a glycol, a polycaprolactone diol is produced; if initiator is a triol, a polycaprolactone triol is produced.

Ricinoleic Acid (12-hydroxy oleic)   89.5%
Dihydroxystearic Acid   0.7
Palmitic Acid   1.0
Stearic Acid   1.0
Oleic Acid   3.0
Linoleic Acid   4.2
Linolenic Acid   0.3
Eicosanoic Acid   0.3

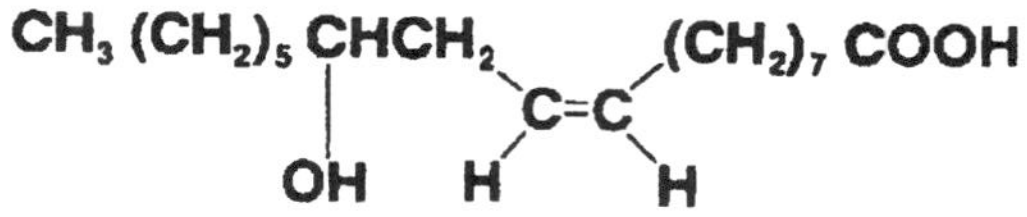

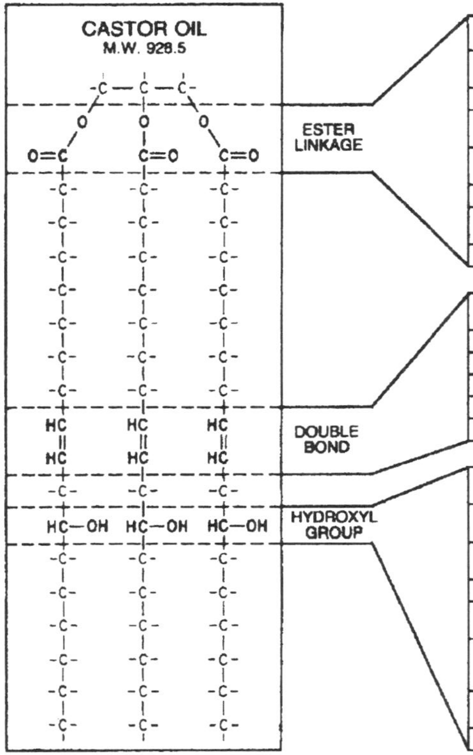

$$CH_3 (CH_2)_5\ CHCH_2 \overset{\displaystyle}{\underset{OH}{|}}\ \ C=C\ (CH_2)_7\ COOH$$

CHEMICAL REACTIONS OF CASTOR OIL

NATURE OF REACTION	ADDED REACTANTS	TYPE OF PRODUCTS
ESTER LINKAGE		
(1) Hydrolysis	Acid, enzyme, or Twitchell reagent catalyst	Fatty acids / Glycerol
(2) Esterification	Monohydric alcohols	Esters
(3) Alcoholysis	Glycerol, glycols, pentaerythritol, etc.	Mono- and diglycerides, monoglycols, etc.
(4) Saponification	Alkalies / Alkalies plus metallic salts	Soluble soaps / Insoluble soaps
(5) Reduction	Na reduction	Alcohols
(6) Amidation	Alkyl amines, alkanolamines, etc.	Amine salts / Amides
(7) Halogenation	SOCl₂	Fatty acid halogens
DOUBLE BOND		
(8) Oxidation, polymerization	Heat, oxygen, crosslink agent	Polymerized oils
(9) Hydrogenation	Hydrogen (moderate pressure)	Hydroxystearate
(10) Epoxidation	Hydrogen peroxide	Epoxidized oils
(11) Halogenation	Cl₂, Br₂, I₂	Halogenated oils
(12) Addition reactions	S, maleic acid	Polymerized oils
(13) Sulfonation	H₂SO₄	Sulfonated oils
HYDROXYL GROUP		
(14) Dehydration / (14a) Hydrolysis, distillation	Catalyst (plus heat)	Dehydrated castor oil / Octadecadienoic acid
(15) Caustic fusion	NaOH	Sebacic acid / Capryl alcohol
(16) Pyrolysis	High heat	Undecylenic acid / Heptaldehyde
(17) Halogenation	PCl₅, POCl₃	Halogenated castor oil
(18) Alkoxylation	Ethylene and/or propylene oxides	Alkoxylated castor oils
(19) Esterification	Acetic-, phosphoric-, maleic-, phthalic anhydrides	Alkyl and Alkylaryl esters / Phosphate esters
(20) Sulfation	H₂SO₄	Sulfated castor oil (Turkey red oil)
(21) Urethane reaction	Isocyanates	Urethane Polymers

CASTOR OIL
M.W. 928.5

FIGURE 5.9   Structure and chemistry of castor oil. *Source:* CASCHEM Technical Bulletin 100.

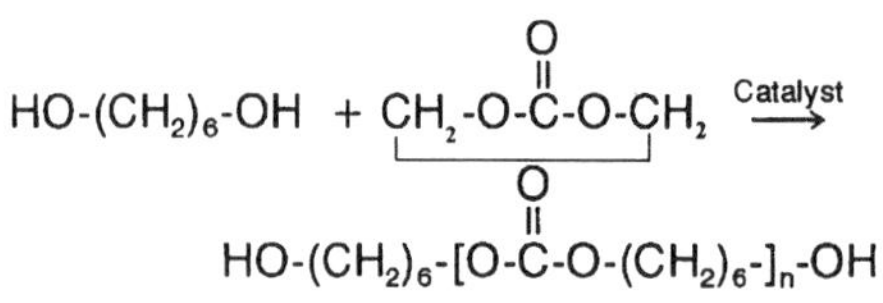

$$HO\text{-}(CH_2)_6\text{-}OH \ + \ CH_2\text{-}O\overset{O}{\overset{\|}{C}}\text{-}O\text{-}CH_2 \ \overset{Catalyst}{\longrightarrow}$$

$$HO\text{-}(CH_2)_6\text{-}[O\overset{O}{\overset{\|}{C}}\text{-}O\text{-}(CH_2)_6\text{-}]_n\text{-}OH$$

Features:
  - Excellent low-temperature properties
  - Good hydrolytic stability
  - Good oxidation stability
  - Excellent weatherability

FIGURE 5.10 Polycarbonates.

5.3 Commercially Available Polyols

In 1994, of the four billion pounds of polyurethanes consumed in the United States, nearly 90% were based on polyethers, 9% on polyesters, and approximately 1% were based on other specialty polyols, demonstrating that polyether-based polyurethanes are preeminent in the industry. The construction, transportation, and upholstered furnishings markets represent the largest consumers of polyurethane materials. These applications predominantly use polyether-polyols.

The structure of the polyol plays a dominant role in the final physical and chemical properties of the resultant polyurethane polymer. While the molecular weight and functionality of the polyol are important, the molecular structure of the polyol chains in equally important. In general, the distinguishing characteristics of the polyols used to make polyurethanes are summarized in Table 5.1.

TABLE 5.1 Distinguishing Characteristics of Polyols

Characteristics	Elastomers, Coatings, Flexible Foams	Elastoplastics, Rigid Coating, Rigid Foams
Molecular weight range (Daltons)	1,000 to 6,500	150–1,000
Functionality	2.0 to 3.0	3.0 to 8.0
Hydroxyl number	28 to 160	250 to 1,000

5.3.1 Polyols Based on 2-methyl-1,3-propanediol and Isophthalic Acid

Polyurethane-based thermosettable resin compositions containing low-molecular-weight polyesters are frequently used as protective coatings for substrates such as steel coil. An ideal protective coating is one that adheres well to the substrate, is simultaneously hard and flexible, and resists solvents, abrasion, and dry heat. A combination of such properties in a polyester-based coating is ordinarily difficult to achieve, because one property can often be enhanced only at the expense of another. For example, excellent coating flexibility is essential during substrate shaping, which normally follows coating application. High hardness is desirable for aesthetic reasons and for greater durability and resistance to stains and solvents. Hardness may be enhanced by increasing the cyclic (i.e., the proportion of aromatic or cycloaliphatic recurring units) content of the polyester. However, a high cyclic content tends to result in inflexible or brittle coatings. Another way to increase hardness is to incorporate significant amounts of a polyol having three or more hydroxy groups into the polyester. The polyester is consequently highly branched rather than linear and tends to decrease the flexibility of the cured coating. Problems with premature gelation are sometimes also observed with such branched polyesters.

Ideally, polyesters are noncrystallizable and dissolve freely in common organic solvents. Resin compositions are normally applied in solution form to lower the viscosity of the neat polyester/aminoplast mixture. Since a polyester typically is stored for an extended period of time prior to application, it is important that the polyester not crystallize or precipitate out of solution. Increasing the cyclic content of a polyester to enhance the hardness of a cured coating, particularly if the dicarboxylic acid component is terephthalic acid, tends to increase the crystallizability of the polyester and to decrease its solubility in organic solvents.

5.3.2 Carbamate-Pendant Polyols

Polyesters are widely used in curable compositions such as coating compositions. These resins offer many beneficial properties, such as good durability, good dispersibility in aqueous systems through incorporation of appropriate ionic or nonionic stabilizing groups, impact resistance, good adhesion, and other physical properties such as stress release.

One area of concern with polyester resins for curable compositions has been the incorporation into the resin of sufficient levels of functional groups to achieve the desired cure performance. Hydroxyl groups are commonly used as functional groups in curable compositions, but polyester resins with pendant hydroxyl groups are difficult to prepare, since any pendant hydroxyl groups would be consumed by reaction with acid groups during formation of the polyester. Hydroxyl functional groups are usually incorporated onto polyester resins by the use of polyol capping agents such as trimethylol propane resulting in terminal OH groups, but no pendant OH groups. Such resins provide only limited cross-link density upon cure. The cross-link density may be increased somewhat by using branched polyesters, which are prepared by the incorporation of trifunctional or higher functional polyols or polyacids in the polyester reaction mixture. However, the degree of branching is often limited due to gelation. Low cross-link density in curable polyester resin systems must often be compensated for by using higher-molecular-weight resins that more closely resemble thermoplastic compositions than thermoset compositions.

Accordingly, Menovcik[1] in U.S. Pat. 5,451,656 directed the invention toward a new method of preparing polyester polymers or oligomers having pendant carbamate groups. The patent also disclosed how to prepare polyester polymers or oligomers while avoiding the formation of side products that could reduce the purity of the final polyester.

According to the invention, a method of preparing a polyester polymer or oligomer consists of the following:

1. Reacting a hydroxyalkyl cyclic carbonate with ammonia, ammonium hydroxide, or a primary amine to form a dihydric alcohol having a carbamate group appended thereto
2. Reacting the dihydric alcohol from (1) and a cyclic anhydride to form a half-ester diacid having a carbamate group appended thereto
3. Reacting a mixture comprising the half-ester diacid from step (2) and a polyol to form a polyester having carbamate groups appended thereto

5.3.3 Amine-Containing Polyesters with Self-Catalytic Activity

Reactive polyoxyalkylene-polyols can be advantageously used in the production of elastomeric or flexible cellular polyurethane products. These polyols are prepared by oxyalkylation of an initiator molecule containing, in bonded form, at least two reactive hydrogen atoms and at least one tertiary amino group bonded via a spacer bridge, comprising an alkylene radical having at least three carbon atoms, by means of at least one alkylene oxide—preferably ethylene oxide and/or 1,2-propylene oxide.[2]

U.S. Pat. 4,144,386 describes a process for the production of flexible PU foams that are extremely soft. Polyoxyalkylene-polyols that are suitable for this purpose have a functionality of from 2–4 and a hydroxyl equivalent weight of from 700 to 2200, and they can be prepared using any difunctional to tetrafunctional alcohols, diamines, or alkanolamines, e.g., pentaerythritol, ethylenediamine, or triethanolamine.

While the use of amino compounds as initiator molecules achieves a significant increase in reactivity of polyoxyalkylene-polyols of low molecular weight (as are used, for example, for the production of rigid PU foams) compared with polyhydric alcohols in the reaction with polyisocyanates. This is not the case for polyoxy alkylene-polyols of relatively high molecular weight, e.g., those having molecular weights of greater than 1800 (as are employed, e.g., for the production of soft PU foams). This different catalytic effect of amine-initiated polyoxyalkylene-polyols can be interpreted as meaning that the tertiary amino groups incorporated into the polyoxyalkylene chain are considerably screened by the high-molecular-weight, highly entangled polyoxyalkylene chains and thus lose their catalytic activity.

Low-molecular-weight amine catalysts that are inert toward isocyanate groups, e.g., diazabicyclo [2.2.2]octane, bis(dimethylaminoethyl) ether or N,N,N'N'-tetramethylalkylene diamines, are frequently relatively volatile, have a strong odor, and leave the polyaddition product slowly. In particular, films and coatings that occur in automobiles due to fogging have been found on analysis to contain amine catalysts as a constituent. The proportion of these volatile constituents in polyisocyanate polyaddition products therefore must be reduced. A target limit for the future is a value of less than 1 mg of condensate, measured by determining the fogging in accordance with DIN 75 201, method B. To allow observation of this limit, the use of volatile amine catalysts should be avoided.

In U.S. Pat. 5,976,969, Hinz discloses high-molecular-weight polyoxyalkylene-polyols containing tertiary amino groups whose catalytic activity is not adversely affected by the molecular-weight, so that the addition of low-molecular-weight tertiary amines as catalysts for the formation of polyisocyanate poly-addition products can be avoided to the greatest possible extent. The polyoxyalkylene-polyols should be suitable for the preparation of flexible, compact, or cellular polyisocyanate polyaddition products; should be readily miscible with the other synthesis components, in particular conventional polyoxyalkylene-polyols containing no amino groups; and should be foamable in the absence of chlorofluorocarbons for the formation of microcellular PU elastomers or PU foams.

The objective is achieved by using selected initiator molecules containing tertiary amino groups in which there is essentially a spatial separation between the tertiary amino groups and substituents containing reactive hydrogen atoms, for the formation of polyoxyalkylene-polyols.

In spite of the self-catalyzing property of these polyols, sometimes it is desired to further accelerate the reaction. To accelerate the reaction of the self-catalyzing polyoxyalkylene-polyol, synergistic catalysts additionally can be introduced into the reaction mixture. Compounds that are highly suitable for this purpose are metal salts such as iron(II) chloride, zinc chloride, lead octanoate, and (preferably) tin salts such as tin dioctanoate, tin diethylhexanoate, and dibutyltin dilaurate, which are usually employed in an amount of from 0.03 to 0.25% by weight, based on the weight of the polyoxyalkylene-polyol.

The cellular or elastomeric moldings of polyurethane products produced from highly reactive poly-oxyalkylene-polyols or mixtures of conventional polyoxyalkylene-polyols are claimed to be readily demoldable, have an improved surface, and have no odor problems. Compact moldings have a lower Shore hardness than products made from conventional polyoxyalkylene-polyols.

The cellular PU elastomers have densities of from about 0.9 to 1.0 g/cm^3, it being possible for the density of filler-containing products to achieve higher values, for example, up to 1.4 g/cm^3 or greater. Moldings made from cellular elastomers of this type are used in the automotive industry as headrests, external parts such as rear spoilers and bumpers, and internal paneling. They are also used as shoe soles.

The soft, flexible PU foams made with self-catalyzing polyols have a density of 0.025 to 0.24 g/cm^3, and in particular from 0.03 to 0.1 g/cm^3. The density of the structural foams ranges from 0.24 to 0.6 g/cm^3. The soft foams and structural foams are used in the vehicle industries (e.g., automotive, aerospace, and shipbuilding), the furniture industry, and the sports equipment industry as, for example, cushioning materials, housing parts, ski boot inners, ski cores, and so on. They are particularly suitable as insulation materials in the construction and refrigeration sectors, for example, as the middle layer of sandwich elements or for foam filling refrigerator and freezer housings.

5.3.4   Polycarbonate-Containing Polyols[*]

When a compound having one or more hydroxyl groups is reacted with a cyclic alkylenecarbonate, the hydroxyl group is converted into a hydroxy-terminated alkylcarbonate group by the addition of the cyclic alkylenecarbonate moiety to the hydroxyl group.

---

[*]U.S. Pat. 5,446,110, Aug. 29, 1995, Nakano, et al.

Present technology employs a two-step process for producing a linear polycarbonate oligomer or polymer having a terminal hydroxyl groups by reacting a glycol with a cyclic alkylenecarbonate. A series of reactions take place in this process. First, the cyclic alkylenecarbonate is ring-opened with the glycol to produce a linear carbonate containing both of the hydrocarbon moieties corresponding to the glycol and cyclic carbonate reactants. Then, a transesterification occurs between the linear carbonate and the glycol, with the formation of an alkylene glycol corresponding to the cyclic alkylenecarbonate as a by-product. The above reaction sequence continues to take place until the linear polycarbonate is chain-extended to a desired length. This process must be carried out under high vacuum to remove the alkylene glycol by-product from the reaction system. Therefore, the cyclic alkylenecarbonate used in the process is comparable, as a reactive derivative of carbonic acid, to phosgene conventionally used in the synthesis of polycarbonate polymers.

Nakano, et al., in U.S. Pat. 5,446,110, have disclosed a new method in which a cyclic alkylenecarbonate may be addition reacted with or addition polymerized to a compound or polymer having one or more alcoholic hydroxyl groups while avoiding the formation of a glycol corresponding to the cyclic alkylen-ecarbonate as a reaction by-product. Therefore, the resulting linear carbonates are terminated with one or more linear hydroxyalkylcarbonate groups corresponding to the cyclic alkylenecarbonate. Since the terminal hydroxyl group is spaced apart from the backbone moiety by the linear carbonate linkage, the hydroxy-terminated linear carbonate is highly reactive and refractory to hydrolysis. Using these unique properties, alcoholic hydroxyl group-containing compounds or polymers used in the polyurethane or coating industry may be modified to have enhanced reactivity and resistance to hydrolysis.

The ring-opening reaction of the cyclic alkylenecarbonate for producing a linear carbonate is per-formed by heating a mixture of a cyclic carbonate and an initiator to an elevated temperature up to 150° C in the presence of a catalyst. Usually, greater than 0.5 equivalents of the cyclic carbonate are used relative to each hydroxyl group possessed by the initiator. The amount of catalyst generally ranges between 1 ppm and 5%, preferably between 5 ppm and 5,000 ppm relative to the reaction mixture. The reaction may be carried out in an aprotic solvent. Examples of usable solvents include aromatic hydrocarbons such as benzene, toluene, or xylene; esters such as ethyl acetate butyl acetate; ketones such as acetone or methyl isobutyl ketone; halogenated hydrocarbons such as dichloromethane or dichloroethane; ethers such as tetrahydrofuran or 1,4-dioxane; and other aprotic solvents such as acetonitrile, nitrobenzene, or nitromethane. The above materials may be charged in a reaction vessel in any desired order but preferably in the order of the solvent, initiator, cyclic carbonate, and catalyst. The reaction end point may be confirmed by measuring the concentration of the cyclic carbonate in the reaction mixture using any conventional technique such as gas chromatography. Normally, less than 1% concentration of the cyclic carbonate may be regarded as the end point. After the reaction, the resulting linear carbonate may be recovered and purified in any conventional technique such as distillation, precipitation, recrystallization, and the like.

Glycol hemiesters such as ethylene glycol monoacetate are used as a solvent. Polycaprolactone or other polyols containing ester linkage are used, for example, in the polyurethane industry. However, the ester linkage is known to be susceptible to hydrolysis. The aliphatic linear carbonate is refractory to hydrolysis and therefore can find use as a raw material for the production of water-resistant (hydrolytically resistant) polyurethanes.

5.3.5 Low-Fluorine Polytetramethylene Ether Glycol

Polytetramethylene ether glycol (PTMEG) is frequently used as a soft-segment component in polyure-thane elastomers. These elastomers are produced by addition polymerization of PTMEG with diisocy-anates (e.g. diphenylmethane diisocyanate).

In these addition-polymerization reactions, the terminal hydroxyl groups of the PTMEG react with the isocyanato groups of diisocyanate. But fluorine is known to be present, albeit in small amounts, on the ends of PTMEG instead of the terminal hydroxyl groups. The terminal fluorine in PTMEG is inert to these reactions, and for the production of high-molecular-weight polymers, it is necessary to use

PTMEG with a low amount of terminal fluorine. Furthermore, the presence of the terminal fluorine lowers the heat resistance of PTMEG-derived elastomers.

One of the most generally used methods for the production of PTMEG is one in which THF is polymerized using FSA as the catalyst, and this is an industrially desirable method due to the high activity of the polymerization catalyst. In this method, SO3F groups are bonded to the ends of the polymer produced by the polymerization. Therefore, after the polymerization reaction has been completed, the resulting polymer is hydrolyzed to eliminate the end-bonded SO3F groups from the polymer.

However, the PTMEG produced by this method still contains fluorine in minute amounts, although the SO3F groups are completely removed by the hydrolysis. (For example, fluorine has been detected at about 200 ppm in commercially available PTMEG having molecular weights of about 1,000.) This fluorine in the PTMEG is assumed to be bonded directly to the terminal carbon of the polymer instead of the hydroxyl groups. This terminal fluorine cannot be eliminated easily by either the hydrolysis after polymerization or the neutralization and washing thereafter.

Murai et al.[3] disclosed a method for the production of polytetramethylene ether glycol wherein polymerization of tetrahydrofuran in the presence of fluorosulfonic acid is followed by hydrolysis of the resulting polymer. According to the method, high-quality polytetramethylene ether glycol with a low fluorine content may be obtained at a high yield without lowering the polymerization temperature. For example, PTMEG may be obtained with 2 or fewer terminal fluorines per 2,000 termini of the resulting PTMEG. As a result, when the resulting PTMEG is used as a starting material for elastomers, it is possible to easily obtain a high-molecular-weight polymer with excellent heat resistance.

The invention teaches a method for the production of PTMEG by polymerization of tetrahydrofuran with fluorosulfonic acid (FSA) as the catalyst to obtain PTMEG with a low fluorine content.

5.3.6 Modified Polyols

There are three main types of modified polyols.

1. Polyvinyl-modified polyethers or *polymer polyols*
2. Polyols containing polyurea dispersions (PHD polyols)
3. Polyols containing polyurethane dispersions (PIPA polyols, poly isocyanate addition)

The graft polyols are ethylene oxidepropylene oxide adducts of propylene glycol, containing ethylene oxide as a cap, and between 15 and 45% styrene, acrylonitrile, or a combination of both as a dispersion. Representative examples of such polyols include Pluracol 973 and Pluracol 994 (sold by BASF), Voranol 4925 (sold by Dow Chemical, USA), and UCC3128 (sold by Union Carbide).

Poly Harnstoff dispersion (PHD) polyols are dispersions of polyurea particles in conventional polyols. These polyols are prepared by the reaction of diamine (hydrazine) with a diisocyanate (toluene diisocyanate) in the presence of a polyether polyol. Representative examples of such polyols are PHD 9151 and PHD 9184, sold by Miles Chemical. PIPA polyols are similar to the PHD polyols but instead contain dispersed particles of polyurethanes formed by the *in situ* reaction of an isocyanate and an alkanolamine, e.g., triethanolamine. In general, the amine is blended into a conventional polyol with, for example, toluene diisocyante under rapid stirring. PIPA polyols are typically made and used by foam-producing companies that have acquired patent licenses.

PHD Polyol

Stable polyol dispersions for use in manufacturing polyurethanes have been available for years. One particular family of such dispersions that has met with substantial commercial success consists of those polyureas and/or polyhydrazodicarbonamide dispersions prepared by reacting an organic polyisocyanate with polyamines containing primary and/or secondary amine groups, hydrazines, hydrazides, or a mixture thereof in the presence of hydroxyl-containing materials. See, e.g., U.S. Pats. 4,042,537, 4,089,835, and 4,324,716. Other patents describing this type of dispersion include U.S. Pats. 3,325,421, 4,092,275, 4,093,569, 4,119,613, 4,147,680, 4,184,990, 4,293,470, 4,305,857, 4,305,858, 4,310,448, 4,310,449,

4,496,678, 4,668,734, 4,761,434, and 4,847,320. While the commercially available dispersions are used to produce polyurethane products having excellent physical properties, it would be desirable if lower durometers could be attained.

The use of relatively low-molecular-weight isocyanate-reactive compounds as so-called chain extenders and cross-linking agents in the preparation of polyurethanes is well known. Certain low-molecular-weight glycol and polyol chain extenders and cross-linking agents have also been reported useful in the preparation of low-viscosity liquid PHD polyol dispersions (for example, U.S. Pats. 4,089,835, 4,324,716, 4,847,320, and 4,855,352) that are suitable for use in commercial foaming equipment. In addition, U.S. Pat. 4,324,716 (columns 13 and 16) indicates that dispersions prepared using predominantly or exclusively low-molecular-weight polyols can be used as a substitute for the generally used chain extenders to obtain highly elastic and transparent polyurethanes.

It is generally accepted that the reaction of polyisocyanates with hydrazines or hydrazides in the presence of hydroxyl-containing compounds involves the preferential reaction of isocyanate groups with NH2 groups. According to U.S. Pat. 4,089,835 (column 2), the reaction of polyol hydroxyl groups also occurs and, in fact, may even be necessary to ensure the stability of the resultant dispersions. Because a given quantity of a low-molecular-weight polyol contains a larger number of hydroxyl groups than the same weight of a higher-molecular-weight polyol, the reaction of hydroxyl groups presumably can be enhanced by using greater proportions of the lower molecular weight polyol. Indeed, U.S. Pat. 4,089,835 (column 9) indicates that up to 60 percent by weight of the base polyol having a molecular weight above 500 can be replaced with lower-molecular-weight glycols and polyols, thereby producing liquid dispersions having low viscosities. The advantage of low-viscosity dispersions is their ease of use in available foaming equipment.

Lucarelli et al.[4] found that the use of relatively small quantities of compounds containing at least two isocyanate-reactive hydrogen atoms and having a molecular weight in the range from about 3 to 700 gives rise to PHD polyol dispersions that are eminently useful for the preparation of urethane-based foams having exceptional load-bearing capacity. When compared to conventional foams made using the same quantity of low-molecular-weight isocyanate-reactive compounds as chain extenders or cross-linkers during the foaming process (rather than as components for preparing the dispersions), foams prepared according to the invention exhibit superior hardness. The invention is therefore directed to a novel process for producing stable dispersions, the dispersions produced by the process, and the use of such dispersions in the manufacture of polyurethane products having exceptionally low hardness.

The dispersions are prepared by the reaction of organic isocyanates, isocyanate-reactive compounds having a molecular weight in the range from 400 to 10,000, relatively low-molecular-weight isocyanate-reactive compounds, and hydrazines or hydrazides. This invention further relates to the dispersions prepared by said process, the use of the dispersions to prepare polyurethane foams having exceptional hardness, and the foams prepared thereby.

Polymer/polyol compositions that found initial commercial acceptance in high-resiliency foams were primarily compositions produced using acrylonitrile.[6] Such compositions were somewhat higher in viscosity than desired in some applications. Furthermore, such compositions were at least primarily used commercially in producing foams under conditions such that the heat generated during foaming is readily dissipated (e.g., the foams have a relatively thin cross section) or under conditions such that relatively little heat is generated during foaming. When polyurethane foams were produced under conditions such that the heat generated during foaming was not readily dissipated, severe foam scorching usually resulted. Later, polymer/polyol compositions produced from acrylonitrile-methylmethacrylate monomer mixtures were commercialized and were convertible to polyurethane foams having reduced scorch.

More recently, polymer/polyol compositions produced from acrylonitrile-styrene monomer mixtures have become available. Use of low ratios of acrylonitrile to styrene in the monomer mixture affords polymer/polyols that do not give rise to a scorch problem. However, it is increasingly difficult to make satisfactorily stable polymer/polyols as the ratio of acrylonitrile to styrene is reduced to the desired levels.

In Pat. 4,357,430, Van Cleve[7] discloses normally liquid, stable polymer/polyol compositions formed by polymerizing, in the presence of a free radical catalyst, an ethylenically unsaturated monomer or monomers dissolved or dispersed in a polyol mixture including a coupled polyol consisting of the reaction product of a polyol having a functionality in excess of 2 reacted with polyisocyanate in such proportion that the ratio of hydroxyl groups to isocyanato groups is greater than 1. In one embodiment, the coupled polyol is made *in situ* in the base polyol by addition to the required amount of the polyisocyanate. The novel polymer/polyol compositions are easily convertible to polyurethane foams and elastomers.

Copolymer Polyahl

Vinyl urethane composite polymers are disclosed in U.S. Pats. 4,098,733 and 4,125,487. These composites are prepared by the simultaneous polymerization of vinyl monomers and polyurethane-forming reactants in which the heat of reaction of the polyurethane-forming reactants initiates the polymerization of the vinyl monomers. Such composites exhibit advantages such as rapid curing and good tensile properties. However, the impact strength and elongation of these composites are not as good as desired for some applications. It is desirable to synthesize a vinyl urethane composite polymer having improved impact strength and elongation properties, and which also largely retains the beneficial properties of previously known vinyl urethane composites.

Ellerbe et al.[5] disclosed vinyl urethane composite polymers having improved impact strength prepared by reacting a polymer polyahl containing dispersed soft polymer particles, a polyisocyanate, and a vinyl monomer. This invention is a vinyl-urethane composite polymer that is a reaction product of a reaction mixture comprising (a) a copolymer polyahl containing dispersed soft polymer particles, (b) a polyiso-cyanate, and (c) a vinyl monomer that has a boiling point above the temperature incurred in reacting said reaction mixture. These reactants are reacted in the presence of catalysts for the free radical poly-merization of said vinyl monomer and for the polymerization of the polyurethane-forming reactants (i.e., those containing isocyanate and active hydrogen containing groups).

The polymer polyahl employed herein comprises a dispersion of discrete soft polymer particles in a continuous polyahl phase. Preferably, the polymer particles are stabilized by being grafted onto at least a portion of the polyahl molecules. The polyahl may be any polyfunctional compound having at least two active hydrogens such as are described more fully in subsequent text.

Typically, the polyahl is a polyol, polyamine, polyamide, polymercaptan, polyacid, or a compound containing a mixture of active hydrogen groups. Preferred are polyols and amine terminated polyethers as are described in U.S. Pat. 3,654,370. Of these, the polyether polyols are most preferred. Such polyether polyols are advantageously prepared by reacting a C2-C6 alkylene oxide, preferably C2-C4, alkylene oxide, more preferably propylene oxide or mixtures thereof with ethylene oxide, with a polyfunctional initiator compound having a plurality of active hydrogen atoms. Preferably, the polyether polyol has a molecular weight about 1,000 to 6,000.

The use of a polymer polyahl provides improvement in elongation and impact strength. In general, the polymer polyahl is employed in an amount such that such improvement is seen. Typically, the dispersed particles of the polymer polyahl range from approximately 1 to 20%, preferably from about 3 to 12%, and more preferably from about 4 to 8% by weight of the combined weight of all reactive components employed in preparing the vinyl urethane composite.

5.3.7 Polytetramethylene Ether Glycol Using a Modified Fluorinated Resin Catalyst

Polytetramethylene ether glycol (PTMEG) is a commodity in the chemical industry, widely used in the manufacture of block copolymers in combination with polyfunctional urethanes and polyesters such as polybutylene terephthalate. It is commonly prepared by the reaction of tetrahydrofuran (THF) with a strong acid catalyst such as fluorosulfonic acid and then quenching and hydrolyzing the product with water.

Dorai et al.[8] disclosed an invention related to an improved process for producing polytetramethylene ether glycol from tetrahydrofuran using a catalyst which is a fluorinated resin containing both sulfonic acid groups and carboxylic acid groups or a blend of fluorinated resin containing sulfonic acid groups and a fluorinated resin containing carboxylic acid groups. An important aspect of the Dorai invention involves activating the sulfonic acid groups containing fluorinated resin prior to its use as catalyst. This activation involves a partial dehydration of the catalyst to contain from not more than 1,000 to 2,000 ppm (0.1 to 0.2 wt. %) water. Further, the catalytic resin may be blended with one or more other resins to control its catalytic activity.

5.3.8 Polycarbonate Polyol

Aliphatic polycarbonate polyols have long been known. They are prepared from nonvicinal diols by reaction with diarylcarbonate, dialkylcarbonate, dioxolanones, phosgene, bischlorocarbonic acid esters or urea. Of the diols described in the literature, only those exclusively or largely based on 1,6-hexanediol have hitherto acquired any real technical significance. Thus, high-quality polyurethane elastomers and coating compositions are prepared from hexanediol polycarbonate.

Outstanding resistance to hydrolysis makes the hexanediol polycarbonates particularly suitable for the production of articles having a long useful life. The hydrolysis resistance of such polyurethanes is known to be far better than that of polyurethanes based on adipic acid polyester as the diol component. Pure hexanediol polycarbonates (molecular weight 500 to 4000) are waxes having a softening point of 45 to 55° C (according to molecular weight). As a result of the tendency toward crystallization of the soft segment, the corresponding polyurethanes tend to harden and lose their flexibility at low temperatures. To eliminate this serious disadvantage, hexanediol polycarbonates were developed.

High-quality polyurethanes (PU) are increasingly used in applications where they are exposed not only to hydrolytic influences but also to attack by microorganisms. This applies, for example, to rollers in printing works or textile factories, to cable sheaths, to spring elements and vibration dampers in machine construction, to coatings for awnings, flat roofs, and garden furniture, and to elastomeric fibers in leisure fabrics. In these fields, polyurethanes based on aliphatic polycarbonates show a susceptibility similar to that of polyurethanes based on aliphatic polyesters.

It is known that polyurethanes based on polyethers are significantly more resistant to degradation by microorganisms. The polymers of tetrahydrofuran, which are the only materials contemplated and actually used for the above-mentioned applications, are in turn attended by other disadvantages. For example, their crystallinity leads to a tendency of the PU to harden at low temperatures, particularly when the desired property spectrum of the PU requires the use of soft segments having average molecular weights of 2000 and higher.

In 1978, Lai and Silvers[9] were awarded a seminal patent relating to a novel process for preparing linear polycarbonates having terminal hydroxyl groups, from glycols with number of carbon atoms >4 and cyclic esters of carbonic acid, hereinafter called alkylene carbonates, by ester interchange reaction with or without catalysts. More particularly, the ester interchange reaction is carried out with distillation under reduced pressure whereby the low-boiling glycol by-product is removed. The amount of the glycol withdrawn controls the molecular weight of the product.

Preferably, polycarbonates high in carbonate and low in ether group concentration are prepared by reacting cycloaliphatic or aliphatic diols containing from 4 to 12 carbon atoms or polyoxyalkylene glycols containing 2 to 20 alkoxy groups per molecule, each alkoxy group containing 2 to 4 carbon atoms, with alkylene carbonates composed of a 5- to 7-membered ring in the mole ratio of 10:1 to 1:10 at temperatures between 100 and 300° C and pressures between 0.1 and 300 mm Hg in the presence of ester interchange catalysts, or without any catalyst, while removing low-boiling glycols by distillation.

According to the invention, the preparation of polycarbonates is conducted in two stages. In the first stage, cycloaliphatic or aliphatic glycols containing from 4 to 40 carbon atoms or polyoxyalkylene glycols containing from 2 to 20 alkoxy groups per molecule, each alkoxy group containing 2 to 4 carbon atoms, are reacted with alkylene carbonates composed of a 5- to 7-membered ring in a mole ratio of 1:10 to 10:1, preferably 1:3 to 3:1, to form a low-molecular-weight polycarbonate. The boiling points of the selected glycol should be at least 5° C above that of the glycol by-product, which is generated from the selected alkylene carbonate during the ester interchange reaction. The lower boiling glycol is removed by distillation at temperatures ranging from 150 to 250° C and under reduced pressure ranging from 50 to 200 mm Hg. If desired, ester interchange catalysts may be used. A fractionating column is used to separate by-product glycol from the reaction mixture.

The by-product glycol is taken off at the top of the column and unreacted alkylene carbonate and reactant glycol returned to the reaction vessel as reflux. A current of inert gas or addition of an inert solvent may be used to aid in the removal of by-product glycol as it is formed. When the amount of by-product glycol obtained indicates the degree of polymerization of the polycarbonate is in the range of 2 to 10, the pressure is gradually reduced to 0.1 to 10 mm Hg, and the unreacted glycol and alkylene carbonate are removed. This marks the beginning of the second stage of the reaction. In this stage, the low-molecular-weight polycarbonate is condensed by distilling off glycol as it is formed at temperatures ranging from 250° C and at a pressure of from 0.1 to 10 mm Hg until the desired molecular weight of polycarbonate is obtained.

5.3.9 Poly THF

Many applications that could utilize the good low-temperature properties of PTMEG require a lower viscosity of the starting polyol. For such use, a "copolymer THF" may be used. The copolyether glycol starting material is one based on tetrahydrofuran (THF) and an alkylene oxide (AO), and is sometimes also referred to as a copolymer of THF and an AO. "AO" as used herein means an alkylene oxide whose ring contains two or three carbon atoms. The AO can be unsubstituted or substituted with, for example, alkyl groups or halogen atoms. Illustrative alkylene oxides are ethylene oxide (EO), 1,2-propylene oxide (PO), 1,3-propylene oxide, 1,2-butylene oxide, 1,3-butylene oxide, 2,2'-bischloromethyl-1,3-propylene oxide, and epichlorohydrin. The copolymers preferred for use are those of THF and EO and THF and PO. The copolymer can also be of THF and two or more alkylene oxides as, for example, a THF/EO/PO copolymer.

In those applications where enhanced chemical resistance is important, a poly(THF) co(carbonate) polyol may be advantageous. These complex products may be properly termed "polymeric carbonate diols of copolyether diols," as disclosed in the Robinson patent.[10] Polymeric carbonate diols are prepared by coupling segments of copolyether glycols with a dialkyl carbonate, with a cyclic carbonate, or with phosgene.

The polyether carbonate diols of the Robinson invention are made by coupling copolyether glycol segments with a dialkyl carbonate, with a cyclic carbonate, or with phosgene. The polyurethanes are made by reacting a polyether carbonate diol with a conventional isocyanate to form a prepolymer and then reacting a prepolymer with a conventional polyol or polyamine chain extender.

The catalyst used can be any of the conventional ester interchange catalysts. Tetrabutyl titanate is preferred. The catalyst may be employed at a concentration of 0.0001 to 0.1% (by weight) of the reaction mass. The reaction can be carried out in bulk, using the copolyether glycol itself as the reaction medium. When a dialkyl carbonate or a cyclic carbonate is used as the coupling agent, the carbonate and the catalyst are added directly to the medium. The reaction can be carried out at a temperature of 120 to 240° C and at atmospheric pressure. By-product alcohol can be removed from the reaction mass by vaporization or by sweeping the reaction zone with nitrogen.

5.3.10 Polydispersivity

It has long been recognized that a narrow molecular weight distribution and/or a numerically low polydispersity (i.e., Mw/Mn, where Mw is weight average molecular weight and Mn is the number average molecular weight) for poly(tetramethylene ether) glycols, PTMEGs, is desirable for a number of end-use applications in specialty polyurethanes. Consequently, various methods of altering or controlling the molecular weight distribution of PTMEG have previously been suggested, with varying degrees of commercial success.

To overcome this deficiency, Dorai et al.[11] disclosed a process for reducing the molecular weight polydispersity of polyether glycols, particularly poly(tetramethylene ether) glycols, by membrane fractionation. The invention involves an improved process for recovering PTMEG characterized as having a narrow molecular distribution and an unusually low polydispersity achieved by batchwise or continuous removal of low-molecular-weight species using a cross-flow ultrafiltration or membrane separation technique.

The Dorai invention may viewed as an improved method for narrowing the molecular weight distribution of poly(tetramethylene ether) glycols by physical separation/fractionation of a polydisperse polymer melt. For example, the improved process is employed to prepare PTMEG having a molecular weight between about 400 and about 4,000, with a polydispersity between about 1.20 and 1.8, starting from a PTMEG having a correspondingly similar average molecular weight but a broader (i.e., numerically greater) polydispersity.

The actual separation of relatively low-molecular-weight PTMEG species from relatively high-molecular-weight PTMEG species, according to the Dorai invention, is achieved by what can be categorically viewed as an ultrafiltration process or a membrane separation process. In either case or perspective, the molten PTMEG being fractionated flows over the permeable or semipermeable surface (i.e., the membrane), thus sweeping across the surface and maintaining the desired flux rate through the membrane. Thus, for purposes of this invention, the term *membrane fractionation* refers to ultrafiltration and membrane separation collectively, while the additional term *cross-flow* is used to emphasize the sweeping of the surface by the molten polyether glycol being fractionated.

5.3.11 Perfluoropolyether–

Heat-processible elastomeric polyurethanes are widely used as structural materials in mechanical, textile, automotive, and biomedical sectors. These materials yield a frequently satisfactory combination of mechanical and elastomeric properties within a rather wide temperature range (of from −40° C up to +100° C) and are used, for example, in the sealing systems of heavy-duty automotive industry.

In these elastomers, when the hard-phase content is high, the products show high tensile strength and elastic modulus, while the chemical resistance of the elastomers in particular their resistance to hydrolysis, and their heat resistance, although being generally satisfactory, sometimes is not high enough to secure a reliable performance under severe operating conditions. The elastomers with a low content of hard-phase, on the contrary, display insufficient mechanical properties and, when submitted to mechanical stresses, can show creep phenomena that render them useless.

The surface properties and chemical resistance are usually not great, so when these materials are used in applications that require them to come into contact with hydraulic fluids or lubricants (e.g., in gaskets or sealing system elbows), a chemical degradation of polymer takes place, with a consequent decay in mechanical properties, which may lead to the break of the polyurethane articles.

In 1994, Ferrari et al.[12] disclosed novel fluorinated polyurethanes and polyurethane-ureas of heat-processible, elastomeric type, endowed with much higher values of resistance to chemicals and surface properties. The fluorinated polyurethanes and polyurethane-ureas are heat-processible, elastomeric types, constituted by rubber-like blocks and hard blocks, comprising segments derived from a perfluoropolyether-diol of formula $HO(CH_2CH_2O)_n-CH_2-Q-CH_2-(OCH_2CH_2)_nOH$ wherein Q is a perfluoropolyether chain, and "n" is within the range of 1 to 5.

One purpose of the Ferrari invention is to provide novel fluorinated polyurethanes and tolyurethane-ureas that combine mechanical properties at least equal to, or even better than, such properties of hydrogenated polyurethanes, with improved chemical and surface properties—in particular, as regards the chemical stability toward aggressive agents such as, e.g., hydrocarbon fluids and chlorinated solvents, oil repellency, water repellency, friction, and wear.

5.3.12 Color Level of Poly(tetramethylene Ether)

One of the several methods of preparing PTMEG is the catalytic polymerization of tetrahydrofuran, using fluorosulfonic acid (FSA) as the catalyst. PTMEG made this way sometimes has an undesirable yellow cast, that must be reduced or eliminated to meet most commercial specifications. It has been found that this yellow cast is due, at least, in part, to the presence of iron sulfites in the FSA. These sulfites are formed during shipment of the FSA in steel tanks by reaction of the steel's iron with the FSA's residual sulfur trioxide.

The yellow cast becomes apparent in the PTMEG when the iron sulfite concentration in the FSA used as catalyst is more than about 1200 ppm. Preventing formation of the yellow cast is therefore a matter of bringing the iron sulfite content of the FSA to below that level before it is used. The iron sulfite content can be brought to below 1200 ppm by conventional methods such as distillation, but it is far easier, and therefore preferred, to prevent iron sulfite formation in the first place by eliminating the FSA's residual sulfur trioxide through reaction with water according to the following equation:

$$SO_3 + H_2O \rightarrow H_2SO_4$$

Kasper et al.[13] were awarded a patent relating to a method for preparing poly(tetramethylene ether) glycol (PTMEG) having a low level of color, for example, an APHA color of less than 10. According to the invention, the color level of poly(tetramethylene ether) glycol made by polymerizing tetrahydrofuran using fluorosulfonic acid as the catalyst is significantly lower if the fluorosulfonic acid contains no more than about 1200 ppm of iron sulfites.

References

1. Menovcik, Gregory L., U.S. Pat. 5,451,656, "Carbamate-functional polyester polymer or oligomer and coating composition," to BASF Corporation, Sept. 19, 1995.
2. Hinz, Werner, et al., U.S. Pat. 5,476,969, "Highly reactive polyoxyalkylene-polyols containing tertiary amino groups in bonded form, their preparation, and their use for the preparation of polyisocyanate polyaddition products," to BASF Aktiengesellschaft, Dec. 19, 1995.
3. Murai, Nobuyuki, Shirato, Masayuki, Takeo, Hiroshi, and Tanaka, Hidetoshi, U.S. Pat. 5,393,866, "Method for the production of polytetramethylene ether glycol," Feb. 28, 1995, Mitsubishi Kasei Corporation, Tokyo, Japan.
4. Lucarelli, Michael A., Keegan, Richard E., Koshute, Mark A., Freitag, Hans-albrecht, Avar, Geza 5,342,855, Aug. 30, 1994. "Stable dispersions of polyhydrazodicarbonamides useful in the preparation of polyurethane products having improved load-bearing capacity," Miles, Inc.
5. Ellerbe, III, Gilbert B., and Parrish, Donald B., U.S. Pat. 4,696,976, Sept. 29, 1987, "Vinyl urethane composite polymer containing dispersed soft copolymer particles," the Dow Chemical Company, Midland, Michigan.
6. Spitler, K.G., Lindsey, J.J., "PHD Polyols-A New Class of PUR Raw Materials," Journal of Cellular Plastics, 43-49, Jan./Feb., 1981.
7. Van Cleve, Russell, U.S. Pat. 4,357,430, Nov. 2, 1982, "Polymer/polyols, methods for making same and polyurethanes based thereon," to Union Carbide Corporation, Danbury, Connecticut.
8. Dorai, S., Pruckmayr, G., Marsi, M., and Quon, W.L., U.S. Pat. 5,118,869, June 2, 1992, "Polymerizing tetrahydrofuran to produce polytetramethylene ether glycol using a modified fluorinated resin catalyst containing sulfonic acid groups," to E.I. Dupont de Nemours and Company, Wilmington, Delaware.
9. Lai, Kwo-Hrong, and Silvers, Harold N., U.S. Pat. 4,131,731, Dec. 26, 1978, "Process for preparing polycarbonates," to Beatrice Food Company, Peabody, MA.
10. Robinson, Ivan M., U.S. Pat. 4,476,293, Oct. 9, 1984, "Polymeric carbonate diols of copolyether glycols and polyurethanes prepared therefrom," E.I. Du Pont de Nemours and Company, Wilmington, Delaware.
11. Dori, S., Goudie, W.W., Ochsenhirt, S.E., Patel, D.V., and Cavall, J.R., U.S. Pat. 5,434,315, July 18, 1995, "Reducing molecular weight polydispersity of polyether glycols by membrane fractionation," to E.I. du Pont de Nemours and Company, Wilmington, Delaware.

12. Ferreri, E., Giavarini, F., Tonelli, C., Trombetta, T., and Zielinski, R.E., U.S. Pat. 5,332,798, July 26, 1994, "Fluorinated polyurethanes and polyurethane-areas, and methods for preparing them," to Ausimont S.p.A., Milan, Italy.

13. Kasper, William W., Quon, Willard L., and Van Domelen, Timothy D., U.S. Pat. 4,590,054, May 20, 1986, "Process for preparing poly(tetramethylene ether) glycol," to E.I. du Point de Nemours and Company, Wilmington, Delaware.

6

Chain Extenders

6.1 Introduction

Chain extenders (or curatives) are low-molecular-weight reactants that produce the familiar elastomeric properties of the polyurethanes. Chain extenders typically have molecular weights in the 40 to 300 Daltons range and can be classified as either hydroxyl-terminated or amine-terminated. Difunctional compounds are considered chain extenders, while compounds with higher functionality are considered cross-linkers. A wide variety of different cross-linkers with different reactivities and functionalities are employed, for example, glycerol, trimethanol propane, glycols, and diamine compounds.

Introduction of high hard-segment content in the polyurethaneurea backbone is important in determining the final properties and performance of the polymer. Hard-segment content of the polyurethane urea usually controls the mechanical properties such as modulus and ultimate strength, in addition to the thermal and hydrolytic stability of the finished product.

Ethylene glycol, 1,4-butanediol and bis-(hydroxyethyl) hydroquinone have attained great commercial significance. 1,4-butanediol is the most important chain extender in elastomeric systems; for elastomers of greater hardness, 2,3-butanediol can be used as a stereoisomeric mixture. Bis-ethoxylated hydroquinone is also used in cast elastomers as well as in thermoplastic elastomers to produce polyurethanes with higher softening temperatures.

The chain extenders provide the *hard segment* to the segmented polyurethane elastomers. Hard-segment content in the polyurethaneurea backbone is important in the final properties and performance of the polymer. Hard-segment content of the polyurethaneurea usually controls important mechanical properties such as modulus and ultimate strength in addition to the thermal and hydrolytic stability of the finished polyurethane. Three popular chain extenders are shown in Table 6.1.

TABLE 6.1 Three Popular Chain Extenders

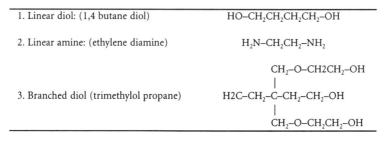

1. Linear diol: (1,4 butane diol)	HO–CH$_2$CH$_2$CH$_2$CH$_2$–OH
2. Linear amine: (ethylene diamine)	H$_2$N–CH$_2$CH$_2$–NH$_2$
3. Branched diol (trimethylol propane)	CH$_2$–O–CH2CH$_2$–OH | H2C–CH$_2$–C–CH$_2$–CH$_2$–OH | CH$_2$–O–CH$_2$CH$_2$–OH

Chain extenders containing hydroxyl groups are frequently slow to react with the polyisocyanate and require one or more catalysts to achieve a sufficiently fast reaction. Such catalysts are typically organometallics such as dibutyl tin dilaurate. The residues of metal-containing catalysts can cause thermal instability in the ultimate polymer. Another potential difficulty with hydroxyl-containing chain extenders is that some of these have a limited solubility in the polyol employed as the primary active hydrogen-containing compound of the composition, thus limiting the amount of chain extender which can be used in the formulation.

Amine-containing chain extenders generally react more rapidly than the corresponding hydroxyl-containing materials but sometimes react too fast; in addition, these amine-containing chain extenders may impart an odor to the resultant polymeric product. Some polyether-amine chain extenders are shown in Table 6.2.

TABLE 6.2 Polyetheramines

Jeffamine T:

$$H_2N \left(OCHCH_2 \right)_n OCH_2CHCH_2O \left(CH_2CHO \right)_n NH_2$$

with CH_3 substituents and side chain:
$$O-CH_2-CH(CH_3)-O-NH_2 \;]_m$$

Jeffamine D:

$$H_2N \left(OCHCH_2 \right)_n OCH_2CH_2O \left(CH_2CHO \right)_n NH_2$$

with CH_3 substituents.

It should be understood that the reaction between a polyol and a polyisocyanate produces a polyurethane. A reaction between a polyamine and a polyisocyanate produces a polyurea, and a reaction involving a polyol, a polyisocyanate, and a polyamine chain extender produces a poly(urethane/urea). Table 6.3a presents a list of some important chain extenders. Table 6.3b lists the trivial name, proper chemical name, and structure of the chain extenders.

6.2 Chain Extender Patents

6.2.1 Chain Extenders Based on Dipiperezinyl Ureas

Polyurethanes are prepared by reacting a compound that contains at least two reactive hydrogen atoms with a polyisocyanate. One of the active hydrogen-containing compounds is typically a polyether or polyester polyol, but it may also be a polyamine containing primary or secondary amino functionalities produced by replacing the hydroxyl functionalities of a polyol with amino groups. The other reactive hydrogen atoms come from low-molecular-weight compounds known as *chain extenders*.

Both the active hydrogen-containing compounds and the polyisocyanate contain at least two reactive functionalities. The polyisocyanate is generally employed in a slight molar excess relative to the total amount of the active hydrogen-containing materials in the composition. By appropriate selection of the particular active hydrogen-containing materials and the particular polyisocyanates, polymers having a wide variety of properties may be produced.

Researchers are always interested in finding new chain extender materials that possess adequate reactivity with polyisocyanate and produce polymers having good physical properties, in particular, improved hardness and modulus. One new chain extender is the subject of U.S. Pat. 5,364,924 by Gerkin, et al.[1]

The Gerkin patent relates to bis-(3,5-dimethylpiperazinyl) urea. These monomers are made by reacting the appropriate cyclic diamines with a reactant such as diphenylcarbonate, urea, a substituted urea such as carbonyldiimidazole, or phosgene.

TABLE 6.3a Chain Extenders

Trivial Name	Chemical Name	Structure
1, 4 BD	1, 4 Butane diol	$HO\text{-}(CH_2)_4\text{-}OH$
CHDM	1,4 cyclohexandedimethanol	$OHCH_2$— (cyclohexane ring) —CH_2OH
Glycerine	1,2,3 propane triol	$HO\text{-}CH_2CHCH_2\text{-}OH$ with OH on central carbon
1, 6 HD	1, 6 hexane diol	$HO\text{-}(CH_2)_6OH$
HQEE	Hydroquinone di (B hydroxyethyl ether)	$HOCH_2CH_2O$—(benzene ring)—OCH_2CH_2OH
MPD	2-methyl-1, 3 propanediol	$HOCH_2\text{-}CH\text{-}CH_2OH$ with CH_3 on central carbon
Quadrol	N,N,N', N' tetrakis (2-hydroxyethyl) ethyl diamine	$(HOCH_2CH_2)_2NCH_2CH_2N(CH_2CH_2OH)_2$
Sorbitol	d-glucitol	$HOCH_2\text{-}CHCH\text{-}CH\text{-}CH\text{-}CH_2OH$ with OH groups
TMP	trimethylolpropane	$HO\text{-}CH_2\text{-}CH\text{-}CH_2\text{-}OH$ with $CH_2\text{-}OH$ and CH_2CH_3 branches

Such monomers are useful as chain extenders in polymeric systems. More particularly, the chain extenders of the present invention can be used to prepare poly(urethane/urea) polymers with very good physical properties, including improved hardness and modulus relative to prior art materials. They are also useful in other polymeric systems such as epoxides and polyamides.

One method of carrying out the polymerization reaction is the *one-shot* process, in which the reactive hydrogen-containing starting material, the chain extender, and any required polymerization catalyst are combined to form a first mixture, and this mixture is then combined with the polyisocyanate and the resulting reaction mixture is allowed to cure.

A second process for carrying out the polymerization reaction is first to form a prepolymer by reacting an active hydrogen-containing material, such as the polyol, and the polyisocyanate to form an isocyanate-terminated prepolymer, and then to react this with the chain extender to produce the final polymeric product.

TABLE 6.3b Chain Extenders

Trivial Name	Chemical Name	Structure
Glycols		
3 BDO	1,3 butanediol	$HO\text{-}CH_2CH_2\overset{\displaystyle OH}{\underset{\displaystyle |}{C}}HCH_3$
Neopentylglycol	2,2 dimethyl-1,3 propanediol	$HOCH_2\text{-}\overset{\displaystyle OH}{\underset{\displaystyle |}{C}}H\text{-}CH_2OH$
PTMEG 250	Poly(tetramethylene ether glycol)	$HO\text{-}[(CH_2)_4\text{-}O]_n\text{-}H$
Amines		
IPDA	Isophoronediamine	
MOCA	4,4'-methylene bis orthochloroaniline	
Piperazine	Diethylene diamine	

It is important to employ the correct stoichiometry in carrying out the polymerization reaction. Typically, the total number of isocyanate groups will be greater than or equal to the sum of the active hydrogen-containing groups in the reacting system. This concept is expressed in the *isocyanate index*, which is usually somewhat greater than 100. In other words, if the isocyanate index is greater than 100, there is an excess of isocyanate groups relative to the active hydrogen-containing groups in the reacting system.

The polymers produced using the chain extending agents of the invention are generally tough and abrasion-resistant materials that find use as cast parts such as solid tires, molded gears, and bushings, as well as other molded parts such as fenders or bumpers of automotive vehicles. Softer polymers containing the chain extenders of the invention find use as components of sealants.

Dihydric alcohols suitable for use as chain extending agents in the preparation of the thermoplastic polyurethanes include those containing carbon chains that are either uninterrupted or that are interrupted by oxygen or sulfur linkages, including 1,2-ethanediol, 1,2-propanediol, isopropyl-a-glyceryl ether, 1,3-propanediol, 1,3-butanediol, 2,2-dimethyl-1,3-propanediol, 2,2-diethyl-1,3-propanediol, 2-ethyl-2-butyl-1,3-propanediol, 2-methyl-2,4-pentanediol, 2,2,4-trimethyl-1,3-pentanediol, 2-ethyl-1,3-hexanediol, 1,4-butanediol, 2,5-hexanediol, 9,5-pentanediol, dihydroxycyclopentane, 1,6-hexanediol, 1,4-cyclohexanediol, 4,4-cyklohexanedimethylol, thiodiglycol, diethylene glycol, dipropylene glycol, 2-

TABLE 6.3c Trade-Named Chain Extenders

Trivial Name	Chemical Name	Structure
Cyanacure	1,2 bis(2-amino-phenylthioethane)	H_2N—⬡—SCH_2CH_2S—⬡—NH_2
Dianol	1,1' isopropylidine-bis-(p-phenylene-oxy)-di-2 ethanol	$HO\text{-}CH_2CH_2\text{-}O\text{-}$⬡... CH_3 / $HO\text{-}CH_2CH_2\text{-}O\text{-}$⬡ C CH_3
Ethacure 100	Diethyltoluene diamine	(structure)
Ethacure 300	Dimethylthiotoluenediamine	(structure)
Polacure 740	1,3 trimethylene glycol bis(p-aminobenzoate)	(structure)
Unilink	Methylene bis N,N dibutylaniline	(structure)

methyl-1,3-propanediol, 2-methyl-2-ethyl-1,3-propanediol, dihydroxyethyl ether of hydroquinone, hydrogenated bisphenol A, dihydroxyethyl terephthalate and dihydroxymethyl benzene and mixtures thereof, 1,4-butane diol, 1,2-ethane diol, and 1,6-hexane diol.

6.2.2 Reacting Unhindered Amines with Prepolymers

Historically, 4,4'-methylene bis (o-chloroaniline) (MOCA) has been a widely used hindered amine. However, MOCA is a suspected carcinogen, so taking the necessary safety precautions substantially increases the cost involved in producing elastomers using MOCA as the chain extender. Another hindered amine that has also seen considerable use is 4,4'-methylene bis (methyl-anthranilate) (MBMA).

Unhindered diamines have not been used for reaction with an isocyanate terminated prepolymer to produce an elastomer. Pot life is short—only a few minutes—even when MOCA and MBMA are used as the diamines. The pot life is only five seconds if one attempts to produce a urethane elastomer from an isocyanate terminated prepolymer and an unhindered diamine, 4,4'-methylene dianiline (MDA).

Frisch and Damusis[2] discovered that an isocyanate terminated prepolymer can be stabilized by reaction with benzotriazole or a tolyl triazole, and that a mixture of the stabilized prepolymer with a diamine has a longer pot life than does a mixture of the unstabilized prepolymer with that diamine. When a hindered diamine is used in producing a urethane elastomer according to the invention, e.g., MOCA or MBMA, the pot life is increased substantially so that the labor required for carrying out the mixing operation can be minimized, and the prepolymer is preferably stabilized with from 0.1 to 0.5 equivalent of the triazole per equivalent of isocyanate in the prepolymer. On the other hand, when the diamine is unhindered, and urethane elastomers (which, so far as is known, had not been made) are the result of practicing the invention, it is usually preferred that the prepolymer be stabilized from 0.5 to 1.0 equivalent of the triazole per equivalent of isocyanate in the prepolymer.

Example

An isocyanate-terminated prepolymer was prepared from 1 mole of a poly(oxytetramethylene) glycol having a molecular weight of 1000 and a melting temperature of 50° C (hereinafter PTMEG 1000) and 2 moles of toluene diisocyanate (TDI). The specific TDI used was an 80/20 blend of 2,4- and 2,6-isomers. The isocyanate (hereinafter NCO) content of the prepolymer, determined by di-n-butyl amine titration, was 6.2 percent. Benzotriazole (hereinafter BT) was then melted, vacuum degassed, and mixed with a sample of the prepolymer which had been preheated to 75° C. The BT was added in the proportion of 0.25 equivalent (mole) thereof per isocyanate equivalent of the prepolymer. The resulting mixture was stirred at 80° C for 1 hr to ensure complete reaction between the BT and the free NCO groups of the prepolymer. The stabilized prepolymer that resulted at 80° C was mixed with MBMA, which had been vacuum degassed at 150° C, mixing temperature 140° C, at an NCO: NH_2 ratio of substantially 1:1. The resulting composition was found to have a pot life of 20 min at 100° C; it was vacuum degassed and poured into a mold preheated to 100° C. The mold, which had a central cavity 6 × 6 × 0.06 inches (length × width × depth), was then covered by a flat plate and placed in a hydraulic press where a compressive force of 10,000 lb/in² was applied to the mold and plate. After approximately 1 hr in the mold at 100° C, the partially cured elastomer was removed from the mold and transferred to an oven, where it was cured for an additional 19 hr at 100° C.

After conditioning at room temperature and 50% RH for one week, the elastomer was found to have the following physical properties:

Test Procedure	
100% modulus, psi	1260
300% modulus, psi, ASTM-D-412	2500
Tensile strength, psi	4380
Elongation at break, percent	440
Shore hardness ASTM-D-2240	92A
Graves tear strength ASTM-D-624	400

6.2.3 Chain Extenders Containing Alkylthio Substituents for Long Pot Life

Polyurethane polymers may be prepared by any of various methods using diols, diamines, or other chain extenders in various polyurethane processes. The cast elastomer process is widely used for the preparation of molded articles. Cast elastomers are produced by conducting the necessary chemical reactions in a mold. Polyurethane cast elastomers are used in tires, grain chute liners, conveyor belts, and in many other applications. It is known to use amines such as 4,4'-methylenebis(2-chloroaniline) (MOCA) as a chain extender in the cast elastomer processes. (See U.S. Pat. 3,752,790.) Presently, the use of MOCA as a chain extender is not favored because of its toxicological properties. Nevertheless, the use of MOCA provides a polyurethane of suitable physical properties. A replacement with roughly equivalent reactivity and physical properties is sought. It would also be desirable to obtain a polyurethane chain extender that, when used in a cast elastomer process, permits operation at a temperature suitable for maintaining the proper viscosity of the prepolymer while still permitting a pot life of suitable length for proper handling.

Nalepa et al.[3] found that use of the polyurethane chain extenders described in the patent is well suited for such preparation of polyurethanes. The invention describes polyurethanes and a process for their preparation with aromatic diamines having at least one aromatic ring, of which at least one aromatic ring has thereupon at least two alkylthio substituents and at least one amino substituent. An isocyanate prepolymer may be combined with a diamine chain extender of the invention and cured at suitable processing temperatures to form elastomers with good physical properties.

According to this delineation of the invention, a separate portion of a polyol or other active hydrogen group-containing compound may also be used as part of the reaction mixture, especially to decrease viscosity such as where a mixture of diisocyanate isomers is used as in U.S. Pat. 4,294,934. Quasi-prepolymer based systems are also within the invention. The aromatic diamines of the invention may be used for the preparation of polyurethanes by any of various methods including reaction injection molding (RIM), cast elastomer processes, foam elastomer processes, spray processes, and other processes known in the polyurethane arts. The diamines of the invention are especially suited for various casting processes, because they permit advantageous pot lives at suitable operating temperatures.

According to the cast elastomer process of the invention, the reaction components are reacted in a casting process employing favorably long casting times. The mixture is subjected to vacuum to prevent holes in the cast molded product. The polyisocyanate and organic active hydrogen group-containing compound, often a polyol, are first prereacted and then cured with a chain extender. The cast elastomer process is ordinarily carried out at ≈ 25 to $150°$ C, and the reaction times are characteristically ≈ 5 to 15 min.

The elastomers are made by pouring a reaction mixture into a mold where they are allowed to cure. Production can be carried out by a one-shot or a prepolymer method, the latter being presently preferred. In that method, an organic active hydrogen group-containing compound, often a polyol, of ≈ 400 to 5,000 molecular weight and of the polyester or polyether type, is reacted with a stoichiometric excess of diisocyanate to produce a prepolymer. Chain extension is then carried out with a chain extender of the type described herein. In the one-shot method, an isocyanate stream is mixed with a stream containing the other components, usually including a catalyst such as dibutyl tin dilaurate.

Factors influencing the final properties are molecular structure, chain entanglement, and secondary binding forces such as the hydrogen bonding and Van der Waals forces. These are controlled by selection of the molecular components employed, as is well known in the art. For example, each backbone exhibits its own advantages and disadvantages; e.g., polyether types display better resilience, lower heat buildup, better hydrolytic stability, good low-temperature properties, and low processing viscosity. The polyesters, however, are tough, abrasion-resistant elastomers with better oil resistance, and they can be produced at lower costs. It should be noted that these comparisons are made with urethanes having the same hardness.

A common technique for producing high-quality castings is by the use of an automatic dispensing machine. The machine degasses the prepolymer either in a batch or continuous mode and meters it along with the chain extender in controlled proportions to a mix head. Here the components are mixed thoroughly, then discharged into a mold. A significant advantage in liquid casting is that less capital expense is required for processing equipment as compared to that needed for the thermoplastic or millable gum elastomers.

6.2.4 1,2-Diaminocyclohexane Cross-Linker to Enhance the Load-Bearing Characteristics

High-resiliency (HR) foams have been known and manufactured for a number of years. These foams may be made by reacting a polyol, preferably a polyether polyol, with an organic polyisocyanate in the presence of a catalyst, a blowing agent such as water or the volatile halocarbons, and a cross-linking agent. As the density of the HR foams is reduced, the load-carrying ability of the foam at between a density of 2 and 2.2 becomes unacceptable, or only marginally acceptable, as measured by a specification such as that of Fisher Body.

Sullivan[4] discovered that improved flexible polyurethane foam can be produced using 1,2-diaminocyclohexane as the cross-linker to enhance the load bearing characteristics. For example, a mixture (preferably made by the one-step method of mixing a polyol of 1000 to 7000 molecular weight, an organic polyisocyanate, blowing agent, catalyst, and 1,2-diaminocyclohexane in cross-linking amounts and allowing the mixture to foam) gives a foam having improved load-bearing characteristics relative to an identical mixture containing an equivalent amount of isophorone diamine.

Stabilizers, for example, sulphonated castor oil and silicones (e.g., polydimethylsiloxanes) may be added to improve the miscibility of the components. However, care should be taken when using silicones of the polyoxyalkylene ether-polydimethylsiloxane block copolymer type to avoid the production of closed cell foams or collapsed foams, i.e., foams having inter alia low sag factors. When manufacturing molded foams, it may be advantageous to include a relatively minor amount of a polydimethylsiloxane surfactant or a silicone of the above described block copolymer type to modify the surface cell structure of the product. If it is desired to formulate a polyurethane foam exhibiting flame retardant characteristics, then, as explained above, it is generally advisable to omit silicones of the block copolymer type or to use only such materials in relatively minor amounts. Optimum amounts of such materials to be used for the particular effect desired may readily be determined by a simple trial run.

Examples

The following table lists formulations of the foams, being produced on standard one-shot foam machines, and physical properties of the cured molded foam.

Formulations	Control	1	2	3
		\multicolumn{ Formula Run Number		
Polyol A	80	80	80	80
Polyol B	20			
Polyol C		20	20	20
Isophorone diamine	1.0			
1,2 diaminocyclo-hexane		1.0	1.0	1.5
DABCO33LB	1.0	1.0	0.5	0.8
Niax A-107	0.5	0.5	0.5	0.5
F-11-630 fluorocarbon	0.03	0.03	0.03	0.03
Q2-5043 silicone	0.8	0.8	0.8	0.8
Water	3.3	3.05	3.3	4.1
90/10 TDI/PAPI index	1.04	1.04	1.04	1.04
80/20 TDI/PAPI	1.04	1.04	1.04	1.04
Physical Properties of 1, 2 Dach Stabilized Foam				
Density (lb/ft^3)	2.0	2.0	1.76	1.53
Tensile, PSI	19.3	22.3	22.0	23.9
Elongation, %	200	207	195	197

6.2.5 Diamine/Diol Chain Extender Blends

In U.S. Pat. 4,549,007, Lin et al.[5] disclosed a chain extending agent for reaction injection molding (RIM) comprising a normally liquid blend of butanediol and a toluene diamine monosubstituted with an electron withdrawing group vicinal to an amino radical and selected from the group of halogen, CN, NO$_2$, acyl, carboxylate, and isomeric mixtures thereof, and wherein the butanediol and diamine are combined in a critical mole ratio of from 4:1 to ≈19:1. The above blend is particularly useful in the preparation of large moldings of polyurethane-polyurea elastomer by the RIM or RRIM (reinforced reaction injection molding) process wherein instant preformed blend is added to a mixture of polyol reactant and catalyst, before polymerization, in the mold with isocyanate, producing a urethane/urea polymer product of improved properties. Accordingly, the invention also concerns an enhanced RIM or RRIM process and improved product by the use of a mixture of the present chain extender blend and polyol reactant maintainable in the liquid state at ambient temperature.

This invention relates to an improved chain extender blend of critical proportion such that its liquid state is maintained at room temperature for direct mixing with a polyol reactant at ambient temperature, and the chain extending profile of the blend is suitable for large-mold reaction injection molding. The invention is particularly directed to the preparation of urethane/urea polymer by the RIM or RRIM process, the basic principles having been developed in Germany by Bayer AG. Typically, this process has been used to produce elastomeric urethanes in molded parts employed in production of fascia for automobiles and recreational vehicles, and the manufacture of shoe soles and chemically resistant coating compositions.

The RIM process involves the production of moldings, having good performance and superior processing efficiency, by a one-step process. Generally, it is carried out by introducing a reactive polymerizable mixture into a mold, with the mixture based on compounds having several reactive hydrogen atoms, such as polyethers and polyesters having terminal hydroxy groups, and isocyanates, such as diphenylmethane diisocyanate, toluene-2,4- and 2,6-diisocyanates or polymeric diphenylmethane diisocyanate or mixtures thereof. The reaction mixture essentially contains the polyol and isocyanate reactants, a chain extender, and catalyst.

The RIM method, wherein the reaction mixture is rapidly polymerized in the mold, is opposed to the substantially slower castable process, which involves a first isocyanate prepolymerization step, relatively long pot life of the prepolymer to afford time for mixing and reaction with polyol before molding, and curing and demolding times of 20 to 40 minutes. Accordingly, chain extenders used in the castable process such as, for example, the polyhalogenated toluene diamines, methylene bis(2-chloro-aniline), must be slow acting to accommodate the requirements of the process.

The rapid mold polymerization that characterizes the recently developed RIM process has many advantages over the well known, time-consuming *castable* process. With the advent of (RIM) techniques, perfected in the mid 1970s, it became desirable to have fast-acting systems. Therefore, it is not surprising that there are many different aromatic diamines known in the art, ranging typically from the slow methylene bis(o-chloroaniline), to the intermediate polymethylene phenyldiamine, to the very fast methylene bis(aniline) and unhindered toluene diamine.

Lin discovered that extended elastomeric polyurethane-polyurea polymers (which are characterized by good hardness, tear, improved modulus properties, and higher demold strengths as compared to prior art extended urethane/urea polymers) can be prepared by using certain aromatic diamine-diol extender blends in which the diamine is the minor component. With the incorporation of these blends, the polymer-forming reactants have reaction profiles fast enough to be suitable for rapid molding and demolding processes as in the RIM or RRIM applications, yet are slow enough to enable the filling of large mold sizes before gelation can occur, thereby avoid the prior art problems noted above. The present blend [1,4-butanediol (B1D) and chlorotoluenediamine] is ideally suited to the large mold RIM or RRIM process due to its high activity with controlled gel time and facilitated incorporation into the reaction mixture at ambient conditions.

6.2.6 Chain Extenders for Polyurethanes

The term *polyurethane* is used generically herein to denote polymers produced by reaction of at least one polyisocyanate starting component and at least one polyfunctional active hydrogen containing starting component. Production of polyurethanes often involves more than one active hydrogen compound. When two or more active hydrogen components are used, at least one is generally a relatively high equivalent weight active hydrogen containing compound. Generally speaking, when incorporated into polyurethanes, relatively high equivalent weight active hydrogen compounds produce segments of polymer called *soft segments,* which have relatively low glass transition temperatures, which are defined as those generally below the temperatures of intended use of a polyurethane.

Relatively low equivalent weight active hydrogen compounds, referred to herein as *chain extenders,* are often used in addition to the relatively high equivalent weight active hydrogen compounds in a polyurethane-forming composition. Chain extenders in polyurethanes generally produce high melting segments called *hard segments,* which are believed to result from an increased intermolecular association or bonding within the polymer. Intermolecular association or bonding can be by covalent or hydrogen bonds. An amount and type of chain extender is generally chosen to achieve preselected processing characteristics, preselected hardness, and other preselected physical properties in a final polyurethane.

In reaction injection molding (RIM), chain extenders are chosen to impart desired physical properties to resulting molded objects and also to achieve a preselected reaction time, which is generally a reaction time sufficiently long to allow complete filling of a mold before gelling of the polymer, yet sufficiently fast to achieve short demold times. Demold times are typically 30 s to 2 min with a shot time, during which the components are injected into the machine, of ≈5 s.

Reaction injection molding is a process for producing and molding polyurethanes, which process has proven especially useful for making large molded objects having resilience, such as furniture items; lightweight building materials; shoe soles and heels; industrial service parts such as rollers, gears, bearing pads, and pump housings; auto body parts such as bumpers, fascia, hoods, doors, and fenders; and so on.

The RIM technique involves filling a mold with a stream of reactive, liquid starting components, which are rapidly injected. The starting components mix by impingement as they are injected into a mixing head from which they flow rapidly into the mold. On mixing, the components quickly begin to react to form polyurethanes, including those having urea bonds. The mixture should remain fluid for a period of time sufficient to fill the mold, which is often of large volume and complex design. The period of time required to fill the mold is the filling time. The period of time between initial mixing of starting components and the first visible reaction in the form of creaminess of the reaction mixture is called *cream time.* The period of time between initial mixing and formation of sufficient gel to solidify the polyurethane is called *gel time.* Shortly after gelling, the polyurethane cures to form a molded object having sufficient dimensional stability that it can be removed from the mold, leaving an empty mold for another injection of starting components. The time from injection until such an object is formed is called the *demold time.* The demold time is but one component of the cycle time, which is a measure of the time from one injection until a RIM machine is ready for the next injection of starting components. Cycle time includes time necessary for mechanical operations such as opening and closing a mold, time required for mold preparations such as removing flash left after demolding, applying external mold release agents, etc.

Starting components enter the mold, gel, cure, and are removed very rapidly. Typically, the mixed stream of components enters the mold at a rate of from ≈20 to ≈1200 lb/min (≈9 to ≈545 kg/m). Typical gel times range from less than ≈1 to ≈15 s. Cycle times are typically from ≈3 to ≈5 minutes and, with high-speed equipment, are often less than ≈2 min. Every stage of the process is preferably optimized to achieve cycle times that are as short as possible.

When RIM is used in a one-shot process of producing polyurethanes, the polyisocyanate starting component is supplied to the mixing head in one stream, and the active hydrogen starting components are supplied in another stream. In a two-shot process, a prepolymer is supplied in one stream while remaining active hydrogen components are supplied in the other. Prepolymers having isocyanate groups are included in the term *polyisocyanate component.* The stream of polyisocyanate component(s) is referred to as the *A side,* or *A component,* while the stream containing the active hydrogen component(s) is referred to as the *B side,* or *B component.* Chain extenders, catalysts, and additives are typically blended and introduced in the *B side,* but in certain cases catalysts and some additives can be mixed into the *A side.* In some instances, a third or fourth stream is required to meter an active ingredient that is incompatible with, prematurely reacts with, or otherwise causes difficulty with one or more of the other components of the polyurethane. The term *reaction mixture* as used herein refers to an admixture of at least one polyisocyanate component and at least one hydrogen component that will form a polyurethane. Additives are optionally included in a reaction mixture.

It is important in RIM that starting components be mixed in desired proportions and that they be mixed intimately. It is also important that inclusion of air in the molded product not be in visible bubbles or pockets. The mold must be filled completely. These and other requirements are met by processes known to those skilled in the art. Such processes include processes disclosed in U.S. Pat. 3,709,640; 3,857,550; 4,218,543; 4,298,701; 4,314,962; and U.S. Pat. 4,582,887, that are incorporated herein by reference. Additional description of RIM processes is found in Prepelka and Wharton, "Reaction Injection Molding in the Automotive Industry," *Journal of Cellular Plastics,* 2(2), pp. 87–98 (1975) and Knipp, "Plastics for Automobile Safety Bumpers," *Journal of Cellular Plastics,* 2, pp. 76–84 (1973).

Chain extenders have important effects on processes for producing polyurethanes, particularly on RIM processes. If a chain extender reacts too rapidly, the polyurethane reaction mixture may gel before a mold is completely filled or before a foam has reached its greatest volume. If the chain extender is too slow, each cycle time will be increased by the time necessary for the reaction mixture to react sufficiently to form a demoldable polyurethane. Those skilled in the art know that the reactivity of the chain extender must be appropriate for use with the other components of a polyurethane formulation to achieve desired reaction times. At the same time, the chain extender must act together with the other components of the reaction mixture to yield a polyurethane having the desired physical properties.

A number of chain extenders have been suggested for use in polyurethane formulations. Aromatic diamines having at least one linear alkyl substituent of one to three carbon atoms in a position ortho to an amine group, such as those disclosed in U.S. Pats. 3,428,610 and 4,218,543, are commonly used as chain extenders. Alkyl substituted methylene dianilines, such as those disclosed in U.S. Pats. 4,294,934 and 4,578,446, may be used alone or in combination with the substituted aromatic diamines. Variations of substituted aromatic diamines, such as the aromatic diamines having an electron withdrawing group disclosed in U.S. Pat. 4,523,004, the vicinal toluenediamine disclosed in U.S. Pat. 4,529,746, the dialkyl aromatic diamines disclosed in U.S. Pat. 4,526,905, and oxyethylated toluenediamines disclosed in U.S. Pat. 4,596,685, have been proposed for use as chain extenders in RIM-produced polyurethanes. Combinations of chain extenders, such as the combinations disclosed in U.S. Pat. 4,269,945, which include at least one primary amine, have also been proposed for use in polyurethanes formed in RIM processes.

Combinations of reactivity and physical properties (e.g., hardness, modulus, load deflection. and the like) imparted to resulting polyurethanes by known chain extenders are limited. It would be desirable, for instance, in situations involving large complex molds, to achieve the hardness attainable with some known substituted aromatic diamines yet to have somewhat slower reaction times to allow complete filling of molds. It is also desirable, in many applications, to produce unfilled polyurethanes having physical properties, such as thermal stability and modulus, generally associated with the use of fillers in polyurethanes produced from formulations including known chain extenders.

The Priester[6] invention is a process for producing polyurethane or polyurethane-polyurea polymers by polymerizing, in a reaction injection molding apparatus, a reaction mixture consisting of

1. at least one polyisocyanate component
2. at least one relatively high equivalent weight active hydrogen component, and
3. at least one chain extender selected from the group consisting of (a) heterocyclic diamines having at least one ring nitrogen, (b) di(aromatic amine) disulfides, (c) alkyl ethers of halogen-substituted aromatic diamines, (d) polysubstituted alkyl diamines having at least one highly electronegative group geminate to each amine group, and (e) alkyl acid esters of halogen substituted diamino phenols and mixtures thereof.

The invention also encompasses the polyurethane or polyurethane-polyurea polymers so produced and motor vehicles having parts thereof formed of said polymers.

6.2.7 Polyurethane Chain Extenders[7]

Polyurethane chain extenders covers certain aromatic amine-amides that comprise the reaction product of an alkylene polyoxypropylene diamine, triamine, or tetramine, and an isatoic anhydride.

R = hydrogen, alkyl, nitro, halo, hydroxy, amino, and cyano

Such chain extenders provide for the production of polyurethane elastomers having good tensile strength, tear strength, and elongation properties.

Background of the Invention

When an organic polyisocyanate is reacted with a polyether polyol to produce a polyurethane composition, various components are introduced into the system to adjust the physical properties of the resulting polyurethane composition. For example, if a cellular product is desired, water or an appropriate blowing agent is added to the polyurethane reaction mixture. Various other additives are used to adjust properties of various polyurethane compositions, such as the tensile strength, elongation, tear strength, flexibility, softness or hardness of the resulting composition, or the color. Often, the addition of an additive to improve one particular property results in the degradation of other properties of the polyurethane composition. For instance, an additive that increases the tensile strength of a solid polyurethane composition, such as various fillers, may result in a decrease in the elongation of the resulting polyurethane composition. Therefore, it is necessary to achieve a balance of properties for a given use.

Solid polyurethane compositions have found usefulness in gaskets, sealants, floor coverings, and the like. More recently, with the advent of molded, rigid plastics, it has become desirable to provide a flexible polyurethane mold for use in the place of the more expensive silicone-type molds currently being used. For a polyurethane composition to be acceptable for this use, it must be soft and flexible yet have good tensile and tear strength so that the mold does not quickly become unusable due to tears or splits in the mold material. Therefore, polyurethane compositions have not been acceptable for this purpose.

Polyurethane compositions generally in use as floor coverings are systems dissolved in a solvent and moisture cured by the atmosphere after application on the floor. These floor coatings have been found to suffer considerably from *bleedthrough*, especially when placed on a substrate that had previously been covered with some other type of floor covering. While there are some single-component floor coatings, i.e., solvent types, they have been found to be lacking in one or more of the desired properties for an acceptable floor coating. To be an acceptable floor covering composition, it is desirable that the elastomer be strong and scuff-resistant yet flexible enough to conform to shifts in the floor.

With the widespread use of foam crash pads in automobiles, it has become desirable to develop a crash pad with a tough scuff-resistant skin that is integral to the foam of the crash pad itself. Previously, it was necessary to line the mold in which the crash pad was to be cast with a decorative coating such as vinyl and the like to achieve the strength and scuff resistance necessary for the pad, and yet maintain an attractive appearance of the crash pad itself. Previous attempts at producing a polyurethane foam crash pad having an integral skin that would meet these qualifications have met with considerable difficulty and disappointing results.

Summary of the Invention

This invention relates to aromatic amine-amide compositions that incorporate a reaction product of an alkylene polyoxypropylene diamine, triamine, or tetramine and an isatoic anhydride. The invention also relates to the production of polyurethane compositions having improved physical properties due to presence of the above compounds, which act as chain extenders.

The chain extender of the invention is incorporated into the reaction mixture of an organic isocyanate and an organic polymeric polyhydroxy compound such as polyester or polyether polyols used for the production of polyurethane compositions, along with a urethane catalyst and various additives frequently used in the polyurethane art.

Detailed Description of the Invention

Particularly preferred chain extenders have the following structural formula:

$$
\underset{\overset{\displaystyle R''_n}{\big|}}{\text{benzene}} \quad \overset{\displaystyle O}{\overset{\|}{C}}\text{-NH-(CH}_2\text{CH}_2\text{O)}_x\text{-R-(OCH}_2\text{CH)}_y\text{NH}_2
$$

with CH_3 groups and NH_2 on the ring

where R = CH_2, s = $[O\text{-}CH_2]$, z = $[O\text{-}CH_2]$, and t = CH_2; R'' is selected from a group consisting of hydrogen, alkyl, nitro, halo, hydroxy, and cyano; n is a number of from 1 to 4; x and y are numbers from 1 to 10; r, s, and t are numbers of from 1 to 6; and z is a number from 4 to 50.

To prepare the above compounds, an isatoic anhydride of the formula

where R'' is selected from the group consisting of hydrogen, alkyl, nitro, halo, hydroxy, and cyano, and n is a number of from 1 to 4, is reacted with an alkylene polyoxypropylene diamine, triamine, or tetramine. The isatoic anhydrides are well known materials, and their preparation need not be discussed in detail. A preferred reactant is isatoic anhydride itself, where R'' is H. Preferred amine reactants are diamines and triamines.

A preferred alkylene polyoxypropylene diamine reactant has the following formula:

$$
\text{H}_2\text{N-(CH}_2\text{CH}_2\text{-O)}_x\text{-R-(OCH)}_y\text{NH}_2
$$

with CH_3 substituents

where R = CH_2, and x and y are numbers from 1 to 10.

A preferred alkylene polyoxypropylene triamine reactant has the following general structure:

$$
\text{H}_2\text{N-(CH-CH}_2\text{-O)}_x\text{-R-(OCH}_2\text{CH)}_y\text{NH}_2
$$

with CH_3 and $(OCH_2CH\text{-}NH_2)$ with CH_3

where x and y are numbers from 1 to 10 and R is an alkyl or aryl residue. Typical compounds would include the propyleneoxide adducts of 1,2,6-hexane triol and pentaerythritol, which are then aminated.

The above amines are well known materials and can be made by a wide variety of techniques. One excellent technique of preparing the diamines is to first provide the corresponding diol compound, such as 1,6-hexanediol. This compound, in turn, is propoxylated with sufficient propylene oxide to provide an adduct. This adduct is then reacted with ammonia or ammonium hydroxide in presence of a suitable reductive amination catalyst to produce the desired diamine.

The same type of sequence of reactions may be carried out to prepare the diamines from propoxylated lower alkyleneglycols. That is, a polyoxy lower alkylenedoil of the formula:

$$
\text{H-(O-CH}_2\text{)}_r\text{-(O-CH}_2\text{)}_{s,z}\text{-(CH}_2\text{)}_t\text{-OH}
$$

where r, s, t, and z are as above, is first reacted with propylene oxide. This adduct is then reductively aminated as set out above. Again, starting diols of the above type, which can be propoxylated, are commercially available and need little elaboration. For example, a typical useful starting material of this type is a polyoxybutylene diol sold under the trademark *POLYMEG™* 1000 by Quaker Oats Co. This particular material has an average molecular weight of approximately 1000.

As a typical example, the first step involves providing as a starting reactant, i.e., a polyoxybutylene diol having the following structural formula:

$$H-(-OCH_2CH_2CH_2CH_2)_z-OH$$

where z is an average number ranging from ≈ 6 to 50. Propoxylating said polybutanediol with sufficient propylene oxide provides an adduct having the structure

$$\underset{\substack{| \\ \text{HO-CH-CH}_2\text{(O-CH-CH}_2)_x\text{-O-(CH}_2\text{CH}_2\text{CH}_2\text{CH}_2\text{-O})_z\text{-(CH}_2\text{CHO})_y\text{-CH}_2\text{CH-OH}}}{\overset{\substack{\text{CH}_3 \quad\quad \text{CH}_3 \quad\quad\quad\quad\quad\quad\quad\quad\quad\quad \text{CH}_3 \quad\quad\quad \text{CH}_3}}{}}$$

where x and y range from 1 to 8. The propoxylation reaction can be conducted using conventional methods and conditions such as temperatures in the range of ≈ 40 to $\approx 200°$ C and pressures from ≈ 0 to ≈ 100 psig. Usually, the reaction occurs under basic conditions established through the use of alkali metals, their hydroxides, oxides, and hydrides and, in some cases, basic amines. Representative alkoxylation procedures may be followed here as described by Schick.[8]

The propylene oxide adduct is then reacted with ammonia or ammonium hydroxide in presence of a suitable reductive amination catalyst to produce the desired polymeric amine. A wide number of known catalysts of this type are useful here. Preferred are nickel- and cobalt-based catalysts, with the most preferable being a nickel-based catalyst, including Raney nickel and nickel in combination with other metals or oxides of metals.

The above-described propylene oxide adducts are reacted with ammonia or ammonium hydroxide (preferably ammonia) in the presence of said hydrogenation-dehydrogenation catalyst at elevated temperatures in the presence of hydrogen to form the amines. Suitable reactors include either a closed autoclave resulting in a batch process or a tubular reactor which can be operated in a continuous manner.

As just noted, the class of useful catalysts here is well known and may include one or more metals including copper, nickel, cobalt, chromium, aluminum, manganese, platinum, palladium, and rhodium, and the oxides of these metals. The metals or their oxides may be employed in combination with normally nonreducible metal oxides such as chromium oxide, molybdenum oxide, and manganese oxide. The amount of the nonreducible oxide employed may be varied considerably, and some catalysts, notably those based on cobalt, require the absence of nonreducible metal oxides.

One preferred catalyst that is very effective for the amination reaction includes the metals or oxides of nickel, cobalt, and chromium. A particularly satisfactory catalyst is one in which the active ingredients consist essentially, in mole percentages on an oxide-free basis, of 60 to 85 percent nickel, 14 to 37 percent copper, and 1 to 5 percent chromium, as produced in accordance with procedures described in U.S. Pat. 3,152,998. As used herein, this catalyst will be referred to as a nickel-copper-chromium catalyst.

The reductive amination reaction is carried out from 160 to 250° C. The reaction pressures are from 750 to ≈ 4000 psig with a hydrogen partial pressure of at least 200 psig. The preferred pressure range is from ≈ 1000 to ≈ 2500 psig and a hydrogen partial pressure from ≈ 200 to ≈ 2000 psig.

The residence times in the reactor to be used to produce the amine reactants are those that would occur at space velocities of ≈ 0.2 to ≈ 3.0 weight of reactants per volume of catalyst per hour, with the preferred space velocity being from ≈ 1.0 to ≈ 2.0. The space velocity herein described is in grams total liquid feed/ml of catalyst/hour, but rates in equivalent units are equally applicable.

The ratio of reactants, i.e., propylene oxide adduct and ammonia can vary over a wide range to produce the amines. The preferred ratio of ammonia to adduct is 10 to 100 moles ammonia/mole adduct.

By following the above-discussed techniques of the invention, substantially all of the hydroxyl groups of the propylene oxide adduct are transformed into primary amine groups.

To make the products of the invention, the above amine and isatoic anhydride reactants are simply mixed together without the necessity of solvent or diluent and heated. When the reaction is finished, the product requires no further treatment or purification. In addition, no catalyst is required to effect the reaction. The products are generally viscous liquids (pourable when warm) rather than crystalline solids. The temperature of reaction may range from ≈20° C to ≈200° C, at a pressure ranging from atmospheric pressure to ≈1000 psig.

When one mole of the isatoic anhydride is added per mole of diamine, only one of the terminal amine groups is reacted to produce a monoamide containing one aromatic and one aliphatic amine group. On the other hand, if two moles of the anhydride are reacted with one mole of the diamine, both terminal groups are reacted to produce a diamide structure containing two aromatic amine groups.

The above *chain-extenders* are particularly useful in preparing improved solid polyurethane compositions useful as sealants, floor coatings, and molds. In addition, when employing the additives of the invention, one may provide an integral skin on a foamed cellular polyurethane composition containing the chain-extenders of the invention. This integral-skinned cellular polyurethane composition produces a product having the desired properties of a foam crash pad in addition to having a tough, scuff-resistant integral skin, thus obviating the necessity of lining the mold with a separate skinning material.

As noted above, in the production of polyurethane compositions, polymeric polyhydroxy compounds such as polyester or polyether polyols are reacted with organic polyisocyanates to produce a polyurethane composition. Polyether polyols are described herein, and polyester polyols are described in U.S. Pat. 3,391,093, for example. This reaction usually occurs in the presence of a catalyst but may occur noncatalytically when a polyol-containing tertiary nitrogen atom is used. In the practice of the invention, the above-described chain extenders are included in this reaction mixture to produce improved polyurethane compositions. When a solid polyurethane composition is produced using the chain extender of the invention, we have discovered that improved tensile strength, tear strength, and elongation results. With the chain extender of our invention, strong yet flexible floor coverings and sealants are possible. In addition, soft, flexible molds can be produced that have improved tear strength but yet sufficient compression strength to withstand pressures produced when the mold made from our polyurethane composition must contain an expanding cellular plastic.

Suitable organic polyisocyanates useful in the practice of our invention are those organic diisocyanates, triisocyanates, and polyisocyanates that are well known in the polyurethane art. Mixed isomers of toluene diisocyanate, which are readily commercially available such as described in U.S. Pat. 3,298,976 and others, may be used. Especially preferred are diisocyanates and polyisocyanates prepared by the phosgenation of the reaction product between aniline and formaldehyde such as 4,4'-diphenylmethane diisocyanate, 2,4'-diphenylmethanediisocyanate, and higher-functionality polyphenylmethylene polyisocyanates, hereinafter called *polyarylpolyisocyanates*. Especially preferred organic polyisocyanates for forming solid polyurethane compositions are diphenylmethane diisocyanate and modified diphenylmethane diisocyanates sold under the trademark of *ISONATE*™ 143L. Polyarylpolyisocyanates that are used in the practice of our invention, particularly to produce cellular polyurethanes, have a functionality of from ≈2.0 to ≈3.3. An especially preferred functionality range is from ≈2.2 to ≈2.9.

Polyether polyols useful in the practice of our invention are those diols, triols, tetrols, and mixtures having a molecular weight from ≈500 to ≈10,000. The diols are generally polyalkylene ether glycols such as polypropylene ether glycol, polybutylene ether glycol, etc., and mixtures thereof. Mixed polyether polyols can also be used such as the condensation products of an alkylene oxide with a polyhydric alcohol having three or four primary hydroxyl groups such as glycerol, trimethylolpropane, 1,2,6-hexanetriol, pentaerythritol, and the like.

As previously mentioned, any suitable polyhydric polyalkylene ether may be used. For example, the condensation product of an alkylene oxide with a polyhydric alcohol. Any suitable polyhydric alcohol may also be used such as ethylene glycol, 1,2-propylene glycol, 1,3-propylene glycol, 1,4-butylene glycol, 1,3-butylene glycol, glycerine, trimethylolpropane, 1,2,6-hexanetriol, pentaerythritol, etc. Any suitable alkylene oxide may be used, e.g., ethylene oxide, propylene oxide, butylene oxide, amylene oxide, their various isomers, etc. Of course, the polyhydric polyalkylene ether polyols can be prepared from other starting materials such as tetrahydrofuran, epihalohydrin, aralkylene oxides such as styrene oxide, and the like. Polyhydric polyether polyols having three or four hydroxyl groups per molecule and a molecular weight of from ≈2,000 to ≈10,000 can be used. The polyol used can be a blend of diols with triols or tetrols to produce a polyol blend having an average molecular weight of from ≈500 to ≈10,000. Blended diols and triols for use in solid polyurethane elastomers are generally discussed in U.S. Pat. 3,391,101. Most preferred for use either alone or blended with a diol are the polyoxyalkylene triols and tetrols having a molecular weight of from ≈2,000 to ≈7,000.

The polyether polyols may have primary or secondary hydroxyl group termination. When the polyhydric alcohol is reacted with an alkylene oxide such as propylene oxide, butylene oxide, etc., the terminal groups are predominantly secondary hydroxyl groups. However, it is within the scope of our invention to use polyether triols or polyether tetrols which have from ≈5 to ≈15 wt. % ethylene oxide added thereto in a final alkoxylation step by the known alkoxylation processes to increase the terminal primary hydroxyl content of the said polyether polyol. The manufacture of ethylene oxide-*tipped* polyether polyols is generally discussed in U.S. Pat. 3,336,242.

As previously mentioned, the polyether polyol and the organic polyisocyanate are reacted to form the polyurethane composition. This reaction may occur noncatalytically when a polyol is used that contains tertiary nitrogen compounds, or it may be carried out in the presence of known polyurethane catalysts. The use of a separate catalyst is preferred. The catalyst employed may be any of the catalysts known to be useful for this purpose, including tertiary amines and metallic salts. Suitable tertiary amines include N-methylmorpholine, N-ethylmorpholine, triethylenediamine, triethylamine, trimethylamine, and N-dimethylethanolamine. Typical metallic salts include, for example, the salts of antimony, tin, mercury and iron—for example, dibutyltin dilaurate phenylmercuric acetate and stannous octoate. The catalyst is usually employed in a proportion of from ≈0.01 to 2% by weight based on the weight of the overall composition.

Various additives can be employed to provide different properties, e.g., fillers such as clay, calcium carbonate, talc, or titanium dioxide. Dyes and pigments may be added for color, and antioxidants also may be used.

When the embodiment of our invention is practiced that involves the production of the self-skinning cellular polyurethane product, a foaming agent is employed that may be any of those known to be useful for this purpose, such as water, the halogenated hydrocarbons, and mixtures thereof. Typical halogenated hydrocarbons include but are not limited to monofluorotrichloromethane, difluorodichloromethane, 1,1,2-trichloro-1, 1,2-fluoroethane methylene chloride, and the like. The amount of foaming agent employed may be varied within a wide range. Generally, however, the halogenated hydrocarbons are employed in an amount from 1 to 50 parts by weight per 100 parts by weight of the polyol used in the production of the polyurethane composition. When water is employed as the blowing agent, it is present in the amount of from 0.1 to 10 parts by weight per 100 parts by weight of the polyether polyol. Halogenated hydrocarbon blowing agents for use in the production of a foamed polyurethane composition are discussed in U.S. Pat. 3,072,582.

When it is desired to practice our invention in producing a floor coating or sealant, it is often desirable to include therein a polyhydric cross-linking agent. Such cross-linking agents include but are not limited to polyhydric alcohols such as glycerol, trimethylolpropane, 1,2,6-hexanetriol or pentaerythritol, or amines such as ethylenediamine, N,N,N',N'-tetrahydroxy propylethylene diamine, etc. These are included in the polyurethane composition that they make up ≈0.02 wt. % to ≈10 wt. %, based on the entire polyurethane composition. The use of such cross-linking agents is well known, and those skilled in the art wil! be able to readily determine the amount and type of cross-linking for use to achieve desired physical properties.

The chain-extending agent of our invention as described above is used in both solid polyurethane compositions and the self-skinning flexible or semiflexible polyurethane foam composition. The amount of the chain-extending agent may be as low as 0.1 wt. % based on the polyol component in a solid elastomer polyurethane composition to ≈50 wt. % of the entire formulation when used in the self-skinning foam polyurethane composition. It may be used either alone as the chain-extending agent or in conjunction with known chain-extending agents such as 1,4-butanediol, diethylene glycol, 4,4'-methylene bis(2-chloroaniline), and the like. However, we have discovered that, whether used alone or in conjunction with known chain-extending agents, the chain extender of our invention improves the tensile strength of the resulting polyurethane composition without detriment to other desired physical properties. When used in solid polyurethane compositions, the amount of 0.1 to ≈15 wt. %, based on the weight of the polyether polyol, and more preferably from ≈0.5 to ≈7 wt. %, is employed.

In the production of the cellular self-skinning polyurethane compositions, the chain-extending agent used in the practice of our invention would be present in the amounts of from 10 to ≈50 wt. % of the polyurethane reaction mixture, with preferred amounts being from ≈15 to ≈35 wt. %.

6.2.8 Method of Producing Polyurethane with Improved Flex Crack Resistance[10]

Polyurethane products having improved flex crack resistance can be obtained by curing a mixture of (1) a polyurethane prepolymer having terminal isocyanate group and (2) at least one curative selected from polyamine and polyol to prepare a polyurethane product, and heat treating the polyurethane product at a temperature of 60 to 180° C for 1 to 30,000 min under an atmosphere having an absolute humidity of at least 2%.

Background of the Invention

Field of the Invention
The present invention relates to a method for improving the flex crack resistance, cut resistance, and tear strength of polyurethane products.

Description of the Prior Art
Polyurethane has high strength, abrasion resistance, and weather resistance, and the use of polyurethane has increased remarkably in recent years. In particular, casting-type polyurethane is widely used due to its processibility in the liquid state. However, polyurethane is insufficient in flex crack resistance, cut resistance, and tear strength, and when a polyurethane product is once flawed, the flaw rapidly grows up under a severe use condition to give a fatal blow to the product. To obviate the above described drawbacks of polyurethane, the following methods have been proposed, i.e., a method wherein a plasticizer is added to a polyurethane prepolymer and a method wherein a curative (chain extender) is added to a polyurethane prepolymer in an amount of not less than equivalent based on the amount of isocyanate groups in the prepolymer. However, in the former method, which uses a plasticizer, the curing and chain-extending reactions of the polyurethane prepolymer are often adversely affected, and the physical properties of the resulting polyurethane always decrease considerably. The latter method, wherein polyamine or polyol is added to a polyurethane prepolymer in an amount of not less than chemical equivalent based on the amount of isocyanate groups of the prepolymer to effect the curing and chain extension of the prepolymer, has been known for a long time as a method of obtaining a polyurethane having an improved flex crack resistance. However, polyamines and polyols, which give a high strength to polyurethane, are expensive. In particular, polyamines have problems related to toxicity to human body, particularly in carcinogenicity. Moreover, the use of a large amount of such chain extenders shortens the pot life at the casting and often causes difficulties in the casting operation. Accordingly, it is desirable to decrease the amount of chain extender as much as possible. This fact is clearly contrary to the commercial demand for obtaining polyurethane having high strength, flex crack resistance, cut resistance, and tear strength.

Summary of the Invention

It is an object of the present invention to solve the above-mentioned drawbacks in the prior art and to produce easily polyurethane products having an improved flex crack resistance. It is another object of the present invention to produce polyurethane products having improved cut resistance and tear strength.

The present invention provides a method of producing polyurethane products having an improved flex crack resistance, which involves curing a mixture of (1) a polyurethane prepolymer having terminal isocyanate groups and (2) at least one curative selected from the group consisting of polyamines and polyols to prepare a polyurethane product, and heat treating the polyurethane product at a temperature of 60 to 180° C for 1 to 30,000 min under an atmosphere having an absolute humidity of at least 2%.

According to the present invention, the flex crack resistance, cut resistance, and tear strength of polyurethane product can be remarkably improved without shortening the pot life and affecting adversely the creep property by using a small amount of curative.

Furthermore, in the present invention, the use amount of curatives such as polyamine and polyol can be decreased while maintaining the high level of the physical properties of polyurethane products, and therefore the present invention is excellent in the low production cost of polyurethane and in the prevention of toxicity to the human body.

Description of the Preferred Embodiments

As the polyurethanes to be used in the present invention, there may be mentioned polyurethanes obtained by a so-called prepolymer process, wherein a polyurethane prepolymer having a terminal isocyanate group is reacted with a curative such as polyamine, polyol, or the like.

The polyurethane prepolymer having a terminal isocyanate group, used as component (a) in the present invention, is a compound having an average molecular weight of 800 to 20,000, preferably 2,000 to 5,000, that is obtained by reacting a polyether, a polyester, or an unsaturated hydrocarbon, each having a terminal active hydrogen, with a molar excess of an organic polyisocyanate.

The polyether having a terminal active hydrogen, polyester having a terminal active hydrogen, and unsaturated hydrocarbon having a terminal active hydrogen mean a polyether, polyester, and unsaturated hydrocarbon, each having, at the terminal, functional group, such as hydroxyl, amino, imino, carboxyl, and mercapto groups and the like, which have active hydrogen reactive with an isocyanate group.

6.2.9   Use of Monol Chain Extenders to Control Molecular Weight

The production of polyurethane compositions by reacting organic polyisocyanates with organic compounds containing hydrogen atoms reactive with isocyanate groups is well known to the art. Polyurethane compositions so produced are employed in a variety of different applications such as, for example, the production of molded articles, and castings. It is difficult, however, to consistently prepare reproducible products having uniform physical properties when polyurethanes are processed near the equivalence point, i.e., at an NCO/OH ratio of from ≈0.90:1 to ≈1.10:1. The variances in product reproducibility and uniformity are due to a number of possible reactions that depend on difficult to control parameters.

One of the most important parameters is the NCO/OH ratio itself, which is controllable as a practical matter only within a range of approx. ±1%. A small change in the NCO/OH ratio results in a large change in rheological properties as measured by the intrinsic melt index or solution viscosity and the content of highly cross-linked particles. These undesirable changes are due to the reaction of the isocyanate groups with urethane groups to form allophanates rather than the normal chain lengthening reaction. Generally, the allophanate reaction, due to its low rate constant, is not detrimental, but near the equivalence point the reaction becomes of importance due to the relative scarcity of hydroxyl groups.

To produce a polyurethane composition of uniform quality and reproducibility, it is necessary to control the chain lengthening reaction so as to achieve the desired extent of polymerization or the desired molecular weight by minimizing the undesirable effects of allophanate formation. The desirability of interrupting the polymerization reaction is suggested in U.S. Pat. 3,310,533, wherein the reaction is temporarily terminated by casting the reaction mixture and cooling to ambient temperature. It is also known to add a monoalcohol, for example, ethanol, to stop the polymerization reaction as taught in Canadian Pat. 888,781. Neither of these methods, however, is selective in minimizing allophanate formation.

Meckel and Britain[11] provided a method of producing polyurethane compositions of uniform quality and reproducibility. Polyurethane elastomers are prepared by reacting organic polyisocyanates with organic polyhydroxyl compounds and a monofunctional secondary hydroxyl group containing a chain terminator such as 2,6,8-trimethylnonanol-4 and 2,4,8-trimethyl-2,4,8-trichloro-6-nonanol. The chain terminator can be added to the polyhydroxyl compound prior to the reaction with the polyisocyanate.

The chain terminators of the invention are monofunctional, secondary hydroxyl compounds wherein the hydroxyl group is sterically hindered, and are significantly more selective than the monofunctional alcohols of the prior art. The increased selectivity of the chain terminators of the invention is due largely to the following properties:

1. The reactivity of the secondary hydroxyl group is lower than the activity of primary hydroxyl groups, thus allowing an undisturbed chain-lengthening reaction while assuring availability of the chain terminator at the end of the reaction.
2. The reactivity of the chain terminators of the invention is higher than that of urethane groups toward isocyanate groups, thus minimizing the extent of allophanate formation.
3. The boiling point of the chain terminators of the invention is sufficiently high to prevent its easy evaporation from the reaction mixture.

Due to equipment limitations, inaccuracy of metering the components, etc., variations in NCO/OH ratio will always be present and, depending on the desired viscosity of the final product, some fluctuation in the NCO/OH ratio can be tolerated. However, as the equivalence point is approached, the tendency toward allophanate formation is enhanced, resulting in increased viscosity due to excessive cross-linking rather than normal extension. Due to the properties of the chain terminator compounds of the invention, they become increasingly reactive toward -NCO groups as the polymerization reaction proceeds to the equivalence point, thus minimizing allophanate formation and assuring a polyurethane product of uniform quality which is more easily reproducible.

The amount of chain terminator employed is a function of the hydroxyl groups available for the polyurethane forming reaction. The chain terminator is used in an amount of from ≈0.1 to ≈3%, based on the molecular equivalent of all available hydroxyl groups and is most preferably used in an amount of from ≈0.2 to ≈2%.

The chain terminators employed in the invention can be any materials having the formula described above, such as materials composed completely of carbon and hydrogen atoms, with the single exception of the one hydroxyl group, as well as compounds containing heteroatoms, i.e., atoms other than carbon and hydrogen, such as halogen substituted compounds. Additionally, heteroatom-containing compounds such as esters and ethers are also suitable. Compounds of this nature are well known and are either readily obtainable commercially or easily synthesized by those skilled in the art. Although structural symmetry is not necessary about the OH group, i.e. both R and R' being the same, and compounds such as 2,6,8-trimethylnonanol-4 and 2,4,8-trimethyl-2,4,8-trichloro-6-nonanol are quite satisfactory, we prefer, however, to employ chain terminators in which R and R' are the same, when R and R' contain heteroatoms in the form of ether or ester groups, and particularly ester groups. Illustrative of such materials are 4-heptanol-2,6-diacetate or 2,6-dimethyl-4-heptanol-2,6-diacetate. These materials can be described as diesters obtained, for example, by the reaction of a carboxylic acid with a triol.

Example

A polyol mixture of ≈100 parts of polybutylene-1,4-adipate (hydroxyl number, ≈56), ≈7.5 parts of 1,4-butane diol, plus ≈0.35 part of diisobutylcarbinol are reacted with 4,4'-diisocyanato-diphenylmethane at a temperature of from ≈90 to ≈140° C. The mixture is stirred for ≈30 s, cast in a preheated mold, cured for ≈10 min, ground, and aged for ≈78 hr at ≈80° C.

6.2.10 Neopentyl Glycol and Trimethylolpropane to Produce Soluble Polyurethane

Vaeth et al.[12] found that polyurethanes that utilize aliphatic branched-chain diols and triols chain extenders are soluble in certain solvents if the polymers are reacted by conventional methods, in the presence of appropriate catalysts, solvents, assistants, and additives.

Vaeth found, surprisingly, that certain polyurethanes not only have a high surface hardness and a high modulus of elasticity without adverse effect on tensile strength and extensibility, but they are also very readily soluble in ethers and ketones, especially in cyclic ethers and ketones such as tetrahydrofuran, dioxan, and cylcohexanone. A further advantage of the process of the invention is that the starting materials for the manufacture of the polyurethanes are not restricted to specific polyester-ols or polyether-ols, e.g., those based on phthalic acids and branched-chain diols according to German Pat. 1,106,958 or polyalkylene glycol ethers of the formula $HO[(CH2)[n]O][x]H$, in which n is from 3 to 6 and x is greater than 7, as in German Pat. 1,112,291. Instead, all linear polyester-ols and/or polyether-ols falling within the stated molecular weight range can be used. Thus, e.g., polyester-ols such as adipates based on glycols, polycaprolactones, or aliphatic polycarbonates, as well as polyether-ols based on alkylene oxides and tetrahydrofuran, can be used for the manufacture of the polyurethanes.

As previously explained, the polyols used for the manufactures of the thermoplastic, elastic polyure-thanes that are virtually free from branching are soluble in ethers and/or ketones and have a high surface hardness and a high modulus of elasticity are polyester-ols and/or polyether-polyols.

It is an essential feature of the present invention that branched-chain aliphatic diols be used. Examples are 2-methyl-1,3-propanediol, 2-methyl-2-ethyl-1,3-propanediol, 2-methyl-2-isopropyl- 1,3-propanediol, 2,2-diethyl-1,3-propanediol, 2-methyl-2butyl-1,3-propanediol, and 2-ethyl-2-butyl-1,3-propanediol. The preferred branched-chain aliphatic diol is 2,2-dimethyl-1,3- propanediol, also known by the trivial name of *neopently glycol*. By using such branched-chain aliphatic diols as chain extenders, numerous hard segments may be introduced into the polyurethane molecule without significantly reducing the solubility of the products. The branched-chain diols may be used as individual compounds or as mixtures.

The hardness of the product can be modified within certain limits by the use of linear diols or triols. Thus, e.g., partial replacement of 2,2-dimethyl-1,3-propanediol by a glycol (e.g. ethylene glycol, 1,4-butanediol or 1,6-hexanediol) gives softer polyurethanes, while partial replacement of branched-chain diols by triols, e.g., trimethylolethane, hexanetriol, and (preferably) trimethylolpropane or glycerol, permits a further increase in surface hardness as a result of cross-linking.

The new thermoplastic, elastic polyurethanes that are soluble in ethers and/or ketones may be man-ufactured in the absence of solvents. Preferably, however, the products are manufactured in solution by the one-shot or prepolymer process, if appropriate in the presence of catalysts and other assistants and/or additives such as dibutyl-tin dilaurate or triethylenediamine.

Preferred solvents are cyclic ethers, e.g., tetrahydrofuran and dioxan, and cyclic ketones, e.g., cyclo-hexanone. Of course, the polyurethanes may also be dissolved in other strongly polar solvents, e.g. dimethylformamide, pyrrolidone, dimethylsulfoxide, or ethylene glycol acetate. The solvents mentioned can equally be mixed with aromatics, e.g., toluene or xylene, and esters, e.g., ethyl acetate or butyl acetate.

The thermoplastic and elastic polyurethanes manufactured according to the invention are very readily soluble in ethers and/or ketones, especially in cyclic ethers and ketones, have melting points above 100° C, preferably at from 120 to 220° C, and have a surface hardness of from 70 to 160 s, moduli of elasticity of from 300 to 2,000 N/mm^2, elongations at break of >250% (especially of >400%), and tensile strengths of >55 N/mm^2, especially of >65 N/mm^2.

The products may be used for coating, e.g., textiles, paper, and plastics; for the manufacture of coverings; for impregnation; and as surface-coating binders. Preferably, however, the products are used as dressings for leather and leather-like materials.

Example

A solution of 250 parts of tetrahydrofuran, 100 parts (0.05 mole) of an ethylene glycol adipate of molecular weight 2,000, and 131.2 parts (0.525 mole) of 4,4'-diisocyanatodiphenylmethane is stirred with 0.02 parts of dibutyl-tin dilaurate for one hour at 50° C. A mixture of 167 parts of tetrahydrofuran, 45.8 parts (0.44 mole) of neopentyl glycol, 0.9 part (0.0067 mole) of trimethylolpropane, and 0.05 part of dibutyl-tin dilaurate is then added in the course of 1 hr to the prepolymer solution, which contains isocyanate groups, at 55° C, while stirring. On reaching a viscosity of 2,000 to 3,000 cp at 55° C, the polyurethane solution is diluted in stages to a solids content of about 20% by weight with 232 parts, 185 parts, and

278 parts of an 0.01% strength by weight dibutyl-tin dilaurate solution in tetrahydrofuran. After the reaction mixture of about 20% strength by weight has again reached a viscosity of from 1,500 to 2,000 cp at 55° C, the reaction is stopped by adding 1.8 parts of dibutylamine, and the reaction solution is cooled. The polyurethane obtained has a viscosity of 3,500 cp as a 20% strength by weight solution in tetrahydrofuran, a Konig pendulum hardness of 125 seconds, and a modulus of elasticity of 1,380 N/mm^2.

6.2.11 Water as Chain Extender

Magnusson and White[13] described a polyurethane-urea that is highly resistant to textile fiber dressings. The composition is a product from the reaction of an ester urethane prepolymer and an excess of water.

This invention relates to novel polyurethane-urea compositions. It is more particularly concerned with novel linear polyurethane-urea compositions having high resistance to textile fiber-dressing solvents and a process for making such polyurethane ureas.

Polyurethanes have been recognized as high-molecular-weight polymeric materials exhibiting excellent resistance to abrasion, cut growth, oxidation and oils. These properties, combined with the fact of easy processibility such as by extrusion or casting, make them well suited for the fabrication of complex parts. The thermoplastic polyurethanes are particularly useful in extrusion operations where high output rates of extrudate are desirable. Because of their ease of fabrication, the thermoplastic polyurethanes have found extensive use in the textile industry as aprons, belts, and the like on texturizing frames. While exhibiting comparatively long life in relation to the common materials found useful in this industry, the polyurethanes have generally suffered from a severe drawback due to their sensitivity to certain textile fiber-dressing solvents. These polymers gradually mechanically degrade in environments where contact with solvents, such as those of the polyethylene carboxylate type, for extended times at higher than ambient temperatures occurs. This mechanical degradation requires the part to be replaced, such replacement resulting in the economically disadvantageous shutdown of the textile machinery.

Unexpectedly, the Magnusson patent teaches thermoplastic polymers are polyurethane-ureas that are linear in nature and have good extrudability and capable of dissolving in dipolar aprotic solvents. Therefore, they can be molded, extruded, or cast. These polyurethane-ureas have high resistance to textile fiber-dressing solvents. These solvents include the glycerol fatty acid esters such as glyceryl monooleate and glyceryl dioleate; the polyoxyalkylene fatty acid esters such as polyoxyethylene, tallow amine acidic phosphate ester of ethoxylated lauryl alcohol, polyoxyethylene sorbitan monooleate, polyoxyethylene hydrogenated castor oil, polyoxyethylene, and tridecyl ether; and the amide-containing fatty acids esters such as lauric myristic diethanolamide. These novel compositions undergo substantially no swelling or other characteristic that would suggest the diminution of physical properties when exposed to the above fiber-dressing solvents at elevated temperatures over extended periods of time. For example, at 50° C and 48 hr, these polyurethane-ureas exhibit little swelling from common fiber-dressing solvents. Comparable polyurethane compositions undergo distortion, swelling, and mechanical property loss under such conditions. The more common nitrile rubber materials, in many cases still being used as texturizing frames, are even more delicate, showing cracking and extensive swelling at 50° C in a 48-hr period.

Prepolymer formation is carried out at temperatures of from ≈40 to ≈100° C, advantageously at ≈60 to 70° C. The prepolymer can be formed in a solvent inert to isocyanate, such as xylene, benzene, toluene, and the like, or advantageously, the prepolymer can be formed in bulk (melt polymerization techniques). The resultant isocyanate-terminated prepolymer composition is subsequently chain extended by reaction with excess water. By *excess of water* is meant sufficient water to act as a chain extension agent, as well as a reaction heat modifier. Surprisingly, when very large excesses of water are used, rather than the common cross-linked polyurethane-urea predictively obtained, linear polyurethane-ureas result. At least a threefold excess of water, based on the total residual NCO content of the prepolymer, should be added in accordance with the present invention.

The chain extension reaction is generally regarded as exothermic, but due to the large amounts of water present, such exotherm is greatly depressed. Thus, the chain extension reaction is conveniently effected at 70° C. Temperatures over ≈100° C should be avoided. The resultant reaction mass is advantageously immediately removed from the reaction vessel, usually in the form of a slurry. The water is allowed to drain from such and, after drying, a polyurethane-urea crumb results.

References

1. Gerkin, Richard M. and Richey, Jr., Froest A., U.S. Pat. 5,364,924, "Substituted ureas containing cyclic amines as chain extenders in polymeric systems" to OSi Specialties, Inc. Nov. 15, 1994.
2. Frisch, Kurt C., and Damusis, Adolfas, U.S. Pat. 4,096,128, June 20, 1978, "Polyurethane elastomer produced by reaction between a triazole-stabilized isocyanate prepolymer and a diamine."
3. Nalepa, Christopher J., Ranken, Paul F., and Wiegand, Karl E., U.S. Pat. 4,595,742, June 17, 1986, "Di(alkylthio)diamine chain extenders for polyurethane elastomers," to Ethyl Corporation, Richmond, Virginia.
4. Sullivan, Carl M., U.S. Pat. 4,269,946, May 26, 1981, "Low density high resiliency foams," to The Goodyear Tire & Rubber Company, Akron, Ohio.
5. Lin, I., Sioun, Gromelski, Jr., Stanley J. Werner, Jesse, Brown, Michael J. and Chakrabarti, Paritosh, M., U.S. Pat. 4,459,007, October 22, 1985, "Diamine/diol chain extender blends for rim process," to GAF Corporation, Wayne, New Jersey.
6. Priester, Jr., Ralph D., Strojny, Edwin J., and Stutts, Debra H., U.S. Pat. 4,931,487, June 5, 1990, to Dow Chemical Company.
7. Marquis, Edward T., and Yeakey, Ernest L., U.S. Pat. 4,180,644, December 25, 1979, "Polyurethane chain-extenders," to Texaco Development Corporation, White Plains, New York.
8. Schick, M.J., ed. "Nonionic Surfactants," New York: Marcel Dekker, 1967, 187–204.
9. *Encyclopedia of Chemical Technology,* vol. 7. Interscience Publishers, Inc., 1951, 257–262.
10. Nakauchi, H., et al., U.S. Pat. 4,086,211, Apr. 25, 1978, "Method of producing polyurethane products having an improved flex crack resistance," Bridgestone Tire Company, Tokyo, Japan.
11. Meckel, W.N., and Britain, J.W., U.S. Pat. 4,071,505, January 31, 1978, "Molecular weight control of polyurethane elastomers," to Mobay Chemical Corporation, Pittsburgh, Pennsylvania.
12. Vaeth, G., Rudolf, F., Hartmann, H., Limburgerholf, H., Limburgerhof, L., August and Roedersheim-Gronau, U.S. Pat. 4,058,506, November 15, 1977, "Thermoplastic, elastic polyurethanes which are soluble in ethers and/or ketones," BASF Aktiengesellschaft, Ludwigshafen, Federal Republic of Germany.
13. Magnusson, Alan B., and White, Howard S., U.S. Pat. 4,049,632, September 20, 1977, "Chain extending polyurethanes with a large excess of water," to Armstrong Cork Company, Lancaster, Pennsylvania.

7

Flexible and Semiflexible Foams

7.1 Introduction

Foams are microcellular structures, produced by gas bubbles formed during the polyurethane polymerization mixture. The process of bubble formation in polyurethane foams is called *blowing*. The blowing reaction is one of several reactions occurring in the final polyurethane mixture while it is foaming. The chemical ingredient in the formulation that provides the gas is called the *blowing agent*. Low-boiling liquids added to physically assist in the foaming process are called *auxiliary blowing agents*.

To the public, as well as to many technologists, the word *urethane* or *polyurethane* is synonymous with flexible foams. With a U.S. consumption of about 2 billion pounds, flexible polyurethane foams are dominant in the bedding, furniture, carpet underlay, and transportation industries. Flexible foams are produced in a variety of shapes by either cutting or molding. Flexible foams are commonly used in upholstered furniture, mattresses, automotive seating, etc.

The largest markets for flexible polyurethane foams are

Furniture and bedding
- mattress cores and pads
- furniture cushions

Transportation
- Seat cushions and trim panels (flexible)
- door panels
- arm rests
- head restraints, and energy absorption (semiflexible)

Carpeting
- Virgin and bonded underlay (flexible)

To this point, *flexible foam* has been used as a fairly general term. Before going any farther, some definitions are in order, although it will be seen that the terminology is still rather general. The American Society for Testing and Materials (ASTM) makes a distinction only as to whether a material is flexible or rigid. ASTM D-883 states, "A cellular plastic is considered flexible if a piece eight inches by one inch by one inch can be wrapped around a one inch mandrel at room temperature." If the foam cannot be wrapped around the mandrel, it is considered to be rigid. Obviously, there are many intermediate types of cellular that defy description in this manner. There are other definitions in which the compression and glass transition temperature are specified, but even these fall short of giving a clear picture of what defines a specific type of foam.

Rather than attempting a universal definition, it may be preferable to describe the foams in terms of stress-strain relationships, as shown in Fig. 7.1. To assist in understanding, consider a flexible foam intended for furniture cushioning, and a rigid foam used for thermal insulation of appliances.

Both foams have a density of 2 lb/ft³. The flexible foam shows relatively low load-bearing properties with high recovery properties, similar to a coiled metal spring, while the rigid foam displays high load-bearing (but with a definite yield point) and subsequent cellular collapse and lack of recovery. A semiflexible foam shows a blend of these characteristics. Semiflexible foams display high load-bearing without a definite yield point, nonlinearity of load deformation with energy absorption, and good recovery qualities. Any cellular product that falls between the values for the rigid and flexible foam curves could be classified as semiflexible. The term *semiflexible* is now generally preferred to *semirigid,* since most commercial uses for this type of foam primarily utilize more of the flexible characteristics.

If other properties, such as tensile strength, elongation, tear strength, and resilience are measured, semiflexible foams will display properties intermediate between flexible and rigid plastics. High-density products, such as microcellular foams used for automotive fascia, do not easily fit into these definitions and will be discussed separately.

Flexible foams are produced in two forms: slab stock and molded. In slab stock (or bun) production, large slabs are made in long, continuous pours often lasting for an hour or more. Long production runs are made to minimize start-up and shut-down waste. Bun dimensions vary according to the intended end use and typically run from 36 to 86 inches in width and up to six feet high. The buns are cut to the desired shape after a period of curing under ambient conditions. Trimming losses due to skin surfaces that are cut off, and from cutting the part, must be minimized by selecting the proper bun size and by making the bun as high as possible. Special machinery options have extended the capabilities for slab production.

Molded foams are made when a large number of relatively small parts are required, when inserts such as wires or fabric are required, when producing complicated configurations, when making composite products to reduce material losses and labor costs, and where a skin surface may be required. By proper

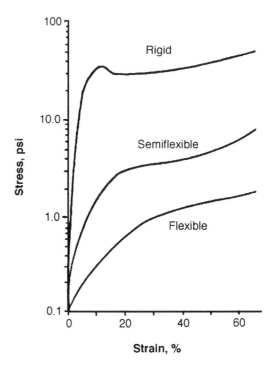

Figure 7.1 Stress-strain diagram for foams.

equipment selection, it is possible to mold multiple-density or load-bearing foams. Conventional (hot-cure) molded foams require high-temperature oven cures to develop satisfactory properties.

The diisocyanate most commonly used in forming flexible foams is an 80:20 isomeric mixture of 2,4 and 2,6-toluene diisocyanate (TDI). Rigid foam can also be made from this diisocyanate if a quasiprepolymer route is chosen. The trend, however, is to use a one-shot process for the production of rigid foams. For one-shot rigid foam production, a modified TDI, MDI, or polymethylene polyphenyl diisocyanate is generally chosen.

The isocyanate serves several purposes. The isocyanate may react with water to form CO_2, a suitable gas for foaming. In many cases, a low-boiling gas, such as trichlorofluoromethane (Refrigerant 11) may be added to provide part or all of the foaming. The diisocyanate also reacts with the functional groups in the resin of polyhydroxyl compound, ensuring that the resin is built in the final polymer molecules.

The principal commercially available polyhydroxy compounds are ethylene and propylene oxide adducts of polyfunctional active hydrogen compounds, such as glycerine, sorbitol, trimethylpropane, sucrose, etc. These compounds are generally referred to as *polyethers,* but since they are primarily polyols, the term *polyether polyols* will be used. The several types of polyether polyols have the generic formula

$$R\text{-}(C_2H_5O)n\text{-}H$$

where R is a polyfunctional active hydrogen compound, and n is the number of ethylene oxide molecules that react. The polyfunctionality (number of reactive hydroxyls or reactive radicals) of the base compound determines the final functionality of the polyether polyols. In contrast, rigid foams require the use of highly functional reactants which are most useful when based on polyalcohols having functionalities of 3 or greater.

The polyether polyol must be liquid at room temperature, since the reactions that take place between the polyhydroxyl compounds and polyisocyanates are so rapid that a completely homogeneous distribution of reactants must be obtained in just a few seconds.

Most polyether polyols are low-cost and highly refined sources of hydroxyls because they are a single species, made to have very low levels of contaminants and uniformly narrow molecular weight distribution. Because of their ether linkages, the polyols and foams in which they are present have sufficient hydrolytic stability for most purposes.

The hydroxyl-terminated polyesters constitute another class of polyhydroxyl compounds. These materials are made by the direct esterification of polybasic acids with polyhydric alcohols, and they offer the chemist a large variety of combinations to obtain specific properties. Their chief advantage lies with their polymeric nature.

However, the polyesters cannot compete with polyethers on a cost basis, and all polyesters suffer from the same basic limitation of having a predominance of ester linkages. The formation of the ester linkage involves a condensation reaction between an acid and alcohol and the elimination of a mole of water per equivalent of acid and alcohol. This, of course, produces higher costs for polyesters than tar polyether polyols, since the direct reactions of a polyol with an isocyanate does not involve condensation loss. Hydrolytic stability and acid and alkali resistance of the ester linkage are generally poor. These failings are multiplied in practice, since the polyester cleavage contains unreacted carboxyl groups that further reduce chemical resistance. Commercially produced polyesters also generally have a greater molecular weight range, as well as some unreacted starting materials and water of esterification. The greatest limitation is probably the high viscosity of the polyester, which requires the manufacture of intermediate prepolymers previous to foaming. The one-shot systems circumvent the need for prepolymers.

In spite of these shortcomings, the hydroxyl-terminated polyesters command a particular advantage in that they, and the foams made from them, can be formulated to provide fire retardance or self-extinguishing characteristics. Several types of halogenated dibasic acids, particularly chlorinated phthalic acids or anhydrides, may be used to provide useful foams, provided the halogen content is sufficiently high.

The halogen content of the polyester obviates the need for various fire retardant agents, such as halogenated plasticizers, which generally tend to degrade the physical properties of the foams. Formulations involving halogenated polyesters are somewhat simpler than those involving halogenated additives, although the halogenated polyesters are more costly. In the final analysis, many polyester-based foams have physical and chemical resistance characteristics that are highly satisfactory, largely because the final properties are a function of the urethane polymer.

7.2 Types of Polyurethane Foam

7.2.1 Flexible Foam

Flexible polyurethane foams, in contrast to rigid types, yield open-cell materials that allow free movement of air throughout the materials when flexed. Flexible foams based on both polyether and polyester polyols are now in general use, and both types can be made in the density range 0.93 to 2.8 lb/ft³. Polyesters are less resilient and less stable to hydrolysis but exhibit higher tensile strength and elongation at break and better *drape* qualities. Load-bearing characteristics are primary controlled by density, but size, shape, texture, method of production, filler content, and variations in the quantities of auxiliary chemicals all affect the values.

Flexible foams exhibit anisotropy, as do rigid foams, but to a much lesser degree, so that it is not necessary to take anisotropy into account when quoting physical test data. Both polyester and polyether types of flexible foams exhibit excellent sound absorbing properties, low thermal conductivity, and good resistance to most solvents and detergents, although some solvents (e.g., acetone and trichloroethylene) cause swelling, from which the foam recovers unchanged upon drying. Flexible urethane foams are rapidly discolored by exposure to ultraviolet radiation and have poor resistance to strong acids and alkalies, especially when hot. The power factor is very low, and volume resistivity is between 10^{10} and 10^{12} Ω-cm. Working temperature can vary between 50 and 100° C (122 and 212° F), depending on the application. Compressive deformation at high temperatures may lead to some degree of permanent deformation. Flexible urethane foams typically remain flexible down to –40° F. Table 7.1 shows typical properties of flexible foams.

TABLE 7.1 Comparison of Typical Properties of Polyether- and Polyester-Based Flexible Foams

Mechanical Properties	Test Method	Polyether Foam	Polyester Foam
Density	—	0.98–2.8	0.94–2.8
Compression set (%)	ASTM 1564		
@ 50% strain		1–6	3–7
@ 75%		2–8	5–10
@ 90%		2–10	6–15
Indentation hardness characteristics (kg)	BS 3667 Pt. II		
@ 25% indentation		3–21	13–21
@ 50%		5–26	16–30
@ 65%		7–38	26–49
Tensile strength (kg/cm²)		0.55–1.6	1.5–2.4
Elongation (%)	BS 3369	100–280	200–450
Tear Strength (kg/cm²)	BS 3379	0.20–0.55	0.5–0.9
Resilience (%)	ASTM 1564	45–65	20–35
Cells per linear inch	ASTM 1564	15–70	15–70
Aging properties	—		
(a) Heat aging, 16 hr @ 140° C, % retained tensile	BS 3379	>70%	>70%
(b) Humid aging, 3 hr @ 105° C, % retained tensile	BS 3379	>70%	>70%
Thermal properties			
Thermal conductivity (BTU/in²/hr/of)		0.22–0.25	0.22–0.45
Flammability			
unprotected	ASTM C177	burns	burns
with flame retardant	ASTM 1692	self-extinguishing	self-extinguishing

Flexible urethane foams are used for upholstering furniture and for automotive cushioning, clothing interlining, sponges, air filters, and packaging. High-density flexible foam containing large amounts of inert filler has cushioning properties similar to those of latex foam.

7.2.2 Semiflexible Foam

Foams of the semiflexible type are manufactured by using suitable combinations of polyesters and isocyanates. They offer high modulus and excellent shock-absorbing qualities. As with flexible urethane foams, considerable variation in hardness and other properties can be achieved with formulation changes.

Figure 7.2 illustrates some of the compressive strength differences between rigid and semirigid urethane foams. It is apparent from this figure that the lower-density rigid foam is stronger at room temperature than at the higher temperature. The modulus of rigidity of the semirigid foam, as represented by the slope of the curves, is about one-third that of the rigid, approximating the condition for flexible foams. Temperature elevation accentuates this difference even more. At 160° F, the semirigid foam cannot carry any load, while the rigid foam is affected relatively little. These factors, although not clearly defining the foams, give an indication of how they may be expected to act in practice.

Cellular polyurethane systems based on castor oil formulations were among the early developments. These foams were essentially of the open-celled, semirigid type and are capable of being produced in a density range of 2 to 20 lb/ft³. Castor-oil based total prepolymer foams are not being made or used in as large quantities as they were in the past.

Semiflexible urethane foams are somewhat thermoplastic and will not melt, but they will become noticeably softer with a moderate increase in temperature. Nonetheless, these foams will not distort under their own weight below 200° F and hence can be used at higher temperatures if not under stress.

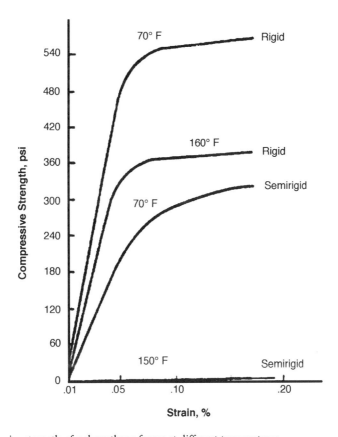

Figure 7.2 Compressive strength of polyurethane foams at different temperatures.

However, in any possible application where severe load-bearing requirements must be met at these elevated temperatures, these foams should be carefully scrutinized. In contrast to rigid foams, where the elastic limit is reached much before or, at the extreme, near the 10% deformation point in compression, these foams can be used while in compressive strengths up to 50% and even up to 75% deformation. Good recovery is found after this high deformation.

One of the outstanding advantages of castor oil-based urethane foams is the excellent bond formed between the foamed-in-place foam and the various materials used to form the sides of the cavities. Prior to the advent of fluorocarbon gas-blown rigid foams, these semirigid foams represented a comparatively low-cost material with good insulation properties. Depending on density, thermal conductivities ranged from 0.20 to 0.30 BTU/hr/ft^2/in °F at 75° F.

Since semiflexible urethane foams are composed essentially of open cells, water can be mechanically absorbed. Moisture from the air is not absorbed, however, and these foams can be considered hydrophobic. Where contact with liquid water is likely, a coating of flexible plastic is desirable. Acoustic insulation is one of the advantages of the open-cellular structure. Semirigid urethane foams also appear to have inherent self-extinguishing properties, apparently caused by localized melting, which quenches the flame. Under similar conditions, rigid foams will burn readily. Furthermore, the flame-retardant properties of semirigid urethane foams are easily enhanced by the usual flame-retardant additives.

The existence of a stress-strain curve without a definite plateau, coupled with an excellent, though slow, recovery after compressive deflection beyond its apparent elastic limit, is evidence that these foams have good energy absorptivity. Indeed, semirigid urethane foams show excellent impact and shock absorption characteristics. Applications include automobile dashboard panels, cushioning, and packaging.

7.2.3 Reticulated Foam

Reticulated urethane foams are highly porous (97% voids) structures that permit the passage of gases and liquids but provide barriers against dirt and dust. They were accidentally discovered when the Scott Paper Company, while attempting to develop a household sponge, subjected urethane foam to a strong caustic solution. The caustic soda dissolved away the membrane-like windows that connected the foam network, leaving only a skeletal structure. The newly developed material did not fulfill the original hope of holding water but was later developed for use as air conditioner filters. These filters have the advantage of being easily cleaned for reuse.

Foams of pore sizes ranging from 10 to 100 pores per linear inch are available in reticulated foams. This offers a considerable advantage to the designer who, for example, can provide three-phase filters merely by combining three different pore-size foams. Reticular urethane foams are illustrated graphically in Fig. 7.3. The polyhedral structure is apparent. On average, the polyhedra are pentagonal dodecahedra.

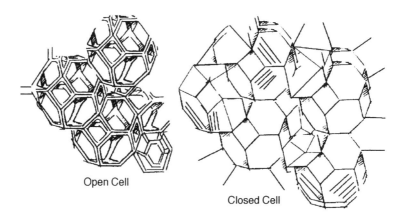

Open Cell

Closed Cell

Figure 7.3 Reticulated foam: (left) open cell, showing its polyhedric three-dimensional structure, and (right) closed cell, before caustic soda treatment.

These foams can be shaped with a reciprocating saw or hot-wire cutting device. They can be nailed, stapled, sewn, glued, laminated, taped to themselves or other materials, framed, or used as filter pads without additional fabrication. The foams resist damage from abrasion, aging, tearing, and most chemicals, and are unaffected within a –50 to +250° F range. They do not to crack, peel, or wear thin. Applications, in addition to air conditioner filters, include filters for warm-air furnaces, automobile carburetors, electronic computers, lawn mower, motors, vacuum cleaners, ventilators, movie projectors, and air compressors.

Reticular urethane foams are used by airlines to separate water from gasoline and jet fuel and as microphone covers to block undesirable noises. They are also used in consumer markets as blackboard erasers, paint brushes, scrub pads, and lint-free wipers. A compressed form of reticular urethane foam is available that is squeezed to as little as 1/15th of its original bulk. It is used for high-efficiency filtering, such as removing invisible oil vapor from the air. Uncompressed foams may also be *flocked* with fibers, combined with cotton, and coated with polyethylene for high chemical resistance.

In the original reticulated urethane foams, caustic solution was used to remove the cell membranes by hydrolysis. While this process effectively produced satisfactory foams, there were a number of disadvantages. The alkali had to be neutralized after application and the reticulated foam washed and dried. Furthermore, this process worked well only with the flexible polyether polyurethane cellular materials. These are inherently more expensive than polyesters. In U.S. Patent 3,175,025, "Process for Bonding and/or Reticulation," assigned to Chemotronics, Inc., a process is described for the reticulation of polyurethane foams by providing a combustible mixture of an oxidizer material (oxygen) and an oxidizable material (natural gas). In the process, air is first removed by applying a vacuum, and a combustible mixture of 2:1 oxygen:natural gas is introduced into the evacuated chamber to a pressure of 1/2 atmosphere absolute. The gas mixture is then ignited with a spark, and the resultant explosion tears out the cell membranes. The explosion reticulation process was exclusively licensed to the Scott Paper Company.

The Scott Paper Company used both the alkali hydrolysis process to produce what is called *Q-Foam*, and the explosion or thermal reticulation process to produce *Z Foam*. Thermal reticulation process provides a cleaner foam, with no residue of incompletely hydrolyzed urethane polymer. To make apparel foam used as interliners for clothing, Scott designed a centrifugal casting machine that continuously spins out a four-foot diameter cylinder of foam which is automatically cut into sections of any desired length. Cutting is accomplished by peeling off a continuous slice from the cast foam. Reticular foams have also been used to reduce the possibility of explosion in aircraft fuel tanks. This is possible because the fuel readily fills the 97% void space and passes easily through the foam structure.

7.2.4 Microcellular Foam

Microcellular urethanes are high-density (22 to 38 lb/ft³) elastomers of cellular composition designed for heavy-duty mechanical applications. They combine the advantages of a cellular material with those of solid cast elastomers. One of these advantages is linear load deformation. Load deformation curves are essentially straight to 40 or 50% deflection, permitting spring responses that are difficult to achieve with other elastomers.

Compressive deflection sufficient to cause an 80% transverse bulge in rubber causes only a 30% bulge in microcellular urethane, a fact that is important where there are space limitations or mounting problems. When loaded rapidly, either one time or cyclically, microcellular urethane exhibits a dynamic load deflection response that is significantly stiffer than the corresponding static response. The shape of the dynamic curve will vary with specific conditions. However, dynamic moduli several times higher than the static modulus are typical. Compression set is 6% or less.

Microcellular urethanes are cast under pressure in closed molds. The interior structures exhibit a combination of open and closed cells—predominantly open. The surfaces of parts made from these materials are denser than the body. These materials are highly resistant to grease, oil, and gasoline. They are not recommended for use with strong acids, bases, and some solvents. Microcellular urethanes can be sawed, sliced, or ground. They cannot be turned or milled. Holes can be cut with a hole saw, sharpened tube, wood augur, or a drill bit with a concave end. Coolant is sometimes required. The material becomes sticky if overheated.

Typical applications for microcellular urethanes include:

- *Springs for vehicles.* Examples include primary springs, auxiliary springs, and jounce bumpers. Such springs offer soft initial response, followed by a smooth "bumpless" transition into a firm final response for overloads. These springs are unaffected by oil and grease.

- *Impact bumpers.* These range from limit stops in delicate instruments to extremely heavy-duty applications, such as exerting more than 500,000 lb peak force in halting the movement of a 200-ton rail crane.

- *Seals and gaskets.* High compressibility, small lateral expansion, and resistance to abrasion and environmental attack make these materials ideal for many seal applications, such as for ball joints in the suspension systems of automobiles, trucks and other vehicles.

- *Vibration isolators.* Microcellular urethanes exhibit exceptional noise and vibration isolating properties. The lower-density grades are ideal for light loads such as instruments, and the higher-density grades for machinery mounts and motor mounts.

- *Wheels and rollers.* Conveyor components made from these materials combine gentle, quiet handling with resistance to cuts, tears, abrasion, and lubricants.

- Other miscellaneous applications include recoil absorbers, footwear components, and orthopedic devices.

7.2.5 High-Resiliency Foam

Molded foams of the high-resilience (HR) type are characterized by high sag factors and improved hysteresis curves, compared to the usual slab foams. These properties make high-resiliency foams the choice for many seating applications. Polyurethane foams are also used as sponges and for other uses that require liquid absorption properties, such as specialty packaging and personal care and hygiene items. Typically, the foams are made in the form of slabs, which are cut to shape, or they are molded to specific needs. Molded foams are commonly of the high-resilience (HR) type and are characterized by high sag factors and improved hysteresis curves compared to the usual slab foams. HR foams are often crushed to give the foams sufficient *breathability,* since HR foams tend to have a high percentage of closed cells.

McGovern[1] disclosed the use of high-resiliency polyurethane foams with improved static fatigue properties in U.S. Pat. 5,157,056, Oct. 20, 1992. The invention relates to the synthesis of high-resiliency (HR) polyurethane compositions that use flame retardant additives to improve static fatigue properties and provide flame retardance.

The McGovern patent disclosed a high-resiliency (HR) polyurethane resin foam composition that contains phosphorus compounds to improve the static fatigue properties thereof. An unusual benefit of using certain phosphorus compounds in HR foams is that improved compression set characteristics are obtained. As might be expected, flammability properties are also improved. Some preferred phosphorus additives include phosphonates, phosphites, and phosphine oxides such as tri(beta-chloroethyl) phosphate, tri(beta-chloropropyl) phosphate, dichloroethyl methylphosphate, tetrakis(2-chloroethyl) ethylene diphosphate, dimethyl methylphosphonate, diphenyl methylphosphate, triethylphosphate, tricresylphosphate (TCP), triphenylphosphite, triethylphosphite, triethylphosphine oxide, triphenyl phosphine oxide, and mixtures.

7.2.6 Flame-Retardant Foam

There are many occasions when polyurethane foams need to be flame retardant, for both technical and legislated reasons. One of the most effective methods of flame retarding polyurethane foams is to incorporate halogenated compounds into the molecular composition. Traditionally, most flame retardants, although efficient in their function of retarding open-flame combustion in a polyurethane foam, promote smolder, thermal degradation (scorch), and hydrolysis. It is essential that one or more of these tendencies be minimized in certain foam types or in certain applications areas.

This is particularly true of polyurethane foam produced from polyester polyols. These polyols tend to degrade under humid conditions and to degrade quickly when thermal decomposition or hydrolysis products of flame retardants are present. It is also true that polyurethane foams for certain applications, such as used in furniture sold in the State of California, need to be resistant to smoldering ignition due to contact with a cigarette or other ignition source. In addition, in the case of polyester-based polyurethane foams, hydrolytic stability of additives is another important criterion that must be met.

Polyurethane foams are used primarily for insulation and cushioning. Since they are thermoset polymers, additives such as flame retardants are generally incorporated during polymerization. Equipment for producing polyurethane foams is generally designed to handle liquids, discouraging the use of solid flame retardant additives.

A recent patent awarded to Favstritsky[2] presents a thorough discussion of this subject. The invention relates to polyurethane foams having improved resistance to scorch, smoldering, and hydrolysis and, in particular, it relates to polyurethane foams incorporating polybrominated higher alkylbenzenes and, optionally, triaryl phosphate flame retardants.

A principal object of the Favstritsky invention is to provide flame-retardant polyurethane foam compositions that do not exhibit scorching or smoldering problems and that exhibit increased hydrolytic stability. Another objective is to provide polyether and polyester type polyurethane foams incorporating a superior flame-retardant agent. Yet a further objective is to utilize polybrominated higher alkylbenzenes as flame-retardant additives in polyether- and polyester-type polyurethane foam compositions.

7.3   Foam Preparation

Polyurethane foams can be prepared by three basic approaches: one-shot, quasiprepolymer, and prepolymer. These variations are discussed below

7.3.1 One-Shot Method

In the one-shot method, all the components (polyol, isocyanate, surfactant, water or volatile solvent, and catalyst) are combined at one time to produce a foam, and the reaction is completed in one shot, as shown in Fig. 7.4. Since the reaction producing the foam is an extremely rapid one, and since the viscosities and other physical relationships between all five materials are not always easy to handle, it is desirable to reduce the number of ingredients being mixed at one time. This is the reason for the development of the quasiprepolymer method.

The possibility of direct contact with the toxic and irritating diisocyanate is fairly strong with this system. Also, one-shot foam reactions produce a very high exothermic temperature foam reactions produce a very high exothermic temperature so that, in many cases, the center or core of the formed part is charred. This is especially true of larger foam parts.

The traditional one-shot procedure has been widely used in making flexible foam buns. The primary objection to the one-shot system in the past has been the expensive equipment and nonreproducibility of the foam. Since the one-shot process is potentially the cheapest process, this process is generally adopted wherever feasible. The process dates to the development of the trimethylene diamine catalyst (DABCO) and was given additional impetus with the development of dialkyltin and polyalkyl organotin compounds that, along with stannous and stannic compounds, were found to be highly favorable in one-step foaming reactions.

Tests have shown that foams prepared by the one-shot system can be made equal or superior to prepolymer foams, where the reaction took from 30 min to 4 hr. This is especially true with regard to density, tensile strength, elongation, load indentation, constant deflection, compression set, and compression load deflection.

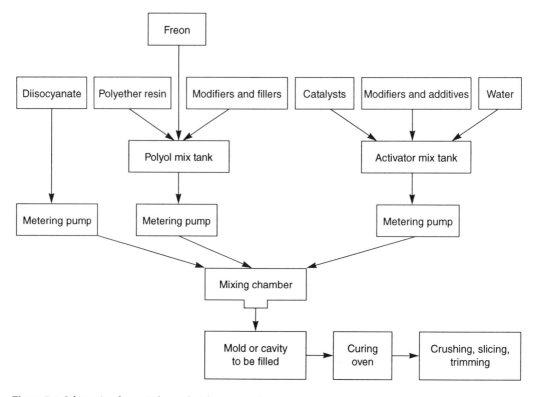

Figure 7.4 Schematic of a typical one-shot foam manufacture sequence.

A typical formulation for a one-shot flexible foam is as follows:

Component	Parts by Weight
Polyether triol, 3000 MW	100
80:20 TDI	38
Stannous octoate	0.5
N-ethyl morpholine	0.5–1.0
Tetramethyl 1,3 butane diamine	0.1
Silicone copolymer surfactant	1.0
Water	2.9

The above components are combined so that they are pumped into the foam machine mix head in four streams as follows:

1. Polyether triol
2. 80/20 TDI
3. Stannous octoate and ethyl morpholine
4. Diamine, water, and silicone copolymer

Today, virtually all large volume production of flexible polyurethane foam utilizes the one-shot approach. Prepolymer or quasiprepolymer steps are eliminated. This approach became possible only after the development of hydrolytically stable organotin catalysts, introduced in 1958. For ease of metering, components that do not react with each other are frequently preblended to cut down the number of streams that must be used. The advantages of the approach are that it is the least expensive chemical method, it is possible to make very low-density foams from a given number of reactants, there is greater versatility, and the resultant foam properties are better. Disadvantages include greater sensitivity to operating conditions, a greater need for metering accuracy in the foaming machinery, and a greater isocyanate hazard.

7.3.2 Quasiprepolymer Method

In this approach, part of the polyol to be used in the formulation is prereacted with all of the isocyanate. The resultant product is NCO terminated as is a prepolymer; however, the free NCO content of the quasiprepolymer is much higher. Foams are prepared by adding water, catalysts, surfactants, and the remaining polyol. Figure 7.5 is a schematic diagram of the quasiprepolymer (adduct) method.

The quasiprepolymer approach finds favor where the isocyanate of choice is either a solid at normal operating conditions or when a modification will result in improved properties or processing of the resultant foam. Quasiprepolymers also have been made with polyols or chain extenders that are difficult to process when used alone. In many applications, the high free NCO-terminated products are handled as though they were basic isocyanates.

7.3.3 Prepolymer Method

In the prepolymer foaming process, used mostly for flexible foams, the hydroxyl compound is reacted with an excess of isocyanate (NCO/OH ratio of about 2:1) to form an isocyanate-terminated prepolymer. The chemical reaction is shown below for the case of a trifunctional resin:

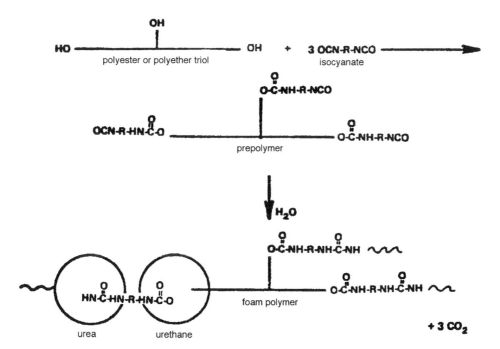

When CO_2 is used as the blowing agent, the prepolymer is further reacted with water to yield the foamed polymer which contains substituted urea groups in addition to urethane groups, as shown above. In this process, the catalyst (water and amine) is mixed into the prepolymer to effect foaming. Additives, e.g., emulsifying agents and colorants, may be added to the catalyst mix. The reaction between the polyol and the isocyanate is generally carried out in the absence of a catalyst. Figure 7.6 is a schematic of a typical prepolymer-based foam manufacture.

In prepolymer reactions, the water content of the polyol-TDI system is very critical because of the preferential reactivity of water with isocyanate groups. When water is added to the urethane prepolymer in the presence of a suitable catalyst, it reacts simultaneously with the prepolymer terminal isocyanate

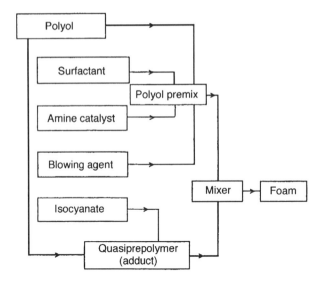

Figure 7.5 Schematic of a typical quasiprepolymer foam method.

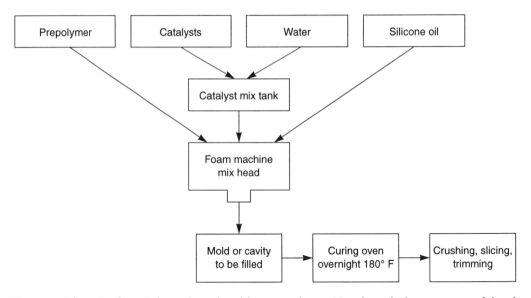

Figure 7.6 Schematic of a typical prepolymer-based foam manufacture. Note that only three streams are fed to the mix head.

groups and the free TDI monomer. The reaction with an isocyanate-terminated prepolymer results primarily in viscosity build-up and cross-linking, while the reaction of the TDI with water is the main source of carbon dioxide, the blowing agent.

In this method, a resin manufacturer markets the partially polymerized diisocyanate resin in which there is very little (approximately 5%) free diisocyanate, and the catalyst mix, which includes surfactants and other additives. Usually, the ratio of catalyst-surfactant mixture to prepolymer is about 2–10:100. By mixing the catalyst and the prepolymer resin, the end user produces the foamed part.

The prepolymer system is relatively unimportant for rigid foam manufacture, with the possible exception of castor-oil based foams. This is mainly due to the fact that relatively high viscosities are obtained in the reaction of low-equivalent-weight polyols with diisocyanate. For instance, a triol having an equivalent weight of about 250 gives a prepolymer with a viscosity of about 100,000 cps at room temperature.

As the equivalent weight of the polyol decreases, the viscosity rises sharply. Although high-viscosity prepolymers can be used, lower viscosities of under 10,000 cps are generally preferred for machine processing. For this reason, the semiprepolymer method has almost completely replaced the complete prepolymer method.

Prepolymers were used primarily with polyesters when less effective catalysts than those presently available were in use. The prepolymers may serve many specific purposes, however, especially with regard to the highly exothermic reactions that take place during final polymerization. Since the polymerization reaction is partially complete with the use of a prepolymer, less heat of reaction is generated in foaming than in the quasiprepolymer or one-shot method. This is particularly important in foams with densities of 6 lb/ft^3 or higher. Cure temperatures of such foams can reach a level high enough to cause internal scorching or charring.

In this technique, all the polyol to be used in the formulation of the foam is prereacted with all of the isocyanate at any period of time prior to the foaming reaction. This generally results in a high-viscosity, high-molecular-weight product dissolved in a slight excess of isocyanate. The reaction may be carried out at room temperature with the addition of a catalyst or by heating under controlled temperature conditions. The prepolymer is foamed by the addition of water, usually in the presence of amine catalysts, surface active agents, and emulsifiers. This approach was the original method of foam preparation. Today, very little foam is made in this way; however, the basic concept is used to make adhesives and coatings. Prepolymers have the advantage of being less sensitive to operating conditions than the other methods, there is less of an isocyanate hazard, and the foam can be made with relatively simple equipment. There are also disadvantages: these foams are more expensive, there is limited stability (since they can react with atmospheric moisture), curing times are long, and there is limited versatility for a given prepolymer.

7.4 Formulating Principles

Table 7.2 lists the various components that may be found in a flexible foam formulation. The exact components and the quantities to be used depend on the grade of foam required; therefore, ranges given are typical. In the table, the polyol set at 100 parts by weight. This is basically standard convention for convenience.

TABLE 7.2 Formulation Basis for Flexible Foams

Component	Parts by Weight (pbw)
Polyol	100
Water	1.5–7.5
Auxiliary blowing agent	0–35
Inorganic filler	0–400
Amine catalyst	0.1–1.0
Organotin catalyst	0–0.5
Silicone surfactant	0–2.5
Isocyanate	25–60
Additives:	Variable
Flame retardant	
Pigment	
Bacteriostat	
Plasticizer	
Antistatic agent	
UV stabilizer	
Cell opener	

7.4.1 Isocyanates for Flexible Foams

The isocyanate moiety is used as the source of the NCO groups that react with the hydroxyl groups from the polyol and water in the formulation. For flexible foams, the most commonly used isocyanate is toluene diisocyanate (TDI). In the United States, the 80/20 isomer ratio of 2,4/2,6 TDI is by far the most commonly used. In contrast, the 65/35 TDI is used for some specific applications and is the more widely used in Europe. 65/35 TDI will give foams with higher load-bearing characteristics, but significant changes are normally required in catalysis, since the reactivity of this isomer ratio is less than that of the 80/20 isomer ratio. The acidity of TDI also affects reactivity. The acidity of normal production grade TDI is 0.002 to 0.005 percent HCl. For prepolymer use, acidity is typically adjusted to 0.008 to 0.011 percent. Higher acidity reduces the reactivity of the isocyanate.

Diphenylmethane diisocyanate (MDI), in various forms has found increasing use with time, particularly in high-resiliency flexible, semiflexible, and microcellular foam. Since pure MDI is a solid at room temperature, it is used either as a crude product (also referred to as polymeric MDI) or as modified products that are liquid and offer processing and/or property enhancements. Table 7.3 lists some of the various types of isocyanate used in specific flexible foam applications.

TABLE 7.3 Isocyanates Used in Flexible Foams

Hot-cure	80/20 and 65/35 TDI
Cold-cure and high-resiliency	TDI, modified MDI, TDI-prepolymers
Semiflexible	Crude MDI, TDI prepolymers, crude TDI
Microcellular	Modified MDI, MDI, MDI prepolymers

At present, the 80/20 ratio is the most popular in the United States. MDI is second in importance to TDI. The pure grade of MDI is used in textiles, elastomers, and coatings. Some of the differences are shown in Table 4.4.

TABLE 7.4 Differences between TDI and MDI Isocyanates

Group 1	Group 2
Toluene diisocyanate	Methylene bis (4-phenyl isocyanate)
(a) Distilled	*(a) 100% distilled*
100% 2,4 isomer	Solid
80% 2,4 isomer: 20% 2,6 isomer	Flakes
65% 2,4 isomer: 35% 2,6 isomer	
100% 2,6 isomer	
(b) Crude grades	*(b) Crude grades (undistilled)*
	PAPI
	"Mondur" series

The diisocyanates are employed in chemical reactions because they will react with and add to any chemical compound containing an active hydrogen. The most important criterion in designing a foam formulation is the stoichiometry involved. The basic value in all urethane calculations is the amine equivalent of a given isocyanate compound. The amine equivalent of any isocyanate compound (pure or derivative) is the measure of its ability to react as an isocyanate; thus, it is the same as its apparent equivalent chemical weight in the ordinary sense. Since the molecular weight of toluene diisocyanate is 174, its chemical equivalent weight is 174/2, or 87, as is its amine equivalent (AE).

The amine equivalent is defined as the number of grams of an isocyanate consumed by 1 g-mol of a secondary amine in the formation of the corresponding urea. A simple chemical titration method can be used for the determination of the amine equivalent of an isocyanate monomer. To find the amount of isocyanate needed for a normal foam formulation, it is necessary to know the quantity of hydroxyl (OH) and carboxyl (COOH), as well as the water content available to react with the isocyanate. This calculation can be done as follows:

$$\frac{(OH + COOH)(AE)}{(10)(56.1)} \text{ (grams of isocyanate per 100 g of material)}$$

The calculation of the quantity of isocyanate compound needed for the water content takes into account the fact that 1 mole of water reacts with two equivalents of isocyanate; hence, water has an equivalent weight of 9. Its requirements are as follows:

$$\frac{(\% \text{ water content})(\text{AE})}{9} \text{ (grams of isocyanate compound per 100 g of material)}$$

These equations, when used for a given resin with a given water content, will determine the theoretical quantity of isocyanate compound (whether pure diisocyanate or derivative) necessary for this resin. Often, an excess of isocyanate, usually 0.5%, is used This basic tape of calculation can be used for all isocyanate systems if the hydroxyl, carboxyl, and water content of each ingredient is taken in consideration.

As a general rule, for any rigid foam system, a small excess of isocyanate is favorable for heat distortion characteristics. Increasing the excess tends to improve the heat distortion but renders the foam more brittle. Not all foams will absorb large excesses of isocyanate; this excess acts as a plasticizer, increasing the density and requiring a longer cure at high temperatures to bind it into a foam. This is not always feasible, and it may be said that the maximum excess possible depends on the resin structure and its water content.

In the U.S. flexible foam field, the 80/20 mixture of 2,4 and 2,6 TDI is by far the most commonly isocyanate used. The 65/35 ratio is quite common in Europe, but MDI has been championed by claims of better dimensional stability of the resulting foams. Additional diisocyanates used in flexible urethane foams are crude grades of TDI and polyisocyanates.

The crude isocyanates are essentially crude grades of TDI and, as such, are less expensive than TDI and produce light brown foams. TDI teams are pure white. Most TDI-based foams are limited to use temperatures in the 200 to 250° F range.

Two polymeric MDI-based polyisocyanates are currently being marketed, trade named *PAPI* and *Mondur MR.* These isocyanate being polymeric have significantly lower vapor pressures and therefore minimize toxicity problems encountered with isocyanates. They also produce foams with enhanced flame retardance. PAPI-based systems can be used to produce foams with good dimensional stability and structural strength at temperatures as high as 300 to 400° F.

The isocyanate/hydroxy ratio (NCO/OH ratio) is of prime importance and is often expressed as the *isocyanate index* ([NCO]/[HO] × 100). The isocyanate index of commercial rigid foams usually ranges between 105 and 110, with 105 being the preferred range for the best all-around properties. Using a larger excess of -NCO usually tends to give higher strength but also a more brittle foam. Generally, the best humid aging properties are obtained at an isocyanate index between 105 and 100. Also, by using a slight excess of NCO, the tendency toward foam shrinkage is lessened.

Certain relationships exist between molecular structure of diisocyanate used in prepolymer preparation and the load-bearing ability of the finished flexible foam. The rigid, bulky diaromatic structures, such as those typified by TDI and 1,5 NDI, produce flexible foams with the highest load-bearing capacity. Intermediate values are obtained with MDI. The benzene type, such as TDI isomers, tend to produce lower load-bearing foams. Thus, the rigidity of the diisocyanate molecule governs the hardness of the foam. The diaromatic rings are believed to produce firmness by causing a hindrance to rotation within the polymer.

An increase in the ratio of 2,6 TDI isomer to 2,4 TDI isomer was shown to give an increase in both the compression set and load-bearing capacity of the finished foams. This relationship of increased load-bearing capacity to increased compression set is exhibited by all foams.

The effect of varying diisocyanate isomer ratios in polyurethane foams is shown in Table 7.5. This table lists some of the physical properties obtained with the formulation shown in the top portion of the table. This formulation was used in a foam machine.

The 100% 2,4 TDI resulted in excessive shrinkage, which prohibited sampling for any physical properties. The 75/25 TDI ratio area resulted in the foam with the highest density (5.6 lb/ft^3). The lowest-

TABLE 7.5 Effects of Varying TDI Isomer Ratios

Component	Parts by Weight					
Multron 118 polyester	90.00					
DABCO catalyst	1.50					
N-ethyl morpholine	1.50					
Non-ionic emulsifier	1.00					
TDI (isomer, as shown below)	28.75					
% 2,4 TDI	% 2,6 TDI	Density (lb/cu^3)	% Compression set	Tear (lb/in)	% Rebound (Schoppe)	% Elongation
65	35	4.3	14.2	5.2	23	223
70	30	4.25	14.5	3.7	43	248
75	25	5.6	19.4	3.1	37	256
80	20	3.7	11.5	4.0	30	245
85	15	3.4	13.4	4.1	29	230
90	10	5.0	16.3	3.0	26	233
100	0	Excessive Shrinkage				

density foams were obtained using TDI containing from 80 to 85% of the 2,4 isomer. The compression set values were also highest in the 75/25 isomer area, with the lowest values obtained in the TDI with 2,4 isomer content of 80 to 85%.

Increasing the 2,6 isomer content of a TDI isomer mixture resulted in increased branching in the polymer. Increased branching leads to an increase in foam firmness of foams. This can be seen in Table 7.3 as the TDI ratio changes from 80/20 to 65/35, as evidenced by the increase in the percent rebound.

The effect of different isomer ratios in rigid urethane foams is slight, as opposed to flexible foams. The isocyanate serves as a di- or polyfunctional reactant, joining polyol molecules together by way of reaction between isocyanate and hydroxyl groups. Thus, in rigid foams, the entire system builds itself into a highly cross-linked polymer. This reaction is strongly exothermic and consequently provides the heat required to vaporize the inert blowing agent, and also provides the heat required to ensure a good cure of the foam. In water-blown foams, the isocyanate reacts with the water to provide CO_2 for the blowing reaction. The polyurea linkages produced by this reaction become part of the polymer chain.

7.4.2 Polyhydroxy Compounds (Polyols)

Although the isocyanates are the key functional groups in the formation of urethane polymers, the polyhydroxy compounds that furnish the hydroxyl group for the reaction crucially influence the properties of the final urethane polymer. The functionality (number of reactive groups) of the polyol determines the character and the degree of cross-linking or branching. The equivalent weight of the polyol determines the physical properties of the foam.

Water is also a source of hydroxyl groups but, because of the low molecular weight, its primary function is to react with the isocyanate to generate carbon dioxide that acts as the blowing agent for the foam. Cross-linking also occurs as a result of the water-isocyanate reaction. Above about 4.0 pbw, care must be taken to avoid high exotherms that can lead to foam discoloration or auto-ignition.

The sources of hydroxyl groups for almost all commercial uses of urethane polymers are:

1. polyethers
2. polyesters
3. naturally-occurring hydroxyl-bearing oils such as castor oil

Of these, the polyethers (mainly polyoxypropylene) are the most important.

Although polyoxypropylene diols have some application, the polyethers primarily used are triols or polyols or higher functionality. They are based on such starter molecules as glycerine, trimethylpropane, sorbitol, methyl glucoside, and sucrose. Polyesters based on dibasic acids, such as adipic acid, and polyols, such as diethylene glycol, have been largely replaced by polyethers, particularly in foam applications.

Nevertheless, polyesters are still important in some applications, such as foam laminates for fabrics. Castor oil used in foams is minor.

The choice of the polyol has a major influence on the physical properties of the resultant foam. The polyol determines whether the foam will be rigid or flexible, brittle or non-brittle, and the extent of its permeability to gas and moisture. Flexible foam is produced from polyols of moderately high molecular weight and low degree of branching, while rigid foams are prepared from lower-molecular-weight, highly branched resins

Present-day flexible urethane foams are formulated chiefly from polyethers. The polyether foams cost less than polyester types and have more desirable properties for seating applications. These foams are based on polyoxypropylene diol of 2000 molecular weight and polyoxypropylene triols with molecular weights up to 4000. These triols generally use glycerine as the precursor molecule. The polyether foams are used for cushioning, crash pads, arm rests, automotive seating, toys, novelties, and other slab or molding applications.

The polyol combined with the isocyanate forms the polymer network that provides the rigid foam with its structure and characteristic properties. The structure can be varied to give rigidity to very flexible foams. The most important characteristics of the polyol are its equivalent weight, functionality, and rigidity or flexibility of chain units.

In general, in rigid foams, the reaction exotherm increases as the equivalent weight of the polyol decreases—hence a greater tendency toward scorching of the foam. The compressive strength and dimensional stability increases with decreasing equivalent weight while the tensile strength decreases. However. as the equivalent weight decreases (i.e., the hydroxyl number increases), the friability of the foams increases. The combined effect of polyether hydroxyl number on the dimensional stability and friability of a typical one-shot polyether foam is shown in Fig. 7.7.

Since it is apparent from Fig. 7.7 that dimensional stability and friability vary in opposite directions with the polyether hydroxyl number, a compromise must be effected to strike a suitable balance of properties. The MW range of hydroxyl number also varies with the specific type of polyether. Decreasing

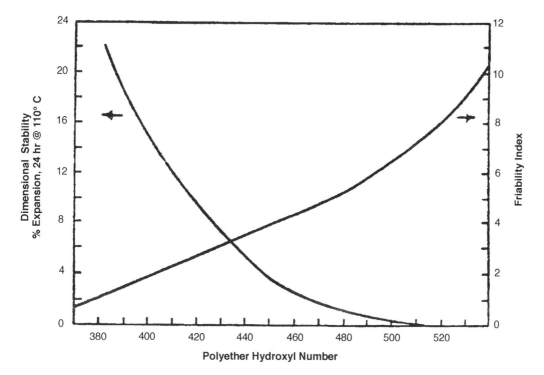

Figure 7.7 Effect of polyether hydroxyl number on dimensional stability and friability of a typical one-shot foam.

the equivalent of the polyol(s) also results in foams with lower water vapor permeability and lower water absorption.

The functionality of polyols also has a profound effect on the properties of rigid foams. Higher functionality favors greater heat resistance and dimensional stability (assuming that the equivalent weight is the same in this comparison). The compressive strength of the foams usually increases with increased functionality, while the tensile strength and elongation tend to decrease. Table 7.6 shows the relationship between the functionality of polyethers and compression/tensile strength properties of the resulting foams.

TABLE 7.6 Foam Strength vs. Functionality of Polyether Polyol

Hydroxyl number	450	490	450
Functionality	4	6	8
Polyol type	pentaerythritol	sorbitol	sucrose
Component "A"			
Polyester	65.0	73.4	88.9
Trichlorofluoromethane	27.0	35.0	28.4
Stannous octoate	0.5	0.8	0.6
Silicone DC-113	0.1	0.01	0.001
	0.5	1.1	0.6
Component "B"			
Quasiprepolymer	100.0	100.0	100.0
Density (lb/cu ft)		1.9	2.1
Compression strength (psi)	32.0	29.8	55.0
Tensile strength (psi)	60.0	46.9	34.0

The rigidity of the polyol component also affects foam properties. For example, in polyether systems, the use of a polyether having a cyclic initiator such as α-methyl glucoside usually leads to better temperature resistance than a polyether of equal functionality and equivalent weight based on pentaerythritol.

In general, polyethers give softer, more resilient foams with better hydrolysis resistance than polyesters, while the polyester-based foams may be expected to have greater tensile strength and better resistance to oils, solvents, and oxidation. With either type of poly, an increase in the degree of branching results in increased load-bearing capacity, lower elongation, and a higher glass transition temperature. A reduction in the equivalent weight of the resin usually leads to higher tensile strength, higher load-bearing capacity, and an increase in the glass transition temperature.

To impart flame resistance to urethane foams, special polyols have been developed containing either halogen or phosphorous or both. In the case of polyesters, the use of chlorendic anhydride produces polyols that result in flame-resistant rigid foams. A number of phosphorous-containing polyols, which can be chemically defined either as phosphates, phosphites, or phosphonates, are commercially available. These polyols may be used in conjunction with polyesters but are more often blended with highly branched polyether polyols.

Polyesters

Although polyesters have been largely replaced by polyethers in most major flexible urethane foam markets in the United States, they are still an important factor in Europe as well as in special applications in the United States. They are primarily adipate esters of diols and triols, with the triol concentration depending on the desired degree of branching. Foams prepared from dimer-acid polyesters show improved hydrolytic stability over those made from adipate polyesters. This is because the former are more hydrophobic in nature and contain a lower weight percentage of ester groups. Flexible foams derived from dimer-acid polyesters have excellent load-bearing capacity at low density, excellent resilience, good low-temperature resistance, and very low compression set.

Hydroxyl-terminated polyesters with varying degrees of branching were the first polyols used in rigid urethane foams. Depending on the amount of triol (e.g., glycerine, trimethylpropane, etc.) in the polyester formulation, foams with varying degrees of cross-linking may be obtained. Polyesters give foams with high strength and a high percentage of closed cells. Depending on the choice of polyol used, they may possess primary hydroxyl groups so that reactions with isocyanates proceeds more rapidly to completion. Polyester foams are usually foamed by the quasiprepolymer and one-shot methods.

The polyesters cannot compete with polyethers in cost, and all suffer the same basic limitation of having a predominance of ester linkages. The formation of the ester linkage involves a condensation reaction between an alcohol and an acid and the elimination of a mole of water per equivalent of acid and alcohol. Alkali resistance, as well as hydrolytic stability, of the ester linkage are generally poor. These failings are multiplied in practice, since the polyester always contains unreacted carboxyls that further reduce chemical resistance. Commercially produced polyesters also have a greater molecular weight range as well as some unreacted starting materials and water of esterification. The greatest limitation is probably the high viscosity of the polyester, necessitating the manufacture of intermediate prepolymers previous to foaming. The one-shot systems that circumvent the need for prepolymers are much more economical.

In spite of these shortcomings, the hydroxyl-terminated polyesters command a particular advantage in that they, and the foams made from them, can be formulated to provide fire retardance or self-extinguishing characteristics. Several types of halogenated dibasic acids, particularly chlorinated phthalic acids or anhydrides, are useful in this manner if the halogen content of the polyester is sufficiently high. The halogen content of the polyester obviates the need for fire-retarding agents, such as halogenated plasticizers, which generally tend to degrade the physical properties of the foam. Formulations are somewhat similar, although such polyesters yield more costly foam.

The polyesters most suitable for the production of polyurethane are essentially saturated, of low acid number, and very low water content. Usually they are liquids of 1000–3000 molecular weigh and terminated in two or more hydroxyl groups. Some commercial polyesters designed especially for urethane foam production are listed in Table 7.7.

TABLE 7.7 Polyesters for Flexible Urethane Foam Production

Trade Name	Hydroxyl Number	Acid Number	Viscosity (cps)	Foam type
Multron R-68	45–52	1.2 max	800–900 @ 73° C	Flexible
Multron R-70	50–57	1.2 max	850–100 @ 73° C	Flexible
Multron R-18	57–63	1.2 max	1000–1100 @ 73° C	Flexible
Formrez 50	49–55	2.0 max	17–22000 @ 25° C	Flexible

Polyethers

With the advent of urethane-grade polyethers in 1956–1957, these raw materials became widely used for the production of flexible foams, and to a lesser degree of rigid foams. For rigid foams, the polyethers often used are the propylene oxide adducts of materials such as a sorbitol, sucrose, aromatic diamines, pentaerythritol, and methyl glucoside. These range in hydroxyl numbers from 350 to 600. For flexible foams, polyethers of hydroxyl numbers ranging from 40 to 160 are used. Examples are condensates of polyhydric alcohols such as glycerine, sometimes containing small amounts of ethylene oxide to increase reactivity.

Side chains are also responsible for a decrease in the hydrophilic characteristics of urethanes. Polyethylene glycol, the most hydrophilic, yield foams that are seriously affected by water and moisture. For this reason, polyethylene glycol based urethanes can be used only in applications where water and moisture are not problems. Polypropylene glycol, on the other hand, forms urethane foams that have a high degree of resistance to moisture. Polybutylene glycol produce urethanes that are even more hydrophobic.

The polyethers constitute by far the latest group of hydroxyl-terminated foam intermediates, and the number of commercially available polyethers grows steadily. For prepolymer systems, the diols or combinations of diols with triols are generally being used. For one-shot systems, the triols, either alone or in combination with diols, are the most frequently used.

The polyether polyol must be liquid at room temperature, since the reactions that take place with isocyanates are so rapid that a completely homogeneous distribution of reactants must be obtained in a few seconds. The addition of a solvent, especially trichlorofluoromethane, to the polyether poly is generally recognized as providing foams with the best insulating properties and is almost universally employed in low-density foams. The selection of this solvent reduces the need for very low polyether polyol viscosities and permits the use of highly functional, high-viscosity polyols, which provide a good balance of properties.

The polyether polyols are a low-cost and highly refined source of hydroxyls, because they are a single species, made to have very low levels of contaminants and uniformly narrow molecular weight distribution. Because of other linkages, the polyols and foams containing them do not have an extremely high degree of chemical resistance but exhibit superior hydrolytic stability.

Commercially available polyether polyols usually range in molecular weights from 400 to 6000. The most widely used polyethers for flexible foams, however, are triols having a molecular weight of about 3000. In general, triols of higher molecular weight than 4000 give rise to foams that have larger cell size, are less resilient, and have slightly higher compression sets.

As a rule, using polyols of increasing functionality gives higher compression modulus but decreased tensile strength, tear strength, and elongation. Decreasing the functionality tends to decrease the compression modulus while increasing the other properties. The most widely used trials of 3000 MW have predominantly secondary hydroxyls. Increasing the primary hydroxyl content by capping the polyethers with ethylene oxide increases the rate of cure and may slightly increase the compressive modulus, tensile strength, tear strength, and elongation.

Effect of Polyol Structure

The polyol structure has a direct bearing on both processing and properties of the resultant foam. Flexible foam polyols may be broadly grouped into the following seven categories:

1. Polyoxypropylene diols.
2. Polyoxypropylene triols.
3. Ethylene oxide *tipped* or *capped* polyols.
4. Copolymer polyols, both random and block copolymers, where the polyols are made with both ethylene and propylene oxide. When the oxides are mixed, the polyols are generally referred to as *hetero* or *heteric* polyols.
5. Blended diol/triol mixtures.
6. Graft or polymer polyols that are stable dispersions of a solid particulate polymeric phase in a liquid polyol phase. The solid phase typically has been a copolymer consisting of a vinyl polymer grafted onto a polyether polyol. Similar products are available where the solid phase is a polyurea.
7. Cross-linkers that are typically short chain, polyfunctional polyols added to increase load bearing of the resultant foam. These polyols often are rigid foam polyols based on amines and, therefore, are quite reactive.

Graft or polymer polyols were developed commercially in the early 1970s to meet the requirements of the automotive industry for improved seating foams. These polyols made it possible to produce foams with better comfort characteristics. It was found that these polyols gave significant load bearing, which broadened the usage spectrum.

By definition, graft polyols consist of a stable dispersion of solid graft copolymer in a polyether polyol. The solid polymer particles serve the same purpose in elastomeric polyurethanes as does carbon black in rubber; that is, they reinforce properties such as load bearing, tensile, and tear strength. An important point is that the solid polymer particles are obtained by the polymerization of vinyl monomer(s) in a polyol. The usual vinyl monomer is acrylonitrile or styrene. There is evidence that there are reactive groups on the solid particles that bond them into the foam matrix. If acrylonitrile or styrene homopolymers or acrylonitrile-styrene copolymers are dispersed polyol, and a polyurethane foam is made from it, the desired improvement in load bearing is not obtained. The foam properties are often degraded, not improved.

While the solid particles do enhance the load-bearing capabilities of the resultant foam, they are not the only reason. When a polyol reacts with an isocyanate, a urea hard phase and a urethane soft phase are formed. The ratio of phases is determined by the polyol-to-water ratio. If a graft polyol is substituted for a conventional polyol, and the same amount of water is used, the base polyol-to-water ratio is increased, and the amount of urea hard phase increases. This also increases load bearing. This effect can be used to advantage in formulating specific types of foams. The various uses of graft polyols are listed in Table 7.8.

TABLE 7.8 Uses of Graft Polyols

Molded high-resiliency foams
High-resiliency slab stock foams
High load-bearing slab stock foams
Semiflexible foams
Foamed elastomers
Coatings

Foam Catalysts

Polyurethane foams are characterized by a wide variety of polymer structures that address a broad range of commercial needs. Both the gelling and the blowing reaction are profoundly influenced by the type and level of catalyst selected. Catalyst selection influences the overall reactivity of the foam system, and the selectivity toward specific reactions. Reactivity of the system is measured by pot-life time for the system, the cure profile, and the demold (cure) time. Catalyst selectivity impacts the reaction balance, type, and order of polymer linkages formed, as well as the flowability of the foam system. Catalysts also affect the ultimate physical properties of the resultant foam.

Amine catalysts primarily affect the reaction between the water and the isocyanate. They are often referred to as the *blowing* catalysts. However, they do have varying degrees of activity in promoting the reaction between the polyol and the isocyanate. This is particularly true when the polyol has high inherent reactivity. In some formulations, the amine catalyst may be the only catalyst used. There are also some non-amine catalysts, such as some antimony compounds, which can act as the blowing catalyst.

The catalysts most commonly used with polyurethane foams are tertiary amines, quaternary amines, amine salts and metal carboxylates (such as SnII, SnIV, or K^+). Tertiary amines display high specificity toward the blowing reaction while also influencing the gelling and cross-linking reactions. Amine salts and thermally sensitive amines such as diazabicycloundecane are used to provide delayed reactions.

Tin catalysts are a family of organotin compounds used to promote the polyol-isocyanate reaction. As such, they are referred to as the *gel* catalysts. As in the case of the amine catalysts, there are formulations that use only an organotin catalyst. The metal carboxylates strongly favor the gelling reaction. Stannous compounds (SnII) offer low cost, but they are hydrolytically unstable. They are typically used in applications that have the capability to meter a separate catalyst stream, such as flexible slab stock. Stannic compounds (SnIV) have greater hydrolytic resistance and can be formulated into systems such as flexible molded foams.

The trimerization reaction is strongly favored by the selection of certain compounds such as quaternary amines, potassium carboxylates, tris(dimethyl aminomethyl) phenol, and 2,4,6 tris [3-(dimethylamino) propyl] hexahydro-s-triazine.

7.4.3 Catalyst Balance

The reaction between an isocyanate and the hydroxyl group in a poly will proceed without a catalyst, but at too slow a rate to be practical. Without a catalyst, a foam may expand, but it may not cure adequately to give good physical properties. The urethane reaction is catalyzed by basic materials. Tertiary amines, such as trimethylenediamine, tetramethyibutane diamine, diethylaminoethanol, and triethylamine, have been widely used.

In rigid urethane foaming systems using the CO_2 formed by the water-isocyanate reaction, a balance of the relative rates between the urea and urethane reactions is necessary. If the urethane reaction is not fast enough, the gas will not be trapped, and no foam will be formed. On the other hand, if the urethane reaction is too fast, the polymer will set up before the gas is formed, and a high-density team will result. The foaming reaction is much less controllable by catalytic action than the urethane reaction.

Tertiary amines can be used alone as catalysts but, for some applications, such as spraying, more speed is desirable. Metal salts, particularly tin salts, accelerate the foaming reactions and can be used alone or in combination with the tertiary amine-type catalysts. Tin catalysts of importance for rigid urethane foaming are stannous octoate and dibutyl tin dilaurate. Stannous octoate will hydrolyze rapidly in the presence of a basic catalyst with loss of activity. Master batches containing stannous octoate and water are stable for only a few hours at room temperature.

Resin master batches containing dibutyl tin dilaurate may stay stable for months. For this reason, this catalyst is preferred for foaming systems packaged for use at other locations or plants where the resin master batch is not used immediately.

Delayed-action catalysts have been made successfully. Buffered amine catalysts, where the activity of the amine has bee reduced by the presence of an acid, have also been used. Acidic materials can be used to retard the urethane reaction. Hydrogen chloride and benzoyl chloride have been used in combination with amine-type catalysts to control reaction rates. A small percentage of acid can increase foaming time from 2.2 to 6 min.

Temperature can also be used to control the urethane foaming reactions; higher temperatures accelerate the foaming reactions. Some delayed-action rigid urethane foaming systems have been made by premixing all of the foaming ingredients at very low temperatures (as low as $-300°$ F). When these systems are heated, foaming of the mass takes place.

The vapors of tertiary amine catalysts are irritating, and contact with the skin can cause dermatitis. The catalysts can produce severe irritations by contact with the skin. Their vapors are also irritating. Care must be taken to ensure that the materials in solid, liquid, or vapor form do not come in contact with the human body.

For CO_2 blown foams, tertiary amines as the sole catalyst are adequate. For solvent-blown foams, however, a more reactive catalyst is necessary because of the cooling effect of the solvent. A synergistic action exists between tin catalysts and tertiary amine. Rigid foam systems, because of their greater degree of cross-linking, build gel strength so rapidly that tertiary amine are adequate for one-shot or prepolymer systems using polyethers or polyesters. The structure of the tertiary amine has a considerable influence on its catalytic effect and also on its usefulness for foam production. The catalytic strength generally increases as the basicity of the amine increases and as steric shielding of the amino nitrogen decreases.

The tertiary amine catalysts provide satisfactory foaming with either the one-shot polyester or the polyether prepolymer systems; both are relatively high in initial viscosity. The density and moldability of foam is greatly influenced by the choice of the catalyst system and concentration, due to the effect of catalysts on the foaming rate. The effect of a tertiary amine catalyst (triethylamine) concentration on the foaming rate, density, and compressive strength of a typical rigid polyester urethane foam is seen in Table 7.9.

The choice of catalyst for the preparation of flexible foams is governed by the type of polyol used. In polyether-based systems, where the effects of low resin viscosity and reactivity must be countered, very potent polymerization catalysts are required. For such systems, the preferred catalysts are the stannous octoate salts of dicarboxylic acids (for example, stannous octoate). Tertiary amines, if used alone, are too inactive for controllable processing of one-shot polyether-based foams. Thus, the tertiary amines are frequently used in conjunction with tin compounds. Generally, polyester-based systems process best with catalysts having a low order of activity; these are more effective foaming catalysts than polymerization catalysts. Tertiary amines, such as n-methyl, ethyl, cocomorpholine, and the dialkyl amine, are used almost exclusively.

In prepolymer systems, the main reaction is that of gas formation, so catalysts that give good control of CO_2 generation are preferred. Again, as in the polyester one-shot systems, tertiary amines are used. With higher catalyst amounts, strength properties of foams generally decrease, while the compression set increases. Load-bearing properties are also affected by the amount and relative proportion of catalysts. In general, softer foams are obtained with increasing catalyst quantities.

TABLE 7.9 Influence of Catalyst Concentration on Foaming Parameters

Formulation	Parts by Weight			
Polyester polyol	56			
Silicone L-521	0.5			
Water	3.0			
Quasiprepolymer	100			
Tertiary amine catalyst	As indicated below			
Triethylamine (Parts by Weight)	Foam Time (min)	Tack-Free Time (min)	Density (lb/ft³)	Compressive Strength (psi)
0.3	4.5	16	2.1	33.6
0.5	3.5	15	2.0	32.4
1.0	2.5	10	1.9	30.8
2.0	1.0	3	1.7	21.5
3.0	0.6	1	1.5	20.0

An increase in catalyst concentration produces an increase in tensile, shear, and compressive strength, as well as a reduction in cure time. There is an upper limit to catalyst concentration, however, above which the foams tend to fissure and crack. An increase in catalyst concentration usually produces a decrease in K factor. This phenomenon can be explained by the fact that a low K factor results from a good retention of fluorohydrocarbon by the foam, which implies a negligible permeation of fluorohydrocarbon through the cell walls. An increased catalyst concentration would be expected to produce a more tightly cross-linked polymer structure and hence reduce the diffusion constant, and possibly the solubility of the fluorohydrocarbon.

Data demonstrating the relative efficacy (and synergism) of a number of catalysts for both the isocyanate-water reaction and the isocyanate-hydroxyl reaction are shown in Table 7.10. A number of catalysts commonly used for flexible foams are listed. Many of these catalysts are also used in rigid systems.

TABLE 7.10 Order of Catalytic Activity

Catalyst	Isocyanate-water reaction		Isocyanate-hydroxyl reaction	
	Conc (%)	Activity	Conc (%)	Activity
Uncatalyzed reaction	—	0.001	—	1
DABCO*	0.1	2.7	—	—
TMBDA**	0.1	1.6	—	—
Triethylamine	0.1	1.5	—	—
N-ethyl morpholine	0.1	1.1	—	—
Stannous octoate	0.1	1.0	0.1	1270
Stannous octoate	0.5	1.5	0.5	1820
Stannous octoate + DABCO	0.1 + 0.1	1.2	0.1 + 0.1	2110
Stannous octoate + DABCO	0.5 + 0.1	1.7	0.5 + 0.1	5140
Stannous oleate	—	—	0.1	25
Dibutyl tin dilaurate	—	—	0.1	280
Dibutyl tin dilaurate	—	—	0.3	610
Dibutyl tin octoate	—	—	0.1	300
Dibutyl tin octoate + DABCO	—	—	0.1 + 0.1	870

*DABCO = 2,2,2-diazabicyclooctane
**TMBDA = N,N,N,'N'-tetramethyl-1,3-butane diamine

Proper catalyst balance is critical to the production of foams. Without proper balance changing the mechanical parameters alone generally will not be able to overcome plant production problems. The function of the amine catalyst is to promote the water-isocyanate or blowing reaction, while the organotin catalyst promotes the polyolisocyanate or gelling reaction.

If the amine catalyst is too high, the following problems may appear:

1. Collapse or splits because the blowing reaction is out of balance with the gelation reaction
2. Excessively fine cell structure
3. Thick skin

With low amine level, some possible problems are

1. Slow skin cure
2. Undercutting in slab foams
3. Increased foam density
4. Loose skin

Conversely, if the organotin level is high, some of the processing problems that may occur are

1. Splits because the gelling reaction is faster than the blowing reaction
2. Foam shrinkage accompanied by closed cells
3. Blow holes from trapped gases
4. Harsh, boardy feel
5. Poor *fingernail* or *elephant hide*
6. Dimpled skin on molded part

With low organotin catalysts, we can expect to see

1. Collapse or splits
2. Slow cure
3. Coarse cell structure

7.4.4 Silicone Foam Surfactants

The word *surfactant* is a contraction of *surface active agent*. Surfactants have also been called stabilizers, cell control additives, foam stabilizers, and cell control agents, among other titles. Surfactants impart stability to the urethane polymer during the foaming process. They help control cell structure by regulating the size and, to a large degree, the uniformity of the cells. The choice of surfactant is governed by factors such as polyol type and method of foam preparation.

Many different types of surfactants have been employed for flexible foams, but usually organic nonionic and anionic surfactants and silicones are successfully used. The choice of surfactant depends on the method of foam preparation and the desired end use (e.g., fine, regular-celled cushions vs. large, irregular sponges). The most commonly used surfactants for polyether-based foams are silicones, such as the polyalkylsiloxanepolyoxyalkylene copolymers. Conventional polydimethylsiloxanes of relatively low viscosity (10 to 100 cps @ 25° C) have been used for polyether systems prepared and foamed as prepolymers.

Silicone surfactants are used to stabilize the foaming reaction, to act as an emulsifier for certain compounds, and to assist in mixing. These surfactants are generally dimethyl silicone fluids or organosilicone/polyoxyalkalene copolymers. These surfactants have varying degrees of efficiency, and the specific product used must be chosen to match the type of foam being produced. Some surfactants act as defoamers if used in the wrong application.

The silicones generally perform differently in polyester foam systems, causing an unstable foaming situation. Silicone oils can be used in very small amounts to enlarge the cell size of polyester teams. Stabilizers used for polyesters are ionic, such as sulfonated castor oil and other natural oils, amine esters or fatty acids, and long-chain fatty acid partial esters of hexatol anhydrides. In polyester systems, nonionic surfactants are used to help modify the viscosity of the polymer during foaming and to provide control of reaction rate.

In the prepolymer method for one-shot polyether flexible foam, the primary role of the silicone surfactant is to lower the surface tension and to provide film (cell wall) resilience. Resilient films prevent the collapse of the foam during rise and continue to stabilize it until the team is self-supporting. A secondary but important role of the silicone surfactant is cell size control. The silicones can be added to the formulation in any of the two to six streams usually fed to the mixing head in the one-shot process. Usually, however, the silicone is metered separately, in combination with the polyol, or added as a water/amine/silicone mixture. It can also be added in the fluorocarbon stream.

7.4.5 Water and Auxiliary Blowing Agents

CO_2 obtained in situ by the reaction of water with TDI is the chief blowing agent for all commercially produced flexible foams. The stoichiometric amount of water and TDI used will determine foam density, provided most of the gas formed is used to expand the urethane polymer. Because water participates in the polymerization reaction leading to the expanded cellular urethane polymer, it has a very pronounced influence on the properties of foams. For better control of the foaming process, most foam manufacturers employ distilled or deionized water.

In addition to water, auxiliary blowing agents may be included in the foam formulation to further reduce the foam density. These agents can be used in addition to, or as partial replacement for, the water in developing special foam properties. An example is the use of methylene chloride or fluorocarbon-11 in either polyether or polyester-based systems for softening the resulting foam.

The amount of water used in flexible foam formulations, along with the corresponding amount of TDI, largely determines the foam density. As the amount of water increases, with a corresponding increase in TDI, the density decreases. If water content is increased without increasing the TDI, foams may be obtained with coarse cells and harsh textures. Lower tensile and tear strengths and compression modulus result, while the compression set tends to increase. Another important effect of too much water is poor aging characteristics. Conversely, too little water will result not only in higher densities than desired but also in slower curing and may cause shrinkage in the foam.

In polyester flexible foams, concentrations of two to five parts of fluorocarbon blowing agents, such as fluorocarbon-11, produce a considerable amount of softening. Good stability in polyester one-shot foams is obtained with concentrations containing from 5 to 15% of fluorocarbon-11. The effect of variable concentrations of fluorocarbon-11 on the softness and other properties of polyether flexible foams is shown in Table 7.11. Importantly, as the amount of fluorocarbon in increased, the silicone and the catalyst content are also increased. In the table, they were set at an artificially high concentration to handle the maximum amount of fluorocarbon-11 added. Another important effect of adding fluorocarbon-11 to the formulation is to reduce the rate and amount of heat buildup in slab form. Too much solvent can have an adverse effects, because it retards the cure, and post curing of the foam is required.

TABLE 7.11 Foam Indentation Load As a Function of Fluorocarbon 11 Concentration

Formulation			
Polyether triol, 3000 MW	70	Stannous octoate	0.35
Polyether diol, 2000 MW	30	Dibutyltin dilaurate	0.075
DABCO	0.10	TDI, 80:20	43.7
N-ethylmorpholine	0.30	Water	3.5
Silicone L-520	2.0	Fluorocarbon 11	Variable
Physical Properties			
Fluorocarbon-11, Parts/100 Polyol	0	5	10	15
Density, lb/ft^3	1.6	1.45	1.35	1.25
Elongation (%)	430	450	380	330
Tensile strength (psi)	21	19	15	12
Tear strength (pli)	4.0	3.5	3.5	2.5
Indentation load, lb (4-in thickness)				
25% deflection	36	30	24	18
50% deflection	51	41	33	27
65% deflection	66	51	45	37

Auxiliary blowing agents can be used to provide and additional degree of foaming to reduce the firmness of the foam. Typically, these blowing agents are low-boiling-point organic liquids such as those listed in Table 7.12. When using an auxiliary blowing agent, certain formulation adjustments are required to ensure the stability of the foaming mixture. When one blowing agent is substituted for another, the relative gas efficiencies have to be considered. Some of these factors are given in Table 7.13. Because of environmental concerns, most of the agents shown in Table 7.12 are being phased out; they are presented here for historical reasons. Newer, environmentally friendly blowing agents continue to be developed and are discussed elsewhere.

TABLE 7.12 Auxiliary Blowing Agents

R-11	Trichloromonofluoromethane
R-11 B	Trichloromonofluoromethane plus a based polyols—a typical stabilizer is allocimine
R-113	Trichlorotrifluoroethane
UCON 11-H	Binary azeotrope of R-11 with 18% methyl
FREON HE	50% R-11, 30% methylene chloride, 20%
MeCl$_2$	100% methylene chloride

TABLE 7.13 Comparison of Blowing Agents

	R-11	U-11H	75/25 R-11/McCl$_2$	McCl$_2$	Freon HE
Boiling Point, °C	23.8	20.0	28.7	40.1	24.1
Mol weight	137.4	111.5	119.0	84.9	97.4
Gas efficiency, %	55	56	60	38	71.5
Relative quantity required	1.5	1.2	1.2	1	1

7.4.6 High-Permeability Foams

The production of low-density flexible foams with a high permeability to air depends on the rate of thin cell wall membrane production, and on the amount of polymer from the membranes that becomes incorporated into the fibrillar struts or ribs at the joint between adjacent membranes. This depends largely on the rate of polymerization compared with the blowing rate. The blowing rate, in turn, varies with the temperature inside the reacting foam mixture as well as with the level and type of catalysis used. The foam structure is also affected by the presence of surfactants that dissipate local expansion stresses and help to prevent premature cellular rupture.

The cell membranes are broken by the increasing internal gas pressure, in combination with a rapid loss in the extensibility of the polymerizing material. Most water-blown, TDI-based, low-density flexible foam systems remain homogenous until the reaction temperature reaches about 80° C, when molecules of substituted and polymeric ureas begin to form a separate phase within the polymer network. This separation, which also plays a part in cell opening by creating areas of stress concentration, represents the beginning of hard block segmentation in the polymer. *Please refer to the section on polyurethane elastomers for an in-depth analysis of segmentation.*

At the time of cell opening, the polymer strength is low with a significant proportion of isocyanate groups remaining unreacted. The temperature at the moment of cell opening is thought to be a measure of cell stability and is related to the chemical composition of the foam system. The temperature continues to rise after the foam has fully expanded, because of the continuing reaction of isocyanate groups with water, with amine end-groups (resulting from the reaction of isocyanate groups with water vapor), and also (depending on the temperature) with the reactive hydrogen in the urea linkages. Some reactive isocyanate groups remain, even when the foam has reached its maximum reaction temperature. This maximum temperature occurs, in large foam blocks, between approx. 30 min to 1 hr after manufacture.

The temperature may remain near this level for 1 to 8 hr (particularly near the center of the foam), depending on the block size, the amount of excess isocyanate, the ambient conditions, and the orientation of the foam block, because orientation affects the rate and direction of flow of convected air through the permeable foam. The *cure* of large foam blocks (i.e., the reactions of those bound isocyanate groups remaining from the initial polymerization and the stabilization of labile bonds) is virtually complete within 48 hr after manufacture. The cure of small foamed moldings may require much longer times, and the mechanism is different. In molded foam cushions, the level of bound but unreacted isocyanate groups falls steadily over several days, and changes in foam properties often continue for about a week after manufacture. MDI-based foams tend to reach equilibrium much faster than those based on TDI. The differing polymerization and curing conditions are reflected in the differing physical properties of slab stock and molded foams. Foams made in large blocks result in stiffer, higher load-bearing polymers with higher tensile strength and lower elongation at break compared with similar foams made on a small scale from the same chemical components.

7.4.7 Miscellaneous Additives

In general, the addition of additives to flexible urethane foams is similar to their use in rigid foams. Plasticizers used include phthalate esters, e.g., dioctyl phthalate. Their use generally does not affect physical properties if the plasticizer is present in amounts of up to 10 parts per 100 parts of polymer, except for an increase in density of the polymer. However, plasticizers are often employed as dispersing media for pigments, dyes, fillers, stabilizers, and catalysts.

Flameproofing polyester flexible foams can be accomplished using either tricresyl phosphate or tris (beta-chloroethyl) phosphate. A maximum of 4.0 parts of these flame retardants per 100 parts of polyester resin can be used; greater amounts adversely affect the humid aging characteristics, especially when tris(beta-chloroethyl) phosphate is used. Flame-retardant additives such as tris(beta-chloroethyl) phosphate and tris(dibromopropyl) phosphate have also been used for polyether-based flexible foams. Vircol 82, a phosphorous-containing diol mentioned above for rigid foams, is a reactive flame retardant used for both prepolymer and one-shot flexible, polyether-based urethane foams.

Fillers have been used in flexible urethane foams to increase load bearing characteristics. One limitation to the use of fillers is that filler-reinforced foams have softened excessively as a result of flexing so that the stiffening benefit from the filler is lost.

Barium sulfate barytes can be used as filler for high-density (2.5 to 4.5 lb/ft³) flexible urethane foams to reduce costs and compete with latex foam products in the 3.25 to 3.75 lb/ft³ range. The physical property most affected by increasing the filler concentration is compression set, with a loss of bun height.

Pigments or dyes may be used to color flexible urethane foams, although the dyes generally are preferred, since they are readily incorporated into the foam during manufacture and produce uniform color throughout the foam. Dyes are satisfactory in fastness both due to light and washing and, for the most part, have only very minor effects on foam properties.

Pigments for a polyether urethane may be introduced in the water stream (water dispersible pigments) or in the polyether stream (non-water dispersible pigments). Different types of pigments are necessary, however, for the two methods. Water-dispersible organic pigments are frequently used because they can be easily dispersed, they permit easy cleaning of equipment, and the colors formed are stronger. Colloidal dispersions of carbon black may also be used. An ideal carbon black dispersion should disperse readily in the water component of the formulation without viscosity increases, have fine particle size, display high tinctorial powers, and produce a medium pH.

Pigments in flexible urethane foams have been used in amounts up to 1% of the weight of the polyether. As with any solid material, when pigments are added, the tensile properties decrease slightly. Pigmented foams are usually slightly softer than unpigmented. Humid aging and dry heat aging properties of pigmented foams are not harmed.

Greater depth of color can be obtained with dyes than with pigments. However, this process requires an extra processing step. Also, the thickness of the foam that can be dyed is limited. About 0.2% dye based on the weight of the foam results in a light color, while medium shades are obtained with 1 to 2%. Up to 7% dye can be used for deeper shades.

Inorganic fillers such as calcium carbonate or barium sulfate are used to modify the density and load-bearing characteristics of the foam. High-quality furniture foams have been a primary market for filled foams because of the feel and comfort characteristics. They were originally developed to compete with high-quality rubber latex foams. Fillers can, however, degrade the strength properties of foams and do increase the flammability unless other formulation modifications are made.

Other additives can be used to modify the characteristics of foams or to improve processing. Among these additives are the following:

1. Flame retardants used to modify the burning characteristics of the foam as measured by standardized tests. These tests do not necessarily relate to actual fire conditions.
2. Pigments most frequently used to identify grades of foam or as masking for the normal yellowing of foams due to exposure to light and air. Important inorganic agents include TiO_2, iron oxide, chromium oxide, cadmium sulfite, mixed oxides, magnesium aluminates, carbon black, etc. Organic pigments are usually based on azo/diazo dyes, phthalocyanines, and dioxazines.
3. Bacteriostats. While foam does not in and of itself support the growth of bacteria, the cellular structure can trap materials that can support that growth. The bacteriostat can also act as a retardant.
4. Plasticizers or organic fillers that can serve the same function as the inorganic fillers mentioned above.
5. Antistatic agents, used in applications where a static build-up can be a hazard, such as in hospitals and power transmission equipment.
6. UV stabilizers, required for certain applications where the natural tendency of foams based on aromatic isocyanates to discolor must be minimized.
7. Cell openers for specialty foams such as those used in the manufacture of synthetic sponges.
8. Mold-release agents such as zinc stearate and soluble amines.

Other additives are antioxidants (which retards the thermal oxidation of polyurethanes). Antioxidants work by stopping the chain-breaking reactions initiated by oxygen, oxygen free radicals. Antioxidants in synergistic mixtures with phosphites or phosphines are particularly effective.

7.4.8 Particulate Additives for Flexible Foams

There are many instances where flexible foams need to be modified to substantially reduce cost by the addition of small particle size modifying agents. One such agent (finely ground rubber) was disclosed by Joubert.[3]

Conventional polyurethane foam products are modified by incorporating into the urethane reaction mixture a junction-modifying particulate material that is selected with a critically controlled particle size relative to certain dimensional parameters of the foamed polymer matrix. This particulate material is selected with a particle size that results in location of the material in the foam cell wall junction regions rather than in the cell walls. The resulting products provide excellent physical properties (such as modulus of compression, tensile strength, elongation, and tear resistance) at lower costs than normally encountered with similar unmodified foams. The preferred junction-modifying particulate materials are organic polymeric materials such as ground rubber or plastics. As a result, the large-particle materials needed to achieve the junction modifying properties can be used without causing harmful abrasion or wear of processing equipment surfaces.

The product of the Joubert invention is a flexible polyurethane foam composition that finds particular utility in the manufacture of furniture, mattresses, carpet underlays, and similar applications. The products of this invention can be formulated to provide products corresponding to any of the normal grade foam products. By the addition of the particulate junction-modifying material of the present invention, however, the properties of these normal-grade foam compositions may be desirably modified. Among the commonly employed foam grades to which the present invention applies are the following:

Foam Grade	Typical Use
A	Firm seating, sheeting, boat seats
B	Majority of lounge seats, mattresses (low-grade), sheeting, and kitchen chair backs
C	Firm, high-quality cushioning
D	Dining seats, footstools, etc., boat bunks, and mattresses
E	High-quality mattress and cushions
F	Low-cost, low-quality, single-use packaging
G	Lounge backs and overlays
H	Seats and mattresses, premium foam

7.5 Colloid Chemistry of Polyurethane Foam Formation

The colloidal aspects of foam formation are of vital importance in understanding the basic phenomena occurring during the different stages through which a foaming system must pass—from the mixing of the components to the curing of the finished foam. The process of foam preparation involves the formation of gas bubbles (derived from the blowing agent) in a polymerizing liquid system, and the growth and stabilization of the bubbles as the foam is formed and cured. In a rigid foam system, the liquid phase undergoes rapid polymerization, increasing in viscosity to a gel phase, which cures relatively quickly to a hard, solid phase. The colloidal nature of the gas-liquid phase exists at least in the initial fluid stages of foam formation, which may persist for only 10 to 30 s after the mixing of the foam components.

In addition to the mechanical mixing function, the mixer must also nucleate the reaction mixture to provide growth-initiation points for the bubbles. Controlled nucleation is essential to control the cell size and structure of the final product. If there are too few nucleation sites, the evolved carbon dioxide forms a supersaturated solution, and suddenly self-nucleation occurs. Then, as the first bubbles to be formed grow very rapidly, the concentration of the carbon dioxide in solution falls precipitously, and no further bubble nucleation occurs. The end result is a foam with large cells or even collapse of the foam. The nucleation of a foam mixture is theoretically possible by using finely divided solids or liquids that are insoluble in the foam mixture. From a practical standpoint, almost all low-density foam manufacture employs nucleation obtained by fine dispersion of a gas, usually air or nitrogen stirred into the reaction mixture. The formation of bubble nuclei is achieved by metering the nucleating gas into the polyol or the isocyanate streams as it flows through the mixer, or by direct injection into the mixer barrel.

The physicochemical principles involved in flexible foam manufacture are complex. Foam reproducibility demands that components (1) be maintained within 5° F of recommended temperature, (2) be dispensed within required stoichiometry, (3) have properly dispersed nucleating agent, and (4) be well mixed prior to exotherm start.

Most of the polyether polyols used in the manufacture of flexible foams are polyether copolymers containing hydrophobic and hydrophilic segments or blocks. The former are chains based on propylene oxide, and the latter are chains based on ethylene oxide. Such polyols are compatible with water, TDI, modified MDI, and the usual catalysts and additives to form homogenous mixtures.

In some foam formulations, the addition of gas is unnecessary to obtain foams with a uniform medium cell size if there is sufficient dissolved gas present in the chemicals, especially in the isocyanate. The dissolved gas results from the storage of the chemicals under pressure, usually about 2 bar, to provide a positive feed to high-speed piston metering pumps. Nucleation of the foam reaction mixture occurs by effervescence of the dissolved gas and is controlled by the pressure in the mixer and by the agitator speed. Where additional gas is needed to obtain very small cell sizes, the gas is usually metered into the polyol stream at a predetermined rate. Mixers for continuous low-density foam production usually employ simple multipin stirrers in plain cylindrical barrels. The nucleation of the foam mixture increases with increasing pressure drop across the mixer pins, and this pressure drop, in turn, increases with increasing agitator speed. The pressure drop depends also on the length and diameter of the mixer outlet nozzle and the throughput of the foam mixture.

The mixed, nucleated foam reaction mixture is dispensed into a mold or continuous conveyor less than 1 s after the start of mixing. There is an induction period before the mixture becomes opaque with visible bubbles. This induction period is known as the *cream time*. During this induction period, the nucleation bubbles formed in the mixer become fewer in number and larger in size. The larger bubbles grow by diffusion of gas through the liquid at the expense of the smaller bubbles that have a higher internal pressure.

Thus, the initial number of bubble nuclei created in the mixer is not the only factor influencing the cell size of the foam. Other factors that influence the cell size of the foam are

1. An increase in the cream time that increases cell size
2. An increase in the foam reaction rate by increasing catalyst level or temperature of the components
3. Addition of surfactants, such as polysiloxanepolyether block copolymers, that reduce the rate of nucleation bubble loss during the induction period, resulting in a foam with a higher cell count

During the initial period, the reactions of diisocyanate with water and with polyol proceed simultaneously. In TDI-based low-density flexible foams, the initial reactions produce an isocyanate-terminated prepolymer that is formed by the reaction of the terminal hydroxyl groups of the polyol with the more reactive four-position isocyanate group, and a diisocyanato diphenyl-urea oligomer from the TDI/water reaction. There is little chain extension and no detectable formation of biuret or allophanate cross-links. Until the reaction mixture reaches the critical temperature of about 80° C, the main reactions are primarily with the isocyanate group of TDI in the para position. Above 90° C, the ortho isocyanate reaches equilibrium rate parity with the isocyanate of the para position, and cross-linking reactions begin to occur.

7.5.1 Anisotropy of Foam Structure

All low-density foams tend to be anisotropic. At first, the small nucleation bubbles are spherical, but as they expand under the increasing pressure resulting from the generation of carbon dioxide, they tend to elongate in the direction of foam rise. The fewer and larger the bubbles, the greater their elongation. Anisotropic foams display physical properties that depend on direction.

In low-density foams, the volume of the expanding gas bubbles quickly becomes much larger than the volume of the polymerizing liquid reaction mixture, and the area of contact between adjacent bubbles forms a membrane. The membrane is planar if the bubbles are of equal size, and the line of contact between three adjacent membranes forms a triangular rib. The gas pressure within the expanding foam remains low until the reaction temperature rises above 90° C, when the hindered ortho position isocyanate groups begin to react at a significant rate. The pressure within the foam increases as gelation proceeds until the thinning wall membranes of the cells rupture and the blowing gas is released.

The formation of a polyurethane-based foam proceeds through several distinct stages. In the first stage, the blowing agent fluorocarbons or carbon dioxide dissolves in the liquid phase until saturation is reached and the liquid phase then becomes saturated with the gas. The gas then separates out of solution and forms bubbles by means of a process called *nucleation*. Nucleation is aided by the presence of *nucleating agents*, which may consist of a second phase, e.g., a finely divided solid or an irregular solid surface.

In most foaming systems, nucleating agents such as silicone surfactants are present. An increase in silicone concentration results in faster nucleation, resulting in fine cells foams. The presence of dissolved gas, such as entrained air, in the reaction mixture may also result in faster nucleation and finer cells. A similar effect may be produced by the addition of finely divided solids such as silica or other inert fillers.

Cell size, in general, depends on the proportion of gas molecules that diffuse through the liquid into existing bubbles (causing the formation of large cells) to those that diffuse through the liquid to a nucleating liquid (causing the formation of new bubbles). Cell size also depends on the concentration and effectiveness of the selected nucleating agents.

To produce foams with fine cell size, the liquid system must permit rapid diffusion of the blowing agent. In addition, a foam stabilizer should be present to form a protective layer around each bubble.

This will also serve to reduce the diffusion rate of the gas from the liquid into
generation of blowing agent and a high level of efficient nucleating agents are rec

An important stage of urethane foam formation is the stabilization of bubbles
Factors affecting foam stability in soap-water systems are similar to those in a u.
For the dispersion of a given volume of gas in a unit volume of liquid, the free energy of the
must increase as follows:

$$\Delta F = \gamma A$$

where γ is the surface tension and A is the total interfacial area. Hence, a larger increase in free energy
is required to produce finer cells than is necessary for the production of large cells. In a liquid foam
system, from an energy point of view, there is always a tendency to reduce the interfacial area, thereby
favoring coalescence of cells and foam collapse. Unless other factors, such as rapid foam curing, come
into play, foam collapse will occur. The addition of surfactants such as silicones will reduce the free energy
increase associated with the dispersion of the gas in the liquid because of the reduction in surface tension.
The silicone surfactant will also cause the production of smaller cells corresponding to a large value of
A. Care must be exercised, however, since too-fine cells may lead to foam collapse. Rupture of foam cells
occurs more readily with thin cell walls.

Another major factor affecting bubble stability is the drainage of liquid in the bubble wall due to
capillary action and gravity. An increase in the viscosity of the liquid due to polymerization reactions
has a stabilizing influence since it will reduce drainage, in this way retarding thinning of the cell walls.

In flexible foams, cell opening occurs to a large extent when the foam rises to its maximum height. It
has been suggested that the membranes of the cells at this time exhibit high viscosity, but not enough
elasticity to expand and relieve the gas pressure. This combination of factors, along with low mechanical
strength of the thin cell walls, leads to rupture of the membranes, i.e., opening of the cells. If, at the time
of maximum gas evolution, the cell membranes did not rupture, closed cells would result.

7.6   CFCs and the Ozone Level

During the early days of the flexible foam industry, foams were made with high water levels, and they
tended to have a harsh, boardy feel and lacked durability. Research showed that by using low-boiling
organic liquids as auxiliary blowing agents, formulated with reduced water content, a much more
satisfactory product could be made. These auxiliary blowing agent had already found a role in rigid
foams, where they contributed significantly to the thermal insulating qualities of the foams. Trichloromonofluoromethane, more commonly referred as F-11 and now as CFC-11, became the product of
choice due to a combination of effectiveness, ease of use, cost, and stability. Since its use in flexible foams
had been patented, alternatives were evaluated, and methylene chloride became the usual alternative.

In the mid-1970s, studies began to suggest that the use of CFCs was responsible destruction of the
ozone layer in upper atmosphere. Ozone serves to screen ultraviolet radiation from the sun, which may
cause skin cancer. There was controversy then, and it continues today, as to whether CFSs are a significant
factor in the observable loss of ozone. Nevertheless, international actions were undertaken to eliminate
CFCs from nonessential uses such as propellants for aerosols.

By 1980, the European Community instituted a moratorium on new production capacities for CFCs,
and in 1985 the Vienna Convention for the Protection of the Ozone Layer was signed by many countries.
The scientific community gradually accepted the results of the studies. On September 16, 1987, the
Montreal Protocol on ozone-depleting materials was signed by the U.S., Canada, and the E.E.C. The
timetable established by the Protocol called for

1. A freeze in usage at 1986 levels on July 1, 1989
2. A reduction to 80% of 1986 levels on July 1, 1993
3. A reduction to 50% of 1986 levels on July 1, 1998

4. A review in 1989–1990 and every subsequent four years, and
5. No imports into the area from July 1, 1990

The impact on the foam industry can be best appreciated by 1988 statistics that showed the usage of CFCs to be

- 28% for production of plastic foam and foam insulation products
- 22% unallocated uses
- 19% mobile air conditioners
- 12% solvents
- 9% refrigeration and air conditioning
- 5% sterilization of medical equipment and instruments
- 4% aerosol propellants and miscellaneous uses

Among the solutions to the CFC problems evaluated have been

1. A less stable molecule
2. Hydrocarbons
3. Chemical blowing agents
4. Polyol modifications
5. Formulation variables
6. Dissolved gases
7. Recovery of the blowing agent, and
8. Mechanical means

Partially halogenated hydrochlorofluorocarbons (HCFCs) degrade in the lower atmosphere and do not have the ozone-depleting potential of the CFCs. Commercialization has come slowly, and they represent only an interim solution. The major efforts have been focused on the rigid foam applications, and only a few of the newer developed products appear to have application to flexible foams. Ideally, a replacement for CFC-11 would be a low-boiling liquid product completely miscible with water and inert to the other formulation reactants. Consideration also has to be given to the materials of construction used in current foam machines to avoid the necessity of major capital expense. These concerns have limited the availability of products.

HCFC-22 has found commercial success. However, it will not be available as a long-term replacement, since HCFC eventually will be banned from the marketplace. Hydrofluorocarbons also are being developed as auxiliary blowing agents. Since these products contain no chlorine, there is no concern over ozone depletion. As in the case of HCFCs, most of the emphasis has been in rigid foam applications. Some evaluations have been done using HFC-134a (1,1,1,2-tetrafluoroethane) as a blowing agent for integral skin foams. This product has a boiling point of −26.2° C and about the same GWP as HCFC-22. It appears to be twice as effective as CFC-11 as a blowing agent in integral skin foams. Cost and availability currently are problems; however, this could change, as it appears that HFC-134a is becoming the refrigerant of choice for air conditioning applications.

Chemical blowing agents (CBAs) have been used in other plastics where the processing temperatures cause breakdown of the solid powder to form gases, such as dialkyl dicarbonates, and anhydrides of carbonic acids. Some of these have theoretical gas yields greater than CFC-11. For practical use, the thermal decomposition by heat of reaction cannot be relied upon; however, amine catalyzation using standard tertiary amine catalysts shows promise of liberating gases at or near room temperature. This approach needs to be further developed before commercialization. Formic acid salts also offer the potential for gas generation through the reaction with isocyanate with the salt to generate carbon dioxide.

Hydrocarbon usage has found a niche as a replacement for CFCs in certain applications. N-pentane (C_5H_{12}) is being used primarily in Europe in integral-skin flexible foams. Pentane is twice as effective as CFC-11 as a blowing agent. There is no evidence that the use of pentane results in poorer performance in standardized flammability tests. Acetone is being used in the U.S. for the manufacture of slab stock foams. There are some concerns about the use of hydrocarbons. For some, it is a matter of flammability and the precautions that are required during the manufacturing steps. Studies to date do not indicate that there is a problem with increased flammability in the finished parts. Development of petrochemical smog from the use of hydrocarbons will be a major obstacle to broader usage.

The route to soft foams by means of polyol modifications certainly is not a new approach, having been described in the literature in the early 1960s and commercialized to some extent. It is being revived due to the CFC issue. Among the various means in use evaluated are

1. High-molecular-weight, high ethylene oxide content polyols
2. High-functionality, high-molecular-weight polyols
3. Amine-terminated propylene oxide polyols (PUREA)
4. Polyols with carboxylic acid groups
5. Monofuctional polyols

Formulation variables by means of additives also have shown some promise. Most of these approaches are proprietary; however, there are commercially available additives. Among the materials mentioned in the literature are ORTEGAL 300 (described as a softener) and UNILINK (described as a blowing agent enhancer).

Isocyanate variations are also possible as a means of giving stability to otherwise unstable foam systems built with new polyols with other formulation and product variables. Isocyanates can be tailored in many ways, particularly pure MDI types that can offer direct ring configurations that affect the processing and properties. Even before the concern over CFCs, there was wide use of TDI and crude MDI blends for the production of high-resiliency foams. The blends were used to impart stability as well as to assist in meeting flammability test requirements.

Since carbon dioxide generated as a result of the water-isocyanate reaction is the primary blowing agent in flexible foams, it is obvious that increasing water level would be a means of replacing CFCs if the objective were only to maintain density. Unfortunately, other foam properties are also affected, some adversely. Combined approaches with polyol changes, catalyst modifications, and chemical modifiers must be used with high-water formulations to obtain the desired results. Foam scorching (discoloration) or auto-ignition is a hazard from the exotherm generated by from high-water formulations. To overcome this hazard, rapid cooling systems have to be used for the manufacture of slab stock foams, where cooling chambers are used to pull cool air through the foam to reduce the foam temperature. Depending on the foam machine operation, the foam block may have to be trimmed or barrier films applied to ensure that the cool air flows completely through the foam.

Adding carbon dioxide as a gas is an effective way of reducing foam density. This is done by dissolving the carbon dioxide in either the polyol or isocyanate component before the reaction. This may be done either in the component tanks or in-line in the machine mixing head. Density reductions in the 10 to 15% range and more can be achieved. Dry air addition to high-density elastomeric foams has been used for years to improve mixing, reduce density, and improve flow in molds. The most typical method is by dissolving the air into the polyol component. Up to 8% air by volume has been achieved with the proper modification of the metering equipment.

Recovery of CFCs gases from the foaming reaction may be recovered in activated carbon beds, then regenerated for reuse. Obviously, this requires some make-up material, since 100% recovery is not achievable. With cessation of CFC production, there is no long-term viability for this process.

For many years, it was recognized that formulations have to be modified to make the same density foams as the altitude of the production plant varies. This is simply a gas law factor. Taking this into account, Foamex has patented the VPF (variable pressure foaming) slab stock foam process, which can produce a wide range of densities of water-blown foams. This is done by adjusting the ambient pressure around the foam as it rises.

In the past, integral-skin foams have depended on the condensation of the blowing gas on the surface of a mold along with mold overpacking to give the desired dense, void-free surface. It has also been found that water-blown integral-skin foams may be successfully produced by rotational molding of the reactant mixture. So far, this has been limited to high-stiffness materials of a rather simple shape.

7.7 Methods of Manufacture

7.7.1 Continuous Slab Production (Free Rise Foaming)

Continuous urethane foam slab is produced by high-volume production techniques, requiring large initial capital outlays for equipment, a considerable amount of floor space for the work area, and storage space for the work area. In addition, a large market must be assured to support economic production.

Flexible slab production is used to make mattresses, cushions, pillows, shock-absorbing packaging materials, peeled goods for laminated clothing, paint rollers, toys, and sponges. Rigid slab is used for panel insulation, flotation, packaging, and numerous other applications.

The process involves continuous delivery of the chemical components, at fixed rates, to a mixing head that moves across a conveyor in a pattern of parallel lines or ribbons. The foam is moved continuously by the conveyor, which is on a slight downgrade to keep the liquid mix from being deposited on top of rising foam. In this way, a uniform end product is maintained. The liquid mix is dispersed into a continuous mold, which is usually roll-paper formed into a trough by the conveyor and adjustable side-boards. From the conveyor, the foam is moved through a curing oven. Flexible foams are usually moved through crushing rolls, although crushing is not necessary with most of the one-shot formulations. However, more uniform load-bearing characteristics are obtained if the foam is flexed. Most one-shot foams can be flexed as soon as they are tack-free (about 15 to 30 min). Foams catalyzed with stannous octoate alone are slow setting and cannot be flexed for at least an hour after pouring. These foams are sometimes held as long as 16 to 24 hr before flexing.

Temperature control of the raw materials is extremely important, because an entire production run can easily be ruined if the temperature of one of the materials is allowed to change, causing a pressure change and off-ratio output. The density and other properties of slab foam depend on the temperature of the foam ingredients. Foam density increases as the temperature of the foam intermediates exceeds 32° C. The temperatures of the ingredients are usually kept between 21 and 32° C. Once temperatures are set for a given formulation, they must be maintained within +1° C.

Continuous mixers are usually used with high- or low-pressure machines to produce outputs of 100 lb or more per minute, with the usual range being 150 to 200 lb/min. After the foam has risen, it is usually passed under curing lamps, heaters, or low-pressure steam to eliminate surface tack.

Exhaust hoods must be properly positioned to remove any irritants that may have volatilized during the foaming process. The slab then passes through a series of auxiliary equipment, which may include cut-off saws, horizontal and vertical trimmers, slicing machines, hot-wire cutters, etc. A variety of machines for cutting and splitting flexible urethane foam are commercially available. The most common processes employed for the splitting of slab stock are centrifugal peeling and horizontal table splitting.

Slab line operations require a considerable amount of floor space, as well as large storage areas for raw materials and end products. A typical slab line may be from 200 to 400 ft long. To minimize scrap loss, particularly in the rigid foams, low surface-to-volume ratios are necessary. The scrap loss in flexible foams may be nearly eliminated by shredding and rebonding.

One of the major losses in slab production occurs as the result of the removal of the rounded *free-rise* crown. To overcome this problem, a novel four-sided conveyor is used to restrict the rise to produce a flat surface, almost completely eliminating this crown effect. In rigid slab production, it is necessary to have the side walls also conveyorized to prevent *jamming* of the foam against the side walls.

The width of the finished foam bun is determined by the spacing of the adjustable side boards that bank the paper trough. The dimension may be up to 80 in and (for economic reasons) is usually some multiple of the desired width of the end product. The height of the bun is a function of throughput and conveyor speed, and it may be up to 40 in.

Interrupted slab operations may be used where it is not economically practical to produce foam on a continuous-slab basis. The resulting slabs or blocks may be 60 to 72 in wide by 80 to 96 in long by 12 to 15 in high. *On-off* or intermittent equipment is used. While in a continuous slab operation the first few feet of material of a run can be sacrificed, such is not the case here. Continuous slab stock of rigid foams is produced in the form of large buns to minimized scrap loss. Usually, the same slab equipment used for flexible foam is used in this application.

7.7.2 Pour-in-Place (Foam-in-Place) Batch Preparation

Pouring in place is the oldest, most versatile and popular technique for the application of rigid urethane foam. In conventional pouring-in-place molding, freshly mixed liquid formulations are charged directly into a cavity in which the foaming reaction is carried out. The cavity is filled upon completion of the foaming reaction. The technique is applicable over the entire density range, from 0.7 to 70 lb/ft^3. Usually, the mold is retained as the exterior of the finished item. Both the semi- or quasiprepolymer and the one-shot processes are successfully used.

The ingredients are mixed either by hand or with a power mixer in any clean containers ranging from a small glass beaker to a large pail or bucket. Batches may be up to 50 lb or more. Ideally, a power mixer should do a thorough job of mixing in 10 to 15 seconds. Some of the higher-viscosity, slower formulations may require up to 30 s. Often, electric drill motors with various blades attached to the shaft are used. The batch system used in poured-in-place processes may result in nonuniform end products, but the equipment required is not costly.

The reactant mixture, after being poured directly into a cavity or mold, flows to the bottom of the mold and then foams to approximately 30 to 40 times its original volume, filling all cracks and corners and forming a strong, seamless core. The elimination of seams, joints, and lines (which tend to fail at low temperatures) is a distinct advantage. Pouring in place differs from molding primarily because adhesion to the mold is desired, while in molding perfect release is sought. Also, during the expansion process, a skin is formed next to the surface of the filled cavity, eliminating the need for vapor barriers.

Pouring in place is ideal for filling defined areas, such as curtain wall sandwich panels, wall cavities, and irregularly shaped voids. The cavity must be capable of supporting foam pressures of up to 3 to 5 psi. The usual increase in volume of 30 to 40 times is cut down somewhat by the frothing process but is not entirely eliminated. The temperature of the foaming reactants is very critical if a low overall density is desired.

The numerous applications for which urethane foam has been successful may be classified into three types, according to density: low, medium, and high ($\leq$1.5, 1.5 to 3.0, and >3.0 lb/ft^3, respectively). Figure 7.8 shows several specific uses where pour in place has been used. The scale on the right indicates the relative volumes of each of these three ranges.

Maximum compressive strength in panels occurs parallel to the panel walls in vertical pours. The high vertical rise is the easiest to handle, as it may be poured through a single opening at the top of the mold. Panels up to 8 ft high have been successfully foamed using only one pour. The foam properties of vertically filled panels are uniform from top to bottom. A low vertical rise may also be used where high panels cannot be conveniently handled. In this case, the foam may be poured through a single hole at one end if the mold is tilted to ensure adequate distribution. Otherwise, it may be necessary to traverse the length of the mold.

In some cases, it is desirable to achieve maximum strength perpendicular to the panel walls. In such cases, the panel can be foamed horizontally. In pouring horizontal panels, one of the panel skins is placed on the bottom of a mold. The mold skins are used to retain the foam as it rises to meet the other skin being held in place by the upper mold section. As with vertical panels, electromagnets and vacuum may be used to hold the top substrate to the upper mold section. The liquid mixture of foam components can be flowed in by a mixing head in a predetermined pattern, traversing the mold area. Skin materials used for insulated building panels include steel, aluminum, melamine laminates, plywood, sheet rock, and paper.

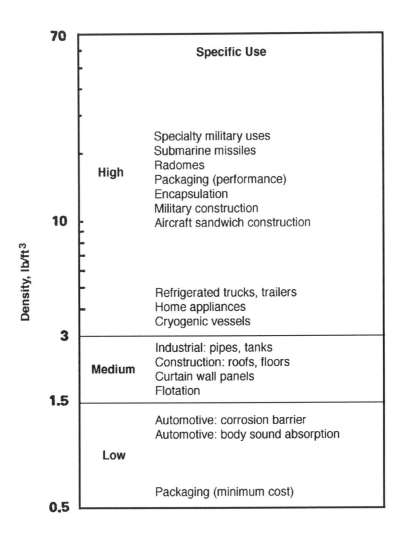

Figure 7.8 Specific uses of pour-in-place method.

Temperatures of 110 to 130° F are desirable with metal molds in one-shot rigid urethane panel molding to ensure low panel densities and optimum K factors. Lower mold temperatures result in high panel density because of the loss of heat to the mold. In addition, a thick skin forms at the foam surface, causing an increase in K factor. Usually, with 110 to 130° F molds, panels may be released from their jigs in approximately 10 min. A postcure at 150 to 175° F may reduce this time to as little as 5 min.

7.7.3 Molding

A number of flexible foam components are molded rather than fabricated from slab stock. For intricately shaped articles, molding results in savings up to 15% compared to slab stock.

In a typical low-pressure molding operation, a mold is preheated and coated with a mold-release agent. A preheated amount of liquid mix is poured into the mold, the mold is closed, and the ingredients are foamed to the mold configuration. After curing and cooling, the part is stripped from the mold. The scrap rate is seldom more than 5 to 10%.

Completely satisfactory methods of molding have never been fully achieved with flexible polyester or polyether prepolymer urethane systems. With the development of improved processing techniques in recent years, molding of flexible one-shot polyether foam has become attractive as a means of reducing materials and labor cost, particularly where large-volume items and standard dimensions are involved. The molding of one-shot foams has also eliminated the somewhat lengthy curing cycles necessary for attaining optimum properties from prepolymer-based items.

Molded items show a better load-deflection curve than slab at high deflections (>50%). The main reason is the load distribution by the molded skin, but random cell variation is also believed to be partly responsible. Mold temperatures are usually 150 to 160° F. The best results are usually obtained when the mold release is spray applied to the clean, warm mold just before each pour. Care should be taken, however, that all solvents from the mold release are removed before the foam ingredients are poured into the cavity.

Curing is usually carried out in two stages: (1) a precure of 15 to 20 min at about 270° F, enabling the item to be removed from the mold, and (2) the final cure of 60 min at the same temperature.

Mold release can be difficult, since uncured polyurethane is an excellent and widely used adhesive. Two basic types of mold release agent are used. The first requires the hot molding to be stripped from the mold. Mixtures of paraffin and microcrystalline waxes are used for this technique, in which hot wax releases from the mold. The molds must be heated and coated with wax before each filling. However, there is a tendency for the paraffin wax to be slowly oxidized by the repeated heating. A release agent containing an antioxidant should be used. The breakdown products formed have no release properties, and it is important that a thin layer of wax be used each time.

The second type of release agent, such as the polyethylene waxes, is used if the mold has to be stripped away when cold. In this case, the foam comes away from the release agent. Probably, the hot-strip technique is better suited to production runs, as it allows the foam to be removed from the molds immediately after they leave the oven. It has been claimed that more than one release can be obtained from some release agents. Mold release problems can be reduced by adding small amounts of dibutyl tin dilaurate, that promotes curing at the surface of the molding and thus improves mold release.

Slab stock foam requires very little curing. A bank of infrared heaters suspended above the conveyor can prove sufficient. In molding, however, the exotherm generated is not sufficient to cure the foam, and external heat must be applied.

Microwave curing permits reduction in curing time from 20 to 4 min. Plastic molds are used with a gel coat of epoxy resin containing iron powder. The plastic molds must be cooled to an even temperature before refilling, however, and this may present problems. It has generally been observed that foam cured with microwaves has properties slightly superior to foam cured by conventional heating, especially in compression set.

Dielectric heating has also been developed for use in curing. Molds with an iron-containing gel are not required. The electrodes which constitute the effective heating area measure 6 × 22 ft and should be capable of curing three front car seats at one time. An important advantage of this method is that a conveyor can be run through the unit, and an oven of the required length therefore can be made by placing units side by side. Dielectric heating reduces cure time in the mold to about one minute.

7.7.4 Laminating

One of the largest uses for polyester-type flexible foam is in foam laminates for clothing interlinings. The first step in preparing the foam for clothing lamination is to take slab stock, cut it into roughly square logs, and peel them as one would peel a wood log for veneer. Thicknesses of about 8/32 and 3/32 in (6.35 and 2.38 mm) are generally used. The foam can easily be butt-joined by heat so that large continuous rolls of foam can be obtained.

There are basically three methods of producing polyurethane foam laminates commercially as described below. The first is the *heat fusion* or *flame lamination process,* which is most popular in the U.S.

Flame Lamination

In the flame lamination process, controlled naked gas flames are allowed to impinge upon a traveling layer of foam. This produces a layer of molten urethane resin by micromelting the upper portion of the foam (about one-third of the thickness). The fabric is immediately pressed against the tacky surface and adheres firmly. The laminate can be rolled up immediately and made ready for use. No additional adhesive is required. Formerly, only polyester foams could be utilized for flame laminations. However, polyether foams are now successfully used for this process. To make them susceptible to flame lamination, organic phosphorous compounds are added. The polyether foams are preferred in preparing soft clothing inter-linings.

In most textile applications, a cloth lining must be sewn against the foam side. A comparatively new adaptation of flame lamination is now being used to obviate this step. In this method, two fabrics form a sandwich with a very thin foam interlining. One fabric is flame laminated to one side of the foam, followed by flame laminating of another fabric to the other side. After flame laminating, so much of the foam is used up that, in reality, it is acting only as an adhesive. In this way, a liner is adhered in place, rather than sewn, with little or no loss in breathability, with gain in warmth, and with practically no gain in weight. Foams as thin as 40 mils can be used. Polyester foams are preferable for this process. Polyether foams are difficult to peel this thin,

Liquid Adhesive Process

The second widely used method for preparing polyurethane foam laminates is the *liquid adhesive* or *wet process*. Here, special adhesives, either in the form of a water emulsion or as a solution in organic solvents, are applied either to the fabric or foam using conventional equipment. The water or solvent is evaporated, and the bond may be set by drying or curing at elevated temperature. One of the adhesives used for this application is based on acrylic interpolymer latices. It is employed in the manufacture of thermal garments, carpet underlay, and insulated bags.

Frothing

The third method, the so-called *frothing* technique, consists of foaming the material directly onto the base materials, such as films or impervious fabrics. The previously unsolved problem of controlling strike-through when foaming on fabrics was solved by simply wetting the fabric with ordinary tap water. Almost any fabric can be used, although the technique is more difficult with the more porous fabrics, which may require the use of thickeners. The fabric is usually immersed in water. After that, the moisture level is controlled by compressing or heating. After the foam has been poured onto the textile, the entire laminate is passed through an oven. This serves the dual purpose of curing the foam and drying out the fabric.

With continuous casting, it is possible to manufacture sandwich structures—fabrics and foam skins enveloping a mesh core—in addition to foam-to-single fabric face laminates. Any external skins can be held to a minimum through the use of belts that cover the foam during its use.

7.7.5 Carpet Backing

TDI-based frothed polyurethanes have an excellent balance of physical properties, including good tensile strength, load bearing, tear strength and resiliency. Previous attempts to duplicate these properties with MDI-based systems failed to achieve this excellent balance. Generally, the MDI systems yield inferior resiliency when formulated to provide load-bearing, tensile, tear and elongation properties equivalent to those provided by TDI-based systems. Resiliency can be improved in these systems, typically by reduction of hard segment levels (decreased use of chain extender), but with this reduction comes a loss of tensile, tear, load-bearing and elongation.

Polyurethane-backed substrates such as attached cushion carpeting can be prepared from a polyure-thane forming composition based on a soft segment prepolymer of MDI or an MDI derivative. The use of the prepolymer provides a backing having good strength properties and good resiliency. In 1992, Jenkins[4] was awarded a patent that related to polyurethane-backed substrates and to a process and composition for making same.

The Jenkins patent permits the use of an MDI-based isocyanate while providing for a combination of physical properties, including tensile strength, tear strength, load-bearing, elongation, and resiliency, which is comparable to that previously achieved only with a TDI-based isocyanate.

7.7.6 Miscellaneous Operations

Slab stock is shaped by band saws to tolerances of 1/32 in or less. Thin flexible sheets are produced in rolls by a peeler. Intricate cuts in flexible foams may be fabricated through the use of cutting profiles to form single, triangle, trapezoid, and other profiles in variable heights and dimensions.

Flexible foams may also be formed by a three-dimensional cutting process. This process uses dies made from wood and metal to compress the flexible foam. Part of the distended foam is cut off by a saw, and when the die is removed, the foam takes the desired shape.

Die cutting is used to form various foam slab shapes from flexible foams, and wood- or metalworking equipment is used for rigid foams. Hot-wire cutting may be used in rigid foams for flat sheets and gentle contours. Rigid urethane foam may be readily postformed by a process also called *plastic stage forming*.

During manufacture, rigid urethane foam goes through three distinct and progressive phases:

- *Expansion* stage
- *Plastic* stage, where foam is pliable and tack-free, with no memory
- *Set* stage, where foam has reached the final physical state

The foam is *worked* while in the second phase. Thin sheets of foam in this phase can be shaped around a mandrel (form) and maintained in the shaped position until *set* has occurred, i.e., completion of the third foam phase. The foam is permanently molded to the desired shape and, upon removal from the forming mold, has no tendency to return to the original flat sheet. The shaped foam is free of densification, because the cells are not ruptured during bending; it exhibits no evidence of internal stress or strain. This technique is commonly used to shape rigid urethane foam into pipe insulation.

7.7.7 Equipment

Independent of the technique used to make polyurethane foams, there are some common elements needed for effective production: material supply containers, metering units, mix heads, temperature control systems, and process control systems.

Material supply includes several containers, including the delivery container from the chemical supplier, in-house storage container, blending container, and the container used with the machine. Sometimes the same container can be used for both purposes. Supplier delivery may be in rail cars, tank trucks, rubber bags in trailers, tote bins, and drums. When the delivery is in large bulk containers, the contents usually are transferred to in-house bulk storage tanks. Handling of components in bulk storage containers depends on chemical requirements—for example, to maintain a specified temperature or for agitation to ensure product uniformity.

In high-output foam machines for slab stock foams, materials may be delivered directly to the metering pumps. Otherwise, there is a transfer from the bulk tank, tote bin, or drum to the metering unit day tank. The specifics of the day tanks vary with the particular machinery manufacturer and include single-walled or jacketed tanks (with or without internal or external temperature control coils or plates), insulation and agitation, means for recirculation, and whether tanks are pressurized or nonpressurized. Materials of construction depend on the component, but tanks typically are carbon steel, coated on the inner surfaces with phenolic or epoxy coatings. Stainless steel tanks also are used when stored components are corrosive. Automatic fill systems for day tanks are used to ensure proper conditioning of components.

Metering units are classified as either low-pressure or high-pressure. There also are some hybrid units sharing characteristics of both types. Metering units should be designed to deliver components to a mix head with a high degree of accuracy, usually ±1 percent. Metering units are available for as many components as necessary to meet the foam system requirements. Output capabilities range from a few ounces per second for applications such as pour-in-place gaskets, to outputs above 1000 lb/min for large molded parts or slab stock production.

Low-pressure metering units generally use high-precision gear pumps. Various drive systems are available including direct drive with dc or ac variable-frequency motors, gear motors, chain drives, gear trains, and power pulley drives. Pressures developed by low-pressure metering pumps typically do not exceed 200 psi. Low-pressure machines are used for open-pour and froth applications and for closed-pour applications when the components arc slow reacting enough to be poured through a pour hole.

High-pressure metering machines are frequently called *reaction impingement mixing* or *high-pressure impingement mixing* machines. Mixing occurs by impingement of the chemical components upon each other at high pressure. Although the most common versions of these machines are two-component units, multicomponent units are also available.

With high-pressure machines, impingement pressures can range from 900 to 3000 psi, depending on the mixing requirements. While high-pressure machines are most frequently thought of as single-shot machines for molding applications, they can also are used as pour machines for continuous production. Two different types of metering systems are used with high-pressure machines, pumps, and cylinders.

Metering pump systems use high-precision axial, radial, or in-line piston pumps capable of developing the required pressures. Because of the close tolerances and materials of construction, these pumps cannot be used with particulate materials, such as glass fibers, or with high-viscosity components. These pumps may be either fixed or adjustable output types. Various direct drive systems are used. Output may be varied manually, through dc or variable-frequency ac drives.

Cylinder metering units use lance or plunger type cylinders that are capable of handling abrasive or high-viscosity materials. Cylinders may be driven independently or with slave hydraulic cylinders. Although operating pressures generally are of the same order as the metering pump systems, special units have been made with impingement pressures to 15,000 psi. While the cylinder units usually are made to accommodate a fixed shot size, tandem units have been produced that provide continuous pumping capability.

With the requirement to change CFC blowing agents used in some formulations, it may be necessary to modify existing equipment to handle the new blowing agents. In many cases, particularly with flexible foams, the equipment currently in use can be employed without modification. Commercially available modification units include those that are capable of handling flammable agents, such as pentane, and those designed to handle gaseous agents, such as carbon dioxide and certain HCFCs and HFCs. Care must be taken to ensure that all materials of construction are compatible with the new blowing agent, since some are particularly aggressive to nonmetallic seals.

Mix heads are either low- or high-pressure types. Low-pressure machine mix heads are chambers with rotating impellers driven by electric or hydraulic motors. The impellers may be low- or high-shear units, depending on requirements. Low-pressure heads give excellent mixing and can be designed for a very wide range of output, including low throughput rates. In mix heads used for shot machines, the components recirculate through the head between shots, and back pressure is controlled by one of several means. Spool valves or ports open when the shot is called. For continuous-pour machines, valving is minimized and, once pouring starts, it continues as long as needed. Formulation changes *on the fly* are possible.

The disadvantage of low-pressure mix heads is the need to purge and flush the head following shots or pours. With shot machines, this may not be necessary after every shot, but it does have to be done on a routine basis. After component flow is stopped, the head may be blown out first with air. Then the chamber is flushed with a suitable solvent. Until recently, this was a product such as methylene chloride, which creates cost and environmental concerns. Recently, for some applications, hot-water flushing systems have been introduced.

High-pressure mix heads work on the principle of impinging materials upon each other at high pressure. The high pressure is dissipated as mixing energy and heat. The heads are opened and closed hydraulically. No mechanical mixers are used, and most modern head designs do not need purging or flushing of the mix chamber. For effective mixing, the mix head chamber diameter must be considered. If output is too low for the diameter, poor mixing results. But if output is too high, splashing can become a problem with open-pour techniques.

A variety of special head designs, including *B* heads with special pins below the impingement zone and *L* heads with self-clean and transverse piston, are available to reduce splashing or to improve mixing efficiency. Particularly with high-density microcellular foams, so-called *aftermixers* are used in closed-mold pouring. These units do not aftermix; they create back pressure in the mixing chamber for more effective mixing in that zone.

Temperature control. Good temperature control is essential in maintaining the consistency of foam production. The requirements are particularly demanding for high-pressure equipment, since recirculating under high pressure generates considerable heat. Several design concepts are available. Systems with plate-type heat exchangers, which offer high efficiency, low pressure drop, and ease of maintenance, have become increasingly popular.

Process control. Systems range from very simple relay logic panels that turn the equipment off and on in response to manual actuation, to highly sophisticated programmable logic systems that monitor, control, and provide data output for statistical process control. Programmable systems frequently feature CRT displays for easy access to process control.

Conveying systems. There are a variety of conveying systems used to produce a range of foams. For slab stock foams, there are tunnel conveying systems where the foam is dispensed onto a paper or plastic film that lines the bottom and sides of the moving conveyor, There are many variations, one of which is designed to increase foam utilization from the bun by generating a flattened top surface instead of the old rounded top.

New developments in equipment for flexible slab stock foam permit the use of very high water content formulations that eliminate or significantly reduce the requirement for auxiliary blowing agents. These systems provide for the rapid cooling of the slab to prevent high exotherms that can cause foam degradation. A chamber is located immediately at the discharge of the slab foam line. Normally, the chamber has three sections for air circulation. The first and third sections pull ambient air through the foam block while the center sections use recirculated air blended with refrigerated air through the block. Conditions are controlled for optimum results. In one version of this equipment, the exhaust gases are passed through a carbon absorption unit to eliminate trace contaminant discharge.

There also are several conveying systems available for molded foam, such as rotary tables, carousels, hanging conveyors, and drag chain conveyors. The type of conveying system chosen depends on such factors as the number of parts required per unit of time, foam cure rate, pour technique, and part complexity, which may call for multiple operations in, or movements of, the mold. It should also be noted that an increasing number of highly successful molding operations use stationary molds. These may have the mix head supported on a long boom or use *ring line,* a system with multiple mix heads operated from a single metering unit.

Double belt laminating lines combine the slab conveyor with a molding operation. Laminating lines sandwich foam between various types of facing layers such as paper, film, aluminum foil, and metal sheet. The laminators have precision metal slat conveyors designed to eliminate marking of the face layers and give precise panel thicknesses. The foam is dispensed on the bottom layer in an open-pour technique. The top layer is guided by the upper slat conveyor. Laminated panels are also made with stationary molds, and it is possible to make long panels with special metering units equipped with a flexible lance that conveys the mix head between the facing layers.

Molds. The variety of foam molds nearly matches the variety of foams. They include

- Simple plywood structures used to form supports for panels
- Silicone rubber laces supported in wood or metal frames for wood simulation
- Epoxy molds for cushions and prototype parts
- Laminated glass fiber-reinforced structures for making a variety of parts
- Epoxy-faced cast aluminum-based molds for making instrument panels
- Cast aluminum for seating or mattresses
- Machined aluminum for specialty parts
- Machined steel used for class-A surface automotive RIM parts

Molds may be self-contained or used in a mold carrier or press. Self-contained molds generally are lower in cost than mold-carrier combinations, but the mold carrier can be used for a different mold over a long lifetime and effectively reduces costs.

Mold carriers. These may consist of a simple clamping frame or a high-tonnage hydraulic press for large automotive parts. While the polyurethane molding process uses low pressure, some techniques increase mold pressure requirements. When components are injected into molds at high rates, the mold experiences a high hydraulic ram effect that tends to force it open. Thus, the design of the mold carrier must match the pressure requirements of the selected molding technique.

7.8   Test Methods

The primary standards for evaluating the physical properties of flexible foams are enumerated in ASTM D-3574, "Standard Methods and Molded Urethane Foams of Testing Flexible Cellular Materials—Slab, Bonded, and Molded Urethane Foams." This standard replaced ASTM D-1564 (slab) and D-2406 (molded), which may still be listed in specifications or the literature. Consolidation of the standards was done to update specific tests and to put most tests in conformance with the International Standards Organization (ISO) tests and nomenclature. For the present discussion, most definitions and generalized testing procedures from ASTM D-3574 will be used.

7.8.1 Density

Foam density is the weight per unit volume of the foam expressed in kilograms per cubic meter (kg/m^3) or pounds per cubic foot (lb/ft^3). This density is *not* a measure or expression of firmness, as is the case with latex rubber foams. Density is an important factor, however, in that for a given firmness foam, the higher-density foam generally results in better quality and performance.

7.8.2 Indentation Force Deflection

Indentation force deflection (IFD) is one measure of load bearing and is expressed in newtons (N) per 323 cm^2 or pounds per 50 in^2 at a given percentage deflection of the foam. To obtain the test value, a 323 cm^2 circular plate is pushed into the top surface of the foam, stopping at a given deflection and reading the force on the test instrument scale. The standard test block is 380×380 mm (15×15 in) and 100 mm (4 in) thick.

As an example, a 25% IFD of 133 N means that it takes a force of 133 N to compress a standard 100-mm sample to a 75 mm thickness. It should be noted that, for some applications, the diameter of the circular plate may be reduced or the shape of the plate changed to a square or rectangle. IFD was formerly known as ILD (indentation load deflection).

The suggested practice for specifying flexible foams is given in ASTM D-3453 by a grade number and a performance grade. The grade number specifies the IFD range and certain other minimum acceptable properties. The performance grade is a letter designation for use categories according to dynamic or static fatigue tests. For example, grade 151, performance grade BD, would describe a foam with a 25% IFD of 151 @ 14 N per 323 cm^2, meeting dynamic fatigue requirements for normal-duty use.

In conducting the IFD test, the foam sample is first preflexed for a given number of cycles to open the cells fully. If this is not done, a slightly closed cell foam will appear to have a higher load bearing than it would exhibit in use. This procedure equalizes foams of varying degrees of cell openness. From the density and IFD measurements, a number of factors can be calculated that are useful in comparing foams or in specifying certain desirable characteristics.

7.8.3 Guide Factor

Guide factor (GF) is the ratio of 25 percent IFD to the density of the foam. This term is useful in determining the relative firmness of foams with different densities. The closer the densities, the better the comparison. When densities are different, the foam with the highest guide factor has a cost advantage, but not necessarily a performance advantage. Another term for guide factor is *normalized IFD*.

7.8.4 Sag Factor

Sag factor is the ratio of the 65% IFD to 25% IFD, and it provides an indication of cushioning quality. A high value indicates a resistance to bottoming out as the load is applied. Foams with low sag factor will often bottom out and give poor performance. Other terms for this ratio factor are SAC factor, support factor, hardness ratio, comfort factor, and modulus.

7.8.5 Recovery

In measuring the IFD, values are normally obtained at 25% deflection, 65% deflection. and again at 25% deflection as the load is removed. When the load is applied and stopped at a given deflection, there is a one-minute wait before the test value is read. This is to take into account the spring response of the foam The recovery is given as the ratio of the 25% deflection on release of the load to the original 25% deflection and is expressed as a percentage. High recovery values are desired for cushioning applications where low recovery would be desirable for shock-absorbing applications. Low-recovery foams are sometimes referred to as *dead*.

7.8.6 Initial Hardness Factor

The initial hardness factor (IHF) is the ratio of the 25% IFD to the 5% IFD. This factor defines the surface feel. Soft and supple foams will have a high value, while boardy or stiff surface foams will have a low value. Another term for initial hardness factor is *comfort* factor.

7.8.7 Indentation Modulus

The indentation modulus is determined by plotting a curve of load vs. deflection between the limits of 20 and 40% deflection. The modulus is the load required to increase deflection by 1%. The slope of the line depends on the resistance of the cell walls to buckling.

7.8.8 Hysteresis Loss

If a curve is developed from 0 to 65% deflection and returned without stopping, the hysteresis loss is determined by dividing the area under the curve for the loading curve into the area under the curve for the unloading curve. This value is sometimes used instead of the recovery value, and there can be some confusion, as the terms have been used interchangeably.

7.8.9 Indentation Residual Deflection Force

Indentation residual deflection force (IRDF) is another measure of load bearing and is expressed as millimeters or inches at a given loading. The same circular plate of 323 cm² area for the IFD test is used, but now the plate is weighed with a given load. The normal loadings are 4.5, 110, and 165 N (2, 50, and 75 lb). The original thickness of the foam must be known to make the value meaningful. As in the case of IFD load, the sample is preflexed prior to testing. Previously this test was known as *indentation residual gauge load (IRGL)*.

7.8.10 Compression Force Deflection

Compression force deflection (CFD) is another measure of load bearing and is generally expressed in pascals or pounds per square inch at a given percentage deflection. The entire sample is compressed in this test, and the values are essentially independent of the foam thickness if the thickness does not exceed the length and width measurements. CLD is used to specify the firmness of certain types of specialty foams and some semiflexible foams. Values are used to determine changes in load bearing under various humid-aging or heat-aging conditions. This test was formerly known as compression load deflection (CFD). Please note that there is no direct correlation between the values obtained for IFD, IRDF, and CFD foams.

7.8.11 Resilience

Resilience is a measure of the elasticity or springiness of a foam. In this test, a small steel ball is dropped from a set height onto the foam, and the rebound is expressed as percent resilience. The test is not highly accurate, because the measurement is made by eye. As with recovery, desirable values depend on the application. With a very soft foam, resilience can be misleading, because the foam bottoms out under the load of the ball. This gives low resilience values, even though the foam is very *lively* or elastic. *Ball rebound* is another term for this property.

7.8.12 Tensile Strength

Tensile strength is a measure of the amount of stress required to break the foam as it is pulled apart, expressed in Pascals (Pa) or pounds per square inch (psi). The sample is die cut using a specified ASTM configuration. Tensile strength can be used as a control strength for quality. One common test is the determination of the change in tensile strength after heat aging.

7.8.13 Elongation

Elongation is generally measured during the tensile strength test procedure. It is a measure of the extent to which the foam may be stretched before it breaks and is expressed as a percentage of the original length.

7.8.14 Tear Strength

Tear strength is a measure of the force required to continue to tear a foam after a split has been started and is expressed in newtons per meter (N/m) or pounds per inch (lb/in). The most common tear test for flexible foams is the *split* or *pant leg* tear of a one-inch square sample. There are other tear test sample configurations stipulated for specific applications. Tear is important in determining the suitability of foams in applications where the material is sewed, stapled or *hog-ringed*.

7.8.15 Compression Set

Compression set is a measure of the deformation of a foam after it has been held compressed under controlled time and temperature conditions. The standard conditions are 22 hr @ 70° C. In this test, the foam is compressed to a thickness given as a percentage of its original thickness. Compression set is most commonly expressed as a percentage of the original compression (Cd). Compression set is also frequently determined following humid aging tests. Compression set finds practical applications in fabrication and use of the foam. Where foam is die cut for automotive trim applications, poor compression set usually results in *edge sealing* when the foam is cut. This renders the foam useless for an application where a sharp profile is required.

7.8.16 Fatigue

Fatigue is a measurement of the loss in load bearing under simulated service conditions and is generally expressed as a percentage of load loss. Two of the most common fatigue tests are static fatigue and dynamic fatigue.

Static Fatigue

In this test, the foam is compressed to 25% of its original thickness for 17 hr at room temperature. IFD losses are calculated as percentages of original values. This test is basically nondestructive, and the foam will generally recover completely with time to give the original IFD value.

Dynamic Fatigue

This test is often referred to as the roller shear fatigue test, and there are a number of variations using the same basic equipment. A roller, longer than the foam sample test width, is rolled back and forth across the foam. The roller is mounted in an offset position to impart a shearing action. The test variations include use of constant deflection settings of the roller weights. Losses are calculated in IFD or IRDF units as specified in the test method.

7.8.17 Air Flow

Air flow is a measurement of the porosity or openness of foam expressed in cubic decimeters of air per second or cubic feet per minute (cfm). Air is pulled through a precisely cut sample of foam held in a fixture. The airflow is measured by a flowmeter.

7.8.18 Steam Autoclave Aging

This test consists of subjecting foam samples to varying conditions of temperature and time in a low-pressure steam autoclave. The common terminology for this test is "humid aging." The most common conditions are 3 hr @ 105° C, or 5 hr @ 125° C. The common property differences evaluated are CFD and compression set.

7.8.19 Dry Heat Aging

This test consists of exposing the foam to dry heat for 22 hr @ 140° C, then measuring the change in properties. The most common property difference determined is in tensile strength.

7.8.20 Pounding Fatigue

This test was developed nearly 20 years ago for the European automotive industry, and has become an ISO standard. It is now used not only for its original automotive use but also for cushioning applications. In this test, the foam is subjected to a reciprocating load of 75 kg (165.3 lb) at a rate of 70 cycles per minute for 60,000 cycles. The apparatus is arranged so that the load essentially free falls on the test sample. The change in load bearing and height is subsequently determined.

7.8.21 Flame Tests

Flexible polyurethane foams are organic materials, and as such they will burn if exposed to a sufficiently intense head source in the presence of oxygen. In this regard, polyurethanes are not different from any other organic material, either natural and synthetic. Some of the burning characteristics of flexible foams can be changed through the use of flame-retardant additives or addition of reactive compounds during foam manufacture. Such foams containing the special materials for modification of burning characteristics still must be considered combustible and be should be handled accordingly. As years pass, an increasing percentage of flexible foams in the United States are subject to federal, state, or other standards regarding flammability. Flammability is a major concern if we consider the following statistics:

- In 1986, there were about 6,000 fire deaths in the United States.
- 4,800 of the deaths were in residential fires.
- 70% of the deaths involved smoking materials.
- Total foams flame retarded in 1970, 4%; in 1985, 40%.

- Furniture foams flame retarded in 1970, 1%; in 1985, 67%.

A number of small-scale tests have been developed that have been used to evaluate the flammability characteristics of foams. It is emphasized that *fire retardant, flame retardant,* and *combustion modified* are relative terms and not intended to indicate hazards presented under actual fire conditions. The results of any individual flame test may be influenced by foreign matter such as foam dust or dirt on the surface of the foam. A foam that passes a given test when clean may fail when it is dirty.

7.8.22 Motor Vehicle Safety Standard No. 302

This federal test method is used for characterizing the flammability of all materials in the interior of a motor vehicle. In some cases, composites are evaluated rather than the individual components. The foam sample is 4 × 14 in by 3.5 in thick and is supported in a U-shaped frame exposing a foam area of 2 × 13 in. The frame and a gas fired burner are enclosed in a specified chamber. Burning rate is determined by timing marks on the foam.

As a general practice, the automotive industry requires that foams be rated as nonburning by this test. This is done to give some assurance that the foam would pass the federal requirement for the foam to not exceed a burning rate of 4 in/min. California has issued a number of tests that evaluate the flammability of foams for specific applications. Before making the tests mandatory, extensive testing was done to ensure that foams could be made commercially that met the requirements at a reasonable cost and with benefit to the consumer. These tests have been widely adopted by other jurisdictions.

7.8.23 California Bulletin 117 Flame Test

This test was a modification of Federal Test Method Standard No. 191, for textile applications. It is used to evaluate resilient materials used in upholstered furniture. Test samples 3 × 12 × 1/2 inches are supported vertically in a special test chamber. A sample is ignited by a gas fired burner at the bottom end. The time of burning and the extent of char are determined.

7.8.24 California Bulletin 117 Smoldering Test

This test simulates the problem of upholstered furniture ignition from cigarettes. Foam is cut into small blocks to fit in a wood test stand to simulate a chair shape. The foam is covered with a standard test fabric. A lit cigarette is placed in the crease between the back and cushion of the chair and is then covered with a specified bed sheeting. The weight loss is determined after all combustion has ceased.

As a final comment regarding flammability, a number of studies have been carried out regarding the presence of toxic gases resulting from burning foams. The major toxic gas usually is carbon monoxide. While other toxic gases such as cyanide may be determined, the quantity is no more than that than may be found from burning synthetic fabrics such as nylon.

International efforts to develop realistic tests have resulted in several full-scale or mock tests. Three of these tests are summarized in Table 7.14.

7.9 Safe Handling of Raw Materials

The diisocyanates constitute the main health hazard in raw materials used in producing polyurethanes The raw materials will be discussed in families.

7.9.1 Isocyanates

The toluene diisocyanate (TDI) isomers are irritants to the mucous membranes, particularly of the respiratory system and eyes. The oral toxicity is quite low, and there is no effect on subcutaneous injection. Application to the skin produces minor irritation, which is most harmful to the eye. The most serious effect found with test animals is the irritation caused by inhalation. Personnel with histories of asthma,

TABLE 7.14 Upholstered Furniture Flame Tests

	California Bulletin 113	Boston FD 1X-10	British BS 58-52
Test Method			
Room	12 × 10 × 8 ft with door	12 × 8 × 8 ft	>20 m²
Specimen	Full-scale end use	Full-scale or mock up	Foam and fabric, steel wire frame
Ignition	Newspaper in metal box	Newspaper in brown bag	Wood crib
Criteria			
Flame out time	X	X	X
Weight loss	X	X	Pending
Temperature	X		Recorded
Smoke obscuration	X		X
Carbon monoxide	X		
Penetration		X	Pending

bronchial lesions, or other respiratory disorders should not be allowed to work with these materials. Adequate ventilation is necessary.

Buildings in which TDI is handled or stored should be well ventilated. A hood should be used where TDI vapors are present in the atmosphere. The hood face velocity should be at least 100 ft/min/ft² of hood opening. In spray applications, adequate ventilation is extremely important, because atomization can produce a high concentration of isocyanate vapors in the atmosphere.

The lowest level of TDI that may be detected by smell is 0.4 ppm. At 0.8 ppm, an appreciable odor may be noted. At 0.5 ppm, irritation of the nose and throat will occur. These levels vary greatly from person to person. A concentration of 0.1 ppm has usually been accepted as a safe working level, but limits as low as 0.02 ppm are generally recommended.

Proper protective equipment, such as goggles, should be worn when TDI is being handled. If the concentration in air cannot be reduced below the suggested levels, air- or oxygen-supplied masks should be employed. Spillage of TDI may be wiped up with alcohols (such as isopropanol). Water is immiscible with TDI and will not readily react with it. Instead, it will tend to spread the diisocyanate and cause more harm. Aqueous solutions of alcohols with soap may be used to destroy and clean away TDI. Spills may also be removed with soda ash or ammoniated saw dust.

The volatility of the diisocyanates seems to be the governing factor that determines their inhalation hazard. The aliphatic diisocyanates such as hexamethylene diisocyanate are usually much more volatile than TDI and, hence, are greater irritants. In addition, these compounds are believed to cause dermatitis. Most or the aromatic diisocyanates are less volatile and hence less irritating than TDI. PAPI, because of its lower vapor pressure, irritates less than TDI, but it is still an irritant. Adequate ventilation should be provided, particularly if the PAPI-based foam is being sprayed.

7.9.2 Polyols

Polyols are generally considered to cause no particular problem. At room temperature, they are thick, nonvolatile liquids, so vapor inhalation is unlikely. Some polyols may cause a mild transient eye irritation, but no lasting corneal injury is likely to occur. Ordinary safety glasses will afford suitable protection. If the eyes become contaminated, they should be flushed with copious amounts of flowing water.

Animal tests have shown that most polyols do not cause significant skin irritation, either intact or abraded. A few polyols are capable of causing at most very mild irritation when in prolonged contact with the skin. Extensive human skin tests show that most polyols are neither primary irritants nor skin sensitizers. Swallowing is unlikely to cause difficulties, since most polyols have low oral toxicity.

7.9.3 Catalysts

Amine and tin catalysts should be handled with caution. Vapor inhalation should be avoided. The amine catalysts are skin irritants to a greater degree than the isocyanates but are generally used at low enough levels that problems will not occur unless a worker is unusually sensitive.

7.9.4 Blowing Agents and Surfactants

Fluorocarbon blowing agents and surfactants have a very low order of toxicity and are not skin irritants.

7.10 Formulating Flexible Foams[5]

In this section, typical formulations are given for various types of flexible foams. Considering the advanced technology of polyurethane foams, it would be impossible to cover all the possible variations. Rather, the objective is to provide a guide for further development. Raw material manufacturers may be consulted for more advanced treatments.

7.10.1 Hot-Cure Molded Foams

Formulations for hot-cure molded foams are based on ethylene oxide capped triols. Diols, generally uncapped, are usually blended with the triols to improve the strength properties of the foams. Capped polyols are used because of their increased reactivity, which helps to overcome the loss of exothermic heat cure. TDI is used as the isocyanate source of choice for hot-cure foams. Table 7.15 shows a typical hot-cure formulation for a seat cushion and the resulting properties. An auxiliary blowing agent would be used to make a foam of lower density and thus a softer back cushion. Increased silicone surfactant would be used in this case. A flame retardant additive would be required to allow this foam to meet flammability test requirements for furniture or automotive applications.

TABLE 7.15 Typical Hot-Cure Molded Foam

Formulation, pbw	
Pluracol 732	100
Water	3.5
DABCO7~33LV	0.45
Catalyst T-9 (stannous octoate)	0.05
L-520 surfactant	1.0
TDI (105 index)	43.3
Properties	
Density, per ft^3	1.97
Tensile strength, psi	19.2
Elongation, %	185
Tear strength, lb/in	2.4
Resilience, %	50.4
IFD, lb/50 in^2	
Part thickness	3.17
25%	26.5
65%	68.9
Sage factor	2.60
Guide factor	13.4
Recovery, %	71.7
Compression set, %	
50%	77.4
Humid aged 5 hr at 250° F	
CFD, % of original	
50%	77.4
Compression set, %	
50%	13.9
Heat aged 22 hr at 284° F	
Tensile strength, % of original	120
Air Flow, cfm	1.15

Note: Pluracol 732 is a blend of a 3000 MW capped triol and a 2000 MW diol.

7.10.2 High-Resiliency Foams

High-resiliency foams were developed in response to the need for improved seating in the automotive industry and was furthered by the fuel crises of the 1970s. Most of the early development was done in the United States and followed two basic approaches:

1. Highly reactive standard polyether polyols and reactive diamine cross-linkers with various modifications of TDI were used as the isocyanate source. The reactive diamines were products like MOCA [4,4'-methylene bis(2-chloroanaline)] and a liquid derivative, LD~13. Because of concerns about the possible carcinogenic nature of these products and impending governmental controls on their use, this approach, while technically good, was abandoned.

2. Highly reactive standard polyethers combined with graft polyols with blends of TDI and MDI as the isocyanate source. Early graft polyols were based on the grafting of acrylonitrile and/or styrene to the polyol backbones. The graft polyols were necessary to meet the required load bearing for the seating applications. Development has continued with variations of graft polyols with different ratios and levels of the monomers to improve specific properties. Because of patent restrictions, alternative technology for graft polyols developed. These include PHD (polyharnstoff dispersions) made by reacting water with isocyanate in a polyol media and PEPA (polyether polyamine) made by terminating a polyether chain with a reactive amine instead of a hydroxyl group.

Component ratios and the specific polyols used are varied to make the foam grade required. Table 7.16 is a formulation for an early high-resiliency molded foam. This formulation used a blended isocyanate, TDI, plus crude MDI to get better foam stability and to impart a degree of flame retardance that would meet the automotive flame test requirements.

TABLE 7.16 Typical High-Resiliency Molded Foam

Formulation, pbw		Properties	
Pluracol 7 Polyol 538, pbw	70	Density, per ft³	2.55
Pluracol 7 Polyol 581	30	Tensile strength, psi	15.4
Water	2.5	Elongation, %	139
DABCO	0.1	Tear strength, lb/in	1.7
NIAX Catalyst A-1	0.12	IFD, lb/50 in²	
Thancat DM-70	0.30	Part thickness	4.06
Fomrez UL-1	0.03	25%	Z3.9
Surfactant DC-5043	1.2	65%	71.5
80 TDI/20 crude MDI (103 index)	32.8	Sag factor	2.99
		Guide factor	9.4
		Recovery, %	83.1
		Compression set, %	
		50%	7.3
		75%	5.6
		90%	6.1
Humid aged 5 hr @ 250° F			
		CFD, % of original	
		50%	77.4
		75%	21.5
		90%	65.4
Heat aged 22 hr @ 284° F			
		Tensile strength	90% of original

Note: Pluracol 538 is a 4500 MW capped polyol, and Pluracol 581 is a graft polyol.

With a continuing interest by the automotive industry to reduce the density of high-resiliency foams, developments continued with emphasis in eliminating CFC blowing agents. Increased water level and the use of TDI only led to processing problems. It was found that the addition of a cross-linker, typically diethanolamine, resolved the processing problems. Initially. a flame retardant additive was needed to pass the automotive requirements, but more recent developments in graft polyols have eliminated the need for these agents. The formulation in Table 7.17 includes a flame-retardant additive.

TABLE 7.17 Low-Density, High-Resiliency Molded Foam

Formulation, pbw		Properties	
Pluracol 538	70	Density, per ft³	1.84
Pluracol 581	30	Tensile strength, psi	16.3
Diethanolamine	1.0	Elongation, %	153
Water	3.4	Tear strength, lb/in	1.25
DABCO 33LV	0.4	IFD, lb/50 in²	
NIAX Catalyst A-1	0.125	Park thickness	4.96
Surfactant DC-5043	2.0	25%	18.8
Fomrez UL-1	0.01	50%	35.2
Fyrol CEF	1.5	65%	53.0
TDI (105 index)	42.7	Sag factor	2.81
		Guide factor	10.2
		Recovery, %	77.0
		Compression set, %	
		50%	9.2
		75%	8.5
		90%	10.2
Humid aged 5 hours at 250° F			
		CFD, % of original	
		50%	76.8
		Compression set, %	
		50%	22.9
		75%	42.4
		90%	81.7
Heat aged 22 hours at 284° F			
		Tensile strength	
		% of original	93.9
		Air flow, cfm	1.10
		MVSS 302 flame test	
		No. not to benchmark	3
Average distance burned, in			0.8
		Burning time, s	15

For high-resiliency foam applications where the foam is poured directly into fabric covers (termed pour-in-place or PIP), the TDI-based systems do have some drawbacks, particularly with bleedthrough. One way to overcome this is to use various MDI based isocyanates. These are pure MDI products compounded to be liquids (prepolymers) or are blends of pure MDI prepolymers and crude or polymeric MDI. MDI-based foams have the following advantages over the standard TDI-based foams:

1. Faster demold time
2. Better humid aging properties
3. More versatility because hardness variations with index changes are greater than with TDI
4. There are no requirements for crushing

Table 7.18 shows a generalized formulation for the MDI based foams. Note that triethanolamine has replaced diethanolamine as the cross-linker. The quantities given are calculated to add to a total of 100 for convenience in calibrating a machine.

New systems have been introduced with modified TDI (prepolymers) which give improved performance and can be used for pour in place applications. Variations also are reported to give better flammability characteristics. Continual development of graft polyol types and catalysts contribute to improve high-resiliency foam properties and performance.

TABLE 7.18 MDI Based High-Resiliency Foams

Formulation, pbw	
35 OH EO capped polyol	87.3
Water	
DABCO 33LV	0.6
NIAX Catalyst A-1	0.3
Triethanolamine	0.8
Silicone surfactant	0.2
Auxiliary blowing agent	8.0

Note: 70/30 blend of pure MDI prepolymer and polymeric MDI to a 105 index

7.10.3 Semiflexible Foams

The early development of semiflexible foams for *crash pad* applications in the late 1950s was based on castor oil/TDI prepolymers cross-linked with glycerine. These foams were quite firm and had a *dead* feel, meaning they were slow to recover. Development of di- and trifunctional polyethers improved foam properties and formulation versatility but, into the 1970s, the prepolymer approach was still followed. Further polyol and catalyst developments led to one-shot formulations based on crude TDI, crude MDI, and high free-NCO adducts that were made to perform like the crude isocyanates. The polyols that were developed and continue to be used today are combinations of 4500 to 6500 MW capped triols with low-molecular-weight cross-linkers such as amine-based polyols originally developed for rigid-foam applications. Compared to the quasiprepolymer systems they have generally replaced, the one-shot systems show advantages in

- Ease of formulation to obtain the desired density and load bearing
- Low sensitivity to temperature-induced load bearing changes
- Low-viscosity isocyanate components
- Excellent flow characteristics
- Rapid cure cycles
- Lower cost

While the one-shot foams generally have lower strength than the quasiprepolymer systems, this has not proven to be a disadvantage in practice. One-shot foams are sufficiently strong to meet the requirements of the automotive industry, both in the original application for instrument panel covers and in later developments in arm rest pads, head restraints, door panels, and center console pads. Table 7.19 shows some typical formulations and properties. Developments in recent years have been to improve the flow of the rising foam in ever increasingly complex pad designs, to improve the adhesion of the foam to both the substrate and cover stock, and to eliminate a problem of staining of the cover by components of the foam.

Graft polyols can be used advantageously in one-shot semiflexible foams. At low levels (1 or 2%), the graft polyol acts as a processing aid by giving a more open cell structure that aids flow and reduces shrinkage. As the level of graft polyol is increased, the expected load bearing increase is obtained. By carrying the level to the extreme of replacing all the conventional polyol with graft polyol, very high load bearing foams can be obtained. Foams formulated this way have utility in energy-absorbing applications such as automotive bumper cores, knee bolsters for instrument panels, and shock absorbers in recreational vehicle seating. Formulations are given in Table 7.20.

The one-shot semiflexible foam approach is energy efficient. Mold temperature is not critical *unless* the foam is part of a composite with a plastic film skin. If this is the case, mold temperature should be uniform, about 100 to 110° F, to avoid processing and part-appearance problems. Otherwise, the temperature can range between 70 and 140° F, and good results will be obtained. As the temperature decreases, the foam density increases. Demold time is in the range of 1 to 5 min, depending on the particular catalysis and mold temperatures.

TABLE 7.19 One-Shot Semiflexible Foams (pbw)

Formulations			
Pluracol 380, pbw	95	95	95
Pluracol 355	5	2.5	
Quadrol polyol	—	2.5	5
Water	1.6	1.6	1.6
Dimethylethanolamine	0.75	0.75	0.75
Polymeric MDI (105 index)	37.1	39.1	41.1
Properties—molded foams			
Density, per ft³ (core)	6.9	6.8	7.0
Tensile strength, psi	21	27	29
Elongation, %	60	60	50
Tear strength, lb/in	0.8	0.9	1.0
IFD, psi			
Part thickness, in	1.0	1.0	1.0
25%	6.9	8.9	10.2
65%	17.5	22.6	26.4
Recovery, %	77	73	69
Compression set, %			
50% original	15	18	20
50% humid aged	13	15	17

Note: Pluracol 380 is a 6500 MW EO capped triol; Pluracol 355 and Quadrol Polyol are amine-based tetrols. Dimenthylethanolamine is a hydroxyl-terminated catalyst.

TABLE 7.20 Graft Polyols in Semiflexible Foams (pbw)

Formulations		
Pluracol 581	93	—
Pluracol 715	—	95
Pluracol 355	7	5
Water	2.0	2.0
Dimethylethanolamine	0.75	0.75
Polymeric MDI (105 index)	45.8	42.9
Properties—molded foams		
Density, per ft³ (core)	6.23	6.25
Tensile strength, psi	48.5	55.1
Elongation, %	33	33
Tear strength, lb/in	2.6	2.7
IFD, lb/1 in²	1.0	1.0
Part thickness, in	1.0	1.0
25%	26.5	27.5
65%	63.3	69.7

Note: Pluracol 581 is a graft triol of styrene and acrylonitrile, Pluracol 715 is a high solid graft polyol, and Pluracol 355 is an amine-based tetrol. Dimethylethanolamine is a hydroxyl-terminated catalyst.

7.10.4 Integral-Skin Foams

Integral-skin foams are generally classified as semiflexible foams because of their applications in the automotive and recreational vehicle industries. Low-density integral-skin foams have found wider use in Europe than in the United States. Most of the integral-skin foam used has been in parts such as armrests, horn buttons, steering wheels, and small trim parts. If left unpainted, instrument panels also have been made. One factor that prompted use in Europe more than the United States is that European consumers accept foams in darker colors (e.g., brown, black, and deep green) that can be obtained with pigmentation and therefore do not need to be painted to mask the yellowing that occurs on aging of natural color foam. United States applications have favored lighter colors, which are obtained by post-painting of the part. The foam may be pigmented to a color close to the paint color to reduce the thickness of the paint and to mask discoloration when the paint film wears. The need for painting limits some applications where styling makes it very difficult to avoid pinholes in the skin. Simulated stitching is a particular problem. In spite of these problems, there has been increased use of integral-skin foams because of potential weight savings or application driven advantages. Customer preferences also come into play, as in the case of automotive steering wheels where the soft feel is preferred. Integral skin has even found use as the core for leather wrapped steering wheels to give the best feel.

The cause of the foam yellowing is from chromophoric groups arising from aromatic isocyanates. On exposure to ultraviolet light, discoloration rapidly occurs. In addition to painting to eliminate discoloration problems, other approaches have been used with varying degrees of success. High levels of UV stabilizers and antioxidants can retard discoloration. Aliphatic isocyanates are inherently more light stable than aromatic isocyanates. Catalytic approaches have been found that make the use of the generally slower-reacting aliphatic isocyanates practical. Both of these approaches have the drawbacks of significantly higher costs, color matching difficulty, and the potential for nonuniform color due to swirls. Greater attention has been given to in-mold coating using light stable urethane coatings. The mold is sprayed with a mold release, then the coating is sprayed on before closing the mold. Use of the coating provides a more economical approach than the other light stable products.

Mold construction for integral skin foams is critical, because the quality of the part is directly related to the quality of the surface of the mold and the tightness of the parting line. Any defect in the tool will be reflected in the part. Heat transfer must also be efficient, since the mold temperature controls the thickness of the skin.

Integral skin formulations in the U.S. developed around the use of reactive diamines such as MOCA and LD-813 with various isocyanate combinations. Polyfunctional short-chain polyol cross-linkers were used in other formulations. Still other approaches used 1,4 butanediol or ethylene glycol as the chain extender.

While the diamine approach is an excellent technical solution, carcinogenic concerns have led to further developments. Improvements in properties have come through the use of pure MDI prepolymers as replacements for the crude MDI products in the early work. To avoid problems with trace quantities of water in the resin component, which could cause defects, molecular sieves were added to some formulations to tie up the water. Table 7.21 lists three of the early approaches. Modifications of these can be made to obtain other desired properties.

TABLE 7.21 Integral Skin Foam Formulations (pbw)

Formulations			
Pluracol TPE 4542, pbw	100	—	—
Pluracol 380	—	80	80
Pluracol PeP 450	—	—	10
MOCA	20		
LD-813	—	20	10
TMEDA	15		
TMBDA	—	0.4	
THANCAT 7 TAP	—	—	1.2
Lead octoate	1.5		
CFC-11	16	20	15
72/28 TDI/PAPI	22.9		
Polymeric MDI, index	—	105	105
Properties			
Core density, per ft^3	8.0	12.9	10.8
CFD, psi			
25%	2.3	8.5	1.5
50%	4.5	18.6	3.7

Note: Pluracol TPE 4542 is a 4500 MW EO capped triol; Pluracol 380 is a 6500 MW. EO capped triol; Pluracol PeP 450 is a 400 MW tetrol; TMEDA, TMBDA, and THANCAT TAP are all amine catalysts.

While it is possible to use some of the alternatives to CFC 11 in integral skin foams, there are some technical problems in doing so. HCFC 22 works well, but it is a gas that must be added under controlled conditions to the polyol. When compounded systems are shipped, there is a problem with maintaining a suitable temperature so that the vapor pressure does not increase. The idea of using water as a blowing agent at first sight seems to be the wrong way to go. Extensive work has been done by a number of researchers to develop water blown systems. This has been successful and commercial production has been achieved. A generalized formulation is given in Table 7.22.

TABLE 7.22 Water-Blown Integral-Skin Basic Formulation

6500 MW EO capped triol, pbw	100
Water	0.5
Ethylene glycol	6.0
Amine catalysts	variable
Pure MDI prepolymer, 28% free NCO	105 index

7.10.5 Filled Foams

Filled slab foams began to be of commercial interest in the early 1960s and were developed primarily as a premium furniture cushioning foam. The increased foam density reduced the tendency of the foam to curl up around a person sitting on the furniture. Seating comfort was improved though higher sag factors. However, strength properties were reduced.

Most of the foams were filled with inorganic mineral fillers. The fillers commonly used were barium sulfate, calcium carbonate, and sand. Barium sulfate was the product of choice because of high specific gravity (which reduces the amount of filler needed) and is low in cost. Calcium carbonate is also low in cost, but its specific gravity is low. Using calcium carbonate alone, and in blends with barium sulphate, was used as a way around the patent on the use of barium sulphate alone. Sand is low in cost and has an intermediate specific gravity but, because of its abrasiveness, requires special equipment design for processing.

In the manufacture of filled foams, the filler is first blended into the polyol in a batch system. Good mixing is required to adequately disperse the filer and, once mixed, constant agitation is required to prevent settling. Generally a polyol-filler blend is not used immediately, since a considerable amount of air is entrained during the blending step. This air would cause metering problems. The water content of the blend must be checked before foam preparation, because an appreciable amount of water is normally introduced with the filler. The water must be accounted for in calculating the isocyanate requirements.

Standard foam machines are suitable for processing foams filled with barium sulphate and/or calcium carbonate. Since the viscosity of polyol–filler blends can be quite high (15 to 20,000 cps), there may be a reduction in machine output unless the feed lines are of a sufficient size to handle the high viscosity.

Filled foams have been made in both slab and molded forms in both conventional and high-resiliency types. Use of inorganic fillers makes it very difficult to flame retard the foams. Since one of the objectives of using fillers was to get improved comfort characteristics, the high-resiliency foam approach that can be adequately flame retarded is a better choice for production.

A typical slab-stock, filled-foam formulation and its properties are shown in Table 7.23.

7.10.6 Colored Foams

Foams may be colored for a number of reasons. These include

1. Color coding of various grades
2. Customer specifications for end uses such as toys or coding for special uses
3. Masking of color imparted by other additives that may give an undesirable color
4. Base color for parts that may be painted
5. Indicator for certain additives to ensure that they are added
6. Masking of yellowing that occurs on aging

TABLE 7.23 Filled Flexible Foam (U.S. Patent 3,296,976) (pbw)

Formulation		
Pluracol 664, pbw	100	
Pluracol 381	—	100
Barium sulfate	100	100
Water	2.2	2.2
Auxiliary blowing agent, CFC 11 equivalent	8.0	8.0
Silicone surfactant	1.3	1.2
DABCO 33LV	0.6	0.4
NIAX A-1	—	0.1
Catalyst T-9	0.35	0.25
TDI		
Index	108	110
pbw	33.2	31.6
Properties		
Density, per ft^3	3.15	3.21
Tensile strength, psi	8.2	8.5
Elongation, %	107	143
Tear strength, psi	1.0	1.4
Resilience, %	45.6	48.7
IFD, lb/50 in^2		
25%	29.4	30.3
65%	71.8	64.6
Compression set, %		
50%	4.0	2.8
90%	3.7	30
Air flow, cfm	1.15	1.85

While dyes can be used to post color foams, the processes are time consuming and costly. Incorporation of dyes directly into the formulations has not been very successful, since many dyes adversely affect the catalysis. Coloring with either organic or inorganic pigments is the preferred method. Dry pigments can be blended in plant with one of the components, usually the polyol. This approach can be messy and requires additional equipment to prepare suitable blends. A word of caution: some pigments may also cause adverse reactions. Certain iron compounds have been cited as causing problems with auto-ignition of foams.

The most satisfactory way to handle pigments is by dispersions. A number of companies supply pigments dispersed in polyols or in water. Typically, these dispersions are on the order of 20% solids. Most have sufficient tinting strength so that only a few parts by weight need be added to obtain the desired results. The dispersions can be added to some other component, or they may be added as separate streams if the viscosity is low enough to permit accurate metering and mixing in the mix head. Separate metering is essential if the foam color must be distinct and has to be changed frequently. Colors are difficult to flush from machine lines. It is important that the pigment dispersion be compatible and stable with all components to which it is added. Generally, no other formulation variables are required when pigment are used.

7.10.7 High Load Bearing Slab Stock Foams

Special graft polyols have been developed specifically for use in slab stock foams. The function of the graft polyol is to increase the load bearing. In general, the graft polyols give significantly increased load bearing and improved tensile strength properties. Elongation is often lower, but this would be expected with the increase in firmness.

Simple substitution of the graft polyol for the standard polyol used in a slab stock formulation will give a firmer foam. A greater advantage for the graft polyol would be to produce a more easily processed foam while maintaining the same load bearing/density relationship. This can be seen in the formulations and properties given in Table 7.24. The graft polyol approach permits the use of lower water levels and lower isocyanate index with an auxiliary blowing agent, all of which make an easier-to-process foam.

TABLE 7.24 GMFT Polyol in Slab Stock Foam (pbw)

Formulation		
Pluracol GP 3030, pbw	100	50
Pluracol 637	—	50
Water	4-5	3-7
Auxiliary blowing agent, CFC 11 equivalent		8.0
Silicone surfactant	1.0	1.2
NIAX A-2	0.36	0.30
Catalyst T-9	0.30	0.25
TDI		
Index	108	108
pbw	57.1	48.7
Properties		
Density, per ft³	1.34	1.26
Tensile strength, psi	19.2	15.6
Elongation, %	330	263
Tear strength, lb/in	2.5	2.9
IFD, lb/50 in²		
25%	34.5	32.4
65%	69.7	63.3

Note: Pluracol GP 3030 is a 3000 MW triol and Pluracol 637 is a graft polyol.

The graft polyols for slab stock foams are more reactive than standard slab stock polyols, permitting the use of lower tin catalyst levels. A reduction of 15% would be expected. Tin catalysts provide broad latitude to permit ease of processing. Foams can be made over a broad range of densities by using graft polyols. The graft polyol used in these formulations is an early-generation high solids product. One of the major applications for high load bearing slab foams is for prime carpet underlay. Newer versions of the graft polyols have been specifically tailored to give outstanding properties in this application.

7.10.8 Superfine-Celled Foam

Superfine-celled flexible foams have a soft supple hand, excellent drape characteristics, and a velvet-like appearance. Specialty uses of this type foam include clothing interlining, protective packaging, anti-mar padding on items such as lamps and bookends, wash cloths or sponges, and decorative items.

Such foams can readily be made by increasing the gas generation rate early in the foam reaction. This is done through the use of very high levels of amine catalyst. With the higher amine catalyst level, the silicone surfactant level is also increased to ensure foam stability. Formulation and operating conditions are shown in Table 7.25, along with data for a standard control foam. It can be seen from the data that the conveyor speed had to be increased; this was due to the very fast creaming of the foam. In fact, the foam mixture exited the mix head in a froth consistency. In adjusting the conveyor speed, the rising foam was often cut by the traversing nozzle. This did not cause splits and the foam bonded without knit lines. The top of the bun was heavily patterned by the traverse and conveyor movements, but the scrap losses would not be excessive.

Foam property differences that appear are slightly lower tensile strength, lower resilience, lower load bearing capabilities, high fatigue losses, and higher compression sets for the superfine-celled foam. For the suggested applications, these differences should not be significant. Strength properties are more than adequate for fabrication and end use. The loss of properties can be attributed largely to the fine cell structure, since there is relatively more cell wall compared to void volume than in a normal foam.

TABLE 7.25 Superfine-Celled Foam

	Superfine	Control
Formulation, pbw		
56 OH triol	100.00	100.00
Blue pigment paste	2.0	2.0
Water	4.0	4.0
Silicone surfactant	2.0	1.0
Dimethylethanolamine	2.5	0.4
Stannous octoate	0.25	0.25
TDI (110 index)	50.2	50.2
Conditions		
Component temperatures, °F	70	70
Conveyor angle, °	6	6
Conveyor speed, fpm	12	7
Properties		
Density, pcf	1.53	1.50
Tensile strength, psi	12.4	15.8
Elongation, %	275	270
Tear strength, pi	2.6	3.0
Resilience, %	32	41
Sag factor	2.06	2.00
Recovery, %	58.0	62.0
Compression set		
50%	13.8	8.7
90%	34.6	8.8
Air flow, cfm	2.2	1.8

While the data are shown for slab stock foams, the basic approach of increasing the catalyst to give a stable foam has been used in some pour in place applications where high viscosity of the foaming mixture prevents bleedthrough of fabrics or stitch lines.

7.10.9 High-Resiliency Slab Stock Foams

High-resiliency slab stock foams have found application in furniture cushioning and bedding. The formulations for molded high-resiliency foams are similar. The use of TDI/MDI blends or prepolymers generally would not be used in slab production, since high-resiliency foam production usually has to alternate with conventional foam. The objective would be to use TDI for all types of foam. A basic formulation is shown in Table 7.26. The ratios of the polyols and cross-linkers are varied to get the desired firmness. Increased levels of flame-retardant additive are used to meet the more demanding specifications.

7.10.10 Flame-Retardant Foams

The terms *flame retardant, fire retardant,* and *combustion modified* are relative terms and are not intended to indicate the hazard presented under actual fire conditions. Tests show differences in the burning character of the foams in a specific test. Large-scale tests are more representative to field conditions, but most foams are tested under controlled laboratory conditions.

The primary approach to adding flame resistance to flexible foams has been through the use of plasticizers containing combinations of phosphorus, chlorine, and/or bromine. Reactive flame retardants used in other plastics, including rigid urethane foams, have not performed well because of processing difficulties. The reactive materials are generally short-chain products and cause reactivity problems that are not easily overcome. The question of permanence of the plasticizer-type retardants has persisted for years. The problem has been recognized, either directly or indirectly, by several specifications and test methods. Specifications for institutional items frequently include requirements for flame testing after washing or autoclaving. High fogging levels in automotive tests often were considered to be an indication of loss in flame retardance. While federal safety standards do not contain aging requirements, the

TABLE 7.26 High-Resiliency Slab Stock Foam (pbw)

Formulation	
Pluracol 538	75
Pluracol 715	25
Flame retardant additive thermolin 101	3.0
Diethylethanolamine	0.8
Silicone surfactant DC 5043	0.9
Water	1.9
DABCO 33LV	0.18
NIAX A-1	0.06
Dibutyltin dilaurate	0.015
TDI (105 index)	27.0
Properties	
Density, per ft^3	2.99
Tensile strength, psi	13.7
Elongation, %	153
Tear strength, lb/in	1.7
Resilience, %	66
IFD, lb/50 in^2	
25%	30.2
65%	71.2
Sag factor	2.36
Guide factor	10.1
Recovery, %	87.3
Compression set, %	
50%	3.6
90%	3.0
Humid aged 5 hr @ 250° F	
CFD, % of original	
25%	61.9
Compression set, %	
50%	7.2
90%	19.2

Note: Pluracol 538 is a 4500 MW EO capped triol, and Pluracol 715 is a high-solids graft polyol.

automotive industry has recognized the importance of permanence through the inclusion of accelerated aging in some material specifications. In the development of tests for upholstered furniture in California, aging characteristics were carefully studied.

Some traditional flame-retardant plasticizer products and products introduced more recently are listed in Table 7.27. Some of these are no longer available in the market, but the data from using them are useful in illustrating differences that occur in testing. Firemaster® LVT23P, commonly known as *TRIS*, is a product with a high level of bromine that was a particularly good product for meeting small-scale tests and was also a widely used flame retardant for children's sleepware. It was found that this product was a suspected carcinogen and is no longer in use. When foams burn, heavy smoke is generally generated. Reogard STG is a smoke suppressant that offers some possibility for reduction of this problem. Expandable graphite is used in patented technology for flame retardant foams that will pass the very demanding tests for aircraft seating. It is used in combination with melamine, ammonium polyphosphate and other additives.

Some approaches to flame retardance appear to be specific to tests, that is, one test may give good results and the other poorer results with a given flame retardant. Table 7.28 shows the results from slab foams made with two different plasticizer type flame retardants. The flame tests were the ASTM D-1692 and MVSS No. 302 methods. As can be seen from the data, both flame retardants gave similar results in the ASTM test, but one was significantly better in the MVSS test. The data in Table 7.29 show that, as density is reduced, it becomes more difficult to meet the flammability test requirements.

TABLE 7.27 Flame Retardants for Flexible Foam

Product Plasticizer Type	Formula or Description
Firemaster LVT23P	$(BrCH_2-CHBr-CH_2-0)3-P = O$
Phosgard 2XC20	$(CICH_2-CH_2-O-P-CH_2)$ 2-C-$(CH_2Cl)2$
	$\|$
	$CICH_2CH_2O$
Thermolin 101	$(CICH_2CH_2O)2-P-OCH_2CH_2O-P$ $(OCH_2CH_2Cl)2$ with two O double bonds
Fyrol CEF	$(CICH_2-CH_2-O)3$ P = O
Firemaster 642	42% Br/7% Cl/6.5% P
Firemaster HP35	36% Br/8% Cl/7.5% P
Antiblaze 315 ~ 345	Proprietary liquids containing Br and P
Fyrol PBR	Mixture of pentabromo-diphenyl oxide and aryl phosphates (≈50% Br)
Reogard STG	Smoke suppressant polyester additive
Expandable graphite	

TABLE 7.28 Flame-Retarded Slab Stock Foam

Formulation		
56 OH heteric triol, pbw	100	100
Firemaster LVT23P	8	
Phosgard 2XC20	—	8
Silicone surfactant	1.3	1.3
Water	3.8	3.8
DABCO 33LV	0.3	0.3
N-ethylmorpholine	0.1	0.1
Stannous octoate	0.18	0.18
TDI (108 index)	47.8	47.8
Properties		
Density, per ft^3	1.60	1.55
Tensile strength, psi	18.0	16.3
Elongation, %	333	298
Tear strength, lb/in	3.6	3.6
IFD, lb/50 in^2		
25%	31.3	32.3
65%	58.0	Sg.2
Compression set, %		
50%	6.2	7.5
90%	6.7	8.2
Airflow, cfm	4.0	3.9
ASTM D-1692 flame test		
Total burn time, s	14.7	22.8
Total burn distance, in	1.24	1.69
Burn rate, ipm	5.06	4.44
MVSS No. 302 flame test		
Burn rate, ipm	*	1.46

*Did not burn past benchmark

Because the plasticizer-type flame retardants do not give the kind of performance desired for more demanding furniture applications, particularly to meet the California Bulletin 117 and 133 tests, developments were made with high-resiliency-type foams with greatly improved flammability characteristics. Primary among these were the Mobay CMHR (combustion modified high resiliency) approach and BASF *Rest Easy* or *Code Red* approach. Both of these approaches utilize fillers that cause the formation of char on the surface of the foam, which will quench a flame. A CMHR formulation is given in Table 7.30, and a Rest Easy formulation is given in Table 7.31.

TABLE 7.29 Density Effect on Flame Retarded Slab Stock Foam (pbw)

Formulation				
56 OH heteric triol	100	100	100	100
PHOSGARD 2XC20	10	10	10	10
Water	2.4	2.4	3.2	4.3
Silicone surfactant	1.0	1.0	1.0	1.0
CFC-11	0	5	5	5
DABCO 33LV	0.22	0.22	0.3	0.4
N-ethylmorpholine	0.06	0.06	0.07	0.1
Stannous octoate	0.15	0.15	0.15	0.15
TDI (110 index)	33.6	33.6	42.2	54.0
Properties				
Density, per ft^3	2.44	1.94	1.56	1.26
Tensile strength, psi	11.8	9.5	11.5	14.7
Elongation, %	180	197	262	285
Tear Strength, lb/in	2.2	1.7	2.3	2.8
IFD, lb/50 in^2				
25%	40.4	25.9	25.4	25.7
65%	78.2	49.0	47.4	46.8
Compression set, %				
50%	3.1	2.9	4.6	8.6
90%	2.7	2.9	4.6	8.6
Airflow, cfm	1.58	1.97	3.02	4.50
ASTM D-1692 flame test				
Total burn time, s	125	121	74	54
Total burn distance, in	5.0	5.0	4.53	5.0
Burn rate, ipm	2.39	2.49	3.67	5.56
MVSS No. 302 flame test				
Time burned past benchmark, s	7	18	41	131
Distance burned past mark, in	0.4	1.5	2.06	10
Burn rate, ipm	—	—	5.03	4.58

7.10.11 Soft and Supersoft Foams

Low load bearing foams are desired for certain applications such as bed pillows and soft toppers for furniture to give a plush feel. The common approach is to use high levels of auxiliary blowing agents. Because of the difficulty in venting of solvent vapors from molds, very soft foams cannot be molded, so they are produced as slab stock. Very strong supersoft foams may be made with high-ethylene-oxide containing polyols. However, they have the disadvantage in being very hygroscopic, i.e., being capable of picking up very significant quantities of water. The foam swells when water is absorbed.

TABLE 7.30 Combustion Modified High-Resiliency Foam (pbw)

Formulation	
Multranol 9238	100
Thermolin 101	30
Hydrated Alumina	120
Decabromodiphenyl oxide	22
Anbmony trioxide	8
Poly (ethylenemaleic anhydride)	2.5
Water	2.6
CFC-11	0–7
NIAX A-1	0–05
Diethylethanolamine-LF	1.2
Dabco 33LV	0.8
Dibutyltin dilaurate	0.15
L-5307 surfactant	1.0
TDI (105 index)	33.5
Properties	
Density, per ft^3	3.5–8
Tensile strength, psi	8–11
Elongation, %	C80
Tear strength, lb/in	0.65–0.9
IFD, lb/50 in^2	
25%	20–65
Oxygen index, % °2	27–29
California 117	Pass
Radiant panel, flame spread index	10–25
Smoke density, flame mode	<200

Note: Multranol® 9238 is a PHD polyol

TABLE 7.31 Rest Easy High-Resiliency Foam (pbw)

Formulation	
Pluracol C-133	100
Pluragard melamine powder	100
Water	2.4
CFC-11	5
NIAX A-1	0.06
Diethylethanolamine	1.2
DABCO 33LV	0.18
Dibutyltin dilaurate	0.1
DC 5043 surfactant	1.4
TDI (110 index)	34.9
Properties	
Density, per ft^3	3.45
Tensile strength, psi	8.8
Elongation, %	98
Tear strength, lb/in	1.1
IFD, lb/50 in^2	
25%	36.0
65%	107.3
Oxygen index % °2	30.5
California 117	Pass
California 133	Pass
Radiant panel, flame spread index	C100
Smoke density, flame mode	200

Note: Pluracol® polyol C-133 is a graft polyol for this application

References

1. McGovern, Michael J. U.S. Pat. 5,157,056, October 20, 1992, "High resiliency polyurethane foams with improved static fatigue properties."
2. Favstritsky, N. U.S. Pat. 4,892,892, January 9, 1990, "Flame retardant polyurethane foam compositions containing polynuclearbrominated alkylbenzene," to Great Lakes Chemical Corporation.
3. Joubert, M.D. U.S. Pat. 4,452,920, June 5, 1984, "Flexible polyurethane foams having junction modifying particulate additives," to Joubert & Joubert Proprietary Ltd., Victoria, Australia.
4. Jenkins, Randall C. U.S. Pat. 5,104,693, April 14, 1992, "Polyurethane carpet-backing process based on soft segment prepolymers of diphenylmethane diisocyanate (MDI)," to Dow Chemical Company.
5. After "Advances in Polyurethane Foams Formulation," by Robert McBrayer, Technomic Publishing Co. Seminars, Section 13-1, 25 (1994).

8

Rigid Polyurethane Foams

8.1   General Foam Properties

Some of the properties common to both polyether and polyester urethane rigid foams are

1. The ability to be foamed-in-place by pouring or spraying
2. Versatility in obtaining a broad range of physical properties
3. Combination of high strength and light weight
4. Good heat-resistance properties
5. Excellent adhesion to metal, wood, glass, and ceramics

The equivalent weight of the polyol or combination of polyols has a considerable influence on foam properties. In general, as the average equivalent weight of the polyol decreases from ≈300 to ≈100 or less, the resulting foam has

1. Higher compressive strength
2. Higher closed-cell content
3. Lower rate of water vapor permeability
4. Lower water absorption
5. Higher heat resistance
6. Lower rate of gas loss in solvent-blown systems
7. Greater tendency towards brittleness
8. Greater tendency to scorch
9. Higher cost per pound of foam

In addition, higher exotherms are developed with lower equivalent weight polyols in both the semi-prepolymer and one-shot methods of preparation.

For a given polyol system, the foam properties can be varied by merely changing the density. Properties such as compressive, tensile, flexural, shear, and impact strength; thermal conductivity; water absorption; and modulus of elasticity depend to a greater degree on foam density. Figure 8.1 shows how the compressive strength and water absorption are affected BV density in a typical rigid polyester-based foam. Figure 8.2 demonstrates the effect of density on the shear strength of a typical rigid polyester foam.

The percentage of closed cells in a rigid urethane foam depends on the degree of cross-linking and the surfactant used during foaming, as well as on the polyol equivalent weight. In spray applications, the degree of mixing is sometimes quite influential on the closed-cell content. For most purposes, a high closed-cell content is desirable. A high closed-cell content is necessary for low water absorption, low moisture permeability, and solvent retention in solvent-blown foams. Most rigid foams have high closed-cell contents, 85 to 95% for a 2 lb/ft^3 foam. For certain specialty uses, such as air filters, a low closed-cell content is required.

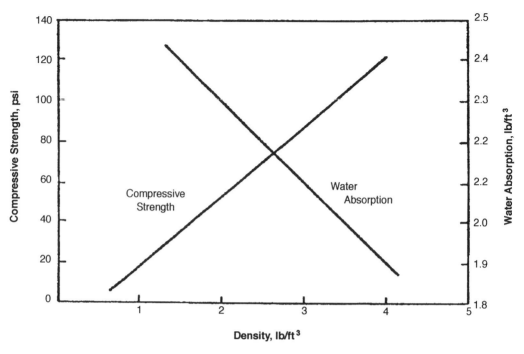

FIGURE 8.1   Variation of compressive strength and water absorption with density of rigid polyurethane foam.

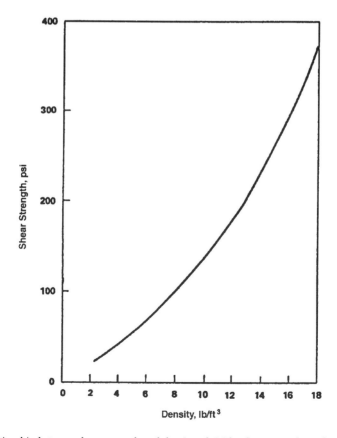

FIGURE 8.2   Relationship between shear strength and density of rigid polyester urethane foam.

8.1.1 Physical and Mechanical Properties

Urethane foams are made in densities as low as 0.5 lb/ft^3 and up to 20 lb/ft^3 or even higher. Foams lower than 1.5 lb/ft^3 are not particularly stable dimensionally. Figure 8.3 shows how density affects some of the more important mechanical properties of rigid urethane foams. This plot covers foam densities up to 7 lb/ft^3. Figure 8.4 shows compressive strength vs. density at increasing temperatures. It is quite apparent that strength increases with increasing density and decreases with increasing temperature, particularly at the higher densities. Figure 8.5 shows the effect of temperature over an extremely broad range, including the cryogenic, on rigid urethane foam.

Table 8.1 shows typical rigid urethane foam properties, including dimensional stability. It is interesting to note here that compression strength perpendicular to foam rise is approximately half that parallel to foam rise. Table 8.2 shows how polyether rigid urethane foams have higher strengths than polyester types. The table includes density variation.

Chemical composition and variables affect the physical properties of high-load-bearing flexible polyurethane foams. In a typical foam system, the following parameters increase load-bearing capacity:

1. Increased isocyanate content
2. Higher functionality of the isocyanate mixture
3. Greater polyoxypropylene triol concentration

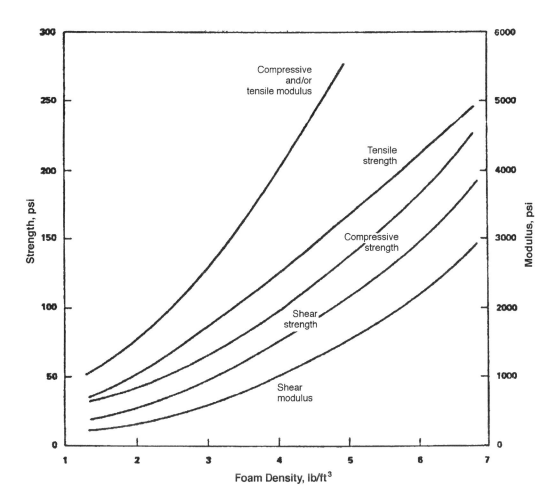

FIGURE 8.3 Effect of density on mechanical properties of rigid urethane foam.

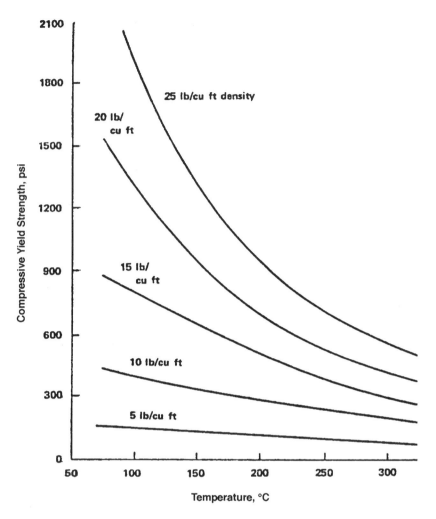

FIGURE 8.4   Relationship between compressive strength, density, and temperature of rigid urethane foam.

TABLE 8.1 Typical Properties of Rigid Urethane Foams

Density (lb/ft³) ASTM D1622	1.5–2.0
Tensile strength (psi) ASTM D1623	30–40
Compression strength at yield (psi) ASTM D1621	
parallel to foam rise	20–45
perpendicular to foam rise	10–25
Compression at yield (%)	5–10
Closed cells (%) ASTM D1940	92–98
Dimensional stability (% volume change)	
@70° C, 100% RH, 2 wk.	7–15
@100° C, 2 wk.	5–10
@–40° C, 2 wk.	0–2

Parameters 2 and 3 cause an increase in the bulk density of the foam, while the density decreases slightly with increasing isocyanate index. Increased quantity of trimerized acids produces a reduction in compressive stress, density and breathability.

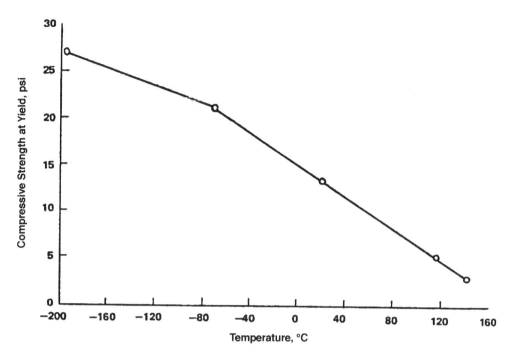

FIGURE 8.5   Compressive strength vs. temperature for a 1.5 lb/ft³ rigid polyether urethane foam.

TABLE 8.2 Compressive Strength Variation with Density for Rigid Urethane Foams

| Density (lb/ft³) | Compressive Strength (Parallel to Foam Rise) | |
	Polyether Types	Polyester Types
2	30–40	25–40
3	65–70	40–50
6	210–220	120–160
9	400	240–280
12	560	400–440

Table 8.3 shows the difference between polyether and polyester flexible urethane foams. The polyester-type foams have greater stiffness and tensile strength than polyether foams of the same density. Generally, properties improve with increasing density, but variations exist among foams of the same type and density.

With regard to compressive strength, most rigid urethane foams have an elastic region in which the stress is nearly proportional to the strain (6.14). They do not entirely follow Hooke's Law, however, because the curve is slightly "S" shaped. In compression, the elastic region varies from 5 to 10% of the initial deflection (strain). In Fig. 8.6, initial stress is nearly proportional to strain, and there is elastic recovery. Beyond the yield point, the foam has little elastic recovery. If a foam is compressed beyond that point, the cell structure is crushed. For low-density foams (<4 lb/ft³), the stress required to crush the foam is about the same as the yield point stress. In this plateau, strain increases with little or no increase in stress. The plateau can extend to 70% compression (strain) in low-density foams. With higher densities, the extent of the plateau decreases, and the stress required to crush the foam becomes greater as the strain increases. The curve in was obtained at a loading rate of 2 in/min on a 1-in sample. At higher loading rates, higher strengths and a smaller plateau may be obtained.

Most low-density rigid foams are anisotropic in that they are stronger in the direction of foam rise. The anisotropic character is generally more pronounced in molded items or panels where the foam rises through a long vertical distance. Under these conditions, a foam may have almost twice as much strength

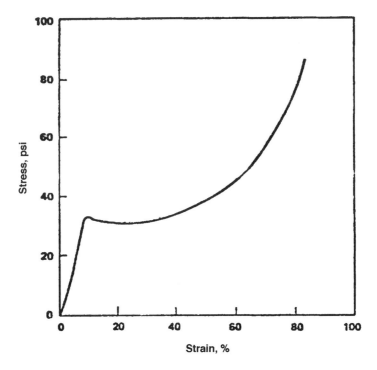

FIGURE 8.6   Stress-strain curve in compression for rigid urethane foam of 2 lb/ft³ density.

TABLE 8.3   Typical Properties of Flexible Urethane Foam (ASTM D-1564).

	Foam Type	
	Polyether	Polyester
Density (lb/ft³)	1.1–2.3	1.6–2.0
Tensile strength (psi)	9–22	25–33
Ultimate elongation (%)	220–310	250–500
Tear strength (lb/linear inch)	2–4	2.3–4.2
Indentation load deflection, 4-in thick sample, Method A, (lb/50 in²)		
@ 25% compression	9–45	—
@ 50% compression	12–60	—
@ 65% compression	16–85	—
@ 75% compression	25–150	—
Compression deflection (psi)		
@ 25% compression	0.1–0.6	0.46–0.48
@ 50% compression	0.2–0.7	0.56–0.80
@ 75% compression	0.4–1.6	1.29–2.09
Compression set, 158° F, Method B (%)		
@ 50% compression, 22 hr	3–4	3.5–6.0
@ 90% compression, 22 hr	6–8	8–20

in the direction parallel to foam rise as in the perpendicular direction. In molded items, the directional properties of the foam can be minimized by overloading the mold, but then a higher-density foam is produced. The use of the frothing process produces a more isotropic foam.

8.2   Insulating Value of Polyurethane Foams

An important way to save thermal energy in commercial buildings is to use proper thermal insulation. Every building material resists thermal transfer to some extent, but rigid polyurethane foams provide the best insulation available. One of the first architectural design considerations is insulation, because of the high cost of air conditioning and heating of buildings, and thus the need for greater insulating efficiency. Polyurethane insulating foams are standard in the fabrication of roofing boards, sheathing, perimeter insulation, sprayed walls/ceilings, industrial tanks, curtain wall panels, etc.

The insulating efficiency of rigid polyurethane foam is unsurpassed. A mere 2 in of rigid polyurethane foam is equivalent in its insulating power to 3 in of polystyrene foam, 3.5 in of mineral wool, 4 in of cork, 6 in of glass fiber matt, 11 in of wood, or 30 in of cemented concrete blocks!

Insulating efficiency is measured by the so-called "K" factor. The K factor measures heat flow in BTUs per hour through one inch thickness of a homogeneous material for a difference in temperature of one degree Fahrenheit. To measure heat flow at other than one inch, we use the "C" factor. Thus, at one inch, the K factor and the C factors are equal.

Mathematically, the K the C factors are defined as:

$$K = \frac{BTU}{hr \times ft^2 \times \frac{°F}{in}}, \; C = \frac{BTU}{hr \times ft^2 \times °F}$$

The R value equals thermal resistance. R is simply the reciprocal of C, i.e., (R = 1/C). C is a convenient mathematical factor because, to determine the thickness of insulation needed to achieve a specific R-rating, all that is needed is to multiply the R by the K factor of the insulation, as T = R × K. The R value determines the thickness of a material needed for a particular insulation need. The higher the R value, the greater the insulation power.

Rigid polyurethane foams are widely used because of their high R value. Rigid polyurethane foams display the highest R value for any given thickness among all the competing insulation materials currently available. For example, one inch of a typical unfaced rigid polyurethane foam has an R value of 6. Therefore, to provide a desired value of R = 19, three inches of a rigid polyurethane foam would be needed, compared to six inches of the best glass fiber matt to provide the same amount of insulating power. Current Industry standards set the following values for rigid polyurethane foams: R = 5 to 6 for unfaced foams, and R = 7 to 8 for faced foams. These values remain unchanged for periods 10 years or greater.

8.3 Formation of Cells in Rigid Foams

In rigid foams, where closed cells are desirable for most applications, the foam system is balanced in such a way that the cell walls do not rupture at the point of maximum gas evolution. The cell membranes require elasticity to permit stretching without breaking. In the preparation of rigid foams, it is necessary to develop sufficient strength to maintain their shape while cooling, or shrinkage will occur because of gas contraction. For this reason, a highly cross-linked structure is most desirable to provide the necessary strength at an early stage of cure. This is shown in Fig. 8.7.

As a general rule, for any rigid foam system, a small excess of isocyanate is favorable for heat distortion characteristics. Increasing the excess tends to improve the heat distortion but renders the foam more brittle. Not all foams will absorb large excesses of isocyanate; this excess acts as a plasticizer, increasing the density and requiring a longer cure at high temperatures to bind it into a foam. This is not always feasible, and it may be said that the maximum excess possible depends on the resin structure and its water content.

8.4 Isocyanates Used in Rigid Foams

The diisocyanates most commonly used in rigid foam production, are modified TDI, MDI, or polymethylene polyphenyl diisocyanate. The isocyanate serves three crucial purposes. First, it may react with water to form CO_2, a suitable gas for foaming. Second, it serves as a di- or polyfunctional reactant,

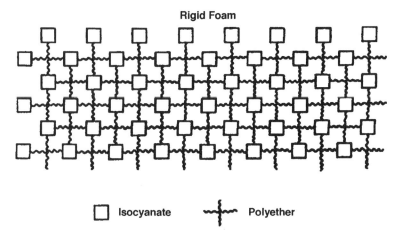

FIGURE 8.7   Highly cross-linked structure provides necessary strength.

joining polyol molecules together by way of reaction between isocyanate and h.ydroxyl groups. Thus, the entire system builds itself into a highly cross-linked polymer. This reaction is strongly exothermic and, consequently, provides the heat required to vaporize the inert blowing agent and also provides the heat required to ensure a good cure of the foam. Third, the isocyanate produces polyurea urea linkages. The polyurea linkages become part of the polymer structure and provide rigidity and thermal resistance to the foam.

Two polymeric MDI-based polyisocyanates are widely used in rigid foams, namely PAPI and Mondur MR. These isocyanates, being polyfunctional, have significantly lower vapor pressures and therefore minimize toxicity problems encountered with isocyanates. They also produce foams with enhanced flame retardance. PAPI-based systems can be used to produce foams with good dimensional stability and structural strength at temperatures as high as 300 to 400° F

The use of lower-activity isocyanates (≈80 to 93% assay) is particularly advantageous in one-shot rigid foams involving highly exothermic reactions during the mixing of the foam components, resulting in scorching of the foam in many cases. Crude isocyanates reduce the incidence of scorching and are lower in cost than the corresponding grades of isocyanates.

The principal commercially available polyhydroxy compounds are ethylene and propylene oxide adducts of polyfunctional active hydrogen compounds, such as glycerine, sorbitol, trimethylpropane, sucrose, etc. Rigid foams require the use of highly functional reactants, which are most useful when based upon polyalcohols having functionalities of three or greater. This is illustrated in Fig. 8.8.

In many cases, a low-boiling gas, such as trichlorofluoromethane (refrigerant 11), may be added to provide part or all of the foaming. The diisocyanate also reacts with the functional groups in the resin of polyhydroxyl compound, ensuring that the resin is built in the final polymer molecules.

8.4.1   Quasiprepolymer Foaming Process (Semiprepolymer)

While the nomenclature is not exact, the term *prepolymer* is used to describe a polyol-diisocyanate adduct with a free isocyanate content of 1 to 12% by weight. The term *quasiprepolymer* is used to describe polyol-diisocyanate adducts with free isocyanate contents as high as 25% by weight. Quasiprepolymers are products with low viscosities and high free isocyanate contents.

The quasiprepolymer technique is really a combination of the prepolymer and one-shot techniques. The polyol vehicle is reacted with diisocyanate to form one component. Water, amine, and additives are mixed with additional polyol to form the second component. The two components are usually mixed in equal quantities. This method is used extensively in the manufacture of rigid foams.

In the quasiprepolymer process, only part of the low-equivalent-weight poly is reacted with all of the di- or polyisocyanate (usually NCO/OH ratio of ≈4:1) to yield a low-viscosity system consisting of a

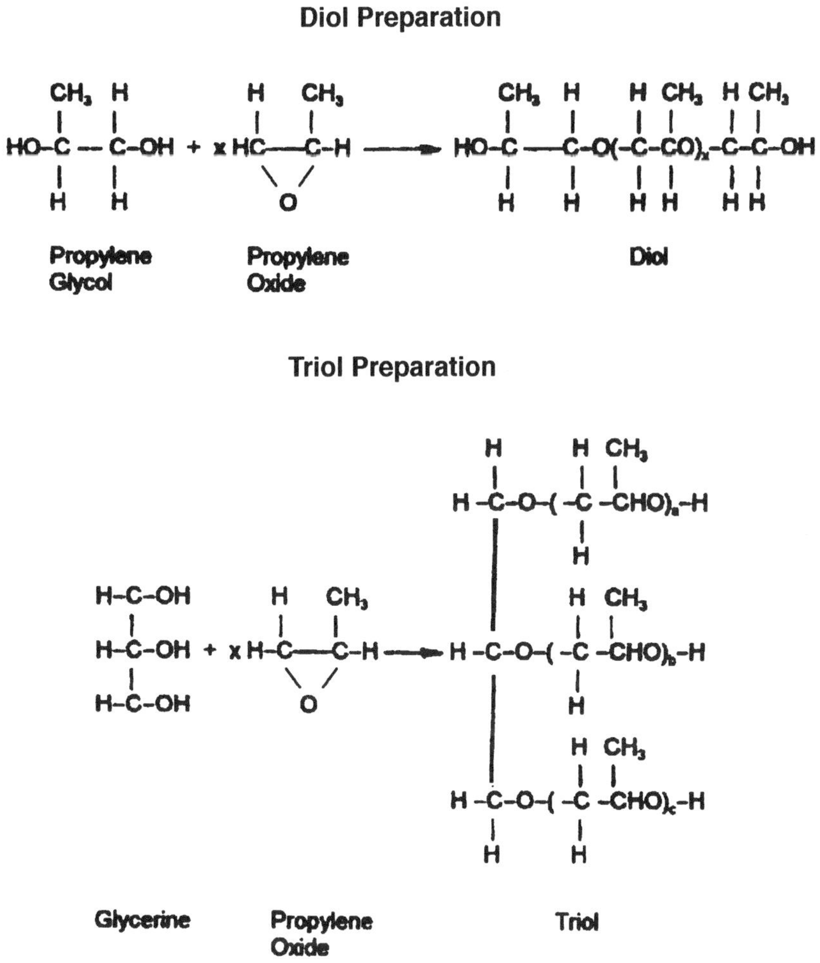

Diol Preparation

Propylene Glycol   **Propylene Oxide**   **Diol**

Triol Preparation

Glycerine   **Propylene Oxide**   **Triol**

FIGURE 8.8   Examples of polyhydroxy compound preparation. In rigid foams, compounds with functionalities of three or greater are most useful.

prepolymer dissolved in the excess of isocyanate. In the foaming step, the semiprepolymer is then reacted with the remainder of the polyol mixed with catalysts, surfactant, and blowing agent. In the absence of water, and using a fluorocarbon blowing agent, the foam polymer should be substantially free of linkages other than urethane. This is a particular advantage, since it has been found that the isocyanate and biuret linkages are the most susceptible to accelerated aging. Depending on the particular hydroxyl compound and final free NCO content, the viscosity will vary from ≈500 cps to as high at 10,000 or even 20,000 cps at 25° C. The semiprepolymer is then combined at the time of forming with an additional amount of hydroxyl compound, a gas producing agent such as water or a volatile solvent, a surfactant, plus a catalyst or combination of catalysts.

The quasiprepolymer system has several advantages. The "A" and "B" components can be made to have very nearly equal viscosities, thus assuring a more rapid and thorough blending of the components. The partial reaction of the TDI with the poly provides lower exotherms when the two components are mixed in the foaming reaction. The TDI is much less noxious when in the form of a quasiprepolymer. Foams of excellent uniformity can be made by this method.

The development of the quasiprepolymer process was brought about to a great extent by the introduction of polyethers, which have largely supplanted hydroxyl terminated polyesters. The lower viscosities of the polyether polyols made the quasiprepolymer practical, since all of the diisocyanate is reacted with part of the polyol. Besides simplification in mixing and meeting, the A component may be heated, and the B component may be kept cool to prevent the loss of the halocarbon. By this means, many other additives, such as fire retardants, extenders, a great variety of catalyst systems, and modifiers, may be incorporated in the B component without regard to reaction with diisocyanate on storage.

Quasiprepolymers may also be formulated into delayed-action, one-part systems by partial blocking of the prepolymer with tertiary-butyl alcohol. Boric acid is included as an auxiliary blowing agent, and the composition also contains surfactant and catalyst. Such blocked adducts are stable at room temperature and produce rigid foams when heated to 150° C.

8.5 Polyols for Rigid Foams

Although the isocyanates are the key functional groups in the formation of urethane polymers, the compounds that furnish the hydroxyl group for the reaction crucially influence the properties of the final urethane polymer. The sources of hydroxyl groups for almost all commercial uses of urethane polymers are (1) polyethers, (2) polyesters, and (3) naturally occurring hydroxyl-bearing oils such as castor oil. Of these, the polyethers are the most important.

Polyethers used are triols or polyols of higher functionality and are based on such starter molecules as glycerine, trimethylpropane, sorbitol, methyl glucoside, sucrose, and certain aromatic derivatives. Polyethers for rigid foams have hydroxyl numbers in the range of 350 to 600, compared to 40 to 75 for polyethers used in flexible foams.

Polyesters, based on dibasic acids such as adipic acid, and polyols, such as diethylene glycol, have been largely replaced by polyethers, particularly in foam applications. Nevertheless, polyesters are still important in some applications, such as foam laminates for fabrics. Castor oil use in foams is minor.

The choice of the polyol has a major influence on the physical properties of the resultant foam. The polyol determines (a) whether the foam will be rigid or flexible, (b) whether it will be brittle or nonbrittle, and (c) the extent of its permeability to gas and moisture. Rigid foams are prepared from lower-molecular-weight, highly branched resins.

The polyol combined with the isocyanate forms the polymer network and provides the rigid foam its structure and characteristic properties. The structure can be varied to give rigid to very flexible foams. The most important characteristics of the polyol are its equivalent weight, functionality, and rigidity or flexibility of chain units.

In general, in rigid foams, the reaction exotherm increases as the equivalent weight of the polyol decreases—hence, there is a greater tendency toward scorching of the foam. The compressive strength and dimensional stability increases with decreasing equivalent weight while the tensile strength decreases. However, as the equivalent weight decreases, i.e., the hydroxyl number increases, the friability of the foams increases. The combined effect of polyether hydroxyl number on the dimensional stability and friability of a typical one-shot polyether foam is shown in Fig. 8.9.

Since it is apparent from Fig. 8.9 that dimensional stability and friability vary in opposite directions with the polyether hydroxyl number, a compromise must be effected to strike a suitable balance of properties. The optimum range of hydroxyl number also varies with the specific type of polyether. Decreasing the equivalent of the polyol(s) also results in foams with lower water vapor permeability and lower water absorption.

The functionality of polyols also has a profound effect on the properties of rigid foams. Higher functionality favors greater heat resistance and dimensional stability (assuming that the equivalent weight is the same in this comparison). The compressive strength of the foams usually increases with increased functionality, while the tensile strength and elongation tend to decrease. Table 8.4 shows the relationship between the functionality of polyethers and compression/tensile strength properties of the resulting foams.

The rigidity of the polyol component also affects foam properties. For example, in polyether systems the use of a polyether having a cyclic initiator such as α-methyl glucoside usually leads to better temperature resistance than a polyether of equal functionality and equivalent weight based on pentaerythritol.

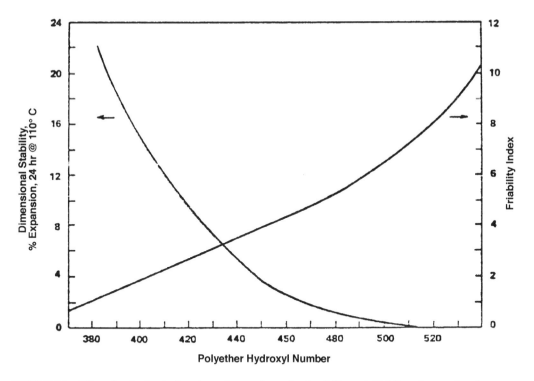

FIGURE 8.9 Effect of polyether hydroxyl number on dimensional stability and friability of a one-shot foam.

TABLE 8.4 Foam Strength vs. Functionality of Polyether Polyol

Hydroxyl number	450	490	450
Functionality	4	6	8
Polyol type	pentaerythritol	sorbitol	sucrose
Component A	65.0	73.4	88.9
Polyester	27.0	35.0	28.4
Trichlorofluoromethane	0.5	0.8	0.6
Stannous octoate	0.1	0.01	0.001
Silicone DC-113	0.5	1.1	0.6
Component B			
Quasiprepolymer	100.0	100.0	100.0
Density (lb/ft3)	1.9	1.9	2.1
Compression strength (psi)	32.0	29.8	55.0
Tensile strength (psi)	60.0	46.9	34.0

In general, polyethers give softer, more resilient foams with better hydrolysis resistance than polyesters, while the polyester-based foams may be expected to have greater tensile strength and better resistance to oils, solvents, and oxidation. With either type of polyol, an increase in the degree of branching results in an increased load-bearing capacity, lower elongation, and a higher glass transition temperature. A reduction in the equivalent weight of the resin usually leads to higher tensile strength, higher load bearing capacity, and an increase in the glass transition temperature.

To impart flame resistance to urethane foams, special polyols have been developed that contain either halogen or phosphorous or both. In the case of polyesters, the use of chlorendic anhydride produces polyols that result in flame-resistant rigid foams. A number of phosphorous-containing polyols, which can be chemically defined either as phosphates, phosphites, or phosponates, are commercially available.

These polyols may be used in conjunction with polyesters but more often are blended with highly branched polyether polyols.

8.5.1   Polyesters

The hydroxyl-terminated polyesters command a particular advantage in that they, and the foams made from them, can be formulated to provide fire retardance or self-extinguishing characteristics. Several types of halogenated dibasic acids, particularly chlorinated phthalic acids or anhydrides, are useful in this manner if the halogen content of the polyester is sufficiently high. The halogen content of the polyester obviates the need for fire-retarding agents, such as halogenated plasticizers, which generally tend to degrade the physical properties of the foam. Formulations are somewhat similar, although such polyesters yield more costly foam.

8.5.2   Polyethers

For rigid foams, the polyethers often used are the propylene oxide adducts of materials such as a sorbitol, sucrose, aromatics, diamines, pentaerythritol, and methyl glucoside. These range in hydroxyl numbers from 350 to 600. Examples are condensates of polyhydric alcohols such as glycerine, sometimes containing small amounts of ethylene oxide to increase reactivity.

The polyether polyol should be a liquid at room temperature, since the reactions that take place with isocyanates are so rapid that a completely homogeneous distribution of reactants must be obtained in a few seconds. The addition of a solvent, especially trichlorofluoromethane, to the polyether poly is generally recognized as providing foams with the best insulating properties and is almost universally employed in low-density foams. The use of this solvent reduces the need for very low polyether polyol viscosities and has permitted the development of highly functional, high-viscosity polyols that provide a good balance of properties.

The polyether polyols are a low-cost and highly refined source of hydroxyls, because they are a single species, made to have very low levels of contaminants, and have uniformly narrow molecular weight distribution. Because of other linkages, the polyols and foams containing them do not have an extremely high degree of chemical resistance but exhibit superior hydrolytic stability.

Quadrol is a special polyol used for making rigid foams to provide increased cross-linking. The tertiary nitrogens provide catalysis for the foaming reaction, and for this reason Quadrol is very useful in sprayed foams requiring a short rise time. Chemically, Quadrol is N,N,N',N',-tetrakis (2-hydroxypropyl) ethylenediamine, and it has the following structure:

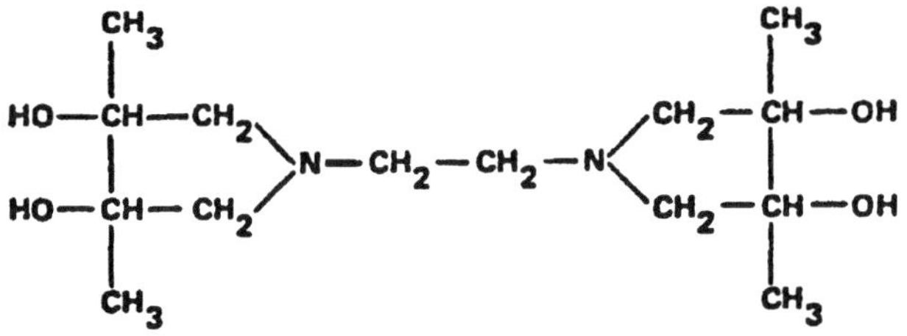

Commercially available polyether polyols usually range in molecular weights from 400 to 6000. The most widely used polyethers for flexible foams, however, are triols having a molecular weight of ≈ 3000. In general, triols of higher molecular weight than 4000 give rise to foams that have larger cell size, are less resilient, and have slightly higher compression sets.

As a rule, using polyols of increasing functionality gives higher compression modulus but decreased tensile strength, tear strength, and elongation. Decreasing the functionality tends to decrease the compression modulus while increasing the other properties. The most widely used triols of 3000 MW have predominantly secondary hydroxyls. Increasing the primary hydroxyl content by capping the polyethers with ethylene oxide increases the rate of cure and may slightly increase the compressive modulus, tensile strength, tear strength, and elongation.

8.5.3 Tertiary Amine Polyols

A rigid polyurethane foam having excellent moldability in mold filling, without the necessity of using a catalyst component, was disclosed by Harada et al. in U.S. Pat. 5,306,735.[1] The process claims to produce a rigid polyurethane foam excellent in mold filling, thermal insulation property, and low-temperature dimensional stability; a spray step for producing a rigid polyurethane foam having excellent mechanical properties and adhesive property; and a process for producing a foamed-in-mold flexible polyurethane foam for use in furniture and automobile cushions. This is accomplished by the use of a tertiary aminoalcohol and having excellent high-temperature moldability at the time of pouring.

Polyurethanes are used in various industrial fields, such as in the formation of elastomers, rigid foam, semirigid foam, flexible foam, and microcellular foam, by virtue of their ease of control of molding density, hardness and other various product properties, and their excellent moldability. In producing these polyurethanes, it is a common practice to use a tertiary amine or an organometallic catalyst as a polyurethane producing catalyst, in addition to a polyisocyanate component, and a polyol component for the purpose of promoting curing or foaming, which enables a polyurethane to be produced on an industrial scale.

Among the polyurethane producing catalysts, tertiary amines are widely used, because they are useful for controlling the balance of the reaction. In many cases, however, they have a strong irritating odor and cause skin irritation (and therefore cause problems in the working environment) and have the drawback that the odor lowers the value of the product. Furthermore, when a rigid polyurethane foam or the like is molded by mold foaming for a use in a refrigerator or a panel, an improvement in the mold filling relating to the fluidity of the resin within a mold is required, so a method for lowering the density in a high yield has been desired in the art.

In recent years, the use of chlorofluorocarbons as foaming agents has been legally regulated for the protection of the ozonosphere, and trichlorofluoromethane (R-11), which has hitherto been used for the production of a rigid polyurethane foam, is among the substances subject to the regulation. This brings about a problem of the necessity of reducing the use of trichlorofluoromethane. In the chlorofluorocarbons-poor formulation wherein the amount of water used as a foaming agent is increased, the increase in the amount of water inevitably accelerates the reaction of water with the polyisocyanate component. This causes the amount of formation of a urea bond derived from the evolution of carbon dioxide to be increased, so the balance between the foaming reaction and the resinification reaction is lost, which causes the mold filling of the polyurethane form to be significantly lowered. The use of 1,1-dichloro-2,2,2-trifluoroethane or 2,2-dichloro-2-fluoroethane instead of trichlorofluoromethane makes it necessary to increase the amount of water use, because the low-temperature dimensional stability, compressive strength, and mold filling are lowered thereby. This, however, causes the mold filling to be further lowered.

The rigid polyurethane foam, produced by a process including a spray step (a spray-type rigid polyurethane foam, hereinafter), is used mainly for the thermal insulation of the internal wall and ceiling of houses, and the thermal insulation of tanks. A special foaming machine is used for the foaming work of the spray-type rigid polyurethane foam. An air spray foaming machine is a system wherein compressed air is introduced into a mixing gun, while an airless foaming machine is a system wherein a feedstock is introduced into a mixing gun through the use of a lightweight compressor, and then sprayed. A liquid mixture comprising a polyol component and an isocyanate component is sprayed on a face of an article through the use of the above-described foaming machines, and a thermal insulation layer, incorporating a rigid polyurethane foam, is formed on that face through the utilization of the properties of the mixture of rapidly thickening, foaming and forming a high-molecular weight polymer.

This spray-type rigid polyurethane foam has met with a great deal of success, and an increase in the amount of use has brought about various problems. One of the problems is that the bonding strength between the foam and the adherend material is so poor that the foam peels off or falls down with the lapse of time to impair the thermal insulation effect, so dewing becomes liable to occur.

Furthermore, the regulation of the use of chlorofluorocarbons such a trichlorofluoromethane has brought about a tendency to increase the amount of water incorporation in the foaming agent, which exacerbates the problems. When the amount of the chlorofluorocarbon subject to the regulation is reduced by increasing the amount of incorporation of water, the agglomeration caused by a urea bond formed by the reaction of water with the isocyanate violently occurs. Furthermore, the boundary between the urethane foam and the adherend or the surface of the foam suffers from less accumulation of the heat of reaction, which brings about drawbacks such as a lack in self-bonding strength (which is the most important property of the spray-type rigid polyurethane foam) and an increase in fragility. This tendency becomes conspicuous in conducting the spraying at a relatively low temperature of 5° C or below.

The flexible, hot-mold foam is produced by blending and sufficiently mixing a polyether polyol, a polyisocyanate, a foaming agent, a silicone foam stabilizer, and a catalyst, pouring the mixture into a mold, and then heating the mixture to allow a reaction to proceed. In this case, after the temperature of the mold is adjusted to 35 to 45° C, a urethane feed-stock is poured into the mold to conduct foaming and cured in a furnace at 160 to 200° C, and the cured foam is demolded. When the temperature of the mold exceeds 45° C, a crack occurs within the foam so that no good product can be obtained. Although trichlorofluoromethane is used in the production of a foam having a low density and a low hardness, it is desirable to reduce or discontinue the use of trichlorofluoromethane. These difficulties have been overcome by the use of the tertiary aminoalcohols of the Harada invention.

Because each of the tertiary aminoalcohols has a tertiary amino group in its molecular skeleton, they exhibit catalytic activity in the reaction of a polyisocyanate compound with an active hydrogen compound. Furthermore, the tertiary aminoalcohol, as such, reacts with an isocyanate group by virtue of the presence of a terminal hydroxyl group and consequently is incorporated in the polyurethane resin skeleton. In addition, since the tertiary aminoalcohol represented by the general formula (I) is a diol type, it neither inhibits an increase in the molecular weight of the polyurethane resin nor deteriorates the final properties. Therefore, unlike the conventional tertiary amine catalyst, the tertiary aminoalcohols are less liable to cause a bad odor, because they have a terminal hydroxyl group and a molecular weight on a certain level. Therefore, although they are incorporated in the polyurethane resin skeleton, neither the polyurethane resin nor the polyurethane foam has a bad odor, so no lowering of the commercial value of the product occurs.

In the production of a rigid polyurethane foam, the balance between the gas evolution rate and the resin cure rate in the reaction is important for improving the mold-filling characteristics of the formulation. When the gas evolution rate is higher than the resin cure rate, no sufficient amount of gas can be entrapped in the resin, and no necessary foam volume is obtained, so the mold filling becomes poor. On the other hand, when the resin cure rate is higher than the gas evolution rate, the resin viscosity becomes so high that the so-called *liquid flow* declines, which causes the mold filling of the rigid polyurethane foam to decline.

Also, when water and trichlorofluoromethane are used in the conventional proportions as a foaming agent, the enhancement in the resin cure rate through a change in the proportion of the polyol, catalyst, or the like, for the purpose of improving productivity or such, causes the balance necessary for the mold filling between the gas evolution rate and the resin cure rate to be lost, which lowers the mold filling of the rigid polyurethane foam. By contrast, the use of the tertiary aminoalcohol as part or the whole of the polyol component, or the use of the tertiary aminoalcohol, promotes the resinification in an early stage of the reaction and the gasification of trichlorofluoromethane, so the balance necessary for the mold filling between the gas evolution rate and the resin cure rate is maintained, and the mold filling is improved.

In the chlorofluorocarbon-poor formulation, wherein the amount of use of trichlorofluoromethane has been reduced, because the amount of water used is increased, a rapid foam-curing reaction lowers the mold filling of the polyurethane foam. In such a formulation, the use of the tertiary aminoalcohol(s), according to the present invention, makes it unnecessary to use the conventional catalyst component and further suppresses the reaction of water with the isocyanate group by virtue of the feature of the tertiary aminoalcohol(s), according to the present invention, so the mold filling of the polyurethane foam is not impaired.

When 1,1-dichloro-2,2,2-trifluoroethane or 2,2-dichloro-2-fluoroethane is used instead of trichlorof-luoromethane, the mold filling lowers due to drawbacks such as a difference in the boiling point between these substances and trichlorofluoromethane, a lowering in the resinification reaction rate due to the dissolution in the resin, and an accompanying delay of the evolution of the chlorofluorocarbon gas. By contrast, the use of the tertiary aminoalcohol(s) according to the present invention enhances the resin-ification reaction rate to prevent the lowering in the mold filling of the polyurethane foam.

In the reaction of the polyol component containing the tertiary aminoalcohol with the isocyanate in the production of the spray-type polyurethane foam, it is possible to complete the reaction through an enhancement in the reaction rate in proportion to the amount of incorporation of the tertiary aminoal-cohol(s), and the reaction can proceed at a low temperature of 5° C or below. Since the reaction can sufficiently proceed at such a low temperature, the necessary mechanical properties and bonding strength of the polyurethane foam can be maintained so that neither peeling nor falling of the thermal insulation layer from the adherend occurs after the spraying.

Furthermore, also in the formulation of the spray-type polyurethane foam wherein the amount of use of chlorofluorocarbons subject to regulation, such trichlorofluoromethane is reduced, and water is used in a large amount of a foaming agent, a desired bonding strength can be obtained, and neither peeling nor falling occurs, even though the reaction proceeds at a low temperature of 5° C or below.

8.5.4 Halogenated Polyether Polyols

A family of halogenated polyether polyols that permit the production of permanently fireproof polyure-thane foams was disclosed by Boulet in U.S. Pat. 4,173,710.[2] Rigid polyurethane foams have many varied uses in industry, particularly in the fields of building and insulation, where resistance to fire is a desirable or even indispensable property. Various means exist for imparting fire-resisting properties to polyurethane foams. A well known process consists of the incorporation in the foams of fireproofing additives such as antimony oxide, or else halogenated and/or phosphorus compounds such as tris(dibromopropyl) or tris(dichloropropyl) phosphates, chlorinated biphenyls, and halogenated hydrocarbons. These additives, which are not chemically bonded to the base polymer, are incapable of providing uniformly distributed permanent resistance to fire. Moreover, they generally have a plasticizing effect on the foam and conse-quently impair its mechanical properties, particularly its compressive strength and dimensional stability. Another means of producing fire-resisting polyurethane foams consists in using halogenated and/or phosphorated polyols.

In French Pat. 1,350,425 of 12.3.1963, in the name of Olin Mathieson Corp., there is described the use of halogenated polyether polyols produced by adding epichlohydrins to monomeric polyhydric alcohols containing at least two hydroxyl groups. This addition reaction yields halogenated polyether polyols having a number of secondary hydroxyl functions equal to the number of hydroxyl functions of the starting hydroxyl reactant. The cellular polyurethanes resulting from the reaction of organic polyiso-cyanates on these halogenated polyether polyols certainly have satisfactory permanent fire-resisting properties, but their dimensional stability is generally inadequate. Moreover, their production is difficult because of the low reactivity of these polyether polyols; this reactivity is still poorer than that of the corresponding nonhalogenated polyether polyols.

The chlorinated polyether polyols according to the Boulet invention are characterized by the presence of alpha-diol groups; they contain primary and secondary hydroxyl groups that are not deactivated by the immediate proximity of chlorine atoms. Because of their special properties, the halogenated polyether polyols, according to the invention, have numerous different applications, such as the production of alkyde resins and additives for epoxy resins. The polyether polyols are also suitable for the production of chlorinated and phosphorated polyether polyols by reaction with organic and/or inorganic phosphorus compounds such as phosphorous, phosphoric, pyro- and polyphosphoric acids, mono- and diphosphonic acids, and their esters.

It has been found that the halogenated polyether polyols are suitable for the production of rigid and semirigid fire-resisting polyurethane foam. The halogenated polyether polyols permit the production of fireproof polyurethane foams possessing mechanical properties similar to, if not better than, those of commercial nonhalogenated polyether polyols.

The halogenated polyether polyols can be used alone or in mixtures for the production of polyurethanes. The relative proportion of halogenated polyether polyols, according to the invention, in the mixture of polyether polyols used may vary within a fairly wide range. The self-extinguishability properties of the resulting polyurethane are obviously the better, the higher this proportion. Self-extinguishable rigid polyurethanes, according to the standard ASTM D-1692, can be obtained by using polyether polyol mixtures containing one or more nonhalogenated polyether polyols and 40%, preferably 70%, by weight of those polyether polyols that have the lowest halogen content and only 20 to 35% by weight of those halogenated polyether polyols which have the highest halogen content.

The rigid and semirigid polyurethane foams are produced by reacting halogenated polyether polyols, or mixtures of nonhalogenated polyether polyols and halogenated polyether polyols, respectively, and organic polyisocyanates in the presence of a foaming agent and of one or more reaction catalysts, optionally water, emulsifying agents, and/or stabilizing agents, filling materials, pigments, etc.

The halogenated polyether polyols are suitable for the production of polyurethane foams by any conventional foaming methods, such as the single-stage (*one-shot*) or methods utilizing a prepolymer or a semipolymer—the so-called *frothing* preexpansion method. The theoretical amount of polyisocyanate necessary for the production of polyurethane is calculated in known manner in dependence on the hydroxyl index of the polyether polyol or polyols and, where applicable, of the water which are present. It is advantageous to use a slight excess of polyisocyanate so as to ensure an isocyanate index of 105 to 120, which improves the hot distortion resistance of the resulting rigid polyurethane foam.

Triethylamine is a particularly preferable catalyst. The amount of catalyst may vary to a certain extent; it affects the mechanical properties of the resulting foam. A range from 0.1 to 3% by weight of catalyst, referred to the polyether polyol or mixture of polyether polyols, is generally used.

The chlorinated polyether polyols forming the object of the invention may be obtained by oligomerization, co-oligomerization, condensation, dehydrochlorination, and hydrolysis, the starting materials comprising, on the one hand, epichlorohydrin, and on the other hand, water or di- or polyhydroxylated compounds that may optionally be halogenated and/or have ether oxide bonds, and/or double bonds capable of being halogenated in a subsequent stage.

8.5.5   Castor Oil

Castor oil, a triglyceride of 1,2-hydroxyoleic acid (richinoleic acid), and some of its derivatives have been the basis of a number of rigid and semirigid foams. Although castor oil is one of the least expensive raw materials for rigid foams, improved strength and heat resistance are generally obtained in the resultant foams, compared to those based on conventional polyesters. Because it contains secondary hydroxyl groups and has a fairly low hydroxyl number, the prepolymer method of foaming is generally favored.

Castor oil and its polyol derivatives have been found to be highly useful in the preparation of low-cost rigid, semirigid, and flexible urethane foams. Foams based on castor oil are resistant to moisture attack, possess excellent shock-absorbing properties, have good low-temperature flexibility, and can be readily flame laminated. Castor oil may be used as the sole polyol in foam systems or as a modifier for polyether-based foams. Castor polyols are also excellent pigment dispersing agents and provide superior color development with colored urethane foams. Castor polyols are currently being used successfully in the manufacture of foam packaging in applications demanding high shock-absorbing properties, in clothing interliners, and in filters.

Foams based primarily on castor oil are semirigid at low densities, becoming more rigid as the foam density increases. They generally (unless special formulations are used) possess an open-cell structure. In the preparation of one-shot flexible foams based on oxidized castor poly (Polycin 120), 99 to 100% open cells are always obtained. This desirable feature is of help in formulating for filtration and interliner application and reduces the necessity for crushing foams to obtain 100% open cells.

Castor poly-based foams, even when formulated to be rigid, possess more "give" than rigid foams based on polyesters and polyethers. In crash-resistant packaging and for shock absorbency, this low resiliency serves to absorb shock rather than allowing the sharp rebound inherent in other team systems. Castor polyols also provide foams with excellent low-temperature properties for packaging.

When foamed in place, castor oil-based foams exhibit good adherence to a variety of substrates. Foaming pressures and exotherms are low. The exotherm, in some cases, is sufficient for complete cure. For uninhibited expansion, the foaming pressure is usually 2 to 3 psi. Approximately 20 psi can be developed during the closed expansion of high-density foams. Shrinkage is encountered during the cure of foams based on castor oil, the precise amount depending on the grade of castor oil used. Often, a finely divided inert material, such as silica, is dispersed in the prepolymer to lessen shrinkage. The electrical properties of properties of foams produced form castor oil polyols vary little with density.

Castor oil foams exhibit excellent heat insulating qualities. There is little change in the K factor with density. This is particularly important in applications where an increase in density is needed to meet strength requirements. The chemical nature of these foams is such that they exhibit thermoplastic tendencies at high temperatures. This limits them to applications where the service temperatures do not exceed 200° F. At the other end of the temperature scale, such materials behave as rigid foams. At temperatures as low as 94° F, the foams will deform but not crumble. Because of their open-cell structure, polyurethane foams based on castor oil should be coated or otherwise protected if they are to come in contact with water. Water can be absorbed into the foams by capillary action. Exposure to high humidity at room temperature may effect an initial increase in compressive strength. This group of foams is inferior to other polyurethane foams in chemical and solvent resistance. Contact with certain solvents may soften and swell the foams: however, upon removal of the solvent, they generally regain their original properties.

8.5.6 Improved Polyether Polyols

An improved method of producing polyether polyol compositions which contain alkoxyalkanol amines by oxyalkylating a polyhydric initiator was disclosed by Klein in U.S. Pat. 4,166,172[3] wherein a polyhydric initiator is oxyalkylated with alkylene epoxides at lower temperatures in the presence of an aqueous ammonia solution. The polyether polyol compositions produced have desirable hydroxyl numbers and low viscosities and are particularly useful in the production of excellent rigid polyurethane foams. In addition to catalyzing the reaction, the nitrogen-containing moiety itself undergoes oxyalkylation during the process such that the polyether polyol compositions produced contain, admixed therewith, various alkoxyalkanol amines. The alkoxyalkanol amines having reactive hydroxyl groups and acting as a mild catalyst for the polyolisocyanate reaction need not be removed from the polyether polyol composition. In addition, the alkoxyalkanol amines increase the blending compatibility of the polyether polyol compositions.

Polyoxyalkylene polyols or polyether polyols are well known. Such polyether polyols are known to be formed by the reaction of a polyhydric compound having from about two to eight hydroxyl groups with a 1,2-epoxide such as ethylene oxide, propylene oxide, or higher alkylene oxide in the presence of a basic catalyst such as aqueous sodium or potassium hydroxide. The polyether polyols produced are useful as reactants with isocyanate containing compounds to form polyurethane material and particularly polyurethane foams.

The above-mentioned method of producing polyether polyols is less than desirable, however, in that the reaction requires a subsequent refining step that includes the neutralization of the caustic alkali catalyst with subsequent removal of the precipitated salts. In addition, the presence of aqueous caustic alkali in the reaction medium is known to facilitate undesirable side reactions. Specifically, the alkylene oxide and water combine to produce diols. These diols tend to decrease the functionality of more desirable higher-functionality polyol compositions.

In an effort to avoid the subsequent refinement step and/or the production of diols, various methods have been proposed. For example, oxyalkylation of the relatively high-melting polyhydric initiators has been proposed where the solid initiators are fused at high temperatures in the presence of an alkylene oxide. This method, while avoiding the disadvantages of the previous method, damages and discolors the final product because of the high temperatures required. Other proposed methods involve the use of nonaqueous solvents with a compatible basic substance; however, most of these methods require catalyst removal and/or solvent recovery prior to using the produced polyols in polyurethane foams. For example, it has been disclosed that certain amine compounds can be used as both a solvent and a catalyst for polyether polyol production. One process, as disclosed in U.S. Pat. 2,902,476, uses lower alkyl tertiary amines, and specifically triethyl-, trimethyl-, and tripropylamines, as a solvent and a catalyst in the reaction of propylene oxide with polyhydric initiators. Water is specifically excluded from the reaction mixture.

While this process eliminates the inherent difficulties encountered with aqueous reaction mediums, it involves the use of expensive, purified solvents. Additionally, trialkyl amines are poor initiator solvents. Thus, large amounts of amine solvent are required to form the desirable single-phase reaction mixture. The presence of large quantities of trialkyl amines in the polyol product is not desirable. Specifically, such substances are highly odoriferous in urethane foam products and strongly catalyze isocyanate-polyol reactions. Therefore, small amounts of these substances must be utilized and/or the solvent must be removed from the polyether polyol composition prior to foam formation, thus requiring a removal step. When small amounts of tertiary alkyl amines are utilized, the amount of solvent is insufficient to form a homogeneous single-phase reaction media. The resulting solid-liquid-gas heterogeneous reaction is difficult to control adequately.

Another such process, disclosed in U.S. Pat. 3,332,934, uses a triethanolamine catalyst-solvent for the reaction of propylene oxide with a polyhydric initiator. As in the previously disclosed process, the reaction proceeds in the absence of water. Pure triethanolamine, like trimethylamine, is relatively expensive. Likewise, triethanolamine is a poor initiator solvent. When those amounts of triethanolamine required to produce polyols of desirable hydroxyl number (i.e., from ≈400 to 600) are used, a heterogeneous slurry of the solid polyhydric material is formed. Thus, the oxyalkylation occurs in a gas-liquid-solid phase reaction. As mentioned previously, such a system is difficult to control with the rate being determined by the solubility of the solid initiator. The time of reaction ranges from 7 to ≈20 hours. If larger amounts of triethanolamine are used, a reformulation of the catalyzed polyol-isocyanate foam reaction is required.

Furthermore, some of the polyether polyols produced using pure triethanolamines exhibit viscosities that render them difficult to ship and use in standard urethane systems. Attempts to use solvents, such as a fluorocarbon, to reduce viscosities limits the use of the polyether polyols in producing low-density foam compositions. Therefore, a process for producing polyether polyols which is compatible with urethane systems, is relatively easy to control, and uses relatively inexpensive starting materials—but does not suffer the inherent drawbacks of caustic alkali catalyzed systems—would be desirable.

Klein, in U.S. Pat. 4,166,172, found that suitable polyether polyols, including those having hydroxyl numbers from ≈400 to ≈650 with viscosities from ≈1,000 to 20,000 centipoise, can be produced in a single process using the relatively inexpensive starting materials of aqueous ammonia and one or more alkylene oxides with a polyhydric initiator. The reactions proceed relatively quickly in a homogeneous reaction media at lower temperatures. The ammonia itself becomes oxyalkylated, producing alkoxyalkanol amines that are compatible with urethane systems and need not be removed prior to the polyol-isocyanate reaction. These alkoxyalkanol amines increase the blending compatibility of the polyol compositions with other polyols. Surprisingly, the production of diols is relatively small, and no refinement step is necessary to remove the nitrogen-containing moiety. Furthermore, it was found that, by varying the amounts (concentration) of the ammonia initially added, the viscosities of the polyether polyols produced can be effectively lowered without materially affecting properties of the foam.

In summary, according to the broader aspects of the Klein invention, polyether polyols are produced by oxyalkylating a polyhydric initiator at lower temperatures with an alkylene oxide in the presence of an effective amount of an aqueous ammonia solution. The nitrogen-containing moiety also undergoes oxyalkylation during the process such that the final polyether polyol contains admixed therewith various alkoxyalkanol amines that contain reactive hydroxyl groups and act as a mild catalyst for polyol-isocyanate reactions.

8.6   Rigid Foam Catalysts

In rigid urethane foaming systems using the CO_2 formed by the water-isocyanate reaction, a balance of the relative rates between the urea and urethane reactions is necessary. If the urethane reaction is not fast enough, the gas will not be trapped, and no foam will be formed. On the other hand, if the urethane reaction is too fast, the polymer will set up before the gas is formed, and a high-density foam will result. The foaming reaction is much less controllable by catalytic action than the urethane reaction.

Tertiary amines can be used alone as catalysts but, for some applications, such as spraying, more speed is desirable. Metal salts, particularly tin salts, accelerate the foaming reactions and can be used alone or in combination with the tertiary amine type catalysts. Tin catalysts of importance for rigid urethane foaming are stannous octoate and dibutyl tin dilaurate. Stannous octoate will hydrolyze rapidly in the presence of a basic catalyst with loss of activity. Master batches containing stannous octoate and water are stable for only a few hours at room temperature.

Resin master batches containing dibutyl tin dilaurate may stay stable for months. For this reason, this catalyst is preferred for foaming systems packaged for use at other locations or plants where the resin master batch is not used immediately.

Delayed action catalysts have been made successfully. Buffered amine catalysts, where the activity of the amine has bee reduced by the presence of an acid, have also been used. Acidic materials can be used to retard the urethane reaction. Hydrogen chloride and benzoyl chloride have been used in combination with amine-type catalysts to control reaction rates. A small percentage of acid can increase foaming time from 2.2 to 6 min. Tables 8.5a and b list some rigid urethane foam catalysts and surfactants.

Temperature can also be used to control the urethane foaming reactions: higher temperatures accelerate the foaming reactions. Some delayed-action rigid urethane foaming systems have been made by premixing all of the foaming ingredients at very low temperatures (as low as −300° F). When these systems are heated, foaming of the mass takes place.

The vapors of tertiary amine catalysts are irritating, and contact with the skin can cause dermatitis. The catalysts can produce severe irritations by contact with the skin. Their vapors are also irritating. Care must be taken to ensure that the materials in solid, liquid, or vapor form do not come in contact with the human body.

For CO_2-blown foams, tertiary amines as the sole catalyst are adequate. For solvent-blown foams, however, a more reactive catalyst is necessary because of the cooling effect of the solvent. A synergistic action exists between tin catalysts and tertiary amine. Rigid foam systems, because of their greater degree of cross-linking, build gel strength so rapidly that tertiary amine are adequate for one-shot or prepolymer systems using polyethers or polyesters. The structure of the tertiary amine has a considerable influence on its catalytic effect and also on its usefulness for foam production. The catalytic strength generally increases as the basicity of the amine increases and as steric shielding of the amino nitrogen decreases.

The tertiary amine catalysts provide satisfactory foaming with either the one-shot polyester or the polyether prepolymer systems, both of which are relatively high in initial viscosity. The density and moldability of foam are greatly influenced by the choice of the catalyst system and concentration, due to the effect of catalysts on the foaming rate. The effect of a tertiary amine catalyst (triethylamine) concentration on the foaming rate, density, and compressive strength of a typical rigid polyester urethane foam is seen in Table 8.9.

The choice of catalyst for the preparation of flexible foams is governed by the type of polyol used. In polyether-based systems, where the effects of low resin viscosity and reactivity must be countered, very potent polymerization catalysts are required. For such systems, the preferred catalysts are the stannous octoate salts of dicarboxylic acids (e.g., stannous octoate). Tertiary amines, if used alone, are too inactive for controllable processing of one-shot polyether-based foams. Thus, the tertiary amines are frequently used in conjunction with tin compounds. Generally, polyester-based systems process best with catalysts

TABLE 8.5a Rigid Urethane Foam Catalysts and Surfactants (*Source:* "Air Products Polyurethane Additives," 1993)

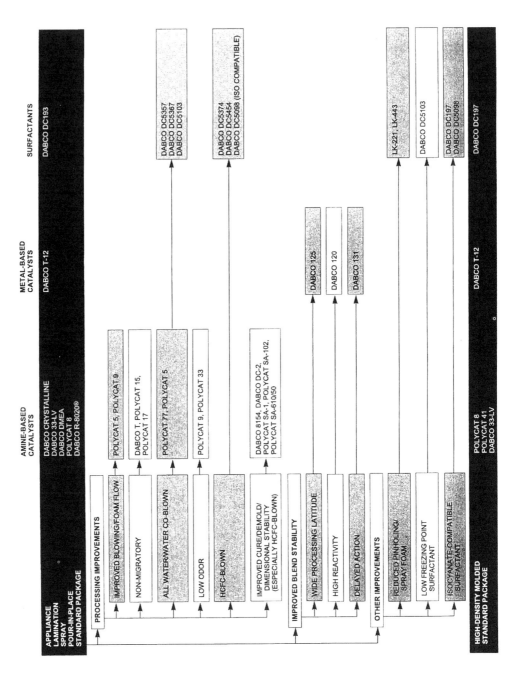

TABLE 8.5b Rigid Urethane Foam Catalysts and Surfactants (*Source:* "Air Products Polyurethane Additives," 1993)

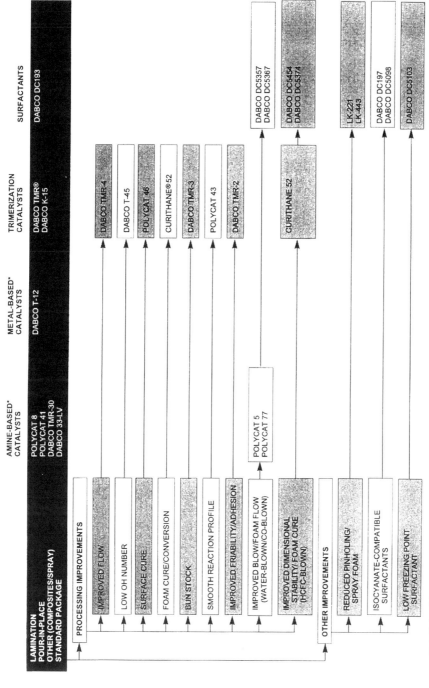

*Substitution of these materials, as depicted on the Rigid Polyurethane chart, will result in similar benefits.

TABLE 8.6 Influence of Catalyst Concentration on Foaming Parameters

Formulation	Parts by Weight			
Polyester polyol	56			
Silicone L-521	0.5			
Water	2.0			
Quasiprepolymer	100			
Tertiary amine catalyst	as indicated below			
Triethylamine (parts by weight)	Foam Time (min)	Tack-Free Time (min)	Density (lb/ft³)	Compressive Strength (psi)
0.3	4.5	16	2.1	33.6
0.5	3.5	15	2.0	32.4
1.0	2.5	10	1.9	30.8
2.0	1.0	3	1.7	21.5
3.0	0.6	1	1.5	20.0

having a low order of activity and, at most, are more effective foaming catalysts than polymerization catalysts. Tertiary amines, such as n-methyl, ethyl, cocomorpholine and the dialkyl amine are used almost exclusively.

In prepolymer systems the main reaction is that of gas formation, so catalysts which give good control of CO_2 generation are preferred. Again, as in the polyester one-shot systems, tertiary amines are used. With higher catalyst amounts, strength properties of foams generally decrease while the compression set increases; load-bearing properties are also affected by the amount and relative proportion of catalysts. In general, softer foams are obtained with increasing catalyst quantities.

An increase in catalyst concentration produces an increase in tensile, shear, and compressive strength, as well as a reduction in cure time. There is an upper limit to catalyst concentration, however, above which the foams tends to fissure and crack. An increase in catalyst concentration usually produces a decrease in K factor. This phenomenon can be explained by the fact that a low K factor results from a good retention of fluorohydrocarbon by the foam, which implies a negligible permeation of fluorohydrocarbon through the cell walls. An increased catalyst concentration would be expected to produce a more tightly cross-linked polymer structure and hence reduce the diffusion constant and possibly the solubility of the fluorohydrocarbon.

Data demonstrating the relative efficacy (and synergism) of a number of catalysts for both the isocyanate-water reaction and the isocyanate-hydroxyl reaction are shown in Table 8.7.

TABLE 8.7 Order of Catalytic Activity

Catalyst	Isocyanate-Water Reaction		Isocyanate-Hydroxyl Reaction	
	Concentration (%)	Activity	Concentration (%)	Activity
Uncatalyzed reaction	—	0.001	—	1
DABCO	0.1	2.7	—	—
TMBDA	0.1	1.6	—	—
Triethylamine	0.1	1.5	—	—
N-ethyl morpholine	0.1	1.1	—	—
Stannous octoate	0.1	1.0	0.1	1270
Stannous octoate	0.5	1.5	0.5	1820
Stannous octoate +DABCO	0.1 + 0.1	1.2	0.1+ 0.1	2110
Stannous octoate + DABCO	0.5 + 0.1	1.7	0.5 +0.1	5140
Stannous oleate	—	—	0.1	25
Dibutyl tin dilaurate	—	—	0.1	280
Dibutyl tin dilaurate	—	—	0.3	610
Dibutyl tin octoate	—	—	0.1	300
Dibutyl tin octoate + DABCO	—	—	0.1 +0.1	870

Note: DABCO = 2,2,2-diazabicyclooctane, TMBDA = N,N,'N'-tetramethyl-1,3,-butane diamine

8.7   Rigid Foam Surfactants

Polyester and polyether-based rigid foams generally require a surfactant, whether expanded with CO_2 from the water-isocyanate reaction or with an inert blowing agent such as fluorocarbon. Without surfactant, the foam may collapse or have a coarse cell structure. Castor oil-based systems generally do not require surfactants, but better results will be obtained if they are used.

Surfactants used in rigid foams range from ionic and nonionic organic types to the silicones. Anionic surfactants, or those that comprise active hydroxyl groups, should not be added to an isocyanate-containing foaming component. The most widely used surfactants are copolymers based on dimethyl polysiloxane and polysiloxane. Some of these silicones are prepared with ethylene and propylene oxides. Some silicones contain Si-O-C linkages and are hydrolytically unstable; others do not contain a silicon-carbon bond and are hydrolytically stable. The higher-viscosity silicone copolymers are more efficient and will produce foams with finer cell structures. Surfactants are used at the 0.5 to 1% level in urethane foams. With too little silicone, foam cell structure is large. Too much silicone does not affect the foam properties but is wasteful.

Surfactants other than silicones used in one-shot rigid foams include Spans (lone chain fatty acid partial esters of hexitol anhydride), Tweens (polyoxyalkylene derivatives of hexitol anhydride partial long-chain fatty acid esters), and Emulphors. These act in a manner similar to the polydimethylsiloxanes. The choice of surfactant to give optimum results will vary with the foam system used, e.g., as the poly or isocyanate is changed. The use of these nonsilicone surfactants usually imparts to rigid polyether-based foams a higher proportion of open cells. In addition to a high content of closed cells, the use of silicone copolymers in rigid polyether foams results in higher strength properties than when other surfactants are used.

8.8   Rigid Foam Blowing Agents

In 1958, halocarbons were first used as blowing agents for urethane foams, providing the first significant advance for the use of rigid urethane foams as thermal and electrical insulators. Until that time, rigid foams had been blown with the CO_2 liberated as a result of the isocyanate-water reaction. The reasons prompting this development are best described by referring to the particular properties of the fluorocarbon blowing agents, originally called refrigerants, listed in Table 8.8.

TABLE 8.8 Properties of Fluorocarbon Blowing Agents

	Fluorocarbon 11	Fluorocarbon 12	Fluorocarbon 113
Chemical formula	$CC1_3F$	$CC1_2F_2$	$CC1_2F\text{-}CC1F_2$
Chemical name	trichlorofluoromethane	dichlorodifluoromethane	trichlorotrifluoromethane
Boiling point, °F	74.78	−21.62	117.63
Freezing point, °F	−168	−252	−31
Thermal conductivity of vapor at 1 atm, 32° F	0.00450	0.00483	

Table 8.9 shows that the CO_2 previously used as a blowing agent had some advantages in thermal conductivity over air, provided the gas did not diffuse through the cell walls and equilibrate with air. In practice, however, this does not occur. The halocarbons display about half the thermal conductivity of CO_2. Furthermore, the thermal conductivities of the halocarbons do not increase in the same proportions as air or CO_2 as the temperature rises.

The effect of aging on K factors of foams with different blowing agents is shown in Table 8.10. The high density (high molecular weight) of the fluorocarbon gas causes it to be a poor conductor of heat. Fortunately, the permeability of a fluorocarbon through the cell walls of common polyurethane foams is extremely slow, so that the fluorocarbon gas and its excellent insulating properties are retained almost indefinitely.

TABLE 8.9 Comparison of Thermal Conductivities of Fluorocarbon Agents, CO_2, and Air

1 BTU/°F(ft²)(hr/in)	Temperature (°F)		
Gas	32	68	86
Air	0.168	0.180	—
CO_2	0.0101	0.117	—
Type 11	0.054	0.057	0.058
Type 12	0.058	0.064	0.067

TABLE 8.10 Thermal Conductivities of Urethane Foams with Different Blowing Agents

Blowing agent	Molecular weight	Wt % to give 2 lb/ft³	Initial K Factor	Final K Factor
Type 11	137	16	0.11	0.14
Methylene chloride *	85	12	0.145	0.24
n-pentane	72	9	0.18	0.20
CO_2	44	1.7	0.21	0.24
Air	29	—	0.24	0.24

*Properties extrapolated from blends with fluorocarbon 11 (foams with 100% methylene chloride cannot be made).

Another critical factor in foam processing is the viscosity of the reactants. Most polyether polyols have high viscosities, and difficulty is experienced in high-speed mixing of them with low-viscosity polyisocyanates in presently available equipment. When halocarbons are added to the polyether polyol component, the viscosity is reduced to that of a thin liquid, thus facilitating pumping, mixing, and metering. The halocarbons also have a high degree of hydrolytic stability and hydrophobicity.

Considerable heat is evolved when foam components are mixed. The vapor provided by type 11 halocarbon, the most commonly used type, occupies 256 times as much volume as its equivalent of liquid at 80° F and atmospheric pressure. This greater volume increase, resulting from the exotherm of the reactants when the halocarbon boils, depends on the halocarbon concentration. Where higher-density foams are desired, the advantages of halocarbons ma be obtained by using type 113. With a much higher boiling point and lower vapor volume, type 113 can provide a distinct advantage in obtaining higher-density foams. This material may also be preferred where ambient temperatures are consistently above 90° F.

Since water is required to react with the isocyanate groups to expand the CO_2-blown foams, it is obvious that the most costly portion of the foam formulation (the isocyanate) must be increased to compensate for the isocyanate, which is required for polymer formation. At densities up to 5 lb/ft³, the halocarbon is much less expensive. Densities above 5 lb/ft³ are more easily obtained with CO_2 blowing. The amount of CO_2 generated is determined by the amount of water used in the foaming composition. After the foam has formed, but before the polymer is completely set, it is very permeable to CO_2, and often the diffusion of CO_2 out of the foam is more rapid than the infusion of air or moisture into the foam, which causes shrinkage of the foam if the polymer structure is not sufficiently rigid. Low-density rigid urethane foams expanded with CO_2 often are friable. Their friability can be reduced by the use of higher equivalent weight polyols and by proper selection of catalysts.

Most rigid foams are produced in the 2 lb/ft³ range. CO_2-blown foams cannot be made with reliably low densities. The lowest practical limit is ≈4 lb/ft³. Halocarbon-blown foams also provide better physical properties than CO_2-blown foam. The greater uniformity of the halocarbon-blown foams is, in part, responsible for their better physical properties. In addition, the polyisocyanate residue from reaction with water is deleterious in several respects. Foaming conditions are less critical with halocarbons because of the heat absorption by the halocarbon.

The type 12 halocarbon is particularly useful in the process called *frothing*. Since its boiling point is very low, it immediately vaporizes when the foam ingredients are discharged from the mixing head. This produces a froth of low density to overcome the pressures exerted by liquid ingredient, which must expand 35-fold to reach densities of ≈2 lb/ft³. Type 11 halocarbon is also included in froth formulations to obtain the final density.

Fluorocarbon blowing agents act as moderating agents and do not produce additional cross-linking in the foam. They are inert and are retained in the polymeric structure of the foam (624). They are also nonflammable and have a very low order of toxicity.

Other blowing agents have been used experimentally in expanding rigid urethane foams. Methylene chloride is of interest because of its low cost. It will soften urethane polymers and cause cracks in rigid foams expanded with it. Methylene chloride is not used in any practical process. Hydrocarbon blowing agents such as propane, butane, and pentane have been tried. Because of their flammability and their value as fuels, they have not gained wide acceptance.

The most pronounced effects of blowing agents in foam formulations are foam density control and thermal insulation effectiveness. The effect or water and fluorocarbon-11 concentration on foam density is shown in Figs. 8.10 and 8.11. The lower K factor of fluorocarbon-blown foams means that the amount needed for any given application may be reduced to as much as 50 percent.

The advantages of fluorocarbon-11 blowing agents over CO_2 are summarized as follows:

- K factor reduced from 0.25 to 0.14 so that one-half of the insulating foam is needed.
- The induction period before foaming is longer, due to latent heat of evaporation of the fluorocarbon blowing agent.
- The gelation rate of expanding foam is decreased, preventing thermal pressure cracks and charring of the foam core in large applications.
- Compressive strength is increased by ≈30%.

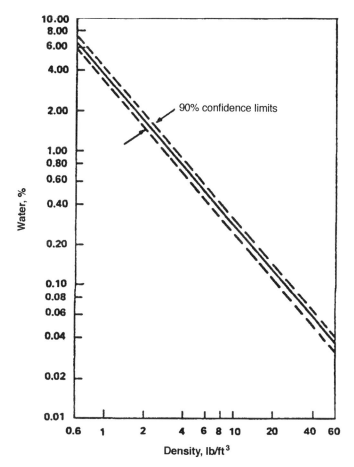

FIGURE 8.10 Effect of water concentration on rigid urethane foam density.

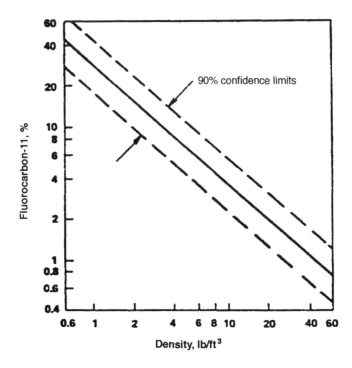

FIGURE 8.11 Effect of fluorocarbon-11 concentration on rigid urethane foam density.

- Moisture vapor transmission is reduced (3.5 perms vs. 5.5 perms for CO_2).
- Adhesion to metal is improved.
- The edge of the foam is less friable.
- There is a higher proportion of closed cells ($\approx$90 vs. 85% for CO_2).
- The foam cost is lower.

As a general rule, thermal conductivities of most materials decrease with decreasing temperature. Polyurethane foams do not adhere uniformly to this rule, however. The thermal conductivity of a typical polyurethane foam blown with Freon-11 will decrease with decreasing temperature until it reaches a temperature of $\approx$50° F. At this point, a temperature decrease results in a limited K factor increase. At still lower temperatures the K factor begins to drop off again. The behavior in the thermal conductivity cure is quite surprising and can be significant when polyurethane foams are used for low temperature insulation.

As indicated above, the thermal conductivity of polyurethane foam is a function of the molecular at of the gases within the cells. Except at relatively high and very low total pressures, the conductivity is independent of the pressure inside the cell. The K factors of rigid polyurethane foams will increase appreciably with exposure to air. Immediately after mixing, the thermal conductivity of C02-expanded foams is $\approx$0.23, as seen in this figure. This value increases slightly an levels at $\approx$0.24.

Because nearly all the rigid foam cells are closed, gases can pass into or out of the foam only by permeating through these walls. In general, polyurethane foam cell walls are permeated rapidly by CO_2, fairly rapidly by air, and extremely slowly by fluorocarbon blowing agents. The thermal conductivity change shown in Fig. 8.12 is caused by CO_2 moving out of the foam and air coming in. CO_2 is a slightly denser gas than air and consequently is a slightly better insulator. The replacement of CO_2 by air is accompanied by a decrease in the insulating properties of the foam. It is important to note that the air does not push out the CO_2, since it actually evacuates the foam at a faster rate than air enters the foam mass.

In the case of fluorocarbon-blown foams exposed to air, as air permeates into the foam, it mixes with the fluorocarbon-11. In this case, the blowing agent fluorocarbon-11 already in the foam is not replaced

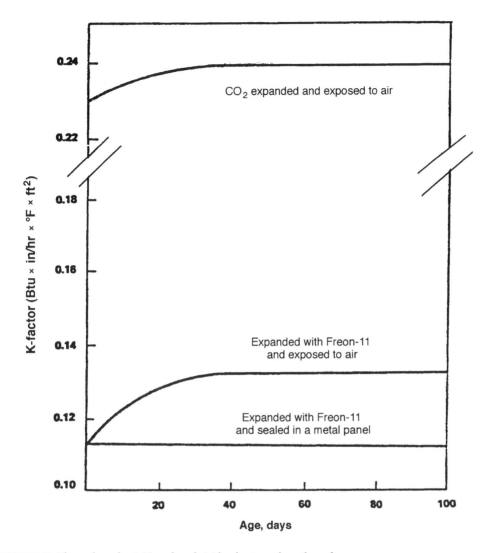

FIGURE 8.12 Thermal conductivities of aged rigid polyester polyurethane foams.

but remains inside the cells. This is because the permeability of the fluorocarbon blowing agent through the cell walls is extremely slow. Before exposure to air, the polyurethane foam is actually under a slight internal vacuum, because the gases have cooled down after the foam attained maximum rise and set to a hard plastic. The pressure inside the cells is ≈0.6 atm (9 psia or 12 in Hg vacuum).

The fluorocarbon-11 does not encourage air entry into the cells. Rather, air permeates the cell wall and achieves a partial pressure of close to 1.0 atm. After equilibrium with the atmosphere is reached, the total pressure in the cell is 1 atm of air plus 0.6 atm of fluorocarbon-11, resuming in a total of 1.6 atm. The gases in the cells are thus exerting a positive pressure, with a resultant slight increase in compressive strength of the foam. This dilution of the fluorocarbon-11 blowing agent with air results in an increase in thermal conductivity, as mentioned above. After the air has reached equilibrium with the outside atmosphere, no additional air permeates, and the rate of fluorocarbon-11 loss is nearly negligible.

Certain bromide-containing fluorocarbons are even more effective fire retardants than are conventional blowing agents such as fluorocarbon-11. These include $CBrF_2CBrF_2$, b.p. 117 ° F (Freon-114B2) and $CBr F_2$ b.p. 77° F (Freon-12B2). Substitution of either of these products for fluorocarbon-11 produces insulating, self-extinguishing polyurethane foams, but at a cost between 15 to 30 times greater.

8.8.1　CFC-Reduced Rigid Foams

In recent years, the use of chlorofluorocarbons as foaming agents has been legally regulated for the protection of the ozonosphere, and trichlorofluoromethane (R-11), which has hitherto been used for the production of a rigid polyurethane foam, is among the substances subject to the regulation.

This brings about a problem of the necessity of reducing the use of trichlorofluoromethane. Examples of the reduction means proposed in the art include one wherein the amount of water used is increased to reduce the amount of trichlorofluoromethane (the so-called "chlorofluorocarbons-poor formulation") and one wherein use is made of 1,1-dichloro-2,2,2-trifluoroethane (R-123) or 2,2-dichloro-2-fluoroethane (R-141b), having ozone destruction factors (ODPs) smaller than that of trichlorofluoromethane.

Physical blowing agents can also be employed in a mixture with water. Suitable liquids are inert toward the organic, modified, or unmodified polyisocyanates (c) and have boiling points below 100° C (preferably below 50° C, and in particular from –50 to 30° C) at atmospheric pressure, so that they evaporate under the influence of the exothermic polyaddition reaction.

Examples of physical blowing agents are hydrocarbons, such as n- and isopentane, in particular technical-grade mixtures of n- and isopentanes; n- and isobutane and propane; ethers such as furan, dimethyl ether, and diethyl ether; ketones such as acetone and methyl ethyl ketone; alkyl carboxylates such as methyl formate, dimethyl oxalate, and ethyl acetate; and halogenated hydrocarbons such as methylene chloride, dichloromonofluoromethane, trifluoromethane, difluoromethane, difluoroethane, tetrafluoroethane, and heptafluoropropane. Mixtures of these low-boiling liquids with one another and/or with other substituted or unsubstituted hydrocarbons can also be used. In addition, organic carboxylic acids, e.g., formic acid, acetic acid, oxalic acid, ricinolinic acid, and other carboxyl-containing compounds.

The blowing agents that can be used for the preparation of the hard foams containing urethane groups or containing urethane and isocyanurate groups preferably include water, which reacts with isocyanate groups to form carbon dioxide. The amount of water that is expediently employed is from 0.1 to 6.5 parts by weight, preferably from 1.0 to 5.5 parts by weight, and in particular from 2.0 to 5.0 parts by weight, based on 100 parts by weight of the high-molecular-weight polyhydroxyl compound.

It is also possible to employ physical blowing agents mixed with water or exclusively physical blowing agents. Suitable compounds are liquids that are inert toward the organic, modified, or unmodified polyisocyanates and have boiling points below 100° C, preferably below 50° C, and in particular from –50 to 30° C, at atmospheric pressure, so that they evaporate during the exothermic polyaddition reaction. Examples of preferred liquids of this type are hydrocarbons such as n- and isopentane; technical-grade pentane mixtures; n- and isobutane and propane; ethers such as foam, dimethyl ether, and diethylether; ketones such as acetone and methyl ethyl ketone; esters such as ethyl acetate and methyl formate; and preferably halogenated hydrocarbons such as methylene chloride, difluoromethane, trichlorofluoromethane, dichlorodifluoromethane, dichloromonofluoromethane, 1,1,1-dichlorofluoroethane, 1,1,1-chlorodifluoroethane, dichlorotetrafluoroethane, tetrafluoroethane, 1,1,2-trichloro-1,2,2-trifluoroethane and heptafluoropropane; and noble gases such as krypton. Mixtures of these low-boiling liquids with one another and/or with other substituted or unsubstituted hydrocarbons can also be used.

The necessary amount of physical blowing agent can be determined in a simple manner, depending on the foam density required, and is from approximately 10 to 30 parts by weight, preferably from 14 to 22 parts by weight, per 100 parts by weight of the high-molecular-weight polyhydroxyl compound, its amount being reduced proportionately if water is also used. It may be expedient to mix the modified or unmodified polyisocyanate with the physical blowing agent and thereby to reduce its viscosity.

8.9　Rigid Foams with Enhanced Compressive Strength

Brown et al., in U.S. Pat. 4,169,922[4], disclosed the development of rigid polyurethane foams with enhanced properties (e.g., compressive strength). Additionally, the proportions of expensive polymerization catalyst and/or surfactant needed to provide a given level of such properties can be lowered by preparing such foams from compositions comprising, in addition to the conventional foam-forming reactants including polyisocyanate and polyol, from ≈3 to ≈10%, based on the weight of the polyol, of a plasticizer selected from essentially halogen-free phosphates and carboxylates containing at least one aryl radical, e.g., isodecyl diphenyl phosphate or butyl benzyl phthalate. The result is a rigid foam with better compressive strength properties at a significantly lower price.

Production of rigid polyurethane is typically carried out by subjecting a mixture of organic polyiso-cyanate and polyol having more than two hydroxy radicals per molecule to polyurethane foam-forming reaction conditions. Usually, the reaction is catalyzed by including an amine or tin compound in the reaction mixture. Also typically included in the mixture are secondary blowing agents to lower the density and surfactants (e.g., silicones) to improve the uniformity of the resulting foam. Such catalysts, surfactants, and blowing agents are normally added in relatively small proportions, but they are expensive. Hence, the quantities in which they must be used have a significant impact on manufacturing cost of the foam.

Accordingly, it would be very advantageous if significant properties of the foam could be improved without need for more of such expensive constituents of the polymerization mixture, or if foams of equivalent properties could be produced while using smaller quantities of such constituents.

In accordance with the Brown invention, the advantages are provided by use of a rigid polyurethane foam-forming composition comprising from ≈3 to ≈10% of a plasticizer selected from essentially halogen-free phosphates and carboxylates containing at least one aryl radical. As used herein, polyurethane foam-forming reactants means polyurethane monomers capable of polymerizing by urethane linkage-forming reactions. The reactants have not yet undergone substantial polymerization to the extent of forming polyurethane "prepolymer," e.g., as disclosed in U.S. Pat. 3,975,316, issued Aug. 17, 1976 to J.L. Villa. Also preferably, the foam-forming compositions of this invention contain essentially no reactants that polymerize under conventional polyurethane foam-forming polymerization conditions. To improve the uniformity of the rigid foam, it has been conventional to include in the reaction mixture a surfactant, normally of the silicone variety, e.g., poly(dialkyl siloxane)s or a silicone-glycol.

Blowing agents (e.g., Freon) are typically employed to lower foam density. Also employed in production of the foams of this invention are conventional polyurethane foam-forming reaction conditions, which normally include uniform mixing of the urethane-forming reactants, and then allowing their exothermic reaction to begin at approximately room temperature and proceed to a desired degree of completion.

Suitable inert blowing agents include hydrocarbons, dialkyl ethers, alkyl alkanoates, aliphatic and cycloaliphatic hydrofluorocarbons, hydrochlorofluorocarbons, chlorofluorocarbons, hydrochlorocarbons, and fluorine-containing ethers. Suitable hydrocarbon blowing agents include lower aliphatic or cyclic hydrocarbons such as pentane, iso-pentane, cyclopentane or neopentane, hexane, and cyclohexane.

Suitable dialkyl ethers to be used as blowing agents include compounds having from two to six carbon atoms. As examples of suitable ethers, there may be mentioned dimethyl ether, methyl ethyl ether, diethyl ether, methyl propyl ether, methyl isopropyl ether, ethyl propyl ether, ethyl isopropyl ether, dipropyl ether, propyl isopropyl ether, diisopropylether, methyl butyl ether, methyl isobutyl ether, methyl t-butyl ether, ethyl butyl ether, ethyl isobutyl ether, and ethyl t-butyl ether. Suitable alkyl alkanoates that may be used as blowing agents include methyl formate, methylacetate, ethyl formate, and ethyl acetate.

The plasticizers used in this invention are essentially halogen-free and selected from phosphates and carboxylates containing at least one aryl radical per molecule. Such phosphates can be triaryl phosphates such as isopropylphenyl diphenyl phosphate, t-butyl-phenyl diphenyl phosphate, tricresyl phosphate, or triphenyl phosphate; alkyl diaryl phosphates such as, C9-C11 alkyl diphenyl phosphate, isodecyl diphenyl phosphate, 2-ethylhexyl diphenyl phosphate, isooctyl diphenyl phosphate, C7-C9 alkyl diphenyl phosphate, or isononyl dicresyl phosphate; or dialkyl aryl phosphates such as, di-isodecyl phenyl phosphate, di-isododecyl t-butyl-phenyl phosphate or di-undecyl cresyl phosphate. Preferred among those phosphates are the alkyl diaryl phosphates, particularly isodecyl diphenyl phosphate, and the triaryl phosphates, particularly t-butyl-phenyl diphenyl phosphate. The carboxylate plasticizers useful in this invention may contain an aryl radical on either or both sides of a carboxyl radical, e.g., dialkyl phthalates, diaryl adipates, or benzyl phthalates. Preferred are dicarboxylates such as phthalates, terephthalates, isophthalates, adipates, glutarates, sebacates, pimelates, azelates, succinates, etc., although mono-carbox-ylates such as benzyl stearate, etc. are also useful. Most preferred are the phthalates, including dialkyl phthalates in which the average carbon atom content per alkyl radical is between ≈6 and ≈12 (e.g. dioctyl phthalate), diaryl phthalates such as dibenzyl phthalate, and most advantageously for many uses, alkyl aryl phthalates such as butyl benzyl phthalate, octyl benzyl phthalate, texanol (e.g., 2,2,4-trimethylpentyl) benzyl phthalates, etc. Also attractively useful herein are isophthalates and terephthalates such as, e.g., bis(2-ethylhexyl) terephthalate. Preferred among the dicarboxylates of aliphatic dicarboxylic acids are diaryl dicarboxylates such as dibenzyl adipate, dicresyl sebacate, etc.

Advantageous foam properties and/or lowered requirements for reaction mixture constituents such as polymerization catalysts, surfactants, and/or blowing agents are provided when the compositions of this invention contain from ≈3 to ≈10% of such a plasticizer, based on the weight of polyol in that composition. In most instances, even more advantageous results are obtained when the composition contains not less than ≈4% and not more than ≈8.5% of the plasticizer, based on the weight of that polyol, and in many instances most advantageous results are obtained when the composition contains not less than ≈5 and not more than ≈7% of the plasticizer, based on the weight of that polyol. Preferably the plasticizer is essentially uniformly dispersed (e.g., by any conventional means) throughout the foam-forming composition employed in this invention.

Various rigid polyurethane foam-forming compositions were prepared by mixing 100 parts of a moderate-viscosity, sucrose-based polyol having a hydroxyl number of 470 mg KOH/gm with a polym-ethylene polyphenyl isocyanate having an NCO content of 31.5% and an amine equivalent of 132, 3 parts of a polymerization catalyst blend composed of 80% dimethylethanolamine and 20% of triethylenedi-amine, 1.6 parts of a low-viscosity silicone-glycol copolymer surfactant having a viscosity of 465 cs at 25°, 35 parts of Freon and 0 or 1 part of distilled water. Separate batches were made using different proportions of the isocyanate such that they had NCO/OH Indices (defined previously) of 96, 108, and 120. In two duplicate sets of such compositions, 5 parts and 10 parts, respectively, of butyl benzyl phthalate were also included. The foam-forming compositions were then thoroughly mixed and subjected to conventional polyurethane-forming reaction conditions beginning at room temperature. Each resulting rigid foam was subjected to measurements of density and compression load deflection by ASTM D-1564-64. Results were that, in the presence of the butyl benzyl phthalate, increased CLD can be obtained despite the use of less silicone.

8.10 Miscellaneous Additives for Rigid Foams

Additives may be incorporated into the foaming system to impart specific desirable properties. These may include air, inert gases, dyes, pigments, plasticizers, flame retardants, synthetic fibers, organic fillers, inert fillers, coal tar, pitch, asphalt, and tall oil.

Air may be used as a nucleating agent to produce fine cell structures when large pours of foam are made by machine mixing. Physical strength properties, K factors, and water pickup are improved. Solids such as carbon black and talc are also used as nucleating agents. Carbon blacks are also used as a means of protection against sunlight.

Urethane color pastes, which are dispersions of color pigments in a polyolplasticizer, are also used. When using color pastes, compensation must be made for the hydroxyl value of the pastes by adding additional diisocyanate. A complete range of colors is obtainable by blending the primary colors (red, blue, and green). Carbon black pastes are easiest to use.

Is is sometimes desirable to color one or both components of the foaming composition for identifi-cation so that a resin component will not be put into a machine used for pumping the isocyanate component. Carbon black and titanium dioxide have also been used as pigments to provide functional protection against sunlight.

Nonreactive liquids are used (a) to reduce viscosity for improved processing and (b) to improve flame resistance. The softening effect of the plasticizer is generally compensated by using a polyol of lower equivalent weight so that a higher cross-linked polymer structure is obtained. These materials will also increase foam density and often adversely affect foam properties.

In general, three methods have been employed to make rigid urethane foams fire retardant.

1. *Incorporation of special flame-retardant additives to the foam components by simple mechanical mixing.* These are organic or inorganic, but occasionally combinations of both types are used. A wide variety of additives are based on phosphorous, sulfur, nitrogen, and halogens. The most commonly used additives are tris(chloroethyl) phosphate (ClCH3–CH,–0)3 P; tris 2,3(dibromopropyl) phosphate diammonium phosphate, various phosphates and phosphonates; and antimony oxide. Also, chlorine-containing flame-proofing agents are used, such as tricresyl phosphate tris(2-chloroethyl) phosphate, tris(2-chloropropyl) phosphate, tris(1,3-dichloropropyl) phosphate, tris(2,3-dibromopropyl) phosphate, and tetrakis(2-chloroethyl) ethylene diphosphate.

   In addition to the above-mentioned halogen-substituted phosphates, it is also possible to use inorganic flame-proofing agents (such as red phosphorus, aluminum oxide hydrate, antimony trioxide, arsenic oxide, ammonium polyphosphate, and calcium sulfate) and cyanuric acid derivatives, e.g., melamine, or mixtures of two or more flame-proofing agents, e.g., ammonium polyphosphates and melamine, and also, if desired, starch and/or expandable graphite to flameproof the hard PU foams. In general, formulators use from 5 to 25 parts by weight of flame-proofing agents or mixtures per 100 parts of polymer.

   These additives usually adversely affect the physical properties of the foam. A decrease in closed cell content and strength properties and an increase in water absorption often occurs. A significant reduction in strength properties and dimensional stability generally occurs on humid aging at elevated temperatures, particularly if the additives are used in excess of 15% by weight.

2. *Addition of flame-retardant compounds containing functional groups, particularly hydroxyls, that become chemically bound in the polymer chain.* This approach is superior to the first. Typical examples of this type are chlorendic acid, which finds application in polyester-based flame-resistant foams. Phosphorous and/or chlorine-containing polyols, among them diol and polyphosphates, phosphonates as well as some chlorinated derivatives of these types of compounds are commercially available.

3. *Coating of the flammable foam by means of flame-retardant materials.* This method is especially useful on spray-applied foams for outdoor applications where low water vapor permeability and good weather protection are desired.

Synthetic fibers have been added to rigid urethane foams to give them better dimensional stability and higher strength properties. However, small amounts of fibers will significantly increase the viscosity of the foaming composition, making it difficult to handle.

Some organic fillers, such as cornstarch, wood flour, etc., contain hydroxyl groups that can react with the isocyanate but, since they are solids, the reaction is very slow, and they act essentially as inert fillers. In general, they increase foam density and sometimes give higher strength. These foams often have reduced resistance to moisture. Bulk fillers, such as phenolic and styrene beads, have also been used, but they produce inferior foams.

Mineral fillers, such as clays and talc, have been used. They behave very much like organic fillers but have little effect on the moisture resistance of the foams. They are more difficult to process than organic fillers, since they are heavy and tend to settle out of the urethane foaming composition. They often increase the apparent viscosity of the foaming composition to such an extent that they are impractical to process. They may also cause excessive wear of pumps and mechanical mixers used for blending. Metal oxide fillers include those used for imparting flame resistance. Heavier metal oxides have been added to modify electrical properties for uses such as radiation shielding and radome applications.

Because of their low cost, there has been a great deal of incentive to add coal tar, pitch, and blown asphalt oil to rigid urethane foams. In every case, foam properties or processing ease are sacrificed, however. Some compositions have been developed using materials satisfactory for vibration damping.

Table 8.11 shows how representative fillers affect the physical properties of a rigid polyurethane foam.

TABLE 8.11 Affect of Fillers on Physical Properties of Rigid Polyurethane Foam

Filler	Properties		
	Density	Compressive Strength	% Closed Cells
Coal tar pitch	increase	decrease	decrease
Hydrated calcium silicate	increase	decrease	decrease
Dixie clay	decrease	decrease	no change
Glass flakes	increase	decrease	decrease
Litharge	decrease	decrease	decrease
Barytes	decrease	decrease	increase
Activated charcoal	decrease	decrease	decrease
Diatomaceous earth	decrease	decrease	decrease

8.11 Types of Polyurethane Foam

Rigid polyurethane foams can be made over a range of densities from ≈ 1.8 to 60 lb/ft³. Some rigid urethane foams have softening temperatures as low as 100° C (212° F), while others do not soften below 150° C (302° F). Low-temperature properties are good. Since they are already in the "glassy" state at room temperature, no further transitions take place at lower temperatures, and there is merely a slow increase in stiffness such that at −200° C (−328° F) the stiffness is approximately doubled.

Applications of rigid urethane foams include thermal and electrical insulation, reinforcement of aircraft parts, sandwich panels, void filling for buoyancy applications, foaming-in-place for package cushioning, salvage by flotation of sunken vessels, etc.

8.11.1 Pour-in-Place (Foam-in-Place) Batch Preparation

Pouring-in-place is the oldest, most versatile, and most popular technique for the application of rigid urethane foam. In conventional pouring-in-place molding, freshly mixed liquid formulations are charged directly into a cavity in which the foaming reaction is carried out. The cavity is filled upon completion of the foaming reaction. The technique is applicable over the entire density range, from 0.7 to 70 lb/ft³. Usually, the mold is retained as the exterior of the finished item. Both the semi- or quasiprepolymer and the one-shot processes are successfully used.

The ingredients are mixed either by hand or with a power mixer in any clean container ranging from a small glass beaker to a large pail or bucket. Batches may be up to 50 lb or more. Ideally, a power mixer should do a thorough job of mixing in 10 to 15 s. Some of the higher-viscosity, slower formulations may require up to 30 s. Often, electric drill motors with various blades attached to the shaft are used. The batch system used in pouring-in-place may result in nonuniform end products, but the required equipment is not costly.

The reactant mixture, after being poured directly into a cavity or mold, flows to the bottom of the mold and then foams to approximately 30 to 40 times its original volume, filling all cracks and corners and forming a strong, seamless core. The elimination of seams, joints, and lines (which tend to fail at low temperatures) is a distinct advantage. Pouring-in-place differs from molding primarily because adhesion to the mold is desired, whereas in molding, perfect release is sought. Also, during the expansion process, a skin is formed next to the surface of the filled cavity, eliminating the need for vapor barriers.

Pouring-in-place is ideal for filling defined areas such as curtain wall sandwich panels, wall cavities, and irregularly shaped voids. The cavity must be capable of supporting foam pressures of up to 3 to 5 psi. The usual increase in volume of 30 to 40 times is cut down somewhat by the frothing process, but it is not entirely eliminated. The temperature of the foaming reactants is very critical if a low overall density is desired.

The numerous applications for which urethane foam has been successful may be classified into three types, according to density, as listed below and shown in Fig. 8.13:

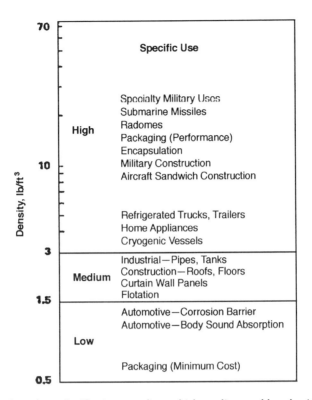

FIGURE 8.13 Polyurethane foam classification according to high, medium, and low density.

- low, up to 1.5
- medium, from 1.5 to 3.0
- high, above 3.0 lb/ft^3

One-shot rigid foam may be poured in place. The molds may be positioned with the walls vertical or horizontal during the foaming process as shown in Fig. 8.14.

Maximum compressive strength in panels occurs parallel to the panel walls in vertical pours. The high vertical rise is the easiest to handle, as it may be poured through a single opening at the top of the mold. Panels up to 8 ft high have been successfully foamed using only one pour. The foam properties of vertically filled panels are uniform from top to bottom. A low vertical rise may also be used where high panels cannot be conveniently handled. In this case, the foam may be poured through a single hole at one end if the mold is tilted to ensure adequate distribution. Otherwise, it may be necessary to traverse the length of the mold.

In some cases, it is desirable to achieve maximum strength perpendicular to the panel walls. In such cases, the panel can be foamed horizontally. In pouring horizontal panels, one of the panel skins is placed on the bottom of a mold. The mold skins are used to retain the foam as it rises to meet the other skin being held in place by the upper mold section. As with vertical panels, electromagnets and vacuum may be used to hold the top substrate to the upper mold section. The liquid mixture of foam components can be flowed in by a mixing head in a predetermined pattern, traversing the mold area.

Skin materials used for insulated building panels include steel, aluminum, melamine laminates, plywood, sheet rock, and paper. Temperatures of 110 to 130° F are desirable with metal molds in one-shot rigid urethane panel molding to ensure low panel densities and optimum K factors. Lower mold temperatures result in high panel density because of the loss of heat to the mold. In addition, a thick skin

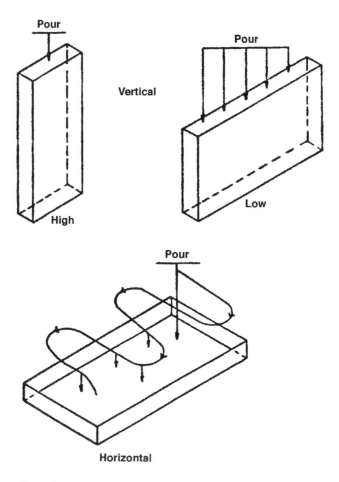

FIGURE 8.14 Pour-in-place techniques.

forms at the foam surface, causing an increase in K factor. Usually, with 110 to 130° F molds, panels may be released from their jigs in approximately 10 min. A post-cure at 150 to 175° F may reduce this time to as little as 5 min.

8.11.2 Spraying

Spray techniques are used for filling molds and panels and for applying foam to plane surfaces. Spraying is particularly useful in applications where large areas are involved, such as tanks or building walls. Sprayed rigid foam coatings provide both physical strength and improved insulation. Spraying is the simplest and least expensive way to produce foam. In addition, spraying equipment is reasonably priced and portable, and foam can be applied without molds or jigs of any kind.

In spray applications, the ingredient mixing is accomplished by atomization of the materials as they leave the nozzle of the spray gun. Resin viscosities in the range of 500 to 1500 cps should be maintained to ensure intimate mixing of the constituents. When necessary, heat may be applied to the polymer to reduce the viscosity to the desired level.

Commercial spray guns will apply a 12-in circle of foam when the gun is held at a distance of 3 ft from the target surface. Larger deliveries and different spray configurations are possible with special guns.

The spray technique may be used for filling molds (spray pouring) by confining the spray pattern as it leaves the gun. This is accomplished by attaching a cone to the gun head. The cone may be from 8 in to 2 ft in length and is attached with the smaller diameter at the outlet end. This device also serves to minimize catalyst loss into the air and confines the foam discharge to a smaller area. With this method, large deliveries of foam material are possible. This technique provides interesting possibilities when filling intricate or large molds. It is relatively inexpensive and highly flexible. Adhesion of the sprayed foam to most surfaces is excellent.

Conventional spraying is usually done with portable metering equipment, which should be light enough to be carried on a small truck or trailer. Most of these units consist of air-driven, dual-action pumps that can be readily calibrated and used with the original 5- or 55-gal shipping containers.

There are two basic types of guns used for spray operations, internal and external mix. The internal mix guns use a number of different methods of mixing. The external mix guns use high-pressure air to bring the components out of the gun. Then, through a swirling action, the air mixes and atomizes the components between the gun and the substrate to be sprayed.

The advantages of an internal mix gun are

- Generally better mixing, especially of high-viscosity materials
- Lower overspray
- Less atomizing air, allowing the operator to get closer to the work
- Usually, better foam properties because of better mixing

The one big disadvantage of most internal mix guns is their tendency to plug. It takes a great deal of care, even with guns having a solvent purge, to keep them in working order.

External mix guns have the advantage of minimal chance of plugging, and they are easily kept clean. Disadvantages are

- Usually, poorer mixing, a disadvantage that is minimized with low-viscosity materials
- Large volume of atomization air, making it difficult to get close to surface being sprayed

To prevent sagging of sprayed urethane foams, formulations with fast reaction rates are used. The faster rates are obtained by using additional catalyst. Additional layers can be sprayed almost immediately. The normal spray rate is ≈4 to 8 lb/min.

The development that has been most responsible for improvement in the spray process is the 1:1 ratio, low-viscosity, one-shot, totally catalyzed system. This system brought about the development of the airless gun. Previously, atomization was achieved by the introduction of high-pressure air into a conventional mixing chamber. Such spray heads are known as *air assisted*.

The rise time for airless sprayed urethane foam is ≈30 s. It can be refoamed for additional thickness after this period. It can be walked on after 3 to 4 min and reaches its full properties in 24 hr. The airless spray gun is held ≈30 in from the surface and moved steadily over it.

Generally, dispensing rates of 4 to 6 lb/min are considered optimum for most spray applications. The surface on which the foam is sprayed must be free of loose scale or grease. The adhesion of urethane foam to steel is essentially equal to the tensile strength of the foam, provided that the surface is clean. Aluminum surfaces, on the other hand, do not provide a good bond unless a primer coat, such as vinyl wash, is used prior to spray foaming.

Figure 8.15 shows a typical stress-strain curve for sprayed urethane foam as measured both with and against the direction of foam rise. The disparity, or degree of nonisotropicity, may be associated with the shape of the foam cell. Since it is necessary to catalyze a spray foam for rapid rise characteristics in order to prevent drooping on vertical and overhead surfaces, the gel time, or that point in the viscosity build-up where there is no longer mobility, is so short that the cell matrix tends to thermoset into a shape similar to an ellipsoid rather than a sphere. Consequently, an increase in compressive strength can be observed when the load is applied parallel to the long axis, while a corresponding decrease can be seen when the load is applied perpendicular to the long axis. In most cases, the high degree of nonisotropicity in spray foam is a significant advantage, since the increased compressive strength is in the direction normal to the foam surface and parallel to the applied loads.

In contrast to the successful techniques developed for spraying rigid and semirigid foams, methods for spraying flexible urethane foams are at a relatively early stage and, at present, rarely used commercially.

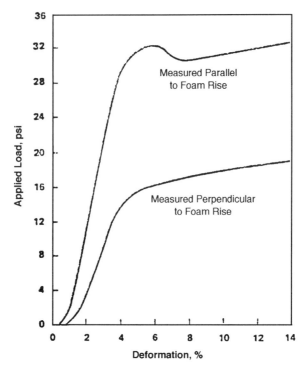

FIGURE 8.15 Stress-strain curves for rigid urethane spray foam.

To date, only polyester systems have been used commercially. Applications are for carpet underlay and textile backing. Water is used instead of fluorocarbon-11 in froth spray resilient urethane foams in the second stage of expansion.

8.11.3 Miscellaneous Operations

Hot-wire cutting may be used in rigid foams for flat sheets and gentle contours. Rigid urethane foam may be readily post-formed by a process also called *plastic stage forming*. During manufacture, rigid urethane foam goes through three distinct and progressive phases:

1. Expansion
2. *Plastic* stage, where foam is pliable and tack-free, with no memory
3. *Set* stage, where foam has reached the final physical state

The foam is "worked" while in the second phase. Thin sheets of foam in this phase can be shaped around a mandrel (form) and maintained in the shaped position until "set" has occurred (i.e., completion of the third foam phase). The foam is permanently molded to the desired shape and, upon removal from the forming mold, has no tendency to return to the original flat sheet. The shaped foam is free of densification, because the cells are not ruptured during bending; it exhibits no evidence of internal stress or strain. This technique is commonly used to shape rigid urethane foam into pipe insulation.

8.12 Rigid Foam Illustrative Patents

8.12.1 Aminoimides as Catalysts

Rigid foam urethane catalysts are employed to promote at least two, and sometimes three, major reactions. These must proceed simultaneously and competitively at balanced rates during the process to provide polyurethanes with the desired physical characteristics. One reaction is a chain-extending isocyanate-hydroxyl reaction by which a hydroxyl-containing molecule is reacted with an isocyanate-containing molecule to form a urethane. This increases the viscosity of the mixture and provides a polyurethane containing secondary nitrogen atom in the urethane groups. A second reaction is a cross-linking isocyanate urethane reaction by which an isocyanate-containing molecule reacts with a urethane group containing a secondary nitrogen atom. The third reaction that may be involved is an isocyanate-water reaction by which an isocyanate-terminated molecule is extended and by which carbon dioxide is generated to blow or assist in the blowing of the foam. This third reaction is not essential if an extraneous blowing agent, such as a halogenated, normally liquid hydrocarbon, carbon dioxide, etc., is employed, but it is essential if all or even a part of the gas for foam generation is to be generated by this in situ reaction (e.g., in the preparation of one-shot flexible polyurethane foams).

The reactions must proceed simultaneously at optimum balanced rates relative to each other to obtain a good foam structure. If carbon dioxide evolution is too rapid in comparison with chain extension, the foam will collapse. If the chain extension is too rapid in comparison with carbon dioxide evolution, foam rise will be restricted, resulting in a high density foam with a high percentage of poorly defined cells. The foam will not be stable in the absence of adequate cross-linking.

It is known that tertiary amines are effective for catalyzing the second cross-linking reaction. However, many amines of this class have a strong amine odor, which is carried over to the polyurethane foam. In still other cases, some tertiary amines impart a color to the product foam known as *pinking*.

Zimmerman[5] disclosed the preparation of a new class of aminoimide compounds, which are useful as polyurethane catalysts. The compounds possess a number of useful characteristics, making them attractive as polyurethane catalysts. For example, they have rapid catalytic activity in the polyurethane foam area. In addition, the compounds are also relatively nonvolatile and possess little if any odor. Also, and most importantly, the compounds do not cause pinking as so often observed when other testing amine catalysts are employed, particularly when polyester polyols are used to make urethanes. The catalysts of the invention are particularly desirable in foaming urethanes in that they provide a sufficient delay in the foaming operation to aid in processing. Yet, the catalysts also give good foams with desirable tack-free times. This delay time is particularly desirable in molding applications to allow sufficient time to situate the prefoam mix in the mold.

8.12.2 1,2,2,2-Trichloroethyl Group as Flame Retardants

Polyurethane foam is normally flame retarded by incorporating phosphorous and/or halogen containing compounds into the formulation. Three such commercial flame retardants, which are chloroalkyl phosphates, are disclosed in U.S. Pats. 3,171,819, 3,817,881, and 3,192,242. Since these flame retardants are nonreactive additives, they are not permanently bound into the polymer and, hence, have the tendency to migrate to the surface of the foam. This results in a loss of flame retardance. Another commercial flame retardant for polyurethane foam is 2,3-dibromo-2-butene-1,4-diol, which is described in U.S. Pats. 3,919,166 and 4,022,718. The disadvantages of this material are its relatively high costs and its poor processibility at high loadings.

An invention by Miano, et al.[6] relates to a flame-retarded polyurethane foam. The flame retardant additives described are effective as flame retardants for hot-cure flexible polyurethane foam, high-resiliency polyurethane foam, rigid polyurethane foam, and rigid polyurethane isocyanurate foam copolymers. These additives are especially effective as flame retardants for high-resiliency (HR) polyurethane foam made from special polyols that allow the foam to cure at ambient temperatures without external heat. This cold-cure process provides for lower molding costs and faster production cycles; it also produces foams with higher load ratios and greater tear strengths than foams made by the conventional hot-cure process.

The compounds of the Miano patent overcome the traditional disadvantages, because they react with the isocyanates used in preparing the polyurethane and thus become permanently bound to the polymer. There is, therefore, no loss in the flame retardance of foams that contain the compounds of the invention due to flame retardant migration or aging.

A high-resiliency polyurethane foam was prepared using the Miano flame retardant, and its flame retardant and physical properties were compared to a foam containing no flame-retardant additives. These results are shown in Table 8.12. The example shows that, without flame retardant additives, the high-resiliency polyurethane foam is totally consumed (5 in) in the ASTM D-1692 test.

Example A shows that when only one part of the new flame retardant is incorporated into a similar 100%-TDI-based high-resiliency formulation, the flame retardance is markedly increased without adversely affecting the physical properties of the foam.

TABLE 8.12 Example of Polyurethane Foam Prepared with Miano Flame Retardant

Composition	Control	Example A
Voranol 4701 (Dow)	60	60
Polyether polyol, molecular weight approx. 5000, functionality 3, primary OH groups		
Niax 34-28 (Union Carbide) polymer polyol	40	40
Bis(1-hydroxy, 2,2,2-trichloroethyl) urea	—	1.0
Silicone surfactant DCF 1-1630 (Dow Corning)	0.04	0.04
Water	2.7	2.7
Diethanolamine	1.5	1.5
Dabco 33LV catalyst (Air Products)	0.3	0.3
Niax A-1 catalyst (Union Carbide)	0.12	0.12
Niax A-4 catalyst (Union Carbide)	0.3	0.3
T-12 catalyst (M & T)	0.03	0.03
Toluene diisocyanate (80/20 mixture of 2,4/2,6 isomers)	35.5	36.1
Isocyanate index	105	105
Physical Properties		
Density (core, lb/ft^3)	2.36	2.46
ILD		
25% (lb/50 in^2)	33	35
65% (lb/50 in^2)	84	88
Sag factor	2.5	2.5
Oxygen index (ASTM D-2863)		
Initial	22.1	23.2
After DHA	21.6	23.2
Extent of burn (inches)		
Initial	5.0	2.9
After DHA	4.0	2.3

8.12.3 Modified Diorganotin Catalysts with Long Pot Lives

Tetravalent tin compounds are effective gel or polymerization catalysts for cellular polyurethanes. Three of the most popular catalysts are stannous 2-ethyl hexoate, dibutyltin di(lauryl mercaptide), and dibutyltin dilaurate. Unfortunately, these catalysts, in addition to most of the other classes of tin compounds, undergo a substantial loss of activity when incorporated into a precursor or *master batch* that is subsequently stored for any considerable length of time before being reacted with the isocyanate component. The master batch contains one or more of the reactants, catalysts, and modifiers, which are subsequently combined with the isocyanate component to form the foam product. These tin-containing catalysts are therefore not suitable for use in any prepackaged two-component systems for preparing rigid cellular polyurethanes.

Hirshman and Treadwell[7] disclosed that diorganotin compounds where at least one of the remaining valences on the tin atom is satisfied by a sulfur atom or a thiocyanate group are unique among diorganotin compounds in that they retain their catalytic activity over extended periods of time in the presence of the precursors or master batches conventionally employed to prepare rigid, cellular polyurethanes.

The invention provides a storage-stable catalyst-containing precursor for rigid cellular polyurethanes. The precursor comprises a polyol containing at least two active hydrogen atoms, as determined by the Zerwitinoff method, and an organotin gel catalyst. The gel catalyst exhibits a formula selected from the group consisting of R2Sn(SCN)2, [R2Sn(SCN)]2O, [R2Sn(SCN)]2S, (R2SnX)2S and R2SnS. R is a hydrocarbon, and X is chlorine, bromine, or iodine.

8.12.4 Delayed-Action Catalysts

Polyurethanes are widely used in preparing rigid and flexible foams, castings, adhesives, and coatings. Typically, the reaction between the isocyanate and the polyol has been catalyzed by using various components such as amines, e.g., tertiary amines, and organometallics, particularly organotin compounds such as stannous octoate, dibutyl tin dilaurate, tin ethylhexanoate, and so forth. The effectiveness of the catalyst is often measured by the cream time, which is the time required for the isocyanate and polyol syrup to turn from a clear solution to a creamy color; the gel time, which is the time required for polymer particles to form in the syrup; rise time, which is the time required for the syrup to rise to its maximum height; and cure time, which is the time to reach a tack-free state.

In some applications for polyurethanes, it is desirable to effect reaction in the shortest time possible; therefore, catalysts having tremendous activity are desired. In some applications, though, as in the molding of intricate parts or large objects, it may be desirable to keep the polyurethane composition in a fluid state for an extended time to permit the composition to completely fill the mold or flow into the cracks and crevices of the mold. Then, once the mold is completely filled, it is desirable to effect polymerization of the polyurethane in the shortest time possible so that the finished parts can be removed and the mold recharged with new materials. In this regard, it is desirable to delay the initial reaction but, after reaction commences, then to catalyze the polymerization rate. To do this, it is necessary to extend the cream time to permit the polyurethane composition to penetrate the cracks and crevices in the mold and to extend the gelation time as the polyurethane foam on gelling becomes intractable and resists molding. However, once the reaction begins, it is desirable to end up with a rise and cure time comparable to those achieved by active catalysts, as this will permit greater productivity.

Bechara et al.[8] were awarded a patent relating to polyurethane compositions containing an effective proportion of a catalyst comprising an amine salt of an amino acid and an organometallic catalyst, and a method for catalyzing the reaction between an isocyanate and a compound having a reactive hydrogen atoms. Broadly, the amine salts of amino acids (DACs) of this invention can be visualized as having at least monofunctionality in terms of amine, and at least monofunctionality (and preferably difunctionality) in terms of the amine salts. Some of these compounds can be formed by reacting ammonia, a primary or secondary amine, and an olefinic unsaturated compound having sufficient reactivity to replace the amine hydrogen atom.

8.12.5 CFC and HCFC-Free Rigid Foams

Since the widespread adoption of the Montreal Protocol, the urethane industry has concentrated efforts directed to eliminating the use of chlorofluorocarbons, such as the widely used CFC-11, from polyurethane foam formulations of all types. The use of HCFCs such as HCFC-22, monochlorodifluoromethane, which have lower ozone depletion potentials (ODPs), has been promoted as an interim solution. However, HCFC-22, an HCFC of choice, is a gas at room temperature, with poor system solubility, and thus extraordinary processing equipment must be used, including, in some cases, pressurized day tanks.

Water has been used for many years in polyurethane and polyisocyanurate foam systems. However, the carbon dioxide generated by the water/isocyanate reaction is markedly inferior to the CFCs and HCFCs with respect to preparing rigid foams having low K factors. To overcome this deficiency, it has been suggested to include perfluorocarbons (PFAs) in water-blown formulations. However, PFAs are quite expensive, despite being used in modest amounts; have exceptionally poor system solubility, often requiring emulsification rather than solution; and moreover offer only a modest advantage over all-water-blown systems.

Low-boiling aliphatic hydrocarbons have been suggested as blowing agents for polymeric foams and are widely used in the expandable and expanded polystyrene industry. However, they have been eschewed by the polyurethane industry due to the flammability of the foams produced through their use, as well as the high K factors obtained in rigid foams, making them undesirable for use in many applications.

Fishback, et al.[8] discovered that polyurethane and polyisocyanurate foams having low flammability characteristics may be achieved with aliphatic hydrocarbon-blown foams. Polyurethane/polyisocyanurate foams with low K factors and enhanced flammability properties, as measured by the Butler Chimney test, are produced through the use of a blowing agent composition comprising aliphatic hydrocarbon and a fully halogenated brominated halocarbon is a mole ratio of from 10:1 to 1:2.

8.12.6 Rigid Foams as Insulators for Water Heaters

The advantage of using rigid polyurethane foam insulation in water heater construction has been recognized for several years. For example, the heat conductivity of polyurethane foam is lower than that of glass fibers, thereby providing superior insulation properties. Thus, it is possible to obtain the same or better insulation properties using a substantially reduced insulation wall thickness, as compared to conventional insulation materials such as glass fibers. This results in more energy-efficient water heaters and reduced size, thereby providing lower packaging and shipping costs.

In addition, the rigidity of the foam insulation, when compared with that of glass fibers, provides improved resistance to dents in the exterior jacket of the tank. This factor permits the use of less sophisticated (and therefore less expensive) shipping containers.

Although the superior insulating properties of expandable foam materials such as polyurethane have been well recognized for many years, the use of foam as an insulating material in water heaters has been quite limited to date. This is due at least in part to the production problems encountered using expandable foam materials, resulting in higher production costs. One of the major problems associated with water heater manufacturing, and particularly the production of foam-insulated water heaters, has been the method by which the foam insulation layer is formed about the tank. Generally, the foam is injected as a liquid that continually expands and eventually matures into a rigid foam layer. Usually, the liquid foam is injected into the annular space between the inner tank and the outer jacket. Unfortunately, the liquid foam has a tendency to leak out of any small openings in the seams of the outer jacket.

West and Marcinkewicz[10] disclosed a method of insulating a water heater with an expandable foam insulating material. The expandable foam insulating material has an initial viscosity of <300 cps, a flow index ≥ approx. 0.7, and a gel index ≥ approx. 0.8, and it generates a maximum pressure of ≤ approx. 1.0 psig. The foam is injected into the annular space between the water tank and the outer jacket. Preferably, the foam is injected into a sleeve, which acts to confine the foam during its most liquid state, positioned within this space. The foam is preferably injected into the annular space between the tank and the jacket while the top cover is removed. In this way, the injected foam can be distributed more evenly around the circumference of the tank.

West et al. discovered that the disadvantages associated with the prior art bags and envelopes may be overcome by using a foam composition having an unusually low initial viscosity, having high flow and gel indices, and which generates lower foaming pressures than hepior art foams. It has also been discovered that, using these new foams, it is no longer necessary to use the foam-restrictive bags and envelopes of the prior art, which were so troublesome from the production and void formation standpoints.

Foams with No Ozone-Reducing Potential

The use of trichloromonofluoromethane (CFC-11) and other chlorofluorocarbons as blowing agents in the production of urethane foams is well known. These CFC blowing agents are also known to have an adverse effect on the ozone layer in the atmosphere. The urethane foam industry is therefore investigating methods for producing foams with good physical properties without using CFC blowing agents.

Initially, the most promising alternatives appeared to be hydrogen-containing chlorofluorocarbons (HCFCs). U.S. Pat. 4,076,644, for example, discloses the use of 1,1-dichloro-2,2,2-trifluoroethane (HCFC-123) and 1,1-dichloro-1-fluoroethane (HCFC-141b) as blowing agents for the production of polyurethane foams. However, HCFCs also have some ozone depletion potential. There is therefore mounting pressure to find substitutes for the HCFCs as well as the CFCs.

Alternative blowing agents which are currently considered to be promising because they contain no ozone-depleting chlorine are fluorocarbons (FCs) and partially fluorinated hydrocarbons (HFCs). The use of 1,1,1,4,4,4-hexafluorobutane as a blowing agent is disclosed in Lambert's "1,1,1,4,4,4-hexafluorobutane, a New Non-Ozone-Depleting Blowing Agent for Rigid PUR Foams," Polyurethanes World Congress 1991 (Sep. 24–26), pp. 734–739.

U.S. Pat. 4,898,893 teaches that a blend of a liquid hydrocarbon and halogenated hydrocarbon is useful as a blowing agent for the production of isocyanurate foams.

The use of mixtures of a chlorofluorocarbon having a boiling point between 74 and 120° F and an alkyl alkanoate having a molecular weight of no more than 88 as a blowing agent for foams is disclosed in U.S. Pat. 4,960,804. HCFC-123 and HCFC-141b are among the chlorofluorocarbons disclosed therein.

U.S. Pat. 5,035,833 discloses the use of a mixture of dichlorotrifluoroethane and at least one paraffin having five or six carbon atoms as blowing agents useful for the production of rigid polyurethane foams.

U.S. Pat. 5,096,933 discloses a process for the production of rigid polyurethane foams in which cyclopentane and/or cyclohexane and optionally a low boiling compound (i.e., boiling point less than 35° C) with no more than four carbon atoms, which is homogeneously miscible in cyclopentane and/or cyclohexane is used.

Azeotropes of HCFCs and various compounds and azeotropes of organic compounds that may be used in combination with HCFCs have also been described in the prior art as being useful blowing agents for the production of foams. U.S. Pat. 4.900,365, for example, teaches that azeotropes of a dichlorotrifluoroethane and isopentane are useful in the production of polyurethane foams. U.S. Pat. 5,106,527 discloses the use of azeotropes of 2-methyl butane and 1,1-dichloro-1-fluoroethane as blowing agents for the production of rigid, closed cell foams.

The azeotropic mixtures discussed in U.S. Pat. 5,166,182 must have boiling points below 50° C. These azeotropic mixtures are formed from organic compounds having surface active properties that enable the blended azeotropic mixture to become miscible with polymer resins. Examples of the organic compounds described as being useful in the production of such azeotropes include: n-pentane, acetone, methyl alcohol, methyl formate, ethyl formate, ethyl alcohol, 2-methyl butane, nitromethane, cyclopentane, 2,3-dimethyl butane, 2,2-dimethyl butane, and dimethyl sulfide. These azeotropes may be used in combination with fluorocarbons, but an azeotrope in which a fluorocarbon is one of the components is not revealed or suggested.

U.S. Pat. 5,227,088 discloses azeotrope-like compositions, which are made up of 1-chloro-3,3,3-trifluoropropane and a hydrocarbon containing five or six carbon atoms.

U.S. Pat. 5,283,003 discloses a blowing agent made up of at least one five-carbon-member hydrocarbon, a chlorinated alkane, and methyl formate. Methylene chloride is the preferred chlorinated alkane.

Azeotropic mixtures in which HCFCs are included are also known to be useful as cleaning solvents. U.S. Pat. 4,055,507, for example, discloses an azeotropic mixture of 1,2-dichloro-1,1-difluoroethane and 3-methylpentane, which is taught to be useful as such a solvent. Japanese Pat. 1,141,995 discloses an azeotropic mixture of 67 to 87% by weight of HCFC-123 and 13 to 33% by weight of 2-methyl butane, which is useful as a cleaning solvent. Japanese Pat. 1,141,996 discloses an azeotropic mixture of HCFC-141b and n-pentane or 2-methyl butane or 2,2-dimethyl butane, which is also said to be useful as a cleaning solvent.

The use of azeotropes formed from specified amounts of 1,1,1,4,4,4-hexafluorobutane and n-pentane as a blowing agent or a cleaning solvent has not, however, been described in the prior art.

The Werner et al.[11] invention relates to novel azeotropic compositions, a process for the production of foams in which these azeotropic compositions are used, and to foams produced using ≈73 to ≈87% by weight of 1,1,1,4,4,4-hexafluorobutane and from ≈13 to ≈27% by weight of n-pentane as a blowing agent in a process for the production of polyurethane foams.

It is an object of the Werner et al. invention to provide an azeotropic composition that contains no chlorine and therefore has an ozone depletion potential of zero. It is also an object of the invention to provide a process for the production of urethane foams in which no chlorine-containing blowing agent is employed.

Blowing Agent Emulsion

The production of cellular polyisocyanate polyaddition products (e.g., cellular polyurethane elastomers and flexible, semirigid, or rigid polyurethane foams) by reacting organic polyisocyanates and/or modified organic polyisocyanates with relatively high-molecular-weight compounds containing at least two reactive hydrogen atoms (e.g., polyoxyalkylene polyamines and/or preferably organic polyhydroxyl compounds having molecular weights of, for example, from 500 to 12,000 and, if desired, chain extenders and/or cross-linking agents having molecular weights of approximately 500 in the presence of catalysts, blowing agents, assistants, and/or additives) has been disclosed in numerous patents and other publications. A suitable choice of the starting components (polyisocyanate, relatively high-molecular-weight compound-containing reactive hydrogen atoms and, if desired, a chain extender and/or cross-linking agent) allows elastic or rigid, cellular polyisocyanate polyaddition products and all modifications in between to be produced by this method.

A review on the production of cellular polyurethane elastomers, polyurethane foams, and polyisocyanurate (PIR) foams, their mechanical properties, and their use is given in Refs. 12 through 15.

Cellular plastics are produced by the polyisocyanate polyaddition process using essentially two types of blowing agents.

1. Low-boiling, inert liquids that evaporate under the conditions of the exothermic polyaddition reaction; for example, alkanes, such as butane, pentane, inter alia, or (preferably) halogenated hydrocarbons such as methylene chloride, dichloromonofluoromethane, trichlorofluoromethane, inter alia
2. Chemical compounds that form blowing gases by a chemical reaction or thermal decomposition

Examples that may be mentioned are the reaction of water with isocyanates to form amines and carbon dioxide, which proceed synchronously with the polyurethane preparation, and the cleavage of thermally labile compounds (e.g., azoisobutyronitrile), which, as a cleavage product, in addition to nitrogen, provides toxic tetramethylsuccinonitrile or azodicarbonamide, used as a constituent of a blowing agent combination.

The last-mentioned method, in which thermally labile compounds (e.g., azo compounds, hydrazides, semicarbazides, N-nitroso compounds, and benzoxazines are usually incorporated into a previously prepared polymer or drum-coated onto the polymer granules and are foamed by extrusion), has remained of minor importance in industry. However, physical low-boiling liquids, in particular chlorofluoroalkanes, are used worldwide on a large scale for the production of polyurethane and polyisocyanurate foams.

The only disadvantage of these blowing gases is environmental pollution. By contrast, the formation of blowing gases by thermal cleavage or chemical reaction gives cleavage products and/or reactive by-products that are included in or chemically bonded to the polyaddition product and can result in an undesired modification of the mechanical properties of the plastic. In the case of the formation of carbon dioxide from water and isocyanate, urea groups are formed in the polyaddition product and can result in an improvement in the compressive strength or even embrittlement of the polyurethane, depending on the amount.

It is furthermore possible to use as blowing agents fluorinated hydrocarbons, perfluorinated hydrocarbons, sulfur hexafluoride, or mixtures of at least two of these compounds. Since these fluorinated or perfluorinated blowing agents are sparingly soluble or insoluble in the starting components for the production of the polyisocyanate polyaddition products, they are emulsified in at least one organic and/or modified organic polyisocyanate, at least one relatively high-molecular weight compound containing at least two reactive hydrogen atoms, or a mixture of at least one relatively high-molecular-weight compound containing at least two reactive hydrogen atoms and a low-molecular-weight chain extender and/or cross-linking agent. This method allows the production of cellular plastics having a uniform and fine cell structure. The only disadvantages of this process are the narrow choice of suitable fluorinated or perfluorinated compounds having a boiling point in the necessary range and the high price of these blowing agents. The production of cellular plastics having the cell structure required in industry is restricted to a narrow choice of mixtures of perfluoropentane and perfluorohexane.

Low-boiling hydrocarbons that can be used as blowing agents are soluble if the starting components for the preparation of the polyisocyanate polyaddition products give cellular plastics having a relatively coarse, frequently nonuniform cell structure and increased thermal conductivity.

The Volkert[16] invention relates to a process for the production of cellular plastics by the polyisocyanate polyaddition process in which the blowing agent used is a low-boiling, fluorinated or perfluorinated, tertiary alkylamine and at least one further physical and/or chemical blowing agent which is sparingly soluble or insoluble in the organic polyisocyanate or the compound containing at least two reactive hydrogen atoms and is therefore emulsified in at least one of these starting components.

It is an object of the Volkert invention to replace all or at least some of the chlorofluorocarbons known as blowing agents for the production of cellular plastics by the polyisocyanate polyaddition process by other, environmentally friendly blowing agents without adversely effecting the fine-celled foam structure, as can be achieved using emulsions based on fluorinated hydrocarbons. It is a further object of the present invention to improve the processing properties of the reaction mixtures containing the blowing agents and to minimize their sensitivity in a variety of foaming equipment.

Acetone as an Auxiliary Blowing Agent

In the manufacture of polyurethane foams, a polyol, an isocyanate, water, and at least one catalyst for the formation of polyurethane foam are mixed to form a liquid mixture, which then foams and solidifies to produce the solid polyurethane foam. Reaction between the isocyanate and the water produces carbon dioxide gas, this production of carbon dioxide normally being accelerated by a catalyst present in the liquid mixture, and the carbon dioxide gas thus produced assists in formation of a foam.

However, many commercial grades of polyurethane foam require low foam densities and high foam flexibilities, and it is not possible to produce a low-density, highly flexible foam using only carbon dioxide as the foam-forming gas. The water-isocyanate reaction that generates the carbon dioxide also produces solid reaction by-products that harden the foam. Therefore, if a high proportion of water is used in the reaction mixture, the resultant foam will be too hard for many applications. Moreover, the water-isocyanate reaction is highly exothermic, to such an extent that, if a high proportion of water is included in the reaction mixture, the resultant foam is produced at a temperature sufficiently high that it may be scorched by atmospheric oxidation; indeed, there have been occasional instances where foams produced from reaction mixtures containing high proportions of water have spontaneously combusted. (Incidentally, the exothermic reactions involved in polyurethane foam manufacture continue for a considerable time, typically several hours, after formation of the solid foam, so heat damage to the foam may not always be visible by observing the foam as it passes along the production line.)

Accordingly, it is conventional to include in the foam-forming mixture an auxiliary blowing agent, that is to say, a blowing agent other than water, this auxiliary blowing agent being a material that is liquid at room temperature but relatively volatile. This means that it can be volatilized during the foam-forming reaction to form a gas, which serves to decrease the density of the polyurethane foam without producing an excessively hard foam. Auxiliary blowing agents also serve to prevent foam becoming too hot, since some of the heat generated by the water is absorbed as latent heat of vaporization in volatilizing the auxiliary blowing agent.

Traditionally, the preferred auxiliary blowing agents for use in commercial polyurethane production were the chlorofluorocarbons (CFCs), especially trichloromonofluoromethane, sold under the trademark Freon-11. The chlorofluorocarbons are highly desirable auxiliary blowing agents because of their high volatility (trichloromonofluoromethane has a boiling point of 23.7° C) and because they are inert and thus do not interfere with the polyurethane-forming reactions. However, the chlorofluorocarbons, which persist for many years in the atmosphere, have recently been discovered to pose a major threat to the environment because they destroy the Earth's ozone layer. Consequently, an international agreement was signed to greatly reduce annual production of chlorofluorocarbons during the next few years. Thus, it is likely that use of chlorofluorocarbons as auxiliary blowing agents in foam production will be outlawed in the near future.

Methylene chloride has been suggested, and indeed is already in commercial use, as a substitute for chlorofluorocarbon auxiliary blowing agents. Methylene chloride is highly volatile (boiling point 39.8° C) and inert in polyurethane-forming mixtures. However, methylene chloride is a suspected carcinogen and has other deleterious effects on workers exposed to it. Accordingly, the concentration of methylene chloride in the air inside a foam plant must be kept low. The American Conference of Governmental Industrial Hygienists recommends that workers be exposed to not more than 50 ppm of the chemical, while the Occupational Safety and Health Administration's Permissible Exposure Limit is 500 ppm. Keeping the levels of methylene chloride in a foam plant below 50 ppm may require additional ventilation equipment, with an associated increase in costs. In addition, the chemical is a recognized environmental pollutant, and both the federal and state governments are beginning to limit releases of the chemical from plants that employ it. Thus, in the near future, plants using methylene chloride as an auxiliary blowing agent may be faced with the substantial additional expense of installing scrubbers or similar equipment to remove methylene chloride from air and/or other gases discharged from the plant. In addition, California has recently proposed that emissions of methylene chloride in that state be subject to a heavy "pollution" tax, and other states are likely to follow a similar course.

Incidentally, although it might be thought that methylene chloride (molecular weight 85) would be more effective as an auxiliary blowing agent than trichloromonofluoromethane (molecular weight 137.4), in practice it is found that, in similar formulations, the two auxiliary blowing agents have essentially the same blowing activity per unit weight. Thus, the activity per unit weight of an auxiliary blowing agent cannot readily be predicted simply from its molecular weight.

Thus, both the auxiliary blowing agents presently used for the commercial production of polyurethane foams cause significant environmental problems, and it appears likely that the industry, within the next few years, will be forbidden to use these two auxiliary blowing agents at all, will be subject to heavy penalties for their use, or will be required to take expensive precautions to prevent their release into the environment. Thus, there is a need for alternative auxiliary blowing agents for use in the production of polyurethane foams.

In theory, the only real requirements for an auxiliary blowing agent are that it be liquid at about ambient temperature (so that it can be introduced into the foam-forming mixture as a liquid), sufficiently volatile that it volatilizes during the formation of the polyurethane foam, and sufficiently inert that it does not interfere with the foam-forming reactions. A wide variety of materials have been proposed for use as auxiliary blowing agents. However, in practice, none other than chlorofluorocarbons and methylene chloride appear to have been used industrially, and some of the proposed compounds have disadvantages so severe that they are totally impracticable for use in industrial plants. Thus, for example, U.S. Pat. 4,546,122, issued Oct. 8, 1985 to Radovich et al., states the following concerning auxiliary blowing agents for flexible polyurethane foams:

> Suitable blowing agents include water and/or readily volatile inorganic or organic substances. Appropriate organic blowing agents are acetone, ethyl acetate, halogen-substituted alkanes such as methylene chloride, chloroform, ethylidene chloride, vinylidene chloride, monofluorotrichloro methane, chlorodifluoromethane, dichlorodifluoromethane; and butane, hexane, heptane, or diethyl ethers. Inorganic blowing agents which may be used are air, CO_2, and N_2O. A blowing effect may also be achieved by adding compounds which decompose at the reaction temperature to give off a gas (e.g., nitrogen, given off by azo compounds, such as azodicarbonamide or azobutyronitrile).

Of the compounds other than CFCs and other haloalkanes in this list, diethyl ether is obviously impracticable as an auxiliary blowing agent under industrial conditions because of its flammability and anesthetic effect, while the use of carbon dioxide or nitrogen requires special equipment for injection of the gaseous auxiliary blowing agents into the developing foam, and consequently these gaseous auxiliary blowing agents are not used in practice for low-density polyurethane foam.

U.S. Pat. 3,121,699, to Merriman, describes the preparation of foamed polyurethane materials using a polyalkylene ether polyol having a mean molecular weight of more than 1500, an organic polyisocyanate, water, a foam stabilizing agent, and a catalyst, and acetaldehyde as an auxiliary blowing agent, the acetaldehyde being vaporized by the heat produced in the exothermic reaction of the ingredients, the acetaldehyde being included upon the ingredients forming the foam and setting of the foam being effected within a mold.

U.S. Pat. 3,165,483, to Gemeinhardt et al., describes the preparation of a skeletal polyurethane foam that is produced using as an auxiliary blowing agent a ketone containing from 5 to 18 carbon atoms. Alternatively, an aldehyde having from 4 to 18 carbon atoms can be used. Among the specific ketones mentioned is methyl n-propyl ketone.

U.S. Pat. 3,179,626, to Beitchman, describes the preparation of polyurethane and polyisocyanate foams using as co-catalyst a mixture of diazabicyclooctane and an aldehyde containing from 1 to 10 carbons atoms, there being from 0.01 to 10 parts of diazabicyclooctane (sold commercially as DABCO) per part of aldehyde. In this patent, the aldehyde appears to be strictly a co-catalyst.

Klesper, in "Application of volatile organic liquids for expanding flexible urethane foam," *Rubber Age,* October 1958, pp. 84–87, describes experiments in which various organic liquids were tested as auxiliary blowing agents for the production of flexible polyurethane foams by the two stage or *two-shot* process, in which an isocyanate is first reacted with a polyol to produce a prepolymer. At some later time, this prepolymer is reacted with water and catalysts to yield the polyurethane foam. This article states that acetone was not very effective in the blowing action because of a too-high boiling point and high solubility in the finished polymer, which also causes the acetone to leave the finished foam slowly. The article further states that pentane alters the surface properties of the foam during rise.

According to Walsmley,[17] a polyurethane foam is produced using acetone as an auxiliary blowing agent. A polyol, an isocyanate, water, acetone, and a catalyst for catalyzing the formation of polyurethane are mixed to produce an unfoamed liquid mixture, which is then retained in bulk form, preferably in an open-topped reaction vessel, for a dwell time during which an exothermic reaction occurs and the liquid foams. Finally, the foaming liquid is spread onto a surface, on which it is allowed to expand and form a solid polyurethane foam material. Significant reductions in the amount of auxiliary blowing agent can be achieved as compared with conventional formulations using methylene or CFCs as auxiliary blowing agents.

References

1. Harada, S. et al. U.S. Pat. 5,306,735, Apr. 26, 1994, "Polyurethane, process for producing the same, and process for producing polyurethane foam," to Kao Corp.

2. Boulet, J.-C. et al. U.S. Pat. 4,173,710, Nov. 6, 1979, "Halogenated polyether polyols and polyure-thane foams produced therefrom," to Solvay & Cie.

3. Klein, H.P. U.S. Pat. 4,166,172, Aug. 28, 1979, "Production of polyether polyol compositions," to Texas Development Corp.

4. Brown, J.H. et al. U.S. Pat. 4,169,922, Oct. 2, 1979, "Rigid polyurethane foam-forming composi-tions" to Monsanto Co.

5. Zimmerman, R.L. U.S. Pat. 4,152,497, "Use of aminoimides as polyurethane catalysts," to Texaco Development Corp., White Plains, New York.

6. Miano, J.D., and S.R. Sandler. U.S. Pat. 4,152,497, May 1, 1979, "Compounds containing the 2,2,2-trichlorethyl group as flame retardants for polyurethanes," to Pennwalt Corp.

7. Hirshman, J.L., and K. Treadwell. U.S. Pat. 4,136,046, Jan. 23, 1979, "Storage-stable precursors for rigid polyurethane foams," to M&T Chemicals, Inc., Stamford, Connecticut.

8. Bechara, I.S., R.L. Mascioli, and P.J. Zaluska. U.S. Pat. 4,115,634, Sep. 19, 1978, "Amine salts of amino acids as delayed action catalysts," to Air Products & Chemicals, Inc., Allentown, Pennsyl-vania.

9. Fishback, T.L., C.J. Reichel, and J.S. Dailey. U.S. Pat. 5,384,338, Jan. 24, 1995, "CFC and HCFC-free rigid insulating foams having low K factors and decreased flammability," to BASF Corp., Parsippany, New Jersey.

10. West, E.L., and R.J. Marcinkewicz. U.S. Pat. 4,904,428, Feb. 27, 1990, "Method of making a foam insulated water heater," to Bradford-White Corp., Philadelphia, Pennsylvania.

11. Werner, J., S.A. Kane, H.P. Doerge, and E.F. Boonstra. U.S. Pat. 5,488,073, Jan. 30, 1996, "Azeotropic compositions of 1,1,1,4,4,4-hexafluorobutane and N-pantane and the use thereof in the production of foams," to Bayer Corp., Pittsburgh, Pennsylvania.

12. Saunders, J.H., and K.C. Frisch. *High Polymers,* Vol. XVI, *Polyurethanes,* Parts I and II. New York: Interscience Publishers, 1962 and 1964, respectively).

13. Vieweg, R., and A. Hochtlen, eds. Kunststoff-Handbuch, Volume VII, *Polyurethane,* 1st ed. Munich: Carl Hanser Verlag, 1966.

14. Oertel, G. Kunststoff-Handbuch, Vol. VII, *Polyurethane,* 2nd ed. Munich: Carl Hanser Verlag, 1983.

15. Piechota, H., and H. Rohr. *Integralschaumstoffe.* Munich, Vienna: Carl Hanser Verlag, 1975.

16. Volkert, O. U.S. Pat. 5,187,206, Feb. 16, 1993, "Production of cellular plastics by the polyisocyanate polyaddition process, and low-boiling, fluorinated or perfluorinated, tertiary alkylamines as blowing agent-containing emulsions for this purpose," to BASF Aktiengesellschaft, Ludwigshafen, Federal Republic of Germany.

17. Walmsley, G.D. U.S. Pat. 5,120,771, June 9, 1992, "Process for the production of polyurethane foam," to Hickory Springs Manufacturing Co., Hickory, North Carolina.

9

Polyurethane Foam
Surfactants

9.1 Introduction

The performance of silicone surfactants in polyurethane foam formulations is principally based on their surface activity and their interaction with the chemical structures of the foam polymer, which is being formed. For this reason, the design and chemical composition of a surfactant should be adjusted to the chemical environment in which it has to develop its efficiency. With the increasingly broadening range of polyurethane foams, to optimize both processing and final foam, it appears essential to select the surfactant by taking into account the relationship between its own composition and the specific characteristics of an actual foam formulation.

9.2 Surfactant Function

Surfactants (or foam stabilizers) are crucial in the manufacture of polyurethane foams. In spite of years of research, their mechanism during bubble formation, especially with polyether foams, is not well understood. The function of foam stabilizers consists of preserving the thermodynamically unstable state of a foam during the time of rising by surface forces until it is hardened. Foam stabilizers act by two mechanisms: (1) they lower the surface tension of the raw material mixture, and (2) they provide emulsification for the entire system.

A specific effect is the formation of surfactant monolayers at the interface. This monolayer is of higher viscosity than the bulk phase; therefore, the elasticity of the surface is increased and the expanding foam bubbles are stabilized. The greater the surface activity of the foam stabilizer, the greater is the expendability of the monolayer, i.e., the more it lowers the surface tension in low concentrations. The stabilization effect at the end of the rise time, when the cells are opened through the drainage of cell membrane material into the cell strands, is especially critical.

For the manufacture of polyester foams, surface active agents containing sulfonic acid groups have proved to be good foam stabilizers. A typical representative is the sodium salt of sulfonated ricinoleic acid or turkish red oil. Mixtures of these compounds with polysiloxanepolyoxyalkylene copolymers, which have an oxyethylene content of at least 75% and have good stabilizing properties.

Stabilizers for polyether foams are predominantly water-soluble polyethersiloxanes. These are reaction products from polydimethylsiloxanes with copolymers of ethylene oxide and propylene oxide. With polyethersiloxanes, which also are designated as polysiloxane-oxyalkylene block copolymer, one differentiates between nonhydrolyzable and hydrolyzable products. This differentiation is derived from the linkage between the polysiloxane and the polyether portion in the molecule. With flexible polyurethane foams, stabilizers have an especially pronounced influence on air permeability as well as compression set and hardness.

The Si-C linkage is hydrolytically stable, while the Si-O-C-linkage is hydrolyzable. For the foam stabilizing influence itself, this kind of bonding is unimportant. However, hydolyzable surfactants may have a deleterious effect on shelf life stability if a mixture of foam stabilizer, tertiary amine catalyst, and water is stored for a period of time. Under these conditions, the Si-O-C-bonding may hydrolyze, and the degradation product loses its foam stabilizing properties.

Fundamentally, both linear and branched polyether-siloxane foam stabilizers can be used. On the basis of their better stabilization properties, the branched types have dominated the market. One differentiates here between the terminally modified types (A) and the products with pendant structures, in which the main chain consists either of a polysiloxane (B) or a hydrocarbon polymer (C).

The polyethersiloxanes are modifiable in three ways: (1) molecular weight, (2) functionality, and (3) in composition of the polyether copolymer. To meet different production requirements, these parameters can be modified in specialized stabilizers. This is necessary because, in practice, numerous polyethers with variable ethylene oxide content and different molecular weights are used, since the activity of foam stabilizers is substrate specific.

9.3   Composition of Surfactants

The composition of silicone surfactants and general chemical structures of these products are shown in Fig. 9.1. Most silicone surfactants belong to one of these two groups of polysiloxane-polyether copolymers. Both contain one polysiloxane segment which can be branched or linear and to which, generally, two or more polyether segments are attached either at the ends or internally. The bonding of the polyether chains can be realized through SiC or SiOC linkages. The SiOC-linked copolymers are stable in neutral or amine basic environment but are gradually hydrolyzed in the presence of lewis acids, such as tin catalysts, and also by mineral acids. The SiC-linked copolymers are chemically stable in both amine basic and slightly acidic environment.

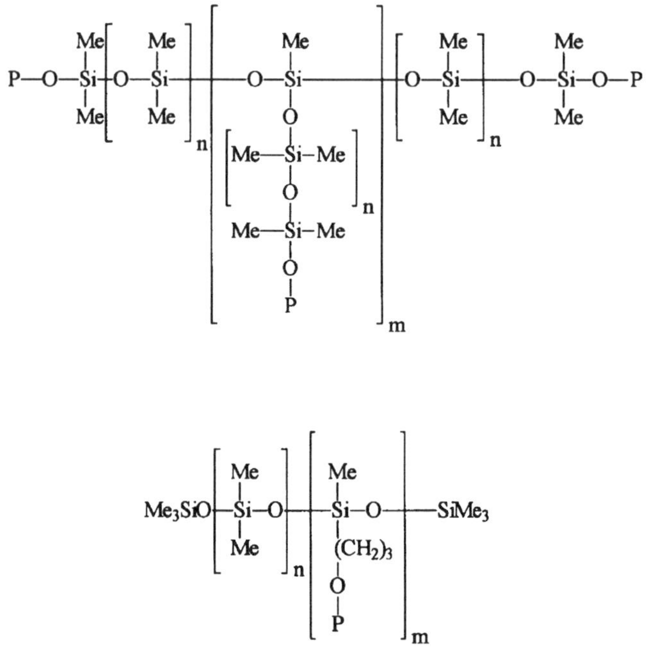

$$P= (C_2H_4)\,x(C_3H_6)R \quad R= \text{alkyl or H}$$

FIGURE 9.1   Chemical structures of silicone surfactants.

Variations in the surfactant properties of these copolymers are obtained by altering the overall polysiloxane-polyether ratio, by varying the ethylene oxide/propylene oxide ratio in the polyether chains, and by the type of end groups by which the polyether chains are capped, which are mainly OH, O-alkyl, or ester groups.

9.4 Surfactant Performance

The basic requirements for a surfactant in rigid foams are high surface activity for nucleation and stabilization of the cells, combined with good emulsifying abilities for the raw materials and the blowing agent.[1] The surface activity is provided by the polysiloxane moiety, and especially by its chain length. Higher chain lengths of the polysiloxane produce higher surface activity. Since polysiloxanes are incompatible with the usual raw materials, the solubility of surfactants and their emulsifying ability have to be regulated by the polyether portion. Increasing the polyether segments improves the solubility in polyols and isocyanate but reduces surface activity. The design of surfactants for different foam systems has therefore mainly to accomplish the best combination between surface activity and emulsifying ability.

Properly designed surfactants are key ingredients for the formation of nearly all polyurethane foams. As the raw materials are considerably different in chemical composition and reactivity, depending on whether a flexible or a rigid foam is intended to be produced, the design and properties of the surfactant have to be adjusted accordingly.

However, in addition to these essential differences, there are some basic similarities. Surfactants have a decisive influence on the nucleation process, i.e., the formation and stabilization of tiny air bubbles. Studies with degassed raw materials showed that self-nucleation does not occur in formulations for flexible or for rigid foam. All the cells of the final foam have to be present as small air nuclei in the reactive mixture before the foam starts rising. Their number and size distribution, and consequently the cell size distribution of the final foam, are determined by the amount of air and the mechanical energy introduced into the system by the mixing procedure, and by the surfactant, which supports physical processes like this due to its surface active potency. This general similarity between rigid and flexible foam systems has to be seen as superimposed and modified by differences in some detail aspects.

The most obvious differences are the compatibility of the raw materials and the morphology of the final foam polymer. The raw materials for rigid foams are incompatible, whereas the basic ingredients for flexible foams are miscible homogeneously. Therefore, the reaction mixture for rigid foams is heterogeneous and represents, including the gas bubbles, a three-phase dispersion consisting of one gaseous and two liquid phases. For this reason, the surfactant likewise has to act as a nucleation supporter and as an emulsifier; otherwise, the structure of a rigid foam would become quite poor.

In formulations for rigid foams, the isocyanate usually represents the continuous phase in which the polyol and the gas are dispersed. As a logical consequence, the whole foaming behavior of a formulation is to expect to be the better, the finer and the more stable this dispersion is. It is furthermore expected that both a fine air dispersion and a good emulsification of the reacting mixture contribute to a superior flowability. Therefore, the physical properties of the rising foam and the final product are determined by the surface activity of the surfactant and its emulsifying ability in the very first moments of the formation of a rigid foam.

One of the most important tasks a surfactant performs during the formation of a flexible foam is to stabilize the system against collapse. The tendency to collapse is much less critical in rigid foams; it could lead to the conclusion that weaker surfactants, which are less complicated to produce than silicone-based ones, are quite sufficient. All our development work in this direction, including tests with commercially available surfactants, produced unsatisfactory results. In some cases, it was possible to obtain acceptable properties of the final foam but, particularly when water was part of the formulation, the results were inferior.

During foam rise of the first spherical bubbles, and after transition to a polyhedral cell structure, the undisturbed growth of the polyhedral cells has to be regulated by the surfactant. As it was found already for flexible foams, increasing the amount of surfactants improves the stability of the rising rigid foam.

A growing rigid foam is much more stable in itself than is a flexible foam at the same life period. This is for several reasons. The sucking or drainage effect is reduced due to an accelerated buildup of viscosity, and there is no danger of defoaming by the separation of a solid phase of polyurea as it occurs during the formation of flexible foam.

Nevertheless, there is difficulty during the stage of expansion, particularly if the rising foam has to fill complicated shape cavities. Due to the growing viscosity and the formation of higher-molecular-weight material, the fluorocarbon representing the major blowing agent becomes more incompatible with the reacting mixture. Therefore, it again needs the emulsifying effectiveness of the surfactant to keep the blowing agent in the system; otherwise, a fine and regular distribution of the blowing agent is no longer guaranteed. Structure defects such as coarser cell-structure and faults and voids at the surface or beneath the skin are the consequence.

9.5   Illustrative Patents

9.5.1   Improved Surfactant Containing an Organic Acid Salt

Surfactants used in each of the several types of foam are generally siloxane-polyether block copolymers that differ from one another depending on the particular types of compositions in which they are to be employed. Such siloxane polyether copolymers are of two classes: *hydrolyzable* siloxane-polyether block copolymers in which the polyether blocks are attached to the siloxane blocks via Si-O-C linkages, and *nonhydrolyzable* siloxane-polyether block copolymers in which the polyether blocks are attached to the siloxane blocks via Si-C linkages.

Siloxane-polyether block copolymer surfactants are very complex mixtures of structures, the precise compositional nature of which are not readily definable. For a given nominal *nonhydrolyzable* siloxane-polyether copolymer, the siloxane block has relatively broad molecular weight distribution. Within the siloxane block molecular weight distribution, there are subdistributions based on the number and spacing of the attachment sites for the polyether blocks. The polyethers attached to the siloxane are also complex distributions of various molecular species, and frequently they are blends of polyethers of different molecular weights and polarities.

The platinum catalyst used to promote hydrosilylation, a reaction commonly employed to prepare such silicone-polyether copolymers by addition of an Si-H moiety across the double bond of an allyl-started polyether, is also poorly defined. Some evidence suggests that this catalyst is homogeneous, and some suggests that it is heterogeneous. Trace impurities in the silicone and allyl-containing polyether reactants have a dramatic impact on the hydrosilylation reaction. When the reactants are relatively free from platinum inhibitors, the hydrosilylation is rapid, and the order in which the polyethers react appears to be substantially random. The presence of certain trace impurities appears to affect both the reaction rate and the selectivity. There are also side reactions such as dehydrocondensation, which causes the loss of polyether attachment sites and cross-links the siloxanes.

Thus, the potential exists for considerable structural variability in the product surfactant if the proper parameters are not precisely controlled during its preparation. The extent to which the above variables affect the quality and consistency of the resulting siloxane-polyether copolymer surfactants is very dependent on the nominal structure of the particular surfactant being produced and the way in which the reactants are prepared and purified.

For each surfactant of a given nominal structure and produced by a given process, there is a practical optimum *ceiling performance* level. Most lots of the surfactant will perform in a given urethane foam at or near this level, but some preparations of the material, ostensibly identical, will perform somewhat less well. For best performance and consistency of the surfactant, it would be very desirable to have the means to modify surfactants to improve their performance. Prior to the present invention, however, such means for improving the performance of siloxane-polyether copolymer surfactants intended for use in the manufacture of conventional flexible polyurethane foam apparently have not been disclosed in the technical or patent literature, and do not appear to be known to the art.

Organic sulfonate or carboxylate salts can be used in combination with certain silicone surfactants for rigid polyurethane foam and polyester foam. In some cases, these organics were used as the sole surfactant. The ability of organic sulfonate or carboxylate salts to function as co-surfactants or as the exclusive surfactant in these two classes of foam has been commonly thought to be due to the high polarity of polyols used for rigid and polyester foams (see U.S. Pat. 3,594,334 for discussion on surface tension lowering).

Anionic organic surfactants do not function as surfactants in conventional flexible polyurethane foam. It is postulated that this is because they are not sufficiently surface active to adequately reduce the surface tension of the less polar polyethers used for conventional flexible polyurethane foam. U.S. Pat. 4,751,251 discloses a surfactant composition for rigid polyurethane and rigid polyisocyanurate foams, this surfactant composition including from about 15 to about 50% each of (1) an organic surfactant, (2) a siloxane-polyoxyalkylene surfactant, and (3) water and alcohol of 1 to 3 carbon atoms, or a mixture thereof.

U.S. Pat. 4,686,240 deals with a process for producing conventional flexible foams and high-resilience polyurethane foams using foam modifiers which are alkali metal or alkaline earth metal salts of Bronsted acids having pKa values greater than 1. Among the several classes of foam modifiers disclosed are salts of alkali and alkaline earth metals and carboxylic acids.

These foam modifiers are used with conventional silicone-polyether surfactants in foam preparations. The anions of the salts can be attached to a polyol, a polyether, a silicone-polyether copolymer, or a hydrocarbon. Generally, the level of the foam modifier employed is a substantial fraction of the sum of the foam modifier and the surfactant. In the foam preparation process, the required ingredients are mixed and immediately foamed. The 4,686,240 patent excludes aryl sulfonate salts as foam modifiers, because the pKa of the parent sulfonic acid is less than 1, which is the minimum value permitted under this patent. There is no indication that the foam modifiers are used in conjunction with the surfactants.

Reedy and Robertson[2] patented a composition related to polysiloxane-polyether surfactants for polyurethane foam manufacture. It contains a minor amount of a salt of an organic acid to aid them in achieving and maintaining a high level of effectiveness.

The Reedy invention provides a method for improving nonhydrolyzable silicone-polyether surfactants intended for use in conventional flexible polyurethane foam production, and improved surfactant compositions. The primary foam properties that are enhanced by this technology are foam height, foam porosity, and uniformity of foam cell structure.

9.5.2 t-butyl Terminated Silicone Surfactants for Enhanced Foam Breathability

High-resilience polyurethane foams are produced by the reaction of organic isocyanates, water, and high-molecular-weight polyols that have greater than 40% primary hydroxyl capping. They are distinguishable, in part, from conventional hot-cure polyurethane foams by the use of such a high percentage of primary hydroxyl groups, as well as by the fact that they require little or no oven curing. Thus, they are often referred to as *cold-cure* foams. High-resilience polyurethane foams are extremely desirable for cushioning applications because of the excellent physical properties they offer, such as high resilience, open cell structure, low fatigue for long life, and high sag factors for good load-bearing capabilities.

The ingredients for high-resilience polyurethane foam are highly reactive. Consequently, there is a rapid buildup of gel strength in the foaming reaction, which sometimes permits the foam to be obtained without use of a cell stabilizer. However, such unstabilized foams typically have very irregular and coarse cell structures, evidenced by surface voids. This problem has generally been addressed by using certain substituted polydimethylsiloxane-polyoxyalkylene or polyphenylmethylsiloxane-polyoxyalkylene copolymers as foam stabilizers.

Polysiloxane-polyoxyalkylene copolymer surfactants for use as stabilizers for high-resilience polyurethane foam are disclosed, for example, in the following: U.S. Pat. 3,741,917 of Morehouse; U.S. Pat. 4,478,957 of Klietsch et al.; U.S. Pat. 4,031,044 of Joslyn; U.S. Pat. 4,477,601 of Battice; U.S. Pat. 4,119,582 of Matsubara et al.; U.S. Pat. 4,139,503 of Kollmeier et al.; and several patents of Kilgour, namely U.S. Pat. 4,690,955, 4,746,683, and 4,769,174. These references variously disclose that the terminal oxygen atom of the polyoxyalkylene portion of the surfactant molecules may bear a hydrogen atom (Morehouse '917, Klietsch '957, Kollmeier '503), an alkyl group of 1-4 carbon atoms (Morehouse '917, Klietsch '957, Kilgour '955, '683, and '174), an alkyl group containing fewer than 10 atoms in total (Joslyn '044), an alkyl group containing a total of less than 11 carbon atoms (Battice '601), or a monovalent hydrocarbon group (Matsubara '582). Methyl capping is commonly used. In addition, several other capping groups are disclosed. These technologists apparently do not believed that there are any advantages to be gained by use of any particular alkyl capping groups.

Surfactants for stabilization of polyurethane foam are evaluated on the basis of several different performance characteristics. Primary among these is the potency or efficiency of the surfactant. The minimal amount of surfactant needed to provide good cell structure in the resulting foam is a relative measure of the potency. Polyurethane foam having good cell structure can be produced using less of a superior surfactant than would be required using a less potent surfactant. The ability to use less material is desirable in the foaming industry to lower the cost of foaming operations.

Of further concern in selecting a surfactant for polyurethane foam stabilization is the breathability or open-celled character of foam. High breathability (more open-celled character) of the foam is desirable, as it provides a greater processing latitude. A narrow processing latitude forces the foam manufacturer to adhere to very close tolerances in metering out the foaming ingredients, which cannot always be accomplished. Furthermore, greater breathability provides foam that is considerably easier to crush, thus avoiding splits that might occur during crushing. This characteristic is particularly desirable in foamed parts that incorporate wire inserts, which are difficult to crush.

In the design and development of surfactants for use as stabilizers for high-resilience polyurethane foam, there traditionally has been a trade-off between increasing the potency (efficiency) of the surfactant and lowering the breathability of the foam produced using it. It generally has been found that the more potent the surfactant, the lower the breathability of the foam made using it. In other words, the more potent surfactants generally afford poorer processing latitudes.

It would be very desirable to have silicone surfactants for stabilization of high-resilience polyurethane foam that afford both good potency and good breathability, thus providing foam manufacturers with relatively low surfactant costs as well as good processing latitude. Such surfactants are the subject of this application.

Kilgour[3] described siloxane-polyether copolymer surfactants for the manufacture of high-resilience polyurethane foam which are capped with t-butyl groups on the ends of the polyether blocks, polyurethane foam compositions containing such surfactants, a method for making polyurethane foam using such surfactants, and polyurethane foam. The t-butyl-capped surfactants afford improved surfactant potencies and foam breathabilities relative to the commonly employed methyl-capped surfactants.

The Kilgour invention provides t-butyl capped siloxane-polyoxyalkylene surfactants for the manufacture of high-resilience polyurethane foam. These surfactants afford higher potency in manufacture of high-resilience polyurethane foam and also produce more open or breathable foam than would be predicted, relative to otherwise similar surfactants in which the polyether portion of the molecule is terminated with a low-molecular-weight alkyl group.

9.5.3   Cell Openers for Urethane Foam

In the stabilization of conventional flexible polyurethane foam, a careful balance of properties must be achieved. Two of the more important surfactant-dependent properties that must be balanced are height of rise and foam porosity. For good heights of rise, the foam bubbles need to maintain their integrity (remain closed) to achieve the full potential of the expanding gases, i.e., low density, with a minimum density gradient from the top to bottom of the foam. The types of foam used for furniture cushions and bedding must also ultimately have a large fraction of open cells.

An excessive number of closed cells will cause the foam to shrink and distort as the gases within the closed cells start to cool. Also, if too many cells are closed or partially closed, the foam will lack resilience; that is, after being subjected to distortion, the foam will not return rapidly to its original configuration.

Thus, the ideal surfactant for polyurethane foam would allow essentially all of the bubbles to stay intact until the foam approached its maximum rise, but at that point it would be desirable for a high percentage of cell walls to open without damaging the skeletal network of the foam. It is also important that there not be densification at the bottom of the foam.

Within the commercially useful concentration range of surfactants, one has historically found an inverse relationship between height of rise and foam porosity. Thus, as air flow (breathability) increases, there is generally a predictable loss in height of rise (i.e., increased density). In some commercially used polyurethane foam formulations, it becomes necessary to increase the surfactant concentration to achieve finer or more uniform cells in the foam. With most commercial surfactants, increasing the surfactant concentration causes an undesirable decrease in foam porosity.

Several examples can be found in the literature of silicone surfactants that have been improved by addition of organic or other surfactants. Most of these were employed in applications other than conventional flexible polyurethane foam, such as rigid polyurethane foam, high-resilience polyurethane foam, and polyester foam. A polyester foam surfactant composition that was a blend of a conventional polyester foam surfactant with an organic sodium sulfonate was revealed in U.S. Pat. 3,594,334, issued in 1971. Subsequently, a sulfonated castor oil was proposed as a co-surfactant for polyester foam in German Pat. 2,615,804, issued in 1977. Blends of cyanoalkyl-containing silicones and more conventional polyether-silicone surfactants have been shown to reduce the flammability of polyurethane foam (see U.S. Pat. 3,935,123, issued in 1976). Improved surfactant compositions for rigid foam have been obtained using a combination of an organic surfactant and a standard silicone surfactant (see U.S. Pat. 4,751,251, issued in 1988).

It would be desirable to be able to increase urethane foam porosity by > 20% without losing more than a few percent in height of rise. It would be very desirable to be able to increase foam porosity by 50% or more with essentially no loss in height of rise.

Farris et al.[4] disclosed silicone surfactants of the type used in manufacturing urethane foams and certain siloxane-oxyalkylene copolymers that act as cell openers when employed in combination with silicone surfactants in urethane foam manufacture. The document covers urethane foam compositions including such cell openers, a method of preparing urethane foams using such cell openers, and urethane foams prepared by the method.

The Farris invention relates to siloxaneoxyalkylene copolymers and certain such materials that, when employed in combination with silicone surfactants used for urethane foam manufacture, produce more open foams than are produced by the traditional silicone surfactants alone.

This need is answered by the *cell opener* materials of the Farris invention. These materials differ from the primary surfactants with which they are employed in that they have a tendency to be less soluble in the urethane foam-forming media than the primary silicone surfactants, and the attached polyether groups are generally somewhat more polar than the average polarity of the polyether groups of the primary surfactants. This is usually because they contain higher percentages of ethylene oxide units and lower percentages of propylene oxide units than are contained, on average, in the primary surfactants.

An additional characteristic of the cell openers of the invention is that the ratio of hydrophilic character (H) to lipophilic character (L) is greater ≥0.3, materials having H/L ratios less than this number being defoamers that cause foam collapse. These terms are defined below for the convenience of the reader.

The term *hydrophil-lipophil* has been used to describe surfactant properties of agents such as ethoxylated linear fatty alcohols. In the Farris patent, a modified version of the concept was employed to aid in distinguishing between silicone-polyether copolymers that are cell openers and other silicone surfactants containing higher percentages of silicone, which are cell closers.

The term *breathability* refers to the ability of a cured foam to permit passage of a gas such as air through it. A *tight* foam has low breathability, while an *open* foam is said to have a high breathability and permits ready passage of gas through it.

The term *air flow* refers to the volume of air that passes through a 0.5-in thick, 2-in dia. circular section of foam at 0.5 in of pressure. A representative commercial unit for measuring air flow is manufactured by Custom Scientific Instruments Inc., of Kearny, N.J.

References

1. Kollmeier, H.J., et al. *Proceedings SPI 26th ANTEC,* 219, 1981.
2. Reedy, James D., and Robertson, Richard T., U.S. Pat. 5,472,987, Dec. 5, 1995, "Surfactant composition for flexible polyurethane foam," to OSi Specialties, Inc., Danbury, CN.
3. Kilgour, John A., U.S. Pat. 5,198,474, Mar. 30, 1993, "Silicone surfactants having t-butyl terminated polyether pendants for use in high resilience polyurethane foam," to Union Carbide Chemicals and Plastics Technology Corporation, Danbury, CT.
4. Farris, David D., et al., U.S. Pat. 5,192,812, Mar. 9, 1993, "Cell openers for urethane foam surfactants," to Union Carbide Chemicals and Plastics Technology Corporation, Danbury, CT.

10

Catalysis of Isocyanate Reactions

10.1 Fundamental Principles

The catalyst plays a key role as the controlling agent of a urethane reaction. Finding the right catalyst may be the biggest challenge of the foam formulator, who must examine the effects it has on the various phases of the foaming reaction.

The formulator of urethane foams must choose raw materials carefully, not only on the basis of the final foam properties but also in accordance with desired processing conditions and the special demands of increasingly sophisticated production equipment. Also gaining in significance are toxicity profiles and cost-effectiveness ratios.

The resulting formulation is a system designed to meet narrowly defined use specifications, with each ingredient playing a particular contributing role. Finding the right catalyst or catalyst combination for a particular formulation can be the most challenging—and rewarding—task of the urethane chemist.

A catalyst is a material that affects the rate of a reaction (most often, a catalyst is added to speed things up) but emerges from the reaction apparently unchanged. The catalyst can be considered as the controlling agent of the reaction; it is the single ingredient that, in effect, makes things happen. In urethane foam production, its primary job is to provide the desired reaction profile, which can be measured by cream, rise, gel, and tack-free times.

To better understand how a catalyst works, we should first review a typical reaction:

$$A + B \Leftrightarrow C + \Delta$$

Raw material A reacts with B to form a product C. Usually, the reaction is accompanied by the liberation of heat. This reaction will proceed on its own if the free energy of product C is lower than the combined free energies of reactants A and B. This is illustrated by the energy-level diagram in Fig. 10.1. However, organic reactions generally do not follow the simple course denoted by path a in the figure.

Instead, the reaction proceeds when A and B combine, via path b, to form an activated complex having a higher energy level than the reactants. The energy level of the activated complex is designated by the symbol E*, and the amount of energy necessary to raise the reactants to this peak is known as the *activation energy.*

The purpose of a catalyst in this case is to lower this peak so that less energy will be required for the reactants to reach the activated state (shown via path b') before the reaction proceeds spontaneously to the product (via path c').

10.1.1 Chemistry of the Urethane Reaction

In applying this information to the preparation of urethane polymers, we must first consider the chemistry of the isocyanate moiety. A polyurethane can be considered to be the product of complex reactions

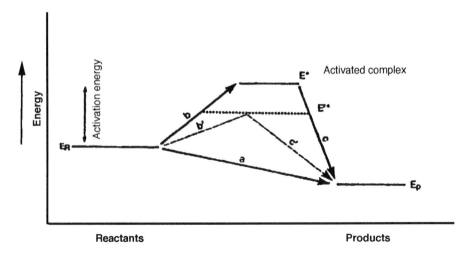

FIGURE 10.1 Polyurethane catalysts affect the energy of activation levels.

centering about this group. A brief look at the electronic structure of the isocyanate moiety helps explain its characteristic behavior. The resonance structures of this group are shown by the following:

The reaction ultimately involves addition to the carbon-nitrogen double bond. In this reaction, the electrophilic (positively polarized), active hydrogen becomes attached to the isocyanate's electron-rich nitrogen. This allows the nucleophilic remainder of the hydrogen-source molecule to attach to the carbon atom of the isocyanate group.

The -NCO group is particularly susceptible to attack by radicals having an active hydrogen in their structure. The active hydrogen compounds of most concern to a discussion of urethanes are hydroxylic materials, such as alcohols (or polyols) and water, and also some amines. Materials of this type will react with the isocyanate to give the product RNHCO-A, where A is the nucleophilic (negatively polarized) portion of the active-hydrogen-containing material.

If both the isocyanate and the active hydrogen compound are difunctional, a linear polymer will be produced. If either the isocyanate or the active hydrogen compound has functionalities greater than two, then a cross-linked polymer will result. Depending on functionality, either an elastomeric, flexible, or rigid material will be produced.

Catalyzing a urethane foam, for example, becomes a matter of choosing a catalyst with the ability to properly balance the complex system of reactions so the gas produced becomes efficiently entrapped in the urethane polymer. This growing polymer, in turn, must have sufficient strength throughout the process to maintain its structural integrity without collapse, shrinkage, or splitting.

While it is true that these reactions may proceed in the absence of catalysts, they generally proceed at rates too slow to be practical. Some of the cross-linking reactions would occur only at very high temperatures, and the proper balance of reactions would not be achieved. However, each of these reactions will proceed readily and in a controllable fashion in the presence of an appropriate catalyst.

The urethane formulator may choose from two major classes of catalysts compounds: tertiary amines and metal salts, primarily those of tin. Since catalysts differ in their activity or ability to affect the isocyanate reaction with other reactants and are selective in their effect, the two types are combined in most applications, not only to provide the desired balance of gelation and blowing but also to tune these reactions to the needs of production equipment.

10.2 Catalytic Effects

In most chemical reactions, there are certain positions on the reacting molecules that are more susceptible than others to attack by added reactants. These positions are therefore more susceptible to undergoing a given reaction. Catalysts characteristically function at these areas. In the case of urethane foams, the catalyst makes it easier for the isocyanate to chemically combine with active hydrogen compounds. It does so by forming an activated complex with the reactants. The ease and extent of the complex formation is largely determined by the catalyst's chemical and physical properties.

10.2.1 Tertiary Amines

An amine by definition is a nitrogen-containing compound having three substituent groups. Table 10.1 shows various types of amines. The catalytic activity of the amine is due to a free electron pair present on the nitrogen atom and is dependent upon the availability of this electron pair for complexation. Two factors affect this availability:

- *Steric hindrance.* The more crowding that exists about the amine nitrogen, such as that caused by branched or bulky substituents, the more difficult it is for the electrons to take part in a reaction.
- *Electronic effects of the substituents.* Some groups tend to push electrons toward the nitrogen, increasing the accessibility of the free electron pair. Other groups tend to withdraw electrons, diminishing accessibility, thus reducing catalytic activity.

Because of their relatively small size and great electron push, methyl groups give good results. Dimethylcyclohexylamine is an example. Since, in general, basicity is a measure of electron accessibility, catalytic strength usually increases as the basicity of the amine increases. Plotting pH vs. catalytic activity generally will yield a straight line.

Some amines such as triethylenediamine do not, however, fall on this line. The substituent groups on the nitrogen atoms of this compound are tied back, thereby reducing steric strain and making the electron pair easily accessible. Thus, the activity of this material is greater than might be expected from its base strength.

Various mechanisms have been suggested to explain the catalytic activity of tertiary amines. Experimental work has shown the formation of an activated complex between the amine and the hydroxyl component of the reaction. The amine tends to separate the active hydrogen from the oxygen of the alcohol to a greater degree than normal, thus greatly increasing its ability to react with the negatively charged nitrogen of the isocyanate.

There is also indication of the formation of a complex between the amine and the isocyanate. In this case, the electron pair of the amine, when complexed to the incipient positive charge on the carbon atom, makes attack by an active hydrogen (on the now more negative isocyanate nitrogen) much easier.

In addition to these electronic effects, other chemical and physical properties play a role in determining the effective activity of a catalyst. For example, materials having low boiling points, such as triethylamine, will be readily volatilized during the exothermic urethane-forming reaction and will consequently be lost from the reaction mixture. Once lost, there can be no further catalytic effect.

Catalysts containing hydroxyl groups, such as dimethylaminoethanol, are generally less active than the corresponding alkyl-substituted derivatives. The hydroxyl group becomes chemically bound to the growing polymer chain. Thus, the catalyst is no longer free to find its way to a reaction site.

TABLE 10.1 Types of Amines Used in Urethane Catalysis

Type	Example	Structure	Comments
Bicyclic	Triethylenediamine		Essentially no steric hindrance
Monoamine	Dimethylcyclohexylamine		Methyl groups cause little steric hindrance
Diamines	Bis(dimethylaminoethyl)ether		More than one available nitrogen, little steric hindrance
Polyamines	N-methyl-N'-(2-dimethyl-aminoethyl)-piperizine		High molar percentage of tertiary nitrogen
Amine salts	DBU-phenate		Heat-activated catalyst

Note: Amines with low molecular weights and high vapor pressures usually flee the system too quickly. These are used as initiators. Larger groups on the nitrogen atom increase steric hindrance, reducing activity.

Tertiary amines have a free electron pair.   N,N-dimethylcyclohexyl-amine   Triethylenediamine

Since the various tertiary amines exhibit different catalytic activities, it is frequently helpful to use combinations of tertiary amines to achieve the desired results. For example, a very volatile or fast-acting catalyst can be combined with a slower or less volatile catalyst to cause the reaction to proceed at a reasonable rate throughout its course. For example, triethylamine initiates a fast foaming reaction and is subsequently lost due to the heat of the reaction, while a less volatile material continues to exert its activity, completing the cure.

10.2.2 Tin Catalysts

While amine catalysts exert some effect on the isocyanate-hydroxyl reaction, organometallic salts favor the isocyanate-hydroxyl reaction almost exclusively and are thus gelation catalysts. Most are based on tin and include stannous octoate, dibutyltin dilaurate, and tin mercaptides. Certain salts of lead, mercury, and antimony have also been used.

Unlike tertiary amines, which usually volatilize and are lost upon foaming, tin catalysts can remain in the foam permanently. However, they may undergo chemical changes. Stannous salts are generally used in conventional flexible foams, although they may hydrolyze or oxidize to the stannic form. Since stannic salts (dibutyltins) tend to promote the thermal oxidative degradation of flexible urethanes, these compounds are primarily used in the catalysis of rigid foams.

The catalytic effect of metal salts is more complex than that of the amine, which complexes with either the hydroxyl or the isocyanate. Tin and other metal catalysts activate both the polyol and isocyanate, first by the formation of a binary complex between the catalyst and polyol and then by the joining of the isocyanate to form a ternary—or bridge—complex of the tin catalyst, polyol, and isocyanate (Fig. 10.2). It should be noted that the bridge complex still allows for the approach of the tertiary amine to exert its effect on the reaction. Amine/tin co-catalyst synergism is thought to be due to the extra stability of the activated intermediate complex of polyol, isocyanate, tin, and amine catalysts.

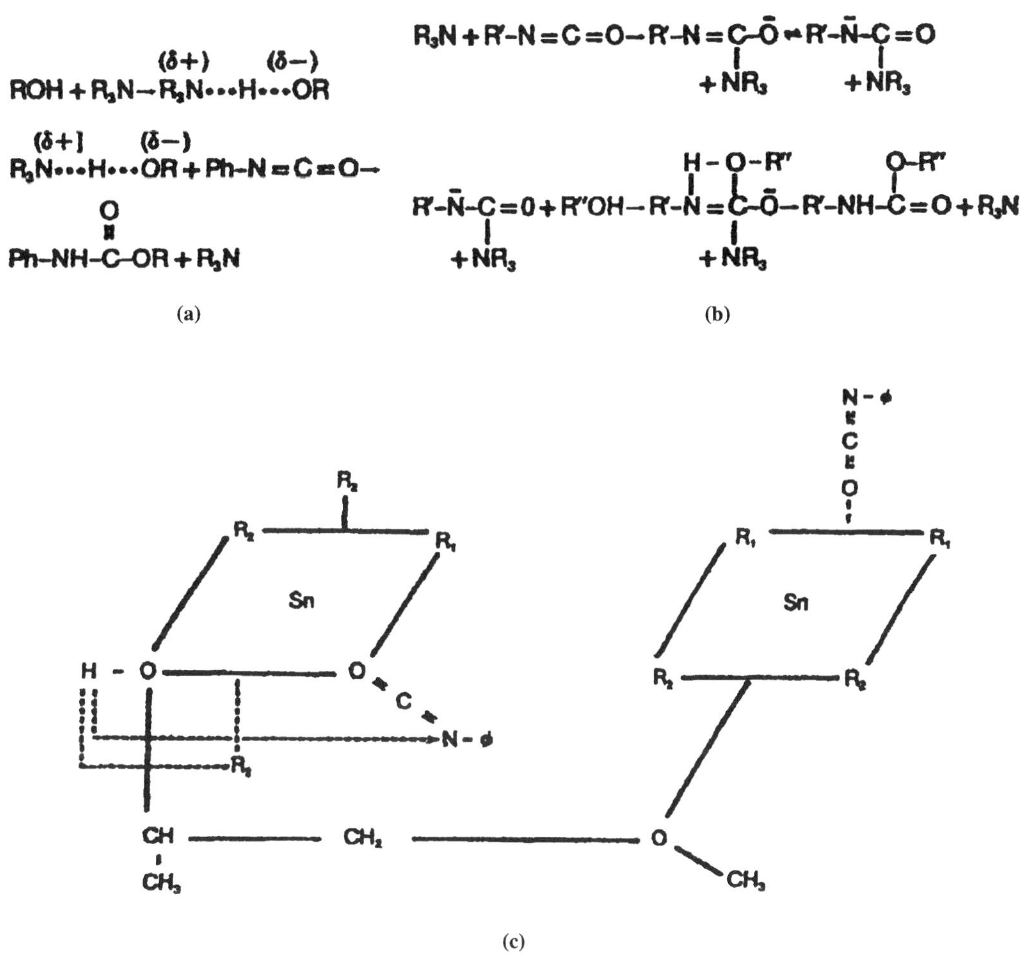

$$R_3N + R'-N=C=O-R'-N=C-\bar{O} \leftrightarrow R'-\bar{N}-C=O$$
$$\underset{+NR_3}{|} \qquad \underset{+NR_3}{|}$$

$$(\delta+) \qquad (\delta-)$$
$$ROH + R_3N \rightarrow R_3N \cdots H \cdots OR$$

$$(\delta+) \qquad (\delta-)$$
$$R_3N \cdots H \cdots OR + Ph-N=C=O \rightarrow$$

$$\underset{\overset{\|}{O}}{Ph-NH-C-OR} + R_3N$$

(a)

$$\underset{+NR_3}{\overset{H-O-R''}{|}} \qquad \underset{+NR_3}{\overset{O-R''}{|}}$$
$$R'-\bar{N}-C=O+R''OH-R'-N=C-\bar{O}-R'-NH-C=O+R_3N$$

(b)

(c)

FIGURE 10.2 (a) Activated complex of amine and hydroxyl, (b) Activated complexes of amine and isocyanate, and (c) bridge complex between 1-methoxy-2-propanol, phenyl isocyanate and tin.

In those reactions, in which both tin and an amine are present as catalysts, the effect of the tin so far overshadows that of the amine on the isocyanate-hydroxyl reaction that the amine is considered to be present in the reaction mixture solely to catalyze the reaction of the isocyanate with water to yield the CO_2 necessary for blowing.

10.2.3 Effects on Processing

In standard or hot-cure flexible urethane foam, the tin and amine must be present in exactly the proper proportions so that the gelation reaction, catalyzed by the tin, is in phase with the gas evolution from the water reaction, catalyzed by the amine (Fig. 10.3).

If the tin level is too high relative to the amine level, a tight foam will be obtained because of the higher number of closed cells produced by too-fast gelation. Carried to a higher level, the foam will shrink badly. On the other hand, if the amine level is high relative to the tin level, the foam will be wide open because of too rapid an evolution of CO_2. It may be so wide open that the foam will collapse from a lack of the gel strength necessary to retain the blowing agent in the foam. The solid line in Fig. 10.4 indicates the theoretical ratio necessary for good foam. However, since existing production machines are not exact, it is essentially impossible to hold the tin and amine levels at this ratio.

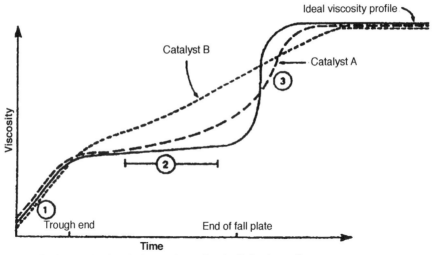

FIGURE 10.3 Representation of viscosity profiles. In a standard foam reaction, the tin/amine catalyst must be selected to ensure proper viscosity profile (gelation reaction).

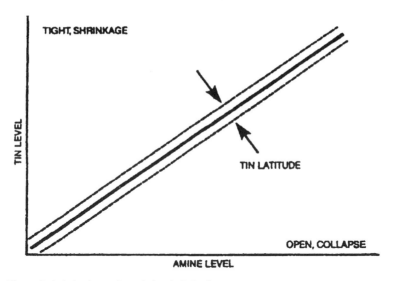

FIGURE 10.4 Theoretical tin/amine ratio and the tin latitude.

Thus, the concept of an amine's tin latitude must be considered by the processor in the formulation of a catalyst blend. An acceptable foam—neither tight nor wide open, shrunken nor collapsed—can be produced within a narrow range on either side of the theoretically perfect ratio. The width of this range is the tin latitude. The wider the tin latitude, the greater tolerance the formulation will have to minor equipment and metering differences during production.

The formulator also has an interest in tailoring the reaction profile of the formulation to the demands of the processing equipment, especially to reduce scrap loss and gain other production economies. Since catalysts are the major controlling influences, their characteristics must be weighed carefully to provide cream, rise, gel, and tack-free times to fully utilize raw materials. For example, newer flat-topping flexible foam machines have throughputs as high as 500 lb/min. Thus, the consequences of error are significant. The formulator will choose to combine with the tin either the amine or the amine mixture that will provide the broadest tolerances.

10.3 Fluorocarbon and Methylene Chloride Blown Foams

Catalyst combinations are often used to meet other types of special processing needs, such as those encountered in the manufacture of methylene chloride blown flexible foams. As stated previously, the primary function of the amine in a flexible-foam formulation is to promote the isocyanate-water reaction, which generates CO_2.

In its efforts to produce foam of extremely low density, the urethane industry has gone to higher and higher water levels. While this does yield lower densities, the greater percentage of urea linkages also results in a rather hard, boardy foam. Because of the higher exotherm, the possibility of scorching the foam is greater in such high-water formulations.

To overcome these disadvantages and still maintain a low density, an auxiliary blowing agent is frequently added. Traditionally, fluorocarbons have been used, although methylene chloride is gaining in acceptance, primarily because processing problems are now being worked out through improved catalysis.

Auxiliary blowing agents are chemically inert. They volatilize as the reaction supplies heat, thus functioning as a blowing agent without entering into the chemical reaction themselves. Use of auxiliary blowing agents provides low densities while maintaining a good "hand" that is neither harsh nor boardy but is rather soft to the touch.

It must be pointed out, however, that the use of auxiliary blowing agents usually requires an adjustment in catalyst levels or a change in the catalyst system. The blowing agent itself may interfere with the catalytic effects of both the amines and tins, which are present in very low concentrations, by hindering the formation of the activated complexes referred to previously. Also, some auxiliary blowing agents, especially in the presence of certain flame retardants, can produce an acidity that will have a negative effect on the amine. The acidity tends to neutralize the effects of amine, reducing its catalytic activity, particularly upon storage. Auxiliary blowing agents also act as heat sinks, absorbing heat much like a sponge soaks up water. The heat evolved during the reactions—which would normally support curing the foam—is absorbed instead by the blowing agents. Thus, to maintain the reactivity and exotherm to properly cure the foam, a higher total catalyst level frequently is needed.

Until recent years, the auxiliary blowing agent most often used was fluorocarbon-11 However, with the possibility of government restrictions on fluorocarbon use, and with possible cost savings as an added incentive, interest in the use of methylene chloride has grown. Much of the impetus has come from Dow Chemical, which has published formulations utilizing as much as 25 phr. Previously, use was restricted to about 6 phr because of foam splitting and other processing problems. However, newer catalysts have been developed to overcome these drawbacks.

High-resilience (HR) foams have also been referred to as cold-cure foams, although this may be a misnomer, at least in molded applications, since the material generally is poured into molds that are first heated to temperatures of up to 140° F. However, it is true that HR foams are produced and molded at temperatures below those associated with standard or hot-cure systems.

High-resilience urethanes are distinguished from conventional foams primarily by the choice of polyol. Conventional urethanes use triols of 3000 to 4000 molecular weight generally prepared from a glycerin initiator and propylene oxide. They therefore have secondary hydroxyl groups at the terminal positions of the polyol chains. HR foams, however, are based on polyols of much higher molecular weight (4500 or more). These polyols also start with initiators such as glycerin, again with much of the chain extension done with propylene oxide. However, the resulting materials are then capped, or reacted at the ends of the chains, with ethylene oxide, providing primary hydroxyl groups as the reactive end of the polyol.

Primary alcohols, $R\text{-}CH_2OH$, react about three times more readily with isocyanates than do secondary alcohols. The rapid and exothermic reaction of primary hydroxyls with isocyanates allows HR molding and slab production to proceed at lower temperatures than with conventional foam polyols.

HR foams are essentially amine catalyzed, since the high inherent activity of their specialized polyols, plus the known tendency of organometallics to promote closed cells if used in excess, could lead to a permanently deformed piece of foam. Additionally, there have been dry-heat aging problems associated with organotin-catalyst foams.

However, dibutyltin dilaurate is often added to HR formulations at levels less than 0.05 phr. At these levels, it is a moot point whether the tin salts exert a large catalytic effect or, as has also been suggested, are in the system simply to improve the compression set of the foam. For this purpose, tin-catalyst suppliers also suggest evaluation of dibutyltin bis lauryl mercaptide, dibutyltin bis isooctylmaleate, dibutyltin bis isooctymercaptoacetate, and the dimethyltin analogs.

Amine catalysts commonly used in HR foams (and other common applications) are indicated in Table 10.2. Usually, several catalysts are blended to maximize the benefits of each. Early HR catalyst packages utilized a mixture of N-ethylmorpholine (NEM), triethylenediamine (TED), and bis(diethylaminoethyl) ether in a ratio of approximately 10:1:1 by weight. The TED generally confers sufficient gelation characteristics to the catalyst mixture, while the other two amines are primarily blowing catalysts. NEM also contributes to a quick initiation of the reaction, although it has an offensive odor, and its vapors may be toxic.

Most traditional catalyst systems were designed to promote release of CO_2 at the same time the fluorocarbon vaporizes, thus obtaining maximum blowing efficiency. With a boiling point of about 75° F, fluorocarbons begin to volatilize at the onset of the CO_2 evolution. Because of its higher boiling point (104° F), methylene chloride does not vaporize until later, during the cross-linking phase of the reaction, causing the foam to split. Therefore, a catalyst system was needed that would delay the cross-linking reaction until the exotherm was great enough to volatilize the methylene chloride, enabling it and CO_2 to evolve together. Such catalysts as Polycat 2 (Abbott Labs) and Dabco WT (Air Products & Chemicals) delayed the gelation but led to some reduction in load-bearing properties. The coaddition of nondelayed catalysts, such as Polycat 77-50 (Abbott Labs) or triethylenediamine, solved the problem.

10.4 Delayed-Action Catalysts

Delayed-action catalysts are of particular interest in molded-foam applications where flow properties are of significance. To permit the complete filling of intricate molds, it is desirable to have the liquid components maintain a low viscosity for as long as possible before gelation takes place. Gel times can be extended by using a less active catalyst. However, this may have adverse effects on the final cure and physical properties. Also, cycle times will be increased.

These problems are alleviated by the use of delayed-action or heat-activated catalysts. The catalysts show low activity at room temperature but increase in activity once the exotherm builds up, or they may be activated by an external heat source such as heat lamps or heated molds. Most delayed-action catalysts are salts of current catalysts.

Flow properties are also important to the formulator of rigid foams, especially in appliance and board-laminate insulation and in high-density molded applications. Better flow properties permit the pouring of wider laminate surfaces. In refrigeration, they reduce the need for overpacking to fill molds. In high-density foams, good flow permits the filling of the most intricate molds, reducing reject rates and thereby improving production economics.

10.5 Catalysis of High-Resilience Urethanes

Within the past 10 years, high-resilience (HR) foams have gained in significance, especially in automotive-seating applications. HR foams, as the name implies, are more resilient than conventional foams of comparable density. They are also ideal for bedding applications, although their higher costs have limited this market.

Dimethylcyclohexylamine, although also having a strong odor, has slightly better blowing characteristics than NEM and significantly better gelation characteristics. Because of its lower odor, Polycat 12 (Abbott Labs) has also been evaluated as an NEM replacement.

Newer catalysts now used to replace one or more of the amines in the HR catalyst package include Niax A-6 (Union Carbide), Polycat 77 (Abbott Labs), and Thancat DD (Jefferson Chemical). Polycat 70, designed for flat topping equipment, is also used in conventional and HR formulations.

TABLE 10.2 Representative Catalysts for Special Foaming Applications

Catalyst	Rigid			Flexible			RIM	Elastomers	Isocyanurate	Polyesters
	Spray	Slab stock	Molded	Slab stock	Molded	HR				
Amines										
Benzyldimethylamine										X
Cetyldimethylamine										X
Curethane[1] 51/52									X	
Dabco[2] R8020	X	X	X							
Dabco[2] T		X	X		X					X
Dabco[2] TMR									X	
Dimethylcyclohexylamine	X	X	X		X	X	X		X[6]	X
Dimethylethanolamine		X		X						
N-cocomorpholine										X
N-ethylmorpholine				X						X
N-methylmorpholine		X		X						X
Niax[3] A-1				X	X	X				
Niax[3] A-6				X	X	X				
Niax[3] ES										X
Polycat[4] 12				X	X	X				X
Polycat[4] 42									X	
Polycat[4] 70				X	X	X				
Polycat[4] 77				X	X	X				
Polycat[4] 77-50				X			X		X	
Quinuclidine	X	X		X		X				
Tetramethylethylenediamine	X	X								
Tetramethylbutane-1,3-diamine		X								
Thancat[5] DD				X	X	X				
Thancat[5] DM-70										X
Thancat[5] DMDEE				X						
Thancat[5] TAP				X	X					
Triethylenediamine	X	X	X	X	X	X	X		X	
Organotins										
Stannous octoate										
Dibutyltin dilaurate	X	X	X				X	X	X	
Dialkyltin dimercaptides	X	X	X				X	X	X	

[1]Trade name of CPL Corp.
[2]Trade name of Air Products & Chemicals.
[3]Trade name of Union Carbide Corp.
[4]Trade name of Abbott Labs.
[5]Trade name of Jefferson Chemical Co.
[6]Urethane modified only.

10.6 Isocyanurate Foams

Under certain conditions, isocyanates will react with themselves, forming cyclic trimers known as isocyanurates. Rigid isocyanurate foams are characterized by higher thermal stability than conventional urethanes.

The isocyanurate reaction requires specialized catalysts. Potassium salts of organic acids, such as octanoic or acetic, are frequently used. Amines used include 2,4,6-tris N,N-dimethylaminomethyl phenol and 1,3,5-tris 3-dimethylamine-propyl hexahydrotriazine.

Because pure isocyanurate foams are very friable, a small amount of polyol is often added to provide some urethane linkages. A 20% urethane-modified isocyanurate foam will have an acceptable balance of physical properties and flammability characteristics. In these foams, the catalyst package should promote both the trimerization and urethane reaction.

10.7 Catalysts for Polyurethane Foams

To accelerate the reaction of the polyoxyalkylene-polyol mixtures, water as the blowing agent, and chain extenders with the organic polyisocyanates and/or modified polyisocyanates, conventional polyurethane catalysts are introduced into the reaction mixture, as shown in Table 10.3.

TABLE 10.3 Some Commercial Catalysts for Urethane Foam

	Flexible		Rigid						
Catalyst	Polyether	Polyester	HR	Spray	Molded	Slab stock	Isocyanurate	Packaging	RIM
Niax[a] A-1	●	●							
Niax A-4		●						●	
Niax A-10	●	●							
Niax A-97		●							
Niax A-200	●								
Niax TMBDA	●			●	●	●			
Niax A-30		●							
Polycat[b] 70	●								
Polycat 91	●								
Polycat 11					●				
Polycat 8				●	●	●			
Polycat 41							●		
Polycat 17					●	●		●	
Polycat DBU	●	●	●	●	●	●		●	●
Dabco[c] 33 LV	●	●	●	●	●	●		●	●
Dabco R 8020	●				●	●			
Dabco X-DM	●	●	●						
Dabco R-595					●	●	●		
Dabco TL	●								
Dabco TMR-2							●		
Thancat[d] TD-33	●	●	●	●	●	●		●	●
Thancat DD	●								
Thancat DM-70	●	●							
Thancat ZR-50				●	●	●		●	
N-methylmorpholin	●	●				●			
N-ethylmorpholine	●	●							

	Flexible		Rigid						
Catalyst	Polyether	Polyester	HR	Spray	Molded	Slab stock	Isocyanurate	Packaging	RIM
Dimethylethanolamine	●			●	●	●		●	
Dimethylcyclohexylamine		●	●	●	●	●		●	●
DMP-30[e]							●		
Catalyst T-9[f]	●								
Catalyst T-10	●								
Catalyst T-11	●								
Catalyst T-12				●	●	●	●	●	●
Catalyst T-1				●		●			
Catalyst T-5					●				
Catalyst 120				●	●	●			●
Catalyst 125				●	●	●			●
Catalyst 131				●	●	●			●
Catalyst T-45							●		
Fomrez[g] UL-1				●	●	●			●
Fomrez UL-2				●	●	●			
Fomrez SUL-3				●	●	●			
Fomrez SUL-4				●	●	●		●	●
Fomrez UL-6				●	●	●			
Fomrez UL-8				●	●	●			
Fomrez UL-22				●	●	●			
Fomrez UL-28				●	●	●			●
Fomrez UL-29									●
Fomrez UL-32				●		●		●	●
Fomrez EC-683							●		
Fomrez EC-686							●		
Fomrez C-2	●								

[a]Trade name of Union Carbide Corp.
[b]Trade name of Abbott Laboratories
[c]Trade name of Air Products & Chemicals Inc.
[d]Trade name of Texaco Chemical Co.

[e]Available from Rohm and Haas Co.
[f]Product of M&T Chemicals Inc.
[g]Trade name of Witco Chemical Corp.

Basic polyurethane catalysts are highly efficient in the production of polyurethane foams. Examples: tertiary amines, such as dimethylbenzylamine, dicyclohexylmethylamine, dimethylcyclohexylamine, N,N,N',N'-tetramethyldiaminodiethyl ether, bis(dimethylaminopropyl) urea, N-methyl-and N-ethylmorpholine, dimethylpiperazine, N-dimethylaminoethylpiperidine, 1,2-dimethylimidazole, 1-azabicyclo [2.2.0]octane, dimethyl aminoethanol, 2-(N,N-dimethylaminoethoxy) ethanol, N,N',N''-tris (dialkylaminoalkyl) hexahydrotriazine, for example N,N',N''-tris(dimethylaminopropyl)-s-hexahydrotriazine, and triethylenediamine. Foams usually require metal salts co-catalysts such as iron(II) chloride, zinc chloride, lead octanoate, and tin salts, such as tin dioctanoate, tin diethylhexanoate, and dibutyl tin dilaurate.

Catalysts are conveniently classified in three groups as shown in Table 10.4. It must also be noted that catalysts display differential reactivities towards different isocyanates. This differential reactivity is shown in Table 10.5.

TABLE 10.4 Catalysts

Group I
Tertiary amines cause moderate rate increases in the reactions of all isocyanates, aromatic and aliphatic.
Group II
Bismuth, tin, and lead are most effective toward the aliphatic isocyanates.
Group III
Zinc, cobalt, iron, antimony, and tin salts activate the aliphatic isocyanates to the point that they may react faster than aromatics toward hydroxyls.

TABLE 10.5 Differential Reactivity (Reactivity of a Polyether Triol with Different Diisocyanates—Gel Time in Minutes, 70° C)

Catalyst, 1%	TDI	MXDI	HDI
None	>240	>240	>240
Triethylamine	120	>240	>240
Dabco	4	80	>240
Dibutyltindioctoate	6	3	3
Lead octoate	2	1	2
Bismuth nitrate	1	0.5	0.5
Stannic chloride	3	0.5	0.5
Zinc naphthenate	60	6	10

Manufacturers market an extensive line of additives for use in all types of polyurethane foam. Included are amine, morpholine, and organometallic catalysts, cross-linkers, and chain extenders. Some important catalysts are discussed below.

- Dabco 33-LV and BL-11, recommended for use in flexible slab stock and flexible molded systems, are tertiary amine catalysts that provide precise control of the gelling and blowing reactions.
- Dabco 8154 and BL-17 are delayed-action catalysts for use in flexible slab stock/flexible molded foam systems.
- Polycat 77 is a 100% active amine catalyst that promotes the gelling and blowing reactions in flexible and semiflexible molded foam.
- Polycat 8 is an industry-standard catalyst for spray systems and pour-in-place rigid applications.
- Polycat 46 catalyst, potassium acetate, and Dabco K-15 catalyst, potassium octoate, are used in isocyanurate lamination systems.
- The Dabco TMR family of catalysts is used alone or in conjunction with polyurethane catalysts to customize the reactivities of polyisocyanurate foam systems.
- Dabco T-9, T-12, and 120 represent three types of tin-based catalysts that are useful in a variety of foam systems.
- Dabco NEM, NMM, and NCM, morpholine-based catalysts, and Dabco B-16 catalyst, are useful in polyester slab stock foam.

AKZO Chemicals Inc.

Stanclere TL dibutyltin dilaurate is said to provide better hydrolytic stability in premixes containing polyol, water, and silicone than stannous catalysts, giving the premix a longer storage life. In molded flexible PU foams, Stanclere TL reportedly promotes faster gassing while delaying gelation of the foam to allow complete filling of the mold before the foam begins to set. Dibutyltin bis IOTG is also offered.

In addition, the AKZO offers liquid potassium acetate and potassium octoate catalysts. These can be supplied at various concentrations in DEG or PEG. These materials are normally more cost-effective than amines for rigid urethane applications and can be blended with amines to achieve various end-point properties.

Anderson Development Co.

The company produces Curene, a series of urethane curatives. They include Curene 442, a standard MBOCA-type material, as well as Curene 185 for reduced hardness and Curene 3005, a liquid curative for polyester-based prepolymers.

BASF Corp.

BASF offers a variety of amine-based catalysts for use in several flexible and rigid applications, including PM-DETA, TM-PDA, TM-HDA, N,N-dimethylcyclohexylamine, N-methylmorpholine, N-ethylmorpholine, 1-methylimidazole, 1.2-dimethylimidazole, dimethylethanolamine, and triethylamine.

Buckman Laboratories, Inc.

Buckman offers Bulab 600 tertiary amine catalysts and TMEDA (tetramethylethylene diamine). Both products are said to be fast, stable catalysts for use in rigid and flexible urethanes.

Calgon Corp., sub. of Merck & Co., Inc.

Metasol 57 is a specially prepared form of phenylmercuric propionate designed to catalyze urethane elastomers and foams. It is supplied as a free-flowing powder or as a 20% solution in dipropylene glycol (Metasol 57/DPG) for use in automatic metering equipment.

Products specifically catalyze the isocyanate/hydroxyl reaction and do not catalyze the isocyanate/water reaction. They are said to be stable, economical, and able to function at low ambient temperatures with rapid gelling to permit fast demolding. Delayed induction period at low viscosity, followed by rapid cure, reportedly is provided, along with good demolding properties.

Cardinal Stabilizers Inc., Cardinal Carolina Corp.

Tin catalysts are available for use in flexible and rigid urethanes, including CC-1 (dibutyltin dilaurate), CC-3 (dibutyltin diacetate), and stannous octoate, including various solutions. A low-freezing-point version of CC-1 is offered for use in colder climates.

Caschem, Inc.

The company's Cocure organomercury compounds function as polymerization or straight-chain catalysts for a wide range of castable two-component PU elastomers. Applications include industrial tires, conveyor belting, shock absorbers, printing rolls, wire sheathing, electrical insulation, seals, gaskets, and sporting goods. Available as liquids and offered in a range of induction periods and cure times, Cocure products reportedly impart excellent elongation and tensile strength, are not affected by humidity, and can give satisfactory cures even at low ambient temperatures.

Cotins, a line of standard and proprietary organotins, including dibutyltin dilaurate and stannous octoate, are offered for urethane elastomers, rigid, and semirigid foam.

Coscat 83, an organobismuth catalyst for two-component urethane elastomer systems, can be used in a wide range of elastomer formulations where reduced catalyst toxicity is desired. It is said to provide an alternative to lead, mercury, or tin-based catalysts.

A complete line of amine catalysts includes Coscat T-33 (triethylenediamine), Coscat DM-8 (dimethylcyclohexylamine), and Coscat DMEA.

Conap, Inc.

Conap offers curatives for use with high-performance, heat-cure polyurethane elastomers.

- Conacure AH-33: MBOCA polyol curative, having an equivalent weight of 280. It functions as a liquid MCOCA alternative in many applications.
- Conacure AH-40: Non-MBOCA liquid aromatic diamine curing agent having an equivalent weight of 133.3.
- Conacure AH-50: Low-viscosity, polyol curative having an equivalent weight of 90.

Elf Atochem North America Inc.

The company produces dimethylaminoethanol (DMAE), which is said to be an effective, economical amine catalyst for flexible and rigid PU foams. Product reportedly is low in cost, reduces viscosity of the polyol mix, offers solvent properties to help disperse polyol-mix components, and provides good latitude in adjusting cream and rise times. When combined with tin catalysts, DMAE is said to give better independent control of the PU foam reactions than other amine catalysts.

Enterprise Chemical Co.

Quincat CQ-330 is a 33% solution of quinuclidene (1,4-ethylene piperidine) in dipropylene glycol. This amine catalyst reportedly shows activity equal to that of triethylenediamine-based catalysts, but at significantly lower use levels and lower cost. Foam properties are said to be comparable to those obtained with standard catalysts. This product is recommended for use in RIM, high-resilience (HR) foams, conventional flexible slab stock and rigid spray, pour-in-place, and refrigeration-type foams.

Ferro Corp., Bedford Chemical Div.

The Cata-Chek line of PU catalysts includes both stannous and alkyl tin systems used in PU foams and elastomers. Stannous octoate in various concentrations is available in the Cata-Chek 860 series. Also available is the standard dibutyltin dilaurate (Cata-Chek 820) and dibutyltin diacetate (Cata-Chek 867).

Kenrich Petrochemicals, Inc.

Kenrich offers titanate coupling agents as urethane catalysts that act via polyol-alcoholysis/ NCO-alkylation mechanism. In a two-component RIM polyurethane, KR 55 titanate will produce a gel time of 16 s, compared with a 27-s gel time for dubutyltin dilaurate. An amino titanate, KR 44 reportedly will gel in 6.3 s and provide an ultimate tensile strength of 21,570 psi at significantly lower cost, compared with 4450 psi for a dibutyltin dilaurate. Tg for urethane cured with KR 44 is 468° F versus 241° F for one cured with tin. Neoalkoxy titanates reportedly provide increased thermal and solvolytic stability.

OM Group, Inc. (Formerly Mooney Chemicals, Inc.)

Organometallic urethane catalysts available include dibutyltin dilaurate, potassium octoate, zinc octoate and lead neodecanoate, bismuth octoate, bismuth neodecanoate, tin neodecanoate, and lithium neodecanoate.

Shepherd Chemical Co.

Shepherd manufactures ferric acetyl acetonate and zinc acetyl acetonate urethane catalysts and offers BiCAT and bismuth versalate formulations for controlled activity and selective polymerizations.

Texaco Chemical Co.

The company's Texacat series of amine catalysts is offered for a broad range of applications in PU foams.

- Texacat TD-33 is a 33.3% solution of triethylenediamine in propylene glycol for use in flexible, semiflexible, and rigid foams and elastomers.
- Texacat TD-33A is a 33.3% solution of triethylenediamine in dipropylene glycol.
- Texacat DME, N,N-dimethylethanolamine, is used with other Texacat catalysts to promote the gas-blowing reaction or to serve as an acid scavenger.
- Texacat TD-20 is a blend of 80% Texacat DME and 20% triethylenediamine for use in a wide range of rigid foams.
- Texacat DM-70 is used in flexible polyethers, polyesters, and HR molded foams. Its advantages are said to include high reactivity, good surface cure, and low odor.
- Texacat NEM, N-ethylmorpholine, is said to promote good top-skin cure in polyester foam molding.
- Texacat NMM, N-methylmorpholine, is for use in high-rise rigid foam panels and polyester foams.

- Texacat M-75 is a unique low-odor catalyst used chiefly in production of polyester slab stock. It features reduced bun discoloration and broader processing latitude, resulting in more open-celled foams.
- Texacat DPA, a low-odor catalyst containing reactive hydroxyl groups, is designed primarily for low-density packaging foams.

The company also offers the following polyether catalysts in the Texacat line:

- ZR-70 is a dimethylaminoethoxyethanol for use in specialty packaging applications and other applications where strong blowing is advantageous.
- ZF-22 is a 70% bis-(2-dimethylaminoethyl) ether diluted in dipropylene glycol, used in flexible polyether slab stock foams and flexible molded foams.
- ZF-23 is a blend of tertiary amines specifically formulated for non-CFC, Maxfoam processes.
- ZF-51 is a formulated amine catalyst used in flexible slab stock to promote higher block yield and is said to be especially useful in methylene chloride-blown foams.
- ZF-52 is a delayed-action, formulated amine catalyst for flexible slab stock foams.
- ZF-54 is a formulated amine for automotive seat molding. Delayed activity allows better flow, and the additive is useful as a co-catalyst with stronger gelation catalysts. Demold time is reduced.

Also offered are the following specialty catalysts:

- DD is a high-boiling tertiary amine offered for use in flexible slab stock and flexible molding.
- DMDEE is a high-boiling tertiary amine with a balanced reactivity to blowing and gelation reactions.
- DMP (N,N-dimethylpiperazine) is a general-purpose catalyst for flexible and semiflexible polyether foam.

Thor Chemicals, Inc.

Thor manufactures proprietary organomercury and organotin catalysts for elastomers and some foams. Thor 535 is a solvent-free, low-viscous, organomercury catalyst containing 35% metal, said to have a long pot life and ambient cure.

Tosoh USA, Inc.

The company manufactures tertiary amine catalyst compounds for rigid, rigid spray, flexible, molded, RIM, and HR polyurethane products. Pure and blended versions of triethylenediamine (TEDA) are offered with various solvent systems, depending on application.

- Toyocat SPF, N-81, and N-31, recommended for flexible and molded foams, are blended catalysts compatible with blended polyols and highly synergistic with organotins.
- Toyocats TF and THN delayed-action catalysts based on TEDA are said to provide excellent retarding effects and gelling characteristics in rigid and RIM systems.
- Toyocats HX4, HX4W, and HX35 reportedly produce good physical properties with no stain in semirigid systems containing PVC.
- Toyocat MR is a strong gelling catalyst; Toyocat DT is a strong blowing catalyst. Both are recommended as replacements for conventional catalysts used in appliances.
- Toyocat F83 and F94 are recommended for CFC-reduced rigid pour-in-place insulations to increase flow and lower K factor in high-water formulations.
- Toyocat TE and DT are recommended for partially CFC-reduced appliance foam systems, which employ high water percentages and a reduced amount of conventional CFC-11.
- Toyocat F2 and F4 are newly developed catalyst systems recommended for CFC-free MDI HR molded foams. Systems assist in reducing foam density while increasing flow and improving foam skin in all MDI-based formulations.

- Toyocat HX63 and HX70 are newly designed catalyst systems recommended for high-quality semi-rigid polyurethane foams, such as instrument panels, headrests, and steering wheels.

Jim Walter Resources, Inc.

The company offers potassium octoate compounds in a diethylene glycol carrier for isocyanurate foams. Percent of potassium by weight ranges from 12.8 to 15%, with respective viscosities in the 5000 and 7000 cps range.

Witco Corp., Organics Div.

Witco offers the Fomrez series of tin-based compounds for various types of urethanes, including rigid, flexible, high-resilience, and microcellular foams, as well as cast and RIM elastomers.

10.8 Illustrative Patents

10.8.1 Catalysis of Hindered Aliphatic Isocyanates

The term *aliphatic isocyanate* refers to compounds in which the isocyanate group (-NCO) is attached to a carbon atom not in an aromatic ring, and hindered isocyanates are those in which the carbon atom to which the isocyanate group is attached is provided with at least one other inert substituent—generally a lower (C1–C8) alkyl group. Primary aliphatic isocyanates, for example, 1,6-hexamethylene diisocyanate, react significantly faster with compounds containing amino, carboxyl, and hydroxyl groups than secondary or tertiary isocyanates.

However, many isocyanates containing secondary and/or tertiary isocyanate groups are known, and some of them provide highly useful polyurethane products. Unfortunately, they usually are exceedingly slow reacting and require catalysis, especially in reactions with hydroxyl-containing compounds.

Reaction injection molding, commonly referred to by the acronym RIM, is a relatively recent advance in polyurethane technology. The process requires fast polyurethane reactions—on the order of seconds at temperatures in the range of about 25 to 40° C. Commonly used polyurethane catalysts, such as lead naphthenate and dialkyl tin dicarboxylates, are not adequate to promote the reaction between hindered aliphatic isocyanates and hydroxyl-containing compounds under these conditions.

Chang[1] disclosed a catalyst composition is for polyurethane production by reaction between aliphatic hydroxyl-containing compounds and secondary or tertiary aliphatic isocyanate group-containing compounds, comprising (a) 1,8-diazabicyclo(5.4.0)-undecene-7, or a salt thereof; (b) an organic tin compound and, optionally, (c) an organic lead compound.

The Chang invention relates to preparation of polyurethanes from hindered aliphatic isocyanates, and particularly to improvements in catalysis of the reaction between hindered aliphatic isocyanates and hydroxyl-containing compounds, and to novel synergistic catalyst compositions for the formation of polyurethanes by reaction between hydroxy-containing compounds and secondary and/or tertiary aliphatic diisocyanates.

Chang discovered that a superior catalyst for polyurethane compositions comprises a combination of (a) 1,8-diazabicyclo(5.4.0)-undecene-7, or its phenolate salt, (b) a tin catalyst, e.g., a dialkyl tin carboxylate, such as dimethyl tin diacetate, or COTIN-222, an organotin carboxylate; and, (c) optionally, but preferably, a lead compound, e.g., an organic lead compound such as lead naphthenate. Such compositions provide rapid curing—unexpectedly so in view of their superiority over lead/tin combinations in this respect. This permits the usage of lead to be reduced, or even eliminated, maintaining the same order of activity and avoiding toxicity problems.

10.8.2 Catalyst Combination for Controlled Shrinkage Resistance

In the production of polyurethane foams, a polyol is reacted with a polyisocyanate in the presence of a polyurethane catalyst and a blowing agent. It is well established that open-cell foams can be produced using water in the reaction mixture to provide a carbon dioxide blowing agent *in situ*, whereas closed-cell foams are typically produced using a chlorofluorocarbon (CFC) blowing agent. Unfortunately, CFC blowing agents have a negative impact on the environment, and alternatives to the use of these blowing agents in the production of closed-cell foams are being sought by the polyurethanes manufacturing community.

Also unfortunately, the use of amine catalysts as supplemental catalysts in combination with water to provide closed cell foams (e.g., rigid and semirigid foams) has heretofore been unsuccessful, largely due to a shrinkage problem associated with the production and storage of such foams. A solution to this shrinkage problem would be highly desirable in the polyurethanes manufacturing community.

The control of shrinkage is therefore a highly desirable in the production of water-blown foams. This particular problem was claimed to be solved by Coppola.[2] The Coppola invention relates to an improved, essentially closed-cell polyurethane foam comprising the reaction product of a polyol with isocyanate in the presence of water to provide a carbon dioxide blowing agent, the improvement being that said reaction is effected in the presence of a catalyst system consisting essentially of N,N,N',N'',N''-pentamethyl-diethylene triamine and triethylene diamine to provide a polyurethane foam characterized by enhanced shrinkage resistance.

10.8.3 Catalyst Composition for Molded Foam with Improved Breathability

Molded polyurethane foams are used extensively in various seating applications, including automobile seats. To provide comfortable seating, the current polyurethane foams must have breathability. The current method of attaining breathability involves the combination of mechanical and/or vacuum crushing with the appropriate silicone surfactants in the polyurethane foam formulation to provide maximum cell openness.

A typical solution to improving the air flow of molded foams is to select less stabilizing silicone surfactants for molded foam formulations. These less effective surfactants will provide more cell opening; i.e., better air flow values; however, more processing problems and overall rising foam instability are offsetting disadvantages. A second practice is to use a mixture of standard silicone surfactant with a dimethylsiloxane fluid. The dimethylsiloxane fluids are normally used as defoamers. These silicone blends work but, again, trade-offs in rising foam stability and processing latitude are necessary.

Catalyst compositions used in making polyurethane foam typically comprise a combination of a blowing catalyst and a gelling catalyst. The blowing catalyst influences the isocyanate-water reaction. Bis(dimethyl-aminoethyl) ether [BDMAEE] is a commonly used blowing catalyst. Polyurethane catalysts are not considered to have cell opening effects. At the levels used in industry to achieve optimum productivity, catalysts are typically blamed for decreasing cell openness by increasing the polymerization rate to polyurethane.

The polyol supplies have worked at producing modifications of their products to improve cell opening. In addition, low levels of selected polyols are sold as cell opening modifiers. These polyols are typically low-molecular-weight polyols used in rigid foam applications.

Petrella[3] disclosed a method for preparing a polyurethane foam which comprises reacting inorganic polyisocyanate and a polyol in the presence of a catalyst composition comprising a blowing catalyst and a gelling catalyst, consisting (a) 25 to 80 wt. % pentamethyl diethylenetriamine and (b) 20 to 75 wt. % bis(dimethylaminopropyl) methylamine. The catalyst composition provides the resulting polyurethane foam with improved breathability.

10.8.4 Catalyst for Isocyanate Trimerization

Since an isocyanurate structure obtained by trimerizing an organic isocyanate brings about an improvement in properties such as enhanced thermal resistance, flame retardancy, chemical resistance, etc. for polyurethane, coating material, and the like, there is a need for newer isocyanate trimerization. For example, various metal salts of carboxylic acids, metal salts of DMF, tertiary amines, alcoholates of metals, etc. are effective catalysts for the above purpose. In addition, an epoxide/pyridine selectively forms a trimer.[4]

However, conventional catalysts for isocyanate trimerization have suffered the following disadvantages:

1. Insufficient activity, which requires severe reaction conditions
2. Insufficient selectivity accompanied by the formation of by-products such as dimers, carbodiimides, etc.
3. Difficulty in eliminating high-boiling additives such as DMF
4. Inevitable side reactions due to the attack by the active species of a catalyst on functional groups such as ester, silyl ether, etc. in the system containing such groups

Endo and Nambu[5] disclosed a process for producing an isocyanate trimer and a process for producing an urethane with a catalyst for isocyanate trimerization or urethane formation, which comprises cesium fluoride or a tetraalkylammonium fluoride. The trimerization is carried out by reacting an organic isocyanate at a temperature of from room temperature to 150° C for 1 min to 2 hr without any solvent by using either 0.005 to 0.02 equivalent of cesium fluoride or a tetraalkylammonium fluoride as the catalyst. After the reaction, an unreacted isocyanate is removed under a reduced pressure, the reaction product is dissolved in methylene chloride, the catalyst is separated by filtration, and the methylene chloride solvent is distilled away to obtain an isocyanurate corresponding to the starting isocyanate.

The urethanization and trimerization accompanied by urethanization are carried out by reacting an organic isocyanate with 0.1 to 1 equivalent of an organic hydroxylic compound under the almost same conditions as those described above to obtain a urethane and an isocyanurate corresponding to the starting isocyanate.

The use of these catalysts according to the Endo invention enables the trimerization of organic isocyanates to proceed under milder conditions than the use of conventional catalysts and thereby to obtain isocyanurate derivatives having a high purity. Furthermore, the catalysts according to the present invention exhibit such a high selectivity for the trimerization of isocyanates that they can be applied to the synthesis of novel isocyanurate derivatives having a variety of reactive groups.

10.8.5 Synergistic Combination of an Organomercuric Compound and a Zinc Salt of an Alkyl Carboxylic Acid

Cellular polyurethane foams are usually prepared by the reaction between a compound containing at least two reactive hydrogen atoms and a polyisocyanate. In producing the foam, there are generally introduced various additives for blowing the polyurethane or emulsifying the urethane being formed, and catalysts are utilized to promote the reaction within desirable time limits. Generally, tertiary amine catalysts are used in combination with an organometallic compound such as stannous octoate or stannous oleate to promote the cross-linking reaction. In preparing such foams, the properties obtained are often dependent upon the water sensitivity in the typical polyisocyanate-polyol reactant mixture.

The polyisocyanate/water reaction leads to the formation of carbon dioxide. Thus, more or less gas is generated, depending on the presence of more or less water being present even in small amounts. For the reaction, it is also known to utilize as catalysts certain mercuric salts of carboxylic acids, for instance, phenylmercuric acetate. These compounds as catalysts for the polyisocyanate-polyol reaction avoid to a substantial degree variations in foam properties as a result of the presence of moisture.

In the preparation of cross-linked, solid, noncellular polyurethane polymers, whether rigid or elastomeric, liquid mixtures of organic polyisocyanates and hydroxy-terminated polyols have been reacted in the presence of certain mercury compounds as disclosed in U.S. Pats. 3,583,945 and 3,592,787, at ambient conditions of temperature without the use of heat or pressure. It is also known to catalyze the reaction of isocyanates with organic compounds utilizing metallic driers as catalysts in accordance with the teaching of U.S. Pat. 2,374,136. Thus, cobalt naphthenate, manganese naphthenate, lead naphthenate, and other metallic driers are known to catalyze the reaction. While other organometallic compounds such as stannous octoate or lead naphthenate are known to function as catalysts for the preparation of polyurethane foams, it is also known that zinc salts (carboxylates) are very poor catalysts for this reaction, as disclosed in U.S. Pat. 3,347,804. Even combining 10% tin octoate with 90% zinc naphthenate provided a catalyst combination that showed an extremely slow foam rise time—too slow to have commercial utility.

Brizgys and Gallager[6] disclosed a reaction of an organic polyisocyanate with an organic compound containing at least 2 active hydrogen-containing groups is catalyzed with a synergistic co-catalyst combination of at least 1 organomercuric catalyst and at least 1 zinc salt of an alkylcarboxylic acid having about 2 to about 22 carbon atoms. In another aspect, the invention relates to a novel catalyst system for the preparation of polyurethanes which system exhibits a faster cure rate at ambient temperature than shown by either component of the catalyst combination when utilized separately.

Brizgys and Gallager discovered that the combination of an organomercuric compound with a zinc salt of an aliphatic carboxylic acid is an efficient co-catalyst that provides a means of accelerating the rate of cure at ambient temperatures for the reaction of an organic polyisocyanate and a compound containing at least two reactive hydrogen atoms, which catalyst combination is more effective than the use of the organomercuric compound when used alone as a catalyst. Useful solid (noncellular) and foamed polyurethane products are obtained, and they are produced at a faster curing rate than with use of either co-catalyst alone.

10.8.6 Mercurial Catalysts

A wide variety of catalysts may be employed to facilitate the polyurethane reaction. The catalysts ordinarily used promote rapid reaction times even at ambient temperatures and, of course, this is considered quite desirable in most instances. On the other hand, there are also instances where it is deemed desirable to prolong or control the initial reaction and yet provide for a fully cured product within a relatively short period. For example, when certain polyurethane reaction products (e.g., polyurethane foams or elastomers) are cast in molds having intricate designs, the reaction mixture must be sufficiently fluid to completely fill all voids in the mold and thus produce a product that, when cured, will be uniformly formed. We have found that a select group of polymercury organic compounds have the unexpected effect of promoting catalysis of polyurethane reactions having a gradual or controlled viscosity buildup and a rapid cure only after heat is applied from some source. This noted effect is quite surprising in view of the known art regarding the catalysis of polyurethane reactions with organomercurials. For example, organomonomercuric catalysts such as phenylmercuric acetate, and other classes of organopolymercury catalysts, rapidly promote the reaction of the isocyanate with, for example, hydroxyl group containing materials at relatively low temperatures.

Fertig and Lederer[7] disclosed polymercury derivatives of organic compounds, structurally distinguished from other polymercurials in that two or more mercury atoms are directly attached to carbon atoms of the same organic nucleus and form part of the same cationic moiety. Accordingly, the organic nucleus and the anion can be a wide variety, the limiting factor being that at least two mercury atoms are attached directly to the organic nucleus.

10.8.7 Chain Extender with Catalytic Properties

The use of catalysts in preparing polyurethane by reacting a polyisocyanate, a polyol, and perhaps other ingredients is well established. To provide polyurethanes with the desired physical properties, the catalyst employed must promote the many reactions that proceed simultaneously and competitively at balanced rates during the polymerization process. One such reaction is between a hydroxyl-containing molecule and an isocyanate to form a urethane. A second reaction is an isocyanate-water reaction by which an isocyanate-terminated molecule is hydrolyzed and CO_2 is generated to blow or assist in the blowing of foam, e.g., in the preparation of one-shot flexible foams.

Tertiary amines have become well established as catalysts for the production of polyurethane foams. They accelerate both urethane-forming and blowing reactions. Many amines of this class, however, have a strong odor that is carried over to the polyurethane foam. Final product odor may be reduced by using volatile catalysts such as N-methylmorpholine, but poor cure often results from too rapid a loss of catalyst in the foaming process. Moreover, high volatility usually means low flash point, high vapor concentration, and handling problems.

Toxicity of some tertiary amines can also be a problem. Tetramethylbutanediamine, for example, is no longer in large-scale use because of its physiological effects, and diethylaminopropionitrile was withdrawn from use in 1978 when it was found to be neurologically active. Therefore, a need exists for low-odor tertiary amine catalysts of low toxicity that will ensure adequate cure.

Tertiary amines containing an isocyanate-reactive group that ties the catalyst into the polymer network offer one solution to the problems of odor, toxicity, and insufficient cure associated with more traditional amine catalysts. Adequate cure is ensured by incorporating the amine into the polymer chain.

Also, the diffusion of the amine from the foam is prevented, which minimizes the odor and possible health hazards associated with free amine. Examples of catalysts containing one isocyanate-reactive group include dimethylaminoethanol, tetramethyl-1,3-diamino-2-propanol, and 1-(2-hydroxypropyl)imidazole. A major disadvantage of these types of catalysts is their action as chain terminators in the polymer forming reaction. An alternative, catalytic alkoxylated amines containing three or more reactive sites such as triethanolamine or "quadrol," a propoxylated ethylenediamine, act as cross-linkers, reducing elastomer or foam flexibility and impact resistance. Tertiary amines containing two isocyanate-reactive groups are often difficult to prepare and isolate in high yield.

Polyurethane elastomers are block copolymers derived from the combination of a polymeric diol, a polyisocyanate, and a chain extending diamine or diol. The resulting copolymer is composed of soft-segment blocks and hard-segment blocks. The polymeric polyester or polyether comprises the flexible soft-segment block, which influences the elastic nature of the product and its low-temperature performance, while the polyisocyanate/chain extender units constitute the hard-segment blocks, which affect modulus, hardness, and tear strength and determine the upper use temperature.

The hard-segment blocks are partially segregated from the soft segment into domains or microdomains. Hydrogen bonding and dipole-dipole interactions between the polar groups provide a pseudo-cross-linked network structure between linear polyurethane chains such that the polymer has the physical characteristics and mechanical behavior of a covalently cross-linked network. Urea linkages that result from the use of diamine chain extenders yield stronger hydrogen bonds than carbamate linkages achieved by use of diol chain extenders. Consequently, the cohesive strength within the hard segments, and thus overall polymer strength, are greater in polyurethanes prepared from diamine chain extenders. Cast elastomer formulations cured with diamine chain extenders generally do not contain tertiary amine catalysts.

Applications using both an aromatic diamine and a tertiary catalyst may include high-resilience foams, semirigid foams, and RIM and microcellular elastomers. As to cast elastomers, the diamines increase the strength and load-bearing capabilities of the polymer product without a significant increase in polymer density.

To produce these polyurethanes, a mixture of an organic polyisocyanate, a polyol and perhaps other ingredients is cured with a derivative of an aromatic diamine, preferably a diaminobenzamide. The preparation of the polyurethane can be accomplished by a one-shot mode or by the preparation of a prepolymer and the subsequent curing of the prepolymer.

The tertiary amine portion of the molecule, which is situated in a side chain of the aromatic diamine, while acting as a polyurethane catalyst, is bound to the polymer network allowing for adequate polymer cure while preventing later diffusion of the catalytic species from the polyurethane thereby eliminating odor and toxicity as problems. The aromatic diamine provides enhanced PU physical properties.

The Casey et al.[8] invention relates to a process for the production of polyurethanes wherein a tertiary amine functionality is incorporated into the side chain of a diaminobenzamide to form a catalytic polyurethane chain extender. These chain extenders are polyurethane catalysts that provide adequate cure without residual odor and safety hazards associated with the more traditional tertiary amine catalysts. When reacted with the isocyanate reactive group the nontertiary amine portion of the chain extender is bound to the polymeric network. Diffusion of the catalyst species from the polyurethane has been prevented, thus negating odor and toxicity as problems. These aromatic diamines provide enhanced polyurethane physical properties.

The Casey invention uses a tertiary amine catalyst that is chemically bound to an aromatic diamine chain extender, incorporated into the polymer chain. Unlike other tertiary amine catalysts, which are bound to the polymer network by an isocyanate reactive group, these catalysts are chain extenders, not chain terminators or cross-linkers.

References

1. Chang, E.Y.C. U.S. Pat. 4,598,103, Jul. 1, 1986. "Catalyst systems for polyurethane compositions," to American Cyanamid Company.
2. Coppola, Pasquale J., and Petteway, Jr., Leny O., U.S. Pat. 5,162,386, Nov. 10, 1992. "Amine catalyst system for water-blown polyurethane foam," to BASF Corporation, Parsippany, New Jersey.
3. Petrella, Robert G., U.S. Pat. 5,039,713, Aug. 13, 1991, to Air Products and Chemicals, Inc., Allentown, Pennsylvania.
4. Jones, J.I., et al. *J. Chem. Soc.*, 4392 (1957).
5. Endo, Takeshi, and Nambu, Yoko. U.S. Pat. 5,264,572. Nov. 23, 1993. "Catalyst for isocyanate trimerization," to Asahi Denka Kogyo K.K., Tokyo, Japan.
6. Brizgys, Bernardas, and Gallagher, James A., U.S. Pat. 4,256,848, Mar. 17, 1981. "Synergistic polyurethane co-catalysts which are a combination of an organomercuric compound and a zinc salt of an alkyl carboxylic acid," to BASF Wyandotte Corporation, Wyandotte, Michigan.
7. Fertig, Joseph, and Lederer, Seymour J., U.S. Pat. 3,642,044, Feb. 15, 1972. "The production of polyurethanes employing organopolymercurial catalysts," to Merck & Co., Inc., Rahway, New Jersey.
8. Casey, Jeremiah P., Clift, Susan M., and Kem, Kenneth M., U.S. Pat. 5,034,426, Jul. 23, 1991. "Aromatic diamine catalytic chain extenders," to Air Products and Chemicals, Inc., Allentown, Pennsylvania.

11

Elastomers

11.1 Introduction

Polyurethane elastomers are a very important group of products, although their consumption is much lower than that of urethane foams. They opened a range of new applications that could not be realized before. The advantage of polyurethane (PU) elastomers is related to their high hardness for any given modulus, high abrasion and chemical resistance, excellent mechanical and elastic properties, blood and tissue compatibility, and also some other specific properties. Another important factor is the possibility of custom tailoring the structure by varying the type and the ratio of starting components during processing, which is usually carried out by the manufacturer.

Many polyurethane elastomers are specifically synthesized to be thermoplastic. Thermoplastic polyurethanes (TPU) can be processed by any standard thermoprocessing technique. TPUs are offered as pellets, ready for processing. Pelletizing may be performed by an underwater pelletizing technique, shown in Fig. 11.1.

11.2 Definitions

Elastomers are a class of materials displaying high reversible deformation. Such behavior requires highly flexible chains, i.e., a low degree of intermolecular interaction, and the presence of cross-links that prevent *sliding* of the chains against their neighbors, causing plastic flow. The cross-links can be of a chemical or physical nature. Both are used in polyurethane elastomers. Physical cross-linking can be obtained through hydrogen bonding and hard domain formation. Chemical cross-linking is introduced via tri- or multifunctional constituents. Once introduced. chemical cross-links cannot be easily destroyed by thermal treatment as is the case with physical cross-links, except in some special cases of labile chemical groups, producing an irreversible network. Thus, physically cross linked polymers allow multiple melting or dissolution of the material which is of great practical importance. Physically cross linked polyurethane elastomers are block copolymers, consisting of alternating rigid and flexible blocks. Due to the different polarity and chemical nature of both blocks they separate into two phases designated as "soft" and "hard." Hard blocks also associate into domains because of rigidity and hydrogen bonding. Hard domains act as physical cross-links.

Segmented polyurethanes are thus two-phase polymers, and their properties are strongly affected by the amount of phase separation. So-called *soft segments* are obtained by reacting polyols having molecular weights between 400 and 6000, although in practice molecular weights of 1000 and 2000 are used almost exclusively. *Soft-segment concentration* (SSC) is defined as the ratio of the mass of polyol chains without terminal hydroxyl groups to the total mass of the polymer, usually expressed as a percentage. The difference, making the total 100%, is *hard-segment concentration* (HSC). Hard segments, defined in this way, consist of all nonpolyol components and may be just the diisocyanate part. Usually, hard segments consist of the diisocyanate and short polyols called *chain extenders*. Hard-segment concentration calculated on the basis of diisocyanate-chain extender blocks may be lower than the difference (100% − SSC).

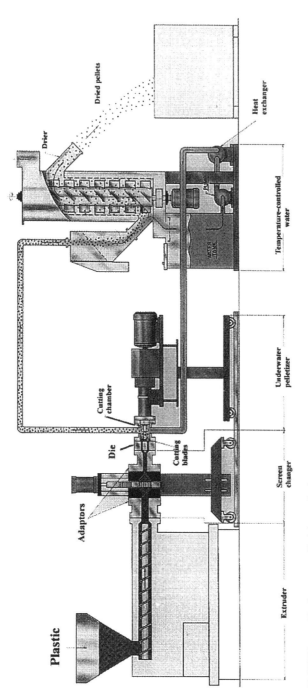

FIGURE 11.1 Underwater pelletizing technique.

One-phase polyurethane elastomers are usually obtained by cross-linking polyols with isocyanates, without the use of chain extenders.

11.3 Chemistry of Polyurethane Elastomers

To understand the structure of polyurethane elastomers, a brief review of the urethane chemistry involved will be given here. As was pointed out earlier, segmented polyurethanes are obtained by reacting polyols and chain extenders with diisocyanates. However, the structure of the polymer will depend on the method of synthesis as well as on the choice of components and their reactivity. Figure 11.2 presents a summary of important definitions and uses of polyurethane elastomers.

The technology of polyurethane elastomers is based on three monomers: (1) an isocyanate source, (2) a macroglycol, and (3) a chain extender. This is shown in Fig. 11.3. To prepare an ideal segmented polyurethane elastomer, most side reactions should be suppressed. Thus, even traces of water must be avoided. That is a difficult task, since commercial polyols contain some minimum amount of water (<0.05%). Allophanates are formed at an appreciable rate at temperatures between 120 and 150° C. This reaction also takes place at lower temperatures in the presence of excess isocyanate. Biuret formation takes place approximately in the same temperature region, except the lower temperature limit is about 20° C below that for allophanate formation. A basic property of allophanate and biuret groups is their low thermal stability, since dissociation into starting components occurs above 150° C.

This fact allows processing of material chemically cross-linked with allophanate or biuret groups; i.e., cross-linking is reversible. Very often, cross-linking is deliberately introduced to improve certain properties or to increase molecular weight. All the above temperatures should be considered as approximate, since these chemical reactions proceed at lower rates at lower temperatures. Thus, time is an important variable when temperature limits are considered.

- Polyurethanes

$$\text{Polymers containing: } -\overset{\displaystyle H}{\underset{\displaystyle |}{N}}-\overset{\displaystyle O}{\underset{\displaystyle \|}{C}}-O-$$

- Elastomers
  Materials that can be stretched repeatedly and return to their original shape.

- Uses
  Applications requiring high mechanical properties (i.e., high tensile strength, elongation, >400 percent chemical resistance)

FIGURE 11.2 Polyurethane elastomers.

Technology based on three monomers:

- Isocyanate source   aromatic
                                                    aliphatic

- Macroglycol
  Polyester
  Polyether
  Polycaprolactone
  Polybutadiene
  Castor oil

- Chain extender   diols
                                          diamines

FIGURE 11.3 Thermoplastic elastomers.

The structure of the segmented polyurethane chain is highly heterogeneous and depends strongly on the reaction conditions as well as on the reactivity of components and use of catalysts. An ideal structure would consist of alternating monodisperse soft and hard segments. In practice, some types of soft segments based on polyether polyols obtained by anionic polymerization are initially monodisperse, whereas polyester polyols have a distribution of molecular weight typical for polycondensation polymers. Hard segments are, on the other hand, almost always polydisperse.

A wider than theoretical distribution of hard segment lengths in the prepolymer technique may be caused by unequal reactivities of monomeric isocyanate and macroisocyanate. The relative reaction rates of the first and second NCO groups of MDI with a particular hydroxyl group are 16 and 8.6, and those of 2.6-TDI are 5 and 2, respectively. If the reactivities of both groups in a diisocyanate are the same, then the ratio of relative reaction rates will be exactly 2, since the concentration of the second NCO group is half that of the original. Due to differences in activation energy, temperature increase may significantly change the relative reactivity of diisocyanates. Thus, temperature may also affect distribution.

A very important factor influencing both molecular weight and distribution is phase transition. During polymerization, the class transition temperature of the mixture increases, and when it reaches the reaction temperature, the reaction stops. Crystallization may be an additional factor in stopping the reaction. Symmetrical diisocyanates like MDI or 2.6-TDI form crystallizable hard segments with symmetrical polyols. To avoid this problem, the reaction temperature should be above the melting point of the crystalline domain. Polyurethanes generally have high melting points, and heating to these temperatures causes degradation.

Thus, to obtain higher regularity of structure, polymerization should be carried out in solution. In dimethyl sulfoxide, dimethylacetamide, tetramethylene sulfone, and dimethyl sulfoxide/4-methylpentanone (50/50), side reactions are minimized. Generally, reaction in solution is "cleaner," with less chance of allophanate formation, which causes branching and molecular weight increase.

11.3.1   Phase Separation in Segmented Polyurethanes

Segmented polyurethanes are two-phase systems, consisting of soft and hard phases. Phase separation strongly affects the properties. These systems may be compared to polymer blends with good adhesion between phases, which is achieved here by chemical bonds, although some secondary interactions such as hydrogen bonding are also present. Properties of the blend from miscible components are usually the average of the properties of the components. On the other hand, blends of immiscible polymers offer some special properties such as the combination of advantages of each component. The soft segment in polyurethanes contributes to the high extension and elastic recovery, while the hard segment contributes high modulus and strength to the composite.

Phase separation depends on the solubility parameters of the phases, their crystallinity, temperature, and previous thermal history. Even in the case of two-phase separation, there is always an interphase layer having transitional properties intermediate between those of the two phases. Thickness of the layer is a measure of the degree of phase separation. It may be assessed using small-angle X-ray scattering (SAXS).

11.4 Hydrogen Bonding in Polyurethanes

The hydrogen bond is the strongest secondary chemical bond, with a strength estimated to be about 20 to 50 kJ/mol. A hydrogen bond is formed between two groups, one of which is a proton donor and the other of which is a proton acceptor. Most frequently, proton donors are hydroxyl or amino groups or their derivatives, e.g., amides. The existence of the hydrogen bond has a great effect on material properties. Formation of the hydrogen bond can be represented by the following scheme:

$$R_1\text{–}X\text{–}H \text{———} Y\text{–}R_2$$

where R–X–H is a proton donating group, and Y–R is a proton acceptor. The urethane bond [–NH–CO–O–] contains both a strong proton donor (N–H group) as well as a carbonyl (C=O) group, which serves as the proton acceptor. Another possible proton acceptor is the alkoxy oxygen (–O–) from the urethane group with its free electron pairs. Thanks to strong hydrogen bonds, urethanes have higher melting points than analogous esters. Hydrogen bonding has a special significance in polyurethanes. Since polyurethanes can have a broad range of chemical structures, only linear and segmented polymers will be considered here.

The existence of hydrogen bonds can be directly observed by infrared spectroscopy. The N–H group participating in the hydrogen bond displays a characteristic absorption band at 3300 to 3300 cm⁻¹. The band of the non-bonded N–H group appears at about 3446 cm⁻¹, but the frequency varies with the bond strength. Higher bond strength is characterized by lower frequency (wave number). A typical Fourier transform infrared spectrum of a segmented polyurethane based on MDI/BD and PTO, having 40% soft segments would display: two carbonyl bands, one at 1707 cm⁻¹ assigned to bonded free C=O groups, and the second at 1731 cm⁻¹ assigned to free C=O groups. A single N-H band at 3327 cm⁻¹ would show that almost all NH groups are hydrogen bonded.

Hydrogen bond strength depends on the type of proton acceptors present. Chlorine atoms in chlorinated solvents are, for example, very weak proton acceptors. Proton accepting ability affects self-association of urethane groups, which is lower in solvents with higher dielectric constant, because of the higher degree of interaction of NH and C=O groups with the solvent. In solvents with carbonyl groups such as ketones and esters, the intensity of NH bands (H-bonded) is stronger than in CCl₄. The band of NH (N–H...O=C) in ketones at 3315 to 3320 cm⁻¹ is located between (N–H...O–) in ethers (3230 cm⁻¹) and NH (N–H...O=C) in urethanes in CCl₄, (3330 to 3350 cm⁻¹). It has been found that ketone carbonyl is a stronger proton acceptor than ester carbonyl. In aromatic hydrocarbon solvents, the proton acceptor is the phenyl group. Judging from the frequency shift of NH and C=O absorption bands, which is related to the interaction strength of proton acceptors with a urethane N-H donor, aliphatic and cycloaliphatic ethers are the strongest acceptors, followed by ketones, esters, urethanes, aromatic ethers, and chlorinated hydrocarbons.

11.5 Effect of Hard and Soft Segment Structure on Tensile Properties

The structures of both diisocyanates and chain extenders have a profound effect on the domain morphology and the strength of polyurethane elastomers. Generally, symmetric isocyanates form crystallizing hard segments with good packing ability, producing higher-strength materials. Usually, diisocyanates have been compared only at a certain specific hard segment concentration, so conclusions presently in the literature may not be general enough. The structure of the diisocyanate affects tensile strength, modulus, and hardness of elastomers. The following diisocyanates are listed in decreasing order of tensile strengths they impart to BD/polyester/diisocyanate elastomers: MDI > p-phenylene diisocyanate > 3,3'-dimethyl-4,4'-diphenylmethane diisocyanate (DMDI) > TDI > NDI > 3,3'-dimethyl-4,4'-diphenyl diisocyanate (TODI) > 4,4'-diphenylisopropylidene diisocyanate (DPDI). Diisocyanates with methyl-1 substituents, such as TDI, DMDF, and DPDI, favor lower modulus than those with an analogous structure but without substituents. The position of individual diisocyanate in the scale given may be altered under different conditions, but the general trend is valid.

Pandya et al.[1] studied the effect of TDI, crude MDI, HDI, and isophorone diisocyanate (IPDI) on mechanical and electrical properties of polyurethanes based on polyester/diisocyanate/BD systems. Polymers were prepared from1 mole of polyol (Mn = 1000), 3.2 moles of diisocyanate, and 2.0 moles of 1,4 butanediol. The highest tensile strength was obtained with TDI, then IPDI, followed by crude MDI and, finally, HDI. Elongation at break was 900% for TDI and IPDI, 800% for HDI, and only 195% for crude MDI. The highest modulus was obtained with crude MDI, followed by, HDI, IPDI, and TDI.

Tensile strength is also highly dependent on the type of chain extender used. Comparison of various diol chain extenders in MDI/polyester systems showed that 1,4-butanediol gives superior tensile strength and rebound elasticity but inferior compression set in comparison with ethylene glycol and hydroquinone bis(2-hydroxyethyl) ether (HQEE).[2] Comparison of the effect of diol chain extenders on tensile strength shows that 1,4-butanediol offers superior properties. The strength of elastomers was found to decrease in the following order for these chain extenders in MDI/polyester polyol polyurethanes: BD > ED > 1,5-pentanediol > HD > 1,3-propanediol. If, however, different polyols are used, then the order of chain extenders according to their effect on strength might be: BD > HD > 1,5-pentanediol > 1,3-propanediol > ED. In all cases, 1,4-butanediol offers the best tensile properties, which is the result of the specific crystalline structure and conformation of the MDI/BD hard segment.

Aromatic diamines as chain extenders offer excellent properties. Tensile strengths decrease in the following order when these diamines are used: 3,3'-dimethylbenzidine > 3,3'-dichlorobenzidine > 3,3'-dichloro-4,4'diaminodiphenylmethane > MDA > benzidine > p-phenylene diamine. Aromatic diamines offer superior tensile strengths compared to aliphatic diamines and diols.

The influence of the structure of the soft segment on tensile strength and elongation is related to the intermolecular association and crystallization during stretching. The effect of the soft segment is pronounced mainly in polymers with a continuous soft phase, i.e., above 50% SSC. Generally, polyurethanes based on polyesters have higher tensile strengths than those based on corresponding polyethers and polybutadiene oligomers. Poly(ethylene adipate) polyester polyurethanes were found to have better tensile properties than poly(ethylene-cobutylene adipate) or PCL polyurethanes, which suggests that crystallization and relative density of ester bonds, i.e., polarity, play decisive roles. Although numerous data have been published on the effect of various components on the properties of polyurethane elastomers, direct comparison of data from different sources is not often possible.

11.6 Effect of Temperature on Strength

Polyurethane elastomers are used in the temperature range limited by Tg of the soft segment (low temperature limit) and Tg of the hard segment (upper temperature limit). Below the Tg of the soft segment, polyurethanes are glassy materials, losing their elastomeric properties. Glassy materials possess higher intermolecular secondary bonds compared to elastomers, having as a result higher mechanical strength. The strength of polyurethane elastomers exponentially decreases as the temperature is increased above the Tg of the soft segment. To vary the Tg of elastomers, either hard segment (or unit) content, which is miscible with the soft phase, may be varied, or a plasticizer may be added. Figure 11.4 shows the thermal behavior of polyurethane elastomers between Tg and Tm.

Concerning elongation at break vs. temperature relationships, a maximum elongation at 20 to 60° C above Tg of the soft segment is obtained. However, in the working range of most elastomers, elongation at break decreases with increasing temperature. The upper temperature limit for practical use of segmented elastomers is related to the softening of the hard segment. Bonart[3] has tried to measure heat distortion temperature (T_{HD}) and relate it to the structure of the hard segment. T_{HD} is measured by gauging the change of length of the sample under small load as a function of temperature. At sufficiently high temperature, when physical cross-links (hard domains) start to break up, a step increase in length is observed. T_{HD} thus characterizes physical cross-linking stability. It depends on the chemical nature of the chain extender, the hard segment length, and domain size. T_{HD} depends also on the nature of the soft segment and phase separation. It has been found that thermal network stability (TNS) of physical cross-links is lower in hard segments with diols having odd numbers of CH_2 groups. However, precise determination of T_{HD} was found to depend on the scale of both temperature and elongation axis as well as on the load, making the method probably unsound for absolute conclusions. An attempt has been made to modify the method and make it independent of the compliance of the soft segment matrix. Extrapolation of the TNS temperature to zero load allows determination of intrinsic thermal stability. For diamine/MDI hard segments, intrinsic TNS values were found to be between 130 and 150° C.

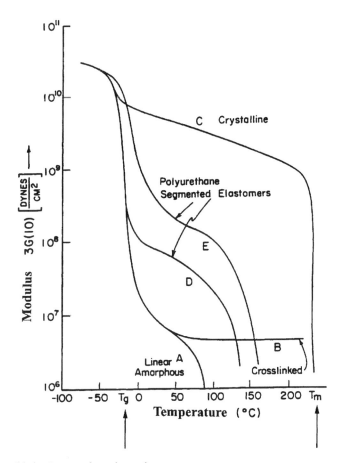

FIGURE 11.4 Thermal behavior or polyurethane elastomers.

11.7 Thermoplastic Elastomers

An elastomer may be defined as a material that may be stretched by 100% and, on release of the stretching force, springs back rapidly. This elasticity was always associated with vulcanized (cross-linked) rubber and is due to the structure of that type of material. The terminology that deals with elastic phenomena is called the *rubbery plateau*, as shown in Fig. 11.5. It should be noted that Tm in a polyurethane elastomer can be modified by increasing the molecular weight of the polymer, as shown in Fig. 11.6. Likewise, slightly cross-linking polyurethane elastomers can increase processing temperature, assuming that the level of cross-linking is held within tightly controlled limits, as shown in Fig. 11.7.

However, before discussing the structure of thermoplastic rubbers/elastomers, it would be as well to explain why this type of material is becoming so popular. TPEs are comparatively new materials, but they already have become significant in terms of tonnage and value. This is because they are creating new markets as well as replacing plastics and traditional rubbers in some of their applications. They provide some of the properties associated with traditional rubbers, but with greater speed and ease of processing. This is because the shaped products do not have to be cured like traditional rubbers; rather, they change from a fluid, processible melt to a solid, rubbery article on cooling.

Other important attributes of TPEs are their ability to be reprocessed, the wide color range possible, and their world-wide availability. The elimination of the compounding stage, essential for traditional rubbers, is also significant. However, at the present time, these materials do not have generally have the high-temperature resistance, compression set resistance, or the oil and solvent resistance of conventional vulcanized rubbers.

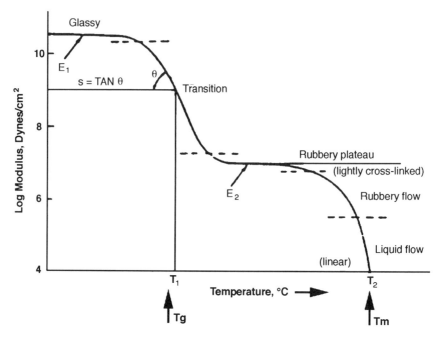

FIGURE 11.5 Modulus-temperature curves for polyurethane elastomers.

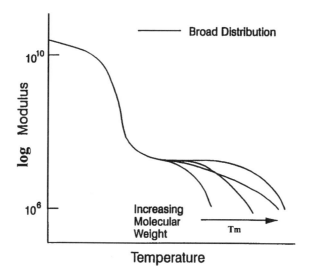

FIGURE 11.6 Polyurethane molecular weight increases the Tm.

11.8 Illustrative Patents

11.8.1 Non-Blocking Polyurethanes

A number of polymeric materials have been investigated for use in the fabrication of medical devices. One of these materials, silicone rubbers, although widely used, are disadvantageous, because large amounts of reinforcing fillers and other additives such as plasticizers, catalysts, and the like are added to the polymer to achieve adequate physical properties (e.g., tensile and tear strength).

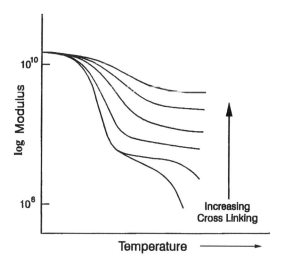

FIGURE 11.7 Cross-linking reduces the thermal sensitivity of polyurethanes.

As a result of several limitations, silicone rubber tubing is not suitable for use in the body for a prolonged period. Certain additives are susceptible to being extracted by body fluids. This is not only a potential source of contamination but, absent the additive, there is a tendency for devices made from silicone rubbers to lose their shape and physical properties. The fillers used in silicone rubber are also a potential source of contamination and have been known to induce thrombosis. A further disadvantage of silicone rubber is that, due to its limited physical-mechanical properties, a relatively large wall thickness is required in medical tubing and the like to achieve adequate strength and kink resistance. Silicone rubber cannot be used to fabricate finer-diameter medical tubing having a narrow wall thickness.

Polyurethanes have arisen as a particularly desirable replacement for silicone rubber in certain applications. One disadvantage of polyurethane resins of the softness desired for many medical devices (e.g., resins having Shore A hardness less than 100) is surface blocking (tack) after extrusion or molding into desired shapes. To avoid this problem, many remedies have been developed in the art, including the use of external mold release agents and the use of various antiblockers or detackifiers in admixture with the polymer. Most antiblocking agents/detackifiers are low-molecular-weight materials that have a tendency to migrate out or to leach (extract) out from the bulk/surface of the polymer. This represents a problem when polyurethanes are used as biomaterials (tubing, prostheses, implants, etc.). The presence of such low-molecular-weight extractables could effect their biocompatibility, hemocompatibility, and surface degradation (fissuring, stress-cracking).

U.S. Pat. 4,057,595, to Rauner et al., teaches the use of other compounds of detackifiers. This patent discloses a method for modifying the physical characteristics of polyurethane elastomers, particularly Spandex fibers, to reduce blocking wherein the polyurethane contains, within the polymer chain, a siloxane-polyoxyalkylene block copolymers.

These silicone-modified polyurethanes contain an Si-O-C linkage that makes them less than completely satisfactory for certain medical applications. The Si-O-C linkage is susceptible to hydrolysis. Upon prolonged exposure to body fluids, these polyurethanes have a tendency to lose their physical characteristics. Accordingly, there is a need for relatively soft polyurethane resins that are nontacky, nontoxic, and thermoplastic.

Zdrahala and Spielvogel[4] disclosed soft, non-blocking, thermoplastic polyurethanes prepared by reacting a long-chain polyester or polyether diol, a short-chain diol, a diisocyanate, and a silicone diol. The polyurethanes of the invention are also advantageous because they can be easily prepared in a one-stage polymerization without solvent. Thus, the invention provides a process for making a soft, non-blocking polyurethane.

11.8.2 Elastomers Produced with Polyols Containing Hydroperoxide Impurities

Thermoplastic noncellular polyurethane elastomers are currently available for use in the preparation of articles by injection molding, extrusion, and similar techniques. These elastomers are generally prepared by the reaction of an organic diisocyanate, a polymeric diol, and a difunctional extender in the presence of a catalyst. Monomer purity is particularly critical both in terms of the effect it can have on the progress of the reaction by way of its effect on the catalyst and the effect it may have on the resultant polymer properties.

Polymeric diols including the types of polyester and polyether diols used in the preparation of polyurethanes are subject to peroxidic impurity formation. Such peroxidic impurities are not limited to polymeric diols but form also in low molecular weight diols of the type used for extending the polyurethane reaction. Peroxide content of diols used in polyurethane elastomer formation must be kept below certain critical levels; otherwise, unacceptable polymer properties or reactions result if, indeed, the polyurethane can be prepared at all. Ideally, elastomer-grade polymeric diols and diol extenders should not have peroxide contents above 10 ppm. (See Ref. 5, which explains an analytical method for determining peroxide content and discusses the addition of various antioxidants to the diols and extenders to inhibit the formation of peroxides.) However, diols and extenders not containing stabilizers or antioxidants, after storage or heating or the like, can easily have peroxide contents ranging up to 200–300 ppm, rendering them literally useless for the preparation of polyurethane elastomers. Heretofore, no method has been disclosed in which such contaminated materials can be used efficiently in the preparation of high-grade elastomer products.

Thus, both the ingredients and the elastomers produced therefrom have been stabilized using various types of antioxidant and scavenging type additives. For example, U.S. Pat. 2,915,496 discloses a method for preparing a heat- and weather-resistant polyurethane elastomer by the addition of an antioxidant material to the urethane forming ingredients. U.S. Pat. 3,205,269 discloses the stabilization of polyethers and polyether alcohols by the addition of certain phosphites, thiophosphites, cyclic phosphites, and the like. The stabilized materials are used, typically, in the manufacture of polyurethanes.

Stabilization of the diisocyanate component is also well known; see, for example, U.S. Pats. 2,950,307 and 3,715,381. Furthermore, certain phosphite and phosphine compounds have been observed to catalyze the formation of integral-skinned polyurethane foams; see U.S. Pat. 4,021,381.

Phosphorus-containing antioxidants, particularly the phosphites (see U.S. Pat. 3,205,269, cited supra), while being recognized as effective antioxidants, are generally avoided because of their tendency to hydrolyze in the presence of trace amounts of moisture. The phosphorus acids thereby produced over an extended period of time are extremely detrimental to the polyurethane forming reaction.

Ehrlich and Smith[6] disclosed an improved process for the preparation of thermoplastic polyurethane elastomers by a one-shot reaction of an organic diisocyanate, a polymeric diol, and a diol extender wherein the diol components, either singly or together, are treated with a trivalent phosphorus compound prior to their reaction with the diisocyanate in the presence of a stannous tin urethane catalyst. The process, when carried out with pure urethane grade diol components, results in very efficient catalysis which, in turn, requires lower levels of stannous tin catalyst than would otherwise be employed. Moreover, the process can be carried out with diol components that are not urethane grade materials but that are contaminated by high levels of peroxidic impurities, which materials would otherwise be useless for the preparation of high-grade elastomer products.

The invention provides a method for the preparation of polyurethane elastomers that can employ polymeric diols or diol extenders containing higher than normally acceptable levels of peroxide impurity, which heretofore would not be considered useful for the preparation of high-grade polyurethane elastomers. It is a further object to effect the polyurethane reaction in the presence of lower levels of the expensive polyurethane-forming catalysts.

These objectives and others are readily accomplished by carrying out the polyurethane reaction in the presence of a particular stannous tin catalyst and using a polymeric diol and diol extender, one or both of which have been treated with a particular phosphorus compound prior to the one-shot polyurethane formation.

11.8.3 Elastoplastics with High Impact and Temperature Resistance (ISOPLAST)

Polyurethanes hitherto available have been used extensively in the fabrication of a wide variety of components, particularly the thermoplastic polyurethane elastomers, which can be fabricated by injection molding or by reaction injection molding (RIM) techniques. However, the use of these materials to prepare components having structural strength properties that match those derived from engineering thermoplastics, such as nylon and the like, has been limited by the need to provide extensive reinforcement using materials such as fiberglass to achieve desirable levels of stiffness, impact resistance, and related properties.

Goldwasser and Onder[7] described polyurethane resins are having high impact strength and other structural strength properties, plus significantly improved resistance to deformation by heat. The properties of these materials are such as to make them comparable to engineering thermoplastics such as nylon and like materials commonly used to fabricate such structural components as automotive parts, equipment housing, sporting goods, furniture, toys, household products, and like consumer goods.

The resins are prepared from organic polyisocyanates, an isocyanate-reactive material (polyol, polyamine, etc.) having an average functionality of at least 1.9, a Tg less than 20° C, and molecular weight of 500–20,000, and one or more low-molecular-weight extenders, the major distinguishing feature being the markedly lower proportion by weight (2 to 25%) in which the isocyanate-reactive material is employed as compared with polyurethane resins conventionally prepared in the art. Depending on the particular combination of reactants employed, the polyurethanes of the invention can be thermoplastic or thermoset and can be prepared in both cellular and noncellular form.

Goldwasser and Onder found that polyurethanes with markedly improved structural strength properties can be prepared by departing significantly from the previous teachings of the art as to the relative proportions of reactants to be employed. The inventors found that very substantial reduction in the amount by weight of the polymeric active-hydrogen containing material (e.g., polyol) employed in the preparation of the polyurethanes is a major factor in producing a highly surprising and dramatic change in the properties of the resulting polyurethanes. The change in properties is enhanced by selection of particular combinations of reactants, as is discussed in detail below. These changes enable us to produce resins that can be employed, without the necessity to incorporate reinforcing fillers and the like, to produce structural components that possess all the desirable impact resistance, stiffness, and other structural strength properties that heretofore have been achievable by the use of other polymers such as nylon and other engineering thermoplastics but not by polyurethanes alone.

This invention encompasses polyurethanes characterized by high impact resistance, high flexural modulus, and a heat deflection temperature of at least 50° C at 264 psi. By the term *high impact resistance* is meant an impact strength at ambient conditions (approx. 20° C) of at least 1 ft-lb/in, and preferably at least 3 ft-lb/in, of notch as measured by the notched Izod test (ASTM D 256-56).

The *heat deflection temperature* is a measure of the resistance of the polymer to deformation by heat and is the temperature (in °C) at which deformation of a specimen of the polyurethane of predetermined size and shape occurs when subjected to a flexural load of a stated amount (e.g., 264 psi). All such temperatures recorded herein were obtained using the procedure of ASTM D 648-56. The term *high flexural modulus* means a flexural modulus under ambient conditions (see above) of at least about 150,000 psi as measured by ASTM-D790.

11.8.4 Polyurethane Elastomers from Polyoxypropylene Polyoxyethylene Block Copolymers

The preparation of thermoplastic polyurethane elastomers from polyester polyols is well known. Such elastomers can be extruded, injection molded, or fabricated in another known manner without suffering any degradation due to momentary exposure to the relatively high processing temperatures (of the order of 400° F) involved in such techniques. In contrast, it has not hitherto been possible to use polyethylene and polypropylene glycols to prepare thermoplastic polyurethanes, which can be molded by techniques involving processing temperatures of the above order. It is highly desirable that such polyurethanes be prepared, since polyethylene and polypropylene glycols are significantly less expensive than polyester polyols and thereby would provide obvious economic advantages. The latter would be in addition to the recognized advantage in hydrolytic stability of the resulting polyurethanes, which hydrolytic stability is associated with the use of polyether polyols as opposed to polyester polyols.

Unfortunately, it has been found hitherto that polyurethane elastomers prepared using polyether polyols, particularly polypropylene glycol, are not capable of withstanding exposure to temperatures of the order of 400° F, even for a brief period such as that required in injection molding and similar techniques.

Bonk and Sha[8] disclosed thermoplastic, recyclable polyurethane elastomers that have increased high-temperature resistance, thus permitting fabrication by injection molding. The elastomers are the product of reaction of 4,4'-methylenebis(phenyl isocyanate), a particular group of polypropylene oxide-polyethylene oxide block copolymers and an extender [straight chain aliphatic diols C2-6 or the bis(2-hydroxyethyl ether) of hydroquinone or resorcinol]. The block copolymers have at least 50% primary hydroxyl groups, a pH in the range of 4.5 to 9, a content of alkali metal ion less than 25 ppm, and a molecular weight of 1000 to 3000. The minimum ethylene oxide (EO) residue content (percent by weight) of the polyether for any molecular weight (MW) is governed by the following equation:

$$\%EO = \left[\left(\frac{MS - 900}{4}\right) \times 3\right]\left[\frac{100}{MW}\right]$$

Bonk and Sha found that, by using a particular group of block copolymer polyether glycols which have not hitherto been regarded as potential candidates for polyols in the preparation of polyurethane elastomers, it is possible to prepare thermoplastic polyurethane elastomers that will withstand temporary exposure for limited periods to temperatures as high as 450° F. These elastomers can be fabricated readily, without degradation, by extrusion, injection molding, and the like. This finding results in marked advantages in terms of reduction in cost as well as the ability to use the more hydrolytically stable polyether-based polyurethane elastomers for fabrication of articles by injection molding and the like.

11.8.5 Prepolymer Formation

The preparation of polyurethanes and polyurethane/urea elastomers by reacting an aliphatic diisocyanate with a polyol and then chain extending with a short chain diol or diamine (e.g., an aromatic diamine) to form the elastomer is well known. Two processes are used, namely, the prepolymer process and the one-shot process, which includes reaction injection molding (RIM). A reactant system widely used in the prepolymer process utilizes a cyclohexanediisocyanate as the isocyanate component of the prepolymer and polytetramethylene glycol and polyethylene adipate glycol as the polyol component. (Such a system is sold under the trademark Elate™ by AKZO N.V. Then, the prepolymer is contacted with an aliphatic diol chain extender, and the formulation is molded.

The prepolymers using cyclohexanediisocyanates, especially trans-1,4-cyclohexanediisocyanate, as the isocyanate component have suffered, because cyclohexanediisocyanates are volatile and toxic, thus requiring special handling procedures. The prepolymers containing unreacted trans-1,4-cyclohexane diisocyanate are difficult to process because of their high melting points and high viscosities at ambient temperatures.

The prior art has produced a variety of prepolymer systems from aliphatic diisocyanates (e.g., cyclohexanediisocyanates and long-chain polyols) and such prior art then includes the following patents and articles.

- U.S. Pat. 3,651,118 discloses a process for the preparation of 1,3-cyclohexylene diisocyanates wherein the corresponding cyclohexanediamines are contacted with hydrogen chloride to form the dihydrochloride salt with subsequent reaction with phosgene to produce the diisocyanate. At column 3, the patentees indicate that the diisocyanates are useful for a variety of applications, and particularly in the preparation of polyurethanes, polyureas, and polyurethane/ureas.

- U.S. Pat. 4,603,189 discloses various araliphatic polyisocyanates which are triisocyanates. The isocyanates are suited for producing polyurethane lacquers and are alleged to have increased resistance to yellowing in comparison to aromatic polyisocyanates, and increased reactivity and hence shorter drying times than lacquers based on aliphatic polyisocyanates. These triisocyanates are methylene-bridged phenyl-cyclohexyl triisocyanates where the phenyl ring has two isocyanate groups. An example is diisocyanato-methylbenzyl-cyclohexylisocyanate.

- U.S. Pat. 4,518,740 discloses moisture-hardening varnishes having an isocyanate resin base of adducts of a mixture of diisocyanates, e.g., 2-methyl-1,5-diisocyanatopentane and 2-ethyl-1,4-diisocyanatobutane. In the background portion of the patent, the patentees point out that cyclic polyisocyanates such as isophorone diisocyanate and methylene-bis-(4-cyclohexylisocyanate) are widely used for preparing polyurethanes, but the isocyanurate and the propanetriol prepolymers form hard, brittle films following moisture hardening. When cyclic polyisocyinates are reacted with higher-molecular-weight polyols, hard or soft films can be produced, but the resulting resins are inactive and must be activated with tin catalysts, etc.

- U.S. Pat. 4,487,913 discloses a polyester polyurethane prepared by reacting trans-1,4-cyclohexane diisocyanate with a mixed polyester polyol. The resulting polyurethane can also contain small amounts of epoxies or a carbodiimide. In the background portion of the invention, the patentee indicates that methylene-bis-(4-isocyanatocyclohexane), sold under the trademark DESMODUR W™ by Mobay Chemical Company, was suited for producing polyurethanes but had poor fuel resistance, even though hydrolytic resistance was acceptable.

- U.S. Pat. 4,487,910 discloses a process for producing polyurethane prepolymers based on mono-cyclic and dicyclic aromatic and aliphatic diisocyanates wherein the prepolymer has reduced residual monomer content. Various polyisocyanates such as toluenediisocyanate or methylene-bis-(phenylisocyanate) and 4,4'-dicyclohexyl-methane diisocyanates are used in preparing the prepolymer.

- U.S. Pat. 4,256,869 discloses the preparation of polyurethanes from trans-cyclohexane-1,4-diiso-cyanate wherein the trans-isomer content is at least 90% of the isocyanate content. It was reported that processes for producing polyurethanes through the use of cyclohexanediisocyanates via a reaction with polyether and polyester polyols were known, but that these resulting polyurethane systems did not exhibit advantageous properties. However, advantages were achieved through the use of the trans-isomer of cyclohexane-1,4-diisocyanate. For example, polyurethanes produced using the trans-isomer of cyclohexane-1,4-diisocyanate had high softening temperatures with low freezing and glass transition temperatures, and they exhibited a remarkable degree of hardness and abrasion resistance over a wide temperature range.

- A series of articles [e.g., Syed et al., "A New Generation of Cast Elastomers," presented at the Polyurethane Manufacturer's Association Fall Meeting-Boston (Oct. 29, 1985); Dieter et al., "Aliphatic Polyurethane Elastomers with High Performance Properties," Polymer Engineering and Science, mid-May, 1987, Volume 27, Number 9 (page 673); and Wong et al., Reactivity Studies and Cast Elastomers Based on trans-Cyclohexane-1,4-Diisocyanate and 1,4-Phenylene Diisocyanate, pp. 75–92] disclose the preparation of polyurethane elastomers based on the trans-isomer of cyclohexane-1,4-diisocyanate and indicate that the diol polyurethanes result in tough elastomers with excellent high-temperature properties in a high-moisture environment and under extreme dynamic stress.

- Quay and Casey[9] disclosed polyisocyanate prepolymer for polyurethane and polyurethane/urea elastomer synthesis and to the resulting elastomer. The prepolymer is formed by reacting a cyclohexanediisocyanate with a long-chain polyol under conditions such that essentially a 2:1 adduct is formed. Excess cyclohexanediisocyanate is removed prior to forming the elastomer.

The Quay and Casey invention relates to a process for producing polyurethane/polyurea elastomers and to the elastomers themselves utilizing a cyclohexanediisocyanate (CHDI)-based prepolymer free of unreacted cyclohexane diisocyanate and essentially free of oligomeric cyclohexanediisocyanate by-products. The prepolymer is prepared by reacting a cyclohexane diisocyanate with a long-chain diol at least 90% by weight where the mole ratio of cyclohexane diisocyanate to polyol is in the range of at least 6 to 20:1, and preferably in the range of from 10 to 16:1, with reaction temperatures ranging from 40 to about 100° C, typically, 65 to 80° C. After formation of the prepolymer, the unreacted cyclohexanediisocyanate is removed from the prepolymer to produce a cyclohexanediisocyanate-free prepolymer—e.g., less than 0.15% by weight. The elastomer is then made by reacting the resultant prepolymer with a diol or diamine chain extender.

Significant advantages associated with the prepolymer of the Quay and Casey and the process for producing such prepolymers are (1) an ability to produce CHDI prepolymers that are low melting and have lower viscosities than conventional CHDI prepolymers (2) a prepolymer system capable of producing elastomers having outstanding high-temperature physical properties, including high hardness and stiffness, (3) a prepolymer that can be used for producing an elastomer having excellent stability to light and to hydrolysis, (4) an ability to produce polyurethane elastomers having excellent modulus, tensile, and shear storage modulus, and (5) a relatively straightforward process utilizing conventional processing equipment and techniques.

11.8.6 Polyurethanes Modified by Siloxane Block Copolymers

Thermoplastic polyurethane elastomers (TPUs) have long been known. They are of commercial significance by virtue of their combination of excellent mechanical properties, with the known advantages of inexpensive thermoplastic processibility. A wide range of mechanical properties can be established through the use of different chemical synthesis components.

Various attempts have already been made to produce TPUs that exhibit the unique surface properties of polysiloxanes such as, for example, water repellency, antiadhesive properties, compatibility with blood, reduced abrasion, and reduced hardness. TPUs with these surface properties are desirable.

One method is to mix TPUs with polysiloxanes. On account of the poor compatibility of polysiloxanes with synthetic polymers, the siloxanes have to be modified. One possibility is to use siloxanes containing vinyl-polymerizable double bonds as described in, for example, Japanese Pat. 59/78236, 1984. According to U.S. Pat. 4,675,361 (believed to be an equivalent to British Patent A 2,140,438), polyurethanes containing polydialkyl siloxanes are used as additives for blood-compatible base polymers.

Another possible solution already been described lies in the use of the soft polysiloxane as a nonpolar soft segment in a thermoplastically processible elastomer (see Japanese Pat. 60/252617, and U.S. Pat. 4,518,758). Due to the high molecular weight of the siloxane components, a considerable quantity of siloxane has to be used. However, this high siloxane content disturbs the phase morphology to such an extent that the mechanical properties of the elastomer deteriorate.

U.S. Pat. 3,562,352 describes polysiloxane-containing block copolymers in which polysiloxane blocks and polyurethane blocks are directly attached to one another by Si-N bonds. However, the materials produced are not thermoplastic polyurethanes.

In addition, U.S. Pat. 4,647,643 describes nonblocking soft polyurethanes containing from 1 to 15% by weight of polyoxyalkylenelsiloxane block copolymers. These materials are also described as nontacky and nontoxic and may be suitable for medical equipment such as tubes, blood bags, and implants.

The hardness of TPUs produced from linear polyhydroxy compounds, diisocyanates, and diols is adjusted through the content of so-called hard phase, which consists essentially of diisocyanate and short-chain diols. In practice, however, the production of TPUs having relatively low hardness is complicated by sticking in the machine and by poor demoldability. TPUs having relatively low hardness generally also suffer from reduced mechanical properties and increased compression set values.

The incorporation of isocyanates consisting of various geometric isomers, for example 4,4'-bis-(isocyanatocyclohexyl)-methane, has a particularly unfavorable effect in soft thermoplastic polyurethanes.

The TPUs produced from these types of isomers generally consist of very slowly recrystallizing hard segments and, accordingly, give products with poor demolding properties.

The most significant thermoplastic polyurethanes used in the medical field are aromatic TPUs. It is suspected that, in the use of these aromatic TPUs, unless processing is optimal, 4,4'-bis-(aminophenyl)-methane (i.e., MDA) is formed after long-term use (i.e., more than 30 days). Although there are also cycloaliphatic TPUs that do not have this problem, they nevertheless present processing difficulties due to the poorly recrystallizing hard segments.

Pudleiner et al.[10] disclosed an invention related to segmented thermoplastic polyurethanes modified with siloxane block copolymers and based on relatively high-molecular-weight dihydroxy compounds, diisocyanates, and difunctional chain extenders Accordingly, the problem addressed by the Pudleiner et al. invention was to economically combine the favorable mechanical properties of a TPU with the particular properties of a polysiloxane to obtain thermoplastically processible products that would be easier to process than the cycloaliphatic TPUs of the prior art.

11.8.7 Polyurethane Elastomers Suitable for Shoe Soles

Elastomeric polyurethane polymer for applications such as shoe soles should exhibit good physical properties, especially including abrasion resistance, flexibility, and durability. Typically, such polymer may be obtained by reaction of a polyester polyol with an isocyanate-terminated polyester polyol-based prepolymer in the presence of water or, alternatively, by reaction of a polyether polyol with a "hard segment" isocyanate-terminated polyether polyol-based prepolymer in the presence of a blowing agent consisting predominantly of a physical blowing agent such as, for example, trichlorofluoromethane. The preparation of polyurethane polymer by such procedures is described, for example, in patent publications E.P. 235,888, E.P. 175,733, U.S. Pat. 3,591,532, U.S. Pat. 3,901,959, U.S. Pat. 4,647,596, and U.S. Pat. 4,757,095.

Isocyanate-terminated prepolymers obtained from low-molecular-weight polyols or diols are frequently identified as a *hard-segment* prepolymers in contrast to *soft-segment* prepolymers generally obtained from high-molecular-weight polyols or diols. The terminology *hard-* and *soft-segment* derives from the morphology of elastomeric polymers, which can contain distinct phase separated regions. Such regions can be detected by thermoanalysis techniques and distinguished by, for example, glass transition temperatures. Generally, soft segments of the polymer can be considered as having glass transition temperatures below room temperature, while hard segments can be considered as having glass transition temperatures above room temperature or even melting points, if a crystallite. It is the current opinion and, hence their classification, that soft-segment prepolymers are associated with the formation of the soft-segment phase of the elastomer and, conversely, hard-segment prepolymers with the hard-segment phase of the elastomer. Structure-property relationships of hard- and soft-segment phases are described, for example, by Redman in Ref. 11. The distinction of the prepolymer type on the basis of molecular weight of the polyol used in the preparation of the prepolymer is arbitrary, but such prepolymers obtained from diols or triols having an equivalent weight of about 150 or less generally are considered to be hard-segment prepolymers.

The use of hard-segment prepolymers, when preparing polyether polyol-based polyurethane polymer, restricts and makes it difficult to substitute or eventually replace all of the physical blowing agent with, for example, water. If water is used as the principal blowing means, the physical properties of the resulting polymer (especially flexibility, abrasion resistance, and hardness) deteriorate. Additionally, processing becomes noticeably inferior with, for example, increased demold times of molded articles. Due to current environmental concern relating to the Earth's atmosphere and ozone levels, it is highly desirable to substitute certain physical blowing agents with alternatives. Particularly, it is desirable to contemplate the use of water as an alternative blowing means. It is therefore desirable to develop a new process for the preparation of polyurethane polymers, particularly microcellular elastomeric polyurethane polymers, which allows for the use of water as principal blowing agent. To this purpose, we have investigated the use of soft-segment prepolymers in the preparation of polyether polyol-based polyurethane polymers.

Use of soft-segment prepolymers in the preparation of polyurethane foam is known from patent publications such as E.P. 22,617 and E.P. 398,304, while use of soft-segment prepolymers in the preparation of elastomeric polyurethane polymers is disclosed in, for example, the patent publications U.S. Pat. 4,190,711, U.S. Pat. 4,532,316, U.S. Pat.4,559,366, and U.S. Pat. 4,374,210. In U.S. Pat. 4,321,333, an isocyanate-terminated prepolymer obtained by reaction of a polyisocyanate with a polyol blend, containing from 15 to 70 weight percent of a polyol having a molecular weight of from at least 1000 and from 85 to 30 weight percent of a polyol having a molecular weight of from 115 to 300, is described. This amount of low-molecular-weight polyol confers hard-segment characteristics to the prepolymer. While such a prepolymer can be used to prepare elastomeric polyurethane polymers in the presence of water, the resulting polymer does not exhibit sufficiently attractive physical properties to meet present commercial demands.

It is therefore desirable to develop alternative, or modified, types of soft-segment isocyanate-terminated prepolymer. It is further desirable that such soft-segment prepolymer can permit the use of water as the principal blowing means and provide for resulting polymers with attractive processing and physical properties.

To overcome these difficulties, Broos et al.[12] disclosed an invention related to a process of preparing polyurethane polymers by reaction of an isocyanate composition having an isocyanate equivalent weight of from 180 to 300 and comprising in from at least 50 weight percent a prepolymer, with an active hydrogen-containing substance in the presence of water which provides for at least 50 mole percent of the total blowing requirement. The prepolymer is obtained by reacting a molar excess of a polyisocyanate (comprising at least 70 weight percent 4,4'-methylene diphenylisocyanate) with an isocyanate-reactive composition containing (a) a branched diol or triol having a molecular weight of from 60 to 300, and (b) a polyoxyalkylene polyol or mixtures thereof having an average functionality of from 2 to 4 and a molecular weight of from 3000 to 12000, wherein the parts by weight ratio of (a):(b) is from 0.01:1 to 0.25:1.

11.8.8   Thermosetting Polyurethane Compositions

Thermoplastic polyurethane materials have many uses as fabricated articles because of their physical properties and their ease of processing. However, as thermoplastic materials, their use at high temperatures is often precluded or limited, they have a tendency to permanent set, and they may be adversely affected by certain solvents. Stable polyurethane elastomeric compositions that can be vulcanized by free radical curing agents, included during or after processing, would be of value in that the disadvantages mentioned above would be at least partially offset.

Schollenberger and Dingsberg[13] disclosed improvements in thermoset or vulcanized polyurethane elastomers obtained by adding to liquid polyurethane reactants prior to reaction to form the polyurethane, organic peroxides having a half-life value of greater than 1 hr at 100° C. Such mixtures may be heated to form the polyurethane article and thereafter thermoset or vulcanized, or the polyurethane containing the unactivated peroxide may be provided in sheet, crumbs, granules, or otherwise, and then formed and heated to thermoset or vulcanize the polyurethane.

Polyurethane elastomers that may be treated as thermoplastic materials during processing are provided in the Schollenberger and Dingsberg invention for use by fabricators in a state of subdivision such as crumbs or granules. This allows easy fabrication into articles by conventional thermoplastic polymer processing operations, which on attainment of a predetermined temperature, activates a peroxide curing agent sufficient to provide the desired degree of vulcanization or cross-linking in the fabricated part. Likewise, liquid polyurethane formulations are provided that may be cast into articles and cured at that time or later. In accordance with this invention, polyurethanes including polyesterurethanes, polylactoneurethanes, polyetherurethanes, polyhydrocarbonurethanes, and the like, containing predetermined amounts of free radical curing organic peroxides having a decomposition temperature greater than the temperature of formation of the polyurethane, are prepared by dissolving the defined peroxide in liquid reactants as the molten macroglycol used to prepare the polyurethane. This technique offers a substantial advantage over conventional methods of adding curing or cross-linking agents by compounding the polyurethane just prior to forming into articles, since the excessive heat often developed limits the peroxides that may be used as curing agents, and it may degrade the polymers.

11.8.9 Use of Diamine to Form Thermoplastics

Polyurethane block copolymers possess an outstanding balance of physical and mechanical properties and superior blood compatibility compared to other polymers such as silicone rubber, polyethylene, polyvinyl chloride, and perfluorinated polymers. As a result, they have come to the forefront as the preferred polymeric biomaterials for fabrication of various medical device components. Some important device applications for polyurethanes include peripheral and central venous catheters, coatings for heart pacemaker leads, and the Jarvik heart.

Polyurethanes are synthesized from three basic components: a polyisocyanate, a polyglycol, and an extender—usually a low-molecular-weight diol, diamine, aminoalcohol, or water. If the extender is a diol, the polyurethane consists entirely of urethane linkages. If the extender is water, aminoalcohol, or a diamine, both urethane and urea linkages are present, and the polyurethane is more accurately and conventionally termed a *polyurethaneurea*. In this disclosure, polyurethaneurea will hereinafter be abbreviated as PUU.

The usual polyglycols are polyethylene glycol (PEG) and polytetramethylene ether glycol (PTMEG). Polypropylene ether glycol (PPG), while providing a polyurethane of desirable high softness, is infrequently used for polyurethanes intended for medical use, because PPG requires a catalyst for reaction with isocyanates. The usual catalysts for polyurethane synthesis, such as dibutyl tin dilaurate, are toxic and contraindicated for medical grade polyurethane synthesis because of the danger of leaching into a patient's body fluid.

Polyurethanes and PUU develop microdomains conventionally termed *hard segments* and *soft segments,* and as a result are often referred to as *segmented* polyurethanes. The hard segments form by localization of the portions of the polymer molecules, which include the isocyanate and extender components and are generally of high crystallinity. The soft segments form from the polyether glycol portions of the polymer chains and generally are either noncrystalline or of low crystallinity. Crystallinity and hard-segment content are factors that contribute to melt processibility.

It is known that PEG is clear viscous liquid at molecular weights below about 900 and is an opaque white solid of increasing hardness as the molecular weight increases above 900. PPG is essentially noncrystalline regardless of its molecular weight, whereas PTMEG develops some crystallinity at higher molecular weight. With PTMEG, the normal chain mobility of the soft segment is decreased as the level of crystallinity increases due to the infusion of crystallites of the soft segment into the hard segment. This, in turn, affects the elastomeric character of the polymer. Nevertheless, polyurethanes made from PTMEG are generally melt processible, but catheters extruded therefrom are less flexible than catheters fabricated from PEG and PPG.

PPG, being totally noncrystalline, gives a polyurethane having maximum phase separation between the hard and soft segments. As a result, PPG derived polyurethanes are soft and elastomeric, and the softness is affected by small changes in temperature. Thus, at body temperature, a typical PPG polymer is about 75 to 90% softer than at room temperature, as compared to a 60 to 75% change shown by a typical PTMEG derived polyurethane.

As is well known in the art, PUU made with diamine extenders is generally not melt processible, regardless of the polyglycol used as the soft segment. For example, a PUU well known as an industrial fiber (Lycra,™ DuPont de Nemours and Co.) has been extensively studied under the trade name Biomer (Ethicon Corp.) for fabrication of various biomedical devices. A review of these studies and the many salubrious properties of PUU has been presented by Phillips et al.[14] However, as stated by Phillips et al., Biomer™ presents some fabrication difficulties that limit production techniques. Biomer has a melt temperature higher than the decomposition temperature of the urethane functionality and therefore can be spun or cast only from solution; i.e., it cannot be melt extruded or injection molded. Severe limitations are thereby imposed on its fabrication latitude. Furthermore, it is essentially insoluble in all solvents except DMAC which, of course, must be completely removed if the product is to be used in a biomedical article.

Onwumere and Solomon[15] disclosed melt-processible polyurethaneurea as prepared from a diisocyanate, a polyglycol, a diol chain extender, and an amine-terminated polyether. Water may be included as a reactant, and the polymer may contain an additive such as a radiopaque material and a coating of an antithrombogenic agent or an antimicrobial agent. The invention includes a one-pot bulk polymerization method for preparation of the polymers.

One aspect of the Onwumere and Solomon invention is a melt-processing PUU prepared by reaction of a polyisocyanate, a polyglycol, a chain extending diol, and an amine-terminated soft-segment component, preferably an amine terminated polyether. Preferred PUUs are prepared from a diisocyanate such as 4,4'-diphenylmethane diisocyanate (MDI) and a polyether glycol, such as PEG or PTMEG. The polyglycol component may include a silicone glycol. Preferred diol extenders are ethylene glycol and 1,4-butanediol (BDO). The preferred amine-terminated polyether is amine terminated polypropylene oxide. Water may be included in the reaction mixture and, in the form of moist air, may also serve to cure the polymer. The polymer may include various additives such as a radiopaque agent.

Another aspect of the Onwumere and Solomon invention is a catalyst-free method to prepare the PUU of the invention. The preferred method is a one-pot reaction in which all the components are combined with efficient stirring. An exotherm takes place during the polymerization reaction, after which the polymer may be transferred to a tray for spontaneous moisture cure at an appropriate temperature.

11.8.10   Melt Processible Polyurethaneurea Copolymers

Polyurethanes develop microdomains conventionally termed *hard segments* and *soft segments* and, as a result, are often referred to as *segmented* polyurethanes. The hard segments form by localization of the portions of the polymer molecules which include the isocyanate and extender components and are generally of high crystallinity. The soft segments form from the polyether glycol portions of the polymer chains and generally are either noncrystalline or of low crystallinity. One of the factors which determines the properties of the copolymer is the ratio of hard and soft segments.

Exemplary of important diol extended polyurethanes are Vialon™ (Becton Dickinson Polymer Research), Pellethane™ (Dow Chemical Co.), and Tecoflex™ (Thermedics Inc.). These products typically have good blood compatibility but, with the exception of Vialon™, generally require processing additives such as antioxidants and detackifiers, a potential disadvantage for use in biomedical articles. They are, however, thermoplastic and therefore may be melt extruded and injection molded.

Diol extended thermoplastic polyurethanes are conventionally manufactured by operationally simple and economical bulk or one-shot polymerization processes wherein all the ingredients are combined, mixed, and reacted. PUUs, although commercially prepared by a two-step procedure described below, have also been prepared by a one-shot continuous process using a catalyst (U.S. Pat. 3,642,964, to Rausch et al.). The catalyst, because it is generally inherently toxic, cannot be present in PUUs to be fabricated into biomedical articles.

The conventional two-step preparation of PUUs is generally carried out by reacting the isocyanate and macroglycol in a solvent to give a prepolymer followed by chain extension with the diamine or aminoalcohol. Exemplary of the two-step procedure is the disclosure of Gilding et al. in U.S. Pat. 4,062,834.

Several disadvantages are encountered in the two-step process. First, the process generally requires a solvent, usually toxic dimethylacetamide (DMAC). Second, as pointed out by Ward et al.,[16] even reagent grade solvents contain enough water as an impurity to hydrolyze a significant portion of the isocyanate groups to amine groups that react with other isocyanate residues to form urea linkages of different structures from those obtained from the diamine extender. The resulting mixed hard segments complicate the structure of the polymers and increase the likelihood of batch-to-batch variations in properties.

Taller et al., in Research Disclosure 12,823, December 1974, and Short et al., in U.S. Pat. 4,522,986, disclose PUU compositions prepared by the two-step prepolymer technique from a diisocyanate, a polyol, and monoethanolamine as extender.

Ward et al. (supra) disclose a new PUU formulation for biomedical use consisting of a blend of PUU and an additive surfactant polymer.

There is a need for a bulk polymerization method to prepare melt-processible PUU having the desirable properties of both diol extended and diamine extended.

To overcome these difficulties, Solon et al.[17] disclosed a melt-processible PUU prepared by reaction of a polyisocyanate, a macroglycol, a chain-extending diol, and a chain-extending diamine. Preferred PUUs are prepared from a diisocyanate such as 4,4'-diphenylmethane diisocyanate (MDI) and a polyether glycol, such as polyethylene glycol (PEG) or polytetramethyleneoxide glycol (PTMO). The macroglycol may be wholly or in part a silicone glycol. Preferred diol extenders are ethylene glycol (EG) and 1,4-butanediol (BDO). Preferred diamine extenders are ethylenediamine (EDA) and 2-methylpentamethylene diamine (MPMD). Water may be included in the reaction mixture and, in the form of moist air, may also serve to cure the polymer. The polymer may include various additives, such as a radiopaque agent.

Another aspect of the Solomon et al. invention is a method to prepare the PUU. The preferred method is a one-pot reaction in which all the components are combined with efficient stirring. An exotherm takes place during the polymerization reaction, after which the polymer may be transferred to a tray for spontaneous moisture cure at an appropriate temperature.

11.9 Hydrophilic Polyurethanes

Flexible polyurethane foams made with TDI (toluene diisocyanate) have been manufactured for many years, especially for cushion and mattress applications. However, for hydrophilic foams used in medical or personal care applications, it is desirable to replace the TDI in the foams with MDI (methylene diphenyl isocyanate) because of the high vapor pressure and relatively high toxicity of TDI, which require special precautionary measures during processing and use.

Furthermore, TDI-based foams can be weakened by hydrolysis during sterilization or storage in a wet package. For example, TDI-based hydrophilic foams can liquefy after a few minutes in a steam autoclave at 120° C. TDI-based hydrophilic foams also swell excessively when wet, such as on the order of more than 100% by volume.

Conventionally, polyurethane foams have been made from MDI. These foams are rigid or semirigid, because MDI imparts crystallinity. In British Pat. 874,430, flexible polyurethane foams are produced by reaction of polyether polyols with at least two hydroxyl groups and a polyisocyanate mixture consisting of diarylmethane diisocyanates and 5 to 10 percent by weight of a polyisocyanate having a functionality greater than 2 in the presence of a small amount of water. A catalyst can be used in optional embodiments. These foams have the disadvantages that they are not hydrophilic and are not made with sufficient quantities of water to allow transport of large amounts of fibers, fillers, antiseptics, or other water-dispersible components into foams used in medical or personal care applications. The term *hydrophilic* as used herein means that the foam product is able to absorb 15 to 20 times its weight of water. A further disadvantage is that, in the case of the optional catalyst, catalyst residue can remain, which is not desirable.

In U.S. Pat. 4,237,240, flexible MDI-based foams with high load-bearing and high energy-absorption capacities are made by reaction of diphenylmethane diisocyanates with polyester polyols or mixtures of polyester polyols and polyether polyols with a polyester polyol content of at least 60 percent by weight of the polyol mixture, and small amounts of water. As set forth in the claims, a catalyst is employed. These foams have the same drawbacks as those of the above-described British Pat. 874,430, including the undesirable catalyst residues in the foam. In addition, they require the use of the more expensive polyester polyols.

In British Pat. 1,209,058, flexible hydrophilic polyurethane foams can be made by reacting a polyiso-cyanate with polyether polyols that contain at least 10% by weight of a block copolymer of ethylene oxide capped with propylene oxide to obtain hydrolytic stability. The method requires using at least one divalent tin salt of a fatty acid and/or at least one tertiary amine as a catalyst. The foam products made by this method, while being hydrophilic, have the drawback of being made with only small amounts of water as well as requiring the use of block copolymers. Moreover, there is no information about the use of MDI, which is hydrophobic, to make hydrophilic foam products, and the resulting foam will contain undesirable catalyst residues.

Guthrie[18] disclosed a flexible hydrophilic MDI based polyurethane foam is produced by mixing together an aqueous phase, which can optionally contain reinforcing fibers and surfactants, and a resin phase comprising a prepolymer derived from a poly (oxy C2-4 alkylene) diol, an MDI containing isocyanate product having a functionality greater than 2.0 made of a mixture of MDI and isocyanate containing derivatives of MDI, and a polymeric poly (oxy C2-4 alkylene) polyol cross-linking agent having 3 or 4 hydroxyl equivalents per mole and a molecular weight of at least 500. The preferred polyol is Poly G176-120. The polymeric polyol cross-linking agent is present so the hydroxy equivalents constitute 5 to 35 mole percent of the total hydroxy equivalents in the diol and the polymeric polyol, while the ratio of the isocyanate equivalents to the total hydroxyl equivalents is in the range of 2.5 to 3.5. The isocyanate-containing product comprises less than 50% by weight of the prepolymer, and Isonate 143L is the preferred isocyanate product. Flexible foams are obtained that are water-absorbing for medical or personal care applications.

11.9.1 Hydrophilic Foam

In the commercial production of polyurethane foams, the chemistry of the aqueous two-stage (prepolymer) process is well known. A urethane prepolymer (the reaction product of an isocyanate and a polyol) is reacted with water to generate carbon dioxide. The carbon dioxide functions as the blowing agent while simultaneous chain extension and cross-linking cures the prepolymer into a polyurethane foam. The aqueous two-stage process has persisted as a foaming technique of significant commercial importance for over three decades.

A particular family of polyurethane prepolymers, derived from methylenediphenyl diisocyanate (MDI) and sold under the trademark HYPOL PLUS™, was developed by W. R. Grace & Company for use in the aqueous two-stage process of foam production. These prepolymers, and the aqueous two-step process foams produced therefrom, are disclosed in U.S. Pat. 4,365,025 to Murch et al., which discloses an isocyanate containing prepolymer in which the isocyanate is a mixture of MDI and polymeric forms of MDI. The prepolymer is foamed by mixing it with an approximately equal amount of water. The resultant flexible foams are characterized by greater hydrolytic stability than those foamed from tolylene diisocyanate (TDI) prepolymers, and the MDI-based foams may be made without the toxin/carcinogen hazards associated with residual TDI in the workplace.

The product is difficult to handle, however, due to the large volume of aqueous reactant necessary and due to the high speed of the reaction. Production equipment must be designed to accommodate the introduction of a substantial quantity of water, must be equipped to evaporate the unreacted portion of the aqueous component (requiring additional energy in the form of heat), and must be capable of producing foamed products larger than needed to accommodate the severe and uneven shrinking that occurs during evaporation. Such production equipment is expensive both to design and to use. In addition, because a maximum of a few minutes is available for fabrication between the time reactants first commingle and the tack-free cure of the final foam, the product is limited to such uses as mold casting and foamed-in-place operations, and it is wholly unsuited to such fabrication techniques as the extrusion and knife-coating processes. For example, none of the MDI-based prepolymers disclosed in U.S. Pat. 4,365,025, to Murch et al., can be mixed with water and fabricated into a thin foam sheet with a Gardner (or other suitable) knife: the foam rises and cures long before a thin sheet material can be cast.

Prior to the development of the prepolymers disclosed in Murch et al., a method was developed for avoiding the handling problems inherent in the two-stage prepolymer process. British Pat. 1,306,372, to Marlin et al., explains a method of mixing an isocyanate with a reactive hydrogen compound in the presence of an organosilicon surfactant, and frothing the mixture with pressurized inert gas. The organosilicon surfactant prevents the foam from curing by imparting chemical and structural stability to the froth until the foam is cured by heating. Only latent metal catalysts may be used in the Marlin et al. process, i.e., those tin and nickel and other metal catalysts for which organosilicon surfactants act as synergistic inhibitors at ambient temperatures without inhibiting catalytic activity at elevated temperatures. In addition to its other chemical properties, the organosilicon surfactant consistently imparts a hydrophobic character to the final cured product.

Accordingly, the Marlin et al. process is wholly unsuited for use in the preparation of hydrophilic foams from the Murch et al. prepolymers not only because the Marlin et al. method requires undesirable elevated-temperature curing and permits only limited uses of surfactants and catalysts, but because the method cannot yield a hydrophilic foam at all in the presence of organosilicon. A need thus remains for a method of producing an MDI-based flexible hydrophilic foam that is well suited for use in all types of fabrication techniques and applications and yet avoids the disadvantages of the aqueous two-stage prepolymer process and the formulational and end-product limitations of the Marlin et al. technique.

Without the need for aqueous reactants or organosilicon surfactants and the limitations they impose, Bowditch et al.[19] disclosed an invention which provides a method for preparing a hydrophilic foam by blowing an MDI-based prepolymer with a substantially nonaqueous blowing agent, such as pressurized air, and polymerizing the prepolymer with a polyoxyethylene polyol having at least two hydroxyl equivalents per mole. The present method permits the hydrophilic foam to be extruded, knife-coated, or cast into sheets, as well as to be fabricated by other known foam preparation techniques, and thus is suitable for use in any flexible foam operation. Hydrophilicity is controlled with an appropriate non-organosilicon surfactant if a surfactant is used at all, and cure may proceed at ambient or elevated temperatures as desired. In addition, because the foam is prepared with polyoxyethylene polyol instead of water, the foam exhibits both superior drape and improved stretch and recovery as compared with prior art MDI-based aqueous two-stage flexible foams.

The Bowditch invention relates generally to and easy-to-handle system for preparing improved hydrophilic flexible foams that demonstrate superior drape, stretch, and recovery. The foams of the invention are particularly suited for use in external biomedical applications.

11.9.2   Catheter Tubing of Controlled In Vivo Softening

Catheterization procedures conventionally include puncture of a patients skin and insertion of a catheter into a body cavity, such as the blood stream, using some type of catheter insertion device. For patient comfort, it is highly desirable that the catheter be of the smallest possible cross-sectional area during insertion. It is nevertheless evident that the catheter lumen must be large enough to achieve the required rate of administration of a medicament solution or drainage of a body fluid through the catheter.

A number of polymeric materials have been investigated for fabrication of catheter tubing. Silicone rubber has been used, but this material, which is soft and pliable, requires inclusion of various additives such as fillers and plasticizers to give sufficient tensile strength. The thick wall needed to prevent collapse due to the pliability requires a large outside diameter to achieve sufficient inside diameter for fluid flow.

Other catheters of the prior art have been made of rigid substantially inflexible polymeric materials. Exemplary of such conventional catheters are the catheters of fluorinated ethylene propylene copolymer (FEP) having stripes of FEP containing a radiopaque agent disclosed by Coneys in U.S. Pat. 4,657,024.

Ostoich, in U.S. Pat. 4,211,741, discloses a catheter having a thin layer of polyurethane laminated on either or both surfaces of a thick polyvinyl chloride layer.

Recently, hydrophilic polymers that absorb water, often termed *hydrogels*, have been disclosed. U.S. Pat. 4,668,221, to Luther, discloses a catheter made of hydrophilic polymer that fits over a stylet for insertion. This catheter, on contact with blood, swells and softens so that the stylet can be removed.

U.S. Pat. 4,883,699, to Aniuk et al., discloses a tubing having a nonhydrophilic polyurethane component and a hydrophilic polyvinyl alcohol component. The tubing is said to absorb water and swell while retaining tensile strength.

Polyurethanes that swell and soften in contact with a body fluid have been disclosed in recent years as an attractive material for catheters. Gould et al., in U.S. Pat. 4,454,309, discloses hydrophilic polyurethane diacrylate compositions that swell in water and may be molded and cured to form shaped products. U.S. Pats. 4,728,322 and 4,781,703, to Walker et al., unveil catheters fabricated of a composition that includes a nonhydrophilic first component and a hydrophilic polyurethane diacrylate second component. When contacted with a liquid, the composition swells and softens due to absorption of the liquid, causing the catheter to increase in cross-sectional area.

Polyurethanes as a class have several advantages as materials for catheters. In general, they have excellent blood compatibility. In addition, they absorb water, soften, and thereby become more pliable. Pliability is a distinct aid in threading a catheter through a tortuous blood vessel to a desired placement.

While a significant improvement in catheter performance has resulted from fabrication using polyurethane, there remains a need for a catheter having the blood compatibility, softness, and pliability of polyurethane that retains sufficient mechanical strength and stiffness for ease of insertion and repositioning if desired. The present invention addresses this need.

Lee et al.[20] disclosed a catheter tubing comprising a stripe or layer of a thermoplastic hydrophobic stiffening polymer encapsulated by a thermoplastic hydrophilic polyurethane base polymer. Preferred stiffening polymers are polyesters, preferably polyester-polyether block copolymers. Preferred polyurethanes are polyetherurethanes. The catheter of the invention may be fabricated by coextrusion, and may include a radiopaque agent.

The catheter of the Lee et al. invention retains the inherent softness and biocompatibility of polyurethane catheters but overcomes the disadvantage of excessive pliability, which may cause difficulty in repositioning conventional polyetherurethane catheters after emplacement in a blood vessel. The degree of stiffening may be controlled by the stripe or layer configuration as well as by the composition and amount of the hydrophobic stiffening polymer. Because the stiffening polymer is completely encapsulated, the blood contacts only the hemocompatible polyurethane surface. If desired, the catheter having a predetermined balance of stiffness for insertion and repositioning and pliability for threading through a blood vessel may include a radiopaque agent as a visualizing aid in emplacement and/or repositioning. The thermoplastic nature of both the stiffening and base polymers retains the economy of manufacture realized by conventional catheter design and extrusion processing.

11.9.3   Hydrophilic Polyurethane

There has been a need for lubricious materials that are not slippery when dry but exhibit lubricious or slippery properties when contacted with aqueous fluids. Applications for such materials are numerous and include fabrication into surface coatings, foams, fibers, films, or solid articles that absorb water, impart wettability, or reduce the coefficient of friction in aqueous environments.

Reaction products of polyvinylpyrrolidones and polyisocyanates are known (U.S. Pat. 3,216,983). It is known to provide hydrophilic coatings that have low coefficients of friction. Such coatings include polyvinylpyrrolidones (U.S. Pat. 4,589,873), polyvinylpyrrolidone-polyurethane interpolymers (U.S. Pats. 4,100,309 and 4,119,094), and mixtures of hydrophilic polyurethanes and polyvinyl pyrrolidones (U.S. Pat. 4,835,003 and published European Patent Application 396,431). Another method of providing low coefficient of friction hydrophilic coatings is to first apply a coating of an isocyanate group containing material to the surface to be coated, and to thereafter apply a coating of a solution containing a polyvinylpyrrolidone (U.S. Pats. 4,585,666, 4,666,437, and 4,729,914). It is also known to prepare hydrophilic, flexible, open-cell polyurethane-poly vinylpyrrolidone interpolymer foams (U.S. Pat. 4,550,126). Blends of thermoplastic polyurethanes and polyvinylpyrrolidones are also known and are described as exhibiting reduced coefficients of friction when wet (U.S. Pat. 4,642,267).

Sarpeshkar et al.[21] disclosed an invention directed to a hydrophilic composition that can absorb more than 50% by weight of water and has a low coefficient of friction when wetted with an aqueous liquid. The invention is also directed to a mixture of the hydrophilic composition with a hydrophobic material.

References

1. Pandya, M.V., Deshpande, D.D., and Hundiwale, D.G. Effect of diisocyanate structure on viscoelastic, thermal, mechanical and electrical properties of cast polyurethanes. *J. Appl. Polym. Sci.*, 32:4959–4966, 1986.

2. Lin, I.S., Biranowski, J., and Lorenz, D.H. Comparison of diol and diamine cross-linkers in castable urethane elastomers, *Adv. Urethane Sci. Tech.*, 8:105, 1981.

3. Bonart, R. Segmentierte polyurethane, *Angew. Makromolek. Chem.*, 58/59, 259, 1977.

4. Zdrahala, Richard J., and Spielvogel, David. U.S. Patent 4,647,643, Mar. 3, 1987, "Soft Non-blocking polyurethanes," to Becton, Dickinson and Company, Franklin Lakes, New Jersey.

5. Siggia. S., *Quantitative Organic Analysis via Functional Groups,* John Wiley & Sons, New York, 1963, p. 255.

6. Ehrlich, Benjamin S., and Smith, Curtis P. U.S. Pat. 4,169,196, Sep. 25, 1979, "Process for improving thermoplastic polyurethanes," to Dow Chemical Company, Midland, Michigan.

7. Goldwasser, David J., and Onder, Kemal. U.S. Patent 4,376,834, Mar. 15, 1983, "Polyurethane prepared by reaction of an organic polyisocyanate, a chain extender and an isocyanate-reactive material of MW 500–20,000 characterized by the use of only 2–25 percent by weight of the latter material," to The Upjohn Company, Kalamazoo, Michigan.

8. Bonk, Henry W., and Shah, Tilak M. U.S. Pat. 4,202,957, May 13, 1980, "Thermoplastic Polyurethane elastomers from polyoxypropylene polyoxyethylene block copolymers," to The Upjohn Company, Kalamazoo, Michigan.

9. Quay, Jeffrey R., and Casey, Jeremiah P. U.S. Pat. 4,892,920, Jan. 9, 1990, "Process for the preparation of cyclohexanediisocyanate-containing polyisocyanate prepolymers and polyurethanes having high temperature performance," to Air Products and Chemicals, Inc.

10. Pudleiner, Heinz, Hugl, Herbert, Dhein, Rolf, and Muller, Hanns-Peter. U.S. Pat. 5,430,121, Jul. 4, 1995, "Thermoplastic polyurethanes modified by siloxane block copolymers."

11. Buist, J.M., ed. *Developments in Polyurethanes-I.* London: Elsevier, 1978.

12. Broos, Rene, Paap, Frans, and Maccari, Bruno. U.S. Pat. 5,418,259, May 23, 1995. "Process for preparing polyurethane elastomer from a soft-segment isocyanate-terminated prepolymer," to The Dow Chemical Company, Midland, Michigan.

13. Schollenberger, Charles S., and Dinbergs, Kornelius, U.S. Pat. 4,255,552, Mar. 10, 1981, "Thermosetting polyurethane compositions," to The B.F. Goodrich Company, Akron, Ohio.

14. Phillips et al., The use of segmented polyurethane in ventricular assist devices and artificial hearts, in *Synthetic Biomedical Polymers,* M. Szycher and W. J. Robinson, eds. Westport, Connecticut: Technomic Publishing Co., 1980, p. 39.

15. Onwumere, Fidelis C., and Solomon, Donald D. U.S. Pat. 4,948,860, Oct. 5, 1993, "Melt processible polyurethaneurea copolymers and method their preparation," to Becton, Dickinson and Company, Franklin Lakes, New Jersey.

16. Ward et al. *Polyurethanes in Biomedical Engineering,* H. Planck, G. Egbers, and I. Syre, eds., Amsterdam: Elsevier Science Publishers B.V., 1984.

17. Solomon, Donald D., Walder, Anthony J., and Hu, Can B. U.S. Pat. 4,948,860, Aug. 14, 1990, "Melt processible polyurethaneurea copolymers and method for their preparation," to Becton, Dickinson and Company, Franklin Lakes, New Jersey.

18. Guthrie, James L., U.S. Pat. 4,384,051, May 17, 1983, "Flexible polyurethane foam based on MDI," to W. R. Grace & Co., New York.

19. Bowditch, W. Raymond, and Rybalka, Borys, U.S. Pat. 4,644,018, Feb. 17, 1987, "Hydrophilic foam," to Norwood Industries, Inc., Malvern, Pennsylvania.

20. Lee, Min-Shiu, Karakelle, Mutlu, Spielvogel, David E., and Taller, Robert A. U.S. Pat. 5,226,899, Jul. 13, 1993, "Catheter tubing of controlled in vivo softening," to Becton, Dickinson and Company, Franklin Lakes, New Jersey.

21. Sarpeshkar, Ashok M., Markusch, Peter H., and Gracik, Charles S. U.S. Pat. 5,283,298, Feb. 1, 1994, "Hydrophilic polyurethane compositions," to Miles Inc., Pittsburgh, Pennsylvania.

12

Reaction Injection Molding

12.1 Background

Reaction injection molding (RIM) has grown in popularity because of its advantages: high flexural modulus, light weight, and low tooling costs. The concept of structural RIM sites is finding use in such activities as prototyping and identification of hundreds of parts for both automotive and nonautomotive applications. Parts vary in size up to and including complete pickup truck beds.

RIM and structural reaction injection molding (SRIM) are popular manufacturing processes because of the variety of end products they produce. Although several materials can be used in RIM and SRIM, the dominant one is polyurethane.

RIM and SRIM parts are the result of a volatile chemical reaction. Compounds containing active hydrogens, namely alcohols in the form of polyols, react with isocyanates in an exothermic reaction to form a polyurethane. Polyol and isocyanate are the basic materials employed. The isocyanates used are either diphenyl methane 4,4-diisocyanate (MDI) or toluene diisocyanate (TDI), MDI being the more common. Polyols include polyesters or polyethers, with polyesters being the more common. In addition to base chemicals, additives such as catalysts, surfactants, and blowing agents are incorporated to propagate the reaction and form a finished product with the desired characteristics.

RIM is an advanced method of molding articles from reactive liquid urethane components at production rates competitive with conventional injection molding. The polyurethane RIM process involves the rapid injection of reactive components into a closed mold through a self-cleaning mixing head mounted directly on the mold.

12.2 Markets

The properties of a final product depend on the combination and percentages of base chemicals and additives used. For example, auto industry parts fabricated by the RIM process are as diverse as steering wheels, horn pad covers, airbag covers, instrument panels, door panels, seat cushions, arm rests, head rests, headliners, floor pans, package trays, body panels, bumpers, and wheel coverings. Another polyurethane market is the rigid segment. Rigid products include household refrigerators, refrigerated tractor trailers, picnic coolers, softballs, water skis, decorative molding, water heaters, and insulation panels.

12.3 Advantages

The advantages of RIM over conventional injection methods are summarized below:

- *Large part production.* Parts weighing up to about 25 lb can be made with standard machinery. Larger parts involve special high-output machines with a marginal increase in capital costs.

- *Low energy use.* The polyurethane RIM process is more energy efficient than competing molding processes.
- *Low equipment costs.* Even for large parts, the process needs only relatively light molds and low clamping pressures, and lightweight mold presses are used. As a result, RIM equipment is less expensive than equivalent thermoplastic injection molding equipment.
- *RIM elastomers* can be produced over a wide range of stiffness.
- *Thin sections* can be produced to meet part performance specifications at optimal material cost.
- *Paintable surface finish.* RIM moldings can produce automotive Class "A" surface finishes.
- *Intricate parts* are routinely produced. The low viscosity of the reaction mixture, coupled the foam expansion, ensures good contact in geometrically complex molds with intricate textured surfaces.
- *Minimal sink marks.* Compared to thermoplastic injection moldings, RIM moldings have a lower tendency to show surface defects around strengthening ribs and inserts.
- *Inserts may be incorporated* during RIM molding. Inserts placed in the mold are wetted by the low-viscosity liquid reaction mixture and become an integral part of the molded article.
- *Stress-free* articles can be produced.
- *Economic production.* This is the result of high production rates and low energy consumption related to RIM.
- *In-mold painting* yields a paint finish that is covalently bonded to the RIM part.

12.4 The RIM Process

A typical polyurethane processing system incorporates a set of tanks that house the liquid chemicals. The chemicals are gravity fed to the metering device, which, in turn, forces the chemicals to the mix head. Along this path, the materials can be processed through filters, pressure gages, and flow meters. Inside the mix head, the material is either mixed and dispensed, or it is recirculated back to the day tanks. As the material is circulated back to the day tank, it usually passes through a heat exchanger where it is temperature conditioned.

Inside the mix head, the two chemical streams collide violently with each other under high pressure, generally 2000 to 3000 psi. This form of mixing is known as *impingement*. Once the streams collide, the flow is very turbulent, and the reaction begins. The stream exits the mix head and is deposited into a mold. At the conclusion of a pour, a piston inside the mix head scrapes the walls of the chambers completely clean so that no reacted foam is left inside.

The "straight-through" mix head was used when polyurethane processing was first introduced. *Straight-through* implies that the chamber inside the mix head is straight rather than bent, as is the case in the L-shaped mix head. The two chemicals collide and exit via a straight path from the chamber. These mix heads are still available, but many processors prefer the features offered by L-shaped mix heads, such as laminar flow to exit the head and an aftermixing action that can be built into the mix head rather than requiring it to be built into the mold.

Key features of a mix head are size and recirculation capabilities. Reduced size offers an advantage in that it is lighter and easier to embed into a clamp—a necessity with SRIM. The ability to recirculate the chemicals through the mix head ensures that both cleaning and temperature conditioning have occurred at the start of pouring. It is important to ensure that the chemicals are continuously maintained at the proper temperature, otherwise the viscosity (and therefore, the throughput) will change, resulting in bad product.

RIM is used to produce a wide range of integral skin parts in flexible, rigid, and semirigid foam, as well as high-density polyurethane parts. A wide range of machinery now available enables RIM to be used to make moldings of all sizes from a few grams to several pounds, depending on the capacity of the metering units. Car bumpers and other external parts for vehicles are being produced with part weights up to 25 lb. The development of systems for stiff, impact-resistant RIM polyurethanes has resulted in lighter-weight auto bumpers and other exterior parts with a high resistance to damage. Polyurethane RIM processes give economical production with short demold times. The minimum demold time is a function of the size, thickness, and complexity of the part and of the design of the mold, since the maximum usable reaction rate of the polyurethane system is governed by the need to obtain a smooth flow of the reaction mixture into the mold.

Thus, the quality of the final product limits the speed at which the mold can be filled satisfactorily. The reaction rate of the polyurethane system may be adjusted to yield the minimum demold time consistent with the speed of filling. Large elastomeric polyurethane RIM parts having thin sections are typically demolded in 30 to 60 s. The use of amine-terminated polyethers and aromatic diamine chain-extenders in high-output machines yields polyurea RIM parts that are demolded in approximately 10 s, depending on their size and shape.

While the MDI-based RIM process is comparable with thermoplastic injection molding in the rate of production and the degree of automation that is possible, RIM requires a relatively low equipment investment. The cost advantage is greatest for large parts, since RIM is performed at low mold-clamping pressures. The surface finish of the RIM molding will provide a negative replica of the mold surface, thus a wide range of surface finishes may be obtained. The thickness of the RIM molding can be varied with minimal risk of surface sink marks.

Additional applications include reinforced reaction injection molding (RRIM) where short fibers and/or flaked glass are incorporated into the reactive mixture, resulting in higher stiffness, dimensional stability, and reduced coefficient of linear thermal expansion.

12.5 RIM Equipment

Structural RIM equipment usually has a preformer, clamp, metering unit, and a mix head. These subjects are discussed below.

12.5.1 Preformer

Many types of preformers are available. They include thermoforming, fiber-directed, knitting, braiding, or a combination of two or more of these processes. The thermoformable mobile heater system was the first generation of preformers and is still readily available today from several manufacturers. In operation, the clamp rapidly closes to form the mat. Once the mat is formed, it is transferred via the side grippers and released. The side grippers then return to the loading station for the next mat.

The main disadvantage of this system is that, when heating occurs in the clamp station, the clamp stays ineffective during the heating stage. Another problem is the length of time it takes to transfer the IR heaters from the clamp to the side position, resulting in mat cooling, and mechanical shock to the IR heaters.

The high-speed thermoformable is a second-generation preformer that uses a stationary heater bank. This permits mat heating while another mat is being formed and unloaded, increasing the efficiency of preformer. This type of preformer offers the latest in technology for preformable, continuous strand mat, and provides the fastest cycle times available. It has five stations:

1. The glass roll station, which can contain 12 or more rolls and can be adjusted to fit rolls of different widths.
2. The loading station, which also can be adjusted to different widths. Mats are on a steel bank table to an approximate length.
3. A heating station. Most preformers have in-line upper and lower IR heaters. The lower banks have metal wires to prevent s fibers from contacting the heaters.
4. The clamp, with a 50-ton closing force, fast-acting hydraulic clamp. The platen size is built to match the maximum size.
5. The unloading station. Unloading is via a one- or two-station table. It is completely automatic.

After the glass mat is cut, it is transferred by the side gripper pins under the IR heaters. After heating, the same grippers rapidly transfer the glass mat to the clamp, simultaneously moving the unload support table. The gripper releases the mat while the part jig frame closes to hold the mat against the unload support table. The clamp upper platen then closes to form the part. The transfer of the mat from the IR heater to the clamp will take about two to four seconds.

The part jig frame is unique to each part and attaches to the clamp platens. The part frame uses the "hold-slip" function for the "flow control" of the mat during forming. When the side gripper releases the heated mat in the clamp station, the side gripper returns to the loading station where the next mat is loaded under the IR heater. Heating occurs while the previous preform is cooling and unloading. After cooling, the upper mold, lower mold, and part frame open. The moving support table then rapidly shuttles out of the clamp station to allow unloading.

An example is a preformer that can make parts as small as 1000×800 mm to as large as 2200×2800 mm. However, a disadvantage of all thermoformable glass mat preformers is the cost of the glass mat and the extreme amount of glass waste on some parts.

The average cost of continuous strand mat is about $1.45/lb. The cost of the final preform is about $2.00/lb. This has led some manufacturers to look at alternatives to manufactured preforms such as fiber-directed preforms that have a bulk roving price of $0.85/lb and a final preform cost of about $1.00/lb.

12.5.2 Clamps

Ideally, a conventional two-cylinder or four-poster clamp with 65/45° upper/lower booking would be best for processing. Unfortunately, some parts are too large for this type of clamp. As a rule of thumb, any part needing a platen surface less than 50×100 in can use a conventional clamp. Any part requiring a larger platen surface needs a shuttle bed RIM clamp for production use.

The shuttle bed RIM clamp has proved ideal for RIM processing because of its ability to book the upper platen 90° for full operator access to the mold. Platen size for a rim clamp is designed for 10×8 ft maximum, with tonnages as high as 600 tons. The clamp can handle total mold weight of 40 tons. The clamp is designed using four cylinder strokes for high-pressure closing via the lower platen. The booking of the upper platen is obtained by two hydraulic cylinders. The bottom shuttle platen is operated by two electrically driven pinions engaging two gear racks fitted on the lower platen's bottom side.

The complete automatic cycle sequence is as follows:

1. The mold is closed.
2. The bottom platen lowers and engages to shuttle.
3. The bottom platen shuttles to service station.
4. The upper platen books 90° to vertical position.
5. The bottom platen shuttles back to clamp station.
6. The bottom platen rises, high tonnage occurs.

Clamps house and coordinate the operation of the molds. Clamps come in a variety of shapes and sizes. Most are custom built. A clamp must move smoothly through its operational sequences; incorrect movements can result in damage to an expensive mold, and improper sequencing can lead to poor product quality.

Nucleation systems inject air into the chemicals. They are used by processors who want to improve flowability of chemicals or who want to improve their final product cell structure. Improved flowability means that more material can be put through the same equipment under greater laminar conditions. Improved cell structure implies that there is a highly consistent, very fine pattern of cells throughout the part.

All turntables are round, which is approximately where their commonality ends. Each table is sized according to the production requirements of the customer. Turntables are used as a production solution when the cure time of the chemicals greatly exceeds the molding-cycle time. This allows optimal usage of the metering unit in supplying the molds. The latest trends have been toward caterpillar drives, which eliminate many of the problems previously incurred with drive systems, and toward robotic manipulation of the mix head or in-mold spray-coating operations, which increases consistency and eliminates human error.

12.5.3 The Metering Unit

The metering unit is responsible for measuring the chemicals and delivering the required amount to the mixing head. Whether the user plans to pour into a bucket, a prototype mold, or a production-scale mold clamp, a metering unit is needed to deliver the chemicals. While research engineers can start with a hand-mix or lab-scale metering unit, it is commonly believed that true results can only be obtained by working on a full-scale metering unit.

A key feature in a metering unit is the ability to process in high pressure to avoid the need to flush the mix head after each shot. A second key feature is modularity. The initial setup operation may be accomplished on a small, economical scale, but once production and sales pick up, the ability to increase the capabilities of the existing equipment is very important. Flow transducers, automatic day tank chemical level control, automatic filling, agitators, and other accessories can all be incorporated as the need arises.

Current trends, with respect to metering units, are heading in two opposite directions. One trend is toward more economical equipment, while the other trend is toward advanced electronics and closed-loop control for pump-type metering units. Economical units are capable of processing quality PU foam yet, at the same time, the economical machine may be limited in its ability to be upgraded as requirements increase. For example, the use of smaller tanks may limit the shot-size capability. Fortunately, as electronic advances are being made, prices of controls are coming down. For this reason, several units are able to incorporate more advanced control technology. On the other hand, advanced electronics and closed-loop control support the processor in his endeavors to enhance capabilities.

There has been some concern in past years regarding the need for specialized RIM metering units compared to standard polyurethane metering units. Depending on the part complexity and size, both can be used. For simple parts with little complexity, a standard pumping unit can be used, having standard positive displacement piston pumps capable of operating with unfilled material.

For higher degrees of complexity, a closed-loop control, variable output cylinder machine that can maintain a constant injection pressure is recommended. A machine of this type offers the manufacturer the ability to control the part injection pressure by controlling the output. Some manufacturers have done optimization injection rate profiles for specific parts to determine the best output injection profiles needed to fill a part.

12.5.4 Mix Head

The injection of the chemical into the mold to get a proper flow distribution is an important part of the RIM process. The RIM mix head must have laminar flow to provide good fill without moving the preform or trapping air. The mix head must be compact in size, allowing for easy center gating of the mold. If the metering unit can vary the output, and the mix head is inaccessible in the mold, the need exists to control the pressure of the injection nozzle. Automatic units are available to keep the pressure constant and allow even filling of the mold.

12.5.5 Mold Construction and Materials

Mold design is critically important in the production of RIM parts. The most important design considerations are discussed below.

The choice of mold material for construction depends on (a) number of parts, (b) the type and quality of the surface finish, (c) the production rate per mold, and(d) the need for future mold modifications. Metals are the most durable materials for mold construction and, because of their high thermal conductivity, are best for use with highly reactive RIM systems. Epoxy resin molds, or metal-filled epoxy molds, can be made strong enough for RIM, and they are easily modified. Their use is restricted to prototype work or small production runs. This is mainly because the surface quality and the ease and speed of release of RIM moldings is a direct function of the quality of the mold surface and the control of the mold surface temperature. Surface temperature is particularly important with high-modulus RIM systems, which produce high exothermic heat of reaction. This heat must be removed from the surface of the molding before it can be cleanly released. Epoxy resin mold surfaces are inferior to metal surfaces in quality, durability, and the ease of temperature control. The low durability of epoxy resin surfaces on heat cycling, combined with their low thermal conductivity, often results in mold-release problems.

Metal molds are made with steel, aluminum, zinc alloys, beryllium/copper, and nickel/copper shell construction. Machined tool steel, finished by polishing or by chromium or nickel plating, is the most durable mold material and gives good release. A highly finished mold surface will allow some polyurea RIM parts to be molded without the use of release agents to the mold surface. Steel molds used in mass production can be used in 100,000 to 200,000 demolding cycles. Machined, forged aluminum is less durable, but it gives good release and easier temperature control than other metals, with the advantage of light weight. Zinc alloy molds are heavy and, because of the relatively low thermal conductivity of the material, are usually cast around closely pitched heat exchanger pipes. They give good release and surface finish and, with reasonable care, have a useful life of >50,000 molding cycles. Zinc alloy molds are easier to repair or refurbish than other metal molds.

Cast aluminum molds are inexpensive but have been found to be unsatisfactory for RIM molding because of surface porosity and surface roughness. Cast aluminum, however, is satisfactory for RIM low-density cushioning foams and for other products that are covered before use. Detailed surface patterns are often reproduced by using electroformed copper-backed nickel shells, which are made by electroplating (or sometimes by vacuum depositing) nickel onto a master pattern. The thin shells, typically 50 to 90 mils of nickel backed with 4 mm of copper, are built into a GRP body, which is usually reinforced with a metal frame. This type of mold is used for the manufacture of low-density integral-skin RIM parts. Mold durability is good for 100,000 molding cycles. Other mold-making techniques include the use of sprayed metal shells backed by GRP or metal and the use of silicone rubbers and polyurethane elastomers. All three techniques will give an accurate reproduction of intricately patterned surfaces at low cost, but the mold surface durability is limited.

The life of these molds depends to some extent on the reactivity of the RIM chemical system, especially on the rate of surface cure. In general, sprayed metal, silicone rubbers, and polyurethane elastomers are not suitable for molding stiff RIM elastomers, and especially polyurea RIM, because the thermal shock from the high exotherm causes breakdown of the mold surface. Sprayed metal molds provide from several hundred to a few thousand self-skinning foam parts, depending to a large extent on the care taken in demolding and in cleaning the mold, as the sprayed metal surface is rather soft and easily damaged. Flexible self-skinning foams can be made in silicone rubber molds, lasting between 100 and 200 molding cycles. Mold linings cast from polyurethane elastomers are more durable than silicone and allow up to 1,000 parts to be made.

12.5.6 Temperature Control

Mold temperature controls product properties, surface quality, and mold release. Both halves of the mold should be constructed with tubes or ducts for the circulation of a heat transfer fluid, usually cooled water. Molds are typically filled at a temperature between 50 to 70° C, depending on the RIM system used, its filler content, and the thermal conductivity of the mold.

Temperatures above 70° C may be needed with some polyurea RIM systems. Integral-skin parts are usually made with mold temperatures in the range from 40 to 50° C. The surface temperature of the mold should be uniform and maintained within 2° C. The RIM process is exothermic; a typical system generates about 50 kcal/kg of material, which, especially in thin parts, is mostly conducted into the mold. Uniformity of mold cooling is therefore crucial to avoid the development of hot spots during production.

The rate of heat removal from the mold will influence the demold time and the rate of production. Fast, uniform cooling is especially important with polyurea RIM.

12.5.7 Mold Venting

Mold venting is essential to allow the air within the mold to escape as the mold is filled with the reaction mixture. Injection times vary from about one to several seconds, depending on the part size and the chemical system, but typical injection times for high-modulus RIM are between 1 and 2 s. Molds are best vented via the parting line at the highest point—remote from the filling gate. This method allows sufficient venting to avoid excessive air pressure in the mold and minimizes polyurethane flashing through the vent, because the reaction mixture has become a viscous polymer by the time it reaches the uppermost parting line. The split line near to the runner and film gate must remain sealed to contain the fast-flowing, low-viscosity reaction mixture during the injection phase.

12.5.8 Mold Release

A mold release agent is an interfacial coating applied between two surfaces that would normally stick together. It causes the surfaces to separate cleanly, smoothly and easily. Mold releases are needed for a variety of molding operations, such as castings, foam manufacture, injection molding, and, of course, RIM.

There is no universal type of release agent for all RIM applications. Molding conditions such as mold material, cycle time, temperature, and secondary finishing must be considered. The properties of the mold release must be matched to the molding conditions.

The mold release (sometimes called a *parting agent*) is best applied by spray coating the open RIM mold. The usual release agents are microcrystalline waxes dispersed in solvents. In addition to microcrystalline and/or polyethylene wax, some proprietary release agent compositions contain other additives including silicone oils. These are valuable in assisting the dispersion of a thin layer of release agent on the mold surface, thus preventing the polyurethane reaction mixture from wetting the surface of the mold. Silicone-based release agents are valuable in obtaining a glossy surface on pigmented moldings, but they should not be used when RIM parts are to be painted. This is because a small part of the release agent coating is transferred to the surface of the RIM molding. Wax-based release agents may be removed from the part before painting, by aqueous detergent washing or by a conventional degreasing solvent, but silicone oils are difficult to remove and will interfere with the paint coating.

12.6 Illustrative Patents

12.6.1 Modified Polyurea-Polyurethane Systems Endowed with Improved Processibility

Polyurethane polymers prepared from an active hydrogen-containing compound and an isocyanate are widely employed in molding processes, particularly RIM processes. RIM articles are finding increased usage as automotive fascia.

The basic polyurethane polymer systems are typically based on an OH polyol component, an OH cross-linker, and an isocyanate component. However, this system suffers from long cream, demolding, and cycle times, greatly increasing the processing time. Modifications to the basic polyurethane system to shorten these processing times have been achieved through substitution of the OH cross-linker with an aminic cross-linking system. Typically, thermosetting urethane polymer compositions comprise an isocyanate component with an excess of isocyanate groups and an aromatic diamine as a chain extender, to form a polyurea-type urethane polymer. Optionally, the polymer composition may also contain additional amounts of a reactive polyol to form a hybrid urea-urethane polymer. Such systems greatly decrease the cream and demolding times, therefore enabling much shorter cycling times in a RIM process.

The use of chain extenders, such as di-alkyl aromatic diamines and, more particularly, di-ethyltoluene diamines and di(alkylthio) aromatic diamines, are often used with isocyanate prepolymers alone or with a polyol component to form a polyurethane/polyurea molding RIM composition (for example U.S. Pats. 4,595,742, 4,631,298, and 4,786,656). While an increase in the flexural modulus is observed through the addition of di-alkyl(thio) aromatic diamines, these compositions are still limited with respect to flexural modulus without observing *cold break* on demolding. Cold break is a brittleness observed in the molded article during demolding. The presence of cold break causes the molded article to fracture on demolding. When trying to achieve a higher flexural modulus (to above 80,000 psi) by increasing the isocyanate content of the isocyanate component (i.e., higher percent NCO), these materials suffer cold break. Alternatively, attempts to increase the flexural modulus by increasing the functionality of the polyol component also result in cold break.

In addition to the mechanical properties of the polyurethane polymer, the processing of the polymer systems plays an important role in the usefulness of a polyurethane system. In RIM processing, a short gel time is desired to increase the productivity of the overall process. However, polyurethane systems based on an OH polyol component, an OH cross-linker, and an isocyanate component had a gel time of from 5 to 8 s and a cycle time of from 3 to 3.5 min. Polyurethane systems using -NH$_2$ cross-linkers and -OH polyols (for example U.S. Pats. 4,595,742, 4,631,298, and 4,786,656) reduced the gel time to about 1.2 s and the cycle time to 1.5 to 2 min. This greatly increased the productivity of RIM processes using these systems, but these -NH$_2$ cross-linkers and -OH polyols systems suffered from an inability to increase the flexural modulus above 80,000 psi without observing cold break, without the addition of fillers.

Polyurea systems based on amine terminated polyether resins and aminic cross-linkers have been developed (U.S. Pats. 4,396,729, 4,433,067, and 4,444,910), that afford superior heat resistance and mechanical properties. Due to the extremely high reactivity of the amine-terminated polyether resin, the gel times are extremely short, in the range of 0.7 s, and the demold times are very short. Such a rapid reaction rate makes these systems very difficult to manipulate and also severely limits the type of RIM technique for which such a composition is suitable.

One of the major problems encountered is the RIM processes in premature gelling of the composition, causing insufficient filling of the mold or limiting the size of the molded article due to the rapidity with which the system gels. In processing by RIM methods, it is necessary that the molding composition maintain a sufficiently low viscosity to completely fill the mold. After the mold is filled, the material must then polymerize very rapidly to reduce the demolding times. The two properties are at opposite ends, since to increase the demolding times by using a rapidly polymerizing system, the mold size has been limited. By decreasing the gel time to accommodate larger molded articles, the demolding times are increased, thereby decreasing productivity of the overall operation.

One solution to increase the mold size without decreasing the rate of polymerization is to increase the output rate of the RIM machine. In this way, more material can be injected into the mold over a short period of time, which allows for the formation of larger articles. However, there are increased costs in high-output RIM machines. Moreover, even this solution has its limitations and is still limited by the viscosity of the material and the rate at which it can be injected.

Another related problem associated with the longer gel times needed to accommodate larger molds is the production of molded articles that exhibit a rubbery feel upon demold due to insufficient curing of the composition. Even though the molded article has sufficient structural integrity to be demolded, the molded article requires further curing to obtain the desired stiffness of the final product. This post-demolding curing adds to the processing time.

Ideally, a polyurethane system that exhibited a low viscosity during the mold filling stage, yet rapidly polymerized and cured after the mold has been filled, is desirable. Such a system would exhibit a nonlinear increase in viscosity initially, followed by a very rapid increase in viscosity at the end.

The reactivity problem becomes more stringent in the case of structural reaction injection molding (SRIM). This technology is not applicable to such systems having short reaction time, resulting in incomplete wetting of the structural reinforcement. Consequently, research continues into systems with excellent mechanical and processing properties.

Smith[1] discovered that the addition of a carboxylic anhydride (I) provides a modified polyurea-polyurethane molding composition with excellent processability (in terms of very long gel times and very short demolding times,) to form articles with high impact strength, flexural modulus, and excellent green strength after demolding.

The process of forming modified polyurea-polyurethane compositions according to the Smith invention is accomplished by conventional RIM, RRIM, or SRIM techniques. The process also allows for spray molding. This includes both open and closed mold processes. This allows for a substitute spray molding composition that gives physical properties as good as or better than conventional polyester resin fiberglass article, yet it avoids the problem of volatile organic compounds. In addition, the composition can be applied using existing spray molding equipment.

In spray molding, it is preferable that the molded article be subjected to a post-mold curing at a temperature of above 200° F for 1 hr, preferably at 250° F for 1 hr. Alternatively, or in addition, it is possible to heat the mold to a temperature of 200 to 250° F to ensure a complete curing.

The invention also allows for the formation of an SRIM-molded article with an A-class surface. This is achieved by RIM molding or spraying a surface layer of the modified polyurea-polyurethane composition of the present invention on either or both surfaces of a mold to form the exterior surfaces, followed by forming an intermediate layer by conventional SRIM methods. The modified polyurea-polyurethane surface layer may be filled or unfilled with conventional fillers and additives. Conventional SRIM articles often suffer the drawback of a surface layer that reveals the fiberglass structural reinforcement. By forming a surface layer of modified polyurea-polyurethane, an A-class surface now can be obtained in an SRIM article. The surface layer of modified polyurea-polyurethane may be from 0.1 to 0.2 in thick, preferably 0.125 in. It is preferable for the intermediate layer be of the same polymer composition as the surface layers to avoid delamination problems due to differences in the coefficient of thermal expansion.

The Smith process also allows for molded articles to contain suitable fillers and reinforcing material and the like, stabilizers, mold release agents, and other additives known to those of ordinary skill in the art as commonly employed in the field of polyurethane molding compositions—especially RIM, RRIM, and SRIM molding compositions.

12.6.2 Method of Using Mold Release Agents

Polyurethane and other plastic parts are used extensively in the manufacture of automobiles, furniture, and industrial service parts and in home construction. In this connection, molds are used to produce the polyurethane article. One constant problem has been that the polyurethane part tends to stick to the mold at the conclusion of the molding operation. To facilitate the release of molded polyurethane articles from the mold, it has been known to coat the mold cavity with a mold release agent (an external mold release agent). As described in U.S. Pat. 4,585,829, external mold release agents have consisted of natural or synthetic compounds such as silicone oils, mineral oils, waxes, fatty acid derivatives, glycols, etc. Generally, these release agents are either dissolved or dispersed in a liquid diluent or carrier and are sprayed into the mold cavity. Such known external release agents require a reapplication of the release agent to the mold after each molding cycle or a limited number of molding cycles.

At the end of each molding cycle, the release agent should enable an easy release of the molded part and leave no visible surface defects on the molded part.

As described in U.S. Pat. 4,473,403, incorporated by reference, release agents have generally contained a blend of several ingredients, including active ingredients that cause the release action, a solvent or diluent, a film-former, a drying or curing agent, and added lubricant to enhance releasibility. Ingredients range from greases, natural waxes, soaps, emulsions, mineral oils, fats, lecithins, metal stearates, silicones, fluorocarbons, synthetic waxes, organo-phosphates, polyaliphatic alcohols, and polymeric resins. See also U.S. Pat. 4,491,607, which is incorporated by reference.

Traditional sacrificial release agents such as waxes and soaps provide molders with limited numbers of releases before recoating is required. One object of this invention is to provide a more durable release coating so that greater numbers of parts may be molded in a particular mold before the release agent must be reapplied.

Traditional sacrificial release agents such as soaps or metal stearates are dissolved, at least in part, in alcohols. These solvents are often flammable. Other objects of the invention are to provide a release coating with lower flammability to the plastic molding market using a solvent that preferably is non-flammable and with the release coating containing no chlorofluorocarbons, and to provide release agents with lower concentrations of volatile organic compounds. Current environmental requirements encourage or require minimization of volatile organic compounds in process aids such as release agents.

To overcome these deficiencies, mold release agents and methods for molding plastics are provided by the Payne and Martin invention.[2] Mold release agents are especially applicable to polyurethane-based RIM systems. The mold release agent includes the lithium, sodium, or potassium salt, or a mixture thereof, of an oligomer of monomeric units, or a mixture of said oligomers, the monomeric units being selected from the group consisting of carboxylic fatty acids having 14 to 24 carbon atoms, and the salt of the oligomer preferably being soluble in water or water/alcohol. Preferably, the Li and Na salts of C36 and C44 dicarboxylic fatty acids are used. Preferred embodiments of the present invention permit the use of water without alcohol in the formulation.

12.6.3 Polyurethane RIM Elastomers Obtained with Hydroxyl-Containing Organotin Catalysts

Polyurethane and polyurethaneurea RIM elastomers require the use of tin catalysts for acceptable processing. In a system based on high-molecular-weight polyether polyols and aromatic diamine chain extenders, it is customary to use dibutyl-tin dilaurate to catalyze the hydroxylisocyanate (OH-NCO) reaction. Without this catalyst additive, it is impossible to make a RIM part with sufficient green strength to be demolded. Yet, it is well known that the presence of tin catalyst in the final elastomer limits the thermal stability of the material, since it can catalyze the unzipping of urethane bonds. The need for high thermal stability in RIM elastomers is the result of high paint-baking temperatures. Physical distortion, cracking, oiling, and blistering can be observed if tin catalyst is present to promote the chemical breakdown of urethane linkages.

One solution to this problem has been the use of polyurea RIM based on isocyanate-reactive components that do not contain OH group, e.g., aliphatic amine terminated polyethers as disclosed in U.S. Pat. 4,433,067. Since the reaction of the amino (-NH$_2$) groups with NCO groups is so rapid, no tin catalyst is needed, although there is still some benefit to their use in polyurea RIM elastomers as disclosed in U.S. Pat. 4,444,910. However, the drawback to this solution is the high reactivity of the aliphatic amine-terminated polyethers, which makes it difficult to achieve high-modulus elastomers with good flowability. Additionally, these products are more expensive than conventional RIM polyols and are not available in "filled" versions such as polymer polyols, grafted polyols, or PHD polyols.

The present invention is a RIM elastomer comprising a cured reaction product of a polyether polyol of greater than 500 molecular weight, an aromatic diamine chain extender, and an aromatic polyisocyanate reacted in the presence of a hydroxyl-containing organotin catalyst, such as, for example, a diorganotin bis(hydroxyl-containing organo) compound. The invention is also a method for preparing a RIM polyurethane elastomer.

The use of a hydroxyl-containing organotin catalyst provides for improved thermal stability of conventional polyether polyol polyurethane and polyurethaneurea RIM systems. In addition, unfilled or filled polyols can be used, and high-modulus elastomers with improved flowability over polyurea RIM systems are obtained.

The RIM elastomer of the invention may be prepared from as few as four ingredients, namely a high-molecular-weight polyether polyol, an aromatic diamine chain extender, an aromatic polyisocyanate, and a hydroxyl-containing diorganotin catalyst.

The Dewhurst and Nichols[3] invention relates to reaction injection molded elastomers derived from high-molecular-weight polyether polyols, an aromatic diamine chain extender, a polyisocyanate, and a hydroxyl-containing organotin catalyst. The RIM elastomers of this invention are useful as automobile body parts.

12.6.4 Polyurethane-Polyisocyanurate Structural RIM Systems with Enhanced Processing

RIM systems produce elastomeric products containing polyurethane-polyurea linkages that have many uses, e.g., as automobile facias. However, such systems have relatively low heat distortion temperatures and lack the flex modulus and tensile strength necessary for many applications. The chemistry of these reactive systems involves the use of a polyisocyanate *A side* (A component) and a *B side* employing a mixture of compounds containing isocyanate-reactive hydrogens. These B-side components generally include one or more hydroxyl-functional polyether or polyester polyols and one or more sterically hindered diamines. The polyol components react with the isocyanate to form urethane linkages, while the amine components react to form urea linkages. Such systems are disclosed, for example, in Weber, U.S. Pat. 4,218,543.

The flex modulus and tensile strength of traditional RIM systems may be improved through the addition of short fiber reinforcement (RRIM). However, the addition of chopped fibers raises the viscosity of the system components as well as seriously affecting the ease of processing. Moreover, this addition of reinforcement has little effect on matrix dominated physical properties such as heat distortion temperature (HDT).

The use of woven or nonwoven (random) fiber reinforcement (structural RIM, or SRIM) can provide parts with yet greater tensile strength and flex modulus. Furthermore, the physical properties may be made directional through the use of unidirectional fibers or combinations of both unidirectional and random fiber orientations. Unfortunately, the use of such reinforcement has not proven practical with traditional polyurethane-polyurea RIM systems due to the high viscosities of such systems. Finally, the low heat distortion temperature of such systems continues to present a detriment to the use of such systems. SRIM systems may be divided into cellular (blown) systems whose resin matrix has a density less then 1.0 g/cm^3 and whose fiber reinforced part has a flexural modulus of less than about 500,000 psi; and high-density systems, which are essentially noncellular, have resin matrix densities greater than 1.0 g/cm^3, and whose flexural modulus is in excess of 500,000 psi.

In U.S. Pat. 4,035,331 are disclosed high-density polyurethane RIM systems employing a polyisocyanate, an amine-initiated polyether polyol, and a liquid modifier having a boiling point above 150° C, all in the absence of a catalyst. The liquid modifier is present in an amount of from 5 to 60 weight percent, preferably from 15 to 40 weight percent of the overall system components. Systems that contain less than about 5 weight percent of modifier are said to produce unacceptable products, subject to swelling, cracking, and distortion. No mention is made of the viscosity of such systems or their suitability as SRIM systems.

In U.S. Pat. 4,709,002, some of the deficiencies of the 4,035,331 patent are overcome through the use of a polyisocyanate and a propylene carbonate adduct of a tertiary amine. The products of these systems are polyisocyanurate polymers that are suitable for use in SRIM systems. However, the alkylene carbonate tertiary amine adduct must be aged for considerable length of time, for example 500 hr, for its catalytic activity to stabilize somewhat. As the catalytic activity continues to change with time, such systems are not storage stable.

It would be desirable to formulate an SRIM system capable of providing a storage-stable, high heat-distortion system yet having a low viscosity suitable for both low-density (microcellular) and high-density (noncellular) applications without the use of large amounts of liquid modifiers.

Low- and high-density structural RIM compositions are disclosed by Kuyzin and Stoll[4] that have low viscosities. These compositions contain a resin B-side component that includes a hydroxyl-functional tertiary amine polyol. The subject invention pertains to the field of fiber-reinforced reaction-injection-moldings. More particularly, the subject invention pertains to reactive systems useful in preparing rigid, fiber-reinforced polyurethane-polyisocyanurate parts especially adapted to the transportation and other industries. Such fiber-reinforced RIM systems and the resulting parts are known as structural-RIM or SRIM.

Kuyzin and Stoll discovered that storage-stable SRIM polyurethane-polyisocyanurate systems having low viscosity, which provide parts having high heat-distortion temperature, may be prepared in the substantial absence of liquid modifiers and yet do not crack or distort, when certain tertiary amine polyols are used as the isocyanate-reactive component. Such systems may be utilized in both low-density, open-pour SRIM and low-density and high-density injection-molded SRIM systems. Finally, low-density systems lend themselves to use in all water-blown systems, eliminating the use of chlorofluorocarbons, although all CFC blown systems and co-blown (CFC and water) are also feasible.

12.6.5 Articles from Flowable Polymerizable Resin-Forming Compositions Employing RIM Techniques

Molding resins are broadly classified as thermoplastic or thermosetting, and molding apparatus and techniques vary considerably according to which of these two categories of materials is being processed. Thermoplastic molding resins (e.g., polyolefins, polyvinyls, polyamides, polycarbonates, polystyrenes, etc.) are highly viscous fluids when heated to the plastic state; e.g., they possess viscosities of from about 10,000 to about 50,000 centipoises or more. Injection pressures of a considerable magnitude (e.g., from about 500 to about 5,000 psia and, in some cases even higher) are required to completely fill the mold cavity and expel the bulk of any air and/or other gas from the mold cavity. Gas expulsion is through vents provided for this purpose—generally small grooves in the parting line of the mold extending from the lip of the mold cavity to its outer edge. The gas, being far less viscous than the thermoplastic, is readily displaced from the mold cavity under the influence of the high injection pressures, whereas the thermoplastic is fully retained in the mold cavity. The high injection pressures typical of thermoplastic injection molding operations also serve to maintain the small quantities of gas that are inevitably present in the molding resins in solution, thereby preventing the formation of bubbles or voids in the molded product.

RIM apparatus and processes have been gaining increasingly wider acceptance for manufacturing shaped articles from thermosetting resins, largely because of the low injection pressures and shorter molding cycles of these resins compared to those for thermoplastic resins, but also due to the fact that the excellent physical and chemical properties of many thermosetting resins (e.g., polyurethane elastomers) make them attractive candidates for numerous engineering applications such as automotive body panels and bumpers, vibration dampeners, gaskets, power transmission belts, conveyor belts, and the like.

A typical flowable polyurethane resin-forming composition is prepared by the high-speed mixing of an isocyanate-reactive liquid polymer such as a polyether or polyester glycol, triol or tetrol, amine-terminated polyether, hydroxyl-terminated polybutadiene, etc. and mixtures thereof; an organic polyiso-cyanate; a chain extender; and, optionally, one or more other components such as catalyst, filler, colorants, etc. to provide a homogeneous liquid of relatively low viscosity—e.g., from ≈50 to ≈10,000 centipoises, and preferably from ≈500 to ≈5,000 centipoises. The composition is then injected into the cavity of a mold, which may or may not be heated, in as brief a time following mixing of the aforesaid components as possible, to prevent any significant degree of gelation from taking place before the full amount of reaction medium required has been introduced into the mold cavity. Injection pressures are generally quite low, usually from about 2 to about 120 psia, since appreciably higher pressures would only force low-viscosity reaction medium through the mold vents.

In practice, it is just about impossible to exclude air or other gas from liquid thermosetting resins due to the chemistry of the polymerization process (gelation followed by curing). Taking the specific case of polyurethane elastomer-forming compositions, there are several sources of gas. all of which can and do prove troublesome in the molding process. All polyurethane reaction media must contain a polyisocyanate component. Polyisocyanates are very reactive toward substances containing active hydrogen, which of course includes water. In the manufacture of polyurethane foams, it is the reaction of polyisocyanate with water that, in a train of chemical events, results in the evolution of carbon dioxide gas. The foaming or blowing activity of carbon dioxide may or may not be supplemented by an auxiliary blowing agent such as a low-boiling halogenated hydrocarbon. In contrast to polyurethane foam reaction media, polyurethane elastomer-forming compositions must scrupulously exclude water and water vapor to avoid the generation of carbon dioxide gas.

When the elastomer is polyester polyol based, it is practically impossible to completely remove all traces of water resulting from the manufacture of the polyester. Another source of gas in polyurethane reaction media is the nitrogen blanket, which is employed for the purpose of preventing contact of each reaction component with atmospheric water vapor during storage. While very little of the gas remains after the reaction components have been passed through a degasser, there may still be enough nitrogen present to significantly contribute to the overall problem of gas entrapment during molding. Still another source of gas results from the manner in which the polyurethane reaction media are injected into the mold. During the injection cycle, air is unavoidably introduced into the reaction media stream. Moreover, any turbulence or "splashing" effect accompanying the filling of the mold cavity can introduce further quantities of air into the reaction media.

The presence of gas bubbles or voids in a molded polymeric article is highly undesirable, for several reasons. Those present at the surface of the article detract from its appearance, a defect that is often sufficient in itself to result in failure of the article to meet minimum quality standards. Those within the interior of the article can significantly diminish its physical and mechanical properties, limiting not only the ability of the article to perform as well as desired but also reducing its useful service life. For example, in the case of a polyurethane elastomer article that experiences frequent compression-decompression cycles (e.g., a vibration-dampening device based on this resin), the *dieseling* effect that can occur within bubbles of entrapped gas can cause localized hot spots within the article, and these accelerate its failure.

Design and process variables in a reaction injection molding system can be controlled to some extent to reduce the amount of air and/or other gas entrapped within liquid resin-forming media. However, since the viscosities of flowable thermosetting reaction media are generally too low to permit the use of injection pressures that would maintain entrapped gas in solution (e.g., the relatively high levels of injection pressure that are commonly used in thermoplastic molding operations), it is not possible to eliminate or suppress the problem of entrapped gas bubbles or voids in reaction injected molded products by resort to high injection pressures.

Graefe[5] disclosed an apparatus for the reaction injection molding of a shaped article from a polymerizable flowable resin-forming composition. Once such composition has attained a predetermined increased level of viscosity as measured *in situ* within the mold cavity, an amount of pressure is applied to the polymerizing composition that is sufficient to maintain any entrapped gases dissolved therein and/or to prevent the composition from pulling away from mold cavity surfaces.

The expression *reaction injection molding* shall be understood herein to refer to any molding system that involves filling a mold with a flowable polymerizable resin-forming composition, which may or may not contain a reinforcement component, the resin-forming composition then undergoing polymerization to provide an article of desired molded configuration. *Reaction injection molding* as used herein is therefore to be regarded as synonymous with, or inclusive of, those operations referred to as *liquid reaction molding (LRM), reinforced reaction injection molding (RRIM), liquid injection molding (LIM),* and *liquid resin molding.*

The expression *flowable polymerizable resin-forming composition* shall be understood herein to refer to any composition that, prior to undergoing a significant degree of polymerization, is a liquid, whether at ambient or elevated temperature, and that following polymerization provides a shape-retaining resin material that may be of the thermoplastic or thermosetting category. Examples of flowable polymerizable resin-forming compositions that can be utilized in the reaction injection molding apparatus and process of this invention include those that provide polyurethanes (both thermoplastic and thermosetting), polyesters and copolyesters, alkyds, polyamides, epoxies, silicones, allyls, amino resins, vinyl resins, interpenetrating polymer networks (IPNs), etc.

12.6.6 Aromatic Diamines as Curing Agents in Polyurethane and Polyurea Manufacture by RIM

As a subclass of commercially available polymers, polyurethane elastomers have several properties whose advantages confer unique benefits on these products. Typically, polyurethanes show high abrasion resistance with high load bearing, excellent cut and tear resistance, high hardness, and resistance to ozone degradation, yet they are pourable and castable. Compared to metals, polyurethanes are lighter in weight and less noisy in use, and they show better wear and excellent corrosion resistance while being capable of less expensive fabrication. Compared to other plastics, polyurethanes are nonbrittle, much more resistant to abrasion, and exhibit a good elastomeric memory. Polyurethanes find use in such diverse products as aircraft hitches, bushings, cams, gaskets, gravure rolls, star wheels, washers, scraper blades, impellers, gears, and drive wheels.

Part of the utility of polyurethanes derives from their enormous diversity of properties resulting from a relatively limited number of reactants. Typically, polyurethanes are prepared on site by curing urethane prepolymers, which are adducts of polyisocyanates and polyhydric alcohols. A large class of such prepolymers are approximately 2:1 adducts of a diisocyanate, OCN-Y-NCO, and a diol, HO-Z-OH, whose resulting structure is OCN-Y-NHCO$_2$-Z-OC(O)NH-Y-NCO. Although Y is susceptible to great variety, usually being a divalent alkyl, cyclohexyl, or aromatic radical, in fact the most available prepolymers are made from toluene diisocyanate (TDI) (most readily available as a mixture of 2,4- and 2,6-isomers), which is rich in the former isomer, or methylene-4,4'- diphenyldiisocyanate (MDI). The diols used display a greater range of variety; Z may be a divalent alkyl radical (i.e., an alkylene group), and the diols frequently are ethers or esters that are the condensation products of glycols with alkylene oxides or dicarboxylic acids, respectively.

The polyurethane elastomers are formed by curing the prepolymer. Curing is the reaction of the terminal isocyanate groups of the prepolymer with active hydrogens of a polyfunctional compound so as to form high polymers through chain extension and, in some cases, cross-linking. Diols, especially alkylene diols, are the most common curing agents for MDI-based prepolymers. Representing such diols with the structure HO-X-OH, where X is an organic moiety, most usually an alkylene group, the resulting polymer has as its repeating unit (-Y-NHCO2ZOC(O)NH-Y-NHCO2-X-O-C(O)NH-). Where a triol or a higher polyhydric alcohol is used, cross-linking occurs to afford a nonlinear polymer.

Although other polyfunctional chemicals, especially diamines, are theoretically suitable, very few have achieved commercial importance as curing agents. The major exception is 4,4'-methylene-di-ortho-chloroaniline, usually referred to as MOCA, a curing agent that is both a chain extender and a cross-linker. TDI-based prepolymers typically are cured with MOCA, and the resulting products account for perhaps most of the polyurethane elastomer market. One reason that polyhydric alcohols generally have gained acceptance as curing agents is that their reaction with urethane prepolymers is sufficiently fast to be convenient, but not so fast as to make it difficult to work with the resulting polymer. In casting polymers, it is desirable that the set-up time be reasonably short, yet long enough for the material to be cast into molds. This property is conventionally referred to as *pot life*. Generally speaking, diamines react with prepolymers, and especially MDI-based prepolymers, so quickly that they are not usable as curing agents. However, primary aromatic diamines with electronegative groups on the aromatic ring, or with alkyl groups ortho to the amino moiety, exhibit sufficiently decreased reactivities with some prepolymers as to afford a desirable pot life, hence the use of, for example, MOCA as a curing agent for TDI-based prepolymers. However, MOCA and others of the aforementioned diamines still remain too reactive to be used, for example, with MDI-based prepolymers.

On the other hand, the advent of RIM provides a means of processing polyurethanes that is well adapted to a short pot life. Reaction injection molding is a process that allows polymerization and cross-linking to take place simultaneously with the forming of a part into its final shape. Because of the rapid curing of polyurethanes, compatible with the fast cycle times of RIM, these polymers seem exceptionally well suited to RIM processing, although epoxies, nylons, and even polyesters have been made by the RIM process.

In RIM, two highly reactive streams of chemicals are brought together under high pressure in a small mixing chamber where the streams are impingement mixed by being sprayed directly into each other before entering the mold. The mixed material flows directly into a mold at 0.35to 0.7 MPa (50 to 100 psi), a low pressure compared to that used in standard injection molding, where the chemical reaction is completed and the part cures. One of the ingredient streams (the first stream) contains the isocyanate, and the other stream (the second stream) contains components having isocyanate-reactive hydrogens, such as polyols and amines, and other components as catalysts, pigments, blowing agents, and surfactants. Much of the technology is currently used in the automotive industry to produce parts such as bumper covers and fenders. Parts are produced on a cycle of 3 min or less, and large urethane parts have been successfully demolded in 30 s or less after injection.

We have found that a large class of N,N'-disubstituted aromatic diamines, mainly alkylated methylene dianilines, are excellent isocyanate-reactive components, or curing agents, for polyisocyanates in the preparation of RIM elastomers. Among the advantages of the curing agents of this invention are that the resulting elastomers can be expected to show excellent compression set, have quite high tensile strength, show greater elongation, and have increased impact properties. The resulting elastomers are thermoplastic or thermosetting polymers, depending on whether a cross-linking agent has been added, or if any of the other components present are capable of cross-linking. Additionally, the curing agents themselves for the most part are liquids at room temperature, facilitating their use at RIM temperature. The curing agents may be used for both TDI and MDI-based polyisocyanates, which give rise to the two largest classes of polyurethane and polyurea elastomers. In short, the unique properties of both the curing agents and the resulting elastomers make each very highly desirable in RIM formulations.

The curing agents of this invention are secondary aromatic diamines, which as a class formerly were not considered as acceptable curing agents for polyurethane and polyurea elastomers. U.S. Pat. 3,846,351 describes the quite narrow use of secondary aromatic alkyl diamines in combination with polyols as catalysts and chain extenders in the production of flexible polyurethane foams. It is important to recognize that such secondary diamines were used only in combination with polyols, preferably at the relatively low level of 0.5 to 5 parts per 100 parts of polyol. In an examination of aromatic diamines as chain extenders in polyurethane elastomers by reaction injection molding, the sole secondary aromatic diamine examined, N,N'-diisopropyl- 4,4'-methylenedianiline, afforded products whose properties were generally unacceptably inferior to those exhibited by polyurethanes made with primary aromatic diamines as curing agents.[6] More recently, we have shown in U.S. Pat. 4,578,446 that contrary to the prior conventional wisdom alkylated methylenedianilines are suitable curing agents for urethane prepolymers, i.e., in elastomer production via non-RIM processes.

To aid in exposition, the isocyanate-reactive components can be classified as either polyols (polyhydric alcohols) or polyamines. Each of these classes has two functionally defined subclasses: backbone polyols (or polyamines) and chain extender polyols (or polyamines). The difference is that, e.g., the backbone polyol, being more reactive, reacts with the isocyanates in the first stream to afford short polymeric segments, and the chain extender polyamine links the short segments to form longer chains. The diamines of this invention act as chain-extender polyamines.

The RIM elastomers that can be made from the amines of this invention are diverse and depend on the nature of the isocyanate-reactive stream. In one variant, the second stream as the isocyanate-reactive component is a mixture of backbone and chain extender polyamines. That is, the second stream may have catalysts, pigments, surfactants, etc., but contain little if any isocyanate-reactive components other than the polyamines. In this variant, the elastomer is exclusively, or almost so, a polyurea.

In another variant, the second stream has as the isocyanate-reactive component a mixture of the amines of this invention, which act as chain extender amines, and various backbone polyols. The mixture will generally have from ≈20 to ≈80% of amine, on an equivalents basis, and more typically contains 30 to 70 equivalent percent of amine. (An equivalent of polyamine or polyol is an amount that furnishes as many amino or hydroxyl groups as there are isocyanate groups in the first stream. As used herein, *equivalents percent* refers to the percentage of amine and/or polyol equivalents relative to isocyanate equivalents.)

Scott and House[7] found that N,N'-disubstituted diamines are effective curing agents in combination with other polyamines and polyols in reaction injection molding for a diverse class of polyisocyanates. The diamines provide a broad spectrum of cure times as well as giving elastomers an interesting and useful diversity of properties. Such diamines may be used as a constituent in a blend of polyamines as the isocyanate-reactive component, in which case the elastomer may be viewed as predominantly a polyurea, or as a constituent in a blend with polyols, in which case the elastomer may be viewed as an elastomer with both urethane and urea segments.

Where the mixture is a blend of backbone and chain extender diamines, the resulting elastomers may be viewed as polyureas. Where the mixture is a blend of backbone polyol and chain extender diamines, the resulting elastomers may be viewed as mixed polyureas-polyurethanes. It should be understood that the elastomers made according to the latter are complex chemical structures having both polyurethane and polyurea segments.

12.6.7 Internal Mold Release Agent for Use in RIM

Products of the RIM process have excellent physical properties. However, the use of the RIM process is limited by the necessity of using sprayed external mold release agents (e.g., waxes, soaps, and the like). It has been necessary to spray these agents onto the mold surface before each shot or every few shots. Recently, internal mold release agents have become available that obviate the need for external release agents. Specifically, internal mold release agents containing zinc carboxylates where the carboxylate group contains from 8 to 24 carbon atoms have met with widespread commercial use. Release agents of this type are described in U.S. Pats. 4,519,965, 4,581,386, and 4,585,803 and British Pat. 2,101,140. In using such zinc carboxylates, it is necessary to mix the carboxylate with a *compatibilizer* that will solubilize the zinc carboxylate so that, when the resultant mixture is mixed with the isocyanate reactive components, the zinc carboxylate will possess improved resistance to precipitation. British Pat. 2,101,140 describes the mixture of a zinc stearate and an epoxidized vegetable oil (such as epoxidized soybean oil).

U.S. Pats. 4,519,965 and 4,581,386 describe the use of compatibilizer selected from the group consisting of nitrogen-containing, isocyanate-reactive acylic compounds and nitrogen-containing, isocyanate-reactive polymers. Preferred compatibilizers include polyether polyamines and amine-or hydroxy-terminated, amine-initiated polyethers. U.S. Pat. 4,585,803 describes the use of compatibilizers that are tertiary amine compounds containing at least one tertiary nitrogen. The tertiary amine compounds described advantageously contain one or more hydroxy groups. Although the combination of the zinc carboxylates and the compatibilizer noted have met with substantial commercial success, the search continues for other satisfactory compatibilizers.

The Dewhurst[8] invention is directed to a mold release, a composition containing the mold release, and a process of using the mold release. The mold release comprises an internal mold release composition consisting of (a) a zinc carboxylate containing from 8 to 24 carbon atoms per carboxylate group, and (b) a compatibilizer incorporating an amidine group-containing compound.

The Dewhurst invention is directed to a novel internal mold release agent for use in a RIM process, to an isocyanate-reactive component containing the mold release agent, and to the use of the mold release agent in a RIM process.

References

1. Smith, Stuart B., U.S. Pat. 5,422,414, Jun. 6, 1995, "Modified polyurea-polyurethane systems endowed with improved processability," to ECP Enichem Polimeri Netherlands B.V.

2. Payne, Jeffrey S., and Martin, David L., U.S. Pat. 5,399,310, Mar. 21, 1995, "Method of using mold release agents," to The Dexter Corporation, Windsor Locks, Connecticut.

3. Dewhurst, John E., and Nichols, James D., U.S. Pat. 5,256,704, Oct. 26, 1993, "Polyurethane RIM elastomers obtained with hydroxyl-containing organotin catalysts," to Air Products and Chemicals, Inc., Allentown, Pennsylvania.

4. Kuyzin, Gregg S., and Stoll, John R., U.S. Pat. 5,073,576. Dec. 17, 1991, "Polyurethane-polyisocyanurate structural RIM systems with enhanced processing," to BASF Corporation, Parsippany, New Jersey.

5. Graefe, Peter U., U.S. Pat. 5,002,475, Mar. 26, 1991, "Reaction injection molding apparatus," to Intellex Corporation, Uniondale, New York.

6. D. Nissen, D., and R.A. Markovs, *Proc. 27th SPI Annual Technical/Marketing Conf.,* 71–78.

7. Scott, Jr., Ray V., and House, David W., U.S. Pat. 4,801,674, Jan. 31, 1989, "Secondary aromatic diamines as curing agents in polyurethane and polyurea manufacture by reaction injection molding," to UOP Inc., Des Plaines, Illinois.

8. Dewhurst, John E., U.S. Pat. 5,420,188, May 30, 1995, "Internal mold release agent for use in reaction injection molding," to Miles Inc., Pittsburgh, Pennsylvania.

13

Polyurethane Adhesives

13.1 Historical

The development of polyurethane adhesives can be traced back more than 50 years to the pioneering efforts of Otto Bayer and coworkers. Professor Bayer extended the chemistry of polyurethanes, initiated in 1937, into the realm of adhesives in about 1940 by combining polyester polyols with di- and polyisocyanates. He found that these products made excellent adhesives for bonding elastomers to fibers and metals. Early commercial applications included life rafts, vests, airplanes, tires, and tanks.[1] Early developments were soon eclipsed by a multitude of new application technologies.

The uses of polyurethane adhesives have expanded to include bonding of numerous substrates such as glass, wood, plastics, and ceramics. Urethane prepolymers were first used in the early 1950s to bond leather, wood, fabric, and rubber composites.[2] A few years later, one of the first two component urethane adhesives was disclosed for use as a metal-to-metal adhesive.[3] And, in 1957, the first thermoplastic polyurethane used as a hot-melt adhesive (adhesive strips) was patented for the use of bonding sheet metal containers.[4] This technology was based on linear, hydroxyl-terminated polyesters and diisocyanates. Additional thermoplastic polyurethane adhesives began appearing in 1958–1959.[5,6]

At the same time, the first metal-to-plastic urethane adhesives were developed. Waterborne polyurethanes were also being developed, with a polyurethane latex claimed to be useful as an adhesive disclosed in 1961 by du Pont.[7] A commercial urethane latex was available by 1963 from Wyandotte Chemicals Corp.[8] The adhesive properties of urethane latexes were further explored by W.R. Grace in 1965.[9]

In the early 1960s, B.F. Goodrich developed thermoplastic polyester polyurethanes that could be used to bond leather and vinyl.[10] In 1968, Goodyear introduced the first structural adhesive for fiberglass reinforced plastic (FRP) that was used for truck hoods.[11]

Polyurethane pressure-sensitive adhesives began appearing in the early 1970s.[12] By 1978, Goodyear made advanced two-component automotive structural adhesives. Waterborne polyurethane adhesives received additional attention at this time,[13] and in 1984 Bostik developed reactive hot melt adhesives.[14]

Polyurethane adhesives are sold into an ever-widening array of markets, where they are known for their adhesion, flexibility, low-temperature performance, high cohesive strength, and cure speeds that can be readily tailored to the manufacturer's demands. Table 13.1 presents the advantages and disadvantages of the most common types of urethane adhesives.

Urethanes make good adhesives for a number of reasons:

1. They effectively wet the surface of most substrates (the energy level of very low-energy surfaces like polyethylene or polypropylene must be raised before good wetting occurs).[15]
2. They readily form hydrogen bonds to the substrate.
3. Their small molecular size allows them to permeate porous substrates.
4. They form covalent bonds with substrates that have active hydrogens.

TABLE 13.1 Advantages and Disadvantages of Urethane Adhesives

Advantages	Disadvantages
Excellent adhesion	Limited thermal stability
Polar polymer	Polyether backbone oxidizable
Chemical reaction with surface	Antioxidants improve performance
High strength	Limited hydrolytic stability
Cross-linkable	Polyesters worse than polyethers
Segmented polymer	Stabilizers available
Rapid cure	Moisture sensitive in bulk
Adjustable with catalyst	Application conditions critical
Good low-temperature properties	Primers
Adjustable hard-segment concentration	
Low Tg	
Many types available	
Cost	

Not only are polyurethanes versatile in their molecular configuration, but they also are compatible with a number of other polymers such phenolics, urea, epoxy, thermoplastic resins and elastomers, that broaden their utility in the area of adhesives and sealants.

Polyurethane adhesives can be classified by several means: by the carrier (solvent or water), the number of component parts, the curing nature (thermoplastic or thermoset), and the physical form of the adhesive (film, powders). Table 13.2 indicates the variety of polyurethanes commercially available for general industrial applications. It should be noted that, because of the high physical strength of polyurethanes (higher than other nonreinforced elastomeric materials), they can be used as thermoplastic adhesives. These adhesives might be either heat or solvent reactivable for specific laminating purposes. Urethanes can also be cured and made into a thermoset polymer by the reaction with water, including moisture in the air. Many urethane adhesive and sealant systems are designed to take advantage of that reaction.

TABLE 13.2 Classification of PU Adhesives and Sealants

Physical form	One component	Two components
Liquid		
Solvent-based	TP or TS	TS
Solvent-free	TS	TS
Waterborne	TP or TS	TS
Solids		
Hot-melt	TP	—
Reactive hot-melt	TS	—
Powder	TP or TS	—

Note: TP = thermoplastic, TS = thermoset

This type of moisture cure technology can be employed in solvent, 100% solid liquids, melts, and other types of systems. Besides the use of water as a curing agent, polyurethanes can also be cured with different chemicals containing an active hydrogen atom resulting in two-component or multicomponent systems. One or more components could be suspended into the other, providing a delayed cure mechanism. Under heat, the total composite fuses, resulting in a chemical-resistant composite.

The selection of the type of adhesive or sealant system depends on the environmental resistance required. Generally, thermoset materials have higher thermal and chemical resistance than their thermoplastic counterparts. The selection of a moisture cure vs. a two-component system is dictated by the application methods, the substrate that is to be bonded (porous or nonporous), and other variables normally found in the utilization of the adhesive. For large laminating areas such as fabrics or plastic films, water- or solvent-carried materials are generally preferred to 100% solids, liquid moisture cure, or two-component systems. Water-based systems can be applicable where porous surfaces are employed. One surface is required to be porous to permit the escape of the water, thereby providing a stronger adhesive film.

Such adhesives typically have been solvent-based lacquer-type products. A lacquer is a solution of a film-forming polymer in a volatile organic solvent which, when applied to a surface, forms an adherent film that solidifies solely by solvent evaporation.

13.2 Consumption and Growth

Polyurethane adhesive consumption has been estimated at a value of over of approximately $355 million. Applications contributing to this volume are shown in Table 13.3. While the packaging market is the fourth largest in volume of urethane adhesives sold, it is substantially larger than the forest products market and the foundry core binder market in terms of dollars.

TABLE 13.3 Market Value of Adhesives Applications

Market Segment	Sales, $Million
Textiles	95
Foundry care binders	75
Forest products	30
Packaging	75
Automotive*	35
Footwear	15
Furniture	20
Recreational vehicles	9
Other	3

*Does not include windshield sealant volumes

Overall, the polyurethane adhesives market grew at approximately 3% per year from 1986 to 1996. Specific market segments like automotive and recreational vehicles easily surpassed GNP growth rate. In the next few years, a number of specific market segments are expected to grow at more than 5% per year, including vehicle assembly (automotive and recreational vehicles), electronics, furniture, and curtain wall manufacture.

13.3 Applications

The textile market has traditionally been the largest consumer of PU adhesives. High-volume applications include textile lamination, integral carpet manufacture, and rebonded foam. Textile lamination occurs through either a solution coating process or flame bonding. Flame bonding textile lamination is accomplished by melting a polyurethane foam by flame and then nipping the foam while still tacky between two textile rolls.

Integral carpet manufacture describes the process by which carpeting is manufactured by attaching either nylon, wool, or polypropylene tufts that are woven through a polypropylene scrim with a urethane adhesive to a polyurethane foam cushion in a continuous process. Rebonded foam is made using scrap PU foam bonded with a urethane prepolymer and is used primarily as carpet underlay. Durability, flexibility, and fast curing speeds are all critical parameters for these applications.

Foundry core binders are the isocyanate-cured alkyd or phenolic adhesives used as binders for sand used to produce foundry sand molds. These sand molds are used to cast iron and steel parts. An economical, fast cure of the sand mold is required under ambient conditions.

Packaging adhesives are used to laminate film to film, film to foil, and film to paper in a variety of packaging constructions. A variety of products are sold to this market with solvent-based, high-solids,

100%-solids, and waterborne adhesives all used. Polyurethane adhesives are considered to be one of the high-performance products offered to this industry. This is because they display excellence in adhesive properties, heat resistance, chemical resistance, and fast curing. Polyurethane adhesives can also be designed to meet FDA approval, a requirement for food packaging applications.

Solvent-borne adhesives represent the majority of the volume in the packaging market, with both one- and two-component systems used. Waterborne polyurethane adhesives are a much smaller segment, driven by environmental considerations. Growth has slowed in recent years because of generally inferior performance compared to solvent-based adhesives, and because most of the major converters have already invested capital in solvent recovery systems.

The forest products industry uses isocyanates to adhesively bond wood chips that are then pressed to form particle board and oriented strand-board. Urethanes are also used to fill knotholes and surface defects in finished plywood boards (*plywood patch*). These filled systems must cure rapidly and be sanded easily.

The transportation market has used polyurethane adhesives for such diverse applications as bonding FRP and SMC (sheet molding composites) panels in truck and car applications, polycarbonate headlamp assemblies, door panels, and weatherstrip flocking.

The construction market for polyurethane adhesives consists of a variety of applications such as laminating thermal sandwich panels, bonding gypsum board to wood ceiling joists in modular and mobile homes, and gluing plywood floors. Early green strength, low shrinkage, and high bond strength are critical properties.

The furniture industry uses both waterborne and solvent-based polyurethane adhesives to bond veneers of various composition to board stock and metal substrates.

Polyurethane adhesives used to attach soles to footwear constitute a sizeable niche. Polyurethane adhesives compete primarily with neoprene-based adhesives and have replaced much of the neoprene volume due to improved performance. The overall market, however, has declined as U.S. manufacturers have moved production overseas.

13.4 Fundamentals Of Adhesion

13.4.1 Adhesive Forces

Adhesion occurs, for the most part, due to attractive forces between molecules. These intermolecular forces include Van der Waals forces, dipole forces, and hydrogen bonding. Van der Waals forces result from temporary dipoles created by imbalances in electron orbits. They are present in all materials but are relatively weak by nature. Van der Waals forces are responsible for virtually all of the molecular attraction between nonpolar materials. Dipole forces are present in polar molecules. Dipole forces tend to create stronger attractions than Van der Waals forces. Hydrogen bonds are a special case of dipole interactions that occur when hydrogen donor and acceptor groups are present. Hydrogen bonds are generally 5 to 10 times stronger than dipole forces.

13.4.2 Surface Wetting

The strength of an adhesive bond is strongly dependent on the ability of the adhesive to wet the substrate. *Wetting* in this case means the tendency of a material to form a broad area of intimate contact with a surface on a molecular level. Intermolecular forces fall off rapidly as the distance between molecules increases. Most surfaces are actually quite rough when examined microscopically. Therefore, it is important to achieve intimate contact between the adhesive and substrate for the attractive forces to take effect.

Wetting is greatly affected by the nature of the attractive forces within the adhesive and adherend. An adhesive exhibiting strong intermolecular forces will resist wetting a surface with lower molecular forces, because the molecules within the adhesive will be more strongly attracted to each other than to the molecules in the substrate. For example, a polyurethane adhesive with strong hydrogen bonds and dipole forces will not wet a polyethylene surface where it can share only weak Van der Waals forces that the polyurethane and polyethylene have in common.

A good measure of the attractive forces in a material is its critical surface energy. Materials with higher surface energy tend to exhibit higher attractive forces than substances with low surface energy. In other words, materials with strong hydrogen bonds tend to have higher surface energies than other polar materials, and polar materials tend to have higher surface energies than other nonpolar materials. Just as materials with strong intermolecular forces resist wetting substances with weaker attractive forces, materials with high surface energies resist wetting low-surface-energy substances. Therefore, to ensure sufficient substrate wetting, the surface energy of the adhesive should be lower than that of the substrate.

13.4.3 Adhesive Joint

An adhesive joint is formed when two substrates are bonded with an adhesive. The joint contains three areas of interest including the substrates, the interfaces between the substrates and adhesive, and the adhesive. Joint failure can occur in any of the three areas. For the purposes of this chapter, the latter two areas are of the most interest; therefore, the discussion will concentrate on the bond interface and the adhesive.

Interfacial adhesion failures occur when the bond strength at the interface is less than the internal strength of the adhesive or the substrate. Failure at the interface is called *adhesive failure* and is indicated by a clean substrate surface upon bond separation. Failures within an adhesive occur when the internal strength of the adhesive is the weakest link in the joint. Internal adhesive failures are called *cohesive failure* and are indicated by a continuous, thick coating of adhesive on both substrates after bond rupture. The thicker the coating, the better. Cohesive failure is generally the preferred mode of failure, as it is more conducive to predictable joint life.

To form a good bond, an adhesive must maximize the interfacial contact between itself and the substrate. This can be achieved through a combination of (1) wetting the substrate while the adhesive is still liquid, and (2) the formation of an intimate physicochemical contact by viscous flow of the adhesive polymer. Thus, herein lies the advantages of using adhesives with a low Tg, such as the polyurethanes. However, to be effective, an adhesive must also resist forces that act to break the bond. That ability rests on a combination of internal cohesive strength and the ability to dissipate and absorb the applied stresses. The relationship between adhesion and cohesion for bonding a generalized substrate pair is shown schematically in Fig. 13.1.

13.4.4 Adhesive Modulus

The strength of the adhesive joint is affected by the intermolecular attractive forces, discussed earlier, as well as the bulk properties of the adhesive. One of the most important bulk properties affecting adhesion

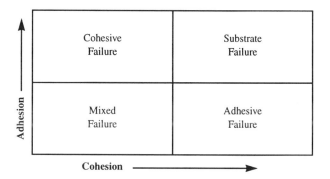

FIGURE 13.1 The adhesion/cohesion balance.

is the modulus of the adhesive material. In general, low-modulus substances make the most useful adhesives. A lower modulus aids adhesion two ways:

1. Low-modulus materials tend to be flexible. This flexibility allows low-modulus substances to conform easily to the adherend, which reduces the distance between the adhesive and substrate molecules. The resulting intimate contact facilitates the intermolecular attractive forces responsible for adhesion.
2. Low-modulus materials also aid adhesion by reducing peak stresses at the bond interface. Brittle, high-modulus materials tend to fail near the adhesive interface, resulting in a very thin coating of adhesive on the substrate upon rupture. By definition, low-modulus materials deform more under the same stresses and thereby create more uniform stress distributions. In a joint, this moves the stress concentration away from the interface and into the adhesive. The resulting thick-film cohesive failure indicates a more durable bond than the thin-film failure typical of high-modulus materials.

Although low-modulus adhesives tend to improve adhesion, it is important to ensure that the modulus does not become too low. Very low-modulus materials tend to have lower strength and tear easily. Therefore, it is important to maintain an adequate modulus to withstand the stresses expected in the adhesive application.

13.5 Bonding Polyurethane

Anyone who has ever dealt with the bonding of an elastomer to a substrate knows that the operation can be very frustrating. A bonding operation can run smoothly for months, but suddenly parts are returned from a customer for poor bond, or the quality control department reports bond failure on a part that was scheduled to ship four days before. Usually, the first reaction to bonding trouble is to call the suppliers of the adhesive and advise them that their product is defective.

The best way to solve bonding problems is to consistently follow a procedure that minimizes the possibility of bond failure. This procedure should be written, with each step explained in detail, including all phases of the bonding operation from incoming materials quality control to final product storage. In other words, learn the basics of bonding, stay with the basics and, if you stray, get back to the basics.

13.5.1 Quality Control on Incoming Raw Materials

A receiving quality control test should be a part of every manufacturer's program. Adhesive suppliers will supply the manufacturer's specifications, which will give the physical properties of any given cement. Visual checks and viscosity checks can be incorporated to determine if the product meets these standards. Manufacturers will certify that products have passed all of their quality control tests, which should give assurance that the product will work for the product.

A good preventive measure to ensure the quality of new lots of any raw material is to run a few parts immediately after opening a new container and carefully inspect the parts to see if they meet specifications. By doing this, you have an immediate measure of the new raw material and do not run the risk of producing a large amount of rejected parts.

Raw material suppliers are provide information on simple tests that can be performed to ensure that their products are as specified in the technical literature. A very strong emphasis must be placed on the use of technical bulletins on any procedure. These bulletins include the typical properties of a bonding agent or any raw material used in a particular operation. They describe the product and its intended use, spray and application data, diluents to be used, and (sometimes) clean-out information. Precautionary information is usually included. Safety data bulletins are also available from each supplier, and these provide more detail than the technical bulletin.

13.5.2 Substrate Preparation

One of the most important and critical steps in bonding a polyurethane polymer to a substrate is the preparation of the substrate. Two prime objectives need to be achieved in substrate preparation:

1. Removal of all contaminants that may be on the surface, such as rust, scale, dirt, oil, mold release, or moisture, which might be detrimental to the achievement of a good bond
2. Development of a surface that will accept a good chemical bond as well as promote the best possible mechanical bond

The most widely used substrate in polyurethane bonding is metal. Several methods are used to treat the metal surface in preparation for bonding. Some of these methods are grit blasting, buffing, chemical etching, wire brushing, phosphatizing, and anodizing. The most common of these methods is blasting. It is simple, fast, can be used on irregular surfaces, and expands the surface area by creating a rough clean surface of peaks and valleys, which enhances bonding. The media normally used in blasting is steel grit, alumina oxide, sand, and glass beads. These materials are available in many sizes and shapes.

In choosing the blasting media, the type of metal substrate should be considered. Number 40 or 50 steel grit is widely used. Alumina grit no. 25 to 50 is preferred for blasting aluminum. Larger grit can cause peening that will imbed contaminants in the bonding surface rather than removing them. The blasting medium needs to be replaced periodically to ensure that it does not become contaminated or worn down by continued use.

Prior to grit blasting, it is important to degrease the metal substrates to remove protective greases and oils. This protects the blasting media from becoming contaminated and helps eliminate the possibility of a contaminated bonding surface.

After blasting, the metal needs to be degreased with clean solvent to remove any contaminants or dust that might remain on the prepared surface. It is normal that some dust and contaminants removed from the metal will remain on the surface. Another possible contamination source that may be involved in the blasting operation is oil and moisture introduced from the compressed air system. If the system is not properly maintained or filtered, oil and moisture can be carried to the blasting media and deposited on the cleaned surface. This can also happen when compressed air is used to blow dust from cleaned parts. Compressed air can be checked by using a piece of glass. A blast of air directed onto a piece of glass for a few seconds will quickly indicate the presence of oil or water contamination.

The selection of solvents for cleaning and degreasing is very important. A solvent that evaporates too rapidly may lower the surface temperature of the metal to below the dew point of the surrounding air, and moisture will condense on the surface. This is especially important in summer months or in tropical areas where the dew point and temperature spread is only a few degrees. For this reason, methyl ethyl ketone, toluene, xylene, or perchloroethylene are better degreasing solvents than acetone or methylene chloride.

13.5.3 Application of Bonding Agent

A freshly prepared bonding surface on metal is a prime surface for rust and oxidation. These processes begin immediately when exposed to atmospheric conditions. To preserve the clean bonding surface, the bonding agent should be applied as soon as possible after final degreasing. This will protect the parts from rust or oxidation. Most bonding agents used in bonding polyurethane will withstand several months of aging without losing any of their capability to achieve a good bond. In storing adhesive coated parts, care must be taken to avoid contamination of the bonding surface. This can be done by storing parts in polyurethane bags or closed containers.

The application of the bonding agent is equally important in achieving a good bond as metal or substrate preparation. It is not possible to overemphasize the importance of the technical data bulletin as a reference to good bonding. The essential information contained in the technical bulletin will help the consumer to use the adhesive properly. The technical bulletin has information on proper application, diluents, spray data, and other information, including safety precautions that are absolutely necessary when using the product.

The methods of application are usually determined by the part itself. Various shapes and sizes may require different methods of application. Most bonding agents used in polyurethane bonding are supplied at brushing viscosity. They can be diluted for dipping; roll application; or air, airless, or electrostatic spraying. The technical bulletin will supply the necessary information for dilution of the adhesive for the particular method of application. For dipping, it is advisable to use agitation in the dip tank to avoid surface filming and settling of solids in the bonding agent.

Adjustment of spray equipment is important to obtain a uniform wet film. Too much atomization pressure will prevent substrate wetting by the adhesive. Too little atomization pressure will result in substrate flooding, causing adhesive running. In the event of cobwebbing, a slower-evaporating solvent should be considered.

The application of a bonding agent may require two coats to get an adequate thickness. It is necessary to dry the adhesive to get rid of all of the solvent prior to applying a second coat. Adhesive films can be air dried or force dried. Force drying can be advantageous in speeding up the drying period as well in preventing moisture contamination during periods of high humidity.

In most cases, a prebake of the adhesive film will improve the bond by drying out any residual solvent in the adhesive film, and it helps set the adhesive to the metal. In cast urethane bonding, it is necessary to bring the metal up to casting temperature. This is comparatively simple for small parts, because it requires only a short time in the oven at 200 to 250° F. On large masses of metal such as large rolls, the cores must be in the oven for 12 to 16 hr to bring the internal temperature of the entire mass to casting temperature prior to casting. Care must be taken to hold the prebake temperature under 250° F to prevent overcure of the adhesive film unless otherwise specified by the adhesive manufacturer.

13.6 Formulating Polyurethane Adhesives

Some of the earlier U.S. patents on diisocyanates disclosed their use as adhesives, especially for the bonding of elastomers to fibers and metals. In particular, methylenebis (4-phenyl isocyanate) (MDI) was utilized during the early stages of World War II in the bonding of elastomers to synthetic fibers for applications such as life rafts and inflatable vests.

Some early work on polyurethane (PU) adhesives was reported by Heiss et al.,[16] who prepared adhesives from castor oil and polyether polyols and diisocyanates. Various methods for applying these adhesives including low- and high-temperature bonding and *carrier* bonding were described.

PUs have found widespread use as adhesives in the various industrial and household environments. They can be used to bond diverse substrates such as metals, plastics, rubber, glass, ceramics, and wood due to their high polarity and the hydrogen bonding between PU and the adherend. The PU adhesive can be designed to meet various needs because of the unique reactivity of the NCO group. PU adhesives, sealants, and binders have proven to be very valuable in different industries such as foundry, construction, automotive, transportation, shoe, paper, textiles, tape packaging, and electronics. PU materials are being accepted as structural adhesives in the automotive industry in bonding composites, plastics, rubbers, glass, and primer-coated metals, particularly for the bonding of structural plastics, which are being used in ever-increasing amounts. PUs are also starting to be used in the manufacture of particle board, replacing more traditional phenolic and urea-formaldehyde adhesives

Other advantages of PU adhesives and sealants over other types of materials are excellent low-temperature resistance, low cure temperature, good flexibility and impact strength, and good wetting to a variety of substrates for good adhesion.

13.6.1 Raw Materials

Isocyanates

The most widely used aromatic isocyanates in adhesive applications are toluene diisocyanates (TDI, usually in an 80/20 mixture of 2,4- and 2,6-isomers), methylene-4,4'-di(phenyl isocyanate) (MDI), and polymeric MDI (PMDI) at NCO functionalities varying from 2.2-3 and 4,4',4''-triphenylmethane triisocyanate (TTI). In addition, adducts of TDI or MDI with trimethylolpropane and other triols or isocyanate derivatives of TDI are used in the formation of adhesives for a variety of substrates.

Where light and color stability are required, aliphatic isocyanates such as isophorone diisocyanate (IPDI), methylene-4,4' di(cyclohexane isocyanate) HMDI (Desmodur W), and the biuret triisocyanate derived from hexamethylene diisocyanate (HDI) with water (Desmodur N), or variations thereof are used. In a similar manner, aromatic isocyanates, adducts, and isocyanate-containing isocyanurates based on HDI or IPDI are frequently employed. The blocking of isocyanates for heat-activated adhesives is used, in particular, for one-component systems. Many different blocking agents can be used, especially phenols, branched alcohols, methylethyl ketoxime, and e-caprolactam.

To avoid the use of monomeric isocyanates with high vapor pressure, NCO-terminated adducts or prepolymers of aromatic or aliphatic diisocyanates with hydroxyl-containing polyesters, polyethers, or polybutadienes are utilized. They also permit wide formulation variations, depending on the required flexibility, chemical resistance, degree of cross-linking, and adhesion to a given substrate.

Polyols

Hydroxyl-terminated polyesters such as the lower alkylene polyadipates are widely used, although other dibasic acid-derived esters, including azelates and dimer acid esters, can be used. Alternatively, polycaprolactones are finding increasing use in a variety of adhesive applications.

Hydroxyl-terminated polyethers for adhesive use consist primarily of low-cost poly (oxypropylene) polyols, which may be modified with oxyethylene units. These polyethers exhibit low viscosity and hence are popular for high-solids or 100%-PU systems.

Poly oxytetramethylene glycols (PTMO or PTMG) have higher viscosity and are higher priced. However, the resulting adhesives exhibit higher-strength values than the corresponding alkylene oxide-based polyether products made with polyols of the same equivalent weight.

Hydroxyl-terminated prepolymers from polyesters or polyethers are also used in combination with polyisocyanates. Hydroxyl-terminated polybutadienes (Poly-bd), as well as their hydrogenated derivatives, are used either alone or in combination with polyethers where more hydrophobicity is required.

Chain Extenders

Many adhesive formulations make use of chain extenders, which form the *hard* segments by reaction with isocyanates. The chain extenders usually consist of short-chain diols such as 1,4-butanediol, 1,6 hexanediol, 1,4-bis(hydroxyethyl) derivative of hydroquinone (HQEE), oxypropylene, and oxyethylene derivatives of bisphenol-A.

In addition, alkanolamines or diamines serve as chain extenders but, while there are no problems in carrying out the chain extension in water (primary and secondary amines react much faster than water), the reaction of isocyanate-terminated prepolymers with amines in solution or in bulk is usually too fast for adhesive applications, unless sterically hindered diamines are used.

Catalysts

Catalysts are employed in the case of both one-component polyisocyanates or NCO-terminated urethane prepolymers, as well as for deblocking of blocked isocyanate based adhesives. Both tertiary amine and organometallic catalysts, such as dibutyltin dilaurate, dibutyltin acetate, dibutyltin dithiocarbonate, and dibutyltin oxide, are used.

In some cases, organomercurial catalysts are used either alone or in combination with organotin catalysts, although their use is restricted in some applications. Combination catalyst systems may also be used, but storage stability, especially for one-component systems, is an important consideration.

Solvents and Additives

Solvent-containing adhesives are formulated with dry organic solvents, including ketones and lower alkyl esters such as acetates, methylene chloride, and trichloroethane. Aromatic solvents such as xylene are frequently employed for binder applications such as sand cores.

Certain flow agents may be added for solvent systems, e.g. cellulose acetate butyrate, vinyl acetate copolymers, etc. Some copolymers, particularly those containing pendant carboxyl groups, may also enhance adhesion to various substrates.

A great variety of fillers can be added to adhesives, sealants, and binders. Commonly used fillers include calcium carbonate (whiting), clays, silica, talc, barytes, and other inorganic fillers. Finely divided polymeric fillers such as polyethylene are also being used. All fillers have to be thoroughly dry to avoid carbon dioxide evolution.

Certain thixotropic materials are employed for adhesion to porous substrates such as paper, textile, leather, concrete, etc. to provide *hold-out* properties. Typical examples are finely divided silicas, for example Cabot Corporation's Cab-O-Sil, polyureas, as well as modified castor oil derivatives.

Plasticizers are often added to adhesives and sealants. The selection of the plasticizer has to be made very carefully to achieve lower hardness and increased extensibility without significant loss of adhesion.

To improve the stability for outdoor exposure to UV and resistance to oxidative degradation, stabilizing systems are frequently used. Certain inorganic pigments such as TiO_2, ZnO, or certain molybdates are used against UV degradation, as is carbon black (if black color is not a factor). Most stabilizing systems for PU adhesives consist of combinations of antioxidants such as hindered phenols and hindered amines (HALS) with organic stabilizers, such as substituted benzotriazoles or benzophenones. In some cases, thioethers and phosphites may also be used in combination with the above-mentioned stabilizing systems.

13.7 Curing Reactions

The curing of urethane polymers and prepolymers involves the reaction of isocyanate groups (either monomers, oligomers, or prepolymers) with compounds containing hydroxyl groups. The functionality of all materials must be >2. If thermoplastic (linear) polyurethanes are desired, a functionality of 2 is required for both the isocyanate and the active hydrogen-containing material. If thermosetting polyurethanes are the goal, functionalities greater than 2 are required for either the polyisocyanate or the active hydrogen compound or both to provide cross-linking. The foregoing holds for all types of urethane products, including elastomers of all types, coatings, sealants, and adhesives.

Suitable active hydrogen-containing materials include the case with hydroxyl groups (polyols) including water, amine groups (primary or secondary), and already-formed urethane and urea groups. Another group that can result in curing of isocyanate-containing material and contains no active hydrogen is the epoxy group. In addition, isocyanate groups can react with themselves and thus participate in the curing of urethane materials with no other active hydrogen-containing compounds.

The curing reaction can utilize isocyanate-terminated prepolymer (NCO/OH = 2) or quasi-prepolymers (NCO/OH >2) or it can be a one-shot approach in which no preformed urethane is involved. Catalysis is often required to obtain complete curing, although the type and amount depend on the type of isocyanate and active hydrogen-containing material.

Polyurethane adhesives (high-modulus elastomers) may be prepared by either the one-shot or the prepolymer approach. In the former, all ingredients (isocyanates polyols with short chain diol or amine chain extender and/or polyamines) are mixed simultaneously, along with catalysts, fillers, plasticizer, and any other additives cast or molded and (generally) thermally cured, although room-temperature curing systems can be prepared by utilizing high levels of mixed catalysts or appropriately fast-reacting isocyanates and active hydrogen-containing materials. Such systems are generally of two components, consisting of an A component, being the di- or polyisocyanate, and a B component, being the polyol(s) and/or polyamine(s), chain extender (generally a low-molecular-weight polyol or polyamine), with any other desired active hydrogen material, catalyst(s) and any other ingredients. Such one-shot systems include reaction injection molding (RIM).

If no polyol is being used and only amine terminated polyethers such as poly(oxypropylene) diols and triols along with a diamine chain extender such as diethyltoluenediamine (DETDA) are employed, a polyurea system is obtained.

When a more structured polyurethane or poly(urethane-urea) is desired, the prepolymer approach is utilized. An isocyanate-terminated prepolymer is first prepared by reacting excess diisocyanate with a polyol. The curing involves the reaction of the prepolymer (A component) with a chain extender (B component), generally a low-molecular-weight polyol or polyamine (functionality depending on desired amount of cross-linking, if any) such as those described earlier.

The isocyanates used are primarily aromatic (toluene diisocyanate or 4,4'-diphenylmethane diisocyanate, pure or modified); occasionally aliphatic isocyanates are employed.

Additional reactions that can occur in the above systems to result in further cross-linking include allophanate formation (reaction of isocyanate with urethane groups) and (biuret formation reaction of isocyanate with urea).

When increased rigidity and/or high temperature performance is desired, further cross-linking to form heterocyclic rings may be accomplished via isocyanurate formation. This trimerization reaction will take place with the employment of excess isocyanate and selective catalysts. Another method for obtaining polyurethanes with high-temperature properties is curing with epoxy-containing materials to form oxazolidones.

Polyurethane elastomers with enhanced energy absorbing abilities (both mechanical and acoustical) represent a more recent application. These materials are made by either under-indexing (NCO/OH < 1), using chain stoppers (mono-functional alcohols), or both. This results in a broad distribution of molecular weights, which gives rise to a very broad glass transition.

13.8 Common Adhesive Types

Table 13.4 presents a summary of the most common types of urethane adhesives. These will be discussed in the following sections.

TABLE 13.4 The Most Common Types of Urethane Adhesives

Two-package reactive adhesives
Low-molecular-weight polyisocyanate, isocyanate prepolymer or mixture in one package, and a low-molecular-weight polyol mixture in the other. When the two packages are mixed, the final, cross-linking urethane linkages are formed in the bondline.

One-package reactive adhesives
Higher-molecular-weight polyisocyanate prepolymers having rather low remaining isocyanate content which react with atmospheric moisture to form the cross-linking bonds in the bondline. The prepolymer might contain urethane groups, but the cross-linking bonds are ureas.

Two-package solvent adhesives
Similar to the 100%-reactive adhesives above but carried in solvent that usually is eliminated after mixing and application but before the bond is closed.

One-package solvent adhesives
A. Similar to the one-package 100% reactive adhesives above but carried in solvent, which is eliminated prior to or during the cross-linking process.
B. Intermediate to high-molecular-weight urethane polymers containing no remaining isocyanate groups. These are not reactive and solidify by solvent elimination.

Thermoplastic urethanes
High-molecular-weight, mostly linear polyurethanes containing no remaining isocyanate groups. These are thermoplastic materials that form the adhesive bonds by cooling from the molten state.

Dispersion or emulsion adhesives
High-molecular-weight, usually nonreactive, polymers dispersed or emulsified in water. The water carrier is eliminated prior to or during use, leaving the precipitated and coalesced polymer to form the adhesive bond.

13.8.1 One-Component Adhesives

The oldest types of one-component polyurethane adhesives were based upon di- or triisocyanates that cured by reacting with active hydrogens on the surface of the substrate or moisture present in the air or substrate. The moisture reacts with the isocyanate groups to form urea and biuret linkages, building molecular weight, strength, and adhesive properties. Prepolymers are also used either as 100% solids or solvent-borne one-component adhesives. Moisture-cured adhesives are used today in rebonded foam, tire cord, furniture, and recreational vehicle applications.

A second type of one-component urethane adhesive is hydroxypolyurethane polymers based on the reaction products of MDI with linear polyester polyols and chain extenders. There are several commercial suppliers of these types of thermoplastic polyurethanes. The polymers are produced by maintaining the NCO:OH ratio at slightly less than 1:1 to limit molecular weight build to the 50,000 to 200,000 range with a slight hydroxy content (approximately 0.05 to 0.1%).

These are typically formulated in solvents for application to shoe soles or other substrates. After solvent-evaporation heat melts the polymer (typically at 50 to 70° C, temperatures at which the polymers reach the soft, rubbery, amorphous state), the shoe upper can be press fit to the sole. Upon cooling, the adhesive recrystallizes to give a strong, flexible bond. More recently, polyisocyanates have been added to increase adhesion and other physical properties upon moisture curing.

The use of waterborne polyurethane adhesives has grown in recent years, replacing solvent-based adhesives in a number of application areas. A number of patents cover the use of waterborne polyure-thanes in shoe soles, packaging laminates, textile laminates, and as an adhesive binder for the particle board industry.[17,18]

Waterborne PU adhesives are environmentally friendly, nonflammable, and have no VOC emissions. They typically can be blended with other dispersions without problems and exhibit good mechanical strength. Water-based systems are fully reacted linear polymers that are emulsified or dispersed in water. This is accomplished by building hydrophilicity into the polymer backbone with either cationic or anionic groups or long hydrophilic polyol segments or, less frequently, through the use of external emulsifiers.

These groups can easily be converted to salts, which, as water is added to the prepolymer/solvent solution, allow the prepolymer to be dispersed in water. The solvent is then stripped, leaving the dispersed product. Variations on this theme allow lower solvent volumes to be used.

Long hydrophilic polyol segments can be similarly introduced into the polymer backbone. Chain extenders with hydrophilic EO groups pendant to the backbone are reacted with the prepolymer to form a nonionic self-emulsifying polyurethane. This reaction is also carried out in a water miscible solvent that can later be stripped from the solvent/water solution.

Blocked isocyanates can also be considered a one-component adhesive. A blocking agent allows the isocyanate to be used in a reactive media that can be heat activated. One-component adhesives based on blocked isocyanates are not amenable to room-temperature curing applications.

One-Component PU Adhesives

One-component polyurethane adhesives include polyisocyanate adhesives and binders, prepolymer-based adhesives, thermoplastic polyurethanes, blocked di- and polyisocyanate adhesives, aqueous polyisocyan-ate and polyurethane adhesives, and the following categories.

Polyisocyanate Adhesives and Binders
This group includes polyisocyanates, either alone or in combination with polymers such as rubber (SBR). Polyisocyanate solutions of 4,4',4"triphenylmethane triisocyanate (TTI), adducts of TDI or MDI with trimethylolpropane, polymeric MDI (PMDI), and trimers of TDI or of other diisocyanates can be used as binders for wood (cellulose) or other cellulose-based products, as well as for primers and adhesion promoters. In addition, polyisocyanates such as TTI are used for bonding rubber to both metals and fibers.

One of the most important applications of polyisocyanate (predominantly PMDI) solutions or aqueous dispersions is for the production of particle board as a substitute for either urea-formaldehyde (UF) or melamine formaldehyde (MF) resins. In some cases, combination systems may be employed by first spraying the wood chips with UF or MF resins and subsequently with the PMDI resin, resulting in increased strength properties.

Di- or polyisocyanates, usually in solution, are also used as *primers* for coatings and adhesives to provide better bonding between the surface layer and the substrate. A primer is a coating applied to a substrate to improve the adhesion, gloss, or durability of a subsequently applied coating.

Many adhesive applications for polyisocyanate solutions are in combination with elastomeric or plastic vehicles. Various types of elastomers including natural rubber, SBR, neoprene, nitrile rubber, etc. are dissolved in either aromatic or chlorinated solvents after being broken down on a Banbury mixer or mill. MDI solutions are added with agitation, resulting in rubber cements that are excellent adhesives for the bonding of elastomers to fabrics. Curing of the composites can be carried out by oven curing, press curing, or by means of air-pressure vulcanizers.

Polyisocyanate solutions are also very effective in the bonding of plastic fibers such as PET to various rubbers, as well as for the bonding of rubber to metals. In this case, the polyurethane is utilized as a binder. (A binder is an adhesive material used for holding particles of dry substances together.)

Prepolymer-Based Adhesives

Isocyanate-based prepolymers for adhesive applications can be applied either with or without a solvent (100% solids) to yield either a one-component moisture-cure adhesive or binder, or reacting the NCO-terminated prepolymer with a hydroxyl-containing compound (e.g., polyester, polyether, or polybutadiene) as a two-component system. TDI- or MDI-based prepolymers are most frequently employed, either alone or with the addition of a small amount of PMDI. The one-component NCO-terminated prepolymer cures on exposure to moist air. Cure can be accelerated by the use of catalysts, although the storage stability has to be taken into consideration. Factors such as percent NCO and type and functionality of the components will greatly affect the strength and environmental properties as well as the adhesion.

One way of applying these systems involves exposing the NCO-terminated prepolymer to air for several minutes after it is applied to the substrate to be bonded to allow for evaporation of the solvent. This step is followed by application of moderate pressure, which is maintained until complete cure is achieved.

Another similar method of application is the curing of the adhesive film to a non-tacky condition. The surfaces to be bonded are then joined and heated under pressure to fuse the adhesive film (175 to 200° C for 1 to 5 min). This method of application is suitable for metals, glass, and other substrates that are not sensitive to high temperatures. Many types of additives may be employed with the PU prepolymers, some of which consist of polymers as modifiers, provided that they do not react with the free NCO groups. Typical properties of a one-component moisture-cured PU sealant with high solids content (>90%) are shown in Table 13.5.

TABLE 13.5 Typical Properties of One-Component Moisture-Cured PU Sealant

Solid content, %	90–95
Viscosity, cps/RT	30,000
Sag index	35/30
Peel strength, lb/in	5.25
Tensile strength, psi	550
Modulus @ 100%	365
Modulus @ 300%	500
Elongation, %	450

A special case of two-component adhesives are certain *anaerobic* PU adhesives. They can be prepared by reacting hydroxyethyl methacrylate (HEMA) with TDI or with an NCO-terminated prepolymer. An organic hydroperoxide is added, and the mixture is stored in an oxygen-permeable container with deliberate inclusion of air to prevent radical initiated polymerization. The adhesive can thus be stored. Only on application of the adhesive to the surfaces of the substrate is access of oxygen prevented, and polymerization of the pendant acrylate groups takes place.

Thermoplastic Polyurethane Adhesives

Fully reacted thermoplastic polyurethanes produced from polyesters such as polyadipates or polycaprolactones and diisocyanates (generally MDI) can be applied as adhesives in a number of ways. They can be dissolved in dry solvents such as THF, DMF, MEK, cyclohexanone, butyl acetate, and chlorinated solvents, with or without addition of a di- or polyisocyanate. Many of these resins are highly crystalline and become amorphous and soft upon heating, thus providing good wetting of the substrate surface.

After cooling, the amorphous forms recrystallize and result in very strong bonds that increase with time. The hardness and peel strength of the adhesives increase with increasing crystallization speed of the hydroxyl PUs.

Many thermoplastic urethane adhesives are compatible with common plasticizers such as certain phthalates, phosphates and adipates, as well as with many resins including PVC copolymers, rosin esters, coumarone resins, etc. Thermoplastic PU adhesives lend themselves readily for use as melt adhesives with or without the addition of plasticizers.

By careful selection of diisocyanates (aromatic or aliphatic), polyols, and chain extenders, as well as plasticizers, hot-melt adhesives can be tailored to meet various industrial requirements. The effect of various diisocyanates and polyols on the lap shear strength of hot-melt elastomeric PU adhesives is shown in Tables 13.6 and 13.7. Unique reactive hot-melt elastomeric PU adhesives exhibiting thermoplastic and thermosetting properties (without and with cross-linker) with excellent adhesion are shown in Table 13.8.

TABLE 13.6 Effect of Aliphatic Diisocyanates on Lap-Shear of Hot-Melt PU Adhesives

Diisocyanates	Lap Shear Strength (Steel/Steel, psi)
HMDI	1,100
CHDI	685
IPDI	740

TABLE 13.7 Effect of Polyols on the Lap Shear Strength of Hot-Melt Elastomeric PU Adhesives

Polyols	Lap Shear Strength (Steel/Steel, psi)
Adipate polyesters	1260
Poly(caprolactone) glycols	1025
Poly(oxtetramethylene) glycol	925
Poly(oxypropylene) glycol	895

TABLE 13.8 Lap Shear Strength of Reactive vs. Nonreactive Hot-Melt Elastomeric PU Adhesives

Polyols	Lap Shear Strength (Steel/Steel, psi)		
	Diisocyanate	Nonreactive	Reactive
Poly(caprolactone) glycol	CHDI	685	>2000
Poly(caprolactone) glycols	IPDI	735	>2000

Blocked Di- and Polyisocyanate Adhesives

The *blocking* of isocyanate groups to produce stable one-package coating and adhesive is widely practiced. Of course, a polyisocyanate may contain more than two isocyanate groups that can be blocked. This reaction is reversible, and generally temperatures of 140 to 160° C will deblock the isocyanate; even lower temperatures and time are needed if a *deblocking* catalyst, such as organotin, zinc salts or tertiary amines.

Thus, blocked polyisocyanates can be used either with polyols, hydroxyl-terminated urethane prepolymers, but also with any active hydrogen-containing polymer, including phenolic resole resins, UF and melamine resins containing methylol groups, hydroxyl- or carboxyl-containing acrylic copolymers with pendant active hydrogen-containing groups, cellulosics, etc. The blocked systems can be applied either in form of 100% solids, solution, or aqueous systems using elevated temperatures to *deblock* the isocyanate, which then reacts with the other component.

Aqueous Polyisocyanate and Polyurethane Adhesives

The trend from solvent to aqueous systems in both the coatings and adhesives industries has intensified over the last decade due to environmental and safety restrictions, as well as the costs of solvents, including the costs of solvent recovery systems.

Although PU dispersions for adhesives use have been produced commercially using extraneous surfactants, most PU dispersions are made from PU ionomas, for example, polyurethanes containing either negative charge groups such as carboxyl or sulfonic acid groups (anionomers) or positive charge groups such as quaternary ammonium groups (cationomers).

Because most PU ionomer dispersions are thermoplastic, various methods can be used to cross-link these emulsions to increase their water and solvent resistance as well as their thermal resistance.

Polyurethane Film and Tape Adhesives

Tape adhesives are generally defined as having a backing with a carrier or reinforcement (which can be paper, plastic, or textiles), while film adhesives are self-supporting.

Both PU tapes and films are produced from thermoplastic PUs, either by solution, casting, extrusion, or calendering. They are used in a great variety of industrial applications. The films are usually applied by using heat and pressure. Applications include bonding of fabrics and bonding with other plastic films, such as polyethylene or polyethylene terephthalate, as well as other polymers. Transparent PU films based on aliphatic isocyanates are being used as interliners for various plastics such as acrylics and polycarbonates.

13.8.2 Two-Component Adhesives

Two-component polyurethane adhesives are widely used where fast cure speeds are critical, such as OEM assembly lines requiring quick fixture of parts, especially at ambient or low bake temperatures. Two-component urethanes are required in laminating applications where no substrate moisture is available, or where moisture cannot penetrate through to the adhesive bond. Two component urethanes are also useful where CO_2 (generated by a one-component moisture cure) or a volatile blocking agent would interfere with the adhesive properties.

Two-component adhesives typically consist of a low-equivalent-weight isocyanate or prepolymer cured with a low-equivalent-weight polyol or polyamine. They may be 100% solids or solvent borne. Since the two components cure rapidly when mixed, they must be kept separate until just before application. Application is followed quickly by mating of the substrates to be bonded.

Efficient mixing of the two components is essential for complete reaction and full development of designed adhesive properties. In line mixing, tubes are adequate for low-volume adhesive systems. For larger-volume demands, sophisticated meter mix machines are required to mix both components just prior to application. Commercial systems for delivering two-component adhesives are segmented based on the viscosity ranges of the components. The ranges can be broken down into low, middle, and high viscosity with, for example, Liquid Control Corp., Sealant Equipment and Engineering Inc., and Graco Inc. supplying equipment for the respective ranges.

High-speed assembly line operations currently use robotically applied adhesives. The adhesive bead is applied quickly and evenly to parts on a conveyor line just prior to fitting. These operations, especially the need to handle the adhered substrates soon after assembly, demand fast-curing adhesive systems.[19]

In two-component PU adhesives or sealants, a wide variety of polyols are used to design for special requirements. Polyester polyols are particularly valuable raw materials for PU adhesives due to the high polarity of the ester groups, which results in high adhesion to many materials—especially to plastics. Sometimes the PU adhesive does not adhere to the surface of the substrate and forms *islands,* a phenomenon commonly referred to as *crawling* or *cratering,* presumably due to the crystallization or high polarity of the PU. This problem can usually be solved by the addition of certain flow agents. Thixotropic agents such as finely dispersed silica are used to prevent the PU adhesive from penetrating into highly absorbent porous materials such as leather, textile, paper, or concrete. The ratio of polyol to polyisocyanate plays an important role in influencing the cross-linking density of the two-component polyurethane adhesive or sealant.

Bonds of the PU adhesive with higher cross-linking density exhibit, in general, a higher lap shear strength and good chemical resistance. With decreasing cross-linking density, a PU adhesive tends to be more elastic but exhibits lower lap shear strength. In general, the resistance of a PU adhesive bond to high temperature, water, solvents, plasticizers, fats, oils, and outdoor weather decreases with decreasing cross-linking density of the adhesive.

Polyisocyanates such as PMDI, TTI, or adducts of trimethylolpropane with TDI and other diisocyanates may be reacted with hydroxyl-terminated polyesters, polyethers, and polybutadienes, as well as hydroxyl-containing urethane prepolymers. In addition, castor oil, castor oil derivatives, and fatty oil derivatives may be employed. Plasticizers, polymeric modifiers, catalysts, solvents, and other additives may be added, which can be applied for curing at room temperature or elevated temperatures.

13.9 Illustrative Patents

13.9.1 One-Component Polyurethane-Type Adhesive[21]

Polyurethane-type adhesives used for general purposes are mostly of solvent types from the standpoint of ease of operation. Polyurethane adhesives are roughly classified into three classes, according to the curing mechanism.

- The first class of the adhesives are two-component polyurethane adhesives in which a first component of a polyisocyanate and/or an isocyanate-terminated prepolymer with a second component of a low-molecular-weight polyol and/or urethane-modified polyol having molecular weight of less than 10,000. They are mixed immediately before application and applied to a base material (adherend) to be cured.

- The second class of the adhesives are moisture-curing type polyurethane adhesives, which contain an isocyanate-terminated prepolymer and cure by reaction with active hydrogen groups of a base material or moisture in the air. This group includes nonsolvent-type reactive hot-melt adhesives that are melted at a temperature of 100° C or higher, applied to a base material and cooled to solidify to exhibit initial adhesive force. They later react with active hydrogens on the base material or moisture in air to be cured by polymerization and cross-linking.

- The third class of adhesives are one-can lacquer-type polyurethane adhesives, which are a solution or dispersion of a high-molecular-weight thermoplastic polyurethane resin in a solvent or water, and the solution is applied to a base material. The high-molecular-weight polyurethane resin having high cohesive energy forms an adhesive layer to exhibit adhesive force by simple evaporation of the solvent or the water.

The respective classes of adhesives have advantages and disadvantages as described below.

The two-component polyurethane adhesive forms an adhesive layer that generally has a cross-linked structure and is superior in heat resistance and durability. However, its pot life is limited, because the liquid formulation of the two components becomes viscous due to the reaction of the isocyanate group with the hydroxy group in the system, and it finally gels.

The moisture-curing polyurethane adhesives form an adhesive layer that gives a final adhesive force that is weaker than that of a two-component type polyurethane adhesive, but they are superior in heat resistance and other properties. Since the adhesive force is usually caused by the reaction with moisture on the base material or in air, the adhesive has disadvantages in that the reaction proceeds slowly, the initial adhesive force is exhibited slowly, and the adhesiveness is affected by external conditions such as low temperature and low humidity in winter, and high temperature and high humidity in summer.

The above two classes of adhesives, namely the two-component type polyurethane adhesives and the moisture-curing type polyurethane adhesives, contain a free polyisocyanate monomer in a slight amount in the system. The free polyisocyanate monomer tends to aggravate working environment, depending on the conditions of application of the adhesive, and frequently requires local ventilation. The one-can lacquer-type polyurethane adhesive, which is a solution of high-molecular-weight polyurethane resin such as a thermoplastic polyurethane solution, has a semipermanent pot life unless the solvent is not evaporated, and it is easily handled. This type of adhesive, however, forms an adhesive layer that has thermoplasticity and a low melting point due to no cross-linked structure and, therefore, the heat-resistance is low since the adhesion strength thereof falls at a temperature exceeding the softening temperature.

From the global environmental problems, volatile organic compound (VOC) regulation is being tightened, and movements toward resource conservation, nonpollution, and higher safety are required in the fields of paints, adhesives, and inks. In view of the decreasing pollution, adhesives and paints of high-solid, powder-coating, waterborne, and hot-melt types are attracting attention. At present, however, no high-solid or nonsolvent type polyurethane adhesive for general purposes has been developed that satisfies both the ease of handling and the excellent characteristics of the solvent-type polyurethane adhesive containing a large amount of a solvent. Development of a polyurethane adhesive that satisfies the VOC regulation and contains no free isocyanate monomer is strongly demanded.

Uretdione group-containing polyisocyanate compounds and related application techniques are known as shown in U.S. Pats. 4,044,171, 4,430,474, 4,442,280, and 4,801,663. The techniques relating to toluene diisocyanate dimer have long been known in the field of powder coating and millable rubber hardeners. Uretdione group-containing polyurethane resins of isophorone diisocyanate type and diphenylmethane diisocyanate type have come to be known recently.

Konishi, et. al., disclosed a one-component polyurethane adhesive including, as a main component, a polyurethane precursor prepared by reaction of a polyisocyanate precursor. The precursor containing a uretdione group and the active hydrogen group at an equivalent ratio of the uretdione group to the active hydrogen groups of from 0.25 to 1.0. The present invention relates to formulation and use of a one-component type polyurethane adhesive that uses a polyurethane precursor as a main component having uretdione groups and active hydrogen groups at a prescribed ratio. It cures by chain extension and cross-linking on simple heating without evolution of a free polyisocyanate monomer, and it is excellent in operability and safety.

The features of the novel polyurethane adhesive of the Konishi invention to meet the above demands are as follows:

1. Sufficient adhesiveness to various base materials and excellent film properties owing to high cohesive energy characteristic of conventional polyurethane adhesives
2. Ease of handling in coating and spraying owing to low viscosity achieved by two-component polyurethane adhesives
3. Excellent heat resistance and excellent durability owing to cross-linking formed by reaction at the time of adhesion
4. Semipermanent storage stability and long pot life that are achievable by lacquer type adhesives and blocked isocyanate systems
5. No liberation or no evaporation of free polyisocyanate monomer during formulation and heating, and reduction of the amount of a solvent to meet the VOC regulation for environment protection
6. High-solid, nonsolvent, or solid types of adhesive

13.9.2 HDI-Type Curing Agent with Low Viscosity and High Solid Content

Coating materials and adhesives have come to be strongly desired that meet the requirements on material saving, low environmental pollution, and safety. Consequently, coating materials and adhesives of high-solid, powder, and aqueous types, and hot-melt type adhesives have come into the spotlight.

Particularly in the United States, high-solid types of coating materials and adhesives are being used to meet the regulation of VOC (volatile organic compounds). In Japan, although the conversion to the high-solid type is slow in comparison to the conversion in the United States at the moment, the use of the high-solid type of coating materials and adhesives is being investigated because of (a) the movement in environment protection in local governments, (b) the present status of improvement of the conventional coating materials with conversion to this type, and (c) the trend toward a unified quality assurance system for U.S.-made cars and Japanese-made cars in the automobile industries, and (d) other related considerations.

On the other hand, in practical coating operation, the coating materials needs to be diluted with a suitable solvent to a viscosity required for an coating method to be applied. When a high-viscosity varnish is used in the coating, a larger amount of the solvent has to be used for thinning, which decreases the coating thickness obtained by one coating operation. It also requires repetition of the coating operation to attain a desired coating film thickness, because a minimum coating film thickness and a minimum hiding property are required to achieve the purpose of protection and coloring of articles with a coating material. In the coating process, excessively rapid evaporation or poor evaporation of the solvent causes pinhole formation in the coating film and sagging of the coating material, which tends to give a poor appearance to the coating film.

One-component and two-component types of polyurethane coating materials and adhesives, which employ polyisocyanate of the type of hexamethylene diisocyanate (HDI) as a curing agent and an acrylic polyol or polyesterpolyol as the polyhydroxyl compound, are superior in weatherability, chemical resistance, abrasion resistance, etc. They are used for exterior coating and adhesion-bonding in automobiles and buildings. In such polyurethane applications, a high-solid type system (namely, a low-viscosity type system) is demanded to decrease the amount of the solvent for dilution to get practical coating viscosity, as mentioned above.

As an HDI type of polyisocyanate curing agent, on polyisocyanurate type is described to be superior in weatherability in Ref. 21. However, commercially available curing agents of this type have high viscosity in the state substantially free from HDI monomer and solvent.

Yoshida et al. disclosed a curing agent for a two-component type polyurethane containing as a main constituent a polyisocyanate derived from hexamethylene diisocyanate, the polyisocyanate having a viscosity of lower than 1,400 mPa/25° C, substantially free from monomeric hexamethylene diisocyanate, and a solvent having a uretidine dione dimer content of more than 10%, and having an isocyanurate cyclic trimer content of less than 60%.[22]

For the purpose of solving the above-mentioned problems in conventional polyurethane coating materials and adhesives, Yoshida et. al. noticed the uretidine dione structure, which is of lower molecular weight than the isocyanurate structure, introducing the uretidine dione structure into the molecule in a large quantity to give a lower molecular weight and a lower viscosity to the curing agent.

13.9.3 Clear, Aqueous PU Dispersions as Universal Household Adhesives

Universal household adhesives, also known as *multipurpose* adhesives, are used for bonding a number of substrates encountered in the home (paper, cardboard, photographs, fabrics, leather, felt, bast, cork, films, metals such as aluminum and iron, china, ceramics, glass, wood, and various plastics such as polystyrene foams). The adhesives in question are expected to produce an adequate adhesive effect on these various substrates, which differ chemically and physically in their surface structure and which are normally subjected to a special surface treatment before bonding.

Compared with the large variety of classes and types of adhesive used in industry and workshops, only a few substances are capable of meeting the stringent demands imposed on the universality of a multipurpose household adhesive. Among those substances, only polyvinyl acetate and its copolymers are widely used—normally in solution or, for gluing wood, in the form of a dispersion.

The demand for universality represents a particularly difficult selection criterion for an adhesive. Ultimately, it means that the adhesive molecules must show equally high affinity for polar and apolar interfaces. Accordingly, the statement that a certain substance is suitable as an adhesive does not indicate to the expert whether it can also be used as a universal household adhesive.

In addition to the universality requirement, there has also recently been a demand for solventless, physiologically safe, clear, aqueous formulations of neutral odor in the field of universal household adhesives. However, these formulations are intended at the same time to lead to adhesives of which the dried films, in turn, have a certain resistance to water. In addition, these water-based adhesives are also intended to be able to bond substrates that are difficult to bond, such as plastics. They are also intended to have a long storage life.

This requirement profile could not be fully satisfied either on the basis of the binders hitherto preferred for universal adhesives, namely polyvinyl acetate and vinyl acetate copolymers, or by such alternatives as nitrocellulose. Although polyvinyl acetate can be produced without solvents in the form of aqueous dispersions, the dispersions obtained are not transparent but milky white. They show good performance properties when used, e.g., as wood glue. The acrylates and styrene acrylates widely used as dispersion adhesives are also not known on the market in the form of transparent household adhesives with the properties mentioned.

Klauck et al. disclosed an invention related to the use of a substantially clear and largely solventless, aqueous, one-component polyurethane dispersion based on the reaction products of a polyol mixture consisting completely or partly of polypropylene glycol, a mixture of polyfunctional isocyanates (TMXDI), a functional component capable of salt formation in aqueous solution, and, optionally, a chain-extending agent as a universal household adhesive.[23] They also refer to a process for its production in the absence of inert solvents.

It is known that specially selected aqueous polyurethane dispersions, known for decades, are suitable as multipurpose universal adhesives. Polyurethane dispersions consist of adducts of polyfunctional isocyanates (isocyanate component) with polyfunctional OH compounds (polyol component) that contain co-condensed units capable of salt formation in aqueous solution. Klauck et al. surprisingly found that polyurethane dispersions in which the polyol component is based on polypropylene glycol, and in which the isocyanate component is based on tetramethyl xylene diisocyanate, are also suitable as universal household adhesives and provide good adhesion values. They also found that dispersions of the type in question can be produced without using inert solvents.

13.9.4 High Temperature Resistant Urethane-Isocyanurate Adhesive[24]

Urethane adhesives are used for bonding automotive fiberglass reinforced parts. Recent adhesives have been developed that have high bonding strength without the need for surface treatment of the fiberglass reinforced part as in U.S. Pat. 4,876,308. Such an adhesive in high-temperature applications, e.g. paint baked at temperatures of 400° F (204° C), degrades and results in fiberglass reinforced plastic assemblies having undesirable bond strength.

U.S. Pat. 4,876,308 relates to a two-component type polyurethane adhesive for fiberglass-reinforced plastics that requires no cleaning, preparation, or treatment of the surface.[24] The curative component is a nitrogen-free liquid polyether or polyester in an amount such that, in association with a urethane prepolymer component containing free NCO groups, the ratio of the free NCO groups in the prepolymer component to the OH groups and NH groups in this curative component is at least 1.2, and desirably at least 1.35.

Wang et al. disclosed a two-component polyurethane composition with good shelf stability, and the adhesive made by curing it has good high-temperature stability and no adhesion loss after high-temperature bake. The base or prepolymer component containing a primary hydroxyl intermediate also has a sufficiently large excess of free isocyanate that the ratio of free NCO groups in the prepolymer component to OH curative groups, plus any amine groups in the curative component, is generally from about 1.2 to about 2.0. The curative component contains a trimerization catalyst so that, upon cure, isocyanurate units are produced. The system also contains molecular sieves as well as phosphorous-type adhesion promoters.

The Wang invention relates to a primerless, high-temperature-resistant urethane-isocyanurate adhesive composition made from a two-component (i.e., a prepolymer component and a curative component) system. The urethane system contains trimerization catalyst, molecular sieves, and phosphorous-type adhesion promoters.

References

1. Dombrow, B.A. 1957. *Polyurethanes*. New York: Reinhold Publishing.

2. U.S. Pat. 2,650,212. 1953. Farbenfabriken-Bayer.

3. U.S. Pat. 2,769,826. 1956. Stoner-Mudge Co.

4. U.S. Pat. 2,801,648. Anderson, J.F. and Fiedler, L.F. 1957. B.F. Goodrich.

5. Schollerberger, C.S., Scott, H., Moore, G.R. 1958. Rubber World. 137,549.

6. U.S. Pat. 2,871,218. 1959. B.F. Goodrich.

7. U.S. Pat. 2,968,575. Mallone, J.E., E.I. du Pont de Nemours and Co.

8. *Experimental Urethane Latex E-204 Bulletin,* Wyandotte Chemicals Corp., 1963.

9. Suskind, Stuart P. 1965. *Journal of Applied Polymer Science* 9, 2451–2458.

10. U.S. Pat. 3,015,650. 1962. B.F. Goodrich.

11. Kimball, M.E. *Adhesives Age,* June 21, 1981.

12. U.S. Pat. 3,802,988. 1974. Continental Tapes.

13. Dieterich, D., and Rieck, J.N. 1978. Aqueous polyurethane adhesives and their possible uses. *Adhesives Age* 21(2):24.

14. Volthenberg, H. Von, 1984. *Eur. Adhesives, Sealants* 1(4):28.

15. Bragole, R.A. *Urethanes in Elastomers and Coatings.* Westport, Conn.: Technomic Publishers, 1973.

16. Heiss, H.L., Saunders, J.H., Morris, M.R., Davis, B.R., and Hardy, E.E., *Ind. Eng. Chem.* 46:1498 (1954).

17. Japanese Pat. 80,08,344.

18. British Pat. 1,250,266.

19. MacIver, G.M. Structural adhesives for composite bonding. *Proc. Adhes. Seal. Council Mini Seminar,* April 1987.

20. Konishi, Shin, Hirayama, Shinji, Hidai, Takao, Morikawa, Yukihiro, Uehara, Koichi, and Hashimoto, Sadako. U.S. Pat. 5,410,011. Apr. 25, 1995. "One-component polyurethane type adhesive, and use thereof," to Inoac Corp, Nippon Polyurethane Industry Co., Ltd.

21. *Polyurethane Resin Handbook.* Nikkan Kogyo Shimbun Ltd., 404–406.

22. Yoshida, Mitsuhiro, Sato, Susumu, Obuchi, Yukio, Konishi, Shin, and Shindo, Masanori. U.S. Pat. 5,354,834. Oct. 11, 1994. "Polyisocyanate curing agent, and coating composition and adhesive composition employing the same," to Nippon Polyurethane Industry Co., Ltd., Tokyo, Japan.

23. Klauck, Wolfgang, Gierenz, Gerhard, Maier, Wolfgang, Hoefer, Rainer, and Gruetzmacher, Roland. U.S. Pat. 5,270,433, Dec. 14, 1993. "Polyurethane-based universal household adhesive," to Henkel Kommanditgesellschaft auf Aktien, Dusseldorf, Federal Republic of Germany.

24. Wang, Chia L., Melby, Earl G., and Cocain, H. William. U.S. Pat. 5,175,228. Dec. 29, 1992. "Two-component primerless urethane-isocyanurate adhesive compositions having high temperature resistance," to GenCorp Inc., Fairlawn, Ohio.

14

Waterborne Polyurethanes

14.1 Introduction

Aqueous PUDs have been gaining increasing importance in a wide-range of applications, due to their excellent properties. These include:

- Adhesion to various substrates
- Resistance to chemicals, solvents and water
- Abrasion resistance
- Flexibility and toughness

Polyurethane (PU) dispersions are especially suitable for painting plastics and wood, and for coating both metallic and mineral substrates. Major markets for waterborne polyurethanes are summarized in Table 14.1.

TABLE 14.1 Markets for Waterborne Polyurethanes

Wood floor coatings and finishes
Currently using oil-modified or moisture-cure urethanes
Factory-applied coatings changing to cross-linked WD urethanes
Floor-care products made of urethane acrylics
Vinyl upholstery topcoats
Automotive base coats
Coatings for plastics
Printing inks
Fiberglass finishes
Leather finishes
Adhesives

Conventional PU resin systems usually contain a high proportion of volatile organic solvents (approx. 40 to 60% by weight). At the end of the 1960s, the manufacturers of polyurethane resins developed processes that permitted the synthesis of low-solvent or solvent-free aqueous PU dispersions. The increased use of PU dispersions is attributable not only to the above-mentioned profile of properties but also to pressures from the coatings industry and environmental legislation for raw materials with a low organic-solvent content (low VOC, volatile organic compounds). Advantages of waterborne polyurethanes are summarized in Table 14.2.

Many improvements have been made in waterborne polyurethanes in the past 10 years. Table 14.3 presents some of the most significant characteristics of state-of-the-art waterborne polyurethanes.

Central to the utilization of polymers in aqueous media is the fact that certain polar functional groups are capable of conferring water solubility or water dispersibility to an otherwise water-insoluble poly-

TABLE 14.2 Waterborne Urethane Advantages

Fast dry
Outstanding flexibility and impact resistance
Excellent abrasion resistance
Nonflammable
Easy water clean-up
Low volatile organic content (VOC) = less pollution

TABLE 14.3 State of the Art in Waterborne Polyurethanes

Most water-dispersible urethanes contain N-methyl pyrrolidone solvent
Solvent used to maintain viscosity of prepolymer intermediate
Some solvent-free WD urethanes made in acetone or MEK, followed by solvent removal
Special equipment required to recover low-boiling solvents
Use of tertramethylxyxylene diisocyanate (TMXDI) produces low-viscosity prepolymers
Solvent-free polyurethanes sometimes exhibit poor water resistance upon air drying
Water resistance can be improved by:
addition of co-solvent
use of aziridine cross-linker
baking the dried coating
using ammonia as the neutralizer

urethane. Best known are carboxylic acid groups, sulphonic acid groups, and tertiary amine groups. The concentration of such functional groups in the polymer is highly influential in determining the solubility or dispersibility in an aqueous environment. Thus, at high concentrations, the polymer may be water soluble, and at lower concentrations the polymer may be water dispersible, provided its molecular weight/viscosity is not excessive. At even lower concentrations, the polar group may be capable of providing charge or steric stabilization to a dispersion of the polymer in water.

Aqueous PU dispersions may be divided into two classes. One class consists of polymers stabilized by external emulsifiers, and the other of those in which stabilization is achieved by the inclusion of hydrophilic centers in the polymer. Such hydrophilic centers, in principle, may be one of three types:

- Nonionic groups: e.g., polyethylene oxide chains
- Cationic groups: e.g., alkylated or protonated tertiary amines
- Anionic groups: e.g., carboxylate or sulphonate groups

Introduction of these hydrophilic groups, which function as *internal emulsifiers,* makes it possible to produce stable aqueous emulsions with a mean particle size in the region of 0.01 to 0.2 μm. These internal emulsifiers are summarized in Table 14.4.

A range of synthetic routes are available for the preparation of waterborne polymer systems. The polymers can differ widely in terms of (1) the nature and concentration of the polar solubilizing group, (2) molecular weight, and (3) the hydrophobic/hydrophilic characteristics of the other units in the polymer chain. As a result, aqueous polymer systems with a wide range of different morphological and physical characteristics can be obtained.

Many such systems are utilized in surface coatings; some are well characterized and understood, while others are not. We will discuss the various types of polymers in terms of their methods of preparation, their physical characteristics, and some of their uses in the surface coatings industry. This classification of aqueous polymers is made difficult by the fact that there is no clear-cut demarcation between the various polymer types.

14.2 Solubilizing Groups

A wide range of polar water solubilizing functional groups are known. These functional groups can be broadly classified into the three categories of anionic, cationic, and nonionic (i.e., nonionizable). The

TABLE 14.4 Internal Emulsifiers Used in Waterborne Polyurethanes

Nonionic types
Require emulsifiers
Need high-shear equipment
Cationic types
Difficult to disperse
Require extensive thinning
Solvents must be removed
Anionic types
Freeze-thaw stable
Good pigment wetting
High gloss
Less co-solvent
Water reducible (anionic colloidal dispersions of fully reacted thermoplastic urethanes)
Commonly referred to as:
Colloidal dispersions (preferred)
Urethane latexes
Urethane emulsions

most commonly utilized functional groups conferring water solubility include weak acids (carboxylic), strong acids (sulphonic), weak bases (amine), strong bases (tertiary and quaternary ammonium compounds), and nonionizable groups (ethylene oxide, vinyl pyrrolidone). In the cases where the functional group is ionizable, the water solubilization effect is greater when the group is in its ionized (salt) form. Therefore, many copolymers containing ionizable water solubilizing groups are water soluble in the pH region over which the group is substantially ionized, and water-insoluble in the pH region over which the group is substantially un-ionized. For example, copolymers containing weak carboxylic acid groups (e.g., from acrylic acid) are often water soluble (depending on the concentration of acrylic acid) at pH values above the pKa of acrylic acid (approximately 4.2) and water-insoluble below this pH value. Conversely, basic polymers containing, for example, tertiary amine groups are soluble (assuming a sufficiently high concentration of functional groups) at low pH values (i.e., below the pKb) and insoluble above this pH. This switch in solubility, with change in pH, is widely utilized as the basis of anodic and cathodic electrodeposition, since large changes in pH are experienced at the anode and cathode.

Similarly, where the ionizable functional group is utilized to provide charge stabilization to a polymer colloid in water, the stabilizing effect is highly dependent on the pH of the system. A change in pH can bring about a total loss in colloid stability. This effect is also used advantageously in electrodeposition.

14.3 Classification of Synthetic Polymers Used in Aqueous Media

It is useful to classify the polymers in two distinct ways:

1. Based on the synthetic procedure
2. Based on the physical state of the polymer in water, e.g., highly water-soluble, water-insoluble dispersion, etc.

In this chapter, both classifications will be discussed, since neither one alone brings out all of the important aspects.

14.3.1 Classification According to Synthetic Procedure

Two main categories can be defined:

1. Polymer synthesized in the presence of water
2. Polymer synthesized in the bulk or in solution in a water-miscible organic solvent and subsequently added to water

In the former category, polymerization can be either by free-radical or step-growth polymerization. Ionic polymerization procedures can be virtually ruled out, because such polymerizations cannot be performed in the presence of water. A recent development is the ruthenium catalyzed ring opening metathesis polymerization, which can be conducted in water resulting in an aqueous polymer dispersion.

14.3.2 Synthesis Performed in the Presence of Water

The main procedures falling within this category are

1. Emulsion polymerization
2. Suspension polymerization
3. Free-radical solution polymerization
4. Dispersion polymerization
5. Inverse emulsion polymerization

Some of the main features of each procedure are briefly discussed below.

Emulsion polymerization is the most widely practiced of the procedures and is characterized by the fact that a free-radical initiator (usually water soluble) is used to polymerize free radically polymerizable monomer(s) to give a water-insoluble polymer. In general, the monomers have a low but finite water solubility, although it is common practice to also incorporate water-soluble acid co-monomers such as acrylic acid and methacrylic acid. Emulsion polymerization differs mechanistically from the other forms of radical polymerizations and exhibits the industrially important characteristic of being able to simultaneously obtain both high-molecular-weight polymer and a high polymerization rate. Furthermore, the viscosity of the resultant emulsion is not molecular weight dependent; hence, the ability of this procedure to prepare very high-molecular-weight polymers can be fully utilized. Emulsion polymerization gives particles in the diameter range approximately 50 to 500 nm, with the ability to accurately control particle size. Particle stability is usually achieved through the use of adsorbed surfactants, often in conjunction with the stabilizing effect of charged polymer end groups (e.g., sulphate groups from persulphate polymerization initiators) and the stabilizing effect of ionic groups arising from the presence of minor proportions of ionic co-monomer (e.g., carboxylate groups from acrylic acid and sulphonate groups from sulphoethyl methacrylate). Emulsion polymers are widely used in waterborne paints.

Suspension polymerization differs from emulsion polymerization in that a monomer-soluble free-radical polymerization initiator is utilized to polymerize essentially water-insoluble monomers. The polymerizations can be regarded as a set of bulk polymerizations taking place in each monomer droplet, and thus molecular weight is inversely proportional to initiator concentration and polymerization rate, as is typical of a bulk radical polymerization. Particle stabilization is usually through the use of adsorbed water-soluble polymers such as partially hydrolyzed polyvinyl acetate and acrylic acid copolymers, and particle size is generally in the range 0.01 to 1 mm. The relatively large particle size of suspension polymers facilitates the isolation of polymer particle (e.g., by centrifugation), and such *beads* are widely utilized in solvent-borne surface coatings and in photocopy toner resins. Suspension copolymers containing sufficient carboxylic acid co-monomer to be alkali-soluble are utilized at alkaline pH in printing inks. As a result of the large particle size, suspension polymers are rarely directly utilized in the form of the aqueous dispersion because settlement and film forming problems would be expected.

Free-radical solution polymerization is characterized by the feature that monomer(s), initiator, and polymer are soluble in the continuous phase, which may be either water or a water/solvent mixture. Examples of appropriate monomers are acrylic acid, acrylamide, N-vinyl pyrrolidone, and 2-acrylamido-2-methyl-propane-sulphonic acid. This procedure is utilized for the preparation of dispersants and thickeners. Associative thickeners can be prepared by utilizing a co-monomer bearing a hydrophobic group capable of associating in water, or by introducing such a hydrophobic group by a subsequent reaction with the water-soluble polymer.

Dispersion polymerization entails the polymerization of monomers that are soluble in a liquid medium to produce a polymer that is insoluble in the liquid medium. If the polymerization is performed in the presence of an appropriate colloid stabilizer, the polymer precipitates to produce a stable polymer dispersion. While nonaqueous dispersion polymerization is the best known example of this process, a free radically initialed dispersion process has been described in which the continuous liquid phase is a miscible water/alcohol mixture and in which a copolymerizable surfactant is utilized.

Inverse emulsion polymerization entails the emulsification of an aqueous monomer solution in a continuous oil medium using a water-in-oil emulsifier, and polymerizing using an oil- or water-soluble polymerization initiator to give submicron water-swollen polymer particles dispersed in oil. This process gives the characteristic benefits of emulsion polymerization previously mentioned, i.e., high-molecular-weight polymers produced at high polymerization rate, high solids, and low viscosities. The most practically important feature is that the addition of excess water causes inversion and rapid dissolution of the water-swollen polymer particles, even when the polymer is of very high molecular weight and might otherwise be very slow to dissolve. Typical monomers that can be used in this process are acrylamide, acrylic acid, and 2-sulphoethylacrylate. This process is widely utilized for the preparation of high-molecular-weight flocculants but is rarely used for making water-soluble additives for surface coatings.

The techniques of aqueous polymerization in monomer mini-emulsions, monomer microemulsions, and monomer vesicles have been extensively studied in recent years; however, little if any practical exploitation in surface coatings has been found.

Preformed Polymer Subsequently Added to Water

The main procedures can be classified as

1. Free-radical addition or step-growth polymerization in the bulk to give a low- or medium-molecular-weight polymer in which the water solubilizing group is incorporated via one or more of the co-monomers, followed by addition to water (or vice versa). In some instances, chain extension is carried out during or subsequent to the addition to water.
2. Free-radical addition or step-growth polymerization in the bulk, with subsequent functionalization to introduce a water solubilizing group, followed by addition to water (or vice versa).
3. Free-radical addition or step-growth polymerization in a water-miscible co-solvent or a water/co-solvent mixture with incorporation of the water solubilizing group via one or more co-monomers, followed by addition to water (or vice versa) and, in some instances, co-solvent removal.
4. Free-radical addition, step-growth, or ionic polymerization in a solvent, with subsequent functionalization to introduce a water-solubilizing group, followed by addition to water and, in some instances, co-solvent removal.

While all of these procedures are industrially practiced in the preparation of binders for surface coatings, procedures (1) and (3) are probably the most widely utilized.

An important feature of this type of approach is that the viscosity of the polymer should be sufficiently low to allow subsequent dispersion or dissolution in water. The necessary low viscosity is achieved by a number of techniques as follows:

1. Limiting the polymer molecular weight
2. Adding the polymer as a solution in solvent (which may or may not be the polymerization solvent)
3. Adding the polymer at elevated temperature to the water

Each of these approaches to reducing viscosity is exemplified in the various routes for preparing aqueous polyurethanes, as described later.

Agitation conditions can be highly important when carrying out such processes, especially when the concentration of water-solubilizing functional groups is low and the desired product is an aqueous dispersion. In general, higher-viscosity/high-molecular-weight polymers can be utilized when high-shear agitation conditions are employed. Both shear and the viscosity of the dispersed *polymer rich* phase are highly influential in determining the particle size of the final dispersion.

Not only is the dependence of attainable particle size on the viscosity of the dispersed *polymer* phase and the shear rate indicated, but the importance of the polymer water interfacial tension is also highlighted. The final attainable particle diameter decreases (an indication of increasing ease of dispersion) as the interfacial tension decreases. Such a decrease in interfacial tension can be obtained by the inclusion of water-solubilizing functional groups or by the addition of an appropriate surfactant. In many instances, it is not necessary to utilize a surfactant for this purpose, as inclusion of the water-solubilizing functional groups causes a sufficient reduction in interfacial tension to achieve the desired reduction in particle size. This situation is frequently encountered with carboxylated urethanes when very low particle sizes (typically 30 to 100 nm) can be achieved in the absence of surfactant and under low-shear agitation conditions. Significantly, it is not only the ability to achieve low particle size that is important, but also the ability to maintain the colloid stability of the low particle size dispersion that has been created. Thus, it is necessary to either select a water-solubilizing functional group that provides adequate colloid stability at the relevant pH and ionic strength or introduce an appropriate surfactant for this purpose.

Another important factor involved in dispersing a preformed polymer in water is the order of addition. When water is added to the polymer (or to a solution of polymer in organic solvent), an inversion process occurs at some stage during the addition. This inversion corresponds to the composition at which the continuous phase switches from being the polymer phase to being the water phase. Such inversion processes can be very critical to manage and generally involve passing through a maximum in viscosity. Aqueous step-growth polymers made by the subsequent addition of a preformed polymer to water are frequently used to make waterborne polyurethanes.

14.3.3 Classification According to Physical State

The precise physical nature of the aqueous polymer system obtained using the various synthetic procedures outlined in the previous section is highly dependent on a number of variables, some of which relate to the polymer itself (molecular weight, concentration, and nature of the solubilizing groups; backbone hydrophobicity; cross-linking), some of which relate to the synthetic process employed (shear and surfactant content) and some of which relate to the aqueous solution/dispersion itself (pH, polymer concentration, and ionic strength).

In this section, an attempt is made to classify the types of systems that can be obtained. The two extremes of a continuous spectrum of physical states are

1. Systems in which the polymer is highly water soluble over the entire pH range, e.g., polyethylene oxide homopolymer
2. Systems in which the polymer is highly water-insoluble, exists as an aqueous dispersion of the polymer throughout the pH range, and has the minimum concentration of water-solubilizing groups and surfactant to ensure colloid stability of the dispersed polymer particles, e.g., emulsifier-free polystyrene latex

Between these two extremes is a whole range of possible intermediate states, many of which are utilized in the surface coatings industry. At one end of the spectrum lies polyacrylic acid, a polymer which, in the normally encountered molecular weight range, is highly water soluble over a wide pH range, even at high ionic strengths. Polyacrylic acid is used as a flocculant (high molecular weights) and as a thickener in aqueous coatings (moderate molecular weights). Solutions of polyacrylic acid in water progressively increase in viscosity as the pH is increased. an effect which at its simplest level can be explained by the progressive increase in ionization and hydration of the carboxyl group, with the mutual ionic repulsion between the ionized groups causing the molecules to adopt a more expanded conformation, with a consequent increase in intermolecular entanglement and hence in viscosity.

At the other end of the spectrum lies a polymer dispersion based on a polymer in which all the constituent co-monomers are highly water insoluble (i.e., the dispersion contains no acrylic acid or other water soluble monomer) such as polystyrene or polybutyl methacrylate, with colloid stability being achieved by the use of an adsorbed surfactant that is effective throughout the pH range and/or initiator-derived, charged-polymer end groups. In such a system, the polymer molecules are likely to adopt a similar conformation and packing to what they would adopt in the bulk.

14.3.4 Associative Thickeners

A fundamental problem with aqueous polymer dispersions is that their rheology is less favorable for surface coatings applications than that of polymer solutions that are more Newtonian. For example, the viscosity of a colloidal dispersion drops very quickly with increasing shear rate, and low shear rate viscosity tends to be too high to allow good flow and leveling. The high shear rate viscosity tends to be too low to give good brush pick-up and transfer from roll coaters. Addition of water-soluble polymer (thickener) is generally vital to increase high shear rate viscosity to an acceptable level for application of the coating. However, in so doing, the low shear rate viscosity of the dispersion is raised disproportionately, with an unfavorable effect on the flow and leveling characteristics. The advent of associative thickeners has undoubtedly been a major advancement, since such hydrophobically modified water-soluble polymers provide the opportunity to independently adjust the high shear rate viscosity and the low shear rate viscosity. This control results, in essence, from the fact that the addition of co-solvent or low-molecular-weight surfactant to a system comprising polymer dispersion and hydrophobically modified water-soluble polymer can reduce the viscosity by reducing the magnitude of the hydrophobic associations between thickener hydrophobes and/or between these hydrophobes and the latex particle surface. This reduction in viscosity is far more pronounced at low shear rates than at high shear rates, because the initial hydrophobic associations present before the addition of co-solvent or surfactant have a large effect on viscosity at low shear rates, but much less effect on viscosity at high shear rates.

14.4 Aqueous Dispersions of Preformed Step-Growth Polymers

The technology for the preparation of aqueous polyurethane dispersions has progressed considerably. In many respects, performance levels can now match that of solvent-borne polyurethanes. Optimized polyurethanes can offer a superior balance of properties over those products of typical emulsion polymerizations, especially in terms of the balance between mechanical properties (abrasion resistance, hardness, and tensile strength) and film-forming temperature.

The ability of aqueous polyurethanes to combine film formation at low temperature with the ability to give high film hardness is in part a result of this water-swollen morphology that aids film formation (i.e., water plasticization). The ability to prepare very low particle size polyurethane dispersions (20 to 100 nm) is also helpful in terms of film formation, since decreasing particle size aids the rate and extent of particle coalescence. The high film hardness that can be achieved with these systems arises from the presence of a hard-block phase.

The more modern processes used commercially for the preparation of aqueous polyurethane dispersions share two common features. In all cases, the first step is the formation of a medium-molecular-weight isocyanate-terminated prepolymer by the reaction of suitable di- or polyol with a stoichiometric excess of di- or polyisocyanate. However, the ways in which this prepolymer is chain extended and the viscosity maintained at a manageably low level differ. The polymer to be dispersed in water is functionalized with water-solubilizing/dispersing groups which are introduced either into the prepolymer prior to chain extension or are introduced as part of the chain extension agent. Thus, small-particle stable dispersions frequently can be produced without the use of an externally added surfactant.

In the solution process, the isocyanate-terminated polyurethane prepolymer is chain extended in solution to prevent an excessive viscosity from being attained. The preferred solvent is acetone, and hence this process is frequently referred as the *acetone process*. The chain extender can be, for example, a sulphonate functional diamine, in which case the water-solubilizing/dispersing group is introduced at the chain extension step. The chain extended polymer is thus more properly described as a *polyurethane urea*. Water is then added to the polymer solution without the need for high-shear agitation and, after phase inversion, a dispersion of polymer solution in water is obtained. Removal of the solvent by distillation yields the desired aqueous polymer dispersion. This is shown in Reaction 14.1.

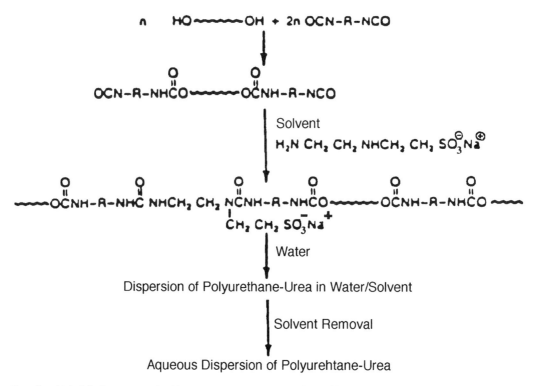

Reaction 14.1 Solution process. In this process, an anionomer is formed by incorporation of a sulfonic salt into the chain.

While the acetone process is effective and reproducible, it is naturally limited to the requirements that the polymer be uncross-linked and acetone soluble, and so the resultant coatings tend not to be very solvent resistant. The need for solvent removal is another obvious process disadvantage.

In the prepolymer mixing process, a hydrophilically modified isocyanate terminated polyurethane prepolymer is chain extended with diamine or polyamine at the aqueous dispersion step. This chain extension is possible because of the preferential reactivity of isocyanate groups with amine rather than with water. To maintain this preferential reactivity with amine, it is necessary to prevent the water temperature from exceeding the value at which significant reactions occur between water and the isocyanate. The choice of isocyanates is clearly important in this respect. The prepolymer mixing process is extremely flexible in terms of the range of aqueous polyurethane ureas that can be prepared. This process has the major advantage of avoiding the use of large amounts of solvent and avoids the need for the final polymer to be solvent soluble.

The ketamine/ketazine process can be regarded as a variant of the prepolymer mixing process. The chain extending agent is a ketone-blocked diamine (ketamine) or ketone-blocked hydrazine (ketazine) that is mixed directly with the isocyanate terminated polyurethane prepolymer. During the subsequent water dispersion step, the ketamine or ketazine is hydrolyzed to generate free diamine or hydrazine, respectively, and thus quantitative chain extension takes place. A claimed advantage of the ketamine process over the prepolymer mixing process is that it is better suited for preparing aqueous urethanes based on the more water-reactive aromatic isocyanates. This process is shown in Reaction 14.2.

The hot-melt process involves the capping of a functionalized isocyanate terminated polyurethane prepolymer with urea at greater than 130° C to form a biuret. This capped polyurethane (which can be solvent free) is dispersed in water at approximately 100° C to minimize viscosity, and chain extension is carried out in the presence of the water by the reaction with formaldehyde that generates methylol groups. These, in turn, self-condense to give the desired molecular weight buildup. This process is shown in Reaction 14.3.

$$OCN-R-NH\overset{O}{\overset{\|}{C}}O \wsim OCNH-R-NH\overset{O}{\overset{\|}{C}}O\ CH_2\ \overset{CH_3}{\underset{CO_2^-\ HNR_3^+}{\overset{\|}{C}}}\ CH_2\ O\ \overset{O}{\overset{\|}{C}}NH-R-NCO$$

Hydrophilic Isocyanate-Terminated Prepolymer
+
Ketimine/ketazine

Reaction 14.2 Ketimine/ketazine process. In this process, a ketimine is preformed by reaction of a ketone with a diamine. This product can be blended with an isocyanate-terminated prepolymer. The mixture is dispersed in water releasing the diamine or hydrazine, which then reacts with the prepolymer to give the desired chain extension.

Hydrophilic Bis-Biuret

Dispersed Polyurethane–Urea

Reaction 14.3 Hot-melt process. In this process, a prepolymer is reacted with urea to form a hydrophilic bis-biuret. The bis-buret is reacted with formaldehyde, which undergoes polycondensation, thus resulting in a dispersed poly(urethane-urea).

Anionic, cationic, or nonionically stabilized aqueous polyurethane dispersions can be prepared. Anionic dispersions are currently the most widely utilized type and are usually either carboxylate or sulphonate functionalized co-monomers, e.g., suitably hindered dihydroxy carboxylic acids (dimethylol propionic acid) or dihydroxy sulphonic acids. This is shown in Reaction 14.4.

Cationic systems are prepared by the incorporation of diols containing tertiary nitrogen atoms that are converted to the quaternary ammonium ion by the addition of a suitable alkylating agent or acid. This is shown in Reaction 14.5. Nonionically stabilized aqueous polyurethanes can be prepared by the use of diol or diisocyanate co-monomers bearing pendant polyethylene oxide chains. Such polyurethane dispersions are colloidally stable over a broad pH range, but a high concentration of polyethylene oxide based co-monomer is required to achieve a low particle size dispersion. This is shown in Reaction 14.6. Thus, combinations of nonionic and anionic stabilizations are sometimes utilized to achieve a combination of small particle size and strong steric stability, without the need to utilize excessively high concentration of nonionic stabilizing co-monomer. Urethane dispersions employ this type of stabilization and can be blended with low pH, acid-containing acrylic latex.

Aqueous polyurethane dispersions are increasingly employed in the surface coatings industry because of the unique property balances that can be obtained by this very flexible chemistry. Important applications include wood lacquers, leather finishing, adhesives, plastic coatings, and glass fiber sizing.

A number of recent developments in aqueous polyurethanes are worth mentioning. Precross-linked urethanes can be prepared by the addition of a multifunctional co-monomer after preparation of the isocyanate-terminated prepolymer and before the aqueous dispersion stage. Such precross-linking need not deleteriously affect film formation, providing that the level of cross-linking is controlled, and it gives rise to an improvement in hardness and solvent resistance. Blocked polyurethanes can be prepared by the reaction of an isocyanate-terminated prepolymer containing acid functionality with, for example, an oxime followed by neutralization and dispersion in water. These water-dispersed blocked polyurethane

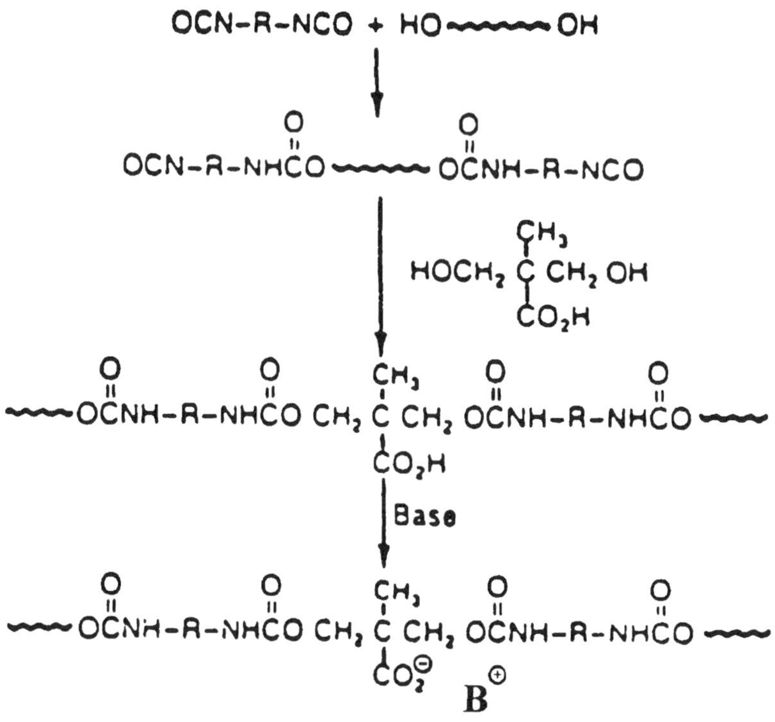

OCN–R–NCO + HO〰〰OH

OCN–R–NHCO〰〰OCNH–R–NCO

CH₃
HOCH₂ C CH₂ OH
CO₂H

〰〰OCNH–R–NHCO CH₂ C CH₂ OCNH–R–NHCO〰〰
CH₃
CO₂H

Base

〰〰OCNH–R–NHCO CH₂ C CH₂ OCNH–R–NHCO〰〰
CH₃
CO₂⊖ B⊕

Reaction 14.4 Preparation of anionic polyurethane ionomers. In this case, the prepolymer is reacted with dimethylolpropionic acid, resulting in a built-in carboxyl group. The product is then neutralized and transferred to water, where spontaneous particle formation occurs.

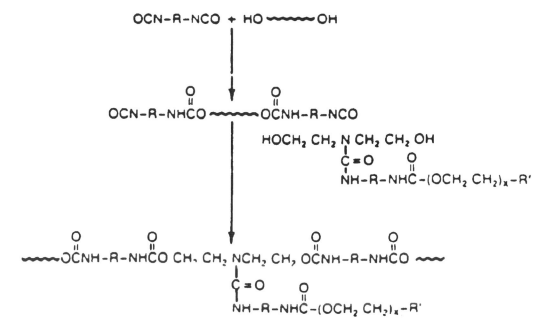

OCN–R–NCO + HO ～～～OH

O   O
‖   ‖
OCN–R–NHCO～～～OCNH–R–NCO

CH₃
HOCH₂ CH₂ N CH₂ CH₂OH

O   O   CH₃   O   O
‖   ‖   ‖   ‖
～～OCNH–R–NHCO CH₂ CH₂ NCH₂ CH₂ OCNH–R–NHCO ～～

Acid
or
Alkylating Agent

O   O   CH₃   O   O
‖   ‖   ⊕   ‖   ‖
～～OCNH–R–NHCO CH₂ CH₂ N CH₂ CH₂ OCNH–R–NHCO～～
A   B⊖

Reaction 14.5 Preparation of cationic polyurethane ionomers, by final reaction with an acid or alkylating agent.

OCN–R–NCO + HO ～～～OH

O   O
‖   ‖
OCN–R–NHCO ～～～OCNH–R–NCO

HOCH₂ CH₂ N CH₂ CH₂ OH
C=O   O
            ‖
NH–R–NHC-(OCH₂ CH₂)ₓ–R'

O   O   O   O
‖   ‖   ‖   ‖
～～OCNH–R–NHCO CH₂ CH₂ NCH₂ CH₂ OCNH–R–NHCO ～～

C=O   O
        ‖
NH–R–NHC-(OCH₂ CH₂)ₓ–R'

Reaction 14.6 Preparation of a polyurethane with nonionic hydrophilic components.

prepolymers can be combined with hydroxy functional acrylic or polyurethane dispersions to provide effective waterborne baking finishes. The choice of both isocyanate and blocking agent is important in controlling deblocking temperature and hydrolytic stability.

Two-pack aqueous polyurethanes based on hydrophilically modified polyisocyanates are another recent development. These reactive two-pack systems comprise a hydrophilically modified polyisocyanate and a co-reactive aqueous polymer, e.g., a hydroxy functional polyurethane dispersion or a water dispersed acrylic polyol. Because of the reaction of isocyanate groups with water, the pot life of such systems is limited, although an increase in pot life up to five to six hours has been obtained by the hydrophilic modification of the isocyanate. These systems are claimed to give very good dry-film properties but do not possess the advantages of the aqueous polyurethane dispersion (previously discussed) in terms of being one-pack systems or free of reactive polyisocyanates during application (spraying).

14.5 Hybrid Systems

Hybrid systems are increasingly utilized in which the binder is a combination of more than one generic class of polymer, e.g., urethane-acrylics and epoxy-acrylics. Many preparative techniques have been reported, especially in the patent literature, and a full review is beyond the scope of this paper.

Probably the most widely utilized technique is to free radically polymerize a combination of monomers in the presence of a preformed polymer that may or may not be intrinsically water dispersible. If the preformed polymer is water dispersible, it can be used directly as a seed for a subsequent free-radical emulsion polymerization, whereas if the preformed polymer is not intrinsically water dispersible, then it can either be emulsified in water using a combination of a surfactant and high shear, or it can be dissolved (if soluble) in the monomers utilized in an emulsion polymerization. In some instances, the preformed polymer can have pendant or terminal functional groups that can participate in the subsequent emulsion polymerization, thus giving grafting between the two polymers.

Hybrid systems prepared by sequential polymerization can provide superior properties to those obtained from the corresponding polymer blends. This is because totally different particle (and hence film) morphology can be obtained in a sequentially polymerized system, since both polymers can exist in the same particle. Furthermore, grafting can provide superior properties (mechanical properties) compared to blends.

Urethane-acrylic aqueous dispersions prepared by performing an acrylic emulsion polymerization in the presence of an aqueous polyurethane can possess a range of advantages over the corresponding blends, e.g., reduced water sensitivity and the ability to prepare in cosolvent-free form. Since the polyurethane is generally more hydrophilic than the acrylic copolymer, the polyurethane concentrates at the particle surface. In a well designed system, the particles coalesce on film formation to give a film with a continuous polyurethane phase. Therefore, the resultant coating behaves more like the parent polyurethane than does the equivalent blend of acrylic latex and polyurethane dispersion.

14.6 Film Formation

The process of film formation is crucial to attaining good coating properties. The film-forming process for aqueous polymer dispersions is much more difficult to control and more critical than for systems that are a solution of polymer in organic solvent or in water. Aqueous dispersions form films by a coalescence process in which the individual polymer particles are forced together as the water is lost during drying, and the particles deform and eventually interdiffuse—ideally on a molecular scale. Incomplete coalescence can give rise to water sensitivity of the coating and, in the case of coatings on ferrous substrates, to a phenomenon known as *early rusting*. The coalescence process is highly dependent upon a number of parameters, being aided by decreasing particle size, decreasing polymer Tg, increasing temperature, cosolvent additions, absence of cross-linking, and increasing contact surface energy. On the other hand, the film-forming process of polymer solutions (aqueous and nonaqueous) is much less critical; i.e., good film formation takes place at room temperature, irrespective of polymer Tg. One reason for this good film formation is that molecular interpenetration is present before drying the coating.

14.7 Polyurethane Dispersions—An Overview for the Printing Industry

Polyurethane dispersions (PUDs) are an increasingly important and very versatile group of waterborne binders for inks, adhesives, and various protective or decorative coatings. Technical quality has improved dramatically during the past decade as a result of many innovations in basic chemistry, dispersion technology, formulations, and application techniques, which have been fueled by consumer awareness, plus current or pending legislation to reduce VOC levels. PUDs are used in, or are at least worthy of investigation for all applications that are currently served by other polymer latices or solvent-borne polyurethanes. The unique chemistry of polyurethanes means that it is possible to achieve endless variation in properties merely by altering the type and relative proportions of monomers used. Thus, PUDs are available that range from extremely soft, highly flexible coatings for textiles and leather to tough, hard coatings for the protection of wood, metal, concrete, and plastic surfaces.

Polyurethane dispersions essentially consist of high-molecular-weight polyurethane or, more precisely, poly(urethane-urea) chains, stabilized as roughly spherical submicron particles. The viscosities are low (typically 50 to 500 cps) compared with homogeneous polyurethane solutions, and PUD formulations often require thickening to achieve a suitable coating rheology. Solids content typically varies from 30% for harder grades up to 50% for softer grades, although some 60% solids versions are available, e.g., for textile coating and dipping applications.

Softer PUD grades are usually free of all solvents. However, harder grades usually contain some co-solvent, which acts mainly as an essential processing aid but will also assist particle coalescence. Strictly minimal quantities of aprotic, polar, and water-miscible solvents such as methyl ethyl ketone, tetrahydrofuran, dimethylformamide, or particularly N-methyl pyrrolidone (NMP) are normally used. While the lower-boiling-point solvents can be substantially distilled out during the final manufacturing stage, the higher-boiling types are retained in the PUD.

Visual appearance ranges from milky white to translucent, with a full range of intermediate opacity, and sometimes with a slight but distinct bluish or golden hue. The actual appearance of a PUD depends on a number of factors influencing particle size, including molecular composition, type of stabilization system used, presence of cosolvent and so on. Their ability to scatter light distinguishes PUDs from true aqueous polyurethane solutions and also facilitates analysis of particle size and size distribution. Translucent grades have very small particle size ($\approx$10 to 50 nm) and other characteristics approaching those of true solutions; consequently, storage stability is good with minimal sedimentation, film coalescence is optimized (which is important, e.g., for edge definition with printing inks), and excellent clarity and gloss levels are possible.

The film-forming properties of PUDs are usually better than those of other polymer latices of equivalent hardness. Another important point is that dried PUD coatings do not redissolve in aqueous or alkaline media, in contrast to other systems such as the alkali-soluble acrylics used in printing inks.

14.8 Classification

Commercial PUDs can be classified on the basis of two characteristics.

1. The internal stabilization mechanism (anionic, cationic, or nonionic), which also determines the usable pH range
2. Their chemical composition, notably the type of polyisocyanate (aliphatic or aromatic) and polyol (polyether, polyester, and polycarbonate) used in their manufacture, which gives some property indications for end applications

High-MW polyurethanes can be stabilized in water solely by using protective colloids or external emulsifying agents, and some of the fundamental research on PUDs followed this route.[2] Unfortunately, the particles of externally emulsified PUDs tend to be coarse (greater than 1–2 μμm), which leads to sedimentation problems and poor coalescence/film-forming properties. Derived coatings are also sensitive to re-emulsification and generally have poor chemical resistance. Virtually all PUDs are now manufactured by various processes that involve internal stabilization; i.e., the provision of self-emulsifying sites evenly distributed along the molecular chains. The active sites have strong hydrophilic character and are usually organic salt groups or nonionic species such as polyethylene oxide chains of low to medium MW.

During the dispersion process, the main hydrophobic portions of the polymer chains agglomerate together at the center of the particle, while the interface with surrounding water is rich in the stabilizing hydrophilic groups. The amount and distribution of stabilizing sites needs to be carefully controlled so that a stable dispersion is obtained, but not at the expense of a dried coating that is oversensitive to aqueous reagents. In practice, for example, as little as 1% w/w of ionic salt groups are required to form stable dispersions (so-called *PU ionomers*). The main advantages of the internal emulsification route are that dispersing processes require lower shear forces and finer particle sizes can be obtained, which in turn leads to better dispersion stability and film formation. In some instances, e.g., manufacture of high-solids PUDs, the internal stabilization mechanism may be augmented by a lower proportion of external surfactant. The main internal stabilization systems are as described below.

14.8.1 Anionic

These are PUDs in which part of the polyol component is replaced by a monomer containing pendant carboxylic acid or sulphonic acid groups. Monomers must be used in which the acid group has low reactivity toward isocyanate, e.g., because of steric hindrance—dimethylolpropionic acid (DMPA) is a preferred reagent for this reason. The acid groups are neutralized with a suitable base (e.g., ammonia, tertiary amines or sodium hydroxide) to generate the ionic centers before the polymer or prepolymer is dispersed in water. Anionic polyurethanes are usually used in formulations maintained in the range of pH 7 to 10 and are normally supplied at pH 7 to 8.5. Raising the pH may increase the viscosity of the PUD, while strongly acidic conditions will normally cause precipitation or coagulation of the polyurethane components. The latter problem may be encountered on reducing the pH of the anionic PUD formulation (especially below pH 5) by shock addition of some incompatible compounding ingredients, by blending with electrolytes of opposite charge such as cationic latices, or even by coating onto low-pH surfaces, e.g., of certain treated papers and leathers. Ammonia- or amine-neutralized salt groups tend to dissociate after the coating has been applied and dried, liberating carboxyl groups and volatile base. The practical effect is that these coatings are much less hydrophilic and therefore have reduced water sensitivity compared with sodium-neutralized analogs. The freed carboxyl groups can be utilized, e.g., as adhesion promoting or cross-linking sites. The majority of PUDs used or intended for printing applications are anionic grades of this type.

14.8.2 Cationic

These PU ionomers are prepared in similar fashion but incorporate monomers containing a tertiary amine group. The ionic centers are formed by protonation with strong acids or by quaternization with alkylating agents. Cationic grades are supplied and used in the lower pH range, e.g., in textile coating/impregnation formulations containing quaternary auxiliaries, such as fabric softeners and water repellents.

14.8.3 Nonionics

These are grades in which the ionic centers are replaced with hydrophilic polyether units, either branching off or terminating on the main polyurethane chains. Polyethylene oxide units (MW 200 to 4,000) are normally used as the dispersing sites. Nonionic grades tend to be more stable to electrolyte addition (of either charge) and strong shear forces compared with ionomers, but they are more sensitive to heat, because polyethylene oxide solubility in water decreases with temperature. Dried coatings retain hydrophilic character and tend to be more susceptible to swelling or whitening by water and aqueous reagents. Some commercial grades now have mixed stabilization systems (ionic and nonionic sites along the same molecular chains) that, to a limited extent, combine the attributes of both PUD types.

Most anionic, cationic, or nonionic PUDs contain poly(urethane-urea) molecular chains, i.e., they are segmented copolymers formed by the multiple addition reactions of polyisocyanates with polyols and polyamines. Manufacture normally involves a two-stage process, in which the urethane and urea linkages *(hard segments)* are formed in consecutive rather than simultaneous reactions. These sequential processes have been reviewed extensively in the chemical literature.[1] Urea and urethane groups have much higher molar cohesive energies compared with ether, ester, carbonate, or methylene linkages; consequently, intermolecular hydrogen bonding increases with hard segment content, and this tends to increase the integrity, 100% modulus (stiffness), hardness, tear strength, elastic recovery, water and solvent resistance of derived films, and coatings. In simpler terms, these properties will tend to improve with increased polyisocyanate content. Other simplified structure/property correlations based on polyisocyanate/polyol types are described in Secs. 14.9 through 14.12.

14.9   Aromatic—Polyether/Polyester

These are PUDs manufactured from relatively inexpensive aromatic diisocyanates (TDI, MDI, and polymeric grades) and various polyol components. They are used in applications where the yellowing tendency is tolerated (e.g., dark pigmented coatings and inks, sandwiched coatings, adhesives, etc.).

14.10   Aliphatic—Polyether

Various aliphatic diisocyanates (IPDI, HDI, HMDI, TMXDI, etc.) are available for imparting improved resistance to yellowing and UV degradation, but they are considerably more expensive than aromatic diisocyanates. Polyether diols impart softness, flexibility, and hydrolysis resistance. The aliphatic-polyether grades are often used as *flexibilizing* agents for hard acrylic latices.

14.11   Aliphatic—Polyester

Numerous polyester diols can be used varying in type and hydrophobic chain length, thus imparting a wide range of properties. The aliphatic-polyester types are considered to be good, multipurpose workhorses of PUD technology and are especially suitable for adhesion to difficult substrates such as polymer films and coatings, especially plasticized PVC.

14.12   Aliphatic—Polycarbonate

Some of the latest PUDs are based on expensive polycarbonate diols that confer high gloss and clarity, plus improved resistance to blocking, heat, hydrolysis, ambient weather conditions, solvents, and other chemicals as compared to equivalent polyester grades. The polycarbonate types are regarded as the doyen of PUDs, which is reflected in their cost, and are often combined with other aliphatic-polyester PUDs and/or acrylic latices for high-performance coatings on a variety of substrates.

14.13   Cross-Linking

High-MW PUDs may be used, either alone or in combination with other compatible polymer latices, as one-pack binders. However, it must be recognized that PUDs containing linear, or perhaps only lightly branched, molecular networks will not compete technically with moisture-cured or two-pack, solvent-borne polyurethanes. Cross-linking will normally improve many properties of concern to the printing industry, including cohesive strength; elastic recovery; heat resistance; water sensitivity; resistance to solvents, oils, and grease; rub-fastness; and, in some cases, adhesion to difficult substrates. An element of cross-linking can be introduced by using a proportion of tri- or polyfunctional monomers (isocyanate, alcohol, or amine) during PUD manufacture. Indeed, many commercial one-pack PUDs, especially the harder grades, contain some intraparticle cross-linking to boost the overall performance and durability of derived coatings. This option is not viable for solvent-based analogs, because cross-linking leads to excessively viscous or gelled polyurethane solutions.

In addition, PUDs can be cured with a number of external cross-linking agents, which are added at the point of application. These two-component PUDs utilize established cross-linking chemistry, which may involve addition, condensation, auto-oxidation, or ion bridging/chelation between the cross-linker molecule and/or receptive sites on the polyurethane chains. Typical function groups in waterborne polyurethanes are located at the chain termini (usually hydroxyl or amine groups), along the main polymer backbone (e.g., urethane, urea, and olefin groups), or as pendant moeities (carboxylic acid and other *active hydrogen* groups). Anionic PUDs, which contain latent acid functionality, are especially amenable to different cross-linkers One further point is that, if PUDs are blended with other polymer latices with complementary cross-linking sites, reaction with external curing agents ought to produce a more fully integrated coating or binder. The extent and ultimate effect of cross-linking on a particular PUD varies from reagent to reagent and also depends on polyurethane composition, distribution of reactive sites along the molecular chains, polymer-to-cross-linker ratio, cure temperature/dwell time profile, presence of catalysts, and other formulation additives and so on. Side reactions of certain cross-linkers with water, which is always present in vast molar excess, may further complicate the issue by inducing short pot lives and effecting poor reproducibility. Not surprisingly, the onus is on PUD manufacturers to develop one-pack PUDs that are technically equal to, or preferably superior to, the present generation of externally-cross-linked PUDs.

Cross-linked coatings and binders can also be obtained from several other waterborne systems utilizing isocyanate chemistry. The types briefly outlined below are usually regarded as separate developments from PUDs and, as such, are outside the scope of this discussion.

14.14   One-Pack, Blocked-Isocyanate

These are waterborne polyurethanes containing latent cross-linking sites that are activated at elevated temperatures (usually above 120° C, depending on type). The active sites are formed by reaction of pendant or terminal isocyanate groups with blocking agents, i.e., low-MW compounds capable of forming thermally reversible bonds during manufacture of the PUD. When the applied coating is heated to the critical temperature, the blocked isocyanate groups decompose quantitatively, yielding free isocyanate groups and the usually volatile blocking agent. Cross-linking is effected through normal isocyanate reactions with urethane or urea groups in the PU backbone at elevated temperatures.

14.15   Two-Pack (Hydroxyl-Terminated Oligomers + Polyisocyanate)

These are true two-component systems based on dispersions of lower-MW, hydroxyl-terminated urethane (or acrylate) oligomers that are mixed with a precalculated amount of water-dispersible polyisocyanate immediately before application to the substrate. A major drawback is restricted pot life, which, in some cases, is only a few hours. These products are waterborne functional equivalents to long-established solvent-borne, two-pack coatings and are intended for similar applications. The main difference is in the NCO:OH ratio (used to calculate the relative proportion of the two reactants), which is higher with the water-based system (2–2.5:1) to counteract side reactions of the polyisocyanate with water, than for the solvent-based analogs (usually 1.1:1).

14.16   One-Pack, EV or EB Cured

Typically, these are colloidal waterborne polyurethanes of varying MW containing unsaturated sites for free-radical addition (e.g., methacrylate functionality). The main advantages of waterborne UV-curable urethanes, compared with similar 100% reactive solids systems, are low coating viscosity (essential for spray application) and significantly lower or zero levels of VOCs and other noxious substances (solvents, free acrylic monomer, reactive diluents, etc.).

14.17   One-Pack, Self-Cross-Linking

Several PUD grades are available that are described as self-cross-linking. Usually, these polymers contain a proportion of free hydrazide groups that are subsequently treated with formaldehyde during manufacture to yield N-methylol groups that are stable in the dispersion. As the coating dries down, self-condensation of the N-methylol groups creates methylene or dimethylene ether bridges between adjacent chains; in theory, this cross-linking reaction proceeds under ambient conditions. The low formaldehyde content (free and released) of these systems is coming under scrutiny, and alternative self-cross-linking mechanisms are being investigated.

14.18   Blends with Other Waterborne Polymers

PUDs are used as the sole binder for some high-performance applications. However, they are relatively expensive compared with other waterborne polymers, which reflects the high cost of raw materials, particularly aliphatic diisocyanates, and processing, which is more time-consuming than polyurethane manufacture in solvents. Wherever possible, coating formulators will look to combine PUDs with compatible (but lower-cost) waterborne polymers to

1. Upgrade the performance of the cheaper binder
2. Retain the superior quality of polyurethane, but at reduced cost

Probably the largest single application for PUDs at present is to modify acrylic or styrene-acrylic latices for inks and coatings. It is important to note that many, but not all, PUDs and acrylic latices are compatible, and that selection is often based on a trial-and-error approach. Simple mistakes such as trying to blend an anionic acrylic latest with a cationic PUD, or vice versa, are easily avoided. Apparently compatible combinations should be thoroughly investigated, e.g., for long-term stability (including elevated temperature and freeze-thaw conditions), binder gloss/clarity, film formation, and acceptance of other formulation additives.

One of the most useful aspects is that the PUD component acts as a coalescence aid by lowering the minimum film-forming temperature (MFFT) of the acrylic latex in the blend. The harder-grade acrylics selected for UV light stability, weather resistance, and water and block resistance tend to have high MFFTs (30 to 60° C) and consequently require a considerable proportion of coalescing solvents (e.g., glycols, glycol ethers) for adequate film formation.

Acrylic latices are most commonly used as the co-binder, but PUDs have also been blended successfully with poly(vinyl acetate) latices, phenolic and MF resins, and waterborne epoxies for specialized adhesives and coatings. In contract, relatively nonpolar polymers, such as polyolefin, natural rubber, and most synthetic rubber latices, are usually thermodynamically incompatible. A wider selection of PUD blends is in prospect, because new dispersion grades of polar polymers, such as polyamides, polyesters, and cellulose ethers (NC and CAB grades) are being introduced that have a proven track record as solvent-borne or 100% solids coatings and printing ink binders. (See Tables 14.5 and 14.6.)

Two major classes of waterborne polymers are in use.

1. Aqueous polymer dispersions prepared by emulsion polymerization
2. Dispersions of preformed polymers in water

TABLE 14.5 Advantages of Polyurethane Dispersions

• Wide variety in molecular composition and, consequently, a wide range of performance characteristics
• Good physical properties (tensile strength, extensibility, elasticity, etc.), impact and abrasion resistance
• Availability of grades with good water/solvent/chemical resistance
• Low MFFTs and Tgs and, consequently, a minimum of coalescers and plasticizers required

TABLE 14.6 Main Additives Used in PUD Formulations

• Reaction solvents, polymerization catalysts, anti-foaming agents, antioxidants, microbiocides, emulsifying agents (added during PUD manufacture to improve processing or shelf life)
• Defoamers, coalescents, multipurpose surfactants, wetting agents, pH modifiers, thickeners, rheology modifiers, flash rusting agents, antioxidants, microbiocides, emulsifying agents (added to assist application to the substrate)
• Other polymer dispersions, coalescents, waxes, adhesion promoters, mar-resist additives, water repellants, weathering preservative (microbiocides, antioxidants, UV inhibitors), cross-linking agents (+ catalyst, one- or two-pack system), reinforcing fillers
• Pigments (+ pigment dispersants), dyes, opacifiers, matting agents, inert fillers, microspheres, flame retardants, intumescent compounds, corrosion-resistant pigments (to improve the appearance/performance of the binder)

Examples of the former type include acrylic copolymers, vinyl acrylics, styrene acrylics, and styrene-butadiene. Examples of the latter class are aqueous polyurethanes and aqueous polyesters. The technology of this class of polymer dispersion, in general, is more recent and is still advancing rapidly. This technology is mainly used for preparing aqueous polymer dispersions that cannot be produced by free-radical polymerization. It generally involves synthesis by step-growth polymerization, followed by dispersion in water.

Emulsion polymers (latices) are well established. They are prepared by free radical polymerization in water using a water-soluble free radical initiator and surfactant(s) to nucleate and stabilize the growing polymer particles. The product is a high-molecular-weight polymer in the form of a well defined colloidal dispersion. The process and products have been studied extensively over the past 30 years.

14.19 Polyurethane Dispersions

Aqueous polyurethane dispersions are finding increasing commercial usage. They offer environmental advantages over both two-pack solvent-borne urethanes and two-pack waterborne polyurethanes in that they are one-pack systems, and the user does not have to handle free isocyanates. Modern polyurethane dispersion technology allows the preparation of the high-molecular-weight polymer dispersions, which can produce surface coatings exhibiting many of the characteristic advantages of polyurethanes (e.g., good abrasion resistance and hardness).

Early attempts at producing polyurethane dispersions employed conventional surfactants to disperse relatively low-molecular-weight polyurethanes under high-shear mixing. Such procedures resulted in large-particle dispersions that exhibited poor colloid stability. Presently, more elegant approaches are used. The polyurethane is rendered water dispersible by the inclusion of a monomer in the backbone or chain extending reagent that contains an internal dispersing functional group, which may be anionic, cationic, or nonionic in nature.

Several synthetic processes for high-molecular-weight polyurethane dispersions are utilized. They include:

- Acetone process
- Melt-dispersion process
- Prepolymer mixing process
- Ketimine process

A common feature of all these processes is that the first step is a conventional polyurethane synthesis in which diols or polyols are reacted with diisocyanates. The next step involves dispersion in water. The necessary low viscosity of the urethane polymer is achieved by one of the following methods:

1. The urethane polymer is low in molecular weight, in which case a chain-extension step is required at the aqueous dispersion stage to achieve the necessary high molecular weight (e.g., ketimine process, prepolymer mixing process).
2. The urethane polymer is dissolved in solvent, in which case the solvent has to be removed after the aqueous dispersion step (acetone process).
3. The urethane polymer is heated to facilitate dispersion in water (melt-dispersion process).

A typical prepolymer mixing process for the preparation of an anionic polyurethane dispersions is as follows. The stoichiometry (isocyanate:hydroxyl ratio) of the condensation reaction is maintained in such a manner that the isocyanate terminated prepolymer formed has workable viscosity/molecular weight to facilitate the aqueous dispersion step. In this case, the prepolymer contains an in-built carboxyl stabilizing group provided by dimethylolpropionic acid (DMPA). The prepolymer is then neutralized and transferred to water, where spontaneous particle formation occurs. Simultaneous chain extension is carried out at this stage, resulting in the formation of a high-molecular-weight polyurethane dispersion. The very nature of the process results in the formation of a polymer colloid that is significantly different from emulsion polymer latices in terms of their colloidal, morphological, and application characteristics.

14.20 Illustrative Patents

14.20.1 Dispersions of Polyurethanes as Dispersing Agents

Nonionic diisocyanate polyaddition products dispersed in polyethers or polyesters are well known. Diisocyanates are reacted with bifunctional primary alcohols in a dispersing medium consisting of a polyether or polyester (MW 500 to 3000) containing at least two (exclusively secondary) hydroxyl groups in the molecule. Compounds containing isocyanate and amino groups undergo a polyaddition reaction *in situ* in a polypropylene glycol ether dispersing agent. Nonionic dispersions of polyurethanes, polyureas, or polyhydrazodicarbonamides in polyvalent, higher-molecular-weight hydroxyl compounds are recommended as thickeners for the textile or dye industry because of their high viscosities, even at low solid contents.

Thus, a 10% (or 20%) dispersion of polyhydrazodicarbonamide in a polypropylene glycol ether has a viscosity of over 10,000 (or 200,000, in the case of a 20% dispersion) cP at 25° C. This amounts to more than 10 times (or 200 times, in the case of a 20% dispersion) the viscosity of the pure dispersing agent. When attempts are made to prepare a 40% dispersion, the reaction mixture solidifies before polyaddition has been completed. The high viscosities that occur at even relatively low solid contents seriously restrict the possibilities of using the products because, in many fields of application, they cannot be dosed with the aid of the usual dosing apparatus. For producing polyurethane foams, for example, a purpose for which such dispersions could be used, the viscosities of the starting materials must be below 2500 cP when conventional high-pressure machines are employed.

Reischl[2] found that stable ionic dispersions having the desired low viscosity can be obtained by reacting polyisocyanates with compounds having salt groups or groups capable of salt formation *in situ*, in dispersing agents consisting of compounds containing hydroxyl groups. In addition to ionic compounds, other nonionic chain-lengthening agents may also be used. The reaction may be carried out continuously in continuous-flow mixers, preferably in the presence of more than 1% by weight of water, based on the total quantity of reaction mixture. Alternatively (and this is preferred because of the simpler dosing and mixing technique and, in many cases, the easier removal of heat of reaction), the reaction is carried out in simple stirrer apparatus (batch-wise reaction in vessels) in the presence of a larger quantity of water, preferably more than 5% by weight, based on the total quantity of reaction mixture.

The ability to use large amounts of water is particularly surprising for several reasons. A water content of 10, 15, or 20% by weight (based on the total quantity of polyether and water), for example, increases the viscosity of a polyalkylene ether glycol at 25° C to 4, 8, and 50 times the original value, respectively (3500, 7300, and over 50,000 cP). If the water content is further increased, the solutions or emulsion originally obtained in many cases separate into two phases. Both the high increase in viscosity and phase separation inevitably led experts to assume that the addition of water would be unsuitable for the commercial production of low-viscosity polyisocyanate polyaddition products in compounds containing hydroxyl groups, particularly since the water might, in addition, interfere chemically with the isocyanate polyaddition reaction.

The Reischl invention relates to a process for the *in situ* preparation of stable dispersions of polyisocyanate polyaddition products in hydroxyl containing compounds as dispersing agents by the reaction of (1) organic polyisocyanates with (2) compounds having primary and/or secondary amino groups and/or primary hydroxyl groups in (3) compounds having at least one hydroxyl group.

Compounds (3) contain secondary hydroxyl groups in cases where compounds (2) contain primary hydroxyl groups. The process is further characterized in that at least one of the components (1) or (2) contains ionic groups or groups capable of salt formation. The components are advantageously reacted in the presence of 10 to 25% by weight of water, based on the total quantity of reaction mixture, the water being subsequently removed in known manner, if desired.

14.20.2 Water-Based Polyurethane Coatings

For over 40 years, waterborne (water-based) polyurethanes have been in existence. The performance properties of these systems have been improved continuously, with literally hundreds of patents being issued during this period in the field of waterborne polyurethanes. There are at least three important reasons why aqueous polyurethanes have become commercially important in the last few years. The first reason is environmental concerns regarding solvents and other volatile organic compounds (VOCs) being emitted into the atmosphere, causing ozone depletion, acid rain, and possibly a chemical imbalance of the Earth's ecosphere. The second reason is economic. Organic solvent systems are expensive, and aqueous polyurethane systems do not bear the extra solvent cost. The third, and perhaps most important, reason relates to the fact that aqueous polyurethanes have been improved to the point that, with regard to performance, they are comparable to or better than the conventional solvent-based polyurethanes for many specific applications.

Typical waterborne polyurethanes are actually polyurethane-polyurea polymers containing both urethane (-NH-CO-O-) and urea (-NH-CO-NH-) groups in a macromolecular chain. These groups are formed by well known polyaddition reactions. Most current waterborne polyurethanes have one manufacturing process in common. In the first phase of production, a medium-molecular-weight *prepolymer,* is synthesized in a reactor at an elevated temperature (60 to 100° C), leaving approximately 2 to 8% free/terminal unreacted isocyanate (-NCO) groups. For this type of polyurethane to have high-performance properties, e.g., flexibility, hardness, acid, solvent, and other chemical resistance, it must be chain-extended in the water phase. The chain-extension phase is a build-up of the prepolymer to a polyurethane having a high molecular weight.

This high-molecular-weight buildup is usually performed by reacting the NCO groups of the prepolymer with amines. Two important problems must be dealt with during the course of this reaction.

1. The control or stabilization of the extremely fast urea formation reaction (e.g., -NCO + NH$_2$- → -NHCONH-)
2. The control or minimization of the ensuing viscosity buildup resulting from the increase in the molecular weight

Because of this, the current state-of-the-art waterborne aliphatic polyurethane dispersions are limited to solid contents of 40% or lower. Moreover, their solvent resistance (as measured by methyl ethyl ketone, or MEK, double rubs) is generally limited to a maximum of 200 to 250 rubs, and their resistance to chemicals such as Skydrol (jet aircraft hydraulic fluid) and jet fuels is very poor.

One attempt to solve the problems related to viscosity (low solids of the dispersion) and high-performance solvent and chemical resistance has been to try to improve waterborne polyurethane dispersions by processing the reaction in a solvent as an intermediate aid to control the viscosity buildup during the critical chain extension phase. Typical of these processes are the so-called *acetone* or N-methylpyrrolidone (NMP) processes.

According to these types of processes, a polyol is reacted with a diisocyanate to form a prepolymer. Then, in the presence of a solvent such as acetone or NMP, the prepolymer is reacted with a chain extender such as a polyamine, e.g., ethylenediamine or diethylenetriamine. The solvent-based extended urethane polymer is then diluted with water, forming an aqueous dispersion of the urethane prepolymer, aliphatic amine chain-extended urethane polymer and solvent. The solvent must then be removed by distillation yielding an aqueous dispersion of urethane containing components ranging from the prepolymer to the highest-molecular-weight aliphatic amine chain-extended polyurethane.

In this process, the solvent must be distilled out of the system, which results in disposal problems with the resulting solvent. This is not a practical solution, and there are relatively few commercial applications of this process.

The most popular process for manufacturing waterborne polyurethanes is the so-called *prepolymer blending process.* This process utilizes hydrophilically modified prepolymers having free terminal NCO-groups, which are more compatible with aqueous systems. These prepolymers, possessing hydrophilicity, are therefore more susceptible to being chain extended with diamines in a water, as contrasted to a solvent phase that helps build up the molecular weight of the extended polyurethane polymers and further enhances the performance properties.

For this hydrophilic prepolymer blending/mixing process to function optimally, the dispersion phase must be performed in as short a period of time as possible, and at temperatures below the critical point where NCO groups rapidly start to react with the water with the formation of carbamic acid groups and the following release of carbon dioxide. To optimize this process, it is often necessary to use 5 to 15% wt. levels of a co-solvent such as NMP to adjust for the viscosity buildup during the chain-extension and/or cross-linking phases. This process utilizes chain extension in the water phase, resulting in the prepolymers being either reacted with the difunctional amines to yield linear, flexible polyurethane-ureas, or cross-linked with polyfunctional amines, which produces cross-linked systems. Waterborne polyurethanes of the cross-linked type contain a combination of ionic and nonionic internal emulsifiers. When compared to films made from the acetone process, the cured polyurethane films from this process exhibit improved solvent resistance when cross-linked with polyfunctional amines. Although this type of process, and variations thereof, is an improved system, the films produced upon curing are almost always inferior to the two-component solvent based aliphatic, fully cross-linked, air-dry polyurethanes. It is believed this is due to the fact that the aqueous-based chain extension is performed in a heterogeneous phase and, therefore, does not proceed as smoothly or as quantitatively as occurs in organic solvent systems, especially the two-component fully cross-linked systems.

Fully cross-linked aliphatic polyurethanes, when prepared from solvent-based, two-component systems, possess performance properties that are generally far superior to any of the current water-based or waterborne prepared polyurethanes. These solvent-based, two-component, fully cross-linked aliphatic polyurethanes can be formulated to provide cured films that resist a 30-day submersion in Skydrol (jet aircraft hydraulic fluid) and also resist more than 1000 double MEK rubs. Such solvent-based systems contain absolutely no added water and have very low levels of moisture content.

It would be desirable to formulate water-based polyurethane dispersions resulting in coatings that do not require the addition of a separate detergent component and in which the polyol phase and amine components could be simplified and more easily prepared and combined.

Hart[3] disclosed a substantially solvent-free, aqueous, two-component coating composition dispersion that, when applied to a substrate, cures to a film having properties of adhesion and bonding that are equal to or superior to films obtained from conventional solvent-based, two-component polyurethane or epoxy systems. The first component is a combined aqueous polyol phase and an amine phase. The polyol phase comprises a member consisting of an acid-containing polyol or a blend of polyols containing an acid-containing polyol such that the average hydroxyl functionality of the polyol phase is at least 1.5 and has an acid number of between about 15 and 200. The amine phase is an amine or blend of amines containing active hydrogens reactive with NCO groups such that the average active hydrogen functionality is at least 1.5. The amine phase is present in a quantity sufficient to substantially neutralize the acid-containing polyols. The second component is a member selected from the group consisting of an aliphatic polyisocyanate, a cycloaliphatic polyisocyanate, and an aromatic polyisocyanate, and adducts and mixtures thereof. The ratio between the isocyanate equivalents and the sum of the active hydrogen equivalents of the polyols and amines is at least 0.5:1 to 1.1:1. The preferential reaction of the isocyanate groups with the active hydrogen moieties from the polyols and amines results in both linear and cross-linking polymerization.

An object of the Hart invention is to provide water-based, no- or low-VOC two-component polyurethane-polyurea dispersions that are fully cross-linked and chain extended in water, and which are free or essentially free of volatile organic compounds or solvents and exhibit superior coating properties. A further object of the invention is to provide water-based, no- or low-VOC two-component polyurethane-polyurea dispersions that do not require the presence of an added detergent.

Another object of this invention is to provide aqueous dispersions of chain-extended polyurethane-polyurea polymers that contain a high solid content but possesses low-viscosity properties without the use of organic solvents.

14.20.3   Water-Dispersible, Cross-Linkable Poly(urethane-urea) Compositions

Solvent-based poly(urethane-urea) compositions made from drying oils have been sold for use as coatings for fabrics, plastics, wood, metal, and the like. These coating compositions offer good properties at a reasonable price. Nevertheless, for a number of reasons, there is a continued need to develop new waterborne poly(urethane-urea) coatings based on drying oils. The reasons include environmental and economical concerns associated with solvents.

One problem, in particular, associated with poly(urethane-urea) water dispersions is the inability of the composition to cross-link at room temperature upon drying. This is desirable, because the composition's properties, such as chemical resistance and mechanical strength, are extended upon the cross-linking. The presence of an polyalkylene ether polyol allows for chain extension upon the inclusion of an amine and subsequent cross-linking of the poly(urethane-urea) upon drying. Although this method represents a valuable advance in the art because it is a one component, cross-linkable poly(urethane-urea) composition, it would be desirable to find alternative methods of preparing cross-linkable, one-component poly(urethane-urea) compositions.

Treasurer[4] disclosed a water-dispersible, one-component, self-cross-linkable poly(urethane-urea) composition that is stable and capable of curing at room temperature. It is prepared by (a) forming a prepolymer comprising an alcoholized drying oil, a diisocyanate, a polyalkylene ether polyol, and a dihydroxy-containing alkanoic acid polyol, (b) neutralizing the prepolymer, and (c) contacting the neutralized prepolymer with water and diamine, wherein the alcoholized drying oil has a sufficient content of monoglycerides to allow for chain extension in step (c) and the poly(urethane-urea) self-cross-links upon drying. The waterborne poly(urethane-urea) composition exhibits good chemical resistance, adhesion, abrasion, and high-gloss properties when used as a coating.

14.20.4   Leather Finishes

A leather finish is the protective layer that is applied to the dried leather after it has been tanned, fat-liquored and/or rendered water repellant. Its purpose is to protect the leather, in particular from soiling and damage, or to modify the surface properties of the leather with regard to properties such as color or gloss. The addition of fat-liquoring agents and/or water repellants to the leather imparts to it the desired softness and the required water-repellant properties.

One of the requirements for a leather finish is that it adhere well to the leather. However, particularly in the case of leathers that have been highly fat-liquored and/or rendered water repellant to a great extent, the adhesion of most leather finishing systems is unsatisfactory. There is, therefore, a need for leather finishing assistants that can be applied to a leather that has been fat-liquored and/or rendered water repellant as a bottoming coat before the actual leather finish, and which impart high adhesive strength to the subsequently applied leather finish without simultaneously substantially reducing the effect of the fat-liquoring and/or water repellant treatment.

Weyland et al.[5] disclosed a finished a leather to which has been applied a fat-liquoring and/or water repellant agent and a particularly formulated polyurethane adhesive, thus improving applied bottoming coat composition. The invention relates to a process for the preparation and use of these formulations as a bottoming coat for finishes on leathers that have been fat-liquored and/or rendered water repellant.

14.20.5 Anionic Polyurethane Dispersions for Coatings and Adhesives

Water-dispersed polyurethanes are generally known. For example, U.S. Pat. 4,066,591, to Scriven and Chang, discloses water-dispersible NCO prepolymers of the type to which the present invention is directed as an improvement therein. Other related patents of interest include U.S. Pat. 4,277,380, to East and Rogemoser; U.S. Pat. 4,576,987, to Crockett and Rimma; U.S. Pat. 4,791,168, to Salatin et al.; and U.S. Pat. 5,023,309, to Kruse, Crowley, and Mardis, all of which relate to water dispersion.

Doshniak[6] disclosed a series of cosolvent-free anionic aqueous polyurethane dispersions containing a special side-chain functionality. The result is lower surface energy of the polymer in both the dispersion state and the dried film state. The new polyurethane dispersions of the invention have enhanced adhesion to plastics and are particularly useful in laminating plastic films as well as providing a general adhesive of improved properties.

Polyurethane-urea polymers in aqueous dispersion are obtained by chain extending an NCO-containing prepolymer in an aqueous medium. The NCO-containing prepolymer is prepared by reacting a polyisocyanate with an active hydrogen-containing compound such as a polyol and dimethylol propionic acid (DMPA) to form a preprepolymer. To this is added a diamine monomer containing a pendant aliphatic to form the prepolymer that is then dispersed in water and chain extended. The improved anionic water-dispersible polyurethane polymers of the Doshniak invention form excellent coatings and adhesives.

The polyurethanes are dispersible in aqueous media. The organic polyisocyanate used is most preferably tetramethyl xylene diisocyanate (TMXDI). An NCO-containing preprepolymer is prepared by reacting an organic polyisocyanate, an active hydrogen-containing compound (such as a dimethylolpropionic acid), and triethylamine. This preprepolymer is then chain extended with a diamine monomer containing a pendant aliphatic chain, as disclosed above, to form a prepolymer having a pendant functionality as defined above, which may be linear or branched. The prepolymer is then dispersed in water and further chain extended with water-soluble organic amine(s) to provide a water-dispersed anionic polymer.

14.20.6 Polyurethane Dispersions for Bookbinding

Bookmaking requires the use of a wide variety of adhesives. Adhesives are used in making hard book covers, casing book blocks into hard covers, applying paper covers to book blocks, and the binding of book blocks themselves. The binding of book blocks is generally the most demanding adhesive application in the industry. Adhesives may be used alone or in cooperation with other mechanical binders such as sewing and staples.

One use of adhesives in binding book blocks is as a primer to wet the spine area of the book block and provide a tie coat for an additional adhesive layer. The additional adhesive layer provides structural support and generally maintains the mechanical integrity of the binding. Priming adhesives are generally aqueous emulsions of film-forming polymers that can readily wet the spine of the book block. The emulsions may be thermoplastic copolymers, acrylates, animal glues, etc.

One of the most widely used binding methods today is known as *perfect binding*. This may be used for either hard or soft cover bindings. In perfect binding, the sewing of signatures is eliminated. The pages are printed on a web of paper, and the book block is formed by folding this web and arranging it into a signature stack. One or more signatures are then clamped and cut, and the page folds are cut and roughened or otherwise prepared for the application of the binding adhesives. The adhesives essentially provide all of the structural support for the book's binding. This requires the adhesives formulator to carefully balance such properties as strength, flexibility, resistance to cold flow, resistance to low temperature embrittlement, etc. The balancing of strength and flexibility has been particularly difficult.

In hard cover or *case* bookbinding, an important structural form of the binding is called the *rounding* of the spine or binding. This is the convex shape of the spine, which distributes the stresses of opening the book across the width of the binding. It is important that a binding maintain its rounding in high-quality books to provide stress distribution for the life of the book. Many current adhesives are incapable of providing years of rounding maintenance. A new development in rounding is disclosed in Carter et al., U.S. Pat. 4,907,822. This reference relates to the use of a polyurethane hot-melt adhesive to bind book block signatures and for the rounding of the bound book blocks. However, there are limitations and difficulties in applying polyurethane hot melts, e.g.,

1. Lack of green strength (3 to 7 days to cure)
2. Need for special equipment
3. Toxicity, requiring increased ventilation due to the presence of free isocyanate

Therefore, bookbinders are constantly demanding improvements in this area.

The bookbinding industry continues to demand increased performance from adhesives manufacturers. In particular, improvements are required in drying characteristics, bond strength, flexibility, shape retention, and versatility of aqueous-emulsion bookbinding adhesives. Therefore, a new and versatile aqueous-emulsion bookbinding adhesive is needed that can be used as a primer by itself to provide permanent rounding and shape retention, excellent bond strength, and acceptable drying characteristics.

Fresonke[7] has disclosed a method of binding book blocks to form books. The method involves applying an aqueous composition to the spine area of a book block and removing the volatile components of the composition. The composition comprises either of

1. An aqueous vehicle, a film-forming polymeric resin, and a polyurethane resin
2. An aqueous polyurethane dispersion or emulsion

The subject of the Fresonke invention is an aqueous adhesive composition useful in bookbinding operations and a method for using this composition in such operations. The adhesive composition comprises an aqueous polyurethane dispersion or emulsion alone or in combination with a latex emulsion that provides increased adhesive structural properties for improved bookmaking. The adhesive may be used as a primer or as a structural adhesive in bookbinding.

14.20.7 PU Dispersion Coatings for Flexible and Rigid Substrates

The preparation of stable, aqueous polyurethane-polyurea dispersions may be subdivided into those that depend on the use of solvents and the so-called solvent-free processes. In the processes that require the use of solvents, high-molecular-weight polyurethanes are synthesized by polyaddition in an organic solution (i.e., in a homogeneous phase) before they are dispersed in water. This method results in exceptionally high-quality polyurethane films that satisfy even the stringent requirements for textile coating compounds.

Among the solvent-free processes, two should be mentioned in particular: the so-called solvent-free dispersion process, which has been outlined in principle (e.g., U.S. Pat. 3,756,992), and the process described in U.S. Pat. 4,108,814, in which a prepolymer containing sulphonate and isocyanate groups is subjected to a reaction with a diamine accompanied by chain lengthening during or after the dispersion process. These aqueous dispersions have all of the advantages of polyurethanes, such as good processing characteristics and excellent physical fastness properties. On the other hand, they often have disadvantages such as poor, rubber-like handle when applied as top coats and insufficient levelling power of the dispersions.

There have been many proposals to obviate these disadvantages by the addition of certain auxiliary agents and additives. Thus, for example, in U.S. Pats. 3,816,168 and 3,823,102, it is recommended to improve the handle of dissolved polyurethanes by the addition of nitrocellulose or other auxiliaries. These methods are substantially limited to the use of organic solutions of the auxiliary agents to modify dissolved polyurethanes. It is recommended that nitrocellulose and/or cellulose acetobutyrate enriched with plasticizers and emulsifiers and dissolved in organic solvents should be added to an organic solution of a polyurethane, and the resulting combination should then be converted into an aqueous dispersion. In this case, the solvent remains in the aqueous dispersion, with the result that the well known disadvantages of systems containing solvents (fire characteristics, environmental problems) are not overcome.

In the case of aqueous dispersions that have been prepared using acetone, substances that are insoluble in water but soluble in acetone may be incorporated. The acetonic solution is converted into a homogeneous dispersion with water, and the solvent is subsequently evaporated off. The problem, however, remained that additives that were insoluble in water and incapable of being dispersed in water without the addition of emulsifiers were required to be incorporated in aqueous polyurethane dispersions without the use of readily inflammable solvents so that such dispersions suitable for storage could be obtained. One particular problem was that auxiliary agents for the dressing of leather (which are neither soluble nor dispersible in water and often are insoluble, or only difficultly soluble in organic media) were required to be incorporated in solvent-free, aqueous polyurethane dispersions suitable for the dressing of leather, in particular as leather finishes, to produce storage stable dispersions containing the aforementioned auxiliary agents.

These problems underlying the present invention could surprisingly be solved by the process according to the invention described below. Wentzel and Dieterich[8] disclosed a process for preparing storage stable, aqueous dispersions of oligourethanes or polyurethanes in admixture with auxiliary agents and additives that are neither soluble nor dispersible in water. The oligourethanes or polyurethanes are rendered dispersible in water by incorporating ionic and/or nonionic hydrophilic groups. For the final product to remain storage stable, the auxiliary agents or additives are added to the oligourethanes or polyurethanes before they are dispersed in water.

The Wentzel and Dieterich invention relates to a novel process for the preparation of aqueous dispersions or solutions of oligourethanes or polyurethanes containing auxiliary agents and additives that are neither dispersible nor soluble in water, and the use of these solutions or dispersions as coating compounds for flexible or rigid substrates.

References

1. Dietrich, D. "Aqueous emulsions, dispersions and solutions of polyurethanes: Synthesis and properties." *Progress in Organic Coatings* 9:281–340, 1981. Elsevier Sequoia, S.A., Lausanne.
2. Reischl, Artur, U.S. Pat. 4,305,858, December 15, 1981, "Process for the preparation of stable ionic dispersions of polyisocyanate-polyaddition products in hydroxyl containing compounds," to Bayer Aktiengesellschaft, Leverkusen, Federal Republic of Germany.
3. Hart, Richard E. U.S. Pat. 5,508,340, April 16, 1996, "Water-based, solvent-free or low VOC, two-component polyurethane coatings," to R.E. Hart Labs, Inc., California.
4. Treasurer, Urvee Y. U.S. Pat. 5,504,145, April 2, 1996, "Water-dispersible poly(urethane-urea) compositions," to The Thompson Minwax Company, Upper Saddle River, New Jersey.
5. Weyland, Peter, Haeberle, Karl, and Treiber, Reinhard. U.S. Pat. 5.401,582, March 28, 1995, "Aqueous polyurethane formulations," to BASF Aktiengesellschaft, Ludwigshafen, Federal Republic of Germany.
6. Dochniak, Michael J. U.S. Pat. 5,354,807, October 11, 1994, "Anionic water dispersed polyurethane polymer for improved coatings and adhesives," to H.B. Fuller Licensing & Financing, Inc., Arden Hills, Minnesota.

7. Fresonke, Flavia M., U.S. Pat. 5,443,674, August 22, 1995, "Polyurethane dispersions for book binding," to H.B. Fuller Licensing & Financing, Inc., Wilmington, Delaware.
8. Wenzel, Wolfgang, and Dieterich, Dieter. U.S. Pat. 4,306,998, December 22, 1981, "Process for the preparation of stable aqueous dispersions of oligourethanes or polyurethanes and their use as coating compounds for flexible or rigid substrates," to Bayer Aktiengesellschaft, Leverkusen, Federal Republic of Germany.

15

Health and Safety

15.1 Health and Safety Factors

Isocyanates are classified as dangerous substances. Isocyanates are generally labeled toxic and should be handled with care. Exposure hazards increase substantially when handling vapors or mists. Isocyanate vapors or mists may be irritating to the nose, throat, and lungs. Even brief exposure may cause irritation, difficult breathing, or coughing. Sensitization may result from excessive exposure. Subsequent exposure to low concentrations has been known to provoke allergic reactions with asthma-type symptoms. Industrially, this is important because MDI-based isocyanates, although having a low vapor pressure, can become airborne during machine flushing and filling procedures. Conversely, toluene diisocyanate (TDI) has a relatively high vapor pressure at ambient temperatures and a vapor density six times that of air. Thus, open containers have the potential to yield high concentrations of TDI vapor. Also, many aliphatic isocyanates have high vapor pressures and therefore must be handled with special caution. Inhalation of aliphatic isocyanates is reported to retard the growth of laboratory mice.

Repeated or prolonged skin contact may cause irritation, blistering, dermatitis, or skin sensitization. Contact with the eye has been reported to cause irritations in testing with rabbits. For these reasons, isocyanates must be handled in well ventilated areas. Respirators should be worn whenever the possibility of vapor exposure exists. If inhalation occurs, the affected person should be moved to a well ventilated area. Chemical goggles should be worn when handling isocyanates. All work areas should be equipped with an eye wash. In the event of eye contact, the eye should be irrigated immediately. The eyes should be held open while flushing with a continuous low pressure stream of water for at least 15 min. In the event of direct skin contact, use a safety shower immediately, removing all clothing while washing. In all cases, call a physician immediately.

The most overlooked hazard and contaminant is water. Water reacts with isocyanates at room temperature to yield both ureas and large quantities of carbon dioxide. The presence of water or moisture can produce a sufficient amount of CO_2 to overpressurize and rupture containers. As little as 30 mL of water can result in 40 L of carbon dioxide, which could result in pressures of up to 40 psi. For these reasons, the use of dry nitrogen atmospheres is recommended during handling. If a plant air system must be used, purification equipment, such as oil traps and drying beds, should be installed between the source and the isocyanate vessel.

Also, the presence of strong bases, even in trace amounts, can promote the formation of isocyanurates or carbodiimides. In the event of gross contamination, the exothermic reaction can sharply increase the temperature of the material. Normally, the trimerization reaction occurs first and furnishes heat for the carbodiimide reaction. The carbodiimide reaction liberates carbon dioxide and forms a hard solid. The liberation of carbon dioxide in a sealed vessel could result in overpressurization and rupture.

Temperature control is important in the handling and storage of isocyanates. Storage at inappropriate temperatures can cause product discoloration, viscosity increases, and dimerization. Handling personnel should consult the technical data sheets for the recommended storage temperature of the specific isocyanate product.

Most commercial isocyanates have a high flash point and are classified as Class IIIB combustible liquids. These materials also burn in the presence of an existing fire or heat source in the presence of oxygen. In the event of an isocyanate fire, it is recommended to use a carbon dioxide or dry chemical extinguisher. For fires covering large areas, use of a protein foam or water spray is recommended. Personnel engaged in fighting isocyanate fires must be protected against nitrogen dioxide vapors and isocyanate fumes. Firefighters should wear approved positive pressure, self-contained breathing apparatus, and fire-resistant clothing.

These data are provided to the reader for reference only. Health professionals are responsible for therapy decisions. Manufacturers should review all current information as well as local regulations and statutes.

15.2 Toluene Diisocyanate (Meta-toluene Diisocyanate)

15.2.1 Overview

Life Support

This overview assumes that basic life support measures have been instituted.

Clinical Effects (OV-CLEF)

Summary of Exposure

- Toluene diisocyanate is an irritant to skin, lungs, conjunctiva and the mucous membranes of the gastrointestinal tract. It also may cause euphoria, ataxia and mental aberrations.

HEENT

- Severe conjunctival irritation and lacrimation from liquid or high vapor concentrations is likely. Lower concentrations may produce a burning or prickling sensation.
- Glaucoma and iridocyclitis have been reported with a splash accident.

Respiratory

- Burning or irritation of the nose and throat, cough, laryngitis, chest pain, and asthmatic syndrome (chemical bronchitis with severe bronchospasm) sensation of oppression or constriction of the chest, bronchitis, emphysema, and cor pulmonale may occur.
- Respiratory symptoms may recur several years after discontinuation of exposure.

Neurologic

- Acute exposures may produce headache, insomnia, euphoria, ataxia, anxiety neurosis with depression, or paranoid tendencies.

Gastrointestinal

- Inhalation of vapor or aerosol may produce vomiting and abdominal pain. Epigastric and substernal pain may be secondary to the paroxysmal or persistent cough associated with inhalation.

Dermatologic

- Irritation and inflammation are common. Dermal absorption is low.

Immunologic

- Elevated specific IgE and IgG antibodies have been noted among sensitized and exposed workers.

15.2.2 Laboratory (OV-LAB)

- Monitor respiratory function closely.

15.2.3 Treatment Overview (OV-TRMT)

Oral/Parenteral Exposure

- Due to the irritant nature of this substance, emesis is not advised.
- *Dilution.* Immediately dilute with 4 to 8 oz (120 to 240 mL) of milk or water (not to exceed 15 mL/kg in a child).
- *Activated charcoal/cathartic.* Administer charcoal slurry, aqueous or mixed with saline cathartic or sorbitol. The FDA suggests 240 mL of diluent/30 g of charcoal. Usual charcoal dose is 30 to 100 g in adults and 15 to 30 g in children (1 to 2 g/kg in infants).
  - Administer one dose of a cathartic, mixed with charcoal or given separately.
- Monitor patient for respiratory distress. If a cough or difficulty in breathing develops, evaluate for respiratory tract irritation, bronchitis, and pneumonia.
- *Allergic reaction.* Sensitized individuals should be cautioned to avoid further exposure, as serious allergic reactions may result.

Inhalation Exposure

- *Decontamination.* Move patient to fresh air. Monitor for respiratory distress. If cough or difficulty in breathing develops, evaluate for respiratory tract irritation, bronchitis, or pneumonitis. Administer 100% humidified supplemental oxygen with assisted ventilation as required.
- Randomized, double-blind crossover studies showed minimal efficacy of bronchodilators. This could be due to the absence of airway hyperresponsiveness in some TDI-induced asthma.

Eye Exposure

- *Decontamination.* Exposed eyes should be irrigated with copious amounts of tepid water for at least 15 min. If irritation, pain, swelling, lacrimation, or photophobia persist, the patient should be seen in a health care facility.

Dermal Exposure

- *Decontamination.* Wash exposed area extremely thoroughly with soap and water. A physician may need to examine the area if irritation or pain persists.

15.2.4 Range of Toxicity (OV-TOX)

- Most data available are of animal origin. The TLV is 0.02 ppm, and the least detectable odor is at 0.4 ppm, so careful monitoring of the atmosphere is essential.

15.2.5 Substances Included (SUBS)

Specific Substances (SPEC)

- Toluene diisocyanate synonyms
  - TDI
  - toluene-2,4-diisocyanate
  - 2,4-tolylene diisocyanate
  - meta-tolylene diisocyanate
  - 2,3-diisocyanatotoluene
  - tolylene diisocyanate
  - CAS 584-84-9
- Methylene diphenyl diisocyanate synonyms
  - MDI

- – CAS 101-68-8
- Hexamethylene diisocyanate synonyms
  - – HDI
  - – CAS 822-06-0
- Isophorone diisocyanate synonyms
  - – IPDI
- Hydrogenated MDI synonyms
  - – Dicyclohexylmethane 4,4'-diisocyanate

15.2.6 Clinical Effects

HEENT (HENT)

Eyes

- *Conjunctivitis.* Severe conjunctival irritation and lacrimation may occur following exposure from liquid or high vapor concentrations (Axford et al., 1976). Burning or pricking sensations from lower concentrations have been reported (Grant, 1986).
- *Iridocyclitis and secondary glaucoma.* Were noted in a workman who accidentally splashed TDI in one eye (Grant, 1986).
- *Rabbits.* A drop of meta- and para-toluene diisocyanate on rabbit eyes caused immediate pain, lacrimation, swelling of the lids, and conjunctival reaction (Grant, 1986).

Respiratory (RESP)

- *Irritation.* Burning or irritation of nose and throat, choking sensation, cough that may or may not produce blood-streaked sputum, laryngitis, retrosternal soreness, and chest pain have been reported (Elkins et al., 1962; NIOSH, 1973).
  - – Depending on length of exposure and level of concentration above 0.5 ppm, respiratory symptoms will develop with a latent period of 4 to 8 hr (Rye, 1973) and, based on the onset of symptoms, asthmatic reactions to isocyanate challenge have been classified as immediate, late, or dual (Fabbri, 1990).
  - – The symptomatology of TDI inhalational exposure is stereotypic (ILO, 1983).
    > At the end of a few days to two months of exposure, lacrimation and irritation of the conjunctivae and pharynx occur and are later coupled with dry nocturnal cough, sternal pains, and dyspnea.
    > The symptoms worsen in the evening and disappear in the morning with minimal mucus expectoration.
    > Symptoms diminish after a few days' rest but recur upon return to work.
  - – The characteristic substernal pain may be due to the paroxysmal or persistent cough often associated with inhalation.
  - – Asthmatic syndrome (chemical bronchitis with severe brochospasm, sensation of oppression or constriction of chest), chronic bronchitis, emphysema, and cor pulmonale have been noted with high exposures (Axford et al., 1976).
- Asthma
  - – Sensitization
    > The onset of symptoms experienced by the TDI sensitized individual may be insidious, becoming progressively more pronounced with continued exposure over days to months. The initial symptoms of dyspnea and cough can progress to severe asthma and bronchitis (ACGIH, 1986; Bruckner et al., 1968; Porter et al., 1975; Weill et al., 1981; Williamson, 1965).

Workers exposed to low TDI levels may also experience minimal or no respiratory symptoms only to experience suddenly an acute and severe asthmatic reaction (Banks et al., 1986).

Late asthmatic reactions have been documented in sensitized workers in association with early elevations of the neutrophils, eosinophils, leukotriene B4, and albumin in bronchoalveolar lavage fluid (Fabbri, 1990; Fabbri et al., 1985, 1987; Zocca et al., 1990).

– Long-term sequelae

Long-term respiratory symptoms with slightly impaired ventilatory function have been reported and, in some, irreversible damage has been documented (Adams, 1970, 1975; Banks et al., 1990; Innocenti et al., 1981; Luo et al., 1990; Mapp et al., 1988; Moller et al., 1986; Paggiaro et al., 1984; Venables et al., 1985; Weill et al., 1981).

Lozewicz et al. (1987) reported 82% of 50 patients they were able to follow, continued to have respiratory symptoms 4 or more years after avoidance of exposure, and nearly one-half of these patients required treatment at least once per week.

– *Fatality.* A 43-year-old male with a 6-year history of toluene diisocyanate-induced asthma developed a fatal asthma attack while mixing two components of a polyurethane paint. Despite advice to change jobs, he continued to work with the paints while taking antiasthmatic drugs at home and work to control his symptoms of asthma (Fabbri et al., 1988).

• *Hemorrhagic pneumonia.* A 34-year-old spray painter presented with hemoptysis, dyspnea, bilateral pulmonary opacities, respiratory failure, and high levels of IgG and IgE antibodies against HDI-HSA (hexamethylene diisocyanate human serum albumin) and TDI-HSA (Patterson et al., 1990). He was declared normal after 2 days of assisted ventilation and 11 days of steroids.

• *Hypersensitivity pneumonitis*

– Hypersensitivity pneumonitis was confirmed by biopsy of a 41-year-old automobile paint sprayer who presented with dyspnea, cyanosis, fever, crepitant rales, reticulonodular radiographic infiltrates, restrictive pulmonary function, and elevated TDI-specific IgG (Yoshizawa et al., 1989). He improved markedly with prednisone and oxygen.

– A 53-year-old steel plant maintenance worker who occasionally glues pipes together presented with cough, fever, malaise, interstitial pneumonitis, eosinophilia, decreased FEF25-75, and elevated IgG antibody levels specific for diphenylmethane diisocyanate (MDI) (Walker et al., 1989).

• Animal exposures

– Tracheobronchitis occurred in animals with 30 exposures to 1 to 2 ppm TDI given for 6 hr each time (ACGIH, 1986).

– Pulmonary edema and hemorrhage were noted on autopsy of rodents following repeated exposure to TDI (ACGIH, 1986).

– Respiratory rate was reduced by 50% after single 6-hr exposures of guinea pigs to 0.18 and 0.5 ppm TDI (Stevens and Palmer, 1970).

– Fibrous lesions were noted in rat lungs after exposure to 0.1 ppm TDI, 6 hours per day, 5 days per week for up to 58 exposures (Niewenhuis et al., 1965).

Neurologic (NEUR)

• *CNS effects.* Firefighters exposed to TDI and possibly other substances experienced neurologic complaints of euphoria, loss of coordination, and loss of consciousness. Long-lasting symptoms of personality change, irritability, depression, and loss of memory were also reported (Le Quesne et al., 1976; McKerrow et al., 1970; O'Donoghue, 1985).

• *Possible impotence.* Firefighters exposed to TDI and possibly other substances suffered from impotence for some time after exposure. This was thought to be due to an indirect neurologic effect rather than to direct toxicity to the male genitalia (Le Quesne et al., 1976).

Gastrointestinal (GAST)

• *Vomiting.* Nausea and vomiting may occur from inhalation of vapor or aerosol (Axford et al., 1976).

- *Abdominal pain.* May be noted from inhalation exposure (Axford et al., 1976).
- *Epigastric pain.* Epigastric and substernal pain may be secondary to the paroxysmal or persistent cough associated with inhalation.

Temperature Regulation (TEMP)

- *Fever.* Occurred following intravenous injection of 0.02 mg TDI/kg (Scheel et al., 1964).

Dermatologic (DERM)

- *Inflammation.* Irritation and inflammation are common but carry a low risk of absorption that would affect internal organs.

Carcinogenicity (CARC)

- *Tumors.* Tumors of the pancreas, liver, mammary glands, and vascular system were reported in rats and mice used in an inadequately described study (National Toxicology Program, 1982).
- *Noncarcinogenicity of chronic exposure.* Rats and mice exposed to 0.05 or 0.15 ppm TDI, 6 hours daily, 5 days per week for 2 years did not develop any tumors (IARC, 1986; Owen, 1980).

Immunologic (IMMU)

- *Positive skin test reactions.* Reactions to TDI-conjugates with human serum albumin and positive TDI-specific IgE and IgG antibodies have been reported but the exact mechanism involved is still unknown (Butcher et al., 1977; Cartier et al., 1989; Finkel, 1983; Karol et al., 1979; Karol, 1980, 1981; Keskinen et al., 1988; Wass and Belin, 1989).
- *Serum chemotaxis factor.* Release of a serum chemoattracting factor for normal neutrophils and activation of asthmatic neutrophils was demonstrated among workers with late asthmatic reaction to TDI (Valentino et al., 1988).

- *HDI-specific IgG antibodies.* These were elevated in a car painter who had three episodes of hypersensitivity pneumonitis-like disease after exposure to acrylic lacquers with hexamethylene diisocyanate (HDI) as the curing agent (Selden et al., 1989).

15.2.7 Laboratory (LAB)

Monitoring Parameters/Levels (MONIT)

Other

- Monitor respiratory function closely.
- RAST testing for IgE antibodies against p-tolyl monoisocyanate antigens is deemed unreliable because of the false positives and false negatives (Proctor et al., 1988).
- When positive, bronchoprovocative challenge with TDI provide a definitive but risky diagnostic tool (Proctor et al., 1988). It has a sensitivity of only 68%, and cross-reactions between TDI and MDI can affect bronchial provocation test results (Banks et al., 1989; Innocenti et al., 1988).

15.2.8 Treatment (TRMT)

Life Support (LIFE)

Support respiratory and cardiovascular function.

Oral/Parenteral Exposure (ORAL-EXP)

Prevention of Absorption

- Emesis
  - Due to the irritant nature of this substance, emesis is NOT advised.

- Dilution
  - Immediately dilute with 4 to 8 ounces (120 to 240 mL) of milk or water (not to exceed 15 mL/kg in a child).
- Activated charcoal/cathartic
  - Charcoal administration
    Administer charcoal as slurry; charcoal slurry may be aqueous, or mixture of charcoal with saline cathartic or sorbitol.
  - Charcoal dose
    The FDA suggests a minimum of 240 mL of diluent per 30 g charcoal. Optimum dose of charcoal is not established; usual dose is 30 to 100 g in adults and 15 to 30 g in children; some suggest using 1 to 2 g/kg as a rough guideline, particularly in infants (FDA, 1985).
  - Charcoal administration, adverse effects
    Refer to the section on activated charcoal/treatment management for further information on administration and adverse reactions.
  - Cathartic contraindications
    Cathartics should not be used in patients who have an ileus. Saline cathartics should not be used in patients with impaired renal function (Gilman et al., 1990).
  - Cathartic administrations/cautions
    Administer in a health care facility, especially in children.
    Monitoring of fluids and electrolytes may be necessary in children.
    The safety of more than one dose of a cathartic has not been established. Hypermagnesemia has been reported after repeated administration of magnesium containing cathartics in overdose patients with normal renal function (Smilkstein et al., 1988).
    Repeated cathartic dosing should be done with extreme caution, if at all.
    Administration of cathartics should be stopped when a charcoal stool appears.
  - Cathartic administration
    Administer *one dose* of a saline cathartic or sorbitol, mixed with charcoal or administered separately.
    Saline cathartic—adult dose. 20 to 30 g per dose of magnesium sulfate or sodium sulfate *or* magnesium citrate, 4 mL/kg per dose up to 300 mL per dose, administered orally (Minocha et al., 1985).
    Saline cathartic—pediatric dose. 250 mg /kg per dose of magnesium or sodium sulfate *or* magnesium citrate, 4 mL/kg per dose up to 300 mL per dose, administered orally (Minocha et al., 1985).
    Saline cathartic administration—adverse effects. Refer to the sections on laxative-saline or magnesium management for further information on administration and adverse effects.
    Sorbitol—adult dose. 1 to 2 g/kg per dose to a maximum of 150 g per dose, administered orally (Minocha et al., 1985).
    Sorbitol—pediatric dose. 1 to 1.5 g/kg per dose as a 35% solution to a maximum of 50 g per dose, administered orally to children over 1 year of age (Minocha et al., 1985).
    Sorbitol administration—adverse effects. Refer to the section on sorbitol management for further information on administration and adverse effects.

Treatment

- Due to the irritant nature of this substance, emesis is not advised. Immediately dilute the ingested substance with milk or water.
- Monitor patient for respiratory distress, if a cough or difficulty in breathing develops, evaluate for respiratory tract irritation, bronchitis, and pneumonia.

- Bronchodilators and oxygen may be resorted to in acute attacks. Several placebo-controlled randomized double-blind crossover studies have been conducted to investigate the efficacy of varying bronchodilators.

   Theophylline (6.5 mg/kg twice a day) has only partial effect (Mapp et al., 1987).

   Verapamil (120 mg twice a day), ketotifen, salbutamol alone, atropine (0.008 to 0.012 mg/kg atropine sulfate administered subcutaneously 30 min before TDI challenge and 90 min after TDI exposure), and cromolyn (20 mg four times daily via a spinhaler) have no protective effect (De Marzo et al., 1988; Mapp et al., 1987; Paggiaro et al., 1987; Tossin et al., 1989)

- Prednisone and aerosolized beclomethasone (1 mg twice daily) have been shown to prevent late asthmatic reactions or increased airway responsiveness in TDI-sensitized patients (De Marzo et al., 1988; Fabbri et al., 1985).

- Non-steroidal anti-inflationary agents did not prevent the reactions associated with late TDI-induced asthma (Fabbri et al., 1985).

- *Allergic reaction.* Sensitized individuals should be cautioned to avoid further exposure, as serious allergic reactions may result.

Inhalation Exposure (INHL-EXP)

Decontamination

- *Decontamination.* Move patient from the toxic environment to fresh air. Monitor for respiratory distress. If cough or difficulty in breathing develops, evaluate for respiratory tract irritation, bronchitis, or pneumonitis.

- *Observation.* Carefully observe patients with inhalation exposure for the development of any systemic signs or symptoms and administer symptomatic treatment as necessary.

- *Initial treatment.* Administer 100% humidified supplemental oxygen with assisted ventilation as required. Exposed skin and eyes should be copiously flushed with water.

- Monitor for allergic reactions.

Treatment

- Treatment should include recommendations listed in the oral/parenteral exposure section when appropriate.

- The specific role of bronchodilators for treatment of severe bronchoconstriction has not been evaluated. Asthma from TDI can be induced in the absence of airway hyperresponsiveness; in these cases, bronchodilator use may contribute little to modifying severity.

Eye Exposure (EYE-EXP)

Decontamination

- Exposed eyes should be irrigated with copious amounts of room temperature water for at least 15 min. If irritation, pain, swelling, lacrimation, or photophobia persist after 15 min of irrigation, an ophthalmologic examination should be performed.

Dermal Exposure (DERM-EXP)

Decontamination

- Wash exposed area extremely thoroughly with soap and water. A physician may need to examine the area if irritation or pain persists after washing.

Treatment

- Treatment should include recommendations listed in the oral/parenteral exposure section when appropriate.

15.2.9 Range of Toxicity (TOX)

Minimum Lethal Exposure (MIN)

- Dermal exposure to 16,000 mg/kg (rabbits) was not lethal.

Maximum Tolerated Exposure (MAX)

- TDI levels of 0.3 to 0.7 ppm were associated with a high incidence of illness, but no cases were observed from concentrations below 0.03 ppm (Hama, 1947).
- The maximum incidence of illnesses occurred when the average concentration of vapor was 0.1 ppm, and very little trouble was reported at 0.01 ppm (Walworth and Virchow, 1959).
- No respiratory symptoms or changes in pulmonary function were noted among workers pouring and molding polyurethane foam and breathing as much as 0.001 to 0.002 ppm TDI (Roper and Cromer, 1975).
- Occasional exposures to TDI beyond 0.02 ppm caused no significant deterioration in lung function (Erlicher and Brochhagen, 1976).
- A dose-response relationship was demonstrated between acute pulmonary function changes and exposure of 112 workers to 0.0035 to 0.06 mg TDI/m^3 (IARC, 1979).
- Exposure of volunteers have shown that 0.05 to 0.1 ppm TDI in the air can cause eye and nose irritation (Grant, 1986).
- A normal age- and smoking-related rate of decline in forced expiratory volume in one second (FEV1) was demonstrated in subjects exposed to 0.001 to 0.0015 ppm TDI, thus negating any effects of TDI at these levels (Musk et al., 1985).
- A daily mean exposure of 0.023 mg/m^3 produced impaired lung function and increased frequency of symptoms among nonsmokers but not among exposed smokers (Alexandersson et al., 1985).

Workplace Standards (WORK)

Systems for more complete information on workplace and environmental standards.

- TLV-TWA
  - *TDI.* 0.005 ppm (approximately 0.036 mg/m^3); STEL of 0.02 ppm (approximately 0.15 mg/m^3) (ACGIH, 1990)
  - *MDI.* 0.005 ppm (approximately 0.051 mg/m^3); no STEL (ACGIH, 1990)
- IDLH value
  - *TDI.* 10 ppm (NIOSH, 1985)
  - *MDI.* 10 ppm (NIOSH, 1985)
- OSHA PEL-Ceiling Transitional Limits (OSHA, 1989)
  - *TDI.* 0.02 ppm Ceiling (approximately 0.14 mg/m^3) (OSHA, 1989)
  - *MDI.* 0.02 ppm Ceiling (approximately 0.2 mg/m^3)
- OSHA PEL Final Rule Limits (OSHA, 1989)
  - *TDI.* 0.005 ppm TWA (approximately 0.04 mg/m^3); 0.02 ppm STEL (approximately 0.15 mg/m^3); no Ceiling Limit
  - *MDI.* 0.02 ppm Ceiling Limit (approximately 0.2 mg/m^3); no TWA or STEL
- Odor Threshold
  - *TDI.* 0.4 to 2.14 ppm (CHRIS, 1990)
  - *MDI.* Not available
  - The least detectable odor is at 0.4 ppm, so careful monitoring of atmosphere is essential.
- Airborne levels of MDI during spray application of polyurethane foam were as high as 0.093 ppm over a 70-min period, well in excess of the TLV or OSHA PEL ceiling limits (Bilan et al., 1989).
- NIOSH REL

- *TDI.* 35 μg/m^3 (TWA) (CDC, 1988)
- *TDI.* 140 μg/m^3 (10-min ceiling) (CDC, 1988)
- *MDI.* 50 μg/m^3 (TWA) (CDC, 1988)
- *MDI.* 200 μg/m^3 (10-min ceiling) (CDC, 1988)
- *HDI.* 35 μg/m^3 (TWA) (CDC, 1988)
- *HDI.* 140 μg/m^3 (10-min ceiling) (CDC, 1988)
- *NDI.* 40 μg/m^3 (TWA) (CDC, 1988)
- *NDI.* 170 μg/m^3 (10-min ceiling) (CDC, 1988)
- *IPDI.* 45 μg/m^3 (TWA) (CDC, 1988)
- *IPDI.* 180 μg/m^3 (10-min ceiling) (CDC, 1988)
- *Hydrogenated MDI.* 55 μg/m^3 (TWA) (CDC, 1988)
- *Hydrogenated MDI.* 40 μg/m^3 (10-min ceiling) (CDC, 1988)
- *Other Diisocyanates.* 5 ppb (TWA) (CDC, 1988)
- *Other Diisocyanates.* 20 ppb (10-min ceiling) (CDC, 1988)

- Persons sensitized will most likely be affected by TDI below the 0.02 ppm level. Rats exposed to 600 ppm for 6 hr died; at 60 ppm no deaths occurred.

LD50/LC50 (LDLC)

- LD50 (ORAL) rat: 5,800 mg/kg (Zapp, 1957)
- LC50 (INHL) rat: 14 ppm/4H (ITI, 1988)
- LC50 (INHL) rat: 600 ppm/6H (Sax and Lewis, 1989)
- LC50 (INHL) Rodent: 12 ppm/4H (Duncan et al., 1962)
- LC50 (INHL) mouse: 9.7 ppm/4H (Duncan et al., 1962)
- LD50 (IV) mouse: 56 mg/kg (Sax and Lewis, 1989)
- LC50 (INHL) guinea pig: 13.9 ppm/4H (Duncan et al., 1962)
- LC50 (INHL) rabbit: 11 ppm/4H (Duncan et al., 1962)
- LC50 (INHL) rabbit: 1,500 ppm/3H (Sax and Lewis, 1989)

Other (TOX-OTHR)

- TCLo (INHL) human: 80 ppb produce nose, eye, and pulmonary irritation (RTECS, 1990; Sax and Lewis, 1989)
- TCLo (INHL) human: 0.5 ppm produce nasal and pulmonary effects (Sax and Lewis, 1989)
- TCLo (INHL) human: 0.02 ppm/2Y cause pulmonary effects (RTECS, 1990; Sax and Lewis, 1989)
- TCLo (INHL) woman: 300 ppt/8H/5D cause pulmonary effects (RTECS, 1990; Sax and Lewis, 1989)

15.2.10   Available Forms/Sources (FORM)

- Toluene diisocyanate (TDI) is usually available in two isomers: 2,4-toluene diisocyanate and 2,6-toluene diisocyanate. Over 95% of the commercially available products used for industry contain 80% 2,4-TDI and 20% 2,6-TDI (ACGIH, 1986).
- TDI is one of the isocyanates most employed in the manufacture of polyurethane foams, elastomers, and coating (ACGIH, 1986).
  - Foams are used in furniture, packaging, insulation, and boat building. Flexible foams are made up of TDI whereas the rigid foams have the less volatile MDI (Finkel, 1983).
  - Polyurethane coatings are used in leather, wire, tank linings, masonry, paints, floor, and wood finishes.
  - Elastomers, which are abrasion and solvent resistant, are used in adhesives, coated fabrics, films, linings, clay pipe seals, and in abrasive wheels and other mechanical items.

• Other synonyms are meta-toluene diisocyanate, tolylene diisocyanate, and 2,4-diisocyanotoluene.

15.2.11 Kinetics (KINET)

Absorption (ABSR)

• The wide distribution of water and other nucleophiles in animal tissue makes it unlikely for toluene diisocyanate to be absorbed and distributed throughout the body (ITIC/USEPA, 1981).

Metabolism (METAB)

Metabolites

• Reaction of TDI with human serum albumin yields mono- or bisureido protein derivatives (ITIC/USEPA, 1981).
• Hydrolysis of both isocyanate groups produce 2,4-toluene diamine, a carcinogen (ITIC/USEPA, 1981).

15.2.12 Pharmacology/Toxicology (PHAR)

Toxicologic Mechanism (TOX-MECH)

• TDI exposure tends to have a cumulative effect on most people. In human toxicology there are two classes of reaction to TDI: (1) primary irritation or pharmacodynamic action, to which all exposed persons are susceptible to some degree, and (2) sensitization reaction or allergic response in those persons who have become sensitized to TDI during earlier exposure (Butcher et al., 1977).

– TDI is a severe irritant to all living tissues with which it comes in contact in liquid or vapor form, especially to mucous membranes of the eyes, the gastrointestinal and the respiratory tract. It also has a marked inflammatory reaction on direct skin contact (Proctor et al., 1988).

A common respiratory system response to inhaled TDI is both acute and chronic diminution of ventilatory capacity, measured by a decrease in FEV1 even in the absence of other overt symptoms (Adams 1970, 1975; Moller et al., 1986; Venables et al., 1985; Weill et al., 1981).

There have been no reports of human ingestion.

Necropsy of rats revealed corrosive action on stomach as well as possible toxic effects on the liver (ACGIH, 1986).

– Respiratory sensitization occurs in susceptible persons after repeated exposure to TDI at levels of 0.02 ppm and below (Elkins et al., 1962). A chronic-like syndrome consisting of coughing, wheezing, tightness, or congestion in the chest and shortness of breath has been characterized with repeated exposures at such low concentrations (NIOSH, 1973).

A sensitized individual, in addition to the aforementioned instant reactions, may be afflicted with marked tissue eosinophilia and acute pneumonitis with inflammatory edema of the lungs (Fabbri, 1985; Fabbri et al., 1987; Zocca et al., 1990).

Some individuals who have been reported to have allergic response have been demonstrated to have circulating antibodies to TDI or to TDI-animal protein conjugates (Butcher et al., 1977; Fabbri et al., 1987; Finkel, 1983; Karol et al., Karol, 1980, 1981).

Further evidence is the demonstration of lymphocyte transformation in TDI-sensitized workers induced by TDI-conjugated proteins.

TDI-induced late asthmatic reactions have been attributed to increased bronchovascular permeability caused by leukotriene B4 levels, which also promote granulocyte adherence and leukocyte migration into tissues (Zocca et al., 1990)

– Because 1 micromole of TDI can stimulate methacholine-induced tracheal ring contraction, the pharmacological effect of TDI is believed to be due to an autonomic imbalance between cholinergic and beta-adrenergic neural control (Borm et al., 1989).

– Epithelial damage, thickening of basement membrane, and mild to moderate inflammatory reaction in the submucosa were demonstrated in TDI-sensitized patients who have ceased work within 4 to 40 months prior to bronchial biopsy (Paggiaro et al., 1990).

15.2.13   Physiochemical (PCHEM)

Physical Parameters (PHYS)

Physical State

- *TDI.* Clear liquid at room temperature, turns straw-colored on standing, with fruity, pungent odor (AAR, 1987; ACGIH, 1986; ILO, 1983)
- *MDI.* Light yellow to white crystals or fused solid (ILO, 1983)

Molecular Weight

- *TDI.* 174.16 (ACGIH, 1986)
- *MDI.* 250.25 (ILO, 1983)
- *HDI.* 168.2 (ILO, 1983)
- *NDI.* 210.2 (ILO, 1983)

Specific Gravity

TDI. 1.22 at 25° C (80:20 mixture) (ACGIH, 1986)
MDI. 1.19 (ILO, 1983)

Other

- A. TDI
    - Freezing point: 11.3 to 13.5° C (80:20) mixture (ACGIH, 1986)
    - Boiling point: 250° C (80:20 mixture) (ACGIH, 1986)
    - Vapor pressure: 0.025 torr (ACGIH, 1986)
    - Flash point (open cup): 266° F (130° C) (ACGIH, 1986)
    - Explosive limits: 0.9% to 9.5% (ILO, 1983)
- B. MDI
    - Melting point: 37.2° C (ILO, 1983)
    - Boiling point: 172° C (ILO, 1983)
    - Vapor pressure: 0.001 mm Hg at 40° C (ILO, 1983)
    - Flash point (open cup): 202° C (ILO, 1983)

Chemical Parameters (CHEM)

Reactivity

- Reacts with water to liberate carbon dioxide (AAR, 1987).
- Polymerizes in the presence of alkali.

Solubility

- TDI is miscible with alcohol, diglycol monomethyl ether, ether, acetone, carbon tetrachloride, benzene, chlorobenzene, kerosene, and olive oil (ITI, 1988).
- When heated to decomposition, TDI emits toxic fumes of NOx (Sax and Lewis, 1989)

References (REFS)

General References (GEN-REF)

1. AAR: Emergency Handling of Hazardous Material in Surface Transportation. Hazardous Materials Systems (BOE). Association of American Railroads, Washington, DC, 1987.

2. *ACGIH: Documentation of the Threshold Limit Values and Biological Exposure Indices,* 5th ed. Am Conference of Govt. Ind. Hyg., Inc., Cincinnati, OH, 1986.

3. *ACGIH: 1990–1991 Threshold Limit Values for Chemical Substances and Physical Agents and Biological Exposure Indices.* Am. Conference Govt. Ind. Hyg., Inc., Cincinnati, OH, 1990.

4. Adams, W.G.F. Long-term effects on the health of men engaged in the manufacture of tolylene diisocyanate (TDI). *Br. J. Ind. Med.* 1975; 32:72.

5. Adams, W.G.F. Lung function of men engaged in the manufacture of tolylene diisocyanate (TDI). *Proc. Roy. Soc. Med.* 1970; 63:378.

6. Alexandersson, R., Hedenstierna, G., Randma, E., et al.: Symptoms and lung function in low-exposure to TDI by polyurethane foam manufacturing. *Int. Arch. Occup. Environ. Health.* 1985; 55:149–157.

7. Axford, A.T., McKerrow. C.B., Jones, A.P., et al.: Accidental exposure to isocyanate fumes in a group of firemen. *Br. J. Ind. Med.* 1976; 33:65–71.

8. Banks, D.E., Butcher, B.T., and Salvaggio, J.E. Isocyanate-induced respiratory disease. *Ann. Allergy* 1986; 57:389–396.

9. Banks, D.E., Rando, R.J., and Barkman, H.W., Jr. Persistence of toluene diisocyanate-induced asthma despite negligible workplace exposures. *Chest* 1990; 97:121–125.

10. Banks, D.E., Sastre, J., Butcher, B.T., et al. Role of inhalation challenge testing in the diagnosis of isocyanate-induced asthma. *Chest* 1989; 95:414–423.

11. Bilan, R.A., Haflidson, W.O., and McVittie, D.J. Assessment of isocyanate exposure during the spray application of polyurethane foam. *Am. Ind. Hyg. Assoc. J.* 1989; 50:303–306.

12. Borm, P.J.A., Bast, A., and Zuiderveld, O.P. In vitro effect of toluene diisocyanate on beta adrenergic and muscarinic receptor function in lung tissue of the rat. *Br. J. Ind. Med.* 1989; 46:56–59.

13. Bruckner, H.C., Avery, S.B., Stetson, D.M., et al. Clinical and immunological appraisal of workers exposed to diisocyanate. *Arch. Environ. Health* 1968; 16:619–625.

14. Butcher, B.T., Jones, R.N., O'Neill, C.E., et al. Longitudinal studies of workers employed in the manufacture of toluene diisocyanate. *Am. Rev. Respir. Dis.* 1977; 116:411–421.

15. Butcher, B.T., O'Neill, C.E., Reed, M.A., et al. Radioallergosorbent testing of toluene diisocyanate-reactive individuals using p-tolyl isocyanate antigen. *J. Allergy Clin. Immunol.* 1980; 66:213–216.

16. Cartier, A., Grammer, L., Malo, J.L., et al. Specific serum antibodies against isocyanates: Association with occupational asthma. *J. Allergy Clin. Immunol.* 1989; 84:507–514.

17. Casarett and Doull. *Toxicology.* Macmillan, New York, 1975, 606–607.

18. CDC. *Recommendations for Occupational Safety and Health Standards.* MMWR 1988; 37 (Suppl. no. S-7):1–29.

19. CHRIS: *CHRIS Hazardous Chemical Data.* U.S. Department of Transportation, U.S. Coast Guard, Washington, DC, 1985.

20. De Marzo, N., Fabri, L.M., Crescioli, S., et al. Dose-dependent inhibitory effect of inhaled beclomethasone on late asthmatic reactions and increased airway responsiveness to methacholine induced by toluene diisocyanate in sensitized patients. *Pul. Pharm.* 1988; 1:15–20.

21. Duncan, B., Scheel, L.D., Fairchild, E.J., et al. *Am. Ind. Hyg. Assoc. J.* 1962; 23:447. As cited in ACGIH: Documentation of the Threshold Limit Values and Biological Exposure Indices, 5th ed. Am. Conference of Govt. Ind. Hyg., Inc., Cincinnati, OH, 1986.

22. Elkins, H.B., McCarl, G.W., Brugsch, H.G., et al. Massachusetts experience with toluene diisocyanate. *Am. Ind. Hyg. Assoc. J.* 1962; 23:265–272.

23. Erlicher, H. and Brochhagen, F.K. Urethane in the Environments. *Proceedings of the Plastic and Rubber Institute Conference,* Sept. 21–22, 1976.

24. Fabbri, L.M. Airway inflammation and late asthmatic reactions. *Eur. Respir. J.* 1990; 3:367–368.

25. Fabbri, L.M., Boschetto, P., Zocca, E., et al. Bronchoalveolar neutrophilia during late asthmatic reactions induced by toluene diisocyanate. *Am. Rev. Respir. Dis.* 1987; 136:36–42.

26. Fabbri, L.M., Chiesura-Corona, P., Dal Vecchio, L., et al. Prednisone inhibits late asthmatic reactions and the associated increase in airway responsiveness by toluene diisocyanate in sensitized subjects. Am. Rev. Respir. Dis. 1985; 132:1010–1014.

27. Fabbri, L.M., Danieli, D., Crescioli, P., et al. Fatal asthma in a subject sensitized to toluene diisocyanate. *Am. Rev. Respir. Dis.* 1988; 137:1494–1498.

28. Fabbri, L.M., Di Giacomo, R., Dal Vecchio, L., et al. Prednisone, indomethacin, and airway hyper-responsiveness in toluene diisocyanate sensitized subjects. *Bull. Eur. Physiopathol. Respir.* 1985; 21:421–426.

29. FDA. Poison treatment drug product for over-the-counter human use; tentative final monograph. *Fed. Register* 1985; 50:2244–2262.

30. Finkel, A.J., ed. *Hamilton and Hardy's Industrial Toxicology,* 4th ed. John Wright, PSG Inc., Boston, MA, 1983.

31. Gilman, A.G., Rall, T.W., Nies, A.S. et al., eds. *Goodman and Gilman's The Pharmacological Basis of Therapeutics,* 8th ed. Pergamon Press, New York, NY, 1990.

32. Grant, W.M. *Toxicology of the Eye,* 3rd ed. Charles C. Thomas, Springfield, IL, 1986.

33. Hama, G.M. *Arch. Ind. Health.* 1947; 16:232. As cited in *ACGIH,* Documentation of the Threshold Limit Values and Biological Exposure Indices, 5th ed. Am. Conference of Govt. Ind. Hyg., Inc., Cincinnati, OH, 1986.

34. IARC. Monographs on the Evaluation of the Carcinogenic Risk of Chemicals to Man. World Health Organization, International Agency for Research on Cancer, Geneva, 1979.

35. IARC. Monographs on the Evaluation of the Carcinogenic Risk of Chemicals to Man. World Health Organization, International Agency for Research on Cancer, Geneva, 1986.

36. ILO. *ILO Encyclopaedia of Occupational Health and Safety,* 3rd ed., Vol. I. Parmeggiani, L., ed. International Labour Office, Geneva, Switzerland, 1983.

37. Innocenti, A., Cirla, A.M., Pisati, G., et al. Cross-reaction between aromatic isocyanates (TDI and MDI): a specific bronchial provocation test study. *Clin. Allergy* 1988; 18:323–329.

38. Innocenti, A., Franzinelli, A., and Sartorelli, E. Long-term study of workers with asthma caused by polyurethane resins. *Med. Lavoro* 1981; 3:231.

39. ITIC/USEPA. *Information Review* 231, Toluene Diisocyanates, 1981.

40. ITI. *Toxic and Hazardous Industrial Chemicals Safety Manual.* The International Technical Information Institute, Tokyo, Japan, 1988.

41. Karol, M.H., Sandberg, T., Riley, E.J., et al. Longitudinal study of tolyl-reactive IgE antibodies in workers hypersensitive to TDI. *J. Occup. Med.* 1979; 21:354–358.

42. Karol, M.H. Study of guinea pig and human antibodies to toluene diisocyanate. *Am. Rev. Respir. Dis.* 1980; 122:965–970.

43. Karol, M.H. Survey of industrial workers for antibodies to toluene diisocyanate. *J. Occup. Med.* 1981; 23:741–747.

44. Keskinen, H., Tupasela, O., Tiikainen, U., et al. Experiences of specific IgE in asthma due to diisocyanates. *Clin. Allergy* 1988; 18:597–604.

45. Le Quesne, P.M., Oxford, A.T., McKerrow, C.B., et al. Neurologic complications after a single severe exposure to toluene diisocyanate. *Br. J. Ind. Med.* 1976; 33:72–78.

46. Lozewicz, S., Assoufi, B.K., Hawkins, R., et al. Outcome of asthma induced by isocyanates. *Br. J. Dis. Chest* 1987; 81:14–22.

47. Luo, J.C.J., Nelsen, K.G., and Fischbein, A. Persistent reactive airway dysfunction syndrome after exposure to toluene diisocyanate. *Br. J. Ind. Med.* 1990; 47:239–241.

48. Mapp, C.E., Boschetto, P., Dal Vecchio, L., et al. Protective effect of anti-asthma drugs on late asthmatic reaction and increased responsiveness induced by toluene diisocyanate in sensitized subjects. *Am. Rev. Respir. Dis.* 1987; 136:1403–1407.

49. Mapp, C.E., Corona, P.C., De Marzo, N., et al. Persistent asthma due to isocyanates. a follow-up study of subjects with occupational asthma due to toluene diisocyanate (TDI). *Am. Rev. Respir. Dis.* 1988; 137:1326–1329.

50. McKerrow, C.T., Davies, H.J., and Jones, P.A. Symptoms and lung function following acute and chronic exposure to tolylene diisocyanate. *Proc. Roy. Soc. Med.* 1970; 63:376–378.

51. Minocha, A., Krenszelok, E.P., and Spyker, D.A. Dosage recommendations for activated charcoal-sorbitol treatment. *J. Toxicol. Clin. Toxicol.* 1985; 23:579–587.

52. Minocha, A., Krenzelok, E.P., and Spyker, D.A. Dosage recommendations for activated charcoal-sorbitol treatment. *J. Toxicol. Clin. Toxicol.* 1985; 23:579–587.

53. Moller, D.R., Brooks, S.M., McKay, R.T., et al. Chronic asthma due to toluene diisocyanate. *Chest* 1986; 90:494–499.

54. Musk, A.W., Peters, J.M., and Bernstein, L. Absence of respiratory effects in subjects exposed to low concentrations of TDI and MDI: a re-evaluation. *J. Occup. Med.* 1985; 27:917–920.

55. National Toxicology program: Carcinogenesis Bioassay of Diisocyanate NTP No 10638W, 1980.

56. Niewenhuis, R., Scheel, L.D., Stemmer, K., et al. Toxicity of chronic low level exposures to toluene diisocyanate in animals. *Am. Ind. Hyg. Assoc. J.* 1965; 26:143–149.

57. NIOSH. *Criteria for a Recommended Standard—Occupational exposure to Toluene Diisocyanate.* DHEW Pub. No. (HSM) 73-11022, 1973.

58. NIOSH. *Pocket Guide to Chemical Hazards.* National Institute for Occupational Safety and Health, Cincinnati, Ohio, 1985.

59. *Occupational Exposure to Toluene Diisocyanate.* U.S. Dept. Health, Education and Welfare, 16–50.

60. O'Donoghue, J.L., ed. *Neurotoxicology of Industrial and Commercial Chemicals.* CRC Press, Inc., Boca Raton, Florida, 1985; Volume I: p 136, Volume II: p 34.

61. OSHA. Department of Labor, Occupational Safety and Health Administration: 29 CFR Part 1910; Air Contaminants; Final Rule. *Federal Register* 1989; 54 (12):2332–2983.

62. Owen, P. The Toxicity ad Carcinogenicity to Rats of Toluene Diisocyanate Vapor Administration by Inhalation for a Period of 113 Weeks. HLE Project 484/1, 1980.

63. Paggiaro., P.L., Bacci, E., Paoletti, P., et al. Bronchoalveolar lavage and morphology of the airways after cessation of exposure in asthmatic subjects sensitized to toluene diisocyanate. *Chest* 1990; 98:536–542.

64. Paggiaro, P.L., Bacci, E., Talini, D., et al. Atropine does not inhibit late asthmatic responses induced by toluene diisocyanate in sensitized subjects. *Am. Rev. Respir. Dis.* 1987; 136:1237–1241.

65. Paggiaro, P.L., Loi, A.M., Rossi, O., et al. Follow-up study of patient with respiratory disease due to toluene diisocyanate (TDI). *Clin. Allergy* 1984; 14:463–469.

66. Patterson, R., Nugent, K.M., Harris, K.E., et al. Immunologic hemorrhagic pneumonia caused by isocyanates. *Am. Rev. Respir. Dis.* 1990; 14:226–230.

67. Porter, C.V., Higgins, R.L., and Scheel, L.D. A retrospective study of clinical, physiologic, and immunologic changes in workers exposed to toluene diisocyanate. *Am. Ind. Hyg. Assoc. J.* 1975; 36:159–168.

68. Proctor, N.H., and Hughes, J.P. *Chemical Hazards of the Workplace.* JB Lippincott Co., Philadelphia, 1978; 483–484.

69. Proctor, N.H., Hughes, J.P., and Fischman, M.L. *Chemical Hazards of the Workplace,* 2nd ed. JB Lippincott Co., Philadelphia, PA, 1988.

70. Roper, C.P., Jr., and Cromer, J.W., Jr. Health Hazard Evaluation Determination Report 74-118-218. HEW, PHS, CDC, NIOSH, 1975.

71. RTECS. *Registry of Toxic Effects of Chemical Substances.* National Institute for Occupational Safety and Health, Cincinnati, Ohio (CD-ROM Version). Micromedex, Inc., Denver, Colorado, 1990.

72. Rye, W.A. Human responses to isocyanate exposure. *J. Occup. Med.* 1973; 15:306–307.

73. Sax, N.I., and Lewis, R.J. *Dangerous Properties of Industrial Materials,* 7th ed. Van Nostrand Reinhold Co, New York, 1989.

74. Scheel, L.D., Killens, R., and Josephson, A. *Am. Ind. Hyg. Assoc. J.* 1964; 25:179. As cited in ACGIH: Documentation of the Threshold Limit Values and Biological Exposure Indices, 5th ed. Am Conference of Govt. Ind. Hyg., Inc., Cincinnati, OH, 1986.

75. Selden, A.I., Belin, L., and Wass, U. Isocyanate exposure and hypersensitivity pneumonitis—A report of probable case and prevalence of specific immunoglobulin G antibodies among exposed individuals. *Scand. J. Work Environ. Health* 1989; 15:234–237.

76. Smilkstein, M.J., Smolinske, S.C., Kulig, K.W., et al. Severe hypermagnesemia due to multiple-dose cathartic therapy. *West. J. Med.* 1988; 148:208–211.

77. Stevens, M.A., and Palmer, R. The effect of tolylene diisocyanate on certain laboratory animals. *Proc. Roy. Soc. Med.* 1970; 63:380–382.

78. Tossin, L., Chiesura-Corona, P., Fabbri, L.M., et al. Ketotifen does not inhibit asthmatic reactions induced by toluene di-isocyanate in sensitized subjects. *Clin. Exp. Allergy* 1989; 19:177–182.

79. Valentino, M., Governa, M., and Fiorini, R. Increased neutrophil leukocyte chemotaxis induced by release of a serum factor in toluene-diisocyanate (TDI) asthma. *Lung* 1988; 166:317–325.

80. Venables, K.M., Dally, M.B., Burge, P.S., et al. Occupational asthma in a steel coating plant. *Br. J. Ind. Med.* 1985; 42:517–524.

81. Walker, C.L., Grammer, L.C., Shaughnessy, M.A., et al. Diphenylmethane diisocyanate hypersensitivity pneumonitis: a serologic evaluation. *J. Occup. Med.* 1989; 31:315–319.

82. Walworth, H.T., and Virchow, W.E. *Am. Ind. Hyg. Assoc. J.* 1959; 20:205. As cited in ACGIH, Documentation of the Threshold Limit Values and Biological Exposure Indices, 5th ed. Am Conference of Govt. Ind. Hyg., Inc., Cincinnati, OH, 1986.

83. Wass, U., and Belin, L. Immunologic specificity of isocyanate-induced IgE antibodies in serum from 10 sensitized workers. *J. Allergy Clin. Immunol.* 1989; 83:126–135.

84. Weill, H., Butcher, B., Dharmarajan, V., et al. NIOSH Contract No 210-75-0006, 1981.

85. Williamson, K.S. Studies in diisocyanate workers. *Trans. Assoc. Ind. Med. Off.* 1965; 15:29–35.

86. Yoshizawa, Y., Ohtsuka, M., Noguchi, K., et al. Hypersensitivity pneumonitis induced by toluene diisocyanate: sequelae of continuous exposure. *Ann. Int. Med.* 1989; 110:31–34.

87. Zapp, L.A., Jr. *Arch. Ind. Health* 1957; 15:324. As cited in ACGIH, Documentation of the Threshold Limit Values and Biological Exposure Indices, 5th ed. Am. Conference of Govt. Ind. Hyg., Inc., Cincinnati, OH, 1986.

88. Zocca, E., Fabbri, L.M., Boschetto, P., et al. Leukotriene B4 and late asthmatic reactions induced by toluene diisocyanate. *J. Appl. Physiol.* 1990; 68:1576–1580.

15.3 Polyamines

Synonyms: polyamine resin, bisphenol A.

15.3.1 Overview (OVER)

Life Support (OV-LIFE)

This overview assumes that basic life support measures have been instituted.

Clinical Effects (OV-CLEF)

Summary of Exposure

- Dermal exposure may result in severe skin irritation or burns. Acute oral ingestion of polyamines results in oral and esophageal burns. Dyspnea, stridor, dysphagia, and shock may be noted. Inhalation of fumes from cured epoxy resins may result in coughing and bronchospasm persisting for several days.

HEENT

- Fumes from cured epoxy resins may result in periorbital edema, facial pruritus and conjunctivitis.

Respiratory

- Fumes from cured epoxy resins may result in coughing, and bronchospasms persisting for several days.

Neurologic

- CNS depression may be noted. Epichlorohydrin ingestion may result in cyanosis, muscular relaxation or paralysis, tremor, seizures, and respiratory arrest.

Gastrointestinal

- Acute ingestion of polyamines may result in burns to the oral pharnyx and esophagus. Dyspnea, stridor, dysphagia and drooling may be noted.

Laboratory (OV-LAB)

- Plasma levels of these agents are not clinically useful. B. No specific lab work (CBC, electrolytes, urinalysis) is needed unless otherwise indicated.

Treatment Overview (OV-TRMT)

Oral/Parenteral Exposure

- *Dilution.* Immediately dilute with 4 to 8 oz (120 to 240 mL) of milk or water (not to exceed 15 mL/kg in a child).
- These products have low oral systemic toxicity. Gastric emptying procedures are seldom warranted.
- If the patient has burns to the mouth or pharynx, or is drooling or has other signs or symptoms of potential esophageal burns without visible burns to the mouth or pharynx, esophagoscopy should be performed.
- Oral and esophageal burns should be managed immediately with dexamethasone (Decadron) 1 mg/kg in children up to 100 mg/dose in adults; followed with doses of 0.1 mg/kg/day in divided doses, administered for 3 weeks and then tapered.
- Strictures may develop and require dilation or esophageal replacement.

Inhalation Exposure

- *Decontamination.* Move patient to fresh air. Monitor for respiratory distress. If cough or difficulty in breathing develops, evaluate for respiratory tract irritation, bronchitis, or pneumonitis. Administer 100% humidified supplemental oxygen with assisted ventilation as required.

Eye Exposure

- *Decontamination.* Exposed eyes should be irrigated with copious amounts of tepid water for at least 15 min. If irritation, pain, swelling, lacrimation, or photophobia persist, the patient should be seen in a health care facility.

Dermal Exposure

- Decontaminate skin or mucous membranes immediately with water and a soft sponge. A physician may need to examine the exposed area if irritation or pain persists after the area is washed.

Range of Toxicity (OV-TOX)

- Insufficient data in the literature to assess the range of toxicity following acute or chronic human exposure.

15.3.2 Substances Included (SUBS)

Therapeutic/Toxic Class (CLASS)

- The term *epoxy* refers to the presence of a three-membered epoxide ring consisting of two carbon atoms and an oxygen atom. Epoxy resins are formed by condensation of an epoxy compound (usually epichlorohydrin) with an alcohol, phenol, or fatty acid to form an ether or ester linkage. The most common type, the *bisphenol A resin* has a glycidyl ether linkage at both ends of the molecule. The resins range in molecular weight from the smallest, diglycidyl ether of bisphenol A (MW 340), to MW 908 and larger resins.

- Once an epoxy resin is formed, it is hardened or *cured* to provide greater stability and strength. Hardeners, such as polyamines, polyamides, and anhydrides, react with epoxy groups on the resin to link several molecules.

- Reactive diluents may be added to high molecular weight epoxy resins to decrease viscosity, in amount as high as 10 to 15%. These consist of low-molecular-weight compounds, such as monoglycidyl ethers, usually containing only one active epoxy group (Anon, 1981). Diluents may be present in epoxy resin products not labeled to contain these substances (Jolanki et al., 1987).

- Epoxy resins systems (resin and curing agent) are widely used as adhesives, molding resins, surface coatings, and reinforced plastics. The resins are usually long-chain polymers produced by condensation of epichlorhydrin and bisphenol A.

Specific Substances (SPEC)

- Epoxy Resin Types
  - diglycidyl ether of bisphenol A (DGEBA)
  - diglycidyl ether of tetrabromo-bisphenol A (Br-DGEBA)
  - tetraglycidyl-4,4'-methylene dianiline (TGMDA)
  - triglycidyl derivative of p-aminophenol (TGPAP)
  - o-diglycidylphthalate
  - 4,4-isopropylidene diphenyl epichlorohydrin
  - vinyl cyclohexene diepoxide

- Hardeners
  - Amines
        triethylenetetramine
        isophorone diamine
        metaphenylenediamine
        xylenediamine
        bis (4-amine-3-methyl-cyclohexyl) methane
        diaminodiphenyl sulphone (DDS)
        methylene dianilene (diaminodiphenylmethane)
        boron trifluoride monoethylamine complex
  - Anhydrides
        dodecenyl succinic anhydride
        hexahydrophthalic anhydride
        phthalic anhydride
        trimellitic anhydride
        tetrachlorophthalic anhydride
        himic anhydride
        maleic anhydride
  - Amides
        dicyandiamide

- Reactive Diluents
        phenyl glycidyl ether
        cresyl glycidyl ether
        butyl glycidyl ether
        allyl glycidyl ether
        1,4-butanediol diglycidyl ether
        neopentyl glycol diglycidyl ether
        1,6-hexanediol diglycidyl ether

Description (DESC)

- Epoxy resins may penetrate through rubber, polyethylene, and PVC gloves. Nitrile and Nitrile-butatoluene-rubber gloves gave protection for 48 hours in sensitized workers (Blanken et al., 1987).

15.3.3 Clinical Effects (CLEF)

Summary (CLEF-SUM)

- The most common adverse effect from occupational exposure to epoxy resins is contact dermatitis. Allergenicity depends on the molecular weight of the resin, and the amount of free oligomer in the *cured* product. The MW 340 oligomer is a potent human sensitizer, and it may be present in trace amounts in products containing primarily larger molecules, resulting in sensitization (Jolanki et al., 1987a). The MW 624 oligomer is an animal sensitizer only, and MW 908 and larger are nonsensitizing (Anon, 1981).
- Irritant dermal reactions may also occur from handling of the alkaline resin hardeners.

HEENT (HENT)

Eyes

- *Conjunctivitis.* Fumes of cured epoxy resins may cause pruritus of face, periorbital edema, and conjunctivitis (Laurberg and Christiansen, 1984).
- *Eyelid Dermatitis.* Contact eyelid dermatitis has been reported in individuals sensitized to epoxy resins (Nethercott et al., 1989).
- *Retinopathy.* Permanent pigmentary retinopathy was reported in a case of accidental ingestion of methylene dianilene. Blurred vision began four days postingestion, and within three weeks visual acuity was limited to perception of light. Gradual partial visual recovery occurred over several months (Roy et al., 1985).

Nose

- *Sneezing.* Nasal pruritus, congestion, rhionorrhea and sneezing have been reported in workers exposed to epoxy resins (Moller et al., 1985; Nielson et al., 1992).
- *Secretions.* Frequency of nasal secretions is significantly increased in workers exposed to epoxy resins (Nielsen et al., 1992).

Throat

- Burns

  Ingestion of polyamines results in burns to lips, tongue, oral mucosa, or hypopharynx. Presence or absence of burns in the mouth may or may not indicate burns of the esophagus. Pain may occur substernally or abdominally.

  Epiglottis, vocal cords, and trachea may be involved with dyspnea and stridor and early shock. Dysphagia may occur with drooling. Spontaneous emesis may aggravate some of these findings.

Cardiovascular (CARD)

- *Bradycardia.* Hypotension, bradycardia, T-wave inversion, and ST segment abnormalities were noted in a case of accidental ingestion of methylene dianilene (Roy et al., 1985).

Respiratory (RESP)

- Coughing

  Fumes of cured epoxy resins may cause coughing, asthmatic attacks, and bronchospasm for several days beyond exposure (Tepper, 1962). A consistent, reversible increase in airway resistance is measured by serial pulmonary function tests has been reported in workers exposed to an amine hardener resin system. This effect was more pronounced in smokers.

2. The prevalence of lower respiratory tract symptoms was higher among shipyard painters exposed to epoxy paints as compared to controls. A linear relation was shown between percentage decrement in FEV(1) and hours of exposure (Rempel et al., 1991).

- Asthma
  - Acute, severe, bronchospasm (occupational asthma) may be provoked by inhalation challenge tests to organic anhydrides and result in both immediate and late asthmatic responses (Bernstein et al., 1982; Zeiss et al., 1977; Howe et al., 1983; Gallagher et al., 1983; Moller et al., 1985; Ward and Davies, 1982; Nielsen et al., 1989; Taylor, 1991).
  - Venables et al. (1990) noted that patients with occupational asthma reported symptoms start late in the shift early in the week, but symptoms occur immediately after coming to work at the end of the week.
- Trimellitic Anhydride
  - Exposure to trimellitic anhydride dust or fumes can produce any of four distinct clinical syndromes (Pien et al., 1988).
    (1) An IGE-mediated asthma-rhinitis syndrome
    (2) A late respiratory systemic syndrome known as the *TMA flu* occurring 4 to 12 hr after exposure, consisting of cough, dyspnea, myalgia, fever, and chills
    (3) Pulmonary disease-anemia syndrome, characterized by cough, dyspnea, hemoptysis, hemolytic anemia, restrictive lung disease, pulmonary infiltrates, and occasionally respiratory failure
    (4) Irritant respiratory effects
  - Other organic acid anhydrides may cause any of the above four clinical syndromes. Methyl tetrahydrophthalic anhydride has been shown to cause an IgE mediated response and bronchial hyperreactivity in an occupationally exposed patient (Nielsen et al., 1989).
- *Asphyxia.* Three workers died of asphyxia after working in a non-ventilated underground water tank with an epoxy resin-based waterproofing paint. A reactive diluent in the epoxy resin, glycidyl ether, produced fumes which displaced oxygen in the tank, thus causing asphyxia (Gavino et al., 1990).
- *IgE.* Strong evidence supports the findings that acid anhydrides may be involved in the IgE antibody combining site. In a study of tetrachlorophthalic anhydride, it was shown that smokers were at a higher risk of developing a specific IgE-mediated bronchial response (Taylor, 1991).

Neurologic (NEUR)

- *CNS depression.* Ingestions of large doses may cause CNS depression.
- *Cyanosis.* Ingestion of epichlorohydrin in laboratory animals produces cyanosis, muscular relaxation or paralysis, tremor, seizures, and death.

Hepatic (HEPA)

- Hepatitis
  - Exposure to an epoxy resin curing agent, methylenedianiline (MDA) produced toxic hepatitis in 84 persons eating contaminated bread (Kopelman et al., 1966; Kopelman et al., 1966a), in 12 workers using the material in a insulating material manufacturing plant via dermal exposure (McGill and Motto, 1974), in 4 of 6 men laying epoxy resin based flooring (Bastian, 1984), and in a case of accidental one-time ingestion in a 28-year-old man (Roy et al., 1985).
  - MDA produces a characteristic mixed cholestatic-hepatocellular jaundice that has an abrupt onset. Symptoms of severe right upper-quadrant pain, high fever, chills, and rash are followed by jaundice with/or without hepatomegaly. The duration of illness ranges from one to seven weeks. In one case, hepatomegaly developed six weeks postingestion, resolving two to three weeks later. Liver transaminases remained elevated one year after the episode (Roy et al., 1985).
- *Biliary Function.* Acute toxicity of methylene dianiline (MDA) on the hepatobiliary function in rats was described. MDA rapidly diminished bile flow and altered the secretion of bile constituents

and was highly injurious to biliary epithelial cells (Kanz et al., 1992). Humans may develop jaundice, cholangitis with cholestasis, or toxic hepatitis.

Hematologic (HEMA)

- *Hemolytic anemia.* May occur in patients exposed to fumes containing trimellitic anhydride. Spontaneous remission may occur within several weeks (Rivera et al., 1981).

Dermatologic (DERM)

Summary

- Epoxy resin exposure may produce severe burns to skin, dermatitis, lacrimation, or sneezing. Some dye bases (i.e., metaphenylenediamine) may turn skin yellow upon oxidation (Cohen, 1985). Cutaneous amine reactions include chemical burns, erythema, persistent itching, severe facial swelling, blistering with leaking of serous fluid, crusting, and scaling.

Scleroderma

- Widespread sclerotic skin changes, erythema, brownish pigmentation, and telangiectasia were reported in two workers engaged in the polymerization of epoxy resins with an amine hardener (Yamakage et al., 1980).

Allergic Contact Dermatitis

- Contact dermatitis is a common occupational illness in workers exposed to epoxy resins. The prevalence in one epoxy resin plant involving 228 workers was 4.4% (Prens et al., 1986), and in another plant involving 135 workers was 18% (van Putten et al., 1984). Hand and arm exudative eczema and purpuric eruptions have been described (Laurberg and Christiansen, 1984).
- While most cases of sensitization occur after exposure to uncured resin, cured resins in the form of sawdust have also been reported to cause dermatitis (Suhonen, 1983; Ward and Davies, 1982).
- Erythema multiforme, exhibiting as an intense pruritic papular eruption, has been reported associated with an allergic contact dermatitis due to an epoxy resin and hardener (Whitfeld et al., 1991).

Eczema

- Acute and chronic eczema, described as an itchy erythematous papular vesicular dermatitis, has been reported from occupational contact epoxy resins and hardeners (Xuemin et al., 1992).

Irritant Contact Dermatitis

- Occupational exposure to epoxy resins may cause irritant dermatitis, believed to be related to the alkaline hardeners (Jolanki et al., 1987a).

Contact Urticaria

- Has been reported after occupational exposure (Jolanki et al., 1987a).

Photodermatitis

- Persistent light reactive photoallergic contact dermatitis was described in eight outdoor pipe repair workers using an epoxy resin (Allen and Kaidbey, 1979).

Endocrine (ENDO)

- *Glycosuria.* Has been reported following methylene dianilene ingestion (Roy et al., 1985).

Carcinogenicity (CARC)

- Epoxy resins have not been associated with human or animal carcinogenicity (Anstadt, 1989).

Immunologic (IMMU)

Summary

- Epoxy resins may cause allergic contact dermatitis, IgE mediated bronchospasm, and IgG formation.

Allergic Contact Dermatitis

- Allergic contact dermatitis is a common occupational illness in workers exposed to epoxy resins (Laurberg and Christiansen, 1984; van Putten et al., 1984; Prens et al., 1986).

Asthma

- Specific IgE and IgG antibodies to tetrachlorophthalic acid and human albumin conjugate was seen in seven patients with tetrachlorophthalic acid-induced asthma (Taylor, 1991).

Immunoglobulins

- IgG against trimellitic anhydride human serum albumen complex has been documented in exposed workers (Gerhardsson et al., 1993).

15.3.4 Laboratory (LAB)

Methods (METH)

- No laboratory determination available.
- The low molecular weight diglycidal ether of bisphenol A is part of the standard contact dermatitis patch test series. Manufacturers should be contacted regarding patch test materials for other types of epoxy resins.

Monitoring Parameters/Levels (MONIT)

Serum/Plasma/Blood

- Monitor liver function tests in patients exposed to methylene dianiline.

Other

- If respiratory tract irritation is present, it may be useful to monitor pulmonary function tests.

15.3.5 Case Reports (REPS)

- A 28-year-old man accidentally swallowed several mouthfuls of a solution of methylene dianilene in potassium carbonate and gamma butyrolactone.
  - Initial signs and symptoms included coma, arreflexia, pallor, miosis, Cheyne-Stokes respiration, bradycardia, hypotension, T-wave inversion, and ST abnormalities. He was alert after two hours and vomited. By the second day, serum transaminases and bilirubin were elevated.
  - On the fourth day, a pigmentory toxic retinopathy developed, which resolved incompletely within a few months. Other complications included hematuria, glycosuria, mild sensory peripheral neuropathy, and erythema multiforme.
  - Pruritus and hepatomegaly developed six weeks after ingestion, and persisted for two to three weeks. Liver function tests remained abnormal one year after the episode (Roy et al., 1985).

15.3.6 Treatment (TRMT)

Life Support (LIFE)

Support respiratory and cardiovascular function.

Oral/Parenteral Exposure (ORAL-EXP)

Prevention of Absorption

- *Dilution.* Immediately dilute with 4 to 8 oz (120 to 240 mL) of milk or water (not to exceed 15 mL/kg in a child).
- *Decontamination.* Decontaminate skin or mucous membranes immediately with water and a soft sponge or soap and water.
- *Emesis/Lavage.* These products have a low order of oral toxicity and emesis or lavage are usually unnecessary.

Treatment

- Esophagoscopy
  - If the patient has burns to the mouth or pharynx, esophagoscopy should be performed.
  - If the patient does not have burns of the mouth or pharynx but is symptomatic (drooling, pain on swallowing), burns of the esophagus may have occurred, and esophagoscopy should be performed.
- Burns
  - If burns have occurred, administer dexamethasone 1 mg/kg in children to 100 mg/kg in adults immediately, with doses of 0.1 mg/kg/day in divided doses given in burned patients for 3 weeks and then tapered off.
- Strictures
  - Strictures may develop and require dilatation or esophageal replacement.

Inhalation Exposure (INHL-EXP)

Decontamination

- *Decontamination.* Move patient from the toxic environment to fresh air. Monitor for respiratory distress. If cough or difficulty in breathing develops, evaluate for respiratory tract irritation, bronchitis, or pneumonitis.
- *Observation.* Carefully observe patients with inhalation exposure for the development of any systemic signs or symptoms and administer symptomatic treatment as necessary.
- *Initial treatment.* Administer 100% humidified supplemental oxygen with assisted ventilation as required. Exposed skin and eyes should be copiously flushed with water.

Treatment

- *Bronchospasm.* If bronchospasm and wheezing occur, consider treatment with inhaled sympathomimetic agents.

Eye Exposure (EYE-EXP)

Decontamination

- Exposed eyes should be irrigated with copious amounts of room temperature water for at least 15 minutes. If irritation, pain, swelling, lacrimation, or photophobia persist after 15 minutes of irrigation, an ophthalmologic examination should be performed.

Dermal Exposure (DERM-EXP)

Decontamination

- Wash exposed area extremely thoroughly with soap and water. A physician may need to examine the area if irritation or pain persists after washing.

Treatment

- Do not cover skin or mucous membranes.

15.3.7 Range of Toxicity (TOX)

Minimum Lethal Exposure (MIN)

- Limited data are available on human toxicity. Epichlorohydrin administered daily for 10 to 20 days at concentrations of 0.1 mL/kg killed all mice in sample population. Application to skin of 0.5 mL/kg daily killed all of a group of rats in four days. A level of 8300 ppm in atmosphere for 30 min was fatal to mice.

LD50/LC50 (LDLC)

- *Diethylenetriamine.* LD50 (Oral) rat, 1,080 mg/kg

15.3.8 Available Forms/Sources (FORM)

- Epoxy resins systems (resin and curing agent) are widely used as adhesives, molding resins, surface coatings, and reinforced plastics. The resins most commonly used are long-chain polymers produced by condensation of epichlorhydrin and bisphenol A.

- Viscosity ranges from low-viscosity liquids to solids, depending on molecular weight. Solvents added to reduce viscosity may include glycidyl ether (phenyl, allyl or butyl), styrene oxide, or styrene epoxide.

- Clay-like epoxy that is solid (parts A and B) has not presented a significant ingestion hazard to this date. Conversion to a solid by cross-linking of the polymers is brought about by the curing agent or hardener.

- Systems requiring external heat for curing commonly employ acid anhydride hardeners, while exothermic or cold curing systems usually involve polyamines (piperazine, di- or triethylenetriamine). Fumes of both the resin and curing agent are emitted during the curing process.

- *Note:* **Nonadhesive products containing the above chemicals are also coded to this management.**

- Small amounts of uncured epoxy resins may be present in plastic medical equipment, such as nasal cannulas (Wright and Fregert, 1983; Toome, 1989), hemodialysis tubing (Mork, 1979), pacemakers (Romaguera and Grimalt, 1981), ostomy bags (Van Ketel et al., 1983; Fregert et al., 1984; Beck et al., 1985), vinyl hospital identification bands (Fisher, 1985), and insulin pump infusion sets (Boom and Vandriel, 1985).

- Uncured epoxy resins may also be found in numerous metallic objects found in household use, such as twist-off bottle caps, film cassettes, metal food and cosmetic packages, signboards, and brass hooks and door knobs (Fregert et al., 1980).

- Other household sources include textile labels (Gregert and Orsmark, 1984), knee patches (Taylor et al., 1983), plastic tool handles (Fischer et al., 1987), and cleaning gloves (Jenkinson and Burrows, 1987).

- Trade names include Epon 825® (contained in the standard patch test tray), and Epon 828®.

- Nonbisphenol A epoxy resins, such as cycloaliphatic epoxies, are used as transformer encapsulating compounds, electrical equipment insulators, coatings for plastics, metals, and paper, and in electron microscopy (Dannaker, 1988).

15.3.9 Kinetics (KINET)

Absorption (ABSR)

- Epoxy resins and polyamines may cause burns to skin in 15 min to 1 hr following exposure. CNS effects may be noted within a few minutes following inhalation or oral exposure.

15.3.10 Pharmacology/Toxicology (PHAR)

Toxicologic Mechanism (TOX-MECH)

- *Caustic.* The polyamine hardeners are extremely basic (pH 13 to 14) and have volatile and caustic alkaline properties. These damage the keratin layer of the skin and remove surface lipids. Ingestion may thus result in severe skin or esophageal burns.

- *Dermal.* Epoxy resin systems are among the most important causes of industrial contact dermatitis and may cause irritation and/or sensitization in up to 43% of exposed workers. Positive patch tests to epoxy resin or triethylenetetramine are found in 66% of cases.

References

1. Allen, H., and Kaidbey, K. Persistent photosensitivity following occupational exposure to epoxy resin. *Arch. Dermatol.* 1979; 115:1307–1310.
2. Anon. *Penetrating* effects of epoxy resin systems. *Occup. Health Saf.* 1981; 50:42–44, 59.
3. Anstadt, G.W. Occupational medicine forum. *J. Occup. Med.* 1989; 31:582–583.
4. Bastian, P.G. Occupational hepatitis caused by methylenedianiline. *Med. J. Aust.* 1984; 141:533–535.
5. Beck, M.H., Burrows, D., Fregert, S., et al. Allergic contact dermatitis to epoxy resin in ostomy bags. *Br. J. Surg.* 1985; 72:202–203.
6. Bernstein, D.I., Patterson, R., and Zeiss, C.R. Clinical and immunologic evaluation of trimellitic anhydride- and phthalic anhydride-exposed workers using a questionnaire with comparative analysis of enzyme-linked immunosorbent and radioimmunoassay studies. *J. Allergy Clin. Immunol.* 1982; 69:311.
7. Blanken, R., Nater, J.P., and Veenhoff, E. Protection against epoxy resins with glove materials. *Contact Dermatitis* 1987; 16:46–47.
8. Boom, B.W., and Vandriel, L.M.J. Allergic contact dermatitis to epoxy resin in infusion sets of an insulin pump. *Contact Dermatitis* 1985; 12:280.
9. Cohen, S.R. Yellow staining caused by 4,4'-methylenedianiline exposure: Occurrence among molded plastics workers. *Arch. Dermatol.* 1985; 212:1022–1027.
10. Dannaker, C.J. Allergic sensitization to a non-bisphenol A epoxy of the cycloaliphatic class. *J. Occup. Med.* 1988; 30:641–643.
11. Fischer, T., Fregert, S., Thulin, I., et al. Unhardened epoxy resin in tool handles. *Contact Dermatitis* 1987; 16:45.
12. Fisher, A.A. Allergic contact dermatitis in early infancy. *Cutis.* 1985; 35:315–316.
13. Fregert, S., Meding, B., and Trulsson, L. Demonstration of epoxy resin in stoma pouch plastic. *Contact Dermatitis* 1984; 10:106.
14. Fregert, S., and Orsmark, K. Allergic contact dermatitis due to epoxy resin in textile labels. *Contact Dermatitis* 1984; 11:131.
15. Fregert, S., Persson, K., and Trulsson, L. Hidden sources of unhardened epoxy resin of bisphenol A type. *Contact Dermatitis* 1980; 6:446–447.
16. Gallagher, J.S., Moller, D.R., Roseman, K.D., et al. In vitro demonstration of specific IgE in a worker exposed to himic anhydride (abstract). *J. Allergy Clin. Immunol.* 1983; 71:157.
17. Gavino, R.R., Salva, E.S., Gregorio, S.P., et al. Occupational fatalities associated with exposure to epoxy resin paint in an underground tank—Makati, Republic of the Philippines. *MMWR* 1990; 39:373–380.
18. Gerhardsson, L., Grammer, L.C., Shaughnessy, M.A., et al. Immunologic specificity of IgG against trimellityl-human serum albumin in serum samples of workers exposed to trimellitic anhydride. *J. Lab. Clin. Med.* 1993; 121:792–796.
19. Howe, W., Venables, K.M., Topping, M.D., et al.: Tetrachlorophthalic anhydride asthma: evidence for specific IgE antibody. *J. Allergy Clin. Immunol.* 1983; 71:5.

20. Jenkinson, H.A., and Burrows, D. Pitfalls in the demonstration of epoxy resins. *Contact Dermatitis* 1987; 16:226–227.

21. Jolanki, R., Estlander, T., and Kanerva, L. Contact allergy to an epoxy reactive diluent: 1,4-butane-diol diglycidyl ether. *Contact Dermatitis* 1987; 16:87–92.

22. Jolanki, R., Estlander, T., and Kanerva, L. Occupational contact dermatitis and contact urticaria caused by epoxy resins. *Acta. Derm. Venereol. Stockholm* 1987a; (Suppl. 134):90–94.

23. Kanz, M.F., Kaphalia, L., Kaphalia, B.S., et al. Methylene dianiline: acute toxicity and effects on biliary function. *Tox. Applied Pharmacol.* 1992; 117:88–97.

24. Kopelman, H., Robertson, M.H., Sanders, P.G., et al. The Epping jaundice. *Br. Med. J.* 1966; 1:514–516.

25. Kopelman, H., Scheuer, P.J., Williams, R., et al. The liver lesion of the Epping jaundice. *Q. J. Med.* 1966; 35:553–564.

26. Laurberg, G., and Christiansen, J.V. Purpuric allergic contact dermatitis of epoxy resins. *Contact Dermatitis* 1984; 11:186–187.

27. McGill, D.B., and Motto, J.D. An industrial outbreak of toxic hepatitis due to methylenedianiline. *N. Engl. J. Med.* 1974; 291:278–282.

28. Moller, D.R., Gallagher, J.S., Bernstein, D.I., et al. Detection of IgE-mediated respiratory sensitization in workers exposed to hexahydrophthalic anhydride. *J. Allergy Clin. Immunol.* 1985; 75:663–672.

29. Mork, N.J. Contact sensitivity from epoxy resin in a hemodialysis set. *Contact Dermatitis* 1979; 5:331–332.

30. Nethercott, J.R., Nield, G., and Holness, D.L. A review of 79 cases of eyelid dermatitis. *J. Am. Acad. Dermatol.* 1989; 21:223–230.

31. Nielsen, J., Welinder, H., Horstmann, V., et al. Allergy tomethyltetrahydrophthalic anhydride in epoxy resin workers. *Br. J. Indust. Med.* 1992; 49:769–775.

32. Nielsen, J., Welinder, H., and Skerfving, S. Allergic airway disease caused by methyl tetrahydrophthalic anhydride in epoxy resin. *Scand. J. Work. Environ. Health.* 1989; 15:154–155.

33. Pien, L.C., Zeis, C.R., Leach, C.L., et al. Antibody response to trimellityl hemoglobin in trimellitic anhydride-induced lung injury. *J. Allergy Clin. Immunol.* 1988; 82:1098–1103.

34. Prens, E.P., De Jong, G., and Van Joost, T.H. Sensitization to epichlorohydrin and epoxy system components. *Contact Dermatitis* 1986; 15:85–90.

35. Rempel, D., Jones, J., Atterbury, M., et al. Respiratory effects of exposure of shipyard workers to epoxy paints. *Br. J. Indust. Med.* 1991; 48:783–787.

36. Rivera, M., Nicotra, B., Byron, G.E., et al. Trimellitic anhydride toxicity—A cause of acute multi-system failure. *Arch. Intern. Med.* 1981; 141:1071–1074.

37. Romaguera, C., and Grimalt, F. Pacemaker dermatitis. *Contact Dermatitis* 1981; 7:333.

38. Roy, C.W., McSorley, P.D., and Syme, I.G. Methylene dianilene: a new toxic cause of visual failure with hepatitis. *Hum. Toxicol.* 1985; 4:61–66.

39. Rudzki, E., and Krajewska, D. Epoxy resin dermatitis. *Contact Dermatitis* 1976; 2:135–138.

40. Sargent, E.V., and Mitchell, C.A. Respiratory effects of occupational exposure to an epoxy resin system. *Arch. Environ. Health* 1976; 31:236–240.

41. Suhonen, R. Epoxy-dermatitis in a ski-stick factory. *Contact Dermatitis* 1983; 9:131–133.

42. Taylor, A.J.N. Acid anhydrides. *Clin. Exp. Allergy* 1991; 21:234–240.

43. Taylor, J.S., Bergfeld, W.F., and Guin, J.D. Contact dermatitis to knee patch adhesive in boys' jeans: A nonoccupational cause of epoxy resin sensitivity. *Cleve. Clin. Q.* 1983; 50:123–127.

44. Tepper, L.B. Hazards to health: Epoxy resins. *N. Engl. J. Med.* 1962; 267:821.

45. Toome, B.K. Allergic contact dermatitis to a nasal cannula (letter). *Arch. Dermatol.* 1989; 125:571.

46. Van Ketel, W.G., Van de Burg, C.K.H.D., and Haan, P. Sensitization to epoxy resin from an ileostomy bag. *Contact Dermatitis* 1983; 9:516.

47. van Putten, P.B., Coenraads, P.J., and Nater, J.P. Hand dermatoses and contact allergic reactions in construction workers exposed to epoxy resins. *Contact Dermatitis* 1984; 10:146–150.

48. Venables, K.M., and Taylor, A.J.N. Exposure-response relationships in asthma caused by tetrachlorophthalic anhydride. *J. Allergy Clin. Immunol.* 1990; 85:55–58.

49. Ward, M.J., and Davies, D. Asthma due to grinding epoxy resin cured with phthallic anhydride. *Clin. Allergy* 1982; 12:165–168.

50. Whitfeld, M.J., and Rivers, J.K. Erythema multiforme after contact dermatitis in response to an epoxy sealant. *J. Acad. Derm.* 1991; 25:386–388.

51. Wright, R.C., and Fregert, S. Allergic contact dermatitis from epoxy resin in nasal canulae. *Contact Dermatitis* 1983; 9:387–389.

52. Xuemin, W., Yingfen, L., Xiafeng, C., et al. 17 cases of epoxy resin dermatitis in Shanghai. *Contact Dermatitis* 1992; 27:202–203.

53. Yamakage, A., Ishikawa, H., Saito, Y., et al. Occupational scleroderma-like disorder occurring in men engaged in the polymerization of epoxy resins. *Dermatologica* 1980; 161:33–44.

54. Zeiss, C.R., Patterson, R., Pruzansky, J.J., et al. Trimellitic anhydride-induced airway syndromes: Clinical and immunologic studies. *J. Allergy Clin. Immunol.* 1977; 60:96.

16

Radiation-Curable Adhesives and Coatings

16.1 Introduction

Radiation curing is the use of electron beam (EB) or ultraviolet (UV) radiation as an energy source to induce the rapid conversion of specially formulated 100% polyurethane reactive liquids to solids. These specially formulated polyurethanes can be coatings, inks, adhesives, sealants, or potting compounds. In the case of adhesives and sealants, the desired conversion may be from a liquid to a semisolid.

Radiation curing is a relatively new, pollution-free process in which a specially formulated coating or adhesive is cured by means of high-energy electrons. Radiation-cured coatings don't pollute the air, cure in seconds, and cut costs. There is no waste because whatever material is applied wet essentially is what remains after curing. Fast-curing is achieved in seconds. Use of radiation-curable adhesives will continue to grow because of several key advantages the technology offers. These include absence of volatile organic components (VOCs), much reduced energy consumption, greater productivity, single-component materials, room temperature cure, and controlled but rapid cure.

At first, radiation curing was limited to flat surfaces. However, recent modifications now allow it to be used to cure coatings on pipe, plastic containers, wire, and electronic circuit boards. Although initial investment is high, a switch to radiation curing often brings about cost savings, because it is much faster than conventional curing, and production can be increased significantly. In addition, properties of cured films often are better than those obtained with solvent coatings, and there are no pollution problems.

16.2 Advantages of Radiation-Cured Polyurethanes

- *Absence of VOCs.* The adhesives are 100% solids, i.e., they contain no solvents and therefore no volatile organic components. As such, their use can meet or exceed clean air regulations.

- *Reduced energy consumption.* Electrical energy is required by both UV and EB equipment. Since most of the UV radiation and electron beam can be sized and directed to the product, the equipment is operated at high efficiency. In comparison to thermal curing systems, only a fraction of energy is consumed by radiation curing systems.

- *Greater productivity.* Production time and labor costs per unit of product are significantly reduced because of the rapid cure of these materials. Cure rate is an insignificant part of the total production time. This fact allows a manufacturer to increase production without additional space for labor requirements.

- *Single-component materials.* Most UV/EB adhesives are supplied as one component, ready-to-use systems, with no blending or open time to worry about. This translates into less waste, because unused material can be saved for the next run.

- *Room-temperature cure*. Both UV and EB curing take place at room temperature, permitting the use of heat-sensitive substrates.
- *Controlled but rapid cure*. This characteristic is particularly useful for UV curing of structural adhesives. The components can be manipulated for as long as necessary, without any changes in the adhesive properties until the item is exposed to radiation. At this point, the adhesive cures instantly, and the product is ready for the next step in the process.

There are three basic types of radiation-curable adhesives, and they represent three different technologies. These include laminating adhesives that are UV or EB curable, depending on the substrates being laminated; pressure-sensitive adhesives (PSAS) that are UV or EB curable; and structural adhesives and sealants that are UV curable.

16.3 Radiation Curing Processes

There are two major radiation-curing processes: ultraviolet (UV) and electron beam (EB). The most commonly used source of energy for UV curing is the mercury vapor lamp. A reflector assembly, called an irradiator, houses the lamp and reflects UV energy onto the substrate. Two other commercial UV lamp systems are available: electrodeless lamps and pulsed xenon. The electrodeless types feature instant on/off capability, whereas the standard mercury and pulsed xenon require a warm-up time for UV output to reach its peak.

Another benefit of UV curing is simplified maintenance. Solvent-borne coating have to be removed thoroughly at the end of each operating period. In contrast, the UV unit is covered with a black plastic film to prevent slow cure from stray light. Only a few minutes are required to allow mercury-vapor lamps to come up to intensity, compared to 30 min to bring the thermal oven up to 350° F. Figures 16.1 through 16.3 present the spectral output of different lamps used in the UV-curing process.

For electron-beam curing, two systems are offered: (1) a linear cathode type with low energy and (2) a scanned-beam type with higher energy. All EB units incorporate an accelerator, controls, a power supply, and a shielded irradiation chamber. Capital investment for electron-beam curing equipment is higher than for conventional systems, and formulation prices are higher, too. But these costs can be offset by higher operating speeds. As for the differences in the two systems, electron beam can cure heavily pigmented materials, whereas coating thickness is limited in ultraviolet systems.

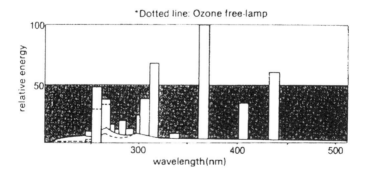

Medium-pressure mercury lamps feature a single-tube design offering extremely high output. These lamps radiate UV rays of 365 nm as a main spectrum and radiate rays of 253.7, 303, and 313 nm with optimal efficiency. Four types are available, with outputs of 50, 80, 120, and 160 W/cm, allowing selection according to the application. Arc lengths of 50 to >2500 mm are available.

FIGURE 16.1 Medium-pressure mercury lamps. *Source:* EYE UV Lamps.

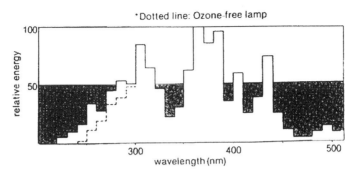

FIGURE 16.2 Medium-pressure metal halide lamps. *Source:* EYE UV Lamps.

Medium-pressure metal halide lamps, available with outputs of 80, 120, and 160 W/cm with arc lengths in excess of 1600 mm, feature comparatively higher UV emission efficacy with particularly high efficacy in the 300–450 nm long UV wavelength region. As a result, curing of thicker inks or paint coatings can be achieved in less than half the time required with conventional curing systems. This makes them particularly useful in screen printing, colored coating, potting, and other fields where thick coatings are used.

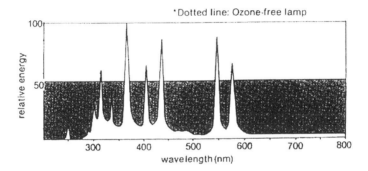

A *spot lamp* features spherical design with arc length of 2 mm and 200 W power. It is primarily used in combination with a focusing cold mirror and optical fiber to allow flexible UV output for spot adhesion of electrical and optical components. A *longitudinal lamp* features single-tube design, housed in a secondary water-cooling tube. Arc length is 300 nm with 100 W/cm. This lamp features very fast start-up time. The extremely compact, water-cooled system is portable and can be transferred easily from one application to another (e.g., tag printing lines).

FIGURE 16.3 High-pressure mercury lamps. *Source:* EYE UV Lamps.

Another difference is that electron beam does not require an inert atmosphere while ultraviolet curing does. Electron beam coating formulations have longer shelf life because, unlike ultraviolet-cured coatings, they do not require an opaque material, like plastic film, during laminating operations. Energy requirements for UV are 1/15 that of solvent-borne coatings; for EB, 1/100.

16.4 UV Versus Hot Melt

Ultraviolet-cured adhesives offer a number of advantages over hot-melt adhesives. UV adhesives can function at temperatures well above 300° F (150° C), compared to 110° F (43° C) for hot melts.

UV-cured adhesives are resistant to mild caustic and acidic water solutions. They resist alcohol and trichloroethylene solvents, but only for a limited time. UV-curable adhesives are finding increased use in pressure-sensitive label stock. They are used on plastic films as well as sheet goods and films for such products as decals, nameplates, printed-circuit boards, and electronic capacitance switches.

When running any UV product, a white or reflective primary substrate is preferred but not mandatory. The reason is that some short-wavelength UV photons will pass through the UV materials without being captured and without performing a curing function. With the primed surface, the photons have a second chance of initiating photopolymerization, because they can be reflected back into the material.

16.5 UV Curing Technology

The UV curing process requires a UV lamp that directs UV light onto the formulated product. Photo-initiators absorb the UV energy from the light source, setting in motion a chemical reaction that quickly* converts the liquid into a solid, cured film.

The bulk of the formulation contains monomers and oligomers. Monomers are low-molecular-weight materials. They can be monofunctional, bifunctional, or multifunctional molecules. These molecules become part of the polymer matrix in the cured coating because of their reactive functional groups. Furthermore, monomers also function as diluents in the formulations. Therefore, monomers are sometimes referred to as *reactive diluents*.

Oligomers, on the other hand, are high-molecular-weight viscous materials. The molecular weight ranges from several hundred to several thousand or even higher. Usually, the type of oligomer backbone determines the final properties of the coating such as flexibility, toughness, etc. These backbones can be polyether, polyester, polyurethane, or other types. The functional groups that provide the linkage between molecules are usually located at both ends of the oligomer molecules. (The most common functionality used is the acrylate functional group.)

A photoinitiator can be categorized in different ways:

According to its *mechanism*:	free radical
	cationic
According to its *form*:	liquid
	solid
According to its *absorption*:	UV light
	visible light
According to its *application*:	clear
	pigmented
	white

The photoinitiator is a critical component of the UV curing process. It is the additive that initiates the polymerization process to quickly reach the final cross-linked product. As UV light energy is emitted, it is absorbed by the photoinitiator in the liquid, causing it to fragment into reactive species. These species can be either free-radical or cationic. The majority of systems are based on free radicals that react with the unsaturated compounds in the liquid and cause them to polymerize. This reaction is almost instantaneous, and discussed below.

Of equal importance to the photoinitiator, however, is the UV light source. Basically, two types of light sources are available: arc light and laser light. The arc-type light includes the medium-pressure mercury lamp and the high-pressure xenon lamp. The medium-pressure mercury lamp is currently the overwhelming industry choice because of its high power (200–700 W/in) and important emission lines that are absorbed by most commercially available photoinitiators. Regardless of the type of light source, however, the emission spectra of the lamp must overlap the absorption spectrum of the chosen initiator.

---

*In most cases, the time is measured in fractions of a second.

The free-radical class of initiators represents greater than 90% of the commercially used initiator chemistry, while cationic curing has been more limited in scope. Free-radical initiators are most typically used with acrylate/methacrylate functional resins and can also be used with unsaturated polyester resins. Cationic curable systems depend on the use of epoxy or vinyl ether functional resins.

16.5.1 Classification of Photoinitiators

Free-radical initiators can be described as either *hydrogen abstraction type* or *alpha cleavage type*. While hydrogen abstraction initiators have their specific uses, especially in three- or four-way photoinitiator blends, alpha cleavage initiators are the most widely used. The alpha cleavage initiators have a generally higher efficiency due to their generation of free radicals via a unimolecular process. Alpha cleavage initiators need only absorb light to generate radicals; hydrogen abstraction initiators, on the other hand, require an extra step: after absorbing light, the excited state photoinitiator must find a hydrogen-donating source to generate the free radicals. It is a bimolecular process.

Beyond the obvious differences in chemical structure, the alpha cleavage initiators are typically differentiated by their absorption profiles. A formulator should be interested in two different characteristics of the photoinitiator's absorption curve. First is determining which wavelengths of light are absorbed, and second is the strength of this absorption (molar extinction coefficients). Photoinitiators developed for curing of pigmented films typically have higher molar extinction coefficients between the wavelengths (from 330 to 400 nm) than those useful for curing clear formulations.

16.5.2 UV Curables Market

The UV curables market has experienced dynamic growth since its introduction in the late 1960s. Market growth through the 1980s was 10 to 15% per year. The value of the 1995 UV curables market was estimated at over $325 million. This is the total value of the formulated coatings, inks, adhesives, photoresists, and printing plates themselves. The value of the converted substrates, of course, is many times greater.

The growth of UV curing technology has two distinct components.

1. Growth of existing markets
2. Adoption of UV curing technology by new markets

The wood and paper coating industries were two of the first markets to utilize UV curing technology. Although growth of UV curing within the wood industry has been modest, growth of UV-curable paper coatings actually accelerated in the late 1980s, making it the largest-volume application for photopolymer technology. Other markets that adopted the UV curing process in the 1980s include structural adhesives, pressure-sensitive adhesives, fiber-optic coatings, and coatings on plastics. The use of coatings on plastics, especially for automotive applications, is expected to increase steadily into the next century.

A number of factors, primarily productivity and product performance, have fueled the growth of UV curing technology. Some of the benefits seen by companies who have adopted UV curing technology to their own manufacturing processes are listed in Table 16.1.

TABLE 16.1 Factors Aiding the Growth of UV Curing

Features	Benefits
Short cure cycle	Higher productivity
Low-temperature process	Allows coating of temperature-sensitive plastics and paper
Energy savings vs. thermal cure	Cost savings
Solventless	Environmentally favored
High gloss	Aesthetically appealing
Highly cross-linked	Abrasion resistance
Novel cross-linking process	Unique end properties

16.6 Industry Trends

Four key trends in the UV curing industry are discussed in this section. A critical factor in each of these trends is the proper choice of photoinitiator.

16.6.1 Trend 1: Higher Line Speeds

Many economists expect productivity to be the watchword of the 1990s. Ultraviolet curing technology fits this need well. However, to continue to be the technology of choice, it will have to meet the demand for higher line speeds. For example, paper coatings are now typically running at 1200 ft/min, but the industry expects line speed to approach 2000 ft/min. Similarly, two-piece can lines are running at 1400 cans/min, but there is an increasing demand that they run at 1600–1800 cans/min. These higher line speeds can be achieved by adding more lamps onto a given line, but space considerations often make this impossible. The capital cost of additional lamps may also be prohibitive. The preferred approach is to develop faster curing coatings by using advanced photoinitiator technology.

16.6.2 Trend 2: Greater Use of Doped Lamps

Doped lamps are those in which the spectral output has been red-shifted (shifted to longer wavelengths). The use of these lamps is growing noticeably faster than the market in general. The reason for this faster growth is their ability to provide better through-cure of both clear and pigmented coatings. The potential use of doped lamps should be considered to select photoinitiators that will make the best use of this longer-wavelength light.

16.6.3 Trend 3: Higher Levels of Pigmentation

The screen printing ink, the offset ink, and the general industrial coatings markets are working to develop thinner and thinner coatings. To achieve thinner coatings and still have acceptable hiding properties, a higher level of pigmentation is required. Photoinitiators with strong absorption at longer wavelengths must be used to properly cure these more heavily pigmented coatings.

16.6.4 Trend 4: Curing in Air

An inert atmosphere, usually formed with nitrogen gas, has been used to overcome the oxygen inhibition effect that is characteristic of free-radical polymerization. The flooring industry, in particular, has found this approach effective. There are, however, two problems associated with this approach that make curing in air a desirable goal. First, the cost of maintaining an inert atmosphere is often higher than expected. In addition, *inerting* is one more process variable that must be monitored. Successful air-cure formulations depend on proper concentration and mixture of photoinitiators.

16.7 Historical Perspectives

In the early 1970s, when UV curing technology was being adopted for industrial purposes, photoinitiator choices were very limited. Benzoin ethers, thioxanthone/amine combinations, or benzophenone/amine combinations were the only photoinitiators available. The commercial introduction of two different benzil ketals in the mid-to-late 1970s accompanied, and greatly assisted, the growth of UV curing technology in the coatings industry. Benzil dimethyl ketal (BDMK) and acetophenone diethyl ketal (ADEK) both brought vast improvements to the shelf life of UV formulations. Poor shelf life had been a major limitation of benzoin ether- and benzophenone/amine-initiated formulations. In addition to improved shelf life, BDMK combined high cure speed and better through-cure of thick coatings. ADEK was an improvement over previous initiators by virtue of its non-yellowing character.

It was not until the early 1980s that two new acetophenone derivatives combined the excellent cure properties of BDMK with the non-yellowing characteristics of ADEK. Hydroxycyclohexyl phenyl ketone

(HCPK) and 2-hydroxy-2-methyl phenylpropanone (HMPP) were readily adopted, in many cases replacing BDMK and ADEK, especially in clear coatings.

In the mid-1980s, the ability to thoroughly cure pigmented inks and coatings at higher line speeds was made possible by the introduction of a morpholino-substituted acetophenone derivative, specifically 2-methyl-1-[4-(methylthio)phenyl]-2-morpholinopropanone (MMMP). This compound's high extinction coefficients, and its ability to be sensitized by thioxanthones, proved highly useful in ink and coating formulations.

In the late 1980s, two new compounds were introduced. Both of these photoinitiators brought additional advances in curing pigmented inks and coatings. One of these compounds was an advancement in the alpha-amino acetophenones, 2-benzyl-2-NN-dimethylamino-1-(4-morpholinophenyl) butanone (BDMB). The other, 2,4,6-trimethylbenzoyldiphenyl phosphine oxide (TPO), was the first commercial compound containing the acylphosphine oxide group. This product was promoted for curing pigmented formulations and thick coatings. Table 16.2 presents a list of some photoinitiators.

TABLE 16.2 Some Photoinitiators

Product	Description
Irgacure 651	This is a general-purpose initiator, useful for curing unsaturated polyester resins.
Irgacure 184	This is best for non-yellowing applications, also has relatively low odor.
Darocur 1173	This is a liquid non-yellowing photoinitiator; good solvency properties make it ideal for making photoinitiator blends.
Irgacure 500	This is a liquid blend of Irgacure 184 and benzophenone, offering good balance of surface cure and through-cure.
Irgacure 907	Strong absorption characteristics make it useful for curing pigmented inks and coatings; has excellent surface cure initiator.
Darocur 4265	This is excellent for curing white inks and coatings.
Irgacure 2959	This offers very low odor and low volatility. It also has a terminal hydroxyl group that may be reacted into a polymer backbone.
Irgacure 369	Strong broad absorption characteristics make it most useful for curing thick pigmented formulations.
Irgacure 261	This cationic initiator is used for curing epoxy resins and is useful for thick film or pigmented coating formulations.
Irgacure 1700	This is the most efficient initiator available for curing inks and coatings containing TiO_2.
Irgacure 784 DC	This visible light indicator responds to all UV and visible light sources, including visible lasers.

16.8 Photochemistry Principles

Photochemical reactions are different from thermal or electrochemical reactions in several respects.

1. Molecules have to absorb incident light to activate photochemical reactions. Thus, molecules without a chromophore cannot undergo photochemical reactions.
2. The electronic and nuclear configurations of the *excited* molecules are usually different from those of molecules activated by chemical or electrochemical means.
3. The products of photochemical processes are usually different from other chemical reaction products. This is due to the large excess energy and the different electronic configuration of the photochemically *excited* molecule compared to the ground-state molecule.

The role of photoinitiators is to produce the free radicals that initiate the chemical reaction, as shown in Fig. 16.4. The photochemically excited molecules may undergo two reactions: alpha cleavage and atom abstraction. These reactions were studied extensively by Norrish in the late 1930s and are named after him.[1] These two important photochemical reactions are called the Norrish type I and type II reactions.[2]

The Norrish type I reaction is a homolytic alpha cleavage reaction between the carbonyl group and the adjacent a carbon atom. The Norrish type II mechanism involves intramolecular hydrogen abstraction from the +carbon. Most often this occurs in the (n,*) state because of its *in-plane* structure. In this structure, a six-membered ring transition state is achieved in which intramolecular hydrogen abstraction can occur more easily (shown in the following reaction diagrams). Figure 16.5 presents the curing mechanisms of several photoinitiators available from Ciba-Geigy.

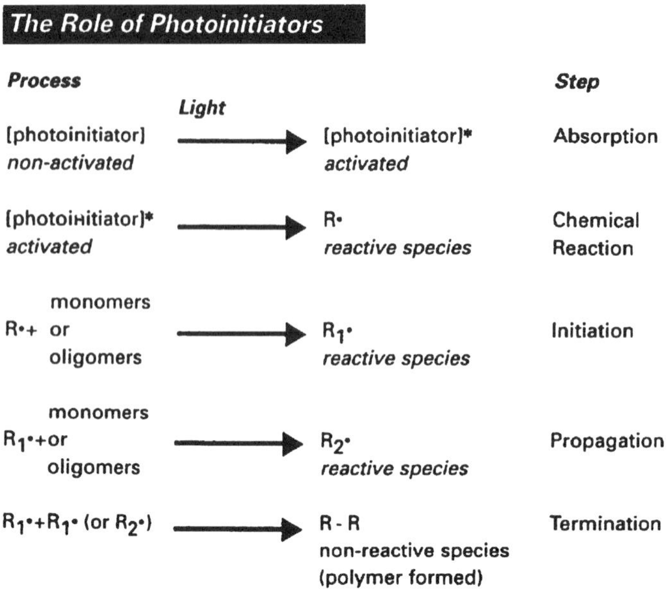

The Role of Photoinitiators

Process   **Step**

Light

[photoinitiator]   ⟶   [photoinitiator]*   Absorption
non-activated   activated

[photoinitiator]*   ⟶   R·   Chemical
activated   reactive species   Reaction

            monomers
R·+ or   ⟶ R_1·   Initiation
            oligomers   reactive species

            monomers
R_1·+or   ⟶ R_2·   Propagation
            oligomers   reactive species

R_1·+R_1· (or R_2·) ⟶   R - R   Termination
                                     non-reactive species
                                     (polymer formed)

FIGURE 16.4 Curing mechanisms of several photoinitiators available from Ciba-Geigy. *Source:* Ciba-Geigy.

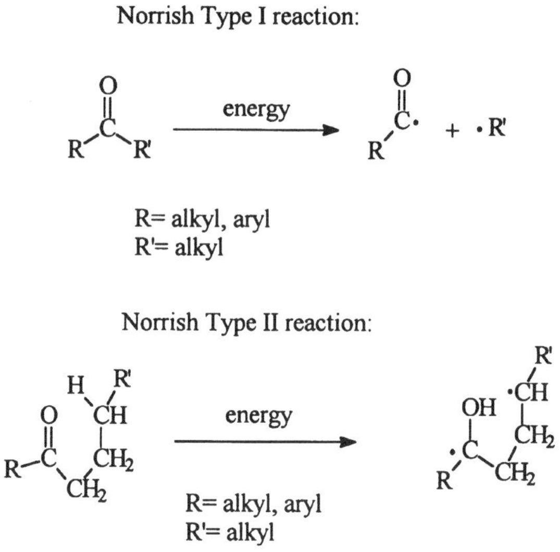

Norrish Type I reaction:

$$R-\overset{\overset{O}{\|}}{C}-R' \xrightarrow{\text{energy}} R-\overset{\overset{O}{\|}}{C}· \; + \; ·R'$$

R= alkyl, aryl
R'= alkyl

Norrish Type II reaction:

R= alkyl, aryl
R'= alkyl

16.9 Mechanisms

16.9.1 Alpha Cleavage

Benzoin Ethers

Benzoin, benzoin ethers, and related compounds are among the earliest patented photoinitiators used in UV curing. Fragmentation of benzoin molecules occurs via the triplet state, whereas benzoin ether decomposes via the singlet state. Both of these classes of photoinitiators undergo Norrish type I photocleavage

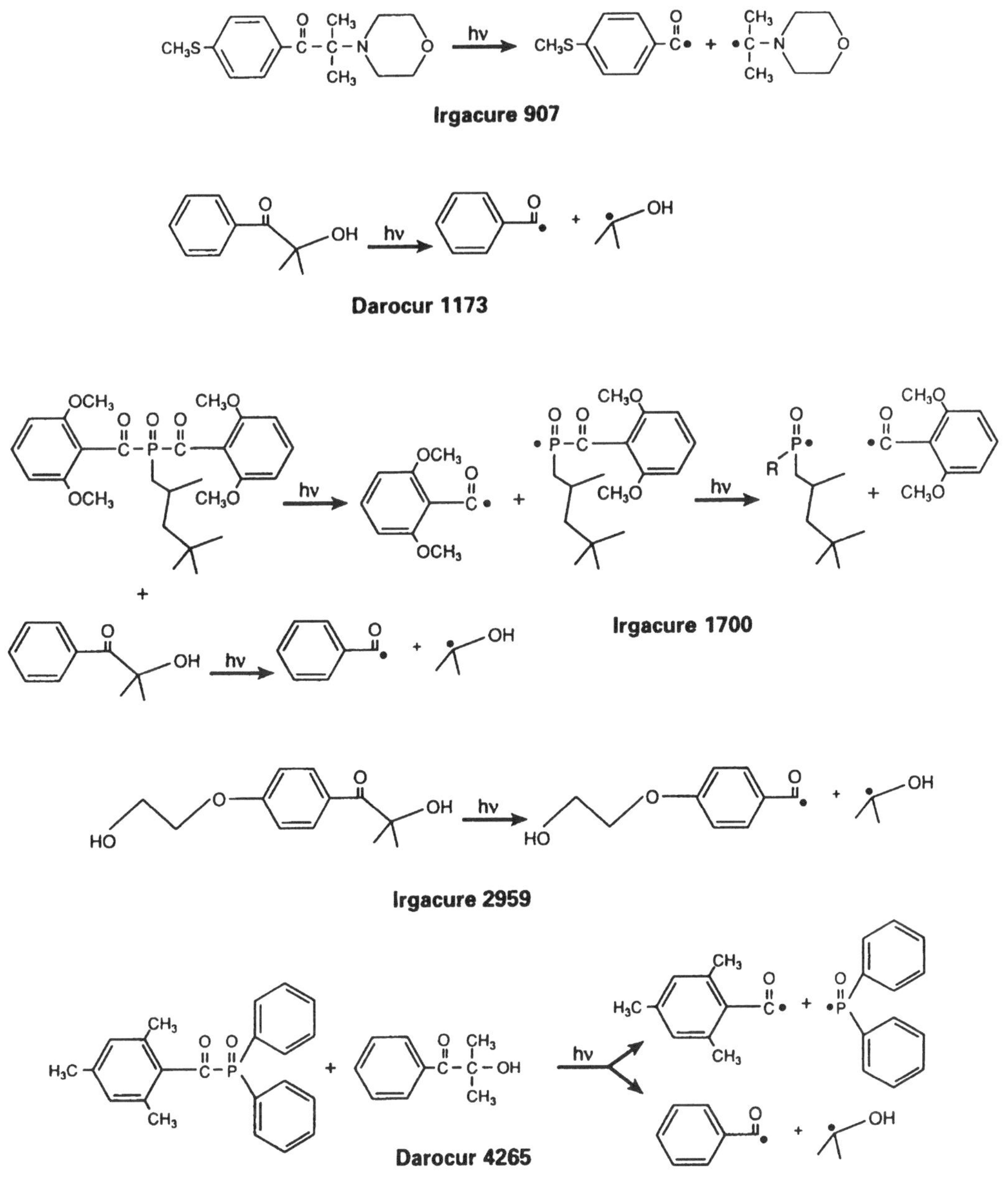

FIGURE 16.5 Cure mechanisms of some photoinitiators. *Source:* Ciba-Geigy.

to yield a benzoyl radical and a substituted benzyl radical. Both radicals are capable of initiating polymerization, but their efficiencies are different. Benzyl radicals usually dimerize, whereas the benzoyl radical besides dimerizing can either initiate polymerization or abstract a hydrogen to form another radical.

The main disadvantage for both of these classes of photoinitiators is the poor dark storage and poor thermal stability of UV polymerizable compositions containing them. This is due to the labile a hydrogen on the benzoin ether, which can be abstracted by peroxy radicals formed during dark storage in air.

Benzil Ketal

To improve the shelf life of formulations utilizing benzoin and its derivatives, the alpha hydrogen was replaced by an alkoxy functional group. The resulting compounds have much better storage stability while retaining high photospeed. Benzil ketals are important alpha cleavage photoinitiators in the UV curing industry. This is especially true of benzil dimethyl ketal (BDMK), as shown in the reactions below:

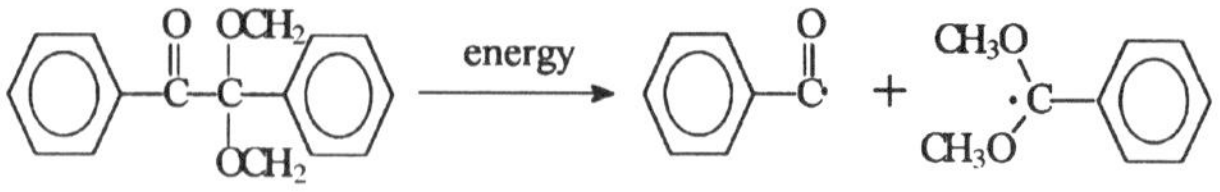

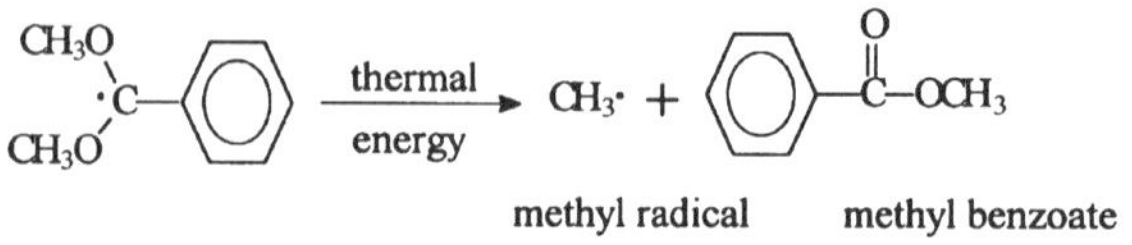

methyl radical   methyl benzoate

The disubstituted benzyl radical can further thermally decompose to form methyl benzoate and a methyl radical. This occurs because the disubstituted benzyl radical is less reactive toward oxygen than the benzoyl radical and is also a longer-lived transient species (i.e., 17 μs vs. 750 ns in degassed solutions). The small methyl radical produced in this reaction has high mobility and reactivity. The UV absorption spectrum of BDMK exhibits absorption beyond 320 nm, where several intense emission lines are available from the medium-pressure mercury lamp. This absorption and the production of reactive benzoyl and methyl radicals makes BDMK an efficient photoinitiator.

TABLE 16.3 Storage Stability of Benzoin Ether in an Acrylate Formulation

Photoinitiator	Conc. (wt. %)	Shelf life at 60° C (days)
BDMK	3	>40
BDMK	5	>40
Benzoin isopropyl ether	3	3

Urethane-acrylate cured with 70 parts polyester acrylate, 30 parts trimethylolpropane triacrylate.

Acetophenones

The following discussion details the four major classes of acetophenone derivatives used in the UV curing industry.

Hydroxyacetophenone

2-hydroxy-2-methylphenyl-l-propanone (HMPP) and hydroxycyclohexyl phenyl ketone (HCPK) are the two major compounds of this class. Both derivatives undergo Norrish type I cleavage to form benzoyl radicals that initiate the polymerization process.

The UV absorption spectra of both hydroxyacetophenone derivatives exhibit a major peak at 250 nm and extend beyond 320 nm.

R, R'= CH₃ (HMPP)

R,R'= (CH₂)₅ (HCPK)

Dialkoxyacetophenones

Unlike most acetophenone derivatives, 2,2-diethoxyacetophenone (DEAP) undergoes both Norrish type I and type II cleavage upon UV irradiation.

As expected, benzoyl radicals are generated via alpha cleavage of either the singlet or triplet state of DEAP. These radicals are responsible for initiation of the photopolymer system. In the second mechanism, a biradical is formed from hydrogen abstraction via the DEAP triplet state. The ratio of Norrish type I to type II products is about 2:1.

Alpha-Amino Acetophenones

In the class of α-amino acetophenones, two compounds have been introduced: 2-methyl-1-[4-(methylthio) phenyl]-2-morpholino propanone-I (MMMP) and 2-benzyl-2-N,N-dimethylamino-1-(4-morpholinophenyl) butanone (BDMB). Both compounds have high extinction coefficients, 18,600 L/mole cm at 320 nm for MMMP and 19,000 L/mole cm at 328 nm for BDMB. Since their absorption spectra extend toward the visible region (more red-shifted for BDMB than MMMP), both compounds are efficient photoinitiators for pigmented coatings. These alpha-amino acetophenone derivatives undergo Norrish type I cleavage via their triplet states, yielding the substituted benzoyl radicals.

Phosphine Oxide

In the early 1980s, acylphosphine oxides and related compounds were reported to be effective photoinitiators for unsaturated resin systems. Laser flash photolysis studies have been used to show their high reactivity toward acrylates and other vinyl monomers This reactivity was attributed to the highly reactive phosphinoyl and phosphinyl radicals generated via the Norrish type I cleavage mechanism.

In the case of diphenyl 2,4,6-trimethyl benzoylphosphine oxide (TPO), both primary radicals resulting from photodecomposition were reported to add to olefinic double bonds. However, the diphenyl phosphinoyl radical is twice as effective as the 2,4,6-trimethyl benzoyl radical. This high reactivity was explained by the tetrahedral structure of the phosphinoyl radical, that was elucidated from the analysis of ESR spectra.

The alpha cleavage quantum yield for several acylphosphine oxides is in the range between 0.3 and 1.0 in various solvents. It was reported in a recent investigation that a tertiary amine acting as a synergist with an acylphosphine oxide would increase the cure speed similar to the synergistic effects found with other photoinitiators. However, after more detailed study, the role of the tertiary amine was shown to be predominantly that of an oxygen scavenger. Another class of phosphine oxides, the bisacylphosphine oxides, has recently been patented (U.S. Pat. 4,737,593). These compounds have properties similar to those of the acylphosphine oxides.

16.9.2 Hydrogen Abstraction

Benzophenone and Related Compounds with Amine

Benzophenone (BP) has long been the focus of numerous photochemical studies. Upon irradiation with UV light, benzophenone is excited to its singlet state followed by intersystem crossing to the triplet state. However, alpha cleavage is not likely, because insufficient energy (E, = 69 kcal/mole) is available for the cleavage process to occur. Free radical species are generated in the presence of a tertiary amine, via a charge transfer complex (exciplex). (See reaction below.)

Amines, especially tertiary alkyl amines, are good hydrogen donors for benzophenone because of their strong affinity for the long-lived benzophenone triplet state. The resulting alpha-aminoalkyl radical is responsible for the initiation reaction; the benzhydril radical will likely dimerize to form a benzpinacol compound. It has also been reported that the resulting aminoalkyl radical can react with oxygen to form a peroxide that can generate another amino radical via hydrogen abstraction. Because of this, formulations containing benzophenone/amine combinations are usually less sensitive to oxygen inhibition. This oxygen consumption effect makes photopolymerization of a thin film coating very efficient utilizing this combination. The most commonly used amine is dimethylethanolamine (DMEA); however, dialkylaminoalkylbenzoates are also very good hydrogen donors and have a lower tendency to yellow in coatings.

Michler's ketone (MK), 4,4'-NNdimethylamino benzophenone, combines both benzophenone and amine groups in one molecule. The MK triplet state can interact with another molecule's amine group to form an exciplex followed by free radical formation and initiation processes. When benzophenone is used with MK, the photoresponse of a UV-curable composition exhibits remarkable improvement as the undesirable MK self-condensation reaction is obviated.

The newest benzophenone derivative, diphenoxy benzophenone (DPB), exhibits photospeed eight times faster than that of benzophenone This has been attributed to a substantially red-shifted absorption peak (about 40 nm compared to the absorption peak of benzophenone) and a higher quantum yield of polymerization (530 moles/Einstein compared to 110 moles/Einstein for benzophenone).

Thioxanthone

Ibioxanthone (TX) and its derivatives, isopropylthioxanthone (ITX) and chlorothioxanthone (CTX), can be regarded as derivatives of benzophenone with a sulfur atom bridge between the two phenyl groups. Because of the sulfur atom, the UV absorption spectra of thioxanthone compounds are red-shifted compared to the spectrum of benzophenone. Thioxanthones are very useful in pigmented coatings owing to their absorption spectrum, which utilizes wavelengths beyond the absorption spectrum of titanium dioxide (greater than 370 nm). The efficiency of this photoinitiator can be further improved with the addition of amines as hydrogen donors because the aminoalkyl radical formed from hydrogen abstraction is more reactive than the thioxanthone triplet biradical. The photochemical mechanism is similar to that of the benzophenone/amine system.

Camphorquinone

Camphorquinone can also be used with amines to generate aminoalkyl radicals. This bicyclic dione compound exhibits efficient photoactivity in both UV and visible light regions owing to its broad absorption profile with a maximum at 470 nm. Camphorquinone is mainly used in visible-light-cured dental impression materials or image-recording compositions because it is not physiologically hazardous.

Bisimidazole

Bisimidazole is an important free-radical photoinitiator for photoimaging applications. Upon irradiation, bisimidazole undergoes homolytic cleavage to form relatively stable imidazolyl radicals, which then abstract a hydrogen atom from a hydrogen donor, generating initiating radicals. Bisimidazole is especially useful in conjunction with various dyes. These dyes can function either as hydrogen donors or as sensitizers. For example, in the bisimidazole-leuco dye system, the initial step is the homolysis of bisimidazole after UV irradiation. The imidazolyl radical then abstracts a hydrogen atom from a leuco dye molecule to form a radical cation species DEAW, a visible-light-absorbing dye, acts as a photosensitizer that can effectively extend the useful wavelength from the UV to the visible region. A proposed photo-initiation mechanism suggests that the formation of an electronically excited dye molecule is the first step after absorption of light. The energy is transferred to the bisimidazole molecules; subsequently, the imidazolyl radicals are formed via homolytic cleavage.

16.9.3 Cationic Photoinitiators

It has been known for years that many metal halides are active initiators for certain olefins. The initiation species are produced by redox reactions and lead to cationic polymerization. One of the attractions of cationic polymerization is that the polymerizable materials are not restricted to those that contain the acrylate double bond but can also include electron-rich double bonds such as in vinyl ether, or molecules containing strained rings such as epoxies and lactones. The reactive intermediates formed during cationic polymerization are insensitive to the presence of oxygen but are sensitive to the presence of nucleophiles such as bases. In addition to metal halides, alkyl metal halides, carbocation salts, protonic acids, ferro-cenium salts, and halogens are all cationic initiators.

16.9.4 Nonmigrating Photoinitiators

Photoinitiators are not usually completely consumed during the UV cure process, and unreacted photo-initiator molecules may eventually migrate to the surface. The development of copolymerizable was undertaken to address these concerns. Copolymerizability is attained or oligomeric photoinitiators by adding an unsaturated functional group (i.e., allyl, vinyl, or acrylic group) onto a conventional photo-initiator moiety or by using a conventional photoinitiator attached to a polymer backbone. Alpha cleavage is the main photolysis pathway for this type of photoinitiator. The number of small molecules produced from fragmentation is greatly reduced. The rate constants for alpha cleavage, however, are usually only one-fifth to one-seventh as fast for copolymerizable or oligomeric photoinitiators as for their small-molecule counterparts. This leads to slower cure in formulated coatings.

16.10 Formulating Considerations

16.10.1 Light Sources

The light source is a critical factor in the UV curing process. In general, two major kinds are used: arc light and laser light. The basic arc light has clear, fused quartz tubing in which a gas is enclosed.

When a direct current is applied through two electrodes to generate an arc, the gas is discharged, and light is emitted. The output of the lamp depends on the pressure and type of gas used. The most common gases are mercury and xenon. A small amount of another gas, such as argon or neon, is also used to aid ionization or to start up the lamp. In a mercury lamp, the gas pressure can be either low (10^{-1} torr), medium (1–2 atm), or high (>2 atm). The medium-pressure mercury lamp is the single most important light source used in the radiation curing industry. This is due to its high power (200–300 W/in) and important emission lines, which are absorbed by most commercial photoinitiators (i.e., 254, 280, 303, 313, 334, and 365 nm). Recently, a 700 W/in medium-pressure mercury lamp was introduced.

High-pressure xenon lamps provide a broad and continuous output from 290 nm in the UV region to >800 nm in the IR region. Both the xenon arc and medium-pressure mercury lamps can be doped with Zn, Cd, or TII_2 to change their output spectra.

In recent years, new lamp designs have been introduced in which the electrodes have been replaced by microwave or other rf (radio frequency) sources. The advantages of electrodeless lamps are longer lamp life, short warm-up times, and highly efficient UV output—40% of the energy can be converted to UV light, compared to about 15–20% conversion for a medium-pressure mercury lamp. This results in less IR output and, in turn, less heat is produced. Lamps of this type with power up to 600 W/in are commercially available.

A laser (light amplification by stimulated emission of radiation) offers the prospect of an excitation source of exceedingly high intensity (one of its inherent properties), compared to classic light sources. Laser output is available in both UV and visible wavelengths. Lasers can be operated in either pulsed or continuous modes. Some lasers offer fixed wavelengths, and others offer tunable wavelengths. The amplifying medium—the medium where population inversion occurs—can be either solid, liquid, or gas. Table 16.4 lists several kinds of lasers and their emissions. Industrial application of lasers will depend on their efficiency and economics. Currently, chemists use them as research tools to probe photopolymerization reactions and photoimaging processes.

TABLE 16.4 Classes and Emission of Lasers

Laser	State	Emission	Type
Ruby	Solid	693.4 nm	Pulsed
ND:YAG	Solid	1064.8 nm	Pulsed/continuous
Ga	Semiconductor	845 or 905 nm*	Pulsed/tunable
Dye	Solution	UV/Vis region	Continuous
Ar + ion	Gas	488.9 nm	Continuous
He-Ne	Gas	632.8 nm	Continuous
Excimer	Gas	UV region	Pulsed

*The wavelength emitted depends on temperature.

The correct choice of lamp system is the first critical step toward success of a radiation curing process. Emission spectra of the lamps must match the absorption spectra of the chosen photoinitiators, sensitizers, or charge transfer complexes. Stable output over a period of time and the intensities of the various emission lines are also important.

16.10.2 Oxygen Inhibition

The oxygen inhibition effect in polymerization has been studied extensively. This inhibition effect is most pronounced in UV-cured coatings owing to the high surface/volume ratio, which allows for optimal oxygen diffusion into the film. Molecular oxygen exhibits natural magnetism; thus, it is a good triplet-state quencher as well as a radical scavenger. This ultimately reduces the rate of polymerization in a system, as peroxy radicals are ineffective in initiating acrylate polymerizations.

(1) Oxygen as a triplet-state quencher

$$PI \xrightarrow{\text{energy}} [PI]^* \xrightarrow{\; ^3O_2 \;} PI \; + \; ^1O_2$$

(2) Oxygen as a radical scavenger

$$R\cdot \; + \; O_2 \longrightarrow ROO\cdot \longrightarrow ROOH + R'\cdot$$

Laser flash photolysis techniques have been used to study the kinetics of the reactions of several carbon-centered radicals with oxygen. The rate constant for the reaction of oxygen with the benzyl radicals was found to be higher than 101/mole-second, leading to less-reactive peroxy radicals (i.e., k, $>>$ k,). Most commercial photoinitiators have very short-lived excited singlet and triplet states, which reduces their tendency to be quenched by oxygen from the surrounding air. In fact, the quantum yield of alpha cleavage in benzoin methyl ether is increased in the presence of oxygen, presumably because part of the photofragment is consumed by the trapping oxygen, thus preventing the recombination of radicals to form the starting photoinitiators. Since the peroxy radicals are less reactive than the alkyl initiation radicals, this decreases the rate of polymerization. Oxygen inhibition results in increased induction period, decreased rate of polymerization, and lower degrees of polymerization. This is shown through differential scanning colorimetry (DSC) and laser nephelometry techniques.

Two major techniques are used to minimize the effects of oxygen: physical and chemical methods.

1. *Physical method.* A UV-transparent film or an inert atmosphere can be used over the coating as an oxygen barrier. Another method is to increase light intensity. It has been reported that the rate of production of initiating radicals in an epoxy/diacrylate system was increased from 0.025 to 0.11 to 20.0 liter/second by using a medium-pressure mercury lamp, an unfocused Ar+ laser, and a focused Ar+ laser, respectively. Increasing photoinitiator concentration will help to generate a high concentration of initiating radicals that will consume most of the oxygen molecules.

2. *Chemical method.* The most common technique is the addition of an oxygen scavenger to the system, such as a highly oxidizable tertiary amine. The effect of amine addition in a urethane acrylate based formulation is shown in Table 16.5. The resulting amine radical can also act as an initiating radical. Another method involves converting triplet ground-state oxygen into singlet oxygen and then removing it by reaction with an adequate acceptor. Methylene blue is usually used to convert oxygen to singlet oxygen via red light irradiation, and then 1,3-diphenyl isobenzofuran is used as a singlet oxygen scavenger. However, this method is seldom used in industry.

TABLE 16.5   Effect of Amine on the Surface Cure Rate of a Urethane Acrylate Coating

Photoinitiator (3 wt. %)	Coinitiator (5 wt. %)	Surface Cure Rate (meters/minute per lamp)
BDMK	—	30
BDMK	Amine	80
DEAP	—	20
DEAP	Amine	50

16.10.3 Pigments

Zinc, antimony, and titanium oxides are the most common filler or extender pigments used in coatings. The cutoff of their absorption spectra is between 380 and 400 nm. Therefore, these pigments compete with the photoinitiators for most of the available UV light. The cure speed of a pigmented system can be changed significantly by changing the pigment concentration, photoinitiator concentration, or film thickness.

Titanium dioxide, a white pigment, provides high scattering of visible light. Because of this, good hiding is obtained in relatively thin coatings with moderate pigment concentrations. The two crystalline forms of titanium dioxide—rutile and anatase—have different absorption characteristics. The anatase grade tends to absorb less radiation energy than the rutile grade; therefore, it is known that the grade of TiO_2 used in a formulation will affect its photospeed. Most UV-cured pigmented systems are initiated by the scattering of the light from TiO_2 crystals within the film. Consequently, crystal size is important in determining which wavelength will be predominantly scattered.

An alternative approach to curing pigmented coatings is to use compounds such as thioxanthones, which have strong absorption in the visible wavelength region. These molecules can also sensitize the photopolymerization reaction through energy transfer to the photoinitiator. Aryl ketone/amine systems that have absorption beyond 380 mn have been reported to be very effective in UV-cured pigmented systems.

Usually, the problem associated with curing a pigmented system is surface wrinkling due to reduced penetration of the UV light through the film. Presumably, the surface of the film cures faster than its interior, resulting in higher stress on the surface. One way to avoid this is to optimize the pigment/binder ratio in the system.

16.10.4 Yellowness

Yellowness in cured coatings can result from the formation of secondary photoproducts. This color can be alleviated through proper choice of photoinitiator. A comparison of yellowness among several commercial photoinitiators is shown in Table 16.6. Initial yellowness indices (YI) show that HCPK and HMPP contribute slightly less color than the other photoinitiators. After 300 hr of QUV exposure, HCPK exhibited the lowest YI value, and BDMK produced the highest YI value.

TABLE 16.6 Comparison of Photoinitiators in an Aliphatic
Urethane Acrylate Coating

	Yellowness Index Upon QUV Exposure	
Photoinitiator (4.75 wt. %)	Time = 0	Time = 300 hr
BDMK	1.8	53.2
HMPP	1.6	44.1
DEAP	1.9	35.7
HCPK	1.5	31.3

16.10.5 Photoinitiator Concentration and Coating Thickness

If photoinitiators have very high extinction coefficients at the major emission lines of light sources, only low concentrations of the photoinitiator are needed to provide adequate film properties. In the absence of oxygen, this principle holds—except with very thin films, because the path length is much shorter. In this case, concentrations of 5% or more of the photoinitiators are needed to generate an adequate radical concentration to effect cure.

In the case of thicker films, photoinitiator concentration can determine both surface-cure and through-cure properties of the coating. In thick films (greater than 1 mil) at high photoinitiator concentration, high radical concentration will be generated near the surface, leading to sufficient surface cure but poor through cure. A low photoinitiator concentration will produce good through cure owing to an even distribution of radicals throughout the film, although at the same time yielding poorer surface cure. Sometimes, blends of photoinitiators are needed to produce the best balance between surface cure and through cure. These theories hold true in the absence of oxygen; however, if oxygen is present, photoinitiator levels must be adjusted.

Usually, if a higher concentration of photoinitiator is used, the film thickness has to be decreased to obtain a completely cured film, because most of the light will be absorbed close to the surface. This is particularly true for compounds with high extinction coefficients. This is called the *screening effect*.[3] If the extinction coefficient of a photoinitiator is low at the major emission lines of the light source, higher concentrations are needed to provide complete surface cure. One exception to this is photobleachable photoinitiators. In this case, high levels of initiator do not prevent through cure, owing to the reduction of photoinitiator absorption during the curing process (see Table 16.7.)

16.11 Coating Applications

16.11.1 Wood Coatings

Coatings on wood, specifically unsaturated polyester wood fillers, represent one of the very first commercial uses of UV curing technology. In a wood filler application, photopolymer is used to fill the voids in a particle-board substrate. This coating is then sanded, giving a smooth surface that readily accepts

TABLE 16.7 Photoinitiator Selection Chart

PI class	Polyester wood coatings	Paper coatings	Clear* coatings	Offset inks	Screen inks	Pigmented coatings	White coatings	Photoresists
Benzoin ethers	✔							
Benzophenones		✔		✔	✔			
Thioxanthones				✔	✔	✔	✔	✔
Ketals	✔		✔	✔	✔			✔
Hydroxy acetophenones	✔	✔	✔					
Phosphine oxide						✔	✔	✔
Titanocenes								✔
Morpholino-substituted acetophenones				✔	✔	✔	✔	✔

*This includes metal, wood, and plastic substrates.

any additional coating layers required. Ultraviolet curing technology has expanded since its initial use with wood fillers and is often used for primary coatings on table tops, chairs, and other wood furniture. Although these coatings are typically roll coated or curtain coated, spray application of photopolymer coatings is an active area of development (see Table 16.8).

TABLE 16.8 Characteristics of UV Market Segments

Application	Film thickness (mils)	Competitive absorbing components	Typical PI concentration (%)
Clear coatings—wood/flooring	0.5–2.0	Talc/fillers	2–4
Paper coatings	0.1–0.3	None	5–10
Offset inks	0.1	Organic pigments	3–6
Screen inks	0.2–0.5	Organic pigments	2–5
Pigmented coatings	0.1–2.0	Dyes	2–6
Photoresists	0.3–2.0	Dyes	2–6
Clear coatings on metal	0.2–0.5	None	3–8

16.11.2 Paper Coatings

Paper coatings, or overprint varnishes, represent the single largest use of photopolymers in the United States. Common uses are on magazines and consumer goods packaging. Ultraviolet cured paper coatings exhibit high gloss and good abrasion resistance. The ability of UV cured coatings to achieve these properties is the reason for the growth of UV cure technology in this market.

Paper coatings are run at very high line speeds, often >200 ft/min per lamp. The coatings must come off the conveyor belt tack-free to allow stacking, and they must also have a certain level of *immediate mar resistance*. In one study, three photoinitiators were evaluated at concentrations of 1.0 to 5.0%. The maximum line speed to cure a coating to a tack-free state was measured for each photoinitiator. As is often true in UV-cured coatings, a mixture (benzophenone with hydroxycyclohexyl phenyl ketone) proved to be more efficient than single-component initiator systems (see Table 16.9).

TABLE 16.9 Performance of Photoinitiators in an Aliphatic Urethane Acrylate System[a]

Photoinitiator (5 wt. %)	Pendulum hardness (seconds)	Percent residual 2-thylhexyl acrylate-
BDMK	17	0.27
DEAP	15	2.53
Benzoin ether mixture	41[b]	9.93

[a]Plaques of 20 mil thickness cured at 100 ft/min under two 200 W/in mercury lamps.
[b]High values are attributed to lubrication of pendulum by unreacted oligomer and/or monomer.

16.12 Commercially Available Photoinitiators

TABLE 16.10 Alpha Cleavage Free-Radical Photoinitiators

Compound	CAS Reg. No.	Trade Name	Supplier
Isobutyl benzoin ether	22499-12-3	Vicure 10	Akzo
2,4,6-Trimethylbenzoyldiphenylphosphine oxide	127090-72-6	Lucirin TPO	BASF
1-Hydroxycyclohexyl phenyl ketone	947-19-3	Irgacure 184	Ciba-Geigy
2-Benzyl-2-dimethylarnino-1-(4-morpholinophenyl)-butane-1-one	119313-12-1	Irgacure 369	Ciba-Geigy
Mixture of benzophenone and	119-61-9	Irgacure 500	Ciba-Geigy
1-hydroxycyclohexyl phenyl ketone	947-19-3		
2,2-Dimethoxy-2-pheylaceto-phenone	24650-42-8	Irgacure 651	Ciba-Geigy
Perfluorinated diphenyl titanocene	Proprietary	Irgacure 784	Ciba-Geigy
2-Methyl-1-4[methylthiolphenyl]-2-(4-morpholinyl)-1-propanone]	71868-10-5	Irgacure 907	Ciba-Geigy
2-Hydroxy-2-methyl-1-phenyl propan-1-one	7473-98-5	Darocur 1173	Ciba-Geigy
4-(2-Hydroxyethoxy) phenyl-2-hydroxy-2-propyl ketone	106797-53-9	Darocur 2959	Ciba-Geigy
Blend of ketones and amines	Proprietary	Darocur 4043	Ciba-Geigy
Blend of aromatic ketones	Proprietary	Darocur 4265	Ciba-Geigy
2,2-Diethoxyacetophenone	6175-45-7	DEAP	First Chemical

TABLE 16.11 Hydrogen Abstraction Free-Radical Photoinitiators

Compound	CAS Reg. No.	Trade Name	Supplier
[4-(4-Methylphenylthio)phenyl] henylmethanone 4-benzoyl-4'-ethyldiphenyl sulfide	83846-85-9	Speedcure BMDS	Aceto Corp.
Ethyl-4-(dimethylamino) benzoate	10287-53-3	Speedcure EDB	Aceto Corp.
Mixture of 2-isopropyl thioxan-thone and 4-isopropyl thio-xanthone	5495-84-1	Speedcure ITX	Aceto Corp.
[4-(4-Methylphenylthio)phenyl] phenylmethanone 4-benzoyl-4'-methyldiphenyl sulfide	83846-85-9	Quantacure BMS	Biddle-Sawyer
2-(Dimethylamino)ethylbenzoate	2208-05-1	Quantacure DMB	Biddle-Sawyer
Ethyl-4-(dimethylamino)benzoate	10287-53-3	Quantacure EPD	Biddle-Sawyer
Mixture of 2-isopropyl thioxan-thone and 4-isopropyl thio-xanthone	5495-84-1	Quantacure ITX	Biddle-Sawyer
d,/-Camphorquinone	10373-78-71	Camphorquinone	Epolin, Inc.
Ethyl d,/-camphorquinone	10287-53-3	EDAB	Hampford Research, Inc.
Mixture of benzophenone and	119-61-9	Photocure 81	Henkel Corp.
4-methylbenzophenone	134-84-9		
Benzophenone	119-61-9	Benzophenone	Marlborough Velsicol
4,4'-Bisdiethylamino benzo-phenone	90-94-8	Michler's ketone	RIT Chemical
4,4'-Bisdiethylamino benzo-phenone ethyl ketone	90-93-74	Michler's ethyl ketone	RIT Chemical

TABLE 16.12 Cationic Photoinitiators

Compound	CAS Reg. No.	Trade Name	Supplier
(n5-2,4-Cyclopentadien-1-yl); (n6-isopropyl benzene)-iron (II) hexafluorophosphate	32760-80-8	Irgacure 261	Ciba-Geigy
Triphenyl sulfonium hexafluorophosphate	57835-99-1	FX-512	3M Co.
Mixed triphenyl sulfonium salts	98452-37-9	Cyracure	Union Carbide
	71449-78-0	UVI-6974	
Mixed triphenyl sulfonium salts	74227-35-3	Cyracure	Union Carbide
	68156-13-8	UVI-6990	

16.13 Illustrative Patents

16.13.1 UV-Curable Adhesives

Increasing concern with energy, environmental protection, and health factors has enhanced the potential of radiation-curable compositions as adhesives. These adhesive compositions typically comprise a polymerizable mixture that can be applied as a thin film to a substrate and polymerized by exposure to a radiation source such as an electron beam, plasma arc, UV light, and such.

Compositions that are curable under the influence of radiation in general, and UV light in particular, are well known. Unfortunately, these known compositions suffer from a number of disadvantages. For example, many of these compositions have insufficient flexibility, which causes them to crack in use when applied to flexible substrates. Other compositions do not adhere sufficiently to substrates such as carbonates and metals, with the undesirable result that the laminated layers become dislodged or peel. Still other compositions require use of solvents that must be evaporated during curing. Evaporation of such solvents increases production time, consumes energy, and creates atmospheric pollution problems.

A need, therefore, continues for radiation-curable adhesive compositions that remain homogeneous, can be readily and uniformly applied to substrates, and that show improved adherence to substrates such as carbonates and metals.

Usifer[4] disclosed UV-curable adhesive compositions that provide improved bonding to substrates such as carbonates and metals. The compositions are formed as the reaction product of a urethane methacrylate and reactive diluents such as N-vinyl caprolactam and isobornyl acrylate, a photoinitiator compound, and adhesion-promoting additives. Urethane methacrylates that may be employed include ester-based urethane methacrylates as well as ether-based urethane methacrylates.

In accordance with the Usifer invention, UV-curable adhesive compositions that provide improved bonding to substrates such as plastics and metals are provided. The compositions are formed as blends of urethane (meth)acrylates, at least one reactive diluents such as N-vinyl caprolactam and isobornyl acrylate, a photoinitiator compound, and optional adhesion-promoting agents. Urethane (meth)acrylates that may be employed include ester urethane (meth)acrylates as well as ether urethane (meth)acrylates. Urethane methacrylates useful in the invention typically have a molecular weight of about 2,000 to about 20,000 and a (NCO:OH) ratio between about 1.1:1 to about 2.0:1. As used herein, the term *(meth)acrylates* encompasses both methacrylates and acrylates.

The adhesive compositions may be employed to bond a variety of materials including, but not limited to, plastics such as flexible and rigid polyethylene, polycarbonate, flexible and rigid polyvinylchloride, stainless steel, polystyrene, polyethylene, tin, cellulose propionate, acrylonitrile-butadiene-styrene, and mylar. In a further embodiment, the urethane (meth)acrylates and reactive diluents are mixed with an epoxy adduct or epoxy methacrylate adduct of carboxy-terminated butylene nitrile (CTBN) to yield adhesives that provide excellent bonding to metals or other nonpolar substrates. The compositions of this embodiment achieve surprisingly strong bonding, good flow, rapid cure rates, and wetting to disparate substrates such as polycarbonate and stainless steel.

Useful photoinitiators include alpha-hydroxy aryl ketone-type compounds such as 1-hydroxycyclohexyl phenyl ketone (Irgacure 184, available from Ciba-Geigy). Other useful initiators include hydrogen abstraction-type compounds such as benzophenone, substituted benzophenone, either alone or in combination with compounds that have labile hydrogens. Included among suitable photoinitiators are 2-benzyl-2,N,N-dimethylamino-1-(4-morpholinophenyl)-1-butanone (Irgacure 369, available from Ciba-Geigy), 2,2-dimethoxy-2-phenylacetophenone (Esacure KB-1, available from Sartomer), or trimethylbenzophenone blends such as those available from Sartomer as Esacure TZT. Still other useful photoinitiators are onium salt-type initiators that generate free radicals and cationic species upon exposure to UV light. Examples of these initiators include, but are not limited to, mixed triarylsulfonium hexafluoroantimonate salts such as UVI-6974, available from Union Carbide, or mixed triraryl sulfonilhexa fluorophosphate salts such as UVI-6990, available from Union Carbide. Based on availability, solubility in the composition of the invention, and efficiency of curing at minimum UV levels, preferred photoinitiators for use in the invention include 2,2-diethoxyacetophenone, benzophenone, and 2-ethylhexyl-2 cyano 3,3-diphenyl acrylate.

In yet another embodiment, organofunctional silanes may be included in the compositions of the invention for adhesion promotion. Such organofunctional silane materials can be employed optionally in the formulation to promote adhesion to various substrates. Examples of useful silanes include, but are not limited to, trialkoxy-functional silane materials such as vinyl trimethoxy silane (A-171, available from Union Carbide) and vinyl trimethoxy silane (A-151, available from Union Carbide). Functionalized trialkoxy silanes also can be employed during the manufacture of the urethane methacrylate resin to bind the silane into the polymer backbone. Examples of useful functionalized trialkoxy silanes include, but are not limited to, gamma-aminopropyl triethoxysilanes such as A-1100 or A-1101, available from Union Carbide; amino functional silanes such as A-1170, available from Union Carbide; gamma-mercaptopropyltrimethoxysilanes such as A-189, available from Union Carbide; and isocyanato functional silanes such as gamma-isocyanatopropyl triethoxysilane (A-1310), available from Union Carbide. These organosilane materials can be employed in amounts of up to about 8% by weight, more preferably about 1 to 3% by weight of the total composition.

16.13.2   Flexible UV-Curable Coatings

Coating compositions that are curable under the influence of radiation in general and UV light, as well as electron beams in particular, are well known. Representative examples of prior coating compositions include those disclosed in U.S. Pats. 3,782,961; 3,829,531; 3,850,770; 3,874,906; 3,864,133; 3,891,523; 3,895,171; 3,899,611; 3,907,574; 3,912,516; 3,932,356; and 3,989,609. Unfortunately, these coating compositions suffer from a number of disadvantages and do not have an in situ UV absorber in the composition. Many of these coating compositions have insufficient flexibility, which causes them to crack when applied to flexible substrates such as those of polyvinyl chloride. Other compositions do not adhere sufficiently to the substrate with the undesirable result that they become dislodged or peel. Still other coating compositions require the use of solvents that must be evaporated during the curing process. The evaporation of such solvents consumes energy and creates atmospheric pollution problems. Other compositions produce coatings that yellow, do not weather well, and have insufficient scratch resistance, stain resistance, abrasion resistance, and/or solvent resistance.

The use of UV absorbers in plastics or coatings to enhance weather resistance is known. The absorbers absorb the radiation and dissipate the energy and thus protect the coating from structural degradation. Considerable economic saving is realized by incorporating the UV absorber on the surface of a plastic article rather than using the UV absorber in conventional bulk application. Conventional surface application, such as the use of a solvent or paint vehicle, is also undesirable in view of the pollution hazard and bulk handling procedures. Radiation curing has made possible production of coating films that are easier to handle, but UV absorbers have consumed the energy from the radiation source, resulting in exceedingly high energy demands in curing or unacceptably slow curing rates. If a small amount of ultraviolet photoinitiator is used to facilitate curing, then addition of use levels of most UV stabilizers would prevent the curing from occurring.

To overcome these deficiencies, Lorentz et al.[5] disclosed an improved coating composition that is substantially free of one or more of the disadvantages of prior radiation-curable coating compositions. Other objects are (1) to provide a coating composition that will produce a coating that is weather resistant, nonyellowing, scratch resistant, stain resistant, abrasion resistant, and solvent resistant, (2) to provide a coating composition that is free of volatile solvents, and (3) to provide an improved process for coating substrates such as those of natural leather, synthetic leather, polyvinyl chloride, polyurethanes, and polycarbonates.

16.13.3 Stain-Resistant Coatings for Floor Covering

The prior art has disclosed polyurethane coating compositions, and vinyl sheet coated therewith, wherein the coating composition consists essentially of low-molecular-weight resinous urethane polymers (often called *oligomers* or *prepolymers*), terminated by isocyanate groups and curable at elevated temperatures by action of moisture. As an improvement on these coating compositions, there are compositions curable entirely by actinic radiation (particularly UV) consisting essentially of photopolymerizable ingredients, including an unsaturated resinous compound (preferably a urethane oligomer) having at least two photopolymerizable ethylenically unsaturated groups per molecule (preferably of the acrylic type), a photo reactive monomer, a multifunctional monomer solvent and cross-linking agent, and a photoinitiator (U.S. Pat. 4,100,318, McCann, et al., for Actinic Radiation Cured Coating for Cushioned Sheet Goods and Method). More particularly, this patent discloses such coating compositions based on acrylourethane oligomer from isophorone diisocyanate and hydroxyethyl acrylate; plus a monomeric acrylate, a glycol diacrylate, and a photoinitiator; and cured by exposure to UV radiation.

U.S. Pat. 4,216,267 of Aug. 4, 1980, to Lorenz et al. for "Flexible Substrates Containing a Radiation Curable Coating Composition," is of interest for its exemplification of radiation-curable acrylourethane compositions derived specifically from isophorone diisocyanate or 4,4'-dicyclohexyl methane diisocyanate.

Boba and Conger[6] disclosed an invention relating to polyurethane coatings, especially upon resilient, embossed vinyl sheet floor covering materials, from compositions that can be rapidly cured using a combination of heat produced, for example, by infrared (IR) radiation and actinic radiation (especially UV radiation) then exposed to the action of isocyanate-reactive substances (such as moisture) to convert a predetermined content of isocyanate groups in the polyurethane into urea groups.

In accordance with the Boba invention, a homogeneous, unitary cross-linked polyurethane coating is provided, combining good retention of gloss as shown in scrubbing tests with good resistance to staining by common staining substances such as tar and asphalt. It also has desirable flexibility, transparency, hardness, adhesiveness, and other desirable properties characteristic of known polyurethane coatings for use on various substrates—particularly on resilient vinyl yard goods. Also, the coatings are useful on resilient tile, wall coverings, upholstery materials, simulated leather, and the like. Also provided are compositions that are curable to form such coatings, the coated vinyl sheets, and methods of producing the sheets.

16.13.4 UV-Curable Automotive Clear Coating Compositions

Clear coat/color coat finishes for automobiles and trucks have been used in recent years and are very popular. Kurauchi et al. (U.S. Pat. 4,728,543, issued Mar. 1, 1988) and Benefiel et al. (U.S. Pat. 3,639,147, issued Feb. 1, 1972) show the application of a clear coat to a color coat or base coat in a *wet-on-wet* application, i.e., the clear coat is applied before the color coat is completely cured.

Productivity of conventional refinish operations using these clear coat/color coat finishes has been lacking in that the conventional refinish clear coating in current use does not dry to a tack-free state in a relatively short period of time, and the vehicle cannot be moved without having foreign particles stick to the clear coat, nor can the clear coat be buffed until completely dried to form a finish with acceptable gloss and smoothness. In a typical refinish operation, after the color coat is applied, the clear coat is applied to the vehicle, and the resulting finish is allowed to dry before the vehicle is moved. Before any further work can be done to the finish, it must be tack-free and must be sufficiently hard to allow buffing to improve gloss or remove minor imperfections. Conventional finishes have long drying and curing times and therefore reduce the productivity of a refinish operation, since the vehicles cannot be moved and worked on quickly after application of the finish.

There is a continued need to use compositions with low VOC (volatile organic content) to meet pollution regulations. It is well known that coating compositions having a reduced VOC also have a high viscosity that reduces the ease of sprayability and application of the composition. To reduce the viscosity of such a composition, the molecular weight of the film-forming polymer of the composition must be reduced. This further lengthens drying and curing times and reduces the productivity of a refinish operation.

Lamb and Simms[7] disclosed a coating composition primarily used as a clear coat of a clear coat/color coat automotive finish. The composition has a film-forming binder solids content of about 30 to 70% by weight containing about

1. Fifty to 80% by weight, based on the weight of the binder, of an acrylic polymer of polymerized monomers of styrene, and two methacrylate monomers and a hydroxy alkyl methacrylate or acrylate, each having 1 to 4 carbon atoms in the alkyl group, with the polymer having a number average molecular weight of about 1,000 to 12,000 and a calculated Tg of at least 40° C

2. One to 20% by weight, based on the weight of the binder, of a polyol component
3. Ten to 49% by weight, based on the weight of the binder, of an organic polyisocyanate

This invention is directed to a coating composition and, in particular, a clear coating composition used for refinishing a clear coat/color coat finish of a vehicle such as an automobile or a truck.

REFERENCES

1. Norrish, R.G.W. *Trans. Faraday Soc.* 33:1521, 1939.
2. Cundall, R.B., and Gilbert, A. *Photochemistry*, Thomas Nelson, London, 1970.
3. Hutchinson J., and Ledwith, A. *Polymer* 14:405, 1973.
4. Usifer, Douglas A. U.S. Pat. 5,426,166, June 20, 1995, "Urethane Adhesive Compositions," to CasChem, Inc., Bayone, New Jersey.
5. Lorenz, Donald H., Shu, T., and Wyman, Donald P. U.S. Pat. 4,216,267, August 5, 1980, "Flexible Substrates Containing a Radiation Curable Coating Composition," to GAF Corporation, New York, New York.
6. Boba, Joseph, and Conger, Robert P. U.S. Pat. 4,393,187, July 12, 1983, "Stain Resistant, Abrasion Resistant Polyurethane Coating Composition, Substrate Coated therewith and Production thereof."
7. Lamb, Douglas M., and Simms, John A. U.S. Pat. 5,286,782, February 15, 1994, "Coating Composition of an Acrylic Polymer, Polyol and Polyisocyanate Cross-Linking Agent," to E.I. Du Pont de Nemours and Company, Wilmington, Deleware.

17

Processing Methods

17.1 Extrusion

Plastic extrusion is a steady-state process for converting a thermoplastic polyurethane raw material to a finished or near-finished product. The raw material is usually in the form of plastic pellets or powder. The conversion takes place by forming a homogeneous molten mass in the extruder and forcing it through a die orifice that defines the shape of the product's cross section. The formed material, or extrudate, is cooled and drawn away from the die exit. The extrudate can then be wound on a spool or cut to a specified length.

In contrast with injection molding, which is a cyclic process, extrusion is a steady-state process. Extruded products are long and continuous and have a cross section that is usually constant with respect to the axis or direction of production. Injection molded products are discrete items with varying cross sections in each axis. Polyurethane thermal processing, from raw material to finished product, is shown in Fig. 17.1.

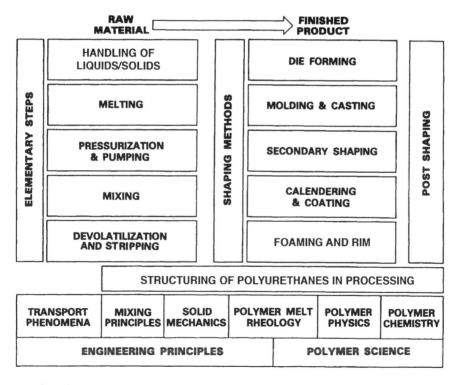

Figure 17.1 Polyurethane processing.

Extruders can be either single-screw or twin-screw in design. Furthermore, twin-screw extruders are either tangential or intermeshing, a shown schematically in Fig. 17. 2. Figures 17.3 and 17.4 show views of twin-screw extruders, one designed to produce tubes and the other to produce sheets. Figure 17.5 shows twin-screw barrels with liners, and Figs. 17.6 and 17.7 show end views of co-rotating and counter-rotating screws, respectively.

17.1.1   Equipment

The major components of an extruding system are the drive, hopper, feed screw, die system, and heating and cooling elements. The drive consists of a motor or belt drive, which should be linked to the extruder through a double-reduction gearbox. Such an arrangement helps to transform the high speed of the motor into the lower speed and high torque required for the extruder. The hopper should hold enough resin pellets or granules to last, at least, two to three hours. When medical products are being extruded, the hopper's flow restrictors should be left fully open for *flood feeding,* which ensures adequate flow of material to the screw.

The feed screw delivers a homogeneous flow of material to the die assembly at a constant melt pressure and temperature. Screw speed is the strongest contributor to raised melt process temperatures, which can result in a weakened extrudate. To ensure good material performance, screw speed should be monitored to keep temperatures as low as possible. A typical screw design for injection molding and extrusion is shown in Fig. 17.8.

Several types of dies are used in the typical product manufacturing, including sheet dies, profile dies, tubing dies, and coating dies. In-line dies (parallel to the extruder path) are commonly used for sheets, profiles, and tubing. Cross-head dies (90° to the extruder path) are used for tubing and coating. Angle dies (any setting other than 90°) are used when more than one extruder is being employed to make a single product (such as striped or multilayer tubing).

Cooling of the extrudate is most often accomplished through the use of an open cooling trough or a vacuum sizer. Other equipment options include the use of an extrudate puller, which can help the manufacturer control the tolerances of completed products. This can be an important factor in medical device manufacturing. An important variant of the thermoplastic extrusion process is film production by the *blown technique* as shown in Fig. 17.9.

17.1.2   Materials

Extrusion techniques can be used to process most thermoplastic polyurethanes. A characteristic that often differentiates extruded from injection-molded plastics is the viscosity of the plastic at normal processing temperatures. Extruded plastics often have a higher melt viscosity, which allows the extrudate to retain the shape imparted to it by the die while the extrudate is in the quenching stages.

Combinations of various resins can be used to gain special physical, biological, or chemical properties. Many additives are used during the extrusion process to enhance processing characteristics of the polymer or to alter product properties. Such additives include lubricants, thermal stabilizers, antioxidants, radio-pacifying agents, and colorants.

Generally speaking, thermoplastic polyurethanes are offered commercially as pellets. Figure 17.10 shows a schematic diagram of the production of pellets by an *underwater pelletizing* method.

17.1.3   Processing Parameters

The parameters important to extrusion processing are similar to those of injection molding processes. Resin temperature, resin pressure, resin moisture content, screw speed, and screw motor amperage are usually controlled or monitored to provide a homogeneous melt at a controlled volumetric rate. Quenching temperature and the rate at which the extrudate is drawn are controlled or monitored to provide a controlled product size. Dimension measurements, using a variety of gauging methods, can be taken of

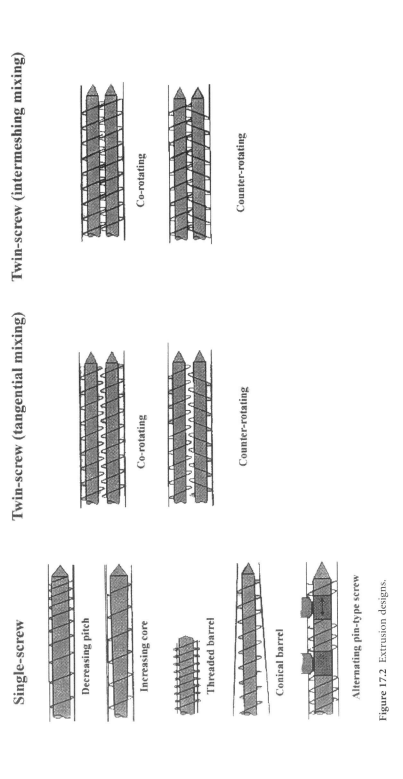

Figure 17.2 Extrusion designs.

Figure 17.3 Micro–18 mm modular/multimode twin-screw extruder with screw pump, downstream tube system, and ECS controls. *Source:* American Leistritz Extruder Corp.

Figure 17.4 ZSE-27 modular/multimode twin-screw extruder and downstream sheeting system. *Source:* American Leistritz Extruder Corp.

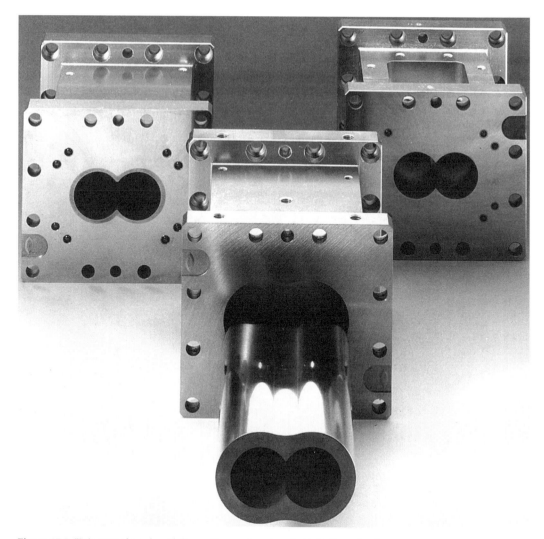

Figure 17.5 Twin screw barrels with liners. *Source:* American Leistritz Extruder Corp.

the extrudate as it is produced. In contrast to injection molding, extrusion can vary the size of the final product without changing the die tooling. Common extrusion production tolerances are within 1% of the nominal measured value.

17.1.4 Design Considerations

Extruded products fall readily into two categories:

1. Those having just one resin in the product cross section
2. Those having more than one

The first category includes tubing with single- or multi-lumen profiles, films for product packaging, and sheets that can be post-formed into fluid containers. The second category includes tubing with encapsulated striping, and multilayer tubing, films, and sheets.

Various materials can also be encapsulated within the extrudate to provide additional properties. Fibers can be braided in to increase burst strength. Stainless steel wires can be added to improve kink resistance or to provide electrical conductivity. Fiber optic bundles can carry images or illumination.

Figure 17.6 End view of co-rotating, intermeshing screw system. *Source:* American Leistritz Extruder Corp.

Figure 17.7 End view of counter-rotating, intermeshing screw system. *Source:* American Leistritz Extruder Corp.

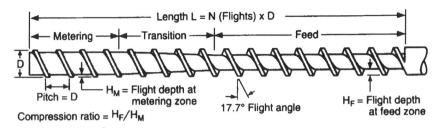

Figure 17.8 Typical screw design for injection molding and extrusion, which includes a helical flight pattern with a constant angle of 17.7°. *Source:* Xaloy, Inc.

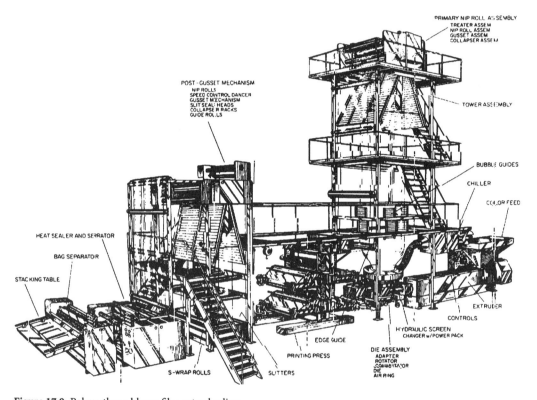

Figure 17.9 Polyurethane blown film extruder line.

17.2   Injection Molding

Injection molding is commonly used to manufacture parts in large quantities with reliable consistency. Understanding all the variables of injection molding and their impact on successful processing is particularly important for manufacturers that require tight tolerances and unique performance requirements.

Equipment design, material performance, process variables, and part design specifics all contribute to the performance quality of any injection molded part. In brief, the injection molding cycle can be broken down into four phases:

- Fill
- Pack
- Hold
- Cooling/plastication

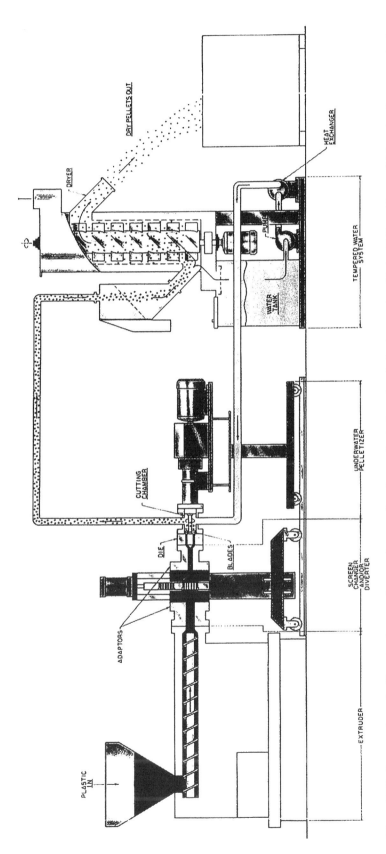

Figure 17.10 Process schematic. The extruder or melt pump forces molten thermoplastic through a die having a series of holes (2.00 to 4.8 mm dia) in a circular pattern. As the polymer emerges from the die holes, it is cut into pellets by rotating blades and solidifies as it passes through the cutting chamber. A tempered water system transports the pellets to a centrifugal dryer where the pellets are dried and discharged. The pellet water is strained, pressurized, cooled, and returned to the slurry, thereby minimizing water use. *Source:* Gala Underwater Systems.

The process begins with the mixing and melting of resin pellets. Molten polyurethane moves through the barrel of the machine and is forced *(injected)* into a steel mold. As the plastic fills and packs the mold, the part takes shape and begins to cool. The molded part is then ejected from the mold, ready for finishing steps and assembly.

17.2.1 Equipment

Several types of injection molding machines are available with different methods for blending, melting, and injecting the polymer into the mold. These are available in a range of sizes, offering choices in clamp tonnage, machine capacity, and screw design, depending on the needs of a particular application. Figure 17.11 shows a typical injection molding machine, and Fig. 17.12 shows a cutaway diagram of a typical injection molding machine.

17.2.2 Materials

Depending on the end-use requirements, manufacturers may select from a broad range of polyurethane thermoplastics. Because processing parameters vary for each material family and resin formulation, the best results are usually achieved by following the handling and processing procedures recommended by the resin manufacturer.

17.2.3 Processing Parameters

While machine selection, material properties, and part design all affect the outcome of injection molding, five processing variables specific to injection molding can have as much or more impact on the success of this process. These variables are as follows:

- Injection velocity
- Plastic temperature
- Plastic pressure
- Cooling temperature
- Time

Control of these variables during each of the four phases of the injection molding process can help improve part quality, reduce part variations, and increase overall productivity.

Phase 1: Fill. In phase 1 the screw advances, and plastic flows into the mold. Flow characteristics are determined by melt temperature, pressure, and shear rate. Injection velocity, the rate at which the ram (screw) moves, is the most critical variable during fill. A polymer flows more easily as injection velocity is increased. However, injection velocity that is too high can create excessive shear and result in problems such as splay and jetting. More importantly, heat from a higher shear rate can degrade the plastic, which adversely affects the properties of the molded part.

The way in which plastics flow during fill is also affected by their viscosity, or resistance to flow. Polymers with high viscosity, such as polyurethanes, are thick and taffy-like; those with low viscosity are thinner and flow more easily. Melt temperature affects viscosity and, to achieve the best results, should be maintained within the temperature range recommended by the supplier.

Plastic pressure, another variable, increases sharply during fill, as shown in Fig. 17.13. The molten plastic, in fact, can be under much greater pressure than is indicated by hydraulic pressure. It is important to understand the flow characteristics during fill of the material being used and to operate the process consistently.

Phase 2: Pack. Phase 2 is when the plastic melt is compressed, and more material is added to compensate for any shrinkage during cooling. Approximately 95% of the total resin is added during fill, with the remaining 5% added during the pack phase. Plastic pressure is the primary variable of concern during

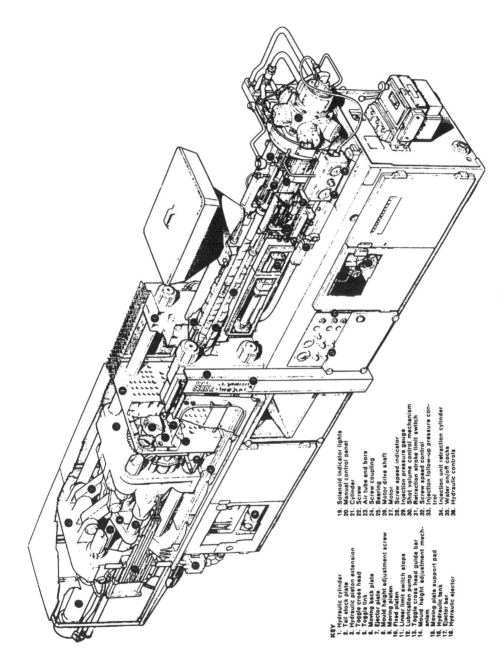

KEY
1. Hydraulic cylinder
2. Tail stock plate
3. Hydraulic platen extension
4. Toggle cross head
5. Toggle link
6. Moving back plate
7. Ejector plate
8. Mould height adjustment screw
9. Moving platen
10. Fixed platen
11. Linear/limit switch stops
12. Lubrication pump
13. Toggle cross head guide bar
14. Mould height adjustment mechanism
15. Moving plate support pad
16. Hydraulic tank
17. Ejector bar
18. Hydraulic ejector
19. Solenoid indicator lights
20. Manual control panel
21. Cylinder
22. Screw
23. Air tube and bore
24. Screw coupling
25. Bearing
26. Motor drive shaft
27. Motor
28. Screw speed indicator
29. Injection pressure gauge
30. Shot volume control mechanism
31. Retraction stroke limit switch
32. Screw speed control
33. Injection follow-up pressure control
34. Injection unit retraction cylinder
35. Water on/off cocks
36. Hydraulic controls

Figure 17.11 Phantom view of a 21/2 in reciprocating screw injection machine.

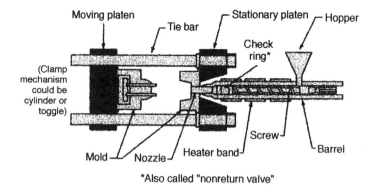

Figure 17.12 Cutaway diagram of a typical injection molding machine.

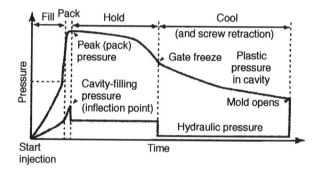

Figure 17.13 Pressure and time in a typical injection molding process.

the pack phase. The screw maintains pressure in the melt, compensating for shrinkage, which can cause sinks and voids. Variations in cavity pressure are a primary cause of deviations in plastic parts.

It is important to completely fill the mold, avoiding overpacking or underpacking, since packing pressure determines part weight and part dimensions. Overpacking can cause dimensional problems and difficulty in ejecting the part, while underpacking can result in short shots, sinks, part-weight variations, and warpage.

Phase 3: Hold. This phase is affected by all five of the process variables described earlier—injection velocity, plastic temperature, plastic pressure, and cooling temperature and time. After the mold is packed, the plastic is held in the mold until it is partially solidified and the gate freezes. The drop in plastic pressure reflects the amount of shrinkage that occurs from cooling. One way to optimize this phase is to decrease the hold time until the part weight changes. At that point, the gate is no longer sealed, and resin backflows out of the mold. If hold continues after the gate seals, cycle time increases, using more time and energy to produce the part. The key is to maintain pressure on the plastic until the gate freezes.

Phase 4: Cooling and Plastication. This is generally the longest part of the molding cycle—up to 80% of the cycle time. Optimizing cooling time can yield substantial gains in productivity. Because the gates are sealed during this phase, cooling temperature and time are the only variables at work. The key to optimizing the cooling phase is to balance the desire to cool quickly against the amount of molded-in stress the final part can withstand.

Injection machines may be either of ram-feed or a screw-feed designs, as shown in Fig. 17.14.

17.2.4   Design Considerations

While designed for functionality, parts should also be designed to maximize overall strength and simplify the manufacturing process. Significant problems in both processing and performance can occur when

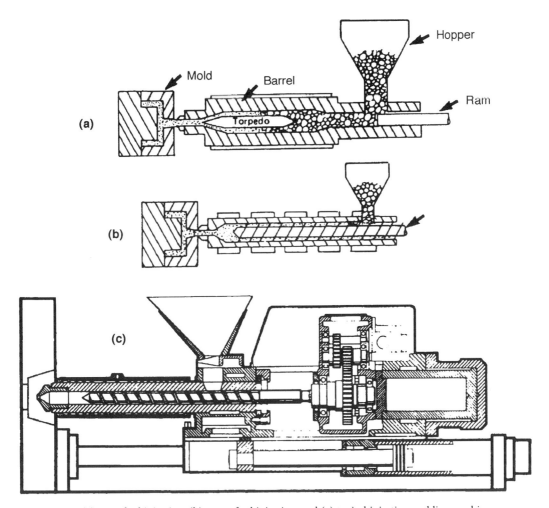

Figure 17.14 (a) Ram-feed injection, (b) screw-feed injection, and (c) typical injection molding machine.

the basic principles of good design are overlooked. Following basic design guidelines for nominal wall, corner radii, holes, projections, draft, and gating, increases the likelihood that the part will process and perform successfully.

Since end-use factors (e.g., sterilization) can also affect material performance, design elements can be used to compensate for certain shifts in material properties. Design factors that increase localized stress should be reviewed with knowledge of the selected material and end-use requirements.

A list of design guidelines shown in Table 17.1 can be reviewed at any time during the product development cycle to focus on fundamentals. While good design does not always guarantee molding success, it does contribute to processing and assembly ease, part performance, and overall productivity.

17.3   Insert Molding

Insert molding is an injection molding process whereby plastic is injected into a cavity and around an insert piece placed into the same cavity just prior to molding. The result is a single piece with the insert encapsulated by the plastic. The insert can be made of metal or another plastic. The technique was initially developed to place threaded inserts in molded parts and to encapsulate the wire-plug connection on electrical cords. Typical applications include encapsulated electrical components and threaded fasteners. Generally, there are few design limitations or restrictions on material combinations.

TABLE 17.1 Design Guidelines for Injection-Molded Parts

The following list offers some basic guidelines for part design. While not all inclusive, it covers design elements common to almost all injection-molded plastic parts.	
Nominal wall	The overall thickness of the part. -Maintain a uniform nominal wall. -Avoid overly thick or thin sections.
Corner radii	The intersection of any two walls. Necessary for part functionality, corner radii act as inherent stress concentrators. *Avoid sharp corners.* -Design rounded inside and outside edges. -Maintain a uniform wall even at corners.
Holes	Openings for attaching components or fasteners, or for providing ventilation or light. Avoid sharp corners, that can localize stress concentrators, cause weld lines and shear the material during fill, and induce polymer orientation.
Projections	Any feature that stand-up off the nominal wall: ribs, bosses, gussets, tabs, and standoffs. Avoid projections that are too thick or too think; they can create problems during processing and impede part performance.
Draft	The degree to which the side walls are tapered or angled. the objective of draft is to make part removal as easy as possible. Ensure adequate draft for any aspect of a part that is oriented perpendicular to the mold so that it can be freed from the mold.
Gating	The opening through which the polymer melt enters the mold. There are several types of gates-sprue, edge, flash, pinpoint, diaphragm, ring, submarine, and tunnel. Consider the several factors that will determine the type of gate used: the number of cavities in the mold, the need for symmetrical filling, the size and shape of the part, and how tight the tolerances are for the part.

Types of bonding occur in insert molding: molecular and mechanical. Molecular bonding can occur when the insert material is the same as or similar to the encapsulating resin. This will yield the best results from the joint, both for physical strength and leak resistance. An example would be molding a polyurethane connector to polyurethane tubing. Mechanical bonding can take place in two ways: by the shrinking of the encapsulating resin around the insert as the resin cools, or by the surrounding of irregularities in the surface of the insert by the encapsulating resin. Although shrinkage always occurs, it is rarely sufficient to produce adequate physical strength or leak resistance of the joint. In general, when insert molding dissimilar materials, the insert should offer some means of mechanical retention such as a sandblasted, flared, or knurled surface.

17.3.1 Equipment

Although insert molding can be performed using a standard injection molding press, doing so can make the critical step of loading and retaining the insert in the cavity a difficult operation, and can thus place restrictions on the design of the part. However, there are some specific molding machine designs that are better suited for insert molding and offer much greater flexibility and productivity. Rotary and shuttle-table-type injection molding machines are excellent for this purpose, because they allow operators to load and unload inserts in the bottom half of one mold while actual molding takes place in another. These machines also lend themselves well to automation of the loading and unloading of inserts and parts. Other machines that offer vertical mold clamping can work well for low volume insert molding, but are generally less productive than rotary or shuttle-table machines.

With a few exceptions, molds for insert molding are generally designed in the same fashion as are molds for injection molding. Molds for a shuttle-table press have two ejector halves, and molds for a rotary press have two to four ejector halves. Tooling costs will be higher due to the additional mold halves. The actual molding cycle time is the determining factor for establishing the number of cavities and mold halves required. For optimum productivity, the time it takes to load the inserts and unload the parts should not exceed the molding cycle time. This consideration is also important when molding resins such as polycarbonate or PVC, for which residence time in the heated barrel of the machine is an important factor. Because the molten plastic is typically injected into the cavity and around the insert at pressures exceeding 1000 psi, when the mold is designed, it is important to determine the exact location of injection gates and how the insert is to be held in place.

17.3.2 Materials

Like injection molding in general, insert molding can be accomplished with a wide variety of materials, including polyethylene, polystyrene, polypropylene, polyvinyl chloride, thermoplastic polyurethane elastomers, and many engineering plastics. The primary factors that restrict the use of insert molding are not process related but are determined by the strength and other properties required for the molded product.

17.3.3 Processing Parameters

One of the chief causes of failure in an insert-molded part is the cleanliness of the insert. It is absolutely imperative that the insert be as clean as possible prior to molding. When molding with large metal inserts, the inserts may need to be preheated to minimize the stresses caused by differential thermal expansion and contraction. When inserts are manually loaded, it is important that the operator maintain a consistent cycle time.

17.3.4 Design Considerations

In general, the basic design rules for insert molding are the same as those that apply to injection-molded parts. However, designers should also be aware of the following elements that may affect the design of their parts:

- The material from which the insert is made
- Pull-and compression-strength requirements of the insert from the plastic
- Leak test requirements
- Torque or axial forces to which the insert will be subjected
- Voltage requirements for electrical applications

Any or all of these elements may establish parameters that can help the designer determine what encapsulating resins will work for the application. They may also create requirements for the type of material preparation that must be performed on the insert to ensure proper performance of the finished part.

17.4 Liquid Resin Casting

Pouring a reactive liquid polyurethane into molds, then allowing it to cure to solid form, describes the fundamentals of liquid resin casting. A technically refined version of this decades-old process is a reliable and cost-effective choice for manufacturers of sophisticated products. With new developments in materials and process controls, liquid resin casting lends itself to many demanding applications, such as cardiac pacemaker encapsulations, hand-held electro-optical surgical devices, and key components of medical imagers and scanners. Product developers use liquid resin casting in two principal ways:

1. For prototyping prior to committing to high-volume production
2. For ongoing low-volume production of 25 to 2000 units per year

The principal advantages of liquid resin casting (comparatively inexpensive tooling, short lead times for tooling and parts, mild processing conditions, and design flexibility) enable the manufacture of highly complex parts with specialized performance characteristics difficult for other technologies to duplicate. For example, unlike molding or machining, liquid resin casting is associated with mild processing conditions that allow delicate components, such as fiber optics or electronics, to be encapsulated directly into the final or near-net shape required.

New tooling starts with a model. A castable material of polyurethane is poured over the model in one or more steps. The material then cures, creating a mold. (Molds may also be machined directly out of aluminum or another suitable material.) Tooling lead times are generally three to six weeks. Once the mold is finished, parts are produced by pouring a resin into it and allowing the material to cure.

For prototyping requirements, plastic patterns often can reduce initial costs and lead time when compared to metal patterns. However, manufacturers that use plastic patterns must be prepared to accept limited options for surface finish and less tolerance control. In addition, if significant design modifications are necessary, a new generation of tooling will probably be required. Production programs are better accomplished with metal patterns, which enable the manufacturer to achieve the best possible results and avoid the limitations of plastic patterns.

17.4.1 Equipment

Special equipment for liquid resin casting includes mixing and dispensing equipment for handling resins, degassing equipment for removing entrapped air within the resin, and ovens for curing materials. The specific equipment needed depends on the kinds of materials being processed and whether they are for a prototype or production activity. The more demanding the application, the more sophisticated the equipment required.

17.4.2 Materials

Thermoset resins most commonly used in liquid resin casting—epoxies, polyurethanes, and silicones—must be in liquid form at approximately room temperature for successful processing. Formulations that will satisfy virtually any application can be developed from these three basic material types. An expanded selection of formulations also is generally available from manufacturers with a production, rather than a prototype, focus.

In general, each type of resin has its own distinct advantages. Epoxies are ideal for high temperatures (up to 450° F) or for highly corrosive applications. Polyurethanes are excellent general-purpose materials for both soft-rubber and hard-plastic applications where exceptional toughness and wear resistance are important. Silicones are best for product applications that require rubber that is soft or of medium hardness over a broad temperature range.

17.4.3 Processing Parameters

Little or no pressure occurs within the liquid resin casting process, but humidity should be controlled during material handling. Polyurethanes are sensitive to moisture and will react to the presence of water in the mold. Release agents can be used on mold surfaces to facilitate part removal and are available in silicone-based and water-soluble formulations.

17.5 Reaction Injection Molding

Reaction injection molding (RIM) creates parts by using impingement mixing to combine reactive liquid intermediates as they enter a mold. RIM differs from traditional injection molding in that it forms solid parts by cross-linking or polymerization in the mold rather than by cooling. The process does not use hot mold cavities to activate the reaction; in fact, RIM molds must often be cooled, because an exotherm forms when the intermediates are mixed in the presence of heat. Completed parts often can be demolded in less than 20 s.

RIM developed from polyurethane foam technology. The needs of the automotive market spurred major growth of the process in the United States, and this market remains the primary application for RIM. Most of the technology's advances have resulted from efforts to improve the materials and processes for use in automobile production. However, there are some significant applications for RIM in other industries, including structural foam cabinet parts, structural parts, furniture, etc.

17.5.1 Equipment

Equipment for RIM was first developed in Germany and is now sold by several manufacturers. The basic elements of a RIM molding system include a conditioning system that prepares the liquid intermediates for use, a metered pumping system that ensures delivery of the intermediates in appropriate quantity and pressure, one or more high-pressure mixing heads in which the liquid intermediates are combined through impingement, and a mold carrier that orients the mold as required and opens and closes it for cleaning and demolding.

Unlike thermoplastic molding, RIM uses liquids that have a low viscosity during mold filling and fills out the part using only internally generated pressure. Consequently, molding pressures in RIM can be as little as 50 psi (compared with 5000 psi or more for thermoplastic molding), making it possible for small machines with very limited clamping force to produce even relatively large parts in large numbers. For the same reasons, RIM molds are typically much less expensive than those used in thermoplastic processing. However, RIM molds made using the criteria for traditional injection molding are often unsuccessful. Molds for RIM have unique requirements related to the low-viscosity liquids they are filled with, and molds from other processes are rarely adaptable.

Low viscosity, low mold pressure, and inexpensive mold costs make RIM attractive for short production runs and prototyping. Selection of equipment appropriate to the anticipated application is critical for successful use of RIM. Key parameters for equipment selection include the type of material to be used (e.g., foam, elastomer), suitability for the size of parts being manufactured, and the desired throughput. As development of the technology has accelerated, corresponding equipment improvements have been made. Options now available include systems that incorporate multiple mixing heads and equipment for a range of processing limitations. Equipment options include various types and sizes of mixing heads, temperature controls for both materials and molds, programmable shot time controls, and process control alarms.

17.5.2 Materials

The earliest use of RIM was with polyurethane, but advancements with the technique now offer opportunities for use with many other materials. Depending on the intermediates used, RIM can be used to produce soft foams, rigid foams, and solid elastomers. For example, RIM technology has produced a reusable foam that would have been impossible to produce with any other technology.

17.5.3 Processing Parameters

Although RIM's use of low-viscosity intermediates is an advantage for productivity, it also has some disadvantages. Handling such reactive or hazardous raw materials requires special equipment and procedures, including spill-cleanup materials. Gowning for operators should include protective coverings, eye protection, and sometimes air filtration masks. Since some components freeze at room temperature, a temperature-controlled environment is required for their shipping and storage.

Gas bubbles can be trapped during filling, and molds can be difficult to seal, which increases flash. Such problems normally can be overcome by careful attention to materials selection, mold design and orientation, shot time, and venting. Because the low-viscosity materials generally penetrate molds, the development of mold release agents for use in RIM has been difficult. Also, the recent exclusion of certain blowing agents (such as chlorofluorocarbons and hydrochlorofluorocarbons) has created the need for extensive research to find suitable replacements. Successful molders must use RIM long enough to learn all the peculiarities of chemical manipulation, mold manufacture, and processing parameters. These typically differ for each project, resulting in a long learning curve for the few companies that choose to offer RIM-produced products.

17.5.4 Design Considerations

Inserts and reinforcement materials can be readily used. Reinforcement materials may include fiberglass, scrap plastics, metal, or wood. Fillers may also be added to improve the flexural modulus of the finished product or reduce its shrink rate in processing.

17.6 Continuous Extruder Polymerization

Thermoplastic polyurethane elastomer (TPE) products are present in important industrial applications owing to their excellent properties and variety of uses. The properties of the products can be tailored to meet specific requirements by selecting the types and ratios of the ingredients. The result is a product that ranges from construction materials of different hardnesses through highly elastic coating materials, fibers, and foils to raw materials for solvent adhesives.

In addition to batch-type processing, an extrusion polymerization continuous process has been developed for the manufacture of TPEs in commercial quantities. For industrial applications, batch-type mixing (agitating) processes are normally used for the synthesis of polyurethane-adhesive and soft thermoplastic elastomers that are made at low reaction speeds. The basic ingredients are fed into a mixer or casting mold and mixed as intensively as possible. The compound remains in the container for final reaction and hardening. Then the castings are tempered, cut, and processed into large chips in a cooled-down grinder. The advantages of mixers over a continuous system are that they ensure long residence times for carrying out the reaction, they require only a small investment for the equipment, and they permit the manufacture of many different chemical recipes.

The basic ingredients are fed into a mixing head and mixed intensively for a very brief time in the low viscous range. Then the reaction mass is transferred to a continuous steel or plastic conveyor belt on which the mixed components react under heat application and the mass sets.

The difference between the continuous conveyor belt system and the continuous conveyor belt extruder system is that, in the first system, the firm plates are removed at the end of the conveyor belt, then thermally treated (e.g., in an oven), ground into chips in a grinder, and perhaps shaped into pellets by extrusion and pelletizing. In the second system, the reaction mass is transferred directly from the conveyor belt to an extruder for final reaction, homogenization, and pelletizing.

Shortcomings in the previously described continuous processes (such as increased costs for personnel and maintenance and the high cost of equipment, as well as the enormous disadvantage of having to do the polymer synthesis and pelletizing in different work processes) have led to the production of continuous extruder polymerization. In this process, the raw materials (diisocyanate, polyol, chain extenders, and additives such as catalysts and stabilizers) are precision metered in one step into the twin-screw extruder, where they are polymerized, and then transformed into pellets of equal particle size.

Co-rotating twin-screw reactors have proved particularly suitable for the production of homogeneous reaction products. These machines meet the processing requirements for successful completion of the reaction by virtue of

- Good conveying action in the low as well as high viscous phase
- Good mixing and homogenization action, even with extremely high viscosities at precisely defined locations within the reaction space, as a result of the variable configuration of the screw elements and kneading blocks
- Narrow residence time spectrum due to the complete and reciprocal cleaning of the screw element surfaces as well as the close tolerance between the screws, and between the barrel sections and screws
- Good exchange of the compound through permanent product transfer from one screw to the other
- Good heat transfer from the barrel walls to the product, and within the product through the intensive exchange of the compound

- Tightly stepped, effective heating and cooling of the product in the reaction space through individually temperature-controlled barrel sections

Screw reactors allow continuous production of soft polyurethane-adhesives, which are made at low reaction speeds, up to the hardest thermoplastic elastomer having Shore-hardnesses >D65, whose synthesis take place in distinct exothermal reaction sequences.

There are two continuous polymerization methods.

1. The two-step *(prepolymer)* process
2. The one-step *(one-shot)* process

The two-step process is a batch manufacturing of isocyanate-terminated (excess of -NCO) prepolymer (viscous liquid). In the second step, a chain extender is added to tie the short molecular chains through free NCO groups into much longer chains of final product. In one-shot process, all components are mixed together and polymerized.

Both methods can be executed on a twin-screw machine. The one-shot process occurs through simultaneous injection of all components into the same feed port. The prepolymer process occurs through split injection of chain extender downstream in the machine.

The steps of the traditional batch process are as follows:

1. Preweighed, premelted isocyanates and polyols are mixed at approximately 150° C for 25 min under vacuum to form a liquid prepolymer. The chain extender is then added and mixed.
2. The mix is quickly cast in molds that are placed for final polymerization into an oven for 24 hr at 110° C.
3. The shredded chips are remelted in an extruder for shaping the material into commercially attractive pellets.

17.6.1 Metering Devices

For reproducible polyurethane synthesis, the proportions of the basic ingredients must correspond exactly to the chemical formulation since, even with a relatively small diisocyanate or NCO deficiency, the required degree of polymerization cannot be reached, and excess NCO will result in branching and cross-linking due to secondary reactions of the residual quantities of NCO. To obtain reproducible production results of a high quality level, the metering devices must be very precise. These requirements can be fulfilled due to the high level of metering technology now available. The continuous metering of the preheated liquid raw materials (polyol, diisocynate, and diol) into the screw reactor is usually carried out gravimetrically using differential-type metering scales. Such metering systems are not just insensitive to viscosity deviations but also allow feedback and recording of the weight actually metered.

The polyol and diisocyanate are metered from heated containers through heated precision gear pumps into the feed barrel of the screw reactor. Diol(s) can be metered the same way as a separate flow into the feed barrel of the reactor or metered at any other point of the processing section and, as a final possibility, even metered premixed with the polyol. Additives such as catalysts, stabilizers, and antioxidants are usually mixed with the polyol in the feed container and metered simultaneously with the polyol stream.

As in batch process, the failure to meter precisely the feed of components will inevitably lead to unacceptable product quality. If, however, in batch process, the polymer mix can be sometimes adjusted if the discrepancy is caught in time, no such possibility exists in continuous process. Therefore, it is important to have top-of-the-line metering systems. The best of them are loss-in-weight liquid systems that are insensitive to viscosity variations. The volumetric gear pumps with mass flow rate control are merely adequate for this purpose.

17.6.2 Discharge and Pelletizing Equipment

The pelletizing of the polyurethane compound depends on its material consistency, such as viscosity or stickiness, as well as on the required type of pellet. It is carried out at the end of the processing section, either by the strand or the hot face-cutting (underwater pelletizing) method. The strand-cutting process is applied to quick-hardening TPE having high melt viscosities, whereas soft types and polyurethane adhesives are pelletized by the hot face-cutting method. Blade drums or underwater pelletizers are used for pelletizing low-viscous and sticky polyurethane types. These pelletizing systems will even process polyurethane adhesives into rice grain size pellets that display good noncaking features and easy separation due to their large surface. When using the hot face-cutting method, it is imperative to discharge low-viscous start-up products through a diverter valve before they reach the pelletizer head to prevent clogging of the cutting tools.

A hard, nonsticky, quick-hardening polymer with high melt viscosity and strength can be pelletized in an inexpensive stranding process. However, softer, sticky polyurethanes with lower melt viscosity are pelletized in more expensive knife-rotor or underwater pelletizing systems wherein the polymer is discharged into the transporting water stream during the cutting. The freshly cut, hot and sticky pellets will agglomerate and clog the system unless the pellet load in the system is reduced to 3 to 20 kg/m³ of water to minimize the chances of pellet collision and agglomeration before they are adequately cooled. To eliminate the stickiness of soft pellets upon discharge from cooling transportation system, they should reach as high a degree of crystallization as possible through increase of their residence time (usually several minutes) inside the cooling systems. That is achieved through the use of long water troughs.

17.6.3   Post-treatment of Pellets

The least complicated and most cost-effective way of pelletizing thermoplastic PU elastomers is the by strand method. As PU strands are discharged from the die head, they are cooled down in a water bath and freed of any remaining water by means of an air stream, and then pelletized.

Much more elaborate equipment must be used when the hot face-cutting method is used for PU adhesives and soft TPEs. These compounds are pelletized by rotary knives as they are discharged from the die plate and directly cooled in the transporting water. The still-soft granulate would form inseparable agglomerates, even at low water temperatures, if the pellets were not transported at high water speeds and at low concentrations in the water. The concentration factor for the pellets depends on the type of pelletizer system applied, the water temperature, and the chemical formulation of the polyurethane elastomers. Concentration is usually between 3 and 20 kg per cubic meter of water.

To eliminate stickiness, the pellets should reach as high a degree of crystallization as possible in the transport or cooling water system. This process requires up to several minutes, especially with PU adhesives. To obtain the required time periods in the flowing water, long troughs or conduits should be used. Owing to the susceptibility of polyesterpolyurethanes to hydrolysis, the pellets should be dried to a minimum moisture content following cooldown.

17.6.4   Layout of Screw Reactors

The layout of extruders depends on the required throughput and the chemical formulation of the PU elastomers to be manufactured. The reactor volume is determined by the structural parameters of the outer screw diameter, and the length-diameter ratio, which defines the length of the processing section of the screw reactor. Table 17.2 lists ZSK-machine sizes plus their respective throughput ranges for various polyurethane formulae. The low throughput volumes are for slow-reacting elastomer types, such as adhesives. The higher throughput volumes are for fast-reacting, harder TPU.

Table 17.2 ZSK-Screw Reactors and Throughput Ranges at L/D = 40

ZSK	Screw Diameter (mm)	Throughput Range (kq/h)
40	40	30–75
58	58	100–250
70	70	150–400
92	92	350–850

The polyaddition-reaction must completed to the highest possible degree within the reaction space. Therefore, the screw speed and the residence will define the reactor volume. It should be pointed out that, in order to increase the residence time with any given volume, the screw speed should not be reduced to such an extent that the mixing intensity is reduced below a critical point. Mixing intensity decisively influences the time the compound is in the reactor. Thus, screw speeds producing shearing rates of $>2,000$ s^{-1} are required in the radial play between the screw crest and the barrel wall to obtain homogeneous reactions and to prevent the buildup of gels or lumps.

To minimize thermal and mechanical disintegration processes, especially high viscous melts of harder TPU should be discharged from the machine directly after transforming the reaction partners. The previously mentioned problem does not exist with PU recipes such as PU adhesive or TPU based on aliphatic diisocyanate, since these products require much higher reaction times for transforming the reactants.

In conclusion, it can be said with regard to the synthesis of linear PU elastomers in screw reactors that extremely short reaction times of between approximately 1 to 3 min are required as compared to mixer and conveyor belt systems. This behavior is not only due to the reaction mass that is processed in screw extruders in accordance with thin-layer technology, in which reaction-accelerating processes such as material and heat exchange are very effective. It is also to the high temperatures of up to 270° C. These reaction temperatures can be realized for PU synthesis in screw extruders on account of the residence times and high pressures.

17.7 Illustrative Patents

17.7.1 Foam Production by Extruder Method

Polymer foams, such as polyisocyanurate and polyurethane foams, have not previously been producible using an extruder, as the foaming material *firms* in the extruder and blocks the equipment.

Richie, in U.S. Pat. 3,466,705, describes apparatus for extruding foamable thermoplastic material such as polystyrene, preferably in tubular form. Richie discloses that the apparatus can be used for thermosetting materials such as polyurethanes, but that steam or hot water would have a damaging effect and therefore are not used. A foam-augmenting gas such as freon is used as the foaming agent. Richie uses an expansion space of transverse cross-sectional area that increases in size gradually in the direction of travel of the extruded product.

Glorioso and Burgess[1] disclosed a method for preparing a thermosetting foam, such as polyisocyanurate or polyurethane foam, that includes introducing polyol and isocyanate to a screw of an extruder and mixing in the extruder screw. Catalyst is then added and mixed with the already formed mixture in the extruder screw. This mixture is extruded onto a conveyor and foamed on the conveyor. Finely divided carbon black, dispersing agent, and/or surfactant may be mixed with polyol before introducing it to the screw of the extruder, or carbon black may be dispersed in polyol in the extruder and a polyol premix made before the remainder of the polyol is added. Foam cell size is decreased when extruder speed is increased. The foaming agent, which is preferably HCFC or water, may be mixed with one of the components, preferably the polyol, before introducing it to the screw of the extruder. The mixture is cooled before extruding onto a conveyor to delay the foaming step until the mixture is outside the extruder. A carbon black dispersion may also be made for future use. The process described may be used for making foam boards or bunstock. The disadvantages of the prior art are overcome by the Glorioso and Burgess invention, as described below.

A method of preparing polyisocyanurate foam using a screw-type extruder for mixing the components of the composition includes introducing isocyanate and polyol to a screw of an extruder and subsequently mixing the isocyanate and polyol, together with foaming agent and, optionally, carbon black and surfactant in the extruder. Catalyst is added and mixed with the already formed polyol/isocyanate mixture in the extruder screw. This mixture is almost immediately extruded onto a conveyor and foamed on the conveyor.

Finely divided carbon black and/or dispersing agent and/or surfactant may be mixed with the isocyanate or with the polyol before introducing to the screw of the extruder. This premix (isocyanate or polyol and/or foaming agent and/or dispersant and/or carbon black) may also be made in the extruder.

The foaming agent, a hydrochlorofluorocarbon (HCFC) or water, is mixed with one of the components, preferably the polyol, before introducing it to the screw of the extruder. If water is the foaming agent, the water is vaporized for foaming the mixture and reacts with the isocyanate to form carbon dioxide, which further foams the mixture.

The mixture is cooled in the extruder to substantially delay the foaming step until the mixture is outside the extruder. Foam cell size is generally decreased when extruder speed is increased. A carbon black dispersion may be made in the extruder using carbon black, dispersant, optional surfactant, and isocyanate or polyol. This dispersion may be used immediately or stored for future use. Use of an extruder provides a better quality dispersion, measuring 8 or more on the Hegman scale, than a dispersion made in a Kady (kinetic dispersion) mill.

It has unexpectedly been found that thermosetting foams (such as polyisocyanurate or polyurethane foams) may be made using an extruder for mixing the foamable mixture before extruding onto a belt conveyor. This allows use of a continuous process in which thermosetting foams may be made at much faster speeds than previously possible, without any decrease in quality.

In making insulation board, the extrusion process described herein allows fast, economical production. The thermal conductivity of insulation board is reduced by adding carbon black, and this advantageously permits reduction of board thickness. The thermal conductivity of polyisocyanurate foams, based on different foaming agents, in btu/ft^3/hr/°F, are as follows:

Isocyanurate/HCFC	0.12
Isocyanurate/carbon black/HCFC	0.09–0.01
Isocyanurate/water	0.14–0.19
Isocyanurate/carbon black/water	0.135–0.17
Isocyanurate/carbon black/HCFC/water	0.09–0.135

17.7.2 Continuous Process for the Making Thermoplastic Polyurethane

It has been proposed to manufacture a polyurethane adapted for processing by thermoplastic shaping methods such as, for example, calendaring, compression molding, extrusion, and the like. The process for making such polyurethanes involves mixing together components that react rapidly with each other and requires careful control of the reaction conditions with interruption of the chemical reaction before a polyurethane has been produced which is no longer processable by thermoplastic methods. Because the reaction has to be controlled very carefully, and the components used in making the reaction mixture must be handled with care prior to mixing, the previously available processes have largely been batch-type processes. It is very difficult to produce a product having the same composition from one batch to another, so the products previously available have varied in physical characteristics and processibility. The net result has been a very complicated and involved process with accompanying high production costs.

Bartel et al.[2] disclosed a process and an apparatus for making thermoplastic polyurethanes continuously. The apparatus combines a storage tank and dehydrating equipment for a polyol and storage tanks for other components of a polyurethane reaction mixture, all connected through conduits to a mixing device that discharges the reaction mixture continuously over the surface of a conveyor belt. The conveyor belt is provided with means for controlling the temperature thereof. The solidified thermoplastic product discharged from the belt is continuously granulated and conveyed to containers.

An object of the Bartel invention to provide a method and apparatus for continuously making a thermoplastic polyurethane. Another object of the invention is to provide an apparatus adapted to dehydrate those components requiring it, and to mix all of the components of a polyurethane reaction mixture together under conditions that produce a solid polyurethane adapted to be processed by thermoplastic shaping methods. A further object of the invention is to provide a method for making a solid thermoplastic polyurethane of constant composition over an extended period of time.

In accordance with this invention, the foregoing objects and others are accomplished, generally speaking, by providing a plurality of storage tanks for the components of a polyurethane reaction mixture, means for continuously charging into a mixing head metered amounts of each component, and means for continuously spreading the resulting reaction mixture over the surface of a conveyor belt provided, plus means for carefully controlling the temperature thereof. More specifically, the invention combines a means for storing, heating, and dehydrating a polyol and for continuously charging this polyol into a mixing chamber, with means for simultaneously charging an organic isocyanate and other components of a polyurethane reaction mixture continuously in metered quantities in the same mixing chamber and with a conveyor belt moving at a constant speed adjacent the discharge opening of the mixing chamber. The conveyor belt is adapted to receive from the mixing chamber the resulting reaction mixture while it is still liquid and will spread as a thin film over the conveyor. Means are provided at that end of the conveyor adjacent the mixing head to heat the reaction mixture spread thereon to provide a temperature that will encourage reaction of the liquid reaction mixture until a solid has been formed.

Means are provided at the other end of the conveyor to cool the solid product to interrupt chemical reaction before the polyurethane is no longer processable by thermoplastic shaping methods. In one embodiment of the process provided by this invention, the polyol is heated to about 130° C, dehydrated at a temperature of at least about 120° C, and then cooled to about 100° C before it is metered into a premixing chamber. In the chamber, it is mixed with a chain extender or other component of the reaction mixture not reactive with the polyol. The premix of liquid components is then mixed with an organic diisocyanate and other components of a mixture that will form a thermoplastic polyurethane such as, for example, fillers, dyes, and the like. The resulting mixture is then spread over the surface of a conveyor where the temperature of which is maintained at chemical reactive levels until the mixture solidifies. Chemical reaction is interrupted after solidification by cooling the polyurethane while it is still processable by thermoplastic methods.

References

1. Glorioso, Sammie J., and Burgess, James H. U.S. Pat. 5,424,014, June 13, 1995, "Method for Extruding Foamable Polymer Material," to Apache Products Company, Jackson, Mississippi.
2. Bartel, G.F., Klawitter, M., and Denker, E. U.S. Pat. 3,620,680, November 16, 1971, "Apparatus for Continuously Making Thermoplastic Polyurethanes," to Die Kunststoffburo Osnabruck Dr. Reuter GmbH, 4531 Lotte, Hannover, Federal Republic of Germany.

18

Compounding
Ingredients

An additive is any substance added to a polymer to alter certain properties. In the polyurethane field, additives are added to the polymers in small amounts to improve processing characteristics, enhance heat stability, provide antioxidant protection, stabilize against UV and gamma radiation, etc. Figure 18·1 presents the universe of polyurethane additives, conveniently divided into process and functional additives. Process additives are incorporated into polyurethanes to improve the manufacture of the polymers, whereas the functional additives are incorporated to enhance certain properties of the finished polymer.

18.1 Internal and External Lubricants for Polyurethanes

The formulation of lubricants appears to have begun 4,000 years ago with addition of metal carboxylates to naturally occurring or derived lipids used to coat chariot axles. Definitions of lubricants have been stated in terms of function, stressing comprehensiveness at the expense of brevity. For the purposes of our discussion, a lubricant is a substance that, when added in small quantities, provides a notable decrease in resistance to movement of chains or segments of a polymer of at least partly amorphous structure, without disproportionate change in observable properties. This definition ignores multifunctional ingre-

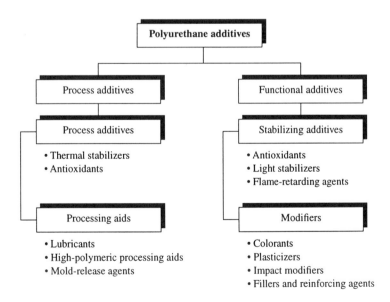

Figure 18.1 Systematic classification of polyurethane additives.

dients. For example, an antidegradant or stabilizer may function also as a lubricant. The antidegradant function, on the other hand, is expected to result in a change in stability disproportionate in effect to its concentration. Figure 18.2 presents the molecular architecture of some important *processing aid* lubricants.

Generally, a distinction is made between lubricants that facilitate movement of one polymer chain with regard to another (internal lubricants) and those intended to increase flow in the vicinity of objects such as the surfaces of process equipment (external lubricants). This division is often not clear. Furthermore, what might be classed as an external lubricant in one system (paraffin in polyurethanes) might well be an internal lubricant in another (paraffin in polyethylene). What is meant, therefore, is that, in a specific composition, an additive functions mainly (although not exclusively) as an internal or external lubricant.

The distinguishing characteristic of internal lubrication is that bulk viscosity is reduced. Unequivocal determination requires apparatus for measurement of bulk viscosity under conditions where perturbation of the composition by the apparatus is controlled and reproducible. This may be done in the laboratory with a variety of instruments: cone-plate and parallel plate viscometers, capillary rheometers, and a number of devices measuring a reflection of bulk viscosity (torque, power draw) in a miniaturization of an actual process such as extrusion or mixing. What is of particular interest is whether viscosity reduction is found under conditions approximating a particular process. Therefore, the apparatus selected should provide a shear rate of the same order of magnitude as the process at similar temperatures.

Although experiments regarding behavior as an internal lubricant are, from a practical viewpoint, referenced to the conditions of a specific mixing or fabrication process, they may be considered independently as simply measurements of the properties of a composition. This consideration may be meaningful to a supplier considering a range of new materials with potential as lubricants. Behavior as an external lubricant must, on the other hand, refer to a particular process.

A common situation for external lubrication relates to the interface between process equipment and the polyurethane compound. The metal surface of the equipment bears a polar oxidized layer and usually a pattern of microcracks and other discontinuities. Polar ends of lubricants are absorbed at such sites. Movement of polymer molecules in the vicinity of the surface then involves interaction between nonpolar ends of lubricant molecules associated with the polymer and those absorbed on the metal surface.

The effects of such lubrication include faster flow through restrictions used to shape the product (higher output, improved fill), lower heat buildup in processes with higher shear rates, and improved surface appearance from easier release and more streamlined flow. Effectiveness of external lubricants should be determined using laboratory or pilot equipment analogous to the actual process, e.g., a small extruder or calender, or a spiral flow mold in the case of injection molding. Then the lubricant concentration can be related to extruder output, appearance, and equipment power draw over a range of process temperature and machine speed, or to mold fill and part appearance vs. injection pressure, time and temperature. These procedures actually evaluate a combination of internal and external lubrication. In practice, it is often convenient to proceed with assessment of both functions at once, particularly when faced with the need for immediate practical problem solving.

The use of laboratory models of production equipment also facilitates evaluation of undesired side effects that may accompany external lubrication. These usually involve the lubricant and some other activity, e.g., as a participant in deposition of materials on metal surfaces, variously known as *plate-out* and *mold fouling*. To minimize such side effects, overlubrication must be avoided. The incompatible part of the lubricant should have low surface energy, i.e., be incompatible, because its solubility parameter is significantly lower (not higher) than that of the polymer. Dipolar materials in which the polymer-incompatible section is more polar than the polymer, can function to improve processability, but these are apt to be tackifiers. There are areas where tackifying agents can serve also as lubricants, but these generally exclude situations susceptible to deposition.

Often, the exact opposite situation is found. The introduction of new or rebuilt machinery having newly finished surfaces is commonly accompanied by a temporary loss in output as external lubricants fight to find a foothold. Attempting to explain to harried middle management that the break-in period on a new machine will far exceed the supplier's estimate, because of a dynamic equilibrium involving a

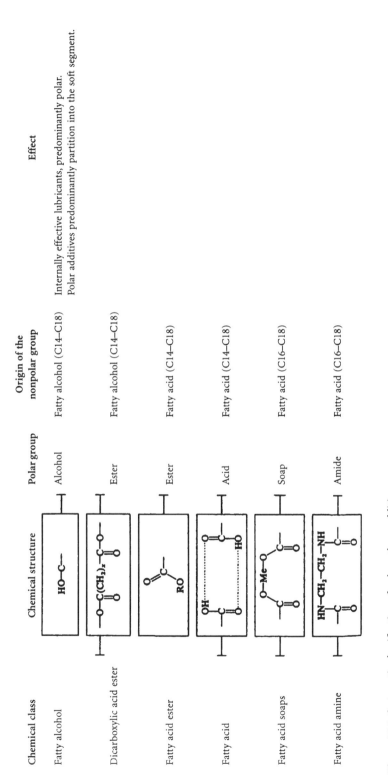

Figure 18.2 Systematic classification of polyurethane additives.

trace additive of incompletely understood function, will strike the technologist as related to rolling rocks uphill. A similar phenomenon is the temporary drop-in output associated with a major formulation change (as one lubricant attempts to displace another, particularly in mature equipment having a high microsurface area). This often registers against adoption of a new recipe that has shown sterling performance in the laboratory (where the small size of the equipment minimizes the effect).

18.2   External Surfaces

Polymer flow in regions adjacent to solid insoluble inclusions presents special problems. It is normally not desired that polymer molecules transport themselves through nips and dies or down runners leaving particles of filler or pigment behind. Occasionally, in service, it is useful if polymeric compositions force a layer of solid particles to the surface, but normally it is expected that dispersed solids will follow the polymer in its travels. In fact, a considerable effort is made to choose fillers and pigments and to disperse them such that extreme magnification is needed to disclose the existence of multiple phases. It is well known to the experienced polyurethane technologist that fractionation of polymer-filler mixtures can nonetheless occur during processing, often at points of sudden pressure change.

At rest, the solid particle functions as a physical cross-link, similar to a point of dipolar or dispersion force interaction between polymer chains. For the composition to flow, polymer-filler bonds must continually break and reform. This is particularly true of particles with high aspect ratios; the consequence of strong interaction (high reinforcement value) is high viscosity. Fillers vary greatly in their ability to absorb dipolar mobile species but, regardless of their surface area or structure, they are almost always more polar than their polymeric binder. Therefore, as the interactive section of the lubricant is made more polar, lubricant-filler interaction becomes more prevalent. This tends to lower process viscosity of a filled composition but may increase overall lubricant requirements. If the lubricant has other functions, e.g., is also a stabilizer, filler competition with the polymer may be detrimental unless compensated.

The lubricants that are effective at the polymer-filler interface are those that are strongly dipolar. They therefore may be classed as *internal* if the polymer is polar (with regard to their polymer-polymer interactive behavior) or *external* if used in nonpolar polymers. This may be altered considerably by use of filters and pigments that have been previously coated or that may be treated during mixing. Such coatings include reagents that bind to the filler surface by ionic, dipolar, or hydrogen bond forces and those that react with surface groups to form covalent links. The net effect is to generate a particle of lower energy surface than existed previously. The ability of lubricant-bearing polymer chains to flow in the vicinity of the particle is thus enhanced.

Important factors in selecting a lubricant are

- The polarity of different sections of a polymer and the volume fractions of these sections
- Other species to be encountered (surface of fillers, pigments, and stabilizers and equipment surfaces)
- Compound behavior without the additive; i.e., what modification is desired

The first of the above factors suggests a donor-acceptor scale for correlating lubricants with polymers. To be generally useful to the applied plastics technologist, a scale should be based on readily available parameters. At this time, the most reasonable choice is the solubility parameter. What is needed is not only the solubility parameter of polymer and lubricant overall, but also estimates of the contributions of the interactive and noninteractive parts of each. These may be calculated by several methods, e.g., from group contributions to the heat of vaporization and to the molar volume. Some values of interest regarding functional groups, polymers, lubricants, and fillers are given in Tables 18.1 through 18.4. *Please note that SI units are used.* Many older tables use Hildebrand units based on calories instead of Joules. These solubility parameters may be converted to SI units by multiplying by 2.05 (the square root of the conversion factor from calories to Joules). This can be a source of confusion at first glance, since 15–16 Hildebrands would denote a very polar material, but 15–16 /MPa$^{1/2}$ (SI units), a nonpolar one.

TABLE 18.1 Solubility Parameters of Some Functional Groups

Functional Group	$1/MPa^{1/2}$
-CH2CH2-	14.3
-CH2-	16.8
-CH,CH(CH,-)CH-*(branch point)*	18.6
-CH,COOH	26.5
-CH2COOH *dimerized*	21.8
-CH2COCH2-	19.7
-CH2CONH-	34.2
-CH2CONH-*hydrogen bonded*	22.5
-CHCl-	19.9

TABLE 18.2 Solubility Parameters of Some Polymers

Polymer	$1/MPa^{1/2}$
Polyolef ins, EP	16–17.5
Styrenics, SBR	18–18.8
Polyurethanes	16–22.5
PVC homopolymers	19–19.5
PVC copolymers, NBR/PVC	18.5–19
Collulosics	20.5–23

TABLE 18.3 Solubility Parameters of Some Polyurethane Lubricants

Lubricant	$1/MPa^{1/2}$
Stearic acid monomer	19.2
Stearic acid dimer	17.8
Glycerol monostearate	17.5
Stearamide	19.8
Divalent stearates	15
Alkali metal stearates	19
PE wax, paraffin	17
Oxidized waxes	18–19
Long chain esters	17.5–18.5

TABLE 18.4 Solubility Parameters of Some Fillers and Colorants

Fillers/Colorants	$1/MPa^{1/2}$
Titanium dioxide	34
Carbon black	25–35
Carbonate fillers	26–28
Sulfate fillers	26–28
Silicate fillers	30–40
Iron oxide	28

Since lubricant function depends on partial compatibility, the first principle is that the solubility parameter of the lubricant must not be the same as that of the polymer (it is then considered to be a plasticizer). At the other extreme, partial compatibility implies that these values be no more than 3.5–4.0/$MPa^{1/2}$ different. The top of this range, if the additive has the lower value, is appropriate to the case of an external lubricant whose exudation is desired. If the additive has a solubility parameter 3.5–4.0/$MPa^{1/2}$ higher than that of the polymer, it may be a filler treatment, a tackifier, a nucleating agent or a partly soluble inclusion of other (or perhaps no) utility.

For internal lubrication the following conditions must be met:

- The overall solubility parameter of the lubricant must be within 3/$MPa^{1/2}$ of the polymer.
- The solubility parameter of the interactive functional group should be a good match for the polar part of the polymer, e.g., ester, ketone, or carboxylate groups for semipolar elastomers; hydrocarbon branch points for nonpolar polymers.

- The solubility parameter of the noninteractive tail of the lubricant should be a poor match for the polymer, $\geq 3/MPa^{1/2}$ lower; for semipolar polymers, a 12-carbon-atom chain normally suffices.

The combination of above requirements places limits on the fractions of polar and nonpolar components. For instance, propionic acid contains both a terminal methyl group and reactive acid dimer. The molar volumes involved, however, generate an overall solubility parameter (21.5–22) that would be useful only in extremely polar polymers. Internal lubrication in common polymers, with additives having typical functional groups, constrains hydrocarbon side chains to the C,,C,, range. If the lubricant has another function, e.g., a lubricating stabilizer or antioxidant, the success of this function is usually dependent on criteria for internal lubrication being met.

For effective filler surface treatment, the functional group of the lubricant should have a higher solubility parameter than that of the polar part of the polymer. This permits the highly polar filler surface to compete with the polymer for lubricant. If a filler is chosen that is less polar than the polymer, the above does not apply. A normally internal lubricant may be of some value if the polymer-noninteractive section is a good match for the nonpolar filler. Filler pretreatment with additives usually does not lower surface polarity drastically enough to put the resultant product in the category of nonpolar fillers.

For external lubrication, the solubility parameter of the functional group of the lubricant should be slightly lower than that of the polar part of the polymer. That of the noninteractive tail should be slightly lower than the nonpolar part of the polymer. Increasing volume fraction of the latter (making the tail longer) increases external function. Except in the case of silicones or fluoropolymers used as trace additives, or other instances where deposition or permanent lowering of surface energy is desired, completely external behavior is neither necessary nor useful. Therefore, the nonpolar fraction of the lubricant cannot be increased beyond certain limits (about 20 times the molar volume of the functional group). Should exudation or deposition become a problem, recompounding should correct the polar-nonpolar balance. One route to this is to decrease the size of the noninteractive section of the lubricant.

18.3 Multifunctional Lubricants

In most processes, a balance among various lubricating functions is sought. In high shear mixing, either with an internal mixer or twin screw extruder, internal lubrication promotes flow of the polymer around other additives and rapid recombination, i.e., enhances distribution. Filler treatment with lubricant (within limits, see above) increases the rate of wetting by the polymer. The combination of these effects should not, however, lower shear stress to the point where dispersion (the reduction of particle agglomerates) is hindered. In such cases, it is preferable to accomplish the needed particle dispersion before lubricant addition. Sufficient external lubrication is required for the compound to remain more cohesive than adhesive and exit cleanly. With high shear mixing, it is therefore desirable that internal lubrication and filler surface treatment occur either very early in the cycle, or immediately after distribution of a filler that requires strong dispersive action, and that external lubrication take place toward the end of the cycle. This demands a lubricant or blend having a polar group of relatively broad solubility parameter and overall size great enough so that total compatibility decreases strongly with increasing temperature.

If, for example, low-molecular-weight PE wax is used as internal lubricant, at the dump temperature of an internal mixer, say 150–165° C, it will also function as an external lubricant (despite being of little use externally at low temperatures, e.g., as a mold release agent). Using this approach, it is feasible to include process release agents in highly adhesive compositions and deliver them more or less cleanly from mixing equipment without compromising their later function. With high shear mixing, the choice lies between a lubricant (or blend) providing sequential function, and the addition of internal lubricant early in the cycle, external lubricant near the end.

A blend to provide balanced lubrication in a given polyurethane should contain components with solubility parameter characteristics discussed above for individual functions. If simultaneous function is desired, the components should have closely similar overall solubility parameters. If sequential function is needed, the components should meet the above criteria for specific behavior but be less compatible among themselves.

It can be seen from the above analysis that the use and selection of lubricants differ little from other areas of formulation; that what is required is principally an accurate statement of the requirements of process and service conditions; that level and choice of lubricant cannot be separated from selection of other ingredients; and that a logical rationale for experimentation, if not perfected, at least is developed sufficiently to help in most cases.

18.4 UV Stabilizers

Protecting polyurethanes from the degradative attack of UV energy is an important requirement in most applications. This section reviews the processes of degradation and stabilization, the functions and major classes of stabilizers, and the products currently on the market.

Annually, 20 billion pounds of plastics go into products that regularly encounter outdoor exposure, and 3 billion pounds go into products exposed to fluorescent light and/or filtered daylight. UV stability is taking on a new urgency, becoming a more important property as plastics compete with more traditional materials. An understanding of why polyurethanes degrade under UV attack provides a starting point for understanding how they can be protected and selecting the product or products to do the job.

18.4.1 Mechanisms of Degradation

The mechanisms responsible for degradation depend on the polymer involved, but the basic culprits are chromophores—chemical groups or structural arrangements of atoms that absorb UV light, which is a source of energy. These chromophore groups may be part of the chemical structure of the polyurethane itself: azo ($N=N$), carbonyl ($C=C$), aromatic, or other double-bonded arrangements. The common denominator is the possibility of a shifting double bond.

On the other hand, there are many polymers that do not contain such groups as part of their chemical structure but may be equally susceptible to UV-induced degradation. Examples of these polymers include polyethylene, polypropylene, and vinyl. In these polymers, chromophoric groups may be present because of impurities in the monomer or other ingredients, aromatic or other double-bond-containing contaminants, residual-monomer groups ($C=C$), hydroperoxide, or carbonyl groups resulting from thermal oxidation during processing.

Whatever the cause or type of chromophore group, it absorbs UV energy. If this excess energy is not rapidly dissipated in some way—or reemitted as fluorescence or phosphorescence—it will begin to break the chemical bonds in the polymer's molecular chains, starting with the weakest link.

The rupture of chemical bonds can lead to two possible reactions.

1. It fragments the molecular chain, and the lower-molecular-weight fragments no longer exhibit the properties of the original polymer.
2. It generates free radicals, initiating and propagating a chain degradation reaction that results in chain scission and/or cross-linking.

Degradation accelerates with the initiation of new free radicals and/or chromophores, resulting in embrittlement, discoloration, and loss of physical properties. While the degradation mechanism is complex, the following is a simplified version of the process: UV energy breaks a bond, creating an excited molecule that forms a free radical.

- Step 1: $RH \rightarrow R^*H \rightarrow R\cdot + H\cdot$
  The free radical $R\cdot$ reacts with atmospheric oxygen to form a peroxyl radical $ROO\cdot$
- Step 2: $R\cdot + O_2 \rightarrow ROO\cdot$
  The free radical $ROO\cdot$ attacks the polymer chain at another point, forming a hydroperoxide $ROO\cdot$ and another free radical, R

- Step 3: ROO• + RH → ROOH + R•
  The hydroperoxide ROOH is unstable under UV and forms new radicals that accelerate the chain reaction.
- Step 4: ROOH → •R + •OH
- Step 5: RO• → chain scission
  In addition to step 4, the hydroperoxides can also form free radicals as:
  2ROOH → ROO• + H_2O

18.4.2 Mechanisms of Stabilization

The process of stabilization involves interrupting the degradation sequence somewhere short of chain scission and formation of other degradation products. Stabilization mechanisms can take several forms, but there are two main routes (perhaps they should be called *roadblocks*) as follows:

1. Inhibition of the initiation shown in step 1 by incorporating additives to screen UV energy, to preferentially absorb it, or to quench the excited state of the chromophore
2. Inhibition of the propagation process by incorporating additives that will undergo chemical reactions with the free radicals and hydroperoxides as soon as they are formed and produce harmless peroxide decomposition and free-radical termination

18.4.3 Absorbers and Screeners

Absorbers, the oldest type of UV stabilizer, inhibit initiation; they act to thwart step 1 of the degradation process. Products in this class compete with the polymer's chromophores for the UV energy, and because their absorptivity is orders of magnitude greater than the polymer's chromophores, they generally win the competition. Thus, the UV energy is preferentially absorbed by the UV absorber, which, in turn, converts it into a nondestructive form, infrared energy, which is dissipated harmlessly as heat. Figures 18.3 and 18.4 show the molecular structure of some light stabilizers/antioxidants.

To be effective in a given polymer, an absorber must have relatively high absorptivity at the UV wavelengths where the polymer is sensitive. It should be noted that transmission is the opposite of absorption. This means that near-zero transmission equates to near-total absorption. Absorbers are more effective in thicker cross sections than in thin cross sections such as film. This may be accounted for by differences in path length because, according to Beer's Law, absorption is proportional to path length as well as absorptivity and concentration of the absorbing species, as follows:

$$A = ecI_o = \log \frac{I_g}{I_t}$$

where A = absorption, e = absorptivity, c = concentration of absorbing species, I_g = path length, I_o = intensity of incident light, and I_t = intensity of transmitted light. This same law also indicates that the surface, where the path length would equal zero, may not be sufficiently protected. Some degradation may occur because of this.

Another problem is that some absorbers may have the secondary function of free-radical termination. To the extent that they are consumed in this process, their absorption activity decreases. Addition of a true free-radical terminator, which is regenerated rather than consumed, can conserve the absorber for its primary function and also protect against any degradation due to insufficient surface protection.

18.4.4 Pigments

In translucent or opaque applications, pigments can act as screening agents, absorbing or reflecting UV light. Carbon black is very effective; it absorbs over the entire UV and visible range, can be used at low

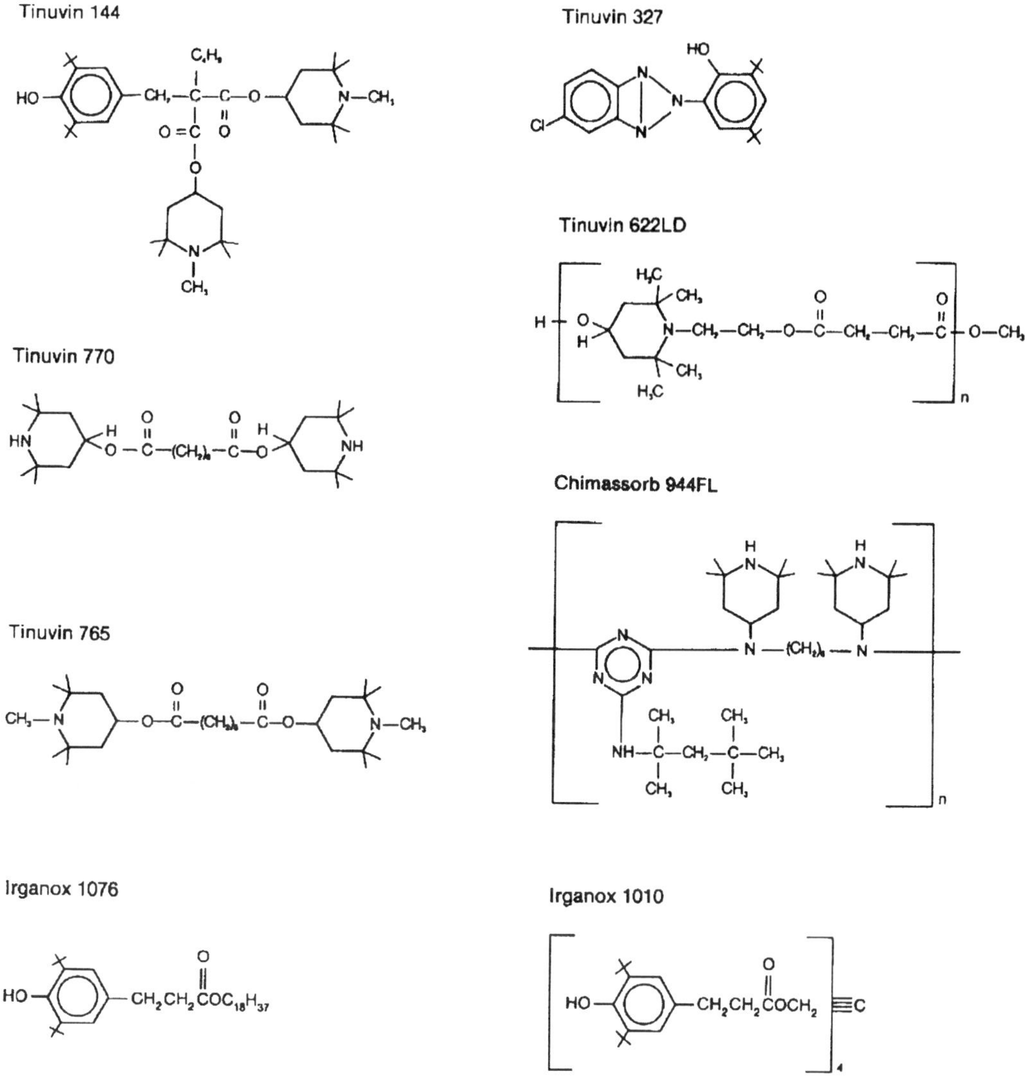

Tinuvin 144

Tinuvin 327

Tinuvin 770

Tinuvin 622LD

Chimassorb 944FL

Tinuvin 765

Irganox 1076

Irganox 1010

Figure 18.3 Light stabilizers and antioxidants. *Courtesy of Ciba-Geigy.*

concentrations (1 or 2%), and may also act as a free-radical scavenger. The black color imparted, of course, limits its application.

18.4.5 Benzophenones

Substituted 2-hydroxy-4-alkoxy benzophenones have moderate to strong absorption in the 230–390-nm range. The type of substituent groups determines both compatibility and volatility, which in turn determine the polymers and processing operations for which these absorbers are suitable. Original products had fairly high volatility. Today, however, higher-alkyl substituents such as octyl (C_8), decyl (C_{10}), and dodecyl (C_{12}) groups are used, which exhibit reduced volatility during processing and increased compatibility with polyurethanes.

The structure below shows a higher-alkyl type. Neglecting the alkyl substituent, the basic structure with a carbonyl group and two aromatic groups gives an idea of how this class competes with a polymer's chromophores to preferentially absorb UV energy. High-molecular-weight monomeric and polymeric

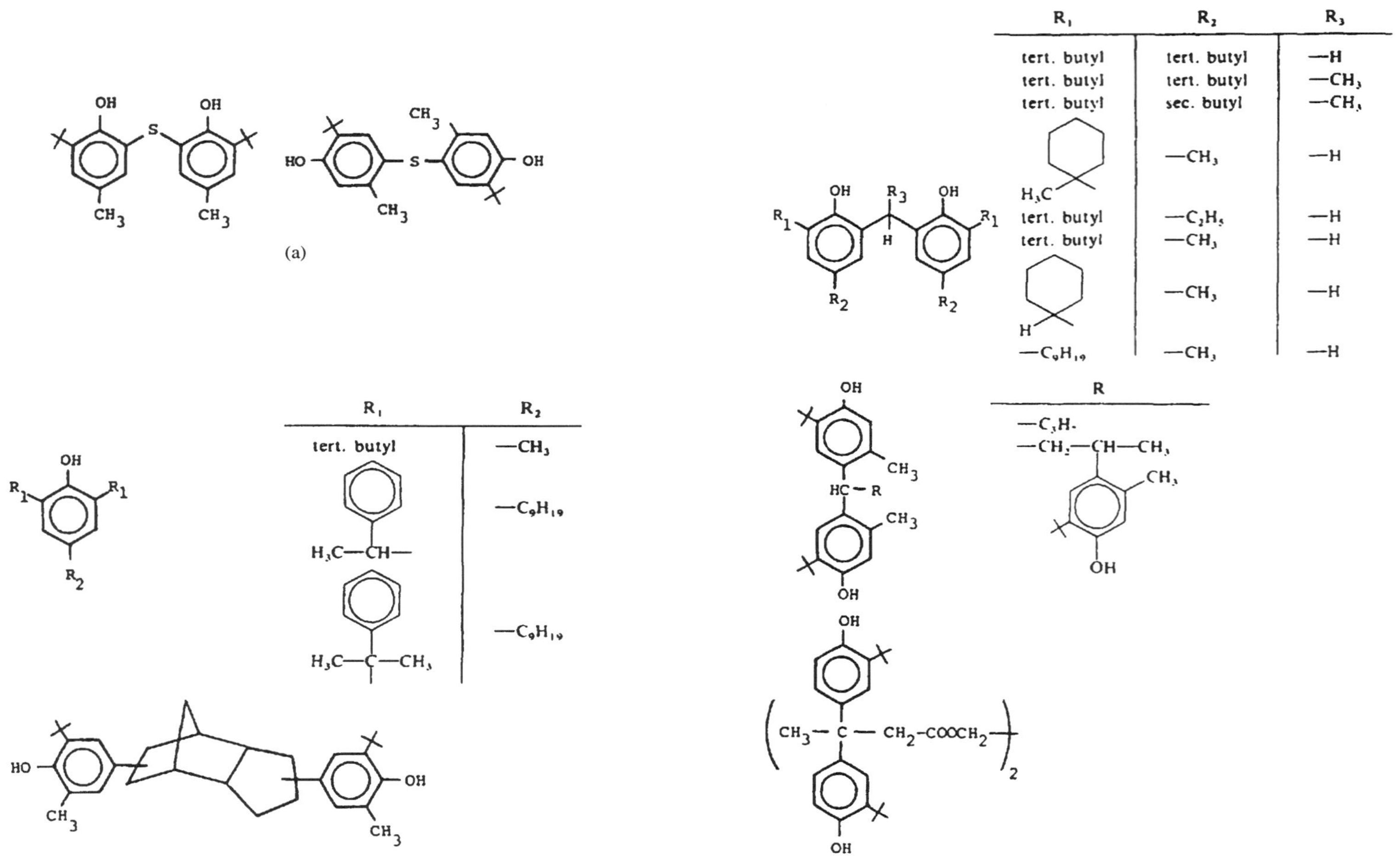
OH
S
CH3
HO
(a)
R1
R2
R3
H
R1 | R2 | R3
tert. butyl | tert. butyl | —H
tert. butyl | tert. butyl | —CH3
tert. butyl | sec. butyl | —CH3
H3C | —CH3 | —H
tert. butyl | —C2H5 | —H
tert. butyl | —CH3 | —H
H | —CH3 | —H
—C9H19 | —CH3 | —H
HC—R
R
—C3H7
—CH2—CH—CH3
CH3—C—CH2-COOCH2
2
R1 | R2
tert. butyl | —CH3
H3C—CH— | —C9H19
H3C—C—CH3 | —C9H19

(c) Alkylphenols

(b) Alkylidene-bisphenols

Figure 18.4 Example structures of important antioxidants *(continues)*

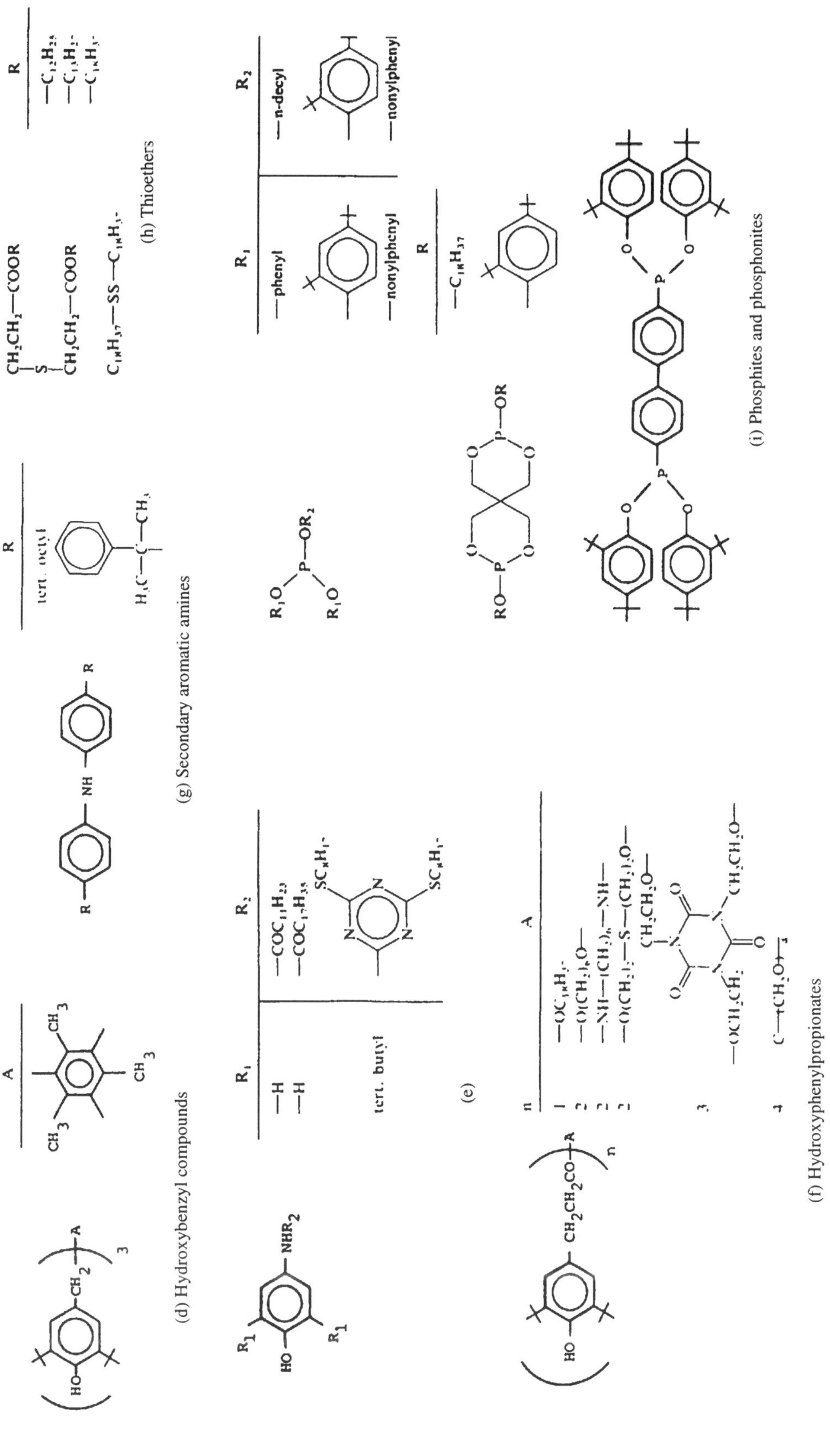

Figure 18.4 Example structures of important antioxidants

versions are now available that are suitable for use with high-temperature-processing engineering thermoplastics.

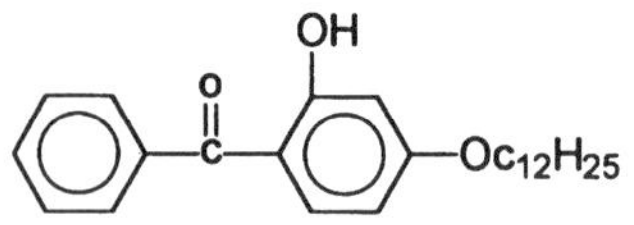

Benzophenone with higher alkyl substituent exhibits reduced volatility, increased compatibility with polyurethanes (4-dodecyloxy-2-hydroxybenzophenone, Eastman Inhibitor DOBP).

Once the UV energy is absorbed, part of the absorber molecule is in an excited state and a rearranged ring structure is formed. It subsequently returns to its original form and gives up the acquired energy as heat. Possible limitations here are that benzophenones are more effective in thick sections than in thin films, and that occasionally objectionable color levels may be developed during compounding, processing, or weathering.

18.4.6 Benzotriazoles

Substituted hydroxy-benzotriazoles are mostly derivatives of 2(2'-hydroxy phenyl) benzotriazole and the 5'-t-octylphenyl analog, as shown below:

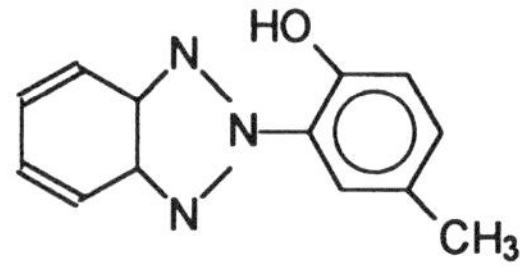

Substituted hydroxyphenyl benzotriazole, resonance hybrid [2(2'-hydroxy-5'-methylphenyl) benxotriazole] Tinuvin P (Ciba-Geigy).

These products exhibit effective UV absorption at 280 to 390 nm (somewhat higher than benzophenones) with a fairly sharp limit close to the visible region (less tendency to discolor the polymer). Again, substituent groups determine compatibility and volatility, and high-molecular-weight monomeric and polymeric versions are available for polyurethane thermoplastics. Benzotriazoles are more expensive than benzophenones, but they exhibit good initial color and long-term color stability, with little tendency to discolor the polymer because of the sharp cutoff of absorption near the visible range. Possible limitations to this category involve thin film and sheeting applications.

18.4.7 Oxanilides

Substituted oxanilides provide absorption in the lower portion of the UV region, from 280 to 320 nm, the range of maximum polymer sensitivity. This absorption range also makes them suitable for use with fluorescent pigments, where they may effect considerable cost savings.

They exhibit very low color and low volatility and are suitable for high-temperature processing applications such as thermoplastic polyurethanes.

18.4.8 Formamidine

This is a new class of absorber, with the structure shown in the structure below. It is a photostable product that provides broad-spectrum protection from both UV-A (320 nm) and UV-B (290 to 320 nm) radiation. This absorber is recommended for ABS, PVC, polyolefins, acrylics, urethanes, and other polymer systems, and it can be used in conjunction with stabilizers having radical-terminating/peroxide-decomposition functionality (hindered amines).

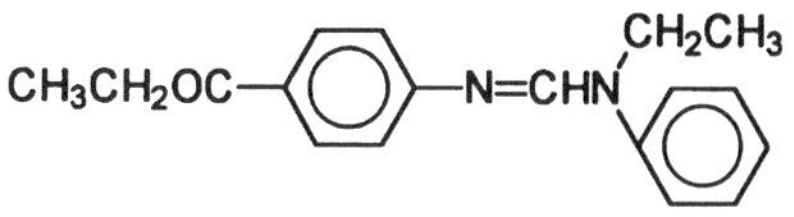

N-(ethoxycarbonylphenyl)-N'-phenylformamidine,
Givsorb UV-2 ultraviolet absorber, Givaudan.

18.5   Radical Scavengers, Peroxide Decomposers

The mechanism of hydroperoxide decomposition, a means of inhibiting propagation, acts on highly unstable hydroperoxide ROOH formed in step 3 of the degradation sequence before it can be decomposed by heat or light to radicals.

Essentially, what this mechanism accomplishes is the prevention of step 4, where the ROOH would form two radicals RO• and •OH, or alternately, where two ROOH molecules would form two ROO• radicals and repeat from step 3. It accomplishes this by reducing peroxides to harmless products (such as ROH); this is similar to the function of secondary antioxidants.

Phosphites and thiodipropionate secondary antioxidants, organic nickel compounds, and carbon blacks have been cited as peroxide decomposers. The antioxidant types would be consumed too rapidly in such service to be used alone. And to the extent that the absorber and quencher types are consumed in such reactions, their primary function suffers.

However, peroxide decomposers, acting after step 3 of the assault, are somewhat late in that radicals have already been formed in steps 1 (R•) and 2 (ROO•). Actually, no type covered so far is able to completely prevent initiation/propagation. Even with absorbers, quenchers, and decomposers used in combination, free radicals are almost always generated. This leads us to the fourth mechanism for controlling UV-induced degradation—free-radical scavenging and termination.

Some absorbers and quenchers may serve in this function. Aryl esters can work to terminate the radicals formed by peroxide decomposition. Benzoates have been reported to be particularly effective in highly pigmented systems in combination with phosphites. Primary antioxidants— phenolics and amines—perform this function, but would be too rapidly consumed for UV service. The best alternatives are the fairly recent hindered amine types.

18.5.1   Hindered Amines

Characteristic of the hindered-amine type is the tetramethyl piperidine structure; a representative hindered-amine structure is shown below. Although hindered amines may function as excited-state quenchers or peroxide decomposers, their major function is free-radical scavenging and termination.

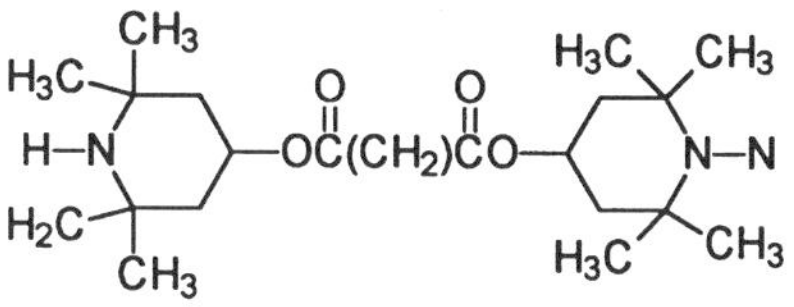

Structure of a hindered amine (Tinuvin 770, Ciba-Geigy)

As radical terminators, hindered amines react with the R• radical formed in step 1 to preclude step 2, and with the ROO• radical formed in step 2 to preclude step 3. The nitroxyl radical may be produced initially through a hydroperoxide decomposition process; it acts as a radical scavenger, being regenerated in the process. This cyclic regeneration represents a real advance in UV stabilization.

Unlike absorbers, the hindered amines provide surface protection and work in thin sections; unlike quencher types, they do not impart color. They can be used alone or in conjunction with absorbers or quenchers to maximize UV stabilization. They are suitable for a wide range of polyurethane applications. The limitation here is that they are not to be used with base-sensitive resins.

Polymeric versions have been introduced that exhibit improved high-temperature processing behavior, better extraction resistance, lower migration, lower volatility, and improved long-term heat resistance.

Selecting a stabilizer for a given polymer and application is not simple. Other ingredients in the formulation, and their concentrations, must be taken into account; the chemical nature of pigments, heat stabilizers, and antioxidants is an important factor. Processing temperatures are another. These considerations are normally covered in suppliers' literature, which should be perused carefully as a source of data on blends of stabilizer types for maximizing performance.

Another area of current emphasis is testing. Evaluating long-term UV stability can quite obviously be a long-term project. Of interest is how it can be accelerated by judicious selection of exposure sites or the use of environmental test chambers. Outdoor exposure of an actual product under end-use type conditions remains the definitive test, but accelerated-aging tests are useful, especially if correlations have been established between the testing device and outdoor exposure for the polymer.

18.6   Antimicrobials

The innovative use of antimicrobials in urethanes can be a benefit in creating new market opportunities. Protection against degradation due to microorganisms can be engineered into many products. This benefit can be marketed as a differentiation among products in the marketplace or between old and new products, all done without a major capital expense.

The differentiation can generate substantial value-added marketing claims such as stain or odor reduction or hygienic claims such as quantitative bacterial-kill results or resistance to spores that cause mold allergies. Figure 18.5 presents the biological response of a typical antimicrobial agent.

18.6.1 Cosmetic Degradation

Structural degradation can be defined as any significant change in the physical, chemical, or electrical properties of a product. But there is another type of degradation-cosmetic degradation. This degradation

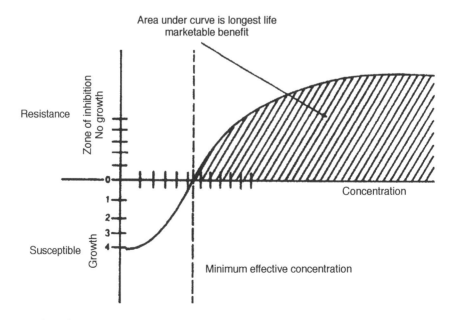

Figure 18.5 Biological response vs. biocide concentration.

is a significant change in the color, odor, or appearance of a product. Such degradation is often a result of attack by microorganisms.

Polyester urethanes are known to be susceptible to biological attack, because the ester linkage in the polymer is easily broken by reactive enzymes. Polyether urethanes were thought to be inert, but the formulating options of the urethane chemist have expanded so that polyether urethanes can now be very susceptible to biological degradation. In dealing with urethane chemistry, cosmetic degradation can be more prevalent than structural degradation.

18.6.2   The Offending Organisms

Algae, bacteria, fungi, and actinomycetes are all organisms that produce polymer deterioration. Green algae, because it produces its own chlorophyll, requires water and sunlight.

Bacteria are not pigmented, and they cannot be seen by the naked eye. Having a ball-like structure, fungal organisms are visible to the naked eye and come in a variety of colors. While bacteria require an aqueous phase, fungi require only moisture.

Finally, actinomycetes are a type of soil organism. Depending on who wrote the biology book, they could be considered very large bacteria or very small fungi. This class of organisms exudes a dye as part of its metabolic by-products. How do these organisms deteriorate plastic parts? As part of their life cycle, organisms excrete enzymes—reactive chemicals that can break up a long-chain molecule into short pieces. These short pieces can be either absorbed directly through the cell wall or be brought back into the cell by other enzymes or proteins. In any case, the cell receives its food, and growth proceeds. In polyurethane formulations, unreacted polyols, silicone surfactants, and mold-release agents can be utilized as food sources.

18.6.3   Testing Methods

Microbiological testing is a very disciplined science. Many methods are used to determine if a material is susceptible to degradation or is resistant to biological attack. The terms *susceptible* and *resistant* are key words on which we need to focus our attention.

Several microbiological test methods are used in the laboratory. For those products that will find their useful life in either fresh or salt water, there is an algal-resistance test (ASTM G29). For such items as pond and pit liners and single-ply roof membranes, soil burial is used in conjunction with physical property tests. Products that will find their useful life outdoors, such as marine upholstery and cushions, can be exposed at commercial test establishments in Florida for an extended period of time.

Many military specifications are built around humidity-cabinet exposure, where a product can be examined in a growth environment ideal for fungal organisms. This type of test is currently run from 30 to 120 days. Extended times are needed for this type of testing, because it takes the organisms longer to get started and to do their dirty work.

The agar-plate method, or petri-dish method, is probably the most widely used test procedure. Under controlled laboratory conditions, it can be determined if a material is susceptible or resistant to the growth of a single or multitude of organisms. This method is relatively rapid, depending on the incubation time set forth in the test protocols, and usually requires only a small sample.

Table 18.5 sets out incubation times of various organisms commonly used with agar-plate methods. The microbiologist is dealing with living organisms and must allow the proper time for their growth. Bacteria grow in 24 hr. Fungal organisms take up to three weeks.

18.6.4   Agar-Plate Ratings

ASTM G21-80, *Standard Practice for Determining Resistance of Synthetic Polymeric Materials to Fungi,* is a mineral-salt (non-nutrient agar) test. There is no carbon source (dextrose) in the agar. If the organisms are going to grow, they must do so on the sample being tested. The untreated sample would be rated as heavy growth. The treated sample is clean, with no growth.

TABLE 18.5 Petri-Dish Test Methods

Organism Time	Method	Incubation
Bacteria		
S. aureus	Qualitative	24 hr
K.pneumoniae	ATCC Method 100	24 hr
	N.Y. State Mattress Test	
Fungi		
A. niger	ASTMG	21 days
P. funiculosum	Mineral salts agar	
C. globosum		
G. virens days	Nutrient agar	14
A. pullulans		
S. reticulum days	Nutrient agar	14

18.6.5 Protection and Prevention

Biological growths begin with contamination. The question naturally arises: *Will a biocide protect a product when it has been contaminated?* To determine the answer, a foam was inoculated with a nutrient inoculum; that is, dextrose was put in with the organisms to give them a head start. The result indicates that the use of a biocide in polyurethane formulations can be either a necessity or an option, as shown in Table 18.6. Microbiological resistance is a necessity if it is truly a performance requirement. As an option, microbiological resistance can be labeled a performance or value-added benefit.

TABLE 18.6 Technical/Marketing Perception of Marketplace

Biocide necessity	Biocide Optional
Carpet underlay	Coated fabrics
Insulation	Flotation/buoyancy
Comfort cushions	Seals/gaskets
Pillows	Filters
Mattresses	Coatings
Impact dissipaters	Adhesives
Vibration damping	Elastomers
Water-borne	Caulking compounds

It should be noted that, if users of the biocide wish to make biocidal claims in their marketing, certain formalities must be followed to comply with EPA regulations. Briefly, claims must be written around the concepts of resistance, prevention, or preclusion of growth of organisms. The biocide supplier makes the bactericidal or fungicidal claims and can assist in proper wording of product claims.

A demonstrable success in this industry has been the profitable introduction of antimicrobials in the rug underlay or carpet-cushion market. There are many other applications where a benefit can be demonstrated in the *top-of-the-line* option applications, and many corporations use these benefits to increase market penetration.

18.7   Colorants

In many polyurethane applications, dyes and pigments are used as colorants. A colorant is generally classified as a dye if it is soluble in the medium used. Pigments, on the other hand, are generally considered insoluble. Dyes are transparent and generally stronger and brighter than pigments. Pigments can vary in transparency/opacity, depending on their particle size and index of refraction difference between the pigment and plastic media. Compared to pigments, dyes are usually less light fast and are more susceptible to migration and bleed problems.

Pigments are classified into organic and inorganic types. Most organic pigments contain the same chromophore groups as organic dyes. Improper processing conditions with organic pigments may result in many of the performance deficiencies characteristic of dyes, namely migration and bleed.

Inorganic pigments are less subject to migration and bleed; however, they are less coloristically intense, but display good opacity, heat stability and light fastness.

At least eight steps must be performed before final chemical colorant type selection can be made on a total colorant system for polyurethanes.

1. The colorant's physical form must be chosen. The forms available are pure colorant as a fine dry powder, color concentrate where a high concentration of colorant has been dispersed in a solid polymer system, color concentrate where the colorant has been dispersed in a compatible solid nonpolymer system, and a color concentrate where the colorant has been dispersed in a liquid vehicle system compatible with the plastic being colored.

2. The physical form of the polymer must be considered. This ranges from powders, flakes, or beads, to pellet shapes of many configurations and sizes such as spherical pellets to small cylinders. The polymer may also exist as a monomer liquid or even an emulsion.

3. Additive materials present other than colorants must be considered for their impact on colorant performance.

4. Colorant/polymer loading, conditioning, and distribution before the dispersion step must be determined. The primary consideration is the pigment loading necessary to achieve the desired color characteristics or the maximum loading possible before colorant and/or polymer properties are adversely affected. A conditioning step may be required such as drying the polymer and/or the colorant to remove moisture. High-shear or high-intensity dry blending may alter the colorant's visual and/or physical properties.

5. The color-compounding or dispersion step is critical. Common basic dispersion techniques are single- and twin-screw extrusion and two- or three-roll milling, depending on whether the polymer is a meltable solid or a liquid. Banburys or continuous mixers may be employed in conjunction with two-roll mills and/or extruders.

6. Part fabrication is a factor in the colorant/polymer mix. Whether the part is injection molded, extruded, blown, compression molded, or heat or chemical reacted has considerable effect on the fabricated part. It is important to note that, in a number of fabrication procedures, steps 5 and 6 can be effectively combined.

7. The end use or environment of the finished product must be considered. A good colorant selection design should provide a system that will equal or most likely outlive the polymer itself.

8. Test considerations and parameters must be clearly determined and stated. These parameters are not only considered during design and fabrication of a colored product, but also must be accounted for during the lifetime of the product. Fabrication considerations include, but are not limited to, the dispersion attained, heat stability required for fabrication, physical abuse sustained by the colorant, and its effect on polymer physicals. Product life considerations include light fastness, heat stability as required by the product, resistance to solvent attack, migration or bleed resistance, and long-term effects on polymer physicals. The last and overriding consideration is one of economics.

General performance and color properties of colorants must also be carefully evaluated. The initial considerations are hue, chroma strength, opacity, and cost. Cleanliness of the hue when blended with other colorants must be considered. The chroma of specific colorants is vital. Inherent color strength will be important for a number of reasons. Tolerance to total loading of the polymer may have to be considered. Dispersibility of colorants is always a factor. Stability of colorants to chemical attack, thermal degradation, and radiation usually are considered as a package. Most colorants exhibiting good chemical resistance will most likely also exhibit good heat stability and light fastness. Migration usually is a problem with partial organic types and is involved with partial solubility of the colorant in some component of the polymer system. Some organic colorants may only become soluble at high processing temperatures. Safety and/or toxicity parameters must be thoroughly analyzed and met from conception, through fabrication, and to the environment of the final product. Compromises may have to be made to arrive at a commercial product.

18.7.1 Opacifying Colorants

Titanium Dioxide

As an opacifier and a white colorant, titanium dioxide (TiO_2) is the most widely used appearance additive. No other material has such brightness with high opacity. Two crystal forms are available, anatase and rutile. The rutile type has the larger market due to its more efficient opacifying power, but it does not show the blue or stark white tone like the anatase type, which is a chalking TiO_2. Chalking keeps the part looking clean and white as the aged outer surface (a microthin layer) erodes and falls away. In most plastics applications, chalking is not a desirable feature. In addition, encapsulated varieties of TiO_2 are available. The encapsulation reduces the potential for surface activity of the pigment particle, resulting in a much more stable pigment. These types find great utility where exceptional light fastness for outdoor use is a necessity.

Carbon Black

Domestic carbon blacks are furnace-type blacks. Channel black, because of environmental problems, no longer is a major commercial item. Furnace blacks' hue range extends from deep-blue, jet-black to a yellow-red, dull-toned black. The jet-black property is produced by very small particles. This size and the tendency to agglomerate usually make them very difficult to wet out and disperse in many plastics. Carbon blacks are the most opaque appearance additives and have a very low cost. These blacks also provide UV stability to many polymers, particularly the polyolefins. Carbon blacks may also affect cure rates of thermoset resins.

18.7.2 Inorganic Colorants

These colorants can be divided into a number of categories that are all bleed resistant.

- Synthetic inorganic complex oxides
- Titanates
- Chromates
- Molybdates
- Encapsulated chromates and molybdates
- Sulfides of cadmium
- Iron oxides
- Chromium oxides
- Ferriferrocyanide

Synthetic Inorganic Complex Oxides

These are colorants with exceptional chemical and physical stability. Their hues range from blue to green, beige to brown and black. These colorants are structurally similar to those used in the ceramic industry. They possess outstanding heat and light stability and bleed and chemical resistance. They are usually opaque but are of lower chroma and are tinctorially weaker than the organic-type colorants. Dispersibility ranges from easy to difficult. These complex oxides have moderate to high cost and are used extensively in architectural products and in some very high temperature engineering thermoplastics.

Titanate

Titanate colorants cover the yellow to beiges and blue to green spectrum. This type of colorant is formed by calcination of TiO_2 with small amounts of other oxides or oxide precursors. These select metal oxides determine the various shades obtainable. The result is a very light- and heat-stable colorant at a low to moderate cost. Inertness to almost any kind of attack provides virtually no restrictions for the use of these titanate colorants.

Chromate

Chromate colorants cover a range of yellows that are clean, intense, and opaque. The chromates contain lead as well as chromium, which may lead to restrictions in their use. However, when used in selected applications, and with some precautions during processing, they fill a need where economy is vital and where light and heat stability and chemical resistance do not pose overwhelming problems.

Molybdate

Molybdate colorants cover a range of oranges that are clean, intense, and opaque. The molybdates contain molybdenum as well as lead and chromium. Their properties and usage are nearly the same as in the chromate-type yellow colorants, except they are moderately more opaque, weaker, and more light fast.

Encapsulated Chromates and Molybdates

These colorants are modified forms of the regular chromates and molybdates. The primary pigment particle is encapsulated in an impervious coating. The resulting pigment exhibits improved performance properties—heat stability approaching 660° F, light fastness much improved over regular types, and improved chemical resistance to easily satisfy conditions only marginally met by regular types. Encapsulated chromates and molybdates are shear sensitive. Processing conditions must be monitored very closely. A fractured or an abraded particle performs only as well as its regular-type counterpart.

Sulfides of Cadmiums

Cadmium sulfides and cadmium sulfoselenides are colorants. Their hues can be modified by the incorporation of zinc, mercury, or selenium metals. In full-tone exterior applications, or where the polymer system protects the colorant from moisture and oxygen, they will exhibit outstanding durability. However, they are not recommended for exterior tint applications. The cadmium sulfide and cadmium sulfoselenides cost ranges from moderate to high.

Iron Oxide

Iron oxide colorants provide a hue range from yellow to red, tan to brown and black. Iron oxides are not very bright or intense, but their opacity is very high. These oxides are low in cost with moderate properties. The heat stability of the yellow oxides limits their processing to under 350° F. The red oxides can be processed about 50° F higher. The iron oxides are available as both synthetic and natural products. The use of natural materials is very low due to the chemical composition varying widely, which influences their performance properties.

Chromium Oxide

Chromium oxide colorants provide blue to yellow shade greens with good opacity at low cost. The blue shades as hydrated products are cleaner and less heat stable than the pure oxide yellow shades. The pure chromium oxide exhibits good weathering properties. Dispersion ranges from poor to good.

Iron Blue

Ferriferrocyanides are the typical iron-blue colorants used predominantly in the ink and coatings industry. These blues are hard to disperse and have low to moderate heat and light stability, but they are low in cost. Their main area of application in plastics is in trash bag films where performance duration is short and economics is of primary importance.

18.7.3 Organic Colorants

Organics are generally very clean, bright, intense, and transparent. Their range of hues encompasses the entire visible spectrum. The properties of these colorants extend over a wide range, and their costs vary considerably. The important categories include

1. Disazo yellows, oranges, and reds
2. Monoazo yellows, oranges, reds, and browns
3. Disazo condensation yellows, oranges, reds, violets, and browns
4. Quinacridone reds and magenta
5. Carbazole dioxazine violet
6. Vat yellows, oranges, blues, violets, red, and browns
7. Phthalocyanine greens and blues
8. Isoindolinones yellow, oranges, and reds

Disazo

This colorant group includes the diarylide yellows and oranges and the pyrazolone oranges and reds. The diarylides that have more complex chemical structures (at a higher cost) find wide use due to improved bleed resistance, heat stability, and light fastness. The simpler, less stable diarylides find limited applications in polyurethanes. The pyrazolone oranges and reds exhibit good light fastness in full tone. The pyrazolone colorants exhibit moderate heat stability.

Monoazo colorants include the Hansa™ (GAF Corp.) yellows and oranges, toluidine and para reds, naphthol reds and browns, nickel azo yellows, red lake Cs, lithols and lithol rubines, 2B reds, and pigment scarlet, plus other more complex types.

Of the monoazo grouping, the Hansa yellows and oranges, and the toluidine and para reds, have little application in thermoplastics resins. They find limited use in thermoset resins. The naphthol reds and brown are noted for low cost and moderate heat stability and light fastness. As an extended colorant, the naphthols may be used for their chemical resistance in floor covering applications. The nickel azo yellow, *green gold,* is the most light-fast azo colorant in both full tone and tint. This yellow also has good heat stability but can be destroyed under acidic processing conditions. Its cost is moderate. The red lake C colorant is a light shade red with good heat stability but poor light fastness. It usually is found in products having a short life or used to fortify and brighten inorganic reds. Its cost is low.

Next in the monoazo colorant group are the lithol reds and the lithol rubines. These reds find limited use in plastics because of modest heat stability. The manganese salt lithol rubine is not recommended due to the detrimental effect the metal salt may have on polymer stability. The lithol reds and the lithol rubines are low in cost, high in strength, and fair in light fastness and bleed resistance. The calcium, barium, and strontium Red 2B colorants are extensively used in plastics. They have moderate cost, high strength, good bleed resistance, and good heat stability. Their light-fast properties in tint are poor. A manganese salt Red 2B exists but is not recommended. Pigment scarlet is a unique, clean, blue shade red with moderate cost and heat stability superior to that of Red 2B. It is moderately difficult to disperse and, unlike many organic colorants, it does not bronze or bleed.

Benzimidazolone

Benzimidazolone pigments range from very transparent yellows through orange to red colorants and even browns. These pigments exhibit very good heat stability, good light fastness, and exceptional chemical resistance. Their chemical resistance leads to their use in polyolefins, vinyls, and styrenics where the end products require great resistance to chemical attack. Costs range from moderate to high.

Disazo Condensation

Disazo condensation yellows, oranges, reds, violets, and browns represent a new generation of colorants that provide clean, intense, bleed-free properties. These are noted for heat and light stability in both full tone and tint applications. Some difficulties may be encountered with their dispersion. The high cost of those colorants may limit their use to premium plastics products. In spite of their cost, the disazo condensation yellows and reds find widespread use in polyolefins, ABS, and vinyls.

Quinacridone

Quinacridone reds and magentas provide a high-cost, clean, intense color. Dispersion may be a problem.

Carbazole Dioxazine

Carbazole dioxazine is one of the strongest violets available. It also provides a clean, transparent, bleed-resistant colorant with light stability and light-fast properties. At very low concentrations, it frequently is used to tone or correct the undertone color of white plastics. Its cost is high.

Vats

The vat yellow, orange, blue, violet, red, and brown colorants are known in the marketplace as

- anthrapyrimidine yellow
- flavanthrone yellow
- perionone orange
- pyranthone orange
- indanthrone blue
- iso violanthrone violet
- perylene red
- bordeaux

They exhibit a broad range of properties; as a result, these colorants and their properties must be closely matched to each specific polymer. All vats show good chemical resistance but vary widely on ease of dispersion.

Phthalocyanine

Phthalocyanine greens and blues provide very clean, intense, transparent colorants. The blues are produced in two crystal forms. The red shade blues are the alpha forms and are unstable at high temperatures. Excessive heat will cause the alpha form to convert to the beta form, which will result in a shift to a green shade. The green shade phthalocyanine blue, the beta form, and phthalocyanine green exhibit good heat stability and light fastness. Economics are good and these colorants are widely used in spite of some dispersion problems. The blues, in particular, must be used with caution in mass tone because they may induce warpage in olefin polymer systems.

Isoindolinone

Isoindolinone colorants are a group of bright and clean yellow, orange, and red colorants. These colorants exhibit heat stability, light fastness, and chemical and bleed resistance. In spite of their high cost, they find significant use in polyolefins, vinyls, cellulosics, and polyester gel coats. Their dispersion properties and higher cost limit their use to premium products.

18.8 Illustrative Patents

18.8.1 Antioxidants

Currently, 2,6-di-tert-butyl-4-methylphenol (*butylated hydroxytoluene, BHT*) is generally used in practice for the above-mentioned purposes, but the improvements in stabilization achieved with it are not satisfactory. Combinations of specific antioxidants have also been proposed, such as mixtures of sterically hindered phenols (see, for example, U.S. Pat. 3,280,049, U.S. Pat. 4,007,230, and U.S. Pat. 3,494,880) or mixtures of sterically hindered phenols with specific diphenylamines (see, for example, U.S. Pat. 4,070,304, U.S. Pat. 4,265,783, U.S. Pat. 4,275,173, and U.S. Pat. 4,021,385). The stabilizers and/or mixtures of stabilizers proposed in those specifications do not, however, meet the stringent demands made on them in practice. Benzofuranone derivatives are already known as stabilizers for various organic materials (e.g., U.S. Pat. 4,611,016).

Michaelis[1] disclosed an invention related to polyether polyol and polyurethane compositions that are protected especially against oxidation and against the undesired phenomenon of core scorching during the preparation of polyurethane foam by the presence of at least one benzofuran-2-one derivative and (at least) one further compound from the group of the phenolic antioxidants and/or from the group of the amino antioxidants of the secondary amine type, as well as to the use of the said derivatives and compounds as additives for the prevention of the said phenomena, and to a process for the preparation of polyurethane using those derivatives and compounds.

Surprisingly, Michaelis discovered that it is possible to prepare compositions that are very effectively protected against oxidation by adding to polyether polyols a combination of at least one benzofuran-2-one derivative with (at least) one further antioxidant from the group of the phenolic antioxidants and/or the amino antioxidants of the secondary amine type, from which compositions it is possible to produce polyurethane foams without the interference of core scorching.

The invention thus relates to compositions comprising (1) a polyether polyol or mixtures of such polyols, and (2) at least one benzofuranone derivative of formula I.

$$\begin{array}{c} -O-CH_2 \\ -OCH_2-\overset{|}{\underset{|}{C}}-CH_2OH \\ CH_2OH \end{array}$$

$$-OCH_2CH=CHCH_2O- \quad \text{or} \quad -OCH_2C\equiv CCH_2O-$$

N-(ethoxycarbonylphenyl)-N'-phenylformamidine,
Givsorb UV-2 ultraviolet absorber, Givaudan.

18.8.2 Internal Mold Release Agents

In an active hydrogen-containing B-side composition for reaction with a polyisocyanate containing A-side composition to make a polyurethane or polyurethaneurea elastomer by reaction injection molding, the improvement that comprises a mold release composition consisting essentially of (1) the reaction product of a mixture of a carboxylic acid, a tertiary amine, and a reactive epoxide, and (2) a carboxy functional siloxane.

Reaction injection molding (RIM) is a versatile process by which elastomeric and thermoset parts can be fabricated. The RIM process involves high pressure impingement mixing of a polyisocyanate stream (A-side) and an active-hydrogen containing isocyanate-reactive stream (B-side) followed by immediate injection into the closed mold. The primary appeal of this process lies in its inherently high productivity. One factor that limits productivity, however, is the necessity to spray the molds with external mold release prior to each injection. This time-consuming task often has a negative environmental impact.

This difficulty can be overcome by the incorporation of an internal release agent into the formulation via one of the two streams to significantly increase the number of molding cycles that can be accomplished between mold release sprayings.

The use of metallic soaps as release agents has been known for a long time. Zinc stearate, in particular, is known to be soluble in aliphatic amines, such as the polyether polyamines and ethylenediamine-initiated polyols. This is the basis for its use as an internal mold release (IMR) agent in RIM. If zinc stearate is simply dispersed as a fine powder in polyol blends, it does not dissolve and does not act as a release agent. Various patents show that zinc soaps can be compatibilized or dissolved in polyol blends with amines, enamines, ketimines, or salts of amidines or guanidines, and that excellent releasability of the subsequent RIM parts will result.

While the IMR approach is commercially applied, there remains significant shortcomings to the currently available IMR systems. The amine-solublized metallic soaps, which are most commonly used in this application, have been implicated in reactivity and/or physical property deficiencies for the RIM elastomers in which they are used. Furthermore, the high melting points and limited solubilities of the metallic soaps make them prone to precipitation in the RIM processing equipment, requiring regular replacement of the piping.

The search for IMR agents that are liquids, without the possibility of solidifying, led to the development of special silicone fluids for this application. U.S. Pat. 4,076,695 discloses certain carboxy-functional silicone fluids as IMR agents for RIM, including Dow Corning's commercial carboxy-functional silicone fluid Q2-7119.

In general, acids have a deleterious effect on the green strength of aryldiamine-extended polyurethaneurea RIM systems due to a general deactivation of the tin catalyst. Thus, higher than normal levels of tin catalysts are needed when acids are present. Due to the sulfur atom, alpha to the carbonyl group, Q2-7119 is a much stronger acid than a typical fatty acid, such as lauric acid.

Therefore, when T-12 (dibutyltin dilaurate) and Q2-7119 are in the same polyol blend, the equilibrium reaction involving the two components leads to a gelled silicone salt. This gelation results from a cross-linking reaction between the trifunctional silicone and the difunctional tin salt. The result is that the system exhibits extremely poor green strength, which cannot be corrected by the addition of more tin catalyst.

Attempts to dissolve this problem include the following:

- U.S. Pat. 4,379,100 discloses the use of a three-stream approach to RIM molding where the Q2-7119 is delivered dispersed in polyol containing no tin catalyst. The other two streams are the normal A and B sides of RIM technology. The A side is isocyanate and the B side is a blend of polyol, diamine chain extender, surfactants, and tin and amine catalysts.

- U.S. Pat. 4,420,570 discloses that the tin catalyst can be placed in the A side. Gelation is avoided, but high levels of catalysts are still needed for adequate green strength. Furthermore, placing the tin catalyst in the isocyanate increases the moisture sensitivity and susceptibility to side reactions, such as allophonate formation, leading to gelation of the isocyanate.

- U.S. Pat. 4,396,729 discloses replacing the polyether polyol and the tin catalyst with polyether polyamines that require no tin catalyst. The result is polyurea RIM, and Q2-7119 can be used with no chemical modification or three-stream approach.

- U.S. Pat. 4,472,341 discloses that the acid groups on Q2-7119 can be converted to amides by reaction with amines or to esters by reaction with alcohols or epoxides yielding nonacidic IMR silicones. These materials have been shown to cause paintability problems. In addition, they have been seen to interfere with polyol nucleation so that low part densities cannot be achieved. In extreme cases, large voids are found in the parts due to coalescence of bubbles.

- U.S. Pat. 4,477,366 discloses that Q2-7119 can be dispersed on the isocyanate side by using a nonisocyanate-reactive silicone as a dispersing and inhibiting agent.

- U.S. Pat. 4,487,912 discloses the use of the reaction products of fatty cyclic anhydrides with primary or secondary amines, including distearylamine as IMR agents.

- U.S. Pat. No. 4,585,803 discloses that salts of Q2-7119 can be made with Group IA, IB, IIA, IIB, aluminum, chromium, molybdenum, iron, cobalt, nickel, tin, lead, antimony, or bismuth. These salts are then compatibilized in the B-side blend with certain tertiary amines. In practice, these salts are extremely viscous or gelatinous in nature and do not disperse well into the polyol.

- U.S. Pats. 4,764,540 and 4,789,688 disclose that salts of Q2-7119 can be made with amidines and guanidines, such as tetramethyl-guanidine, to yield neutralized forms of the silicone that would not gel tin catalysts. Waxy amidines, such as the imidazolines from stearic acid and ethylenediamine derivatives, were cited as particularly efficacious for release.

- U.S. Pat. 4,040,992 discloses the use of N-hydroxyalkyl quaternary ammonium carbonylate salts as catalysts in the production of polyisocyanurates and polyurethanes. Among the exemplary preferred catalysts are N-hydroxypropyl trimethyl ammonium salts of carboxylic acids such as those of formic and acetic acids and of fatty acids such as hexanoic and octanoic acids and the like.

The Dewhurst[2] invention is directed to a method for making a polyurethane, polyurethaneurea, or polyurea elastomer in which a reactive mixture is formed in a mold cavity and cured. The reactive mixture contains polyol, organic polyisocyanate, urethane catalyst, an optional diol and/or diamine chain extender, and a mold-release additive. The present invention provides as the internal mold release (IMR) additive a composition consisting essentially of (1) the reaction product of a mixture of a carboxylic acid, a tertiary amine, and a reactive epoxide, and (2) a carboxy functional siloxane. The resulting compositions function as IMR agents that do not gel tin catalysts. In some cases, normal tin catalyst levels can be used.

18.8.3   HCFC as a Sole Blowing Agent

In the production of polyurethane foams, a polyol is reacted with a polyisocyanate in the presence of a polyurethane catalyst and a blowing agent. Unfortunately, certain blowing agents, namely chlorofluoro-carbons (so-called *CFCs*), are hazardous to the environment, specifically the ozone layer of the atmosphere. Hence, alternatives to the use of CFCs are being actively sought by the polyurethanes community.

The use of a portable foaming apparatus employing a foam forming formulation to provide on-site generation of polyurethane foam is well known in the art. By way of illustration, U.S. Pat. 3,882,052 discloses the use of such an apparatus to provide non-froth polyurethane foam. For other applications, frothed foam is suitably produced using the portable foaming apparatus by incorporating into the foam forming formulation an auxiliary CFC blowing agent, such as dichloro-difouoromethane, commercially available as Freon 12 as a product of the Du Pont Company. When injected into the foam-forming mixture, the auxiliary blowing agent serves to augment the function of the static mixer in bringing about thorough blending of the foam forming ingredients. Since the auxiliary CFC blowing agent is hazardous to the environment, alternatives to the use of CFCs in the production of frothed foam would be highly desired by frothed foam manufacturers.

The Wishneski[3] invention relates to an improved process for producing, by means of a portable foaming apparatus having a static mixer, a polyurethane foam by reacting a reaction mixture composed of a polyol, an organic isocyanate, a foaming/frothing agent, and a reaction catalyst, the improvement being the employment of monochlorodifluoromethane as the sole blowing/frothing agent in said reaction to produce a CFC-free, essentially closed-cell, rigid or semirigid polyurethane foam.

Any polyurethane foam forming composition that is suitable for processing and dispensing by means of a portable foaming apparatus may be employed in preparing the polyurethane foam in accordance with the process of the Wishneski invention, provided that monochloro difluoromethane is employed as the sole blowing/frothing agent. The composition typically comprises a polyol reactant, an organic isocyanate reactant, a foaming/frothing agent, a reaction catalyst and, preferably, a surfactant.

Immediately after completion of mixing of the components, the foam reaction mix is dispensed directly into the appropriate mold, and foaming is allowed to take place in the mold in accordance with procedures well recognized in the art for the molding of polymer foams.

18.8.4   Silicone Surfactants

High-resiliency polyurethane foams are produced by the reaction of organic isocyanates, water, and high-molecular-weight polyols that have greater than 40% of primary hydroxyl capping. They are distinguishable, in part, from conventional hot-cure polyurethane foams by the use of such a high percentage of primary hydroxyl groups, as well as by the fact that they require little or no oven curing. Thus, they are often referred to as *cold cure* foams. High-resiliency polyurethane foams are extremely desirable for cushioning applications because of the excellent physical properties they offer, such as high resiliency, open cell structure, low fatigue for long life, and high sag factors for good load-bearing capabilities.

The ingredients for high-resiliency polyurethane foam are highly reactive; consequently, there is a rapid buildup of gel strength in the foaming reaction, which sometimes permits the foam to be obtained without use of a cell stabilizer. However, such unstabilized foams typically have very irregular and coarse cell structures, evidenced by surface voids. This problem has generally been addressed by using substituted certain polydimethylsiloxane-polyoxyalkylene or polyphenylmethylsiloxane-polyoxyalkylene copolymers as foam stabilizers.

Polysiloxane-polyoxyalkylene copolymer surfactants for use as stabilizers for high-resilience polyurethane foam are disclosed, for example, in the following: U.S. Pat. 3,741,917, Morehouse; U.S. Pat. 4,478,957, Klietsch et al.; U.S. Pat. 4,031,044, Joslyn; U.S. Pat. 4,477,601, Battice; U.S. Pat. 4,119,582, Matsubara et al.; U.S. Pat. 4,139,503, Kollmeier et al.; and several patents of Kilgour, U.S. Pats. 4,690,955, 4,746,683, and 4,769,174. These references variously disclose that the terminal oxygen atom of the polyoxyalkylene portion of the surfactant molecules may bear a hydrogen atom (Morehouse '917, Klietsch '957, Kollmeier '503), an alkyl group of 1-4 carbon atoms (Morehouse '917, Klietsch '957, Kilgour '955, '683, and '174), an alkyl group containing fewer than 10 atoms in total (Joslyn '044), an alkyl group containing a total of less than 11 carbon atoms (Battice '601), or a monovalent hydrocarbon group (Matsubara '582). Methyl capping is commonly used. In addition, several other capping groups are disclosed. Those skilled in the art apparently have not believed that there are any advantages to be gained by use of any particular alkyl capping groups.

Surfactants for stabilization of polyurethane foam are evaluated on the basis of several different performance characteristics. Primary among these is the potency or efficiency of the surfactant. The minimal amount of surfactant needed to provide good cell structure in the resulting foam is a relative measure of the potency. Polyurethane foam having good cell structure can be produced using less of a superior surfactant than would be required using a less-potent surfactant. The ability to use less material is desirable in the foaming industry to lower the cost of foaming operations.

Of further concern in selecting a surfactant for polyurethane foam stabilization is the breathability or open-celled character of foam. High breathability (more open-celled character) of the foam is desirable, as it provides a greater processing latitude. A narrow processing latitude forces the foam manufacturer to adhere to very close tolerances in metering out the foaming ingredients, which cannot always be accomplished. Furthermore, greater breathability provides foam that is considerably easier to crush, thus avoiding splits that might occur during crushing. This characteristic is particularly desirable in foamed parts that incorporate wire inserts, which are difficult to crush.

In the design and development of surfactants for use as stabilizers for high-resilience polyurethane foam, there traditionally has been a trade-off between increasing the potency (efficiency) of the surfactant and lowering the breathability of the foam produced using it. It has generally been found that the more potent the surfactant, the lower the breathability of the foam made using it. In other words, the more potent surfactants generally afford poorer processing latitudes.

It would be very desirable to have silicone surfactants for stabilization of high-resilience polyurethane foam that afford both good potency and good breathability, thus providing foam manufacturers with relatively low surfactant costs as well as good processing latitude. Such surfactants are the subject of this application.

The Kilgour invention[4] provides t-butyl capped siloxane-polyoxyalkylene surfactants for manufacture of high-resilience polyurethane foam. These surfactants afford higher potency in manufacture of high-resilience polyurethane foam and also produce more open or breathable foam than would be predicted, relative to otherwise similar surfactants of the prior art in which the polyether portion of the molecule is terminated with a low-molecular-weight alkyl group.

The Kilgour invention relates to polyurethane foam, and more particularly to polysiloxane-polyoxyalkylene copolymer surfactants containing the t-butyl group in the terminal position of the polyoxyalkylene portion of the copolymers, for use in manufacturing high-resilience polyurethane foams.

18.8.5 Cycloaliphatic Diamines as Additives in High Water

The preparation of high-resiliency polyurethane foams characterized by high strength and load-bearing properties by the reaction of an organic polyisocyanate with a polyol in the presence of a cross-linking agent is well known in the art. Commercial polyurethane foam molders are always trying to optimize their formulations in a cost-effective fashion so as to obtain the maximum set of properties from a given set of materials and processing conditions. It is desired to improve the physical properties at equivalent density or maintain physical properties at reduced density.

Shrinkage occurs in a molded article when the trapped hot gas within unopened cells cools and causes the cells to contract. Currently, methods have been developed (foam crushing) to open the closed cells upon demolding of the article. Generally, the use of cross-linker molecules increases the closed cell content of a molded polyurethane article relative to a control formulation.

The use of aromatic diamine cross-linkers has been suggested to provide benefit in molded polyurethane foams. However, such materials have demonstrated only marginal physical property improvements when compared to control formulations. In addition, aromatic diamines and their derivatives have controversial toxicological properties and thus require special handling.

Previously, increased water levels have been used to reduce foam density and maintain physical properties. The currently used water levels cannot, however, be increased further because of processing limitations.

- U.S. Pat. 4,210,728 discloses a reactive polyol composition useful in the preparation of highly resilient polyurethane foams comprising a polyol and from 0.5 to 5.0 wt. % of a reactive cycloaliphatic diamine.

- U.S. Pat. 4,248,756 discloses thermosetting two-component coating compositions containing a prepolymer having ketoxime-blocked aromatic isocyanate groups and, as a cross-linking agent, certain diamino dicyclohexyl methanes.

- U.S. Pat. 4,108,842 discloses a process for the production of polyurethane ureas from a prepolymer and a mixture of hardeners containing amino groups.

- U.S. Pat. 3,849,360 discloses polyurea-urethanes made by reaction of an isocyanate-terminated polyurethane prepolymer with a diamine component containing 4,4'-bis(aminocyclohexyl) methane and one or more diamines having a single alicyclic ring.

- U.S. Pat. 4,293,687 discloses biscyclohexylamine derivatives.

The Vratsanos invention[5] provides a high-water, high-resiliency polyurethane foam having improved elongation and tear and increased stiffness while being noncollapsing. High-water, high-resiliency polyurethane foam compositions comprise conventional polyol, polymer polyol, an optional diethanolamine, silicone surfactant, water, and a catalyst system comprising a blowing catalyst and a gelling catalyst. The present invention provides for the addition to this foam composition of 0.25 to 10 parts per 100 parts polyol of a cycloaliphatic diamine.

References

1. Michaelis, Peter. U.S. Pat. 5,22,415, June 6, 1995, "Polyether Polyol and Polyurethane Compositions Protected against Oxidation and Core Scorching," to Ciba-Geigy Corporation, Ardsley, New York.

2. Dewhurst, John E. U.S. Pat. 5,312,845, May 17, 1994, "Rim Polyol Blends Containing Acidic Siloxane Internal Mold Release Agents and Tin Catalyst," Air Products and Chemicals, Inc.

3. Wishneski, Todd W. U.S. Pat. 5,264,464, November 23, 1993, "On-Site Generation of Polyurethane Foam Using an HCFC as a Sole Blowing Agent," to BASF Corporation, Parsippany, New Jersey.

4. Kilgour, John A. U.S. Pat. 5,198,474, March 30, 1993, "Silicone Surfactants Having t-butyl Terminated Polyether Pendants for Use in High Resilience Polyurethane Foam," to Union Carbide Chemicals and Plastics Technology Corporation.

5. Vratsanos, Menas S. U.S. Pat. 5,173,516, December 22, 1992, "Cycloaliphatic Diamines as Additives in High Eater, High Resiliency Polyurethane Foams," to Air Products and Chemicals, Inc., Allentown, Pennsylvania.

19

Copolymers and Polyblends

19.1 Polyblends of Polyolefins and Polyurethanes

The blending of polyolefins such as polyethylene or polypropylene with thermoplastic polyurethanes has been a long sought endeavor. If nothing else, one of the primary goals has been the availability of economically attractive polymer blends with properties close or equal to the best of the polyurethane components. Unfortunately, it has long been observed that polyurethanes and polyolefins, with polyethylene in particular, are largely incompatible except at the lowest levels of one in the other. Quite obviously, such incompatibility results in their so-called delamination or separation when formed into molded articles or films, which results in very poor physical properties.

U.S. Pat. 3,272,890 discloses blends of basically 15 to 25 wt. % of polyurethane in polyethylene. This is achieved by first melting and fluxing the polyethylene in a Banbury mixer, to which is added the polyurethane. In a series of U.S. Patents (3,310,604; 3,351,676; and 3,358,052), there are disclosed polyurethanes having dispersed therein 0.2 to 5 wt. % polyethylene.

U.S. Pat. 3,929,928 teaches that blends of 80:20 to 20:80 weight ratio of chlorinated polyethylenes with polyurethanes and containing 1 to 10 pph of polyethylene result in improved processability, particularly in the manufacture of films or sheets by milling or calendering. Such blends are more economical than the polyurethane alone.

U.S. Pats. 4,410,595 and 4,423,185 disclose soft resinous compositions containing 5 to 70 wt. % thermoplastic polyurethanes and 30 to 95% of polyolefins modified with functional groups such as carboxyl, carboxylic acid anhydride, carboxylate salt, hydroxyl, and epoxy. One of the features of the disclosed blends is their adhesion to other polymeric substances such as polyvinyl chloride, acrylic resins, polystyrenes, polyacrylonitriles, and the like. This property leads to their prime utility in the coextrusion, extrusion coating, extrusion laminating, etc. of polymer laminates.

U.S. Pat. 4,525,405 discloses polyurethane polymer forming compositions consisting of polyols, polyisocyanates, and catalysts, which compositions also contain at least one thermoplastic polymer such as chlorinated polyethylene, polyethylenes, polyethylene vinyl acetate, and mixtures thereof. The disclosed compositions are employed to form a polyurethane secondary carpet backing on a primary synthetic backing having the carpet fibers tufted therein. This prevents an edge curl problem in the resulting carpets.

There still remains the need for thermoplastic polyurethane/polyolefin compatible blends wherein the latter component forms a substantial proportion, and the resultant polymer products have good physical properties.

Zabrocki[1] disclosed novel thermoplastic compatible compositions comprising a polyolefin, a thermoplastic polyurethane, and a compatibilizing amount of at least one modified polyolefin. The polymer blends are soft, flexible, resinous compositions of high tensile and tear strengths. The compositions overcome the barrier of incompatibility between polyethylene and thermoplastic polyurethanes. This invention relates to thermoplastic resinous compositions and is more particularly concerned with compatible blends of polyolefins with thermoplastic polyurethanes and a compatibilizing proportion of a modified polyolefin comprising a polyolefin, a thermoplastic polyurethane, and a compatibilizing amount of at least one modified polyolefin.

The term *polyolefin* means polyethylene inclusive of high-density polyethylene (HDPE), low-density polyethylene (LDPE), linear low-density polyethylene (LLDPE), polypropylene, polybutylene, copolymers of ethylene/propylene, ethylene/butylene, propylene/butylene, and the like. The term *compatibilizing amount* as used herein means an amount sufficient to cause the polyolefin and polyurethane to mix together without objectionable separation so that delamination or derivation problems do not occur in the resulting thermally formed products such as molded, extruded, or film-formed parts as described below.

The term *modified polyolefin* means a random, block, or graft olefin copolymer having in a main or side chain thereof a functional group such as carboxylic acid; C1 to C8 carboxylate ester such as carbomethoxy, carbethoxy, carbopropoxy, carbobutoxy, carbopentoxy, carbohexoxy, carboheptoxy, carboctoxy, and isomeric forms thereof; carboxylic acid anhydride; or carboxylate salts formed from the neutralization of carboxylic acid group(s) with metal ions.

19.2 Polyblends

Thermoplastic polycarbonate resins are readily molded at elevated temperatures to make a wide variety of articles. Exemplary of such articles are automotive parts, tool housings, structural components, and the like. The use of polycarbonate on its own for molding purposes is limited, as the polycarbonate has a number of deficiencies including

- Sensitivity of impact toughness to the ambient temperature and, more particularly, thickness of the molded article
- Susceptibility to degradation by solvents including water and hydrocarbons

Correction of the deficiency of polycarbonate resins is known by blending the polycarbonates with other polymeric additives, as disclosed in U.S. Pat. 3,431,224, and addition of other compounds and stabilizers, as disclosed in U.S. Pats. 3,498,946 and 3,742,083. Polycarbonate resins have been modified by blending with other polymers including polyethylene, polypropylene, copolymers of ethylene and an alkyl acrylate, polyamide, polyvinyl acetate, alkyl cellulose ether, and polyurethane elastomer.

In U.S. Pat. 4,034,016, a ternary blend consisting of a polycarbonate, a polybutylene terephthalate, and a thermoplastic polyurethane (hereafter referred to as TPU) is disclosed that has an improved impact strength at critical thickness. U.S. Pat. 4,179,479 discloses a ternary blend of a TPU, a thermoplastic polycarbonate, and an acrylic polymer, with the latter functioning as a processing aid to confer uniformity of melt flow properties. U.S. Pat. 4,350,799 discloses a ternary blend containing a TPU, a polycarbonate, and a polyphosphate, and the blend displays reduced flammability. Ternary blends of TPU, polycarbonate, and rubbers as impact modifiers are disclosed by EP 125739 and U.S. Pat. 4,522,979.

The preparation of binary TPU polycarbonate blends has been little studied due to the inherent problems of compatibility between polycarbonate and TPU. These include, for example, large differences in melt viscosities, processing temperatures, and thermodynamic solubilities. These differences are especially prominent with polyether-based TPUs.

Accordingly, it would be desirable to provide a binary polycarbonate TPU blend wherein the incompatibility difference of the two polymers has been minimized and molded articles prepared therefrom show improved resistance to hydrocarbon solvents and improved impact resistance and toughness compared to polycarbonate alone.

Skochdopole[2] disclosed an invention relating to binary thermoplastic polyblend that comprises a thermoplastic aromatic polycarbonate and a polycaprolactone polyol-based TPU. The polyblends exhibit improved hydrocarbon solvent resistance and melt flow properties. The polyblend consists of a thermoplastic aromatic polycarbonate polymer from up to 70% by weight of the combined weights of the thermoplastic aromatic polycarbonate and the TPU present. The TPU is present in the polyblend in from at least 5%, and preferably up to 45%, by weight of the combined weights of the extant thermoplastic aromatic polycarbonate and the TPU.

Suitable thermoplastic aromatic polycarbonate polymers that can be used in the practice of this invention are those aromatic homopolycarbonates and aromatic copolycarbonates that advantageously have a molecular weight of about 15,000 to about 100,000. In addition, the polycarbonate advantageously has a melt flow rate of at 12 g/10 min to18 g/10 min at 300° C with 1.2 kg weight as measured by the ASTM Procedure D-1238.

19.3 Interpenetrating Polymeric Networks

Interpenetrating polymeric networks (hereinafter referred to as IPNs) of chemically similar polymers such as the vinyl type polymers are known in the art. For example, Millar[*] has disclosed IPNs of styrene and divinylbenzene prepared by the imbibition of a styrene-divinylbenzene monomer solution into a conventional divinylbenzene-styrene resin and subsequent polymerization of the monomers. This is also shown in British Pat. 728,508 and U.S. Pat. 3,078,140, where preparation of ion exchange resins having IPNs is disclosed.

The IPNs of this invention are composed of chemically dissimilar cross-linking polymer chains that have substantially no chemical bonding between them. The existence of these compositions is surprising. The mixing of chemically different polymers in the solid state usually results in multiphase polyblends. Even the mixing of two different polymer solutions generally produces two liquid phases. It has been found, however, that this difficulty can be overcome, without resorting to chemically bonding the two or more networks, by producing an IPN. By cross-linking a polymer in the presence of a cross-linked network, an IPN can be produced composed of several separately cross-linked polymers held together by permanent chain entanglements.

A cross-linked polymer can be thought of as a linear molecule containing macrocycles of various sizes along its chain. The cross-linking of another polymer in the presence of the first cross-linked polymer causes a statistical number of the macrocycles of the first cross-linked polymer to be threaded through by those of the second cross-linked polymer to form an IPN. The properties of such a network reflect those of the constituent polymers. The mechanical properties, such as tensile strength of the IPN, reflect those of the stronger network, while the swelling characteristics of the IPN approach those of the more readily swellable network. Thus, IPNs from the chemically dissimilar polymers are new and desirable materials that are not limited by chemical incompatibility.

Frisch et al.[3] disclosed synthetic resins having a topologically interpenetrating polymeric network comprising macrocyclic structures of chemically different, cross-linking polymers that are permanently entangled with each other, with the several macrocyclics between the individual networks. This invention relates to a composition composed of interpenetrating polymeric networks. More particularly, the interpenetrating polymeric networks of this invention are composed of at least two chemically different, cross-linking polymers that do not interreact chemically.

In accordance with this invention, IPNs are prepared from chemically dissimilar cross-linking polymers that do not interreact chemically, i.e., which are cross-linkable by different mechanisms under the process conditions employed by means of polymer specific cross-linking agents. Polymer-specific cross-linking agents are those cross-linking agents that, under the process conditions employed, act to cross-link one type of the dissimilar polymers but not the other. The polymers are intimately mixed, so as to avoid complete phase separation, in combinations of two or more where at least two of the cross-linking polymers are chemically dissimilar, along with their respective polymer specific cross-linking agents and stabilizers, and subsequently cured to form the IPNs. The cross-linking polymers employed in the IPNs of this invention typically can be olyfunctional polyurethanes. The cross-link density of the foregoing

---

J. Chem. Soc., 1311 (1960).

cross-linking polymers must be such that the resulting macrocyclic rings have at least 20 ring atoms, and preferably at least 50 ring atoms. The relative amounts of the chemically dissimilar cross-linking polymers that are present can vary over a wide range, depending on the desired properties of the ultimate resin. For example, in an IPN made up of two macrocyclic structures, the relative amounts of each polymeric chain can be in a weight ratio as high as 9:1 and higher, or as low as 1:9 or lower.

The IPNs of the present invention can be prepared by several methods. For example, emulsions or solutions of two types of chemically dissimilar polymers can be admixed, along with their respective cross-linking agents where necessary, and then simultaneously cured, i.e., cross-linked, thereby producing two entangled macrocyclic structures. Alternatively, liquid-curable polymers can be sorbed by an already cross-linked polymeric structure and then cured so to produce the desired entanglement. The particular curing conditions in any given instance are determined by the type of polymer that is to be cross-linked.

19.4 Polyblends

Clear polyurethane resins, produced by reaction of an organic polyisocyanate and a low-molecular-weight active hydrogen-containing compound such as an aliphatic glycol, have been known in the art for a considerable time.[*] These materials are rigid and relatively brittle, i.e., they demonstrate a low order of impact resistance when tested in accordance with standard tests such as the Notched Izod impact test (ASTM D256-56). Accordingly, they have found relatively little application in the fabrication of structural components such as automobile body parts, equipment housings, mechanical goods, and the like.

Various methods for modifying the properties of such resins have been suggested. Illustratively, U.S. Pat. 4,076,660 describes the production of rapid-setting rigid polyurethanes by reacting an organic polyisocyanate, an aliphatic monomeric triol, and a low-molecular-weight propylene glycol derivative in the presence of a modifier that is a liquid carboxylic acid ester. The resulting products are said to be moldable by casting to form machine parts and the like. U.S. Pat. 4,076,679 shows making similar compositions but using, as the modifier that is incorporated into the polyurethane-forming reaction mixture, a polymer derived from one or more ethylenically unsaturated monomers, which polymer is in the form of a liquid or a grease meeting certain specifications.

U.S. Pat. 4,342,847 describes a process for the preparation of thermoplastic materials by introducing a thermoplastic polymer into an extruder at a first inlet at a temperature such that the polymer melts, and then adding polyurethane forming reactants through a second inlet. The blend of the thermoplastic polymer and the polyurethane is discharged from the extruder in finished form. The resulting polymer blend is said to possess high impact resistance. Examples 1(a), (b), and (c) and 2(b) of the reference show the use of an ABS polymer as the thermoplastic polymer and butanediol and various diisocyanates (but no polyol) as the polyurethane forming reactants. The minimum amount of ABS polymer used in these various examples is 39.3 percent by weight based on total weight of the ABS polymer and the polyurethane.

Goldwasser[4] found that polymers having properties comparable to those of engineering plastics such as nylon can be obtained by blending a clear polyurethane plastic with a minor amount of certain impact modifiers that are incompatible with the polyurethane. This invention comprises polymer blends that are characterized by high impact resistance, high flexural modulus, and a heat deflection temperature of at least 50° C at 264 psi. They comprise a blend of (a) from 3 to 30 parts by weight, per 100 parts by weight of said blend, of a polymeric impact modifier that is incompatible with the major component of the blend.

19.4.1 Blends of Polycarbonates and Polyurethanes

Aromatic polycarbonate resins are a well known class of synthetic polymeric resins, generally prepared by the reaction of a polyhydric phenol with carbonate precursor. See, for example, U.S. Pat. 3,989,672.

---

[*]See, for example, O. Bayer, *Angewandte Chemie*, 59, No. 9, pp. 255–288, September 1947.

For the most part, these resins are thermoplastic and readily molded under heat to fabricate a wide variety of articles such as automotive parts, tool housings, and similar structural components. Also well known are the molding technique and the advantageous moldable compositions comprising the base polycarbonate in admixture with a wide variety of fillers, reinforcing agents, stabilizers, melt-viscosity modifiers, strength and impact modifiers, fire retardants, pigments, and like additives. Representative of such moldable compositions are described in U.S. Pats. 3,431,224 and 4,481,331.

In spite of the numerous thermoplastic polycarbonate molding resin compositions known and available, there has remained a need for thermoplastic resin compositions meeting specific needs and from which articles of specific physical properties may be molded. For example, there is a need for polycarbonate resins that may be molded into engineering structural components having certain thickness specifications, impact resistance, and resistance to hydrocarbon solvent degradation. These articles are of particular need in the automotive industry as automobile components.

However, polycarbonate-based molding compositions have had restrictions on their use in molding thick-walled articles due to two inherent disadvantages. The first disadvantage is the low critical thickness values of polycarbonates, i.e., the thickness at which a discontinuity in Izod impact values occurs. These low critical thickness values tend to limit wall thickness of molded polycarbonate to a thickness below the critical thickness. Polycarbonates exhibit notched Izod impact values that are dependent on the thickness of the polycarbonates. Thus, for example, while typical notched Izod impact values for a 3.2 mm thick polycarbonate test specimen are generally in the range of about 850 J/M. Typical notched Izod impact values for a 6.4 mm thick polycarbonate test specimen are generally in the range of about 160 J/M. The high Izod values of the 3.2 mm thick polycarbonate test specimen are due to the fact that these specimens are thinner than the critical thickness of the polymer and, therefore, upon impact, a ductile break occurs. The low Izod impact values of the 6.4 mm thick polycarbonate test specimens are due to the fact that these specimens exceed the critical thickness of the polymer and, therefore, upon impact, a clean or brittle break occurs.

It has been proposed that this thickness sensitivity can be reduced by adding a small amount of a polyolefin, e.g., polyethylene, to the polycarbonate resin. While the addition of polyethylene has proven to be somewhat effective, in the main thick section notched Izod strengths are still not comparable to thin section strengths.

Boutni[5] disclosed that thermoplastically moldable polycarbonate resins are improved by the blending therewith a thermoplastically moldable polyurethane characterized by high impact resistance, high flexural modulus, and a heat deflection temperature of at least 50° C at 18.6 kg/cm². The compositions of the present invention solve this particular problem of the prior art in that they are moldable into articles having a thickness both above and below the critical thickness with useful impact resistance properties.

19.4.2 Polyurethane/Polyolefin Thermoplastic Polymeric Blends

Thermoplastic polyurethane polymers are linear noncross-linked polymers produced by coreaction of difunctional reactants comprising hydroxyl terminated polyesters or polyethers, glycols, and diisocyanates. Thermoplastic polyurethanes vary in hardness depending on the ratio of glycol (hard) to polyester or polyether (soft). They produce soft thermoplastics known as elastomers or harder thermoplastics known as plastics. Although high hardness and rigidity are desirable properties for molded plastic products, such products tend to exhibit poor low-temperature properties such as poor resiliency, brittleness, and low impact resistance and toughness, especially for outdoor applications at temperatures of approximately –30° C.

Polyethylene and polypropylene polymers are nonpolar polymers, whereas polyurethane polymers are polar polymers. It has been generally accepted that thermoplastic polyolefin polymers are incompatible with thermoplastic polyurethane polymers. Only additive amounts of polyolefins consisting of less than three percent by weight of the polymeric blend could be utilized.

For instance, U.S. Pat. 3,929,928 indicates that mill blending of a thermoplastic polyurethane with polyethylene results in severe plate-out due to incompatibility of the two polymers Researchers have reported, in *Organic Coatings Plastics Chemistry,* Vol. 40, page 664 (1979), that it was impossible to prepare, with a roll mill, useful test specimens at any polyurethane/polyethylene blend ratios. Similarly, *Walker's Handbook of Thermoplastic Elastomers,* Section 5.4.17, reports that low-density polyolefin modifications of polyurethane polymers must be maintained below three percent to avoid adverse effects due to incompatibility of the two polymers. Although U.S. Pat. 3,272,890 purports useful blends of polyolefin and soft polyurethane polymers, such blends are polyolefin based containing less than 25% by weight polyurethane polymer, where polyurethane polymer content above 25%, are incompatible and cannot be molded into useful plastic containers. Crystalline high-density polyethylene or polypropylene polymeric blends are even more difficult to prepare due to incompatibility of the crystalline polyolefins with polyurethanes. Useful blends of thermoplastic polyurethane elastomers containing less than 15% by weight neutralized ethylene/carboxylic acid copolymers are disclosed in U.S. Pat. 4,238,574 to provide elastomeric blends useful in blow-molding operations.

A patent by Lee[6] reveals polymeric blends of thermoplastic polyurethane elastomers and crystalline or predominately crystalline polyolefin thermoplastics are produced by high-shear mixing of melts of the respective polymers. The stabilized mechanically compatible blend of the ordinarily thermodynamically incompatible polymers comprises from about 3.5 to 20 parts by weight of a polyolefin homopolymer or copolymer with 100 parts by weight of polyurethane. The mechanically compatible polymeric blend is particularly useful for molding plastic articles by injection molding, extrusion, calendering, or similar process for molding thermoplastic articles.

Lee found that up to about 20% by weight of crystalline polyolefin polymer can be successfully blended with high-hardness thermoplastic polyurethane elastomer by high-shear mixing of melts of the respective polymers to produce a uniform, mechanically stabilized compatible polymeric blend of the two polymers. In accordance with the present invention, from about 3.5 to 20 weight parts crystalline polyolefin homopolymer or copolymer can be blended with 100 weight parts thermoplastic polyurethane polymer to produce a mechanically stabilized compatible polymeric mixture. High-shear blending of the polymeric blends can be effected by high-shear mixers such as single- or twin-screw extruders or a Buss-kneader at elevated temperatures above the melt temperatures of the respective polymers for time sufficient to produce the mechanically compatible polymeric mixtures. It has been found that conventional low-shear mixers, such as two-roll mills or a Banbury mixer, provide inadequate shear and resulting macroscopically phase-separated incompatible polymeric blends.

The improved polyolefin-modified polyurethane polymeric blends of the Lee invention unexpectedly provide highly desirable compositions that are useful for molding plastic articles exhibiting considerably improved physical properties, especially low-temperature impact properties as well as improved hardness, resiliency, rigidity, and toughness. Still further advantages of the invention are obtained in the high-shear processing step where improved processability, due to reduced melt viscosity and reduced pressure fluctuation during processing, are obtained by high-shear processing of the polyolefin-modified polyurethane blend. The polymeric blend is particularly useful for molding films, sheets, hoses, exterior automotive body parts, and similar plastic articles that are often subjected to low temperatures of, e.g., −30° C.

19.5 Polyurethane/Polyvinyl Chloride Interpenetrating Polymeric Networks

According to the literature, "IPNs are polymer alloys consisting of two or more distinct cross-linked polymer networks held together by permanent entanglements with only accidental covalent bonds between the polymers."[7]

Pernice et al., make an IPN by reacting (1) a mix of poly(oxypropylene) triol, ethylene oxide capped and grafted with acrylonitrile, and a mixture of short chain diols with (2) an isocyanate (a *modified MDI*) in admixture with (3) a preformed epoxy. A PU network is said to be formed within the epoxy network.

Mixing or blending of preformed PU with various preformed polymers [epoxies, polyacrylates, poly(acrylonitrile-butadiene), polystyrene, etc.] are known. See, for example, U.S. Pat. 4,302,553, to Frisch et al. *Chemical Abstracts* 103(4) discloses a blend of preformed PU/PVC. A physical blend of PU/PVC is also disclosed in Garcia, D., *Journal of Polymer Science*, Part B: Polymer Physics 24(7), 1577–1586.

Thermoplastic PUs melt-blended with PVC in a Banbury mixer are reported by K.R. Gifford, D.R. Moore, and R.G. Pearson in *Chemical Abstracts* 94(12). See also P.M. Khachatryanl, *Chemical Abstracts* 105(16). According to Blank, et al.,[8]

> Interpenetrating polymer networks (IPNs) can be made by admixing polyols, polyisocyanate, and a poly(vinylchloride) plastisol, followed by heating to complete the cure. The resulting polyurethane/poly(vinylchloride) IPN has superior properties as sealant, especially for automotive parts. A polyurethane (PU) network is formed that contains, within its interstices and reticula, discrete particles of a poly(vinyl chloride) plastisol (PVC). This resin composite is heated to cure the PVC, thereby forming a PVC network that interpenetrates the PU network. The result is a PU/PVC IPN.

Blank et al. start with two components:

- One component ("component A") comprises one or more polyols.
- The other ("component B"), is a polyisocyanate.

These two ingredients are selected such that, upon bringing them together in the presence of a PU catalyst, a PU will immediately begin to form. Either component can contain the catalyst. As regards the PVC plastisol, either component can contain this ingredient. Blank et al. prefer that it be admixed into the polyol composition.

The A and B components are mixed (e.g., by spraying or extruding), and the formation of the IPN begins. It forms in two steps.

- Step 1 is where most (typically 90%) of the PU network forms. This happens immediately after mixing. At this point, the PVC plastisol lies dispersed as discrete prepolymeric particles within the reticula of the PU network, substantially unchanged from their original form. This initial PU network is crucial to the invention. It forms a firm framework, or skeleton, that ensures a *uniform* configuration or dispersion of the forthcoming PVC network. At this point, the IPN achieves sufficient initial strength to form a sealant/adhesive coating layer.

- Step 2, the PU/PVC mix is heated at a temperature and for a time necessary to cure the PVC plastisol, i.e., convert it from a sol to a gel. In this second step, the particles of the PVC plastisol join up with each other in the known way from a sol to a gel, to form their own network, and the result is two interpenetrating polymers—a PU/PVC IPN. Any residual polyol and isocyanate also react during this curing operation, thereby completing PU network formation and giving the IPN its full ultimate mechanical strength. As with any IPN system, during curing, a great deal of branching and cross-linking take place, both of which may be aided (if desired) by addition of conventional chain extenders and cross-linkers.

19.5.1 Polyurethane/Acrylic Interpenetrating Polymeric Networks Having Improved Acoustic Damping Characteristics

The acoustic damping properties of viscoelastic polymeric materials render them most effective in their glass transition temperature range where the material changes from hard and glass-like to a soft, rubbery consistency. For a particular polymeric material, the glass transition temperature range is centered about a characteristic temperature for that material. For most polymeric materials, the glass transition temperature range is on the order of 20° C. This temperature range is where the polymeric material provides its maximum acoustic damping. However, it frequently occurs at temperatures that are either lower or higher than the temperature range in which a high degree of acoustic damping is desired from an applications standpoint. Efforts have therefore been made to broaden the glass transition temperature range and to shift it to a designated temperature range such that a high degree of acoustic damping is achieved at temperatures at which acoustic damping ordinarily is low.

Interpenetrating polymer networks having improved acoustic damping characteristics are known. U.S. Pat. 3,833,404 discloses interpenetrating polymer networks to be used for surface layers or coatings for damping vibrations or noise-emitting surfaces. The material consists of polyethylacrylate cross-linked with a polyglycol dimethacrylate, and polystyrene cross-linked with divinylbenzene.

U.S. Pat. 4,302,553 discloses a number of interpenetrating polymer networks that have improved tensile strength and swelling properties in presence of solvents, including combinations of polyurethanes with polyacrylates, polyepoxides, polyesters, styrene-butadiene polymers, and polydimethyl siloxanes.

U.S. Pat. 4,342,793 describes curable resin compositions for protective surface coatings consisting of interpenetrating polymer networks prepared from saturated polyols, acrylate and methacrylate esters, and polyisocyanates, by radiation and thermal curing.

U.S. Pats. 4,618,658 and 4,719,268 describe polymer modified epoxy resin compositions comprising the copolymerization product of an epoxy resin wherein part of the epoxide groups have been modified to provide polymerizable ethylenic unsaturation, vinyl-terminated urethane oligomer, and a polymerizable ethylenically unsaturated compound such as styrene or an acrylate ester.

U.S. Pat. 4,742,128 discloses compositions for molded products consisting of an interpenetrating polymer network comprising a polyamide and a polyurethane.

U.S. Pat. 4,752,624 describes an interpenetrating polymer network for selective permeation membranes comprising a hydrophilic and a hydrophobic polymer component. The hydrophylic component is made from hexamethylene diisocyanate and polyethylene ether glycol, and cross-linked with trimethylolpropane. The hydrophobic polymer component is polystyrene cross-linked with divinylbenzene.

U.S. Pat. 4,766,183 discloses a heat-curable composition comprising a urethane/epoxy/silicone interpenetrating polymer network.

U.S. Pat. 4,824,919 describes vinyl ester/styrene composition flexibilized by the addition of a small amount of polyurethane.

U.S. Pat. 4,902,737 discloses a resin having improved impact properties comprising an aromatic carbonate resin/polyester blend modified by the addition of a first elastomeric phase of cross-linked polyacrylate, and a second phase of cross-linked styrene/acrylonitrile.

Sorathia et. al.[9] disclosed improved acoustic damping materials that comprise interpenetrating polymer networks having a soft polymer component and a hard polymer component. The soft polymer component, constituting from 50 to 90% by weight of the material, is made by polymerizing an aromatic diisocyanate with a polyalkylene ether glycol, and the hard polymer component is an acrylic polymer made by polymerization of the alkyl esters and alkylene diesters of acrylic and/or methacrylic acid, e.g., n-butyl methacrylate and tetramethylene glycol dimethacrylate. The curing of the mixture is carried out at room temperature.

The Sorathia, et al. invention relates to polymer compositions having improved acoustic damping properties, and methods of preparing them. The invention provides a viscoelastic material interpenetrating polymer network having a broadened glass transition temperature range and, therefore, improved acoustic damping over a broad range of temperatures. Furthermore, by adjusting the relative weight percentages of the polymer components, the temperature range over which maximum acoustic damping is achieved can be varied to a desired temperature level.

References

1. Zabrocki, Vincent S. U.S. Pat. 4,883,837, November 28, 1989, "Compatible blends of polyolefins with thermoplastic polyurethanes," to The Dow Chemical Company, Midland, Michigan.
2. Skochdopole, Richard E. U.S. Pat. 5,162,461, November 10, 1992, "Thermoplastic polyblends of aromatic polycarbonates and thermoplastic polyurethanes," to The Dow Chemical Company, Midland, Michigan.

3. Frisch, Harry L., Frisch, Kurt C., and Klempner, Daniel. US Pat. 4,302,553, November 14, 1981, "Interpenetrating polymeric networks."
4. Goldwasser, David J., and Oertel, III, Richard W. U.S. Pat. 4,567,236, January 28, 1986, "Novel polymer blends," to The Dow Chemical Company, Midland, Michigan.
5. Boutni, Omar M. U.S. Pat. 4,743,650, May 10, 1988, "Thermoplastic molding blends of polycarbonates and polyurethanes," to General Electric Company, Mt. Vernon, Indiana.
6. Lee, Biing-Lin. U.S. Pat. 4,990,557, February 5, 1991, "Mechanically compatible polyurethane/polyolefin thermoplastic polymeric blends," to The B.F. Goodrich Company, Akron, Ohio.
7. Pernice, J., Frisch, K.C., and Navare, R. *Cellular Plastics*, 18(2), 121–128.
8. Blank, Norman E., Hartwig, Richard C., and Vu, Cung, U.S. Pat. 5,091,455, February 25, 1992, "Polyurethane-poly(vinylchloride) interpenetrating network," to W.R. Grace & Co., Connecticut and New York.
9. Sorathia, Usman A., and Dapp, Timothy L. U.S. Pat. 5,328,957, July 12, 1994, "Polyurethane-acrylic interpenetrating polymer network acoustic damping material," to the United States of America as represented by the Secretary of the Navy, Washington, District of Columbia.

20

Polyurethane Coatings

20.1 Introduction

The unique combination of the properties of toughness, flexibility, abrasion resistance, and solvent resistance, plus the ability to be *custom tailored,* has led to the widespread and continually increasing use of urethanes. Typical uses include leather coatings, fabric coatings and adhesives, industrial maintenance and corrosion-resistant finishes, floor varnishes, seamless flooring, marine finishes, magnet wire coating, and concrete sealing.

Urethane coatings are classified as belonging to one of five ASTM classifications, which are chiefly related to the curing mechanism:

1. *Oil modified.* These are reaction products of isocyanates and alcoholysis products of drying oils. The process is analogous to that used in alkyd manufacture and, as with alkyds, produces an upgraded drying oil. Curing is accomplished by oxidation of the unsaturated oil.
2. *Moisture cure.* Isocyanates are polymerized with diols and triols such as polyethers, polyhydric alcohols, and castor oil products. These *prepolymers* are designed with unreacted terminal isocyanate groups that react with atmosphere moisture to form the finished cross-linked polymer film.
3. *Blocked.* These are variations of two-component systems. The adduct is *blocked* by reaction with certain polyols, most commonly phenol, making it unreactive at room temperature. Thus, it can be packaged in one can with polyols, pigments, and additives. Heating to temperatures of 250° F (121° C) or higher unblocks the adduct, producing the ingredients of a two-package system.
4. *Prepolymer plus catalyst.* These are essentially the same as type 2, above, but are provided with a separate catalyst to accelerate the cure.
5. *Two-component.* Isocyanates are reacted with relatively low-molecular-weight polyols such as alcohols to form *adducts.* These adducts then form one part of a two-can system; chain extension and curing are obtained from the polyol component.

20.2 Coating Technology

Traditionally, polyurethane coating technology is discussed as

1. Oil-modified
2. One-package prepolymer
3. Two-package systems

20.2.1 Oil-Modified Urethanes

Manufacture

The oil-modified urethanes are most similar to conventional finishes. Their manufacture is analogous to that of alkyds: a drying oil is reacted with a polyhydric alcohol such as pentaerythritol or glycerine to form an alcoholysis product. This alcoholysis product has terminal hydroxyl groups introduced from the alcohol which enable reaction with an isocyanate, typically toluene diisocyanate, to form the urethane-oil resin. The ratio of isocyanate to hydroxyl is adjusted to leave no reactive NCO groups in the finished resin. The majority of oil modified urethanes have been made from linseed oil and, to a lesser extent, soya, safflower, and tall oils.

20.2.2   One-Package Prepolymers

General

The one-package prepolymers are frequently termed *moisture cure* urethanes since reacting the terminal isocyanate group with water vapor is the common method of curing this type of resin. However, cure may also be effected by heat, so the label *moisture cure* is not necessarily strictly true.

Many classifications distinguish between *moisture cure* and *prepolymer plus catalyst* as two different types of urethane coatings. If desired, a separate catalyst with limited pot life after addition may be used with any one-package prepolymer. However, film curing can be achieved without such a catalyst, and this use typically affects only the rate of cure and is a matter of choice. These resins are then truly stable one-package prepolymers whether moisture cured or heat cured.

Manufacture

These urethane prepolymers are excellent examples of planned creation of resins with specific properties. This is especially true with those based on polyethers because of the great variety of intermediates that are available.

Toluene diisocyanate, as a mixture of 80% of 2-4 isomer and 20% 2-6 isomer, is the basis of the majority of these prepolymers because of its low cost, ease of handling, good reaction rates, and water white color. Diphenyl methane 4,4' diisocyanate is used to a lesser degree to obtain some enhanced properties such as better abrasion resistance, adhesion, and chemical resistance.

Polyethers, mixtures of polyethers and polyhydric alcohols, castor oil derivatives, and, to a lesser extent, polyesters, are used to formulate prepolymers. These materials all have two or more hydroxyl groups per molecule, allowing the formation of polymeric molecules. Table 20.1 lists several polyethers, glycols, and castor oils that are used in commercial coatings resins. Many more are available, but these illustrate the variety of molecular construction units the urethane polymer chemist may use.

TABLE 20.1   Typical Polyols Useful for the Production of Prepolymers

Description	Type	MW	Hydroxyl No.	Use
Polypropylene glycol	Diol	1000	110	Prepolymers
Polypropylene glycol	Diol	2000	56	Flexible prepolymers
Polyoxypropylene	Triol	615	274	Prepolymers
Castor oil	Triglyceride	—	164	Prepolymers
Trimethylol	Triol	134	1250	Adducts and prepolymers

Predictable generalities include the following:

- Higher-molecular-weight diols increase flexibility and abrasion resistance at the expense of hardness and chemical resistance.

- A higher ratio of triol imparts toughness through cross-linking, but the ratio must be carefully balanced to avoid brittleness.

Such formulas are illustrated by A and B in Table 20.2. They are typical for one-package prepolymers. Both are mixtures of diol and triols, and each secures a good combination of hardness, flexibility, and toughness. By increasing the average molecular weight of the diol, the resin becomes softer and more flexible.

TABLE 20.2 Formulas Typical for One-Package Prepolymers

	A	B
Polypropylene glycol (1000 MW)	1 Equiv.	2 Equiv.
1, 3 butylene glycol	1	—
Trimethylol propane	2	2
80/20 TDI	8	8
Sward hardness	60	26
Tensile strength	5000 psi	4200 psi
Elongation at break	40%	150%

Solvents required for these prepolymers that are typically available at 40–60% solids are ketones, esters, aromatic hydrocarbons, or mixtures. Practical considerations of coating leveling, rapid dry, and acceptable flash point limit the choice to such compounds as methyl isobutyl ketone, butyl acetate, ethylene glycol monoethyl ether acetate, xylene, and toluene. Methyl ethyl ketone and ethyl acetate are used to a limited extent for rapid drying.

Special *urethane grade* ketones and esters are required with low values of water content and other hydroxyls. Use of such high purity enables good viscosity stability of the finished resin without side reactions with the solvent impurities.

Curing Properties

While the prepolymers will cure as made to usable films either at ambient conditions or elevated temperatures, effective application requires catalysts, typically organo-metals or amines. The effectiveness of such catalysts is shown in Table 20.3, with a typical polyether-based prepolymer.

TABLE 20.3 Effect of Catalysis on One-Package Prepolymer Dry Time

Catalyst	25° C	70° C	120° C	Pot Life
None	14 hr	2.5 hr	7 min	Indefinite
0.5% methyl diethanolamine	3.5 hr	13 min	2 min	24 hr
0.7% cobalt naphthenate	11 hr	13 min	1 min	33 hr
0.7% lead naphthenate	10 hr	5 min	1 min	27 hr
0.5% dibutyl tin dilaurate	6 hr	5 min	1 min	37 hr

Careful formulation can build such catalysts into prepolymers that will secure rapid cure rates along with excellent can stability. Film properties of the urethane prepolymers are outstanding, especially for the remarkable combination of such properties found in a single coating.

Packaging

Isocyanate-terminated prepolymers are *moisture* sensitive and therefore require a certain amount of care in packaging to ensure good shelf stability. The techniques are based on common sense to minimize sources of moisture or other factors that would accelerate premature polymerization of the resin. These precautions include the following:

1. Drums and pails should be closed bead, unlined steel.
2. Smaller containers should have screw caps and, if tin-plated, should be outside soldered.
3. Larger containers should be factory sealed and empty to prevent condensation.
4. Cans should be baked just prior to filling.
5. Filling of small containers should be done in a low-humidity area.
6. All filled containers should be topped with an inert gas.

Material packaged as outlined above is typically stable for a year or more, depending on the sensitivity of the formulation. A further precaution to be observed is to avoid prolonged storage at high temperatures. Ten days at 130° F (54° C) is sufficient to cause gelation of some resins.

The majority of one-package urethane coatings are clear finishes, with special pigment dispersions available as second packages that have limited pot life after mixing. This is because of the moisture invariably absorbed on the finely ground pigments, which is sufficient to cause gelation if added to an isocyanate-terminated prepolymer. Special techniques may be used to incorporate pigments into the urethane. One such is *slurry grinding,* which consists of prereacting the pigment with TDI so that no moisture is left to affect the prepolymer. Another incorporates desiccants into the formulation to adsorb the moisture.

Application

The one-package prepolymers may be applied by all the conventional techniques with regard for its sensitivity to moisture. Brush, roller, mohair applicator, spray, airless spray, and even notched trowels have been successfully used in different applications.

Because air-cured coatings depend on humidity to form a film, the rate of cure for a given formulation is a direct factor of the humidity. This is shown in Table 20.4 for a typical moisture-cure coating.

TABLE 20.4 Effect of Relative Humidity on Dry Time

3 Mils Wet Film @ 70° F (21° C)	
R.H., %	Dry Time, Hr
30	6
45	3
60	1.5
75	0.75

Applicators have taken advantage of the above effect by using humidifiers in enclosed areas to accelerate drying. Because of the fast early hardness of these coatings, recoating is best done as soon as possible after hard dry—usually within 24 hours of application. Otherwise, the fast hardness pickup will require sanding to secure intercoat adhesion.

Dilution, of course, requires the same grade of solvent used in manufacturing the resin. Many manufacturers have special *urethane-grade* solvents such as acetates and ketones available, and most aromatic hydrocarbons from major suppliers are suitable.

These typical prepolymer solvents do require care in the surfaces that can be coated by urethanes. Many previously coated or painted surfaces will be attacked by these solvents, requiring their removal before application. Following manufacturer's or formulator's instructions will result, as with other specialty coatings, in good results. These include such simple practices as diluting prime coats to penetrate porous substrates and using suitable primers over metals. Invariably, the special care required to ensure a good coating is amply rewarded by the excellent results.

20.2.3 Two-Package Systems

Manufacture

The adducts intended for two-package coatings differ from the single-package prepolymers chiefly in their molecular weight. They are typically reaction products of diisocyanates with polyhydric alcohols. The resulting compound is a relatively small molecule and not a film former, so the properties of the coating are derived from the particular *second package* used to react with the adduct.

A prototype reaction is that between toluene diisocyanate and a mixture of trimethylol propane and 1-3 butylene glycol. With approximately 2 moles of TDI per mole of polyol, one isocyanate group theoretically forms a urethane link with the polyol leaving the other free for further reaction. A typical formulation is shown in Table 20.5, and solution properties are shown in Table 20.6.

Free residual TDI monomer is minimized by carrying out the reaction at as low a temperature as possible to take advantage of the difference in reactivity of para- and ortho- groups of the diisocyanate. Temperatures below 60° C effect a fair compromise between low-residual monomer and reasonable reaction time, giving free TDI levels of 3–5% at 75% nonvolatile.

Special techniques make possible very low levels of free monomer—below 1%. This limit is frequently specified by formulators as essentially eliminating any need for special precautions because of the irritating

TABLE 20.5 A Typical Formulation of Toluene Diisocyanate, Mixture of Trimethylol Propane, and Butylene Glycol

	Wt.%
TDI	68.4
Trimethylol propane	8.3
1-3 butylene glycol	8.3
Ethyl acetate	25.0

TABLE 20.6 Solution Properties

% solids	75%
Viscosity	Z-2 Gardner-Holt
NCO	12.7%

properties of the TDI. Such adducts typically have their counterparts at higher free-TDI levels and are somewhat more costly.

A special variety of adducts are the so-called *blocked* resins. These are the same basic adducts as described above reacted with a *blocking* agent, typically phenol, in stoichiometric ratio so that all isocyanate is completely reacted. At higher temperatures the phenol or other blocking agent is broken away from the adduct, leaving the isocyanate available for reaction. Thus, a stable coating can be made of the blocked adduct and polyol in solution with pigments and additives, if desired, that can be applied and then heat cured to a finished film.

Curing of blocked resins offers a different picture in that a *threshold* temperature must be reached before any cure takes place. In the most commonly used phenol-blocked resins, this is approximately 280° F (138° C). Again, the rates of cure may be greatly speeded with appropriate catalysts, although without effect on stability at room temperature. With proper catalysts cure rates of 1 to 2 min are possible at about 450° F (232° C). Commonly used solvents are cresylic acid, cellosolve acetate, and diacetone alcohol with toluol, xylol, etc. used as diluents.

20.3 Specific Applications

20.3.1 Wood

The decorative and protective coating of wood surfaces is a natural use for urethanes. Their excellent abrasion resistance, cleanability, and flexibility, combined with exceptional gloss and clarity to enhance the natural wood beauty in clear finishes, have won them exceptional acceptance. While some finishes are pigmented as outlined in the previous sections, the majority of applications for urethane finishes on wood are as clear coatings.

Urethanes are also very well suited for uses as floor finishes. The most frequently used urethanes for floor varnishes are the relatively *long-oil,* oil-modified types and the one-package moisture cures. An example of a gymnasium finish is shown in Table 20.7.

TABLE 20.7 Example of a Gymnasium Finish

Linseed oil-modified urethane	70 lb
Amsco odorless solvent 450	30
Cobalt naphthenate, 6%	0.07
Solution Properties	
Nonvolatile	42%
Viscosity	A2
Weight per gallon	7.3 lb
Sward hardness	40
Hard dry	4.5 hr
Set-to-touch	1.5 hr

The ultimate in wood floor finishes are the moisture-curing prepolymers that give maximum wear, mar resistance, and beauty. These are typically completely ready from the resin manufacturer but may be secured at higher solids to formulate to special order.

Urethane-based coatings are commonly available oil paint shelves as the *premium* finish for general-purpose wood finishing. These include high-gloss and satin finishes ranging in applications from kitchen cabinets to boats to furniture. Recently available are urethane-based stains enabling a complete urethane finish.

Priming of either type of finish on wood is best done by diluting the finish itself to A or less viscosity. A vinyl butyral sealer may be used for enhanced holdout, build, and gloss.

20.3.2 Concrete

Any urethane coating of appropriate hardness may be used on concrete with all the characteristic urethane properties available for these typically severe service areas. An oil-modified coating should be chosen where ease of handling is important; a moisture cure offers the toughest one-package finish; and possibly a two-package system would best be able to secure specific chemical resistance properties.

Possibly even more than for other substrates, the preparation before coating concrete with urethanes will spell the difference between success and failure. On old surfaces, all grease and oil must be removed, including that which has penetrated the surface. This is accomplished by solvents, detergents, mechanical removal of some of the surface, or a combination of these.

New concrete can present several complications. Consulting with the contractor can avoid extremely smooth, troweled surfaces, which are difficult to penetrate. Such consultation can also prevent the use of *curing* agents that may contain materials that interfere with adhesion. Alkali effects are minimized by allowing the concrete to cure a few weeks and by etching with dilute (10%) hydrochloric acid. The acid also serves to roughen the surface as an aid to bonding. Thorough rinsing with water is, of course, necessary.

20.3.3 Flexible Coatings

Earlier portions of this chapter discussed the variation in polymer property that may be had by changing the chain length and the degree of cross-linking. Moving to the long chain and lesser cross-linking end of this spectrum automatically produces coatings that are sufficiently flexible (Sward hardness of 5–30) for soft substrates such as rubber, leather, and textiles. Rubber coating with urethanes is done on an extensive scale for products such as golf balls and automobile weather strips.

20.3.4 Leather

The use of coatings on leather is somewhat susceptible to fashion's whims but has achieved excellent acceptance for producing scuff-resistant finishes ranging from glossy to dull. The coating of fabrics for such special purposes as conveyor belts and rain wear is increasingly employed. Here, such special considerations as *hand*, strikethrough, and rapid cure at temperatures safe for the fabric are important. While one- and two-part urethanes as previously discussed are widely used, increasing use is being made of nonisocyanate-terminated urethane polymers that cure essentially by solvent evaporation as lacquers.

20.3.5 Seamless Flooring

One of the fastest-growing applications for urethane coatings is *seamless flooring*. This is the name given to floors that are created by spreading coating materials on a floor substrate using varied decorative schemes embodied in the coating resin.

The most popular of these systems uses moisture-cure urethanes as the chief resin. Typically, a coat of urethane is applied to the floor over a primer, colored plastic chips are embedded in the urethane, and top coats of urethane are applied. The prime coat may be either the urethane itself, or special primers may be used to act as barriers against solvent attack and to provide both extra adhesion to difficult surfaces and extra resilience.

Special catalysts have been developed for rapid cure to allow economical multilayer application; these are typically proprietary to the seamless flooring marketer. Because of the highly decorative effect of the floors, protection against UV coloring is frequently provided. This is accomplished either by the use of either UV absorbers or *non-yellowing* urethanes based on aliphatic isocyanates.

20.4 Illustrative Patents

20.4.1 Microporous Structures

Many processes for the production of microporous sheet structures have been described in the art. Most of them involve coagulation methods. Observing certain modes of action and starting from suitable compounds, polyurethanes may be precipitated from their solutions in such a way that microporous sheets are obtained.

This may be effected either by (a) immersing a substrate that is coated with the solution of a polyurethane into a coagulation bath containing a liquid or a mixture of liquids that do not dissolve the polymer (see, for example, U.S. Pats. 3,100,721; 3,536,639; 3,595,685 and 3,622,526) or by (b) evaporating selectively the solvent from a polymer/solvent/nonsolvent mixture. According to U.S. Pat. 3,553,008, the polyurethane is dissolved in a mixture of low-boiling solvent and high-boiling nonsolvent from which the solvent evaporates selectively (the evaporation number of the nonsolvent ought to be at least twice that of the solvent), and the polymer thus is precipitated. As a common feature of these two processes, the solvent in the polyurethane solution is replaced gradually by a nonsolvent so that the initially colloidal solution is transformed to a gel and, finally, to a stable microporous structure.

U.S. Pats. 3,582,396 and 3,595,732 disclose a process whereby the starting components for the polyurethane (NCO-prepolymer and chain-extending agent or polyol, polyisocyanate and chain-extending agent) are dissolved in a liquid or mixture of liquids that is a nonsolvent for the final polyaddition product. It is the object of these two patents to synthesize the microporous sheetings directly, without any additional coagulation step. As reaction proceeds, the growing polymer molecules become less and less soluble. First a dispersion, and then a gel, is formed. Finally, one obtains an interconnecting, stable microporous structure from which the organic liquid is evaporated. It was expected that it would be possible to obtain a microporous sheet structure only if the organic liquid consisted of a nonsolvent, or a mixture of nonsolvents, and a volatile solvent for a polymer that has a lower evaporation number than the nonsolvent (in this case, the solvent preferentially evaporates during reaction so that only the nonsolvents are remaining). The disclosures of U.S. Pats. 3,553,008, 3,582,396, and 3,595,732 therefore explicitly advise against using a polyurethane solvent evaporating at a higher temperature than the nonsolvents present to avoid a collapse of the microporous structure after the bulk of the nonsolvent has been removed by evaporation.

Traubel et al.[1] disclosed microporous sheet structures and a method for their production that comprises reacting an organic polyisocyanate with an organic compound containing at least two hydroxyl or amino hydrogen atoms reactive with NCO groups as starting materials. This produces a polyaddition product that, in the homogeneous state, exhibits a shore A hardness greater than 40 and a softening range above 100° C. The said reaction is effected in an organic liquid that dissolves the starting materials, does not dissolve the polyaddition product, boils at below 250° C, and contains 1 to 300% by weight, based on the polyaddition product to be formed, of a polar solvent that is inert to the starting materials. Said polar solvent has a greater evaporation number than the components of the aforesaid organic liquid (in the following called *nonsolvent*), the nonsolvent being such that, while the polyaddition product forms, it gradually becomes increasingly insoluble in the mixture of polar solvent and nonsolvent. This mixture has no significant swelling effect on the polyaddition product.

Traubel et al. surprisingly found that it is also possible to add certain amounts of highly polar solvents with a greater evaporation number than the nonsolvents present, although one had to expect that such difficultly volatile liquids, being very good solvents and softeners for polyurethanes, would give rise to formation of a homogenous, impervious foil or coating. On the contrary, the process and the microporous sheet structures obtained are even improved considerably when the reacting mixture contains small quantities of polar compounds with a high evaporation number.

20.4.2 Microporous Garments

A well known waterproof, breathable textile laminate of commerce, sold under the brand name GORE-TEX™, is technically based on the use of a membrane of hydrophobic, microporous, expanded polytetrafluoroethylene (hereinafter *PTFE*) as an essential functional component thereof. For most purposes, the microporous PTFE membrane of the laminate is sandwiched between inner and outer fabric layers, the membrane generally, although not necessarily, being continuously bonded and/or adhered to one or both fabric layers. Such hydrophobic microporous PTFE membranes and the preparation thereof are described in U.S. Pats. 3,953,566, Robert W. Gore, issued Apr. 27, 1976, and 4,187,390, Robert W. Gore, issued Feb. 8, 1980. The preparative method broadly comprises the uniaxial or biaxial stretching of an unsintered highly crystalline PTFE sheet, prepared by paste extrusion, at a rate exceeding 10% per second while maintaining the temperature thereof at between about 35° C and the crystalline melt point of the polymer.

The resulting membranous PTFE product of this process has a microstructure characterized by nodes interconnected by fibrils and is possessed of an interesting and useful combination of properties that befits it for use in the preparation of waterproof, breathable textile systems. More specifically, the hydrophobic, microporous, expanded PTFE membranes of the above Gore patents are sufficiently hydrophobic, and the micropores small enough, that the membrane can function as a barrier to the passage of bulk water at significant hydrostatic pressures. On the other hand, due to the microporous character thereof, said membranes also possess the capacity to allow diffusion of water vapor. Thus, these desirable properties of waterproofness and water vapor transmissibility or *breathability* are imparted to textile laminates in which such a membrane constitutes a component layer.

Despite its relative success in the marketplace, the waterproof, breathable textile laminate systems based on the hydrophobic microporous expanded PTFE membrane of the above-identified Gore patents are, nevertheless, possessed of certain deficiencies.

1. Such expanded PTFE membranes are relatively expensive.
2. Because waterproof breathable textile laminates utilizing such microporous expanded PTFE membranes are necessarily constructed by some form of physical lamination of the previously prepared PTFE membrane to at least one previously prepared fabric layer, equipment and operative techniques must be provided to handle the membrane and the fabric, to properly index them, and to secure the PTFE membrane layer to the fabric layer. Thus, the techniques used to prepare such laminates are generally substantially more complex, arduous, and expensive as compared to conventional liquid coating and curing techniques known in the fabric coating art.
3. PTFE materials are generally known to be adhesion resistant. Therefore, where it is desired to prepare a waterproof breathable textile laminate by means of a continuous bonding of the PTFE membrane layer to a fabric layer, such as by interposition of a continuous adhesive layer between the microporous expanded PTFE membrane layer and the fabric layer, the resulting bond strength, albeit usually adequate, is generally relatively low, and the selection of a suitable adhesive that can accomplish the necessary bonding without substantial adverse affect upon the breathability of the system can be a problem.

Another problem associated with the use of the microporous expanded PTFE membranes in textile laminate systems resides in the finding that such membranes, over a period of use, can crack sufficiently so as to provide sites for bulk water leakage or seepage. Apparently, the many micropores necessarily embodied in the polymer matrix can act as crack propagation loci. This, coupled with the crystalline nature of the polymer, results in a membrane whose flexure life is somewhat limited. Finally, the expanded microporous membranes of the Gore patent are apparently susceptible to significant loss of their waterproof characteristics when contacted with such surface active agents as are inherently contained in human perspiration. Obviously, this can be a serious detriment where textile laminates employing these membranes are employed as garment materials. This problem, as well as a solution for it, is disclosed in U.S. Pat. 4,194,041, Gore et al., issued March 18, 1980. The disclosed solution resides in the interposition of a continuous, hydrophilic, water vapor transmissible layer between the microporous hydrophobic membrane surface and the surfactant source. Thus, with respect to garments fabricated with a waterproof, breathable textile laminate prepared in accordance with the Gore et al. patent, the microporous hydrophobic membrane layer faces the exterior of the garment, while the continuous hydrophilic layer faces the interior of the garment.

In this role, the continuous hydrophilic layer functions as a barrier to the surfactant contaminants in human perspiration and prevents its contact with the hydrophobic microporous PTFE membrane, thereby preserving the waterproof character of the membrane. As disclosed, the continuous hydrophilic layer of the construction can be in the nature of a hydrophilic polyurethane based on a reactive cross-linkable prepolymer having an isocyanate-terminated branched polyoxyethylene backbone. Attachment of the continuous hydrophilic layer to the microporous hydrophobic membrane can be discontinuous, such as in the nature of sewing or adhering the edges of the respective layers by thread or adhesive. The use of a continuous adhesive bond between the hydrophobic and hydrophilic layers is apparently believed by the patentees to be potentially detrimental to the water vapor transmission properties of the laminate and so is not discussed or disclosed. Where continuous bonding of the respective layers of the laminate is desired, patentees disclose a technique whereby direct bonding of the one layer to the other is achieved. Said technique involves the casting of the hydrophilic layer directly onto the surface of the microporous hydrophobic membrane layer with application of enough hydraulic pressure to force the hydrophilic layer into the surface voids of the hydrophobic layer. Thus, while the Gore et al. patent may provide a second-generation solution for the surfactant contamination problem disclosed to exist with respect to garment applications involving the first generation hydrophobic microporous expanded PTFE membranes disclosed in U.S. Pats. 3,953,566 and 4,187,390, it is obvious that the solution is achieved at the expense of added complexity and, of course, cost.

Krishnan[2] disclosed polyurethane compositions specifically adapted to produce nonporous membranes exhibiting waterproof and water vapor transmissible characteristics. The membranes can be produced as free-standing products or as coatings on porous substrates to confer similar properties to such substrates. Also disclosed are coated fabrics and fabric laminates utilizing the membranous coatings of the invention and exhibiting waterproof and water vapor transmissible characteristics. Such coated fabrics and fabric laminates find utility in the fabrication of tenting, rainwear, and other garments where waterproofness, coupled with breathability, are important features.

In accordance with the Krishnan invention, there is provided a waterproof, breathable polyurethane membrane whose waterproofness and breathability properties do not depend on the presence of microporosity and in which at least several of the problems related to the known microporous membranes of the prior art have been solved or substantially ameliorated. In another aspect of the invention, the polyurethane membranes are disposed over porous substrates, particularly as coatings applied to fabrics, thereby conferring waterproofness and breathability to such substrates.

20.4.3 Spandex Fibers

A variety of elastomeric polymers are known in the art. By way of illustration, polyurethane elastomers made from polyols prepared using a double metal cyanide complex catalyst are known, as illustrated by U.S. Pat. 5,096,993, which discloses the production of thermoplastic elastomers made using DMC-prepared polyether polyols. These elastomers are disclosed in the 1993 patent as having excellent physical and chemical properties.

One particular elastomeric polymer in fiber form, namely spandex, is a well known component of clothing, particularly sportswear, which adds elasticity to the clothing. Spandex is, by definition, a hard-segment/soft-segment-containing urethane-containing polymer composed of at least 85% by weight of a segmented polyurethane(urea). The term *segmented* refers to alternating soft and hard regions within the polymer structure.

Formation of the segmented polymer structure takes place in several discrete steps. The first step involves the formation of linear polyester or polyether macroglycols typically having molecular weights of between about 500 and about 4000 and having isocyanate reactive hydroxyl group at both ends of the macroglycol molecule. The next step involves the reaction of the macroglycol with an excess of an isocyanate, typically in about a 1:2 molar ratio, to form an isocyanate-terminated soft-segment prepolymer. The hard segments are formed next by reacting the isocyanate-terminated prepolymer with low-molecular-weight glycols or diamines. The resulting hard segments provide sites for hydrogen bonding and act as tie points in the segmented polyurethane responsible for long-range elasticity. As an alternative to the prepolymer method, a *one-shot* method is suitably used to produce the desired elastomeric polymer for subsequent processing to produce spandex fibers by melt spinning.

Spandex fibers are suitably prepared by extruding the urethane or urethane precursor through a spinnerette into a diamine bath where filament and polymer formation occur simultaneously, as disclosed, for example, in U.S. Pat. 4,002,711. The current polyol of choice in the manufacture of spandex is poly(tetramethylene ether glycol) (PTMEG), and the use of PTMEG in the preparation of spandex is disclosed, for example, in U.S. Pat. 5,185,420. Unfortunately, PTMEG is more expensive than might be desired, and the physical properties of the resulting spandex leave room for improvement, particularly with respect to strength and thermal and hydrolytic stability properties.

The Smith and Connor[3] invention relates to the production of spandex fibers made from segmented polyurethane(urea)s that are prepared from low unsaturation-containing polyols made by polymerizing alkylene oxides in the presence of a double metal cyanide complex catalyst. With the low unsaturation-containing polyols, spandex fibers can be prepared that have better physical properties than fibers prepared with PTMEG polyols. This invention relates to the utilization of elastomeric polyurethanes made using low ethylenic unsaturation-containing polyols that are essentially free of mono-ol in the production of spandex fibers. The spandex fibers made using these polyols exhibit excellent physical properties.

In one aspect, the invention relates to a process for enhancing the strength and hydrolytic or thermal stability of spandex fibers from a polyurethane or polyurethane precursor that comprises, prior to fiber formation, fabricating the polyurethane or polyurethane precursor by reacting a polyisocyanate (preferably a diisocyanate) with a polyol (preferably a diol) prepared using a double metal cyanide complex catalyst, and chain extending said polyurethane precursor with a diol or diamine chain extender, said polyol having a molecular weight between 2000 and 10,000 and a low level of terminal ethylenic unsaturation that is, generally, less than 0.02, more preferably less than 0.015, milliequivalents per gram of polyol.

20.4.4   Elasthane Fibers

Elasthane filaments consist of at least 85% by weight segmented polyurethanes. The elastic and mechanical properties are achieved by using polyurea polyurethanes of aromatic diisocyanates. Elasthane filaments are produced by wet spinning or dry spinning of solutions. Suitable solvents are polar solvents, for example, dimethylsulfoxide, N-methylpyrrolidone and, preferably dimethylformamide or dimethylacetamide. Elasthanes have to be stabilized against light and harmful gases and therefore contain stabilizers that are either introduced into the spinning solution as an additive or are incorporated in the segmented polyurethane. Where additives are used, they should not be volatile under the spinning conditions.

Various stabilizers and stabilizer combinations of phenolic antioxidants and phosphites have already been proposed with a view to improving the stability of elasthanes. However, the stabilities of these described elasthane filaments do not satisfy current technical requirements because, on the one hand, the stabilizing compounds used must not adversely affect the mechanical and elastic properties while, on the other hand, none of the processing steps from production of the filaments to the finished article, such as corsets or bathing apparel, should reduce the effectiveness of the stabilizer systems used.

The comparison tests in EP-A 137 408 with phosphites having the structure of pentaerythritol (Weston 618) and the above antioxidants clearly show poorer results than the aryl-alkylphosphites of phenol and secondary alcohols, which are claimed in EP-A 137 408.

The Kausch et al[4] invention relates to elasthane fibers that are protected against harmful environmental influences, particularly sunlight and combustion gases, by a synergistic stabilizer. The invention relates to elasthane filaments and fibers containing a phenolic antioxidant and a phosphite stabilizer.

20.4.5 Process for Dry Spinning Spandex

In dry-spinning conventional spandex, the combination of properties that can be obtained in the spandex is highly dependent on spinning speed. A particular combination of elongation and setting properties is needed in spandex intended for incorporation into women's hosiery. To ensure that such hosiery will have a satisfactory appearance and appropriate elastic characteristics, the maximum spinning speed at which the spandex can be produced is limited. Higher speeds are needed for more efficient and economical production. However, at higher spinning speeds, typical conventional spandex compositions have significantly poorer break elongations and setting characteristics than are desired for spandex intended for use in hosiery.

In the conventional manufacture of women's hosiery, spandex yarns usually are knit into the hosiery along with nylon yarns. After knitting, the hosiery usually is subjected to a *boarding* treatment in which the hosiery is placed on a form and treated with steam at a temperature of about 115° C. The treatment removes wrinkles from the hosiery, sets its final shape, and provides it with a much more elegant initial appearance than untreated hosiery would have.

Spandex yarns generally require higher setting temperatures than do nylon yarns. In boarding operations, excess temperatures or overheating of the hosiery can cause formation of undesirable *board marks* or lines. The undesirable marks correspond to the location of the edge of the form used in the boarding operation. Also, knit hosiery that contains spandex yarns generally does not retain the shape of the boarding form as well as does hosiery that contains no spandex, because of the strong elastic recovery properties of the spandex. These problems of spandex in hosiery are exacerbated with spandex of very low decitex. Nonetheless, over the years, the percentage of women's hosiery that contains spandex yarns has increased greatly because of the better recovery from stretch, the better fit retention, and the better wear life possessed by such spandex-containing hosiery, compared to 100% nylon hosiery. Thus, there is a need for an improved spandex that can be produced at high speed and can be used satisfactorily in women's hosiery.

Although not concerned with the above-described problems of spandex in hosiery, Dreibelbis et al., in U.S. Pat. 5,000,899, disclose a spandex that can be spun at speeds as high as 945 m/min. The spandex is set by a dry heat treatment at a temperature of about 190° C. Such dry heat-setting conditions are suitable for heavyweight warp-knit fabrics. However, under the milder steam heat-setting temperatures typically used for boarding lightweight hosiery fabrics, the Dreibelbis et al. products have very poor set. The Dreibelbis et al. spandex is a polyurethaneurea formed from an isocyanate-capped tetrahydrofuran/3-methyltetrahydrofuran copolymer glycol that was chain extended with a diamine mixture, preferably of ethylenediamine (EDA) and 2-methylpentamethylenediamine (MPMD). Generally, the amount of coextender in the diamine mixture is in the range of 20 to 50 mol %. Among possible combinations of co-extenders disclosed in passing by Dreibelbis et al. (i.e., at column 3, lines 35–42) is a mixture of EDA and 1,2-diaminopropane (PDA) coextender. A similar spandex yarn, formed from a capped glycol of a copolymer of tetrahydrofuran and 3-methyltetrahydrofuran that was chain-extended with an EDA/MPMD diamine mixture, is disclosed by Bretches et al. in U.S. Pat. 4,973,647. However, the spandex of Bretches also suffers from shortcomings similar to those of the Dreibelbis spandex.

The use of mixtures of EDA and 1,2-diaminopropane (PDA) and as chain extenders for polyether-based polyurethaneureas formed into spandex is also disclosed in Japanese Patent Application Publication 3-279,415, and Japanese Patent Application Publication 58-194915. The first of these publications discloses such spandex made with EDA/PDA mixtures containing 10–30 mol % of PDA (only 20% is exemplified) and then spun at a maximum speed of 300 m/min. The second Japanese publication discloses such spandex formed from mixtures of diols that were reacted with a diisocyanate and then chain extended with an EDA/PDA mixture containing 1–40 mol % of PDA. The resultant polymer is then spun at a maximum speed of 210 m/min. Neither of these Japanese publications concerns the problems encountered in boarding spandex-containing hosiery.

Houser at al.[5] provide a process for dry spinning a segmented polyurethaneurea at speeds as high as 900 m/min. The process includes chain-extending an isocyanate-capped polyether glycol with a mixture of ethylenediamine and 1,2-diaminopropane, the 1,2-diaminopropane amounting to 8–17 mole % of the diamine mixture. The resultant spandex is particularly suited for use in women's hosiery.

The Houser et al. present invention relates to a method for dry spinning spandex at high speeds. More particularly, the invention concerns such a process wherein the spandex is dry spun and wound up at speeds as high as 900 m/min or higher from a solution of a polyurethaneurea derived from a polyether-based glycol that was capped with diisocyanate and then chain extended with a particular mixture of specific diamines. The spandex of the invention is especially suited for use in women's hosiery.

The invention provides a dry spinning process for making polyether-based polyurethaneurea spandex that is particularly suited for inclusion in women's knitted hosiery. The process comprises (1) reacting methylene-bis(4-phenylisocyanate) with a poly(tetramethyleneether) glycol to form a capped glycol, with a capping ratio being in the range of 1.65–1.95 and the glycol having a number average molecular weight in the range of 1750–2250, (2) chain extending the capped glycol in an organic solvent with a mixture of ethyenediamine and 1,2-diamino propane, the 1,2-diaminopropane amounting to in the range of 8–17 mol % of the mixture of diamine chain extenders, to form a polyurethaneurea solution, (3) adjusting the concentration of the polyurethaneurea in the solution to be in the range of 32–38 weight % of the total solution, (4) dry spinning said solution through orifices to form filaments of textile decitex, and (5) winding up the resultant filaments at a speed of at least 550 m/min, preferably at least 700 m/min, most preferably at least 850 m/min.

20.4.6   Spandex Formed with a Mixture of Diamines

The art (for example, Wittbecker, U.S. Pat. 3,507,834, and McMillin et al., U.S. Pat. 3,549,596) discloses spandex filaments derived from a polyalkaneether glycol (e.g., polytetramethylene ether glycol, referred to herein as P04G) that was capped with tertiary aralkyl diisocyanate (e.g., alpha, alpha, alpha', alpha'-tetramethyl-p-xylylene diisocyanate, referred to herein as p-TMXDI) and then chain extended with a diamine (e.g., hydrogenated m-phenylenediamine, also known as 1,3-diaminocyclohexane, referred to herein as HmPD). Such spandex filaments have good tensile, whiteness, and degradation resistance properties but are deficient in heat settability and in resistance to swelling by certain solvents. Wittbecker and McMillin et al. disclose that a plurality of chain extenders may be employed. However, neither discloses the particular combination of the three diamine chain extenders that are required in the practice of the present invention.

Frazer et al., U.S. Pat. 2,929,803, disclose polyether-based spandex filaments made by reaction of a polyether glycol, a diisocyanate, and a secondary amine. Poly(tetramethylene oxide) glycol is disclosed. A list of secondary diamines and a list of primary diamines with which the secondary diamines may be mixed are disclosed by Frazer et al. in column 9, lines 55–74. Among the listed primary diamines are p-xylylenediamine and 1,4-diaminocyclohexane. Furthermore, Frazer et al. disclose that these diamines may be substituted with halogens, among other groups. However, tetrachloro-xylylenediisocyanates that are required in the spandex required for the present invention are not disclosed by Frazer et al.

Although not concerned with polyether-based spandex polymers, Altau et al., in U.S. Pat. 3,994,881, disclose polyester polyurethane-urea spandex filaments having improved resistance to degradation by ultraviolet radiation and chlorine-containing bleaches, in which the hard segments contain ureylene groups that are joined to *functionally nonaromatic* radicals, at least 25 mol %.

The polymers are prepared from a polyester glycol, tetra-halogenated diisocyanates and tetra-halogenated diamines. These tetra-halogenated diamines have been found by the present inventor to be useful as a minor constituent of the three-component diamine mixture employed as chain extenders for the spandex of the shaped article of the present invention.

The Lodoen[6] invention relates to a shaped article of a polyurethane-urea spandex polymer that includes soft segments derived from a polyalkaneether glycol and hard segments derived from a certain hindered tertiary aralkyl diisocyanates. In particular, the invention concerns such an article made from spandex polymer that is formed by capping a polyalkaneether glycol with an alpha, alpha, alpha', alpha'-tetramethyl-p-xylylene diisocyanate and then extending the isocyanate-capped glycol with a specific mixture of diamines.

The Lodoen invention provides an improved shaped article of spandex polymer derived from a polyalkaneether glycol that was capped with a tertiary aralkyl diisocyanate and then chain extended with diamine, wherein the improvement comprises the chain-extending diamine being a mixture consisting essentially of 25–80% hydrogenated m-phenylene diamine, 10–50% of hydrogenated p-phenylene diamine, and 10–30% of tetrachloro-p-xylylenediisocyanate, all percentages of the amines being mole %. The chain-extending diamine mixture preferably consists essentially of 35–55% of HmPD, 30–45% of HpPD, and 10–25% of TClpXD. Preferably, the spandex is in the form of a shaped article, and most preferably a fiber or film.

References

1. Traubel, Harro, and Konig, Klaus. U.S. Pat. 3,920,588, November 18, 1975, "Microporous Sheet Structures," to Bayer Aktiengesellschaft, Leverkusen, Germany.
2. Krishnan, Sundaram. U.S. Pat. 5,283,112, February 1, 1994, "Waterproof Breathable Fabric Laminates and Method for Producing Same," to Surface coatings, Inc., Wilmington, Massachusetts.
3. Smith, Curtis P., and O'Connor, James M. U.S. Pat. 5,340,902, August 23, 1994, "Spandex Fibers Made Using Low Unsaturation Polyols," to Olin Corporation, Cheshire, Connecticut.
4. Kausch, Michael, Buysch, Hans Josef, Schroer, Hans, and Suling, Carlhans. U.S. Pat. 4,877,825, October 31, 1989, "Elasthane Fibers Stabilized against Environmental Influences," to Bayer Akteingesellschaft, Leverkusen, Germany.
5. Houser, Nathan E., Bakker, Willem, and Dreibelbis, Richard L. U.S. Pat. 5,362,432, November 8, 1994, "Process for Dry Spinning Spandex."
6. Lodoen, Gary A. U.S. Pat. 4,798,880, January 17, 1989, "Spandex Formed with a Mixture of Diamines," to E.I. Du Pont de Nemours and Company, Wilmington, Delaware.

21

Castables, Sealants, and Caulking Compounds

21.1 Introduction

Cast polyurethane elastomers are tough, flexible materials designed to achieve demanding requirements not met by conventional rubbers and plastics. They consist of essentially three components:

- diisocyanate
- polyol
- curative

each of which is available in a variety of chemical forms. By varying the type and ratio of these three components, the mechanical properties can be tailored over an broad range. Polyurethane cast elastomers are used in a variety of critical applications including electrical encapsulants; abrasion-resistant linings for pumps, chutes and conveyor belts; rollers; wiper blades; seals and gaskets; and wheels for casters, roller and in-line skates.

Castable polyurethane systems are all liquid non-foam urethane polymers, 100% (or nearly 100%) solids, reactive polymer systems. The worldwide market for urethane liquid elastomers is about $1 billion. The market is usually divided into three major segments:

- adhesives and sealants
- cast elastomers
- coatings of comparable size

The adhesives and sealants segment accounts for about 36% of the market. The overall growth rate of the market is about 4–6%.

Castable polyurethane systems are extremely diverse. The urethane systems in the market include durometer hardness from 45 Shore A to 75 Shore D. One particular attraction of urethanes is their ease of processing on low-capital-cost machinery. Urethane polymer systems are processed by pouring, casting, spraying, troweling, injecting, brushing, and other means. Typically, urethane liquid polymer systems comprise an A and a B side, one being an NCO-terminated prepolymer and the other an amine, diol, or polyol curative system. Another common system is the *one-shot* or *inverted* system where the prepolymer is an alcohol or amine functional oligomer, and the curative is a chain-extended, polymeric, or trimerized isocyanate. Urethane polymer systems are also available as one-component systems where curing occurs by the reaction of moisture with available isocyanates or by the reaction of blocked isocyanates with available amine or alcohol. Urethanes are selected because of their excellent toughness, abrasion resistance, broad service temperature range, environmental and fluid resistance, and excellent gloss and colorability.

Applications for urethane liquid polymers are extremely diverse. Many areas appear to be maturing, especially industrial rolls, wheels, bushings, and bearings. Other areas such as urethane coatings have seen high growth in select areas such as window laminates and abrasion-resistant coatings for hoppers, hopper cars, pumps, and piping. RIM experienced rapid growth in the early 1980s, but the growth rate has begun to level off. Urethane adhesives have seen slow, steady growth, whereas urethane sealants have grown rapidly in several areas such as automotive and architectural glazing.

21.1.1 Environmental, Health and Safety

Both governmental regulation and a generally improved overall environmental and health consciousness have significantly affected the urethane industry. In 1972, the Occupational Health and Safety Administration (OSHA) proposed regulations for 4,4'-methylene-bis-(3-chloroaniline) (MOCA, MBOCA) curative, which threatened to virtually cripple major sectors of this rapidly growing industry. Much development work through the 1970s and 1980s was directed toward finding replacements for MOCA. Regulation governing isocyanate exposure, especially tolylene diisocyanate (TDI), have fueled movement to methylene-di-p-phenyl diisocyanate (MDI) development of low-TDI prepolymers, and the use of chain extended or trimerized isocyanates in one-shot systems.

Making polyurethane parts with high quality and consistency is complex. Market demands, competition, and technological advancements have led to increasing improvements in the quality and performance of urethane parts. Lower residual monomer, narrower molecular weight distribution in polyols and prepolymers has led to lower viscosities, easier processing, and improved properties. Purer monomers and improved stabilizers have led to improved color, light, and environmental stability. Improvements in processing machinery and instrumentation have provided better ratio and temperature control and methods of quality measurement.

21.1.2 New Materials Technology

Isocyanates

In the area of isocyanates the developments have included:

- 1, 4 cyclohexylene diisocyanate (CHDI)
- p-phenylene diisocyanate (PPDI)
- m-trimethyl xylene diisocyanate (TMXDI)
- New trimerized and chain extended isocyanates
- Better understanding of the uses of MDI
- New blocked isocyanates

PPDI and CHDI are high-performance diisocyanates. These two isocyanates have demonstrated exceptional dynamic and thermomechanical properties in castable and TPU systems.

Among the aliphatic isocyanates for coatings, there has been continued growth in PPDI and trimerized HDI, along with TMXDI. The development of new liquid MDI systems as polymeric chain-extended MDI and MDI quasiprepolymers has increased the prospects for high-performance one-shot systems.

Polyols

The workhorses of the castable industry continue to be adipate esters. Caprolactone esters have established a strong following in select areas, although their true distinction from butylene adipates may in large part be their high and consistent quality. Poly(tetramethylene ether) glycols have continued to maintain market share against polyesters where superior low-temperature, dynamic, and environmental (microbial/hydrolytic) resistance is important.

Ethylene oxide (EO) capped propylene oxide ether glycols (PPGs) have established a small base in the cast elastomer market, providing moderate properties compared to other urethanes but still greatly improved toughness and abrasion resistance compared to conventional rubber. One particular strength of EO capped PPGs comes to light in one-component moisture-cure systems. Due to their high hydrophillicity compared to PTMEG or polyesters, sealants, coatings, and adhesives containing EO-capped

PPG moisture cure much faster. A new development in the area are the new high-molecular-weight, low unsaturated PPGs and EO-capped PPGs. Targeted first for the sealants market, these 2,000–6,000 molecular weight diols provide improved elasticity and toughness with good extensibility with fillers. Polycarbonate polyols continue to make inroads as high-performance coating systems, due to their excellent combination of UV resistance, toughness, and flexibility compared to acrylics.

Curatives

Curatives are driven by market demands for improvements such as

- Low-toxicity, noncarcinogenic, environmentally safe products
- Liquid state at room temperature
- Broad stoichiometric tolerance
- Manageable gel times with rapid cures
- Competitive cost/performance

A number of new amine curatives have been offered over the past 10 to 15 years. Some have carved out niches, and some have been dropped. Among the amine curatives available today are

- *MOCA.* Due to its long gel time and broad process tolerance, MOCA remains the dominant curative for TDI prepolymer systems above 80–85A Shore A hardness.
- *Ethacure.* These are room-temperature liquid substituted aromatic diamines. Ethacure 100 has very short gel times; consequently, its greatest utility is in RIM systems. Ethacure 300 is a moderately fast curative that provides good to excellent properties with both MDI and TDI prepolymers.
- *Unilink.* These are room-temperature liquid secondary aromatic diamines with a bulky alkyl substituted on the amine, which results in moderate reactivity and reduced hardness. These curatives are used alone and in conjunction with other curatives finding utility in flexible elastomer systems, coatings, and adhesives.
- *Polacure/polamine.* These are p-amino benzoate esters of various chain length diols. These materials have relatively low toxicity in urethane systems and are a Federal Drug Administration (FDA) approved ingredient for certain applications. The Polamine 740 has the disadvantage of a high melting point but produces urethanes of high durometer with excellent properties. The Polamine 250 is a low-melting solid and 650 is a liquid. These curatives produce moderate to high durometer systems with good to excellent properties. Higher-molecular-weight Polamines 1000 and 2000 are used for soft coatings and adhesives or may replace the prepolymer in *inverted* systems, i.e., systems in which the polamine acts as an amine terminated prepolymer, and liquid MDI acts as a curative.
- *Others.* Many aliphatic diamine and amino alcohols are used to cure the much less reactive aliphatic isocyanate systems, i.e., piperazine, propylene diamine, isophonone diamine, mono-ethanol amine and di-ethanol amine.
- *Diol curatives.* Butane diol, hydroquinone di-(B-hydroxy ethyl ether) (HQEE), and butanediol (BDO), together with minor amounts of 2-ethyl-2-(hydroxy methyl)-1,3 propane diol (TMP), are the main curatives for moderate and high durometer MDI prepolymer systems. Together with MDI, these systems provide superior resilience, environmental stability, and toughness. For flexible systems, a broad range of diols, triols, and polyols, along with a catalyst, are formulated into proprietary cure systems for a range of prepolymers.

Prepolymers

Two trends stand out in the castable market. First, due to their low NCO vapor pressure and excellent properties with non-MOCA curatives, MDI-based prepolymers continue to grow against TDI. Due to regulatory and health concerns, suppliers of TDI systems began moving slowly toward low free TDI prepolymers. However, as the market now begins to recognize the performance benefits of the combination of low free TDI and low molecular weight distribution, the move to low free TDI systems is accelerating. 1,4 trans cyclohexylene diisocyanate (CHDI) prepolymers have been marketed on a very limited basis. Here, the cast elastomer industry came to know what the coatings industry recognized all along; that is, the outstanding chemical resistance and hydrolytic stability of urethanes made from aliphatic isocyanates. CHDI, however, also provides mechanical properties equal to or better than the best aromatic isocyanates. CHDI's unique combination of excellent chemical resistance, toughness, and high resilience is of particular interest for rolls in wet paper processing, and for down-hole oil applications. Following the successful start-up of their PPDI business, Du Pont is soon expected to bring CHDI to market. Du Pont has introduced p-phenylene diisocyanate (PPDI) as a commercially available product. This has quickly been followed by the commercialization of PPDI-based prepolymers by several of the high-performance prepolymer suppliers. PPDI is a very compact, rigid, and symmetrical molecule.

Its first benefit is its ease of handling. It is available in a free-flowing flake form. Due to its non-symmetrical reactivity (i.e., the first NCO is 12 times more reactive than its second), PPDI readily produces low free PPDI prepolymer with low oligomer content. Cast elastomers and TPUs with exceptional dynamic mechanical properties, thermomechanicals, and cut growth resistance have recently been produced.

PPG-based prepolymers that provide the good mechanical properties of urethanes, but at a lower cost differential compared to conventional elastomers, are slowly establishing a market share. New technology in the area of PPG and EO PPG polyols may significantly enhance the performance of elastomers made from these polyols. As the market for urethane coatings has grown, so has interest in prepolymers based on aliphatic isocyanates. These are, however, low-volume specialties.

One-Component Systems

One-component systems are typically either moisture-cure systems or heat-activated blocked isocyanate systems. Moisture-cure systems are used extensively for adhesives, coatings, and sealants. In these systems, curing occurs by available NCO groups reacting with ambient moisture. Heat-activated block isocyanate systems are used extensively in baked coatings and in a variety of heat-cured castable systems.

With one-shot or inverse systems, the prepolymer is replaced by hydroxy or amine functional oligomer, and the curative is a quasiprepolymer, a trimerized isocyanate, a chain-extended isocyanate, or a polymeric isocyanate. This approach to formulating a liquid urethane system completely sidesteps the issue of high-melting-point amine curatives, residual free isocyanate, and viscous prepolymers.

Growth in this area has been enhanced by the development of new chain-extended MDIs and the availability of high-NCO quasiprepolymers. Because they are simple mixtures and not reactor products, one-shot systems are often economically formulated to meet exact requirements of a particular end use.

The development of Polamine 650, 1000, and 2000 has provided new opportunity for one-shot systems. These are amine functionalized polyols where the chain end is p-amino benzoate. The inductive effects of the para ester provides an aromatic amine functionality with sufficiently low reactivity to be used with liquid MDIs. This combination provides a room-temperature processible system with manageable gel times. The resulting systems have very good dynamic and mechanical properties over a range of hardnesses.

Applications

Design and styling changes in cars and trucks, and the rapid growth in recreational vehicles, have provided excellent opportunities for polyurethane foam, RIM, elastomers, and coatings. Emphasis on cost and productivity, plus better recognition of performance, have provided major inroads in mining, agriculture, and related materials handling. The recent boom in office machinery has provided opportunities for elastomeric belts, rolls, guides, and RIM housings. In the rapidly growing health care market, tubing, catheters, transdermal drug delivery systems, wound dressings, surgical drapes, and cardiovascular repair parts have all made major inroads.

Rapid growth in leisure and sports, along with rapidly changing trends, have created a wide range of opportunities. Notably, footwear, skate blades, motor sports, and skiing continue to be areas of high growth of rapid change.

21.2 Polyurethanes for Roll Covering

Polyurethanes are established as roll covering materials in a number of roll applications. Polyurethanes are being used in graphic rolls, offset printing rolls, industrial rolls and in paper mill rolls. One of the main driving forces behind their selection is outstanding toughness and wear resistance; particularly in hard roll covering. Polyurethanes also offer versatility in formulation; in fact, polyurethane rollers are being formulated from 10 Shore A hardness to 75 Shore D hardness. Although polyurethanes are more expensive than many rubbers used in roll covering they offer a favorable performance/cost ratio. For instance, in paper milling operations, a shutdown due to roll failure is extremely costly in downtime; therefore, a more durable roll that is longer lasting will more than compensate for high cost.

The dominant type of polyurethane used in roll covering is polyurethane made by casting from liquid urethane prepolymers; in particular, prepolymers made from toluene diisocyanate (TDI). Only these prepolymers will be described in this article. Prepolymers made from methylene-bis diphenyl isocyanate will not be described, although they find some use in hard roll applications.

21.2.1 Manufacture of Liquid Prepolymers

Table 21.1 lists some of the commercially available liquid urethane prepolymer manufacturers and their respective products. These prepolymers are available from each supplier in a large variety of types and hardnesses. Prepolymers are manufactured by capping low-molecular-weight aliphatic polyethers or low-molecular-weight aliphatic polyesters with approximately a one mole excess of toluene diisocyanate. The product is a low-molecular-weight prepolymer that is liquid at processing temperatures; however, at room temperature, it will often crystallize to a low-melting wax-like solid.

TABLE 21.1 Commercially Available TDI-Based Prepolymers

Name	Supplier
Adiprene	Uniroyal, Inc.
Andur	Anderson Development
Conahane	Conap
Cyanaprene	American Cyanamid
Multrrathane	Bayer, Inc.
Polathane	Air Products
Vibrathane	Uniroyal, Inc.

The molecular chain length of the polyester or polyether determines the hardness of the resultant polyurethane. For example, an increase in the molecular chain length, at the same mole ratio of TDI, will decrease the hardness of the polyurethane. The isocyanate-terminated prepolymers are quite reactive with a variety of active-hydrogen compounds. Among these are aromatic diamines, aliphatic diols, and even water. In the context of casting a roll covering, reaction with water is an undesirable side reaction. Maximum care, therefore, is taken to exclude water during the processing of prepolymers. Reaction with aromatic diamines or with polyols is the main method of chain extension or cure of these prepolymers.

21.2.2 Casting of Liquid Urethane Prepolymers

The prepolymer must be melted if solid at room temperature, or warmed (even if liquid) to reduce the viscosity to permit pumping into prepolymer storage tanks. The prepolymer is then maintained at 70 to 100° C. The total heat history imparted to the prepolymer is kept to a minimum to preclude branching reactions within the prepolymer. The liquid curative is maintained at 100 to 120° C, depending on type of curative used.

Mixing of the two liquid components, the hot prepolymer, and the hot diamine or polyol is usually done by metering through high-speed mixing machines. However, hand-mixing is sometimes used in preparing small rolls. The hot liquid mixture is cast onto a metal core in a heated mold. The covered roll is post-cured, usually overnight at 100 to 120° C, to develop the full mechanical properties of the polyurethane.

The hydrogen attached to the nitrogen of the urethane group is reactive with isocyanates although at a relatively slow rate. Therefore, a liquid urethane prepolymer has the capacity to react with itself. The result is formation of an allophanate branch. The consequence of this branching reaction, which results from an excessive heat history, is decay of the prepolymer and poor polyurethane properties.

Table 21.2, lists some effects of excessive heating on liquid urethane prepolymers. The isocyanate content is prematurely used up by the branching reaction with urethane. Branching also causes an increase in prepolymer viscosity, which affects processability. Significant decreases are brought about in the hardness of the polyurethane, rubber elasticity, and other mechanical properties. In the case of hard polyurethanes, excessive branching affects the semicrystalline morphology of these polyurethanes and therefore their properties. In processing liquid urethane prepolymers, the control of time/temperature before the mixing stage is critical if the maximum benefits of polyurethane toughness is to be realized.

TABLE 21.2 Effects of Branching in Prepolymers

Depletion of the isocyanate content
Increase in prepolymer viscosity
Decrease in polyurethane hardness
Loss of rubber elasticity
Loss of static and dynamic properties
Significant change in the morphology of hard polyurethanes

21.2.3 Hard Polyurethanes for Roll Covering

Hard roll covering is made by curing liquid urethane prepolymers with aromatic diamines. The predominant diamine used is methylenbis (orthochloroaniline), commonly known as MOCA. The polyurethanes formed with MOCA range from 70 Shore A to 75 Shore D hardness. The polyurethanes formed have a molecular chain structure of alternating flexible and rigid segments.

Hard polyurethanes do not require reinforcement with fillers as do other elastomers. Polyurethanes of this type are self-reinforcing because of a semicrystalline morphology; that is, a dispersion of semicrystalline domains in a rubbery matrix performs the analogous role of carbon black in rubber.

Table 21.3 shows some typical mechanical properties of polyether-based polyurethane cured with MOCA. A general indication of toughness in rubbery materials is a combination of high modulus, high tensile strength, and high elongation. These hard polyurethanes possess all of these characteristics. The tear strength shown here increases to very high values with increasing hardness. Stiff polyurethanes of this type have a high compression modulus and, although not shown here, a high dynamic modulus. Compression set is also quite low. In roll covering, particularly paper mill press rolls, low compression set and high dynamic modulus are very important.

TABLE 21.3 Hard Polyether Polyurethane Roll Covering Cured with MOCA

Shore Hardness	80A	90A	50D	60D	75D
Tensile modulus, 300%, MPa	7.9	1.9	28	–	–
Ultimate tensile strength, MPa	22	31.7	40	45	55
Ultimate elongation, %	500	380	350	270	180
Compression modulus, 10% deflection, MPa	1.8	3.8	4.4	14	30
Compression set, B, % 22 hours at 70° C	29	30	31	30	–
Tear strength, die C, kN/m	60	77	95	119	175

The outstanding abrasion resistance of polyurethane is shown in Table 21.4. In the Thelin Abrasion Test,[1] polyurethane and steel samples with free-flowing silicone carbide powder between them are abraded against each other. Fresh silicone carbide powder is periodically injected between the abrading surface. Sample weight loss is measured in both polyurethane and steel.

TABLE 21.4 Thelin Abrasion Resistance of Polyester Polyurethane

Shore Hardness	Grams Lost Per 20,000 Cycles (Silicon Carbide Powder)	
	Polyurethane	Steel
80A	0.0012	0.392
90A	0.0037	0.200
60D	0.0210	0.192
70D	0.0359	0.175

21.2.4 Soft Polyurethane Roll Covering

In making roll covering of soft polyurethanes, curing is accomplished by reacting prepolymer with a triol curative such as trimethylolpropane. A prepolymer is used, such as Cyanaprene A-9 urethane prepolymer, which normally yields a 90 Shore A hardness when cured with MOCA. When A-9 is cured with triol, a Shore A hardness of about 55 Shore A is obtained. A plasticizer is used to further reduce hardness level. These polyurethanes are heavily cross-linked and have an amorphous morphology.

Table 21.5 shows some properties of soft compositions. These soft polyurethanes have a lower range of mechanical properties than those polyurethanes cured with MOCA. However, these soft roll coverings are used in graphic arts, offset printing, and other industrial roll applications where resistance to solvents, combined with low hysteresis and low compression set, is important. The high level of cross-linking in these compositions is conducive to achieving these properties.

TABLE 21.5 Soft Polyester-Based PU Roll Covering; Prepolymer Cured with Triol Curative

Hardness, Shore A	1.5	20	30	40	50
Tensile modulus, 300%, MPa	0.4	0.8	1.0	2.4	3.4
Ultimate tensile, strength, MPa	2.1	5.9	11.3	12.7	19.7
Ultimate elongation, %	670	700	650	620	600
Compression set, B, % 22 hours at 70° C	1.0	1.0	1.0	1.0	1.0
Tear strength, Die C, kN/m	4.8	4.4	4.4	6.1	8.8

21.2.5 Environmental Resistance

Polyurethanes can be used under many environmental conditions, but not all conditions can be tolerated. Table 21.6 summarizes the response of polyurethanes, natural rubber, and neoprene to some environmental conditions. The absence of unsaturation in polyurethanes makes them resistant to ozone.

Generally, polyurethanes are more resistant than rubber to aliphatic solvents and oils. However, polyurethanes are less resistant to aqueous conditions, particularly at elevated temperatures. The upper service temperature of polyurethanes is about 100° C, but this depends on type of polyurethane. The polyether-polyurethanes are inferior to the polyester-polyurethane in resistance to heat and oxidation.

21.2.6 A Comparison of Polyester to Polyether Polyurethanes

Since the soft flexible segment, i.e., the polyester or polyether constitutes a major portion of the polyurethane elastomer, it is not surprising that the soft segment type will have a very significant effect on properties. Table 21.7 shows a comparison of properties between polyester-polyurethane and polyether-polyurethane. In making comparisons, it should be noted that the inferior polyurethane type still finds some use in applications where it is not called for because some compromise often must be made in some performance property.

At a given hardness, the polyester polyurethanes are the tougher and more wear resistant material. Nevertheless, the compression modulus and/or dynamic modulus in both types of polyurethane are approximately the same when compared at the same hardness. Polyester-polyurethanes can withstand

TABLE 21.6 Environmental Resistance Derived from TDI Prepolymers

Environment	Polyester polyurethanes	Polyester	Natural rubber	Neoprene rubber
Heat	G	F	F	G
Ozone	E	E	P	F
ASTM #1 oil	E	E	P	G
ASTM #3 oil	E	P	P	G
Aliphatic solvents	E	F	P	G
Chlorinated solvents	F–G	P	G	P
Aromatic solvents	F	P	P	F
Dilute acids	P	F	G	G
Dilute alkali	P	F	G	G

E = excellent; G = good; F = fair; P = poor

TABLE 21.7 Comparison of Polyester vs. Polyether Polyurethane Roll Covering

Property	Polyester	Polyether
Tensile strength	+	−
Toughness, wear resistance	+	−
Resistance to water	−	+
Resistance to fungus	−	+
Resistance to acid/base	−	+
Resistance to oil, solvents	+	−
Resistance to dry heat	+	−
Hysteresis	−	+

higher service temperature and are generally more resistant to oils and solvents. Thus, they are preferred in offset printing.

The polyether-polyurethanes are significantly superior in humid conditions and in acidic or base environments. They also exhibit lower hysteresis and therefore do not heat up under cycle loading as easily as the polyester type.

In summary, polyurethanes made from liquid urethane prepolymers can offer several advantages to roll covering. These advantages are

1. High toughness and wear resistance
2. Versatility of formulation
3. Good environmental resistance
4. High dynamic modulus in hard rolls
5. A high performance/cost ratio

However, to realize the outstanding mechanical properties of polyurethanes, strict control of the thermal history of the liquid urethane prepolymer is critical.

21.3 Illustrative Patents About Sealants

21.3.1 A-99 Used to Speed Windshield Sealant Cure with MDI and 20% Primary Alcohol

One-part moisture-curable polyurethane compositions have been used as adhesives, coatings, and sealants (see, e.g., U.S. Pats. 3,380,950, 3,380,967, 3,707,521, and 3,779,794). They provide convenient application and good ultimate physical properties. For example, most U.S. automobile manufacturers utilize one-part moisture-curable polyurethane sealants to bond front and rear windshield glass to automotive passenger car bodies. The resultant bonded windshield assemblies become integral structural parts of the car body and contribute to the roof crush resistance, thereby assisting the car manufacturer in meeting the requirements of DOT specification No. 216 (see 49 CFR 571.216).

For any adhesive, coating, or sealant composition, the ultimate physical properties available after cure (e.g., tensile strength, shear strength, weathering resistance, flexibility, etc.) are of great importance to the user. However, ultimate physical properties are not the only parameters by which such compositions can be evaluated. Most adhesive, coating, and sealant compositions provide a gradual buildup of physical properties during cure. It is highly desirable for such compositions to provide not only a high level of ultimate physical properties but also rapid attainment thereof. For example, a windshield sealant for use on cars manufactured by General Motors Corporation, not only must pass certain physical property tests specified by the automaker, but must pass such tests within six hours after sealant application. Similarly, adhesives, coatings, and sealants used in structural applications (e.g., building construction, general manufacturing, and the like) not only must provide good physical properties, (e.g., high bond strength) but also should provide such properties as rapidly as possible to speed assembly times and reduce fixturing costs. In general, for an adhesive, coating, or sealant composition with a given level of physical properties, the faster the rate of property buildup, the better.

An additional important parameter by which such compositions are evaluated is the *tack-free time*, that is, the time required for a sample of the composition to become non-tacky to the touch after exposure of the composition to ambient air. Compositions having rapid tack-free time enable parts bonded therewith to be subjected to subsequent operations (e.g., grinding, drilling, handling, packaging, and the like) that could contaminate parts joined with a tacky material. In general, for an adhesive, coating, or sealant having a given level of physical properties, the faster the tack-free time, the better.

A further important parameter by which such compositions are evaluated is the shelf life thereof, that is, the amount of time the compositions can be stored under typical storage conditions without significant loss of handling properties when uncured, physical properties when cured, or cure characteristics during cure. Unfortunately, compositions having rapid physical property buildup or rapid tack-free times typically also have attenuated shelf life, thus making it very difficult to obtain rapid physical property buildup, rapid tack-free time, and long shelf life in a single composition.

Schumacher[2] disclosed an invention related to adhesives and sealants for use in bonding articles to substrates. Also, this invention relates to a method for bonding articles to substrates and to cured assemblies thereof. The adhesive, coating, or sealant compositions containing prepolymer(s) are derived from MDI [or derivative(s) thereof], and polyol(s) containing primary hydroxyl groups, along with bis [2-N,N-dialkylamino) alkyl]ether(s).

The Schumacher invention provides, in one aspect, compositions having excellent ultimate physical properties combined with rapid physical property buildup, rapid tack-free times, and long shelf life, suitable for use as adhesives, coatings, or sealants. It also provides adhesive, coating, or sealant formulations for glass, comprising the above-described compositions and silane-containing primers, such primers being applied to the glass as a separate layer to which the formulations are subsequently applied, or being incorporated into the formulations as a component thereof, or both.

21.3.2 Bismuth Octoate Speeds Cure

Inasmuch as fast room temperature curing, single component sealants and adhesives are desirable and useful, particularly in original equipment manufacturing. It is not surprising that a number of one component elastomeric sealants are now available in the marketplace. Such sealants include various polymer bases such as polysulfides, mercaptan terminated polyethers, polysiloxanes and polyurethanes.

Certain industries need elastomeric adhesives or sealants cure by exposure to ambient conditions, and that will develop a high tensile strength. Applications of this type include sealing automobile windshields that are often intended as structural components in design. For these applications the elastomeric sealant or adhesive must not only have high tensile strength but should achieve such strength in a matter of a few hours so that the automobile may be safely driven shortly after installation of the windshield.

Of the various liquid elastomers available today, cured polyurethanes in general have the highest mechanical strength and therefore are the polymers of choice as a windshield sealant or adhesive—provided that the adhesive or sealant can cure rapidly under ambient conditions without exhibiting other problems such as foaming, storage instability, depolymerization, etc.

An example of a one-component, room-temperature, moisture-curing polyurethane sealant is disclosed in U.S. Pat. 3,779,794 wherein a polyurethane sealant in combination with a particular type of silane primer is described. In that patent, the polyurethane sealant is an isocyanate-terminated polyethylene ether diol-polypropylene ether triol combination having from 1.2 to 1.5% free isocyanate terminals. These terminals are blocked with a volatile blocking agent which evaporates when exposed to air, and the moisture in the air cures the polyurethane. According to the patent, the polyurethane sealant disclosed in this patent, in combination with the silane primer, cures to a tensile strength of 40–60 psi after a 6-hr exposure at 77° F and 30% relative humidity. Although such sealants are satisfactory in terms of ultimate elongation characteristics and the like there nevertheless is a need for the development of a sealant having higher early strength, i.e., 100 psi or more in a 6-hr period, with equivalent ultimate elongation.

Hutt and Blanco[3] disclosed a polyurethane sealant including a polyurethane having terminal isocyanate groups and containing a dual curing catalyst of organic bismuth and organic tin, said polyurethane being the reaction product of a liquid poly (lower) alkylene polyol having a molecular weight of 6,000 or more, and three to five hydroxyl groups, and a sterically unhindered aromatic diisocyanate. The polyurethane sealant may be used in conjunction with a silane-based primer and, when applied to glass surfaces, provides a high strength and rapidly curing seal.

It is an object of the Hutt and Blanco invention to provide polyurethane sealant systems for use in bonding glass and/or metal in combination with a silane-based primer, said polyurethane sealant systems having extremely fast cure rates when exposed to ambient moisture to produce high-strength seals within six hours without sacrificing elongation characteristics and other desirable features.

The present invention is based on the surprising discovery that one-component, room-temperature curing, stable polyurethane sealants having extremely rapid cure rates and high early strengths are produced by incorporating a dual-catalyst system. This system employs certain organic tin and bismuth salts in a polyurethane sealant produced by using sterically unhindered aromatic isocyanates to terminate liquid poly (lower) alkylene ether polyols having functionalities between three to five [i.e., poly (lower) alkylene ethers with three to five terminal functional hydroxyl groups per molecule] and molecular weights above about 6,000, along with a volatile blocking agent to block the isocyanate groups. A three-fold improvement in 6-hr ambient tensile strengths are found, compared with the revelations of U.S. Pat. 3,779,794.

The polyurethanes produced with the dual-catalyst organic tin and bismuth salts have very high tensile strengths, i.e., 100 psi and greater, within 6 hr after exposure to ambient moisture and temperature. Because of the ultimate higher strength, the polyurethane sealants of this invention allow higher extensions with plasticizers, thereby reducing the cost, yet still meet all the ultimate performance characteristics of automobile manufacturers.

21.3.3 Breakdown of Ketimine Due to Catalysis by Minor Amounts of Water

Sealant and coating compositions desirably have a combination of properties that render them particularly suitable for their intended applications. Such compositions should be able to be packaged in sealed containers or cartridges and stored for relatively long periods of time without objectionably *setting up* or hardening (as a result of cross-linking). When applied as a caulking sealant or coating composition, they should form a relatively tack-free surface soon after being applied and exposed to atmospheric moisture, and they should cure without the formation of bubbles within an acceptable time period. They should adhere tenaciously in the cured state to a wide variety of surfaces, such as to glass, aluminum, concrete, marble, and steel. The sealant or coating in the cured state should have sufficient elasticity and flexibility to withstand expansions and contractions of panels, etc., with which it is associated during temperature variations experienced as a result of climatic changes and to withstand wind forces that cause the associated panels to flex or twist.

Various sealant compositions have been proposed heretofore.

- U.S. Pat. 3,248,371 is directed to a blocked isocyanate-terminated urethane coating composition formed by reacting an isocyanate-terminated polyether-based urethane intermediate with a hydroxy tertiary amine.
- U.S. Pat. 3,267,078, the disclosure thereof being incorporated herein by reference, pertains to a coating composition that contains a blocked isocyanate-terminated polyether-based urethane intermediate and a diimine prepared by reacting a diamine with a carbonyl compound such as a ketone or aldehyde.
- U.S. Pat. 3,445,436, the disclosure thereof being incorporated herein by reference, discloses poly-urethane polyepoxide compositions formed by reacting a polyester or polyether triol, with or without a minor proportion of a diol material blended therewith, with a polyisocyanate to produce a liquid polyurethane prepolymer material, the liquid polyurethane prepolymer then being reacted with glycidol or a 2-alkyl glycidol (such as 2-methyl glycidol or 2-ethyl glycidol) to form the polyurethane polyepoxide. The polyurethane polyepoxide compositions can be cross-linked by the addition of an organic polyamine.
- U.S. Pat. 3,627,722 describes a sealant composition formed by the reaction of an isocyanate-terminated polyurethane prepolymer with a trialkyloxysilane such as N-methylaminopropyltri-methoxysilane.
- U.S. Pat. 3,632,557 discloses reacting an isocyanate-terminated polyurethane prepolymer with a trifunctional organosilicon compound such as gamma-aminopropyl trimethoxysilane. A stoichi-ometric excess of the aminosilane is used.
- U.S. Pat. 3,309,261 discloses the addition of an aminosilane to a polyurethane adhesive to improve the adhesive's lap shear strength and peel strength.
- U.S. Pat. 3,372,083 describes the use of a bituminous material that has been reacted with a polyisocyanate in combination with a polyurethane prepolymer.
- U.S. Pat. 4,067,844 describes a one-part polyurethane sealant composition formed by preparing an isocyanate-terminated polyurethane prepolymer and then reacting a minor proportion of the NCO terminates with specified aminosilane materials or the reaction product of a mercaptosilane with a monoepoxide or the reaction product of an epoxysilane with a secondary amine.

Barron and Wang[4] disclosed a sealant or coating composition comprising

1. A blocked isocyanate-terminated prepolymer in which essentially all NCO groups are blocked
2. A multifunctional imine essentially free of any amine functionality formed by the reaction between a primary multifunctional amine with a ketone or aldehyde
3. From 0.1 to 10 parts by weight of an organosilane based on 100 parts by weight of the blocked isocyanate-terminated prepolymer

The invention relates to curable one-part blocked isocyanate-terminated polymeric compositions that are useful as caulking sealants and coating compositions.

In accordance with the Barron and Wang invention, a one-part sealant or coating composition is provided that comprises

1. A blocked isocyanate-terminated prepolymer in which essentially all (desirably at least 90 percent) NCO groups are blocked
2. A multifunctional imine essentially free of any amine functionality formed by the reaction between a primary multifunctional amine with a ketone or aldehyde
3. From 0.1 to 10 parts by weight of an organosilane based on 100 parts by weight of said blocked isocyanate-terminated prepolymer

Sealants or coating compositions of the present invention are moisture curable compositions that adhere tenaciously to a wide variety of surfaces, form tack-free surfaces after being exposed to the atmosphere for only a short period of time, and form a sealing cure within a matter of hours after being applied.

21.3.4 Process Combining an Aldimine and Oxazolidine

Traditionally polyurea polymers and polyurethane polymers are prepared by mixing polyisocyanates with polyamines and polyhydroxy compounds, respectively. If polyamines or polyhydroxy compounds are used as curing agent or component of a curing agent, then the mixtures of polyisocyanates and said curing agents are stable only for very short periods, and they have to be used immediately after mixing.

It is known that oxazolidines, like aldimines and ketimines (i.e., condensation products of keto compounds with amines, generally named Schiff bases), do not react or react only very slowly. Polyaldimines, polyketimines, and polyoxazolidines are described in the art, which also documents their use as components of mixtures with polyisocyanates. Reference is made to British Pat. 1,064,841. Such mixtures are stable for longer periods if humidity is absent. When they are contacted with water or environmental humidity, a polyamine is liberated by the hydrolysis, and it reacts with the polyisocyanate of the mixture, forming urea bonds.

Zabel et al.[5] patented new compounds having one, two, or three aldimine groups and one, two, or three oxazolidine groups in their molecule. They are curing agents for organic polyisocyanates. Mixtures of said new compounds with polyisocyanates are stable for long periods if humidity is absent; if, however, they come into contact with water or environmental humidity, they rapidly cure, resulting in elastic or hard polymers. The new compounds are prepared by reacting a diamine or polyamine with up to six amino groups with an epoxy compound, yielding a polyamino alcohol, which is thereafter reacted with an aldehyde, e.g., formaldehyde, an aliphatic, cyclic, or heterocyclic or aromatic aldehyde.

Surprisingly, Zabel et al. found that new compounds having one or more aldimine groups and one or more oxazolidine groups, if mixed with polyisocyanates, result in mixtures having an extremely long shelf life, provided that the access of humidity is prevented. If, however, said mixtures are contacted with water, the polymerization proceeds very quickly, resulting in hard or elastic polymers. The so-cured polymers have a good resistance against chemicals and solvents. It is an object of the present invention to provide new compounds having one or more aldimine groups and one or more oxazolidine groups in their molecule.

21.3.5 Fibers Entangle Each Other to Form a Network

Since 1965, polyurethane joint sealants have been used extensively in the elastomeric sealant industry because of their quality and economy as compared to other sealants. These have functioned satisfactorily in joints whose movement is small. However, joint widening and closing is a serious problem when the variation in width of the joint is substantial, such as 20–25% or more, because the large tensile forces applied to the sealant can cause adhesive failure or failure of the sealant material itself.

To avoid such failure in joints subject to large variations in width, particularly those frequently changing width due to temperature changes and the like, it is necessary to provide a soft, low-modulus elastomer with excellent tensile and elongation properties and excellent weathering properties. This is less of a problem in highway joints where the sealant can be poured in place, because non-sag properties are less important. In vertical joints between the concrete panels of a modern building, the problem is very serious because of the importance of sag resistance. If a caulking gun is used, the problem is even more difficult because of the need for good flow properties.

A simple approach to the problem of sealing joints subject to severe variations in width is to employ a non-sag, elastomeric-type silicone caulking composition that produces a soft, low-modulus elastomeric product. One-component silicone joint sealants are suitable for this purpose but are relative expensive and energy wasteful, have poor tear resistance, tend to collect dirt, and cannot be painted.

There is a one-component moisture-curing polyurethane joint sealant made by a Swiss company (SIKA) that employs polyvinyl chloride to achieve non-sag properties without an undue increase in the durometer hardness of the cured sealant. This type of sealant that is disclosed in U.S. Pat. 4,059,549, has poor package stability and poor adhesion to water-soaked masonry, and is more difficult to prepare than more conventional polyurethane sealants. (See Table 21.1.)

There is also a two-component epoxy-type polyurethane joint sealant. It has been on the market for many years under the name *DYMERIC*, as disclosed in U.S. Pat. 3,445,436, filed June 14, 1966. This two-component sealant has non-sag properties and, when cured, has a low modulus (e.g., a Shore A durometer hardness of 45 or less) comparable to that of the SIKA sealant but has poor adhesion to water-soaked masonry and requires mixing on the job.

It is preferable to employ one-component sealants and to avoid the extra time and expense involved in mixing two components at the point of use. One-component, non-sag, moisture-curing polyurethane caulking compositions have been very popular for simple joint sealing operations because of the economy of application, but they present an extremely difficult problem to the compounder. Prior to the present invention, it has not been possible to provide a low-modulus, one-component polyurethane sealant of the desired softness with, at the same time, the desired sag resistance and flow characteristics that could be easily manufactured and that meets the federal specifications for Class A, cold-applied elastomeric-type joint sealants. Such federal specifications were developed more than a decade ago by the National Bureau of Standards and are set forth in Federal Specifications TT–S–00227B and TT–S–00230C.

Except for the above-described sealants of U.S. Pats. 3,445,436 and 4,059,549, there was no non-sag polyurethane joint sealant available prior to this invention that could meet such federal specifications for a Class A rating. The known polyurethane sealants could not meet the Class A requirements because they had excessive durometer hardness (e.g., a Shore A durometer hardness of about 60 or greater) when compounded to provide the necessary non-sag properties. Manufacturers of caulking compositions unable to obtain such Class A rating were often unable to sell their product because contracts involving Government construction projects required Class A sealants meeting the federal specifications.

Conventional compounding techniques make it possible to vary the elastomeric properties and flow properties of a polyurethane sealing composition, but it has been impossible to achieve the desired combination of extrudability, non-sag, and low-modulus merely by judicious choice of compounding ingredients or fillers.

No solution to the problem is provided by conventional fillers and reinforcements for polyurethanes and plastics, such as silica, talc, calcium silicate, Wollastonite, Asbestine, kaolin, barium sulfate, graphite, hydrated alumina, chrysotile, serpentine, pearlite, vermiculite, mica, crocidolite, zirconium silicate, barium zirconate, calcium zirconium silicate, magnesium zirconium silicate, glass beads, fiberglass, titanium dioxide, PMF mineral fiber, nylon fiber, polyester fiber, alpha cellulose fiber, polypropylene fiber, and the like. Satisfactory fillers for a joint sealant do not provide the desired combination of low-modulus and non-sag properties when used alone or in combination with other conventional fillers. It therefore has been generally accepted that mixing of different fillers provides no synergistic result and is of no particular consequence when seeking such a combination of properties. This is also true of the conventional fibrillar fillers such as Wollastonite, fibrous talc, asbestos, and processed mineral fiber (PMF).

Fibers used for reinforcing polyurethane products, such as glass fibers, nylon fibers, polyester fibers, polypropylene fibers and other synthetic fibers, are generally unsatisfactory for polyurethane joint sealants, and particularly for one-component caulking compositions. They have poor compatibility, do not disperse properly, and produce a coarse surface that is unattractive and unacceptable. For these reasons, they have not been used in joint sealants.

Another fiber material that has not been used in sealants is synthetic wood pulp fiber. For many years, synthetic fibrillar polyethylene and polypropylene fibers have been produced as a substitute for cellulose fibers with various fiber lengths from 0.7 to 2.5 mm or more. These synthetic wood pulp fibers, sold under the name SWP, have been used for various products such as wallpapers, packaging papers, electrical paper, cigarette filters, photographic paper, molded products, masonry cement, body-molding compounds, roofing compounds, traffic paints, plaster repair, non-woven fabric, texture paints, and the like.

Evans and Leonard[6] found an improvement in the properties and capabilities of elastomeric polyurethane joint-sealing compositions by incorporating a fibrillated polyolefin of high surface area that is rendered compatible by combining it with effective stabilizing cofillers such as titanium dioxide, calcium carbonate, carbon black, fibrous talc, serpentine, kaolin, or various other metal silicate fillers. Such cofillers are capable of improving compatibility and stabilizing the sealant mixture so that unacceptable sweat out or exudation of liquid is prevented, and so that the outer surface of the applied sealant has an acceptable surface quality after curing.

The invention relates to cold-applied elastomeric-type polyurethane joint-sealing compounds for sealing, caulking, and glazing operations in building construction, and more particularly to non-sag sealants that provide soft low-modulus elastomers suitable for vertical joints, and other joints subject to extreme variation in width.

When 2–8% by weight of fibrillated polyolefin of macrofibrillar structure, preferably containing macrofibrils with diameters from 1–10 microns, is incorporated in a polyurethane joint sealant along with 8–30% of stabilizing cofillers, it becomes possible to provide revolutionary new joint sealants with remarkable properties, such as Boeing sag values of 0.1 and below for caulking compositions applied to extremely wide joints. These unique non-sag joint sealants can be compounded to form strong, soft, low-modulus elastomers ideally suited for vertical joints subject to extreme cyclical movement as is common in modern building construction.

Prior to this invention, SWP fibers have been tried in polyurethane joint sealants and other polyurethane compositions and found unsatisfactory, particularly because of poor compatibility, excessive sweat out, and unacceptable surface quality in the cured product. The fibers tend to project beyond the surface of the cured sealant and provide a coarse surface that is unacceptable.

21.3.6 Silane-Terminated Prepolymer and Use of Clay

U.S. Pat. 3,632,557, granted Jan. 4, 1972 to Brode et al., reveals silicon-terminated organic polymers that are curable at room temperature in the presence of moisture. The patent indicates that such polymers, particularly after the inclusion therein of fillers that are conventional for incorporation into elastomeric compositions, can be used for coating, caulking, and sealing.

Indeed, U.S. Pat. 3,979,344, granted Sept. 7, 1976 to Bryant et al., and U.S. Pat. 4,222,925, granted Sept. 16, 1980 to Bryant et al., describe sealant compositions, curable at room temperature in the presence of moisture, comprising the organosilicon polymers of the Brode et al. patent in combination with specific additive materials. Thus, the first of the Bryant et al. patents describes the addition of N-(beta-aminoethyl)-gamma-aminopropyltrimethoxy silane to a sealant composition containing such a silicon terminated organic polymer. The second of the aforementioned Bryant patents relates to the inclusion of a particular carbon black filler in such compositions. These patents reveal the utility of the silicon terminated organic polymers, or of sealant compositions containing the same, for forming tenacious bonds to nonporous surfaces, particularly glass.

The silicon-terminated organic polymers of the present invention are similarly moisture-curable at room temperature and can similarly be formulated into sealant compositions comprising fillers and other additives. As do the prior art materials, the polymers and compositions of the present invention show particularly good adhesion to nonporous surfaces such as glass and are particularly characterized by a very rapid cure rate that facilitates the use of the polymers and sealant compositions compounded with them in industrial applications—for example, as sealants for automotive glass.

The silicon-terminated organic polymers of the Brode et al. patent are prepared by reacting a polyurethane prepolymer having terminal isocyanate groups with a silicon compound containing alkoxysilane groups, and having a mercapto group or a primary or secondary amino group reactive with isocyanate groups. Upon reaction of the mercapto or amino group with the terminal isocyanate groups of the polyurethane prepolymer, a moisture-curable polymer having terminal hydrolyzable alkoxysilane groups is formed. These terminal alkoxysilane groups, in the presence of atmospheric moisture, react to form siloxane (-Si-O-Si-) groups, possibly by way of intermediate silanol formation. The formation of the siloxane linkages not only cross-links and cures the moisture-curable polymer but also promotes adhesion of the polymer to nonporous surfaces such as glass surfaces. Combined with the hydrolyzable alkoxysilane groups, they form particularly tenacious bonds in the presence of atmospheric moisture.

Rizk et al.[7] disclosed the method of making a moisture-curable silicon terminated organic polymer which comprises reacting a polyurethane prepolymer having terminal active hydrogen atoms with an isocyanato organosilane having, at least, one hydrolyzable alkoxy group bonded to silicon, silicon terminated organic polymers so produced, and moisture-curable sealant compositions comprising such a silicon terminated polymer. The invention relates to moisture-curable silicon terminated organic polymers, to methods of making the same, and to moisture-curable sealant compositions comprising such moisture-curable polymers.

According to the Rizk invention, a polyurethane prepolymer is similarly reacted with an organosilane compound having one or more hydrolyzable alkoxysilane groups. However, according to the invention, the polyurethane prepolymer has terminal active hydrogen atoms, present in groups such as hydroxy groups, mercapto groups, or primary or secondary amino groups. These active hydrogen atoms are reacted with an isocyanate group present in the organosilane compound. As in the prior art, urethane, thiourethane, or urea groups are produced by the reaction, but the nature in which these linking groups bond the terminal alkoxysilane groups to the polyurethane prepolymer differs from that known in the prior art, and it accounts for the improved properties of the claimed polymers and sealants.

References

1. J.H. Thelin. *Rubber Chemistry and Technology* 43, 1503, 1970.
2. Schumacher, Gerald F. U.S. Pat. 4,511,626, April 16, 1985, "One-Part Moisture-Curable Polyurethane Adhesive, Coating, and Sealant Composition," to Minnesota Mining and Manufacturing Company, St. Paul, Minnesota.
3. Hutt, Jack W., and Blanco, Fernando E. U.S. Pat. 4,284,751, August 18, 1981, "Polyurethane Sealant System."
4. Barron, Larry R., and Wang, Pao Chi. U.S. Pat. 4,507,443, March 26, 1985, "Sealant and Coating Composition," to The B. F. Goodrich Company, Akron, Ohio.
5. Zabel, Lutz D., Widmer, Jurg, and Sulser, Ueli. U.S. Pat. 4,504,647, March 12, 1985, "Compounds Having One or More Aldimine and Oxazolidine Groups, Process for Their Preparation and Their Use as Curing Agents for Polyisocyanates," to Sika A.G., Kaspar Winkler & Co., Zurich, Switzerland.
6. Evans, Robert M, and Leonard, Thomas M. U.S. Pat. 4,318,959, March 9, 1982, "Low-Modulus Polyurethane Joint Sealant," to MAMECO International, Inc., 4475 E. 175th Street, Cleveland, Ohio.
7. Rizk, Sidky, D., Hsieh, Harry W.S. and Prendergast, John J. U.S. Pat. 4,345,053, August 17, 1982, "Silicon-Terminated Polyurethane Polymer," to Essex Chemical Corporation, Clifton, New Jersey.

22

Medical Applications

22.1 Background

Medical applications of polyurethane elastomers contribute significantly to the quality and effectiveness of the nation's health care system. These products range from nasogastric catheters to the insulation on the leads of electronic pacemakers implanted in many cardiac patients.

The Food and Drug Administration (FDA) estimates there are now 2,600 different kinds of medical devices and another 2,000 diagnostic products. Many critical components of these devices and diagnostic products are fabricated of polyurethane elastomers, since these polymers offer an unsurpassed combination of biocompatibility, performance and ease of manufacture.

The Pharmaceutical Manufacturers Association estimated 1996 sales of medical devices and diagnostic products at about $25 billion. Yet, many chemists or chemical engineers are unaware of the technical revolution in medicine brought about by the utilization of biomedical polymers. This chapter presents the story of one family of biomedical polymer, the polyurethane elastomers, and how these elastomers are helping physicians diagnose disease and save lives of critically ill patients.

22.2 High-Performance Polyurethanes

Among the highest-performing biomedical-grade elastomers are the polyurethanes, block copolymers containing blocks of high-molecular-weight polyols linked together by a urethane group. These have the versatility of being either rigid, semirigid, or flexible. These materials, in general, have excellent biocompatibility, outstanding hydrolytic stability, superior abrasion resistance, excellent physical strength, and high flexure endurance. They have been described as resistant to gamma radiation, oils, acids, and bases.

Thermoplastic polyurethane elastomers consist of essentially linear primary polymer chains. The structure of these primary chains comprises a preponderance of relatively long, flexible, soft chain segments that have been joined end to end by rigid, hard chain segments through covalent chemical bonds. For medical applications, the soft segments are diisocyanate coupled, low melting polyether chains. Hard segments are formed by the reaction of diisocyanate with the small glycol chain extender component.

22.3 Peritoneal Dialysis Applications

In end-stage kidney disease (ESKD), the kidneys cease to perform their function of cleansing and purifying the patient's blood. Systems for treatment of ESKD provide a substitute for kidney function by removing water and toxic waste materials from the blood. The alternative to these substitute waste removal systems is kidney transplantation. However, this cannot be considered as a viable medical treatment for the millions affected by kidney disease, because of limited availability of donor kidneys and numerous medical complications inherent in this method of treatment.

Approximately 200,000 patients throughout the world are presently receiving treatment for ESKD. According to published data, approximately 65,000 patients receive treatment in the United States, and approximately 110,000 receive treatment in Japan and Western Europe, with the remaining patients in various other countries.

Financial assistance is sustained primarily by government reimbursement programs. Almost all dialysis patients require financial assistance provided by government programs to pay for dialysis services and materials. The cost of treatment in the United States is approximately $25,000 per year if the patient is treated at a kidney center, and approximately $18,000 per year if treatment is performed at home. Due to the high cost of treatment, the Department of Health and Human Services has promulgated regulations to limit the level of reimbursement for dialysis in the United States. These regulations generally provide for equal rates of payment to hospitals and kidney centers, regardless of whether treatment is provided in a clinic or at the patient's home. Due to the lower overhead and personnel costs associated with dialysis carried out in the home, these regulations may have the effect of encouraging home dialysis.

Hemodialysis is currently the predominant treatment method for ESKD. Hemodialysis requires the continuous external circulation of blood through a dialyzer (artificial kidney) to remove water and toxins during the treatment and is performed in a hospital, clinic, dialysis center, or in the patient's home. Treatments are usually performed three times a week, with each treatment lasting from four to six hours.

Alternative methods of treatment to hemodialysis include hemofiltration and peritoneal dialysis that could be performed at home. Hemofiltration cleanses the blood by using a filter to separate fluids containing waste products from the blood. These fluids are then replaced with other sterile solutions. Peritoneal dialysis that could be done at home consists of the periodic infusions of sterile fluids into the patient's abdomen through a surgically implanted catheter and the use of the peritoneum (a membrane that surrounds the abdomen and its contents) as a path to remove toxic waste from the blood. Tecoflex® tubing has received FDA approval for use as peritoneal dialysis catheters.

Enteral feeding means, literally, *to provide nutrients through the gut.* Total enteral feeding is gaining acceptance in several clinical states such as post operative care, anorexia, burn patients, trauma, head and neck cancer, and long-term convalescent care, where swallowing may be painful or impossible to the patient. Until recently, nutritional support to impaired patients could be done only intravenously; however, the intravenous route is highly prone to infection. Many patients develop an allergic reaction (anaphylactic shock) to the large-scale delivery of nutrients directly to the blood stream, resulting in fever, convulsions, and other undesirable conditions.

Now, with the advent of enteral feeding, nutritional modular supplements allow physicians the flexibility of customizing a nutritional care plan specifically suited for each patient's need. Enteral feeding tubes are constructed of elastomeric, biomedical-grade elastomers that are nonirritating, flexible, and comfortable. The tubes are designed with a weighed bolus and a stylet, to allow easy transnasal insertion into the esophagus, and on into the stomach or small intestine. Once in place, the stylet is removed, and feeding is initiated at desired intervals.

In the manufacture of enteral feeding tubes, a special grade of polyurethane is compounded with 20% by weight of barium sulfate. The inclusion of radiopaque barium sulfate allows physicians to observe the entry and final position of the catheter tube under X-ray confirmation during passage through the gastroesophageal and pyloric sphincters.

22.4 Cardiac Pacemaker Leads

Cardiac pacing systems are utilized for patients with some impairment of the heart's natural electrical system, which limits the heart's ability to pump blood throughout the body at a rate suitable to fulfill the body's needs. Under normal circumstances, the rhythmic contractions of the heart are stimulated by small electrical signals emitted by the sinus node, the heart's natural pacemaker. These signals are conducted downward along nerve fibers to the four chambers of the heart, first to the two upper chambers or atria, which serve to prime the two lower chambers or ventricles. These, in turn, perform the principal pumping function of the heart.

When the heart's natural pacing mechanism is impaired, a pacemaker may be used to remedy the problem by electrically stimulating the heart to restore its regular rhythmic muscular contractions. Pacemakers are generally implanted and connected to the heart by means of insulated wire leads and electrodes called lead/electrodes. Since the right and left sides of the heart contract in unison, it is necessary to pace only one side, and only the right side is commonly paced. Most modern pacemakers are hermetically sealed, usually in titanium metal cases containing a long-life lithium-based battery and appropriate electronic circuitry to generate pulses and control their characteristics. Pacemakers currently on the market range from approximately 35–100 g (1.25–3.5 oz) in weight, and from approximately 15–50 cm^3 in volume. The useful lives of these pacemakers typically range from 8–15 years.

The cardiac pacing industry began in the 1960s, with the development of relatively simple, single-chamber pacing devices providing a continuous stream of electrical pulses at a fixed rate, generally to the right ventricle of the heart. By the early 1970s, the industry had almost entirely converted to more advanced single-chamber *demand* pacemakers capable of sensing or monitoring the heart's natural rhythm, and providing stimulation only when the natural rhythm is inadequate.

In the mid-1970s, programmable pacemakers came into increasing use, offering the ability to modify, without surgery, several operating parameters of the pacemaker, including pulse rate and pulse energy, simply by activating a programming unit communications head held against the skin over the implanted pacemaker. The ability to adjust certain operating parameters enabled the physician to regulate the pacemaker to best meet the patient's needs, to extend the expected life of the pacemaker under certain conditions, and to eliminate the need for surgical replacements, in some cases, where the physiological condition of the patient's heart had changed. Later in the 1970s, such external control was extended in multiprogrammable pacemakers, enabling the physician to program additional operating parameters.

Another important development during the late 1970s was the creation of pacing systems that enabled bidirectional telemetry between the implanted unit and an external programmer. This improvement enabled the physician, for the first time, to interrogate the implanted unit and to receive back information confirming that the pacemaker had received and responded correctly to the programming signals, and to ascertain the operating characteristics of the implanted pacemaker. Such telemetry systems also permitted transmission of important physiologic and operating data on how the heart and pacemaker function together.

The next advance in pacing systems was the development of dual-chamber devices, the first practical versions of which were released commercially in the early 1980s. Unlike single-chamber systems, which stimulate only the right ventricle of the heart through a single lead/electrode, dual-chamber systems are capable of synchronized stimulation of both the right atrium and right ventricle through two separate lead/electrodes. This synchronized stimulation can increase blood flow with less strain on the heart. As a result, physicians are recommending the use of dual-chamber systems for increasing numbers of patients.

Recently, more advanced dual-chamber systems have been developed. These devices not only provide synchronized stimulation of both chambers but are also capable of sensing and tracking in both chambers as well. When dual-chamber pacemaker senses that there is a regular natural atrial rhythm, it will ensure that the ventricles track this rhythm in synchrony. When a regular atrial rhythm is not present, an atrial pulse is delivered by the system followed by a synchronized ventricular pulse. Because dual-chamber systems more closely simulate the heart's actual physiological rhythm, thereby more effectively meeting the individual physiological needs of each patient, these pacemakers are expected to increase from 20% in 1984 to >50% of all implanted pacemakers by 2000.

Based upon estimates from industry analysts, the world-wide pacemaker utilization reached 300,000 units in 1996, representing a $1.5 billion market, of which approximately 60% was implanted in the U.S. Until recently, silicone elastomers were the polymers of choice for single-chamber pacing; however, with the introduction of dual chamber pacing, a stronger and more slippery polymer became necessary.

Compared to standard silicone rubbers, polyurethane elastomers offer several advantages including higher tensile strengths, significantly higher tear resistance, and excellent abrasion resistance. In the manufacture of cardiac pacing leads, these advantages have resulted in the introduction of polyurethane leads with significantly reduced wall thicknesses. These thinner-wall leads have resulted in easier surgical insertion, less traumatic introduction of multiple leads when inserted into single veins (for dual-chamber pacing), and greater elasticity for the implanted lead.

For dual-chamber pacemaker lead insulation, a thermoplastic urethane polymer appears to offer the best combination of properties. Polyurethanes display high tensile strength (<5000 psi), are highly elastic, are exceptionally blood compatible, and soften significantly one hour after insertion into the human body. This property facilitates the insertion of a semirigid pacemaker lead, which then becomes compliant and less prone to damage the delicate lining of the heart, blood vessels, and other sensitive surfaces of the human body.

22.5 Infusion Pumps

New drug delivery systems are under intense development activity for all routes of drug administration. Statistically, it is likely that buccal/intestinal absorption of drugs will remain the predominant form of delivery. Although other routes of administration have included occular, nasal, vaginal, and rectal mucosa as well, in past years there has been a growing interest in novel methods for controlled drug delivery to ambulatory (nonhospital) patients.

Controlled drug delivery is a relatively new concept in pharmaceutical product development, offering safer and more effective means of administering drugs. It involves releasing a drug into the bloodstream or delivering it to a target site in the body at predetermined rates over an extended period of time. The mechanisms for controlled drug delivery are varied, but in all cases the goal is to provide more effective drug therapy and patient compliance while reducing or eliminating many of the side effects associated with conventional drug therapy. One method of eliminating side effects, increasing patient compliance, and treating chronic disease is to deliver constant intravenous therapy by means of infusion pumps.

Infusion pumps consist of a piston, roller, or peristaltic pump that introduces fluids into a patient in a controlled and predictable manner. In a typical application, the pump is primed with a drug solution and connected to a vein via an intravenous line involving a needle and catheter. Once the intravenous connection has been accomplished, the pump is turned on, and drug delivery is initiated.

Constant intravenous therapy on an ambulatory basis has been achieved by the development of infusion pumps that deliver minute amounts of liquid at extremely constant rates over long periods of time. The pump's major use would be to deliver small amounts of highly potent water-soluble medications. Other types of compounds, e.g., antibiotics, might seem to be useful applications. However, the amount of antibiotic used would require extremely high concentrations that might not be achievable because of solubility limits. It also would involve the risk of causing considerable pain and inflammation at the site of entry into the vein.

Another major application could be the treatment of insulin-dependent diabetes. Hormones of this type are very soluble, reasonably but not entirely stable, and potent in very low concentrations, so a large reservoir of material could be placed internally and infused without interruption for weeks. Techniques for refilling these infusion pumps without their complete removal also exist. Currently, it is believed that close regulation of diabetes will prevent many of the long-term complications of the disease. One way to achieve close regulation would be the capability of constantly infusing insulin. The danger would be that unexpected periods of little or no food intake could lead to relative insulin excess and shock. The ultimate technology will be the constant monitoring of the blood sugar levels and the capacity to adjust the infusion rate to the blood sugar level.

Insulin-dependent (Type 1) diabetics, at over 1.4 million in the U.S. alone, constitute the single largest patient population for which evidence has been accumulating rapidly of improved efficacy of drug administration via an ambulatory, external, or wearable syringe pump, as they are variously called. Other large U.S. markets in which interest is quickly rising in formulating pharmaceuticals to enhance clinical effectiveness of chronic infusion include cancer chemotherapy (850,000 new cases per year), chronic or intractable pain syndromes (1.8 million), and chronic antibiotic therapy (0.5 million). The rapidly escalating interest in this field is, likewise, reflected by the variety of pumps either commercially available or slated for market entry—25 models from 17 different manufacturers.

Product development in infusion pumps has advanced and promises to continue to advance at an astounding rate. From purely experimental devices (heavy, complicated to operate, fragile) major strides have been made toward redesign of pumps that are less obtrusive, more reliable, easier to operate, and less sensitive to the wear and tear of normal living.

Table 22.1 lists some of the drugs, chemicals, hormones, and related diseases for which frequent or chronic administration has been suggested to improve therapeutic effectiveness. The drug list in Table 22.1 is not exhaustive, but it is clear from this partial list that the potential impact of chronic infusion pumps in the practice of internal medicine (particularly diabetology, oncology, endocrinology, and neurology) over the next five to eight years may well become profound.

TABLE 22.1 Therapeutic Applications Improved by Infusion Pump Administration

Disease	Drug
Acute cardiac failure	Nitroglycerine
Asthma	Theophylline
Cancer	Antineoplastic agents
Cardiac arrythmia	Lidocaine
Chronic infection (i.e., osteomyelitis)	Antimicrobial/antineoplastic agents
Chronic intractable pain	Morphine, opiates, endorphins
Depression	Lithium
Diabetes insipidus	Vasopressin
Diabetes mellitus	Insulin
Glaucoma	Pilocarpine
Growth disorders, infertility, etc.	Hormone replacement 9GH, LH/RH)
Hemophilia	Factor VIII
Hypertension	Beta blocker
Hypertension	Clonidine
Menopause, female hypogonadism	Estradiol
Parkinson's disease	Dopamine
Respiratory distress	Scopolamine
Rheumatoid arthritis	NSAIDs
Seizure disorders	Dilantine, phenobarbital, etc.
Thalassemia	Deferoxamine
Thrombosis	Heparin
Ulcerative colitis	Azulfidine, steroids

Several forces are driving this newly emerged technology-linked-to-therapy market. In the broadest, simplest sense, when a drug or chemical is best administered slowly and chronically by reason of limited bioavailability or metabolic half-life, and control over variable delivery rates in the range of microliters to tenths of milliliters per hour is required, ambulatory drug infusion pumps are rapidly gaining recognition as a treatment or drug administration modality of choice. Drug infusion devices are providing a new level of precision over delivery rates, which will allow therapeutic pharmacologists and medicinal chemists to pay attention, as never before, to the implications of pharmacokinetics and pharmacodynamics in new drug development.

The alternative of lying in a hospital bed with an intravenous (IV) line or another route of administration apparatus at upward of $500 per day (assuming only medication and room costs) is not a happy state of emotional or financial affairs for most people. Therein lies a major market force—cost containment. The home health care industry experienced early, explosive growth on the basis of total parenteral and enteral nutrition services alone. Recent projections put this market at >$300 million per year by 1995 plus another $50 million for insulin delivery systems and $50 million for other drug delivery systems including chemotherapy and antibiotics.

For the 1.4 million insulin-dependent diabetics, continuous insulin infusion means an end to daily cycles of hyper- and hypoglycemia and, with normalized blood sugar levels, a far greater chance of avoiding the devastating micro- and macrovascular side effects of intermittent insulin replacement therapy. Similarly, on any given day, approximately 300,000 people undergo cancer chemotherapy in this country alone. Add antibiotic and chronic pain therapy, plus others as suggested in Table 22.1 and, given clinical evidence of improved efficacy with chronic, low-level drug therapy, one cannot escape seeing an impressive picture of large and growing markets representing a major change in medical and pharmaceutical technology. Already, the ethical pharmaceutical industry has responded by reformulating trademarked drugs for wearable pump use, especially those about to expire from patent life, thereby extending the drug's proprietary life.

It is clear that the pump manufacturers and drug houses have just begun to find ways to jointly respond and shape a market force of major dimensions. Approximately 20 companies manufacture, or are clinically testing, wearable/implantable infusion pumps. While the 1982 market for these innovative pumps did not exceed $10 million, their versatility and clinical usefulness are expected to gain rapid acceptance and approach $700 million by the year 2000.

Two principal types of intravenous catheter placement systems currently are being marketed: over-the-needle systems and through-the-needle systems. In an over-the-needle system, a catheter is placed over the needle by molding or shrinking an external plastic sheath over it. These over-the-needle catheters are made of a large fitting that remains on the skin after needle withdrawal, and a stiff catheter of polyethylene or Teflon piercing the skin. In some cases, the intravenous fluids are infused by means of a very stiff stainless steel needle penetrating the skin. Either way, discomfort, insulin aggregation, frequent clogging (particularly with pulsatile pumps), and (worse for the patient) daily catheter changes are the rule. Even though the patient is wearing a pump, frequent needle changing is necessary for patient hygiene and to prevent clogging. Therefore, much of the inconvenience and pain of intermittent insulin therapy remains.

The newest advance in IV/infusion pump catheter systems features a completely removable needle, with the smallest flexible polyurethane catheter available. For instance, a Neonatal Central line is composed of a tiny, soft radiopaque Tecoflex catheter with an outer diameter of only 0.015 in, inside of a breakaway needle. Once the needle has been removed, the pliable catheter remains in the fragile vein, reducing trauma, and ensuring long-term patency. In addition to neonatal care, these systems are also utilized for pediatrics, geriatrics, critical care, long-term control of diabetes, and the treatment of cancer.

The economic cost of diabetes in terms of medical care and losses due to disability or premature death is colossal—$10 billion annually. But just as important is the cost of treating the disease's major complications, which can lead to blindness, kidney failure, heart disease, and peripheral vascular disease.

Insulin infusion devices will be able to control blood glucose levels much more closely than traditional once-a-day injections. There is great hope that this better control will eliminate the long-term problems caused by diabetes. If this happens, cost savings to the health care system could be in the billions of dollars.

Perhaps the most significant contribution that these drug delivery systems are making today is in the treatment of cancer, specifically liver cancer. Side effects are the limiting factors in cancer chemotherapy. In conventional chemotherapy, large doses of medication are injected into the circulatory system, carrying the toxic chemicals not only to the cancer site but to the rest of the body. Since the same medications that are lethal to the tumor are toxic to normal body cells, undesirable side effects are inevitable.

With a medication delivery system, however, the drug is directed through a catheter into a diseased organ's own blood supply. This reduces the dose rate needed to kill tumor cells by confining the toxic effects to the target organ, and diluting the amount delivered to healthy tissue. Thus, side effects can be minimized and sometimes eliminated. Patients are already benefiting from reduced hospital stays and costs and from being able to spend more time at home.

22.6 Cardiovascular Catheters

First, the good news. After half a century of steady increases in heart disease fatalities in the United States, the rate now appears to be on the downswing. The death rate due to all types of heart disease slipped from 369 per 100,000 in 1960 to 362 by 1970 and to 343 by 1980, and it continues to decline.

The bad news is that heart disease is still by far the nation's leading killer, claiming almost a million lives in 1981—nearly twice as many as cancer and accidents combined. About 1.5 million Americans will suffer heart attacks this year, and 500,000 of them will die as a result. Of those, approximately half won't survive long enough to reach medical care.

Cardiologists admit they aren't sure why the death rates are falling. Just as heart disease is a multifactorial process, so is its decline. Most of them agree, however, that biomaterials technology is aiding the process. Biomaterials technology is aiding in the form of better diagnostic techniques, utilizing plastic catheters that permit the visualization of coronary arteries under radiography (angiograph), plastic catheters used to dissolve clots (thrombolysis), and the newest technique, transluminal angioplasty. We will discuss each of these important applications for polyurethanes in sequence.

Angiography Catheters. During the early part of this century, when infectious disease was the leading killer, patching up the patient was the most logical approach. The introduction of antibiotics, in the 1940s and 1950s, virtually eliminated infections as major killers in the United States. Since then, heart and blood vessel diseases have soared into the number-one position, claiming nearly one million lives in 1983. The important difference is that the development of heart disease (like that of cancer, the second-place killer) is almost always accompanied by presymptomatic clues and warning signs.

As recently as the early 1960s, the most common early warning of heart disease was the chest pain called angina pectoris—the signal that a portion of the heart muscle (myocardium) is receiving inadequate supplies of blood. All too often, even that sign was absent. In any event, the cardiologist could do little more than speculate about the origins of the pain and provide the patient with pain-relieving drugs.

The development of coronary angiography (or arteriography) changed that. By snaking a thin, flexible catheter through a patient's artery to the aorta and injecting a small amount of radiopaque dye, the coronary arteries can now be seen in brilliant detail on a conventional fluoroscope. In most cases, so can the blockages responsible for the angina—usually plugs of waxy cholesterol, accumulations of tiny blood clots, or calcified combinations of the two.

More than 400,000 angiograms are performed in the United States yearly, at an average cost of $1,500 each. The angiographic catheters utilized in these techniques must be exceptionally blood compatible so as to not to create life-threatening blood clots during the procedure. Tecoflex catheters, because of their inherent blood compatibility, are particularly useful in this application.

Thrombolysis Catheters. Heart attacks come in different sizes. There are big ones and small ones, depending on how much heart muscle is lost once the coronary artery supplying it with blood becomes blocked. In recent years, intense interest has focused on efforts to limit heart muscle destruction once a heart attack has begun, thus reducing the likelihood of long-term disability in survivors. The type of intervention that is now receiving the greatest attention is called *thrombolysis*, which means *dissolving clot*. Thrombolytic therapy is far from standard practice in any hospital at this time, but enough is being learned to warrant a review of what is involved and how the therapy works and when it works.

The central event in a heart attack is blockage of a coronary artery. As a result, blood flow to a portion of the heart muscle is stopped, and death of that tissue results. About 85% of the time, the block is due to formation of a blood clot within a coronary artery (which is usually already narrowed by cholesterol accumulation in its walls). Once the blockage occurs, the blood-deprived heart tissue dies gradually. (Indeed, muscle fibers at the outskirts of the area that has been deprived may get a small amount of blood from neighboring vessels and be able to live for a number of hours.) If the blocked artery can be opened before the injured heart muscle reaches its *point of no return*, a substantial amount of heart muscle can be salvaged. A substance called *streptokinase* can dissolve the clot if it is directed into the clogged vessel. This can be achieved by the technique of coronary artery catheterization which, in this instance, is put to use at the very onset of a heart attack. A thin catheter is threaded (through an arm or leg blood vessel) to the coronary arteries. The site of blockage is determined by squirting a small amount of contrast material through the catheter and taking X-rays. Then, the same catheter provides the route through which streptokinase is fed to the clot, as shown in Fig. 22.1. Polyurethane catheters can be utilized advantageously in this technique, since they can be made stiff initially, thereby allowing proper placement within the small coronary arteries. Subsequently, the substance softens significantly once it has reached body temperature and absorbed approximately 0.75% by weight of water. Since these catheters are left in place for many hours or days until all the

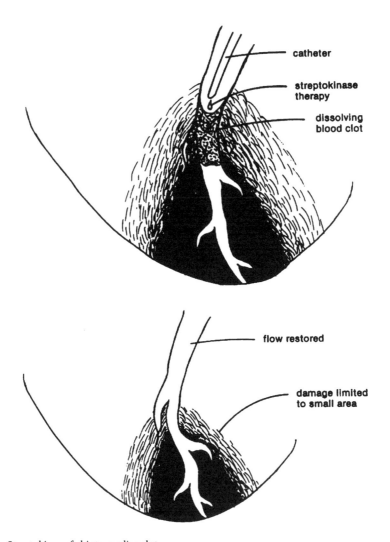

Figure 22.1   Streptokinase fed into cardiac clot.

clot has been dissolved, this softening reduces trauma to the delicate inner lining of the coronary arteries.

Transluminal Angioplasty Catheters.   The normal adult human heart is a muscular, four-chambered organ about the size of a medium grapefruit and weighing less than a pound. It is located almost exactly in the center of the chest (although its lower end, or apex, is tilted slightly toward the left). For all its complexity, it has just one purpose—*to provide the rest of the body with oxygen-carrying blood, at the same time carrying away the waste products for disposal.* The task is carried out by two main pumping chambers, the right and left ventricles, separated by a muscular septum and a series of one-way valves.

Oxygen-depleted blood from throughout the body enters the heart via two large veins, the superior and inferior vena cavae. Both veins empty first into a small chamber called the *right atrium* and then through the *tricuspid valve* into the *right ventricle.* With each contraction of the heart muscle, blood is forced from the right ventricle through the pulmonary valve to the lungs via the pulmonary arteries. In the lungs, the blood exchanges its carbon dioxide for fresh oxygen and returns to the heart through the pulmonary veins.

As these veins empty into the heart's *left atrium,* the oxygenated blood flows through the *mitral valve* into the powerful *left ventricle.* Contractions force the blood through a fourth and final valve (the aortic) into the aorta, the body's main artery. The aorta separates into various arteries to the upper and lower extremities; here the blood's hemoglobin gives up oxygen to the cells, takes a fresh load of carbon dioxide, and begins its journey back to the right atrium.

The heart's constant activity, consisting of approximately 70 contractions per minute, every minute of our lives, exacts a high price—gram for gram, its need for steady oxygen supply exceeds that of any other body organ. The demand is met by a series of coronary arteries, branching from the aorta and encircling the heart like a crown (in Latin, *corona,* hence *coronary*). These arteries carry freshly oxygenated blood directly to the heart muscle, or myocardium.

In more than one million hearts per year in the United States, these arteries begin to narrow to an internal diameter of only a millimeter or two. The arteries become choked with fatty deposits, and once the flow is finally dampened below a critical point, the symptoms of angina pectoris develop. Until recently, pain-reducing drugs (such as nitroglycerine) or coronary bypass surgery were the only available treatments.

Now, there is an alternative to bypass surgery. In the mid-1970s, Swiss cardiologist Andreas Gruntzig developed an ingenious method of unclogging arteries using a small balloon. In angioplasty, now performed on about 12,000 patients a year, a narrow tube (catheter) is threaded into the diseased artery until it reaches the clogged area. At that point, a tiny balloon at the tip of the catheter is repeatedly inflated so that it flattens the deposits against the arterial wall and widens the channel, as shown in Fig. 22.2.

Coronary angioplasty is now being used with increased success in patients with single-vessel disease and provides a low-cost, lower-risk alternative to coronary bypass surgery. With angioplasty, symptoms are relieved immediately, patients are out of the hospital within two days, and the majority are back to work within one week.

In coronary angioplasty catheters, the most crucial part of the device is the balloon. The balloon must be repeatedly inflated, without any appreciable change in dimensions, since an overexpansion would rupture the coronary artery wall. In this case, polyurethane is particularly advantageous, since fabric reinforced balloons can be easily fabricated of the substance while still retaining the blood-compatible properties of this elastomeric polyurethane.

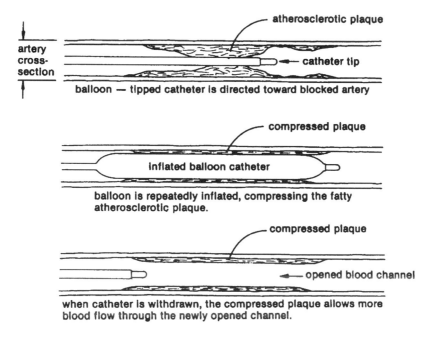

Figure 22.2   Catheter inflation widening arterial wall.

22.7 Illustrative Patents

22.7.1 Polycarbonate-Based Polyurethane

In 1992, a second-generation biomedical-grade thermoplastic polyurethane elastomer was introduced by Szycher and collaborators in U.S. Pat. 5,254,662, "Biostable Polyurethane Products." This polyurethane is considered to be a second-generation elastomer since it is a carbinate-based polymer. This type of linear, segmented urethane is a rubbery reaction product of an aliphatic isocyanate, a high-molecular-weight polyol, and a low-molecular-weight chain extender. Polycarbonate-based polyurethanes represent a substantial advance over older, conventional ether-based polyurethanes since the polycarbonate-based polyurethanes are far more durable in the highly corrosive environment of the human body. In this context, *biomedical-grade* means *a systemic, pharmacologically inert substance designed for implantation within living systems.*

22.7.2 Wound dressings

Burns and other related wounds, such as donor sites and the like, present a serious problem in that they tend to produce large amounts of exudate that can cause conventional dressings to become saturated, or to stick to the wound, or even to become infected. One method of covering such wounds has been to cover the with a material into which new epithelial or fibroblast growth can penetrate. Dressings of this kind are disclosed in U.S. Pats. 3,526,224, 3,648,692, and 3,949,742.

However, such dressings can be extremely painful to remove and often require surgical excision. A fundamentally different approach, requiring a fundamentally different type of dressing, is to employ materials that are designed to reduce the propensity to adhere to the wound. Dressings of this kind are disclosed in U.S. Pat. 3,543,750. One more recent attempt at nonadherent dressings is U.S. Pat. 3,709,221, which discloses a dressing having an outer microporous, liquid repellent fibrous layer, an inner macroporous fibrous layer, and an absorbent intermediate layer, which was also envisaged as normally being fibrous. To reduce the tendency of this material to adhere to the wound, the inner layer had to be treated with an agent to render it non-wetted by body liquid. It is now realized that it would be desirable to provide a dressing in which the wound facing layer does not require special treatment.

Lang[1] disclosed an invention relate to an absorptive wound dressing suitable for use on burns or other wounds, which dressing has a reduced tendency to adhere to the wound and can act as a bacterial barrier. The present invention also relates to the manufacture and use of such dressings.

Lang discovered that, by avoiding fibrous materials, it is possible to produce a dressing with reduced tendency to adhere to wounds without the need for special treatments. An attempt at producing an absorbent dressing is described in U.S. Pat. 3,888,748, which describes a dressing fabricated from at least four sheet materials. The wound facing part of the dressing apparently consists of a grid or scrim coated with polyethylene in such manner that the polyethylene surrounds the filaments of the grid and collects any loose thread or particle that may be present in the core material. It is now realized that it is desirable to avoid the use of wound facing layers that can allow such penetration of the central layer to the wound surface. It has also been realized that it would be desirable to provide a material that was highly conformable to the wound so that it is possible to minimize the quantity of exudate between the wound surface and the dressing. U.S. Pats. 3,709,221 and 3,888,248 disclose materials that are bonded along their edges that may reflect a desire to improve conformability. The dressing of the present invention allows for bonding over the whole of the operative area while retaining flexibility.

Accordingly, the Lang invention provides a low-adherence wound dressing that comprises a wound facing layer, an intermediate absorbent layer, and an outer layer. The wound dressing is distinguished in that the wound facing layer consists of a conformable elastomeric apertured film, the intermediate absorbent layer is made up of a conformable hydrophilic foam, and the outer layer is a continuous moisture vapor transmitting conformable film.

22.7.3 Thermoplastic Medical-Grade Polyurethane

Extensive investigations have been undertaken over many years to find materials that will be biologically and chemically stable toward body fluids. This area of research has become increasingly important with the development of various objects and articles that can be in contact with blood, such as artificial organs, vascular grafts, probes, cannulas, catheters, and the like.

Synthetic plastics have come to the forefront as preferred materials for such articles. However, these materials have the major drawback of being thrombogenic. Thrombogenicity has conventionally been counteracted by the use of anticoagulants such as heparin. Exemplary of procedures for attachment of heparin to otherwise thrombogenic polymeric surfaces are the disclosures in U.S. Pat. No. 4,613,517 to Williams et al., and U.S. Pat. 4,521,564, to Solomon et al.

In general, the most blood-compatible plastics known are the fluorinated polyolefins, such as poly-tetrafluoroethylene, and the silicone polymers. However, while being basically hemocompatible, silicone polymers do not have the desired mechanical strength for most blood-contacting applications. One approach to improving the mechanical properties of silicone polymers has been addition of appropriate fillers and curing agents. Such additives, although providing strength, are usually themselves thrombogenic so that the improved physical strength is offset by the reduced blood compatibility.

Another approach has been to combine the blood compatibility of the silicone with the excellent mechanical properties of polyurethane. U.S. Pat. 3,562,352 discloses a copolymer consisting of about 90% polyurethane and 10% polydimethylsiloxane. This material, under the trade name Cardiothane® (Kontron Cardiovascular, Inc., Everett, Mass.), has been widely used in blood-contacting applications but has the major drawback that it is not thermoplastic and cannot be melt processed.

Thermoplastic polyoxyalkylene polyurethanes having up to 15% of a soft segment formed from a polysiloxane devoid of oxygen atoms bonded to both silicon and carbon are disclosed by Zdrahala et al. in U.S. Pat. 4,647,643. Polyurethanes prepared from 1,3-bis(4-hydroxybutyl) tetramethyl disiloxane are reported by Yilgor et al. in *American Chemical Society Polymer Preprint* 20, 286 (1982) and are suggested to have possible utility in the biomedical field.

Silicone coatings have been achieved by plasma polymerization of silicon-containing monomers onto various polymeric base materials. Preparation and hemocompatibility studies of such materials are described by Chawla in *Biomaterials* 2, 83 (1981).

Ward, in U.K. Pat. GB 2,140,437B, disperses up to 5% of a silicone-containing additive in a polymeric base material by mixing the components as a melt or in a solvent. Biomedical devices are prepared therefrom by conventional techniques such as injection molding and by homogeneous extrusion.

While significant advances have been made toward blood-compatible surfaces for fabrication of medical devices, further improvements are needed. In particular, materials having surfaces are needed that are essentially nonthrombogenic for use in devices that will be in contact with blood for prolonged periods. It is toward fulfillment of this need that this invention is directed.

The Hu et al.[2] invention discloses an article having a hemocompatible surface. In the present disclosure, the term *hemocompatible* describes a surface that does not induce thrombosis or changes in blood cells, enzymes, or electrolytes; does not damage adjacent tissue; and does not cause adverse immune responses or toxic reactions. Preferred articles are medical devices such as catheters.

The Hu et al. invention includes a thermoplastic polymeric base material having thereon a layer of copolymer having urethane and silicon-containing segments. In preferred articles, the base material is a polyurethane, and the silicon-containing segment is a siloxane. Particularly preferred articles have siloxane segments in the copolymer which are hydroxyalkyl terminated and which thereby have outstanding stability when in hydrolytic environments.

22.7.4 Fluorinated PU

PEU compositions develop microdomains conventionally termed *hard-segment domains* and *soft-segment domains*. They are (AB)n-type block copolymers, A being the hard segment and B the soft segment, and they are occasionally termed *segmented* polyurethanes. The hard-segment domains form by localization of the portions of the copolymer molecules that include the isocyanate and extender components, whereas the soft-segment domains form from the polyether glycol portions of the copolymer chains. The hard segments are generally more crystalline and hydrophilic than the soft segments, and these characteristics prevail, in general, for the respective domains. One disadvantage of polyurethane resins of the softness desired for many medical devices, e.g., resins having Shore A hardness less than abut 100, is surface blocking (tack) after extrusion or molding into desired shapes. To avoid this problem, many remedies have been developed in the art, including the use of external mold release agents and the use of various antiblockers or detackifiers in admixture with the polymer. Most antiblocking agents/detackifiers are low-molecular-weight materials that have a tendency to migrate or leach out of the polymer. This represents a problem when the polyurethanes are to be used as biomaterials (tubing, prostheses, implants, etc.). The presence of such low-molecular-weight extractables can affect the biocompatibility of the polyurethanes and lead to surface degradation such as fissuring or stress cracking.

Fluorine-containing polyurethanes are known. Kato et al., in *Progress in Artificial Organs,* 1983, p. 858, disclose polyurethanes synthesized from fluorinated isocyanates. Yoon et al., in *Macromolecules* 19, 1986, p. 1068 discloses polyurethanes synthesized from fluorinated chain extenders. Field et al., in U.S. Pat. 4,157,358, discloses a randomly fluorinated epoxyurethane resin.

Although some progress has been made toward providing a thermoplastic polyurethane that (a) is nonblocking without additives, (b) provides a desirable balance of stiffness in air and softness in liquid, and (c) is suitable for blood contact, further improvement is needed. This invention is directed toward fulfillment of this need.

Zdrahala et al.[3] disclosed nonblocking, hemocompatible, thermoplastic, fluorinated polyetherurethanes and a method for their preparation from fluorinated polyether glycols, isocyanates, chain extenders, and a nonfluorinated polyol. The method includes two steps in which the fluorinated glycol is reacted initially with the diisocyanate to give a prepolymer having terminal isocyanate groups, and the prepolymer is then reacted with the extender and nonfluorinated polyol. Medical devices are fabricated from the fluorinated polyetherurethane.

References

1. Lang, Stephen M. U.S. Pat. 5,445,604, August 29, 1995, "Wound Dressing with Conformable Elastomeric Wound Contact Layer."
2. Hu, Can B., Solomon, Donald d., and Wells, Stanley C. U.S. Pat. 5,059,269, October 22, 1991, "Method of Making an Article Having a Hemocompatible Surface," to Becton, Dickinson and Company, Franklin Lakes, New Jersey.
3. Zdrahala, Richard J., and Strand, Marc A., U.S. Pat. 4,841,007, June 20, 1989, "Fluorinated Polyetherurethanes and Medical Devices therefrom," to Becton, Dickinson and Company, Franklin Lakes, New Jersey.

A

Polyurethane Suppliers and Manufacturers

ACRO PRODUCTS, INC.
2701 Dwenger Ave.
Fort Wayne, IN 46803 USA
Finished Goods Manufacturer: Elastomer; RIM

ACROTECH, INC.
Industrial Court North
Lake City, MN 55041 USA
Finished Goods Manufacturer: Elastomer; Cast

AMERICAN FINISH & CHEMICAL CO.
1012 Broadway
Chelsea, MA 02150 USA
Finished Goods Manufacturer: Coating; Adhesive

AMERICAN FOAM PRODUCTS, INC.
753 Liberty St.
Painesville, OH 44077 USA
Systems Manufacturer: Rigid Foam Systems; Flexible Foam

3M INDUSTRIAL CHEMICAL PRODUCTS DIVISION
Building 223-65-04 3M Center
St. Paul, MN 55144 USA

7 SIGMA, INC.
2843 26th Ave. S.
Minneapolis, MN 55406 USA

A & L FOAM INSULATION, INC.
2033 Durham
Irving, TX 75062-3546 USA
Markets: Aerospace; Construction; Defense/Military; Industrial; Refrigeration; Transportation

A-R PRODUCTS, INC.
11807-7/8 Slauson Ave.
Santa Fe Springs, CA 90670

A & D MOLDED PRODUCTS
1 Front St.
East Rochester, NH 03870 USA

A. SHULMAN, INC.
3550 W. Market St.
Akron, OH 44333 USA

Markets: Aerospace; Appliances; Automotive; Carpet; Chemical; Clothing; Construction; Defense/Military; Electronic/Electrical; Footwear; Furniture; Industrial; Lumber; Marine; Material Handling; Medical; Mining/Drilling; Packaging; Paper; Pipe/Plumbing; Printing; Recreational; Refrigeration; Transportation; Wheels/Tires

A.M.E. CORP.
33 Jacksonville Rd.
P.O. Box 195
Towaco, NJ 07082 USA
Finished Goods Manufacturer: Elastomer; Flexible Foam; Integral Skin; Silicone Molded; Cast

A.W. CHESTERTON COMPANY
9 Forbes Rd.
Woburn, MA 01801 USA
Finished Goods Manufacturer: Elastomer; Cast
Markets: Industrial; Mining/Drilling

A.W. HUBER CO.
10291 Bach Blvd.
St. Louis, MO 63132 USA
Finished Goods Manufacturer: Elastomer; Cast

AASR, INC.
2219 McKinney St.
P.O. Box 1068
Houston, TX 77251 USA
Finished Goods Manufacturer: Flexible Foam; Rigid Foam
Other: Fabricator

ABBA RUBBER CO., INC.
6240 Prescott Ct.
Chino, CA 91710 USA
Finished Goods Manufacturer: Elastomer; Cast

ABBOT INDUSTRIES, INC.
211 E. Page Ave.
P.O. Box 2317
Gastonia, NC 28052 USA
Finished Goods Manufacturer: Flexible Foam; Coating

ABLE R & S ROLLER CO.
Triangle Marketing & Research, Inc.

212 Decatur St.
Doylestown PA 18901 USA
Finished Goods Manufacturer: Coating

ACADIA POLYMERS, INC.
1420 Coulter Dr. NW
Roanoke, VA 24012 USA

ACCEL PLASTIC PRODUCTS CORP.
2701 W. Virginia Ave.
Phoenix, AZ 85009 USA
Systems Manufacturer: Rigid Foam Systems; Flexible
    Foam Systems
Other: Distributor

ACCESSIBLE PRODUCTS
2122 W. Fifth Place
Tempe, AZ 85281 USA
Finished Goods Manufacturer: Flexible Foam
Other: Fabricator
Markets: Industrial; Pipe/Plumbing

ACCUMETRIC, INC.
Meter-Mix, Inc.
350 Ring Rd.
Elizabethtown, KY 42701 USA
Equipment or Machinery Supplier/Manufacturer:
    Dispensing Equipment; Tanks; Pumps

ACCURATE FOAM CO.
819 Fox St.
Laporte, IN 46350 USA
Finished Goods Manufacturer: Flexible Foam; Rigid Foam
Other: Fabricator; Manufacturer *Markets:* Aerospace;
    Appliances; Automotive; Electronic/ Electrical;
    Industrial; Packaging; Recreational

ACCURATE PLASTICS, INC.
18 Morris Place
Yonkers, NY 10705 USA
Manufacturer of elastomer sheets, rods and gasket seals.

ACLA USA, INC.
109 Thomson Park Dr.
Mars, PA 16046 USA
Markets: Aerospace; Automotive; Construction; Defense/
    Military; Electronic/Electrical; Footwear; Industrial;
    Lumber; Marine; Material Handling; Mining/Drilling;
    Packaging

ACME FISHER TANK LININGS
728 S. I 3rd St.
Louisville, KY 4021 0 USA
Equipment or Machinery Supplier/Manufacturer: Tanks
Finished Goods Manufacturer: Coating
Markets: Chemical; Defense/Military; Industrial;
    Material Handling; Mining/Drilling; Transportation

ACME FOAM CORP.
900 Dean St.
Brooklyn, NY 11238 USA

ACME MACHELL CO., INC.
2000 Airport Rd.
P.O. Box 1617
Waukesha, WI 53187 USA

ACOUSTICAL SPRAY INSULATORS, INC.
Box 625
Allentown, PA 18105 USA
Markets: Chemical; Construction; Pipe/Plumbing Spray
    Foam Contractor.

ACUSHNET CO./TITEIST/FOOT-JOY WORLDWIDE
Division of Acushnet Co.
P.O. Box 965
333 Bridge St.

ADAMS ELEVATOR
5640 W. Howard St.
Skokie, IL 60077 USA
Finished Goods Manufacturer: Elastomer; TPU *Markets:*
    Wheels/Tires

ADAN PLASTICS, INC.
7949 Forrest Hills Rd.
Rockford, IL 61111 USA
Finished Goods Manufacturer: Elastomer; RIM; Non-
    automotive; Cast; Flexible Foam; Integral Skin; Rigid
    Foam

ADFOAM
6110 Lamb Rd.
Wyoming, NY 14591 USA
Finished Goods Manufacturer: Flexible Foam
Other: Fabricator

ADHESIVE & SEALANT COUNCIL
1627 K St. NW
Suite 1000
Washington, DC 20006 USA
Other: Trade Association

ADMIRAL COMPANY
Division of Maytag
1 Admiral Dr.
Galesburg, IL 61402 USA
Finished Goods Manufacturer: Rigid Foam
Markets: Appliances; Refrigeration

ADMIRAL EQUIPMENT CO.
Subsidiary of Dow Chemical
305 W. North St.
Akron, OH 44303 USA
Equipment or Machinery Supplier/Manufacturer: Molds;
    Laminators; Dispensing Equipment; Tanks; Pumps;
    Automated Controls; Processing Equipment

ADOBE URETHANE SYSTEMS
P.O. Box 100
Peralta, NM 87042 USA
Other: Equipment Service

ADVANCE RUBBER CO.
3334 Washington Ave. N.
Minneapolis, MN 55412 USA
Finished Goods Manufacturer: Elastomer

ADVANCED FOAM & PLASTIC CO.
3431 W. 140th St.
Cleveland, OH 44111 USA
Finished Goods Manufacturer: Flexible Foam; Integral
    Skin; Rigid Foam

ADVANCE LATEX PRODUCTS, INC.
International Molders, Inc.
3579 Hayden Ave.
Culver City, CA 90230 USA
Finished Goods Manufacturer: Elastomer; TPU; Flexible
   Foam

ADVANCE FOAM PLASTICS, INC.
5250 N. Sherman St.
Denver, CO 80216 USA

ADVANCED INSULATION
123 E. Lake St.
Suite 301
Bloomingdale, IL 60108 USA
Finished Goods Manufacturer: Rigid Foam

ADVANCED PROCESS TECHNOLOGY, INC.
200 Egel Ave.
Middlesex, NJ 08846 USA
Equipment or Machinery Supplier/Manufacturer:
   Dispensing Equipment; Metering; Mixing

AERO SPECIALTIES CORP.
20 Burt Dr.
Deer Park, NY 11729 USA
Finished Goods Manufacturer: Flexible Foam

AEROL CO., INC.
3235 San Fernando Rd.
Los Angeles, CA 90065 USA
Finished Goods Manufacturer: Elastomer; Cast

AG RIM, INC.
2268 Reum Rd.
Niles, MI 49120 USA
Raw Material Supplier/Manufacturer: Isocyanate; Polyol;
   Polymer; Prepolymer; Resins

AIA PLASTICS/ANDERSON ASSOC
508 Wyman Rd.
Box 674
Toledo, OH 43964 USA
Finished Goods Manufacturer: Elastomer; Cast

AIN PLASTICS, INC.
P.O. Box 151
249 E. Sandford Blvd.
Mount Vernon, NY 10550 USA
Markets: Industrial

AIR PRODUCTS AND CHEMICALS, INC.
Polyurethane and Performance Chemical Division
7201 Hamilton Blvd.
Allentown, PA 18195-1501 USA
Raw Material Supplier/Manufacturer: Catalyst;
   Surfactant; Prepolymer; Release Agents;
   Intermediates; Curatives; Cross-linkers; Chain
   Extenders

AIR-O-PLASTIK
150 Fieldcrest Ave.
Edison, NJ 08837 USA
Finished Goods Manufacturer: Flexible Foam
Markets: Construction; Medical

AIREX RUBBER PRODUCTS CORP.
100 Indian Hill Ave.
Portland, CT 06480 USA
Finished Goods Manufacturer: Elastomer; Cast

AIRMEC, INC.
2102 Vanco
Irving, TX 75061 USA
Finished Goods Manufacturer: Elastomer; Cast

AIRTEX INDUSTRIES, INC.
3558 Second St. N
Minneapolis, MN 55412 USA
Finished Goods Manufacturer: Flexible Foam
Markets: Construction; Industrial; Packaging

AJAX FLOOR PRODUCTS CO., INC.
P.O. Box 161
Great Meadows, NJ 07838 USA
Finished Goods Manufacturer: Coating
Markets: Construction; Flooring

AK RUBBER PRODUCTS CO., INC.
P.O. Box 468
248 P Centralia St.
Elkhorn, WI 53121 USA
Finished Goods Manufacturer: Elastomer; Millable Gums

AKRO CORPORATION, IM
Ravena Division
225 W. Lake St.
P.O. Box 472
Ravena, OH 44266 USA
Systems Manufacturer: Urethane Rubbers

AKRON EXTRUDERS, INC.
1119 Milan St.
Canal Fulton, OH 44614 USA
Equipment or Machinery Supplier/Manufacturer:
   Extrusion Equipment

AKRON PORCELAIN & PLASTICS CO.
2739 Cory Ave.
P.O. Box 3767
Akron, OH 44314 USA
Finished Goods Manufacturer: Flexible Foam

AKZO CHEMICALS, INC.
5 Livingstone Ave.
Dobbs Ferry, NY 10522 USA
Raw Material Supplier/Manufacturer: Flame Retardants

AKRON INSULATING CO.
1985 Manchester Rd.
Akron, OH 44314 USA
Other: Installers and Applicators of Formed-in-Place
   Polyurethane

AKRON POLYMER LABORATORY
1080 S. Main St.
P.O. Box 1764
Akron, OH 44309-1764 USA
Consultant: Testing; Research Organization

AL-BE INDUSTRIES, INC.
16633 Minnesota Ave.

P.O. Box 500
Paramount, CA 90723 USA
Equipment or Machinery Supplier/Manufacturer:
Extrusion Equipment

ALACRA SYSTEMS
Race and Ridge Streets
Ambler, PA 19002 USA
Finished Goods Manufacturer: Flexible Foam

ALADDIN INDUSTRIES, INC.
703 Murfreesboro Rd.
Nashville, TN 37210 USA
Finished Goods Manufacturer: Flexible Foam

ALASCO RUBBER & PLASTICS
617 Mountain View Ave.
Belmont, CA 94002 USA
Finished Goods Manufacturer:. Elastomer; Cast

ALBERT TROSTEL PACKINGS LTD.
Specialty Elastomers Group
999 Wells St.
Lake Geneva, WI 53147 USA
Finished Goods Manufacturer:. Elastomer; TPU; Cast

ALBION INDUSTRIES, INC.
P.O. Box 411
Albion, MI 49224 USA
Equipment or Machinery Supplier/Manufacturer: Wheels

ALBRIGHT & WILSON, INC.
P.O. Box 26229
Richmond, VA 23260 USA
Raw Material Supplier/Manufacturer: Flame retardants

ALDRICH CHEMICAL CO.
1001 W. St. Paul Ave.
Milwaukee, WI 53233 USA
Raw Material Supplier/Manufacturer: Isocyanate, Polyol;
Catalyst; Prepolymer

ALEXANDER RUBBER PRODUCTS
25613 Dollar St.
Building #9
Hayward, CA 94544 USA
Finished Goods Manufacturer: Elastomer; Millable Gums

ALFA III CORP.
1225 Lakeview Dr.
Chaska, MN 55318 USA
Finished Goods Manufacturer: Flexible Foam

ALL AMERICAN ENTERPRISES
Manufacturing Division
301 Eubank SS
Albuquerque, NM 87123 USA
Finished Goods Manufacturer: Flexible Foam; Rigid Foam

ALL-STATE BELTING CO.
1824 Industrial Circle
West Des Moines, IA 50265 USA
Finished Goods Manufacturer: Elastomer; Cast; Flexible
Foam

ALLEN FOAM CORP.
175 E. Manville

Compton, CA 90220 USA
Finished Goods Manufacturer: Flexible Foam
Other: Fabricator
Markets: Automotive

ALLIED FOAM CORP.
1530 W. Winton Ave.
Hayward, CA 94545 USA
Markets: Furniture; Packaging; Bedding

ALLIED PLASTIC SUPPLY CO.
1928 East River Dr.
Lake Luzerne, NY 12846 USA
Finished Goods Manufacturer: Elastomer; Cast

ALLIED SIGNAL, INC.
Fluorocarbons Division
P.O. Box 1053
Morristown, NJ 07962 USA
Raw Material Supplier/Manufacturer: Blowing Agent

ALLIED-BALTIC RUBBER, INC.
One Factory St.
P.O. Box 267
Baltic, OH 43804 USA
Finished Goods Manufacturer: Elastomer; Cast; Flexible
Foam

ALLSTATE PLASTICS, INC.
237 Raritan St.
South Amboy, NJ 08879 USA
Finished Goods Manufacturer: Elastomer; Cast

ALM, INC.
1058 S. Vandeventer Ave.
St. Louis, MO 63110 USA

ALMAC PLASTICS, INC.
4742 37th St.
Long Island City, NY 11105 USA
Finished Goods Manufacturer: Flexible Foam; Rigid Foam

ALPHA CHEMICAL DIST CORP.
P.O. Box 514
San Juan, PR 00936 USA
Finished Goods Manufacturer: Rigid Foam

ALUCHEM, INC.
One Landy Lane
Reading, OH 45215 USA
Raw Material Supplier/Manufacturer: Flame Retardants

ALUMA SHIELD INDUSTRIES, INC.
405 Fentress Blvd.
Daytona Beach, FL 32114 USA
Markets: Construction; Refrigeration

AMANA REFRIGERATION, INC.
Highway 220
Amana, IA 52204 USA
Systems Manufacturer: Rigid Foam Systems
Finished Goods Manufacturer: Rigid Foam
Markets: Appliances

AM AMERICAN ACOUSTICAL PRODUCTS
9 Cochituate St.

Natick, MA 01760 USA
Markets: Industrial; Sound Deadening

AMERICAN BARMAG CORPORATION
1101 Westinghouse Blvd.
Charlotte, NC 28273 USA
Equipment or Machinery Supplier/Manufacturer: Pumps

AMARK INC.
1511 Columbia Circle
P.O. Box 1467
Merrimack, NH 03054 USA
Finished Goods Manufacturer: RIM; Non-automotive

AME CORP.
33 Jacksonville Rd.
P.O. Box 195
Towaco, NJ 07082-0195 USA
Finished Goods Manufacturer: Elastomer; Cast; Flexible
Foam

AMERICAN URETHANE INC.
1320 Defense Highway
Gambeills, MD 21054 USA
Raw Material Supplier/Manufacturer: Prepolymer
Manufacturer; Elastomer Systems
Finished Goods Manufacturer: Elastomer; Cast

AMERICO CORPORATION
25120 Trowbridge
Dearborn, MI 48124 USA

AMERICAN TECHNICAL COATINGS, INC.
400 W. Central Blvd.
Cape Canaveral, FL 32920 USA
Systems Manufacturer: Elastomer Systems
Finished Goods Manufacturer: Elastomer; Coating
Other: Consultant; Installation Contracts-Polyurethane;
Foam Roofing
Markets: Aerospace; Construction; Defense/Military;
Industrial; Commercial Buildings

AMERICAN THERMOSET SYSTEMS
P.O. Box 3125
Brentwood, TN 37024-3125 USA

AMERICAN BILTRITE, INC.
57 River St.
Wellesley Hills, MA 02181 USA
Finished Goods Manufacturer: Elastomer; TPU; Coating

AMERICAN COATING, INC.
P.O. Box 19706
Greensboro, NC 27419-9706 USA

AMERICAN RUBBER PRODUCTS CORPORATION
P.O. Box 69
Staples, MN 56479 USA
Finished Goods Manufacturer: Flexible Foam

AMERICAN SEATING CO.
901 Broadway NW
Grand Rapids, MI 49504 USA
Other: Fabricator

AMERICAN PRECISION PRODUCTS
1 314 Buford St.

Huntsville, AL 35801 USA
Finished Goods Manufacturer: Flexible Foam; Rigid Foam

AMERICAN ROLLER COMPANY
2223 Lakeside Dr.
Bannockburn, IL 60015 USA
Finished Goods Manufacturer:. Elastomer; Cast

AMERICAN POLY-FOAM CO., INC.
1580 Atlantic St.
Union City, CA 94587 USA

AMERICAN POLY-THERM CO.
2000 Flightline Dr.
Lincoln, CA 95648 USA
Finished Goods Manufacturer:. Flexible Foam; Rigid
Foam

**AMERICAN LEITSTRITZ EXTRUDER
CORPORATION**
l69 Meister Ave.
Somerville, NJ 08876 USA
Equipment or Machinery Supplier/Manufacturer:
Extruding Equipment

AMERICAN LEWA, INC.
132 Hopping Brook Rd.
Holliston, MA 01746 USA
Systems Manufacturer: Pumps

AMERICAN FURNITURE MFGS. ASSN.
P.O. Box HD-7
High Point, NC 27261 USA
Industry Organization

AMERICAN GFM
1200 Cavalier Blvd.
Chesapeake, VA 23323 USA
Mgr Compforms

**AMERICAN FOAM RUBBER PRODUCTS
CORPORATION**
4909 Lakawana
Dallas, TX 75247 USA
Markets: Automotive; Furniture; Industrial; Packaging

AMERICAN FUEL CELL & COATED
FABRICS
124 S. Jackson
Magnolia, AR 71753 USA
Finished Goods Manufacturer: Coating

AMERICAN EXCELSIOR COMPANY
850 Ave. H E.
Arlington, TX 76011 USA
Finished Goods Manufacturer: Flexible Foam

AMERICAN FIBRIT, INC.
The Becker Group, Inc.
1950 Concept Dr.
Warren, MI 48091-1385 USA
Finished Goods Manufacturer: Automotive; Flexible
Foam; Adhesive; Rigid Foam

AMES RUBBER CORP.
2347 Ames Blvd.
Hamburg, NJ 07419 USA

Finished Goods Manufacturer: Elastomer; Cast; Flexible
   Foam; Coating
Markets: Automotive

AMETEC INC.
P.O. Box 9
Nesquehoming, PA 18240 USA
Finished Goods Manufacturer:. Elastomer; Cast

AMSPEC CHEMICAL CORP.
Foot of Water St.
Gloucester City, NJ 08030 USA
Raw Material Supplier/Manufacturer: Catalyst; Additives,
   Flame Retardants

ANDERSON DEVELOPMENT CO.
1415 E. Michigan St.
Adrian, MI 49221 USA
Raw Material Supplier/Manufacturer: Catalyst;
   Prepolymer
Systems Manufacturer: Prepolymer Manufacturer;
   Elastomer Systems

ANGUS CHEMICAL COMPANY
1500 E. Lake Cook Rd.
Buffalo Grove, IL 60089 USA
Raw Material Supplier/Manufacturer: Moisture
   Scavenger Reactive Diluent

ANGUS FIRE ARMOUR. CORP
Box 879
Angier, NC 27501
Finished Goods Manufacturer: Elastomer; TPU

APACHIE HOSE & BELTING
P.O. Box 1719
Cedar Rapids, IA 52406 USA
Finished Goods Manufacturer: Elastomer

APACHE PRODUCTS CO.
905 23rd Ave.
Meridian, MS 39302 USA
Finished Goods Manufacturer: Rigid Foam

APEX MOLDED PRODUCTS CO.
3574 Ruth St.
Philadelphia, PA 19134 USA
Finished Goods Manufacturer: Elastomer; Millable Gums

APOLLO RUBBER CO.
P.O. Box 9462
Tulsa, OK 74157 USA
Finished Goods Manufacturer: Elastomer; Cast

APPLE RUBBER PRODUCTS, INC.
310 Erie St.
Lancaster, NY 14086 USA
Finished Goods Manufacturer: Elastomer; Cast

APPLI-TEC, INC.
1 51 Essex St.
Haverhill, MA 01831 USA
Systems Manufacturer: Encapsulants; Adhesives

APPLIED PRODUCTS, INC.
10428 Product Dr.

Rockford, IL 611 11 USA
Finished Goods Manufacturer: Flexible Foam

AQUA GLASS CORPORATION
Highway 22-Industrial Park
P.O. Box 412
Adamsville, TN 38310 USA

APPLIED EXTRUSION TECHNOLOGY
P.O. Box 582
Middletown, DE 19703 USA
Other: Fabricator

APPLIED METEROLOGY
900 S.W. Freeway
Suite 400
Houston, TX 77074 USA
Other: Consultant; Distributor

ARBONITE CORP.
3826 Old Easton Rd.
P.O. Box 888
Doylestown, PA 18901 USA
Finished Goods Manufacturer: Elastomer; Coating

ARCO CHEMICAL COMPANY
3801 W. Chester Pike
Newtown Square, PA 19073 USA
Raw Material Supplier/Manufacturer: Polyol *Systems
   Manufacturer:* Flexible Foam Systems

ARCO FLUID PRODUCTS
Division of Ingersol Rand
One ARCO Center
Bryan, OH 43506 USA
Equipment or Machinery Supplier/Manufacturer: Pumps

ARCTIC SEAL URETHANE INSULATION SYSTEMS
7116th St.
Jesup, IA 50648 USA
Systems Manufacturer: Rigid Foam Systems; Flexible
   Foam

ARGENT
41131 Vincenti Ct.
Novi, MI 48375 USA
Finished Goods Manufacturer: Adhesive

ARI-THANE FOAM
P.O. Box 347
Etiwanda, CA 91739 USA
Markets: Construction

ARISTECH CHEMICAL CORPORATION
600 Grant St.
Room 1066
Pittsburgh, PA 15219 USA

ARISTOCRAT DEVELOPMEENT
Foam-Kote
727 N. China Lake Blvd.
Ridgecrest, CA 93555 USA
Finished Goods Manufacturer: Elastomer; Coating; Rigid
   Foam; Roofing

AZ FOAM & SPRAY
P.O. Box 1422

Mesa, AZ 85211 USA
Finished Goods Manufacturer: Rigid Foam

ARMALYFOAM
Armaly Brands
1900 Easy St.
P.O. Box 611
Walled Lake, MI 48390 USA
Finished Goods Manufacturer: Flexible Foam; Polyester
and Polyether Sponge Material

ARMATURE COIL EQUIPMENT, INC.
7300 Manufacturing Rd.
Cleveland, OH 44135 USA
Equipment Machinery Supplier/Manufacturer: Heat
Cleaning Ovens

ARMSTRONG HOSE DIVISION
P.O. Box 850
96 Stokes Ave.
Trenton, NJ 08605-0850 USA
Finished Goods Manufacturer: Elastomer; TPU

ARNCO
5141 Firestone Place
South Gate, CA 90280 USA
Systems Manufacturer: Elastomer Systems; Fast Cast
Systems

ARNESON PRODUCTS, INC
2450 S. Watney Way
Fairfield, CA 94533 USA
Finished Goods Manufacturer: Elastomer

ASC, INC.
American Sunroof Co.
P.O. Box 30
Bowling Green, KY 42102 USA
Finished Goods Manufacturer:. Rigid Foam

ASHBY CROSS CO.
418 Boston St.
Topsfleld, MA 01983 USA
Equipment or Machinery Supplier/Manufacturer:
Dispensing

ASHLAND CHEMICAL, INC.
FRP Supply
55 Scotland Blvd.
Bridgewater, MA 02324 USA
Other: Distributor

ASHLAND CHEMICAL, INC.
5200 Blazer Parkway
Dublin, OH 43017 USA
Finished Goods Manufacturer: Elastomers

ASHTABULA RUBBER CO.
2751 West Ave.
Ashtabula, OH 44004 USA
Finished Goods Manufacturer: Flexible Foam

ASSOCIATED RUBBER PRODUCTS
P.O. Box 605
St. Clair, MO 63077 USA

Finished Goods Manufacturer: Elastomer; TPU; Millable
Urethane

ASSOCIATED RUBBER, INC.
115 S. Sixth St.
Quakertown, PA 18951 USA
Finished Goods Manufacturer: Elastomer; Flexible Foam

ASTEC CO.
P.O. Box 9
Hawthorne, NJ 07507-0009 USA
Finished Goods Manufacturer: Flexible Foam

ASTRO MOLDING, INC.
Runyon Rd.
R.D. #1
Old Bridge, NJ 08857 USA
Finished Goods Manufacturer: Elastomer

ASTRO-VALCOUR, INC.
18 Peck Ave.
P.O. Box 148
Glens Falls, NY 12801 USA

ASTROFOAM, INC.
120 Industrial Dr.
Holden, MA 01520 USA
Finished Goods Manufacturer: Flexible Foam

ATHOL MANUFACTURING CORP.
P.O. Box 105
C and 22nd Streets
Butner, NC 27509 USA
Finished Goods Manufacturer: Coating

ATLANTIC POLYMERS & PRODUCTS
P.O. Box 790
Oakhurst, NJ 07755 USA
Finished Goods Manufacturer: Elastomer; Cast; Flexible
Foam; Coating; Microcellular

ATLANTIC RUBBER CO., INC.
50 Parker St.
Unit 2
Newburyport, MA 01950 USA
Finished Goods Manufacturer: Elastomer; TPU

ATLANTIC THERMO PLASTICS CO., INC.
P.O. Box 457
Blackstone, MA 01504 USA
Finished Goods Manufacturer: Elastomer; RIM; Non-
automotive; Cast; Flexible Foam; Integral Skin

ATLAS ENERGY PRODUCTS
1775 The Exchange
Suite 160
Atlanta, GA
Finished Goods Manufacturer: Rigid Foam

ATLAS FOAM PRODUCTS
12836 Arroyo St.
Sylmar, CA 91342 USA
Markets: Medical; Packaging

ATLAS INDUSTRIES
6 Willows Rd.

Ayer, MA 01432 USA
Raw Material Supplier/Manufacturer: Resins

ATLAS MINERALS & CHEMICALS, INC.
Farmington Rd.
Mertztown, PA 19539 USA
Finished Goods Manufacturer: Elastomer; Coating;
   Adhesive; Sealant

ATLAS ROOFING CORPORATION
802 Highway 19N
Meridian, MS 39307 USA
Finished Goods Manufacturer: Rigid Foam

AUBURN MANUFACTURING
Stack St.
Middletown, CT 06457 USA
Finished Goods Manufacturer: Elastomer; TPU

AUSIMONT USA
44 Whippany Rd.
Morristown, NJ 07960 USA
Finished Goods Manufacturer: Elastomer

AUSTIN URETHANE, INC.
P.O. Box 921
Americus, GA 31709 USA
Finished Goods Manufacturer: Flexible Foam

AUTOCAMIONES IND. DE PUERTO RICO
P.O. Box 2522
Bayamon, PR 00619 USA

AUTOMA INTERIOR SYSTEMS
19700 Haggerty Rd.
Livonia, MI 48152 USA
Finished Goods Manufacturer:. Flexible Foam

AUTOMATIC PROCESS CONTROL, INC.
1123 Morris Ave.
Union, NJ 07083 USA
Equipment or Machinery Supplier/Manufacture:.
   Dispensing Equipment; Tanks

AUTOMATION PRODUCTS, INC.
DYNATROL(TM)
3030 Max Roy
Houston, TX 77008 USA
Equipment or Machinery Supplier/Manufacturer:
   Dispensing Equipment; Automated Controls

AUTOMATIC CHEMICAL PRODUCTS
P.O. Box 130
240 Cambria St.
Randolph, WI 53956 USA
Finished Goods Manufacturer: Coating

AUTORIM
18 Johns Lane
Topsfield, MA 01983 USA
Equipment or Machinery Supplier/Manufacturer:
   Dispensing Equipment

AVON RUBBER COMPANY
2001 Belgrave Ave.
Huntington Park, CA90255 USA

Finished Goods Manufacturer: Elastomer; Millable
   Urethanes

AWOODS RUBBER CO., INC.
2401 Pilot Knob Rd.
Mendota Heights, MN 55120 USA
Finished Goods Manufacturer: Elastomer; Cast

AXEL PLASTICS RESEARCH LABORATORIES, INC.
Box 770 855
Woodside, NY 11377 USA
Raw Material Supplier/Manufacturer: External Release
   Agents; Internal Mold Release Agents; Mold Cleaners;
   Mold Conditioners; Barrier Coat/Release Agents

AXIOM INDUSTRIES, INC.
P.O. Box 1147
Tualatin, OR 97062 USA
Finished Goods Manufacturer: Elastomer; Cast

AZON USA, INC.
2204 Ravine Rd.
Kalamazoo, MI, 49004 USA
Raw Material Supplier/Manufacturer: Resins

BALLY ENGINEERED STRUCTURES, INC.
P.O. Box 98
Bally, PA 19503 USA
Finished Goods Manufacturer: Rigid Foam
Markets: Construction

BALSAM CORP.
727 Goddard Ave.
Chesterfield, MO 63005 USA
Finished Goods Manufacturer: Elastomer
Markets: Recreational

BANDO AMERICAN, INC.
1149 W. Bryn Mawr
Itasca, IL 60143 USA
Finished Goods Manufacturer: Elastomer

BOS BROS.
24 Federal Plaza
Bloomfield, NJ 07003 USA
Finished Goods Manufacturer: Flexible Foam

BARNHARDT MANUFACTURING
P.O. Box 34276
1100 Hawthorne Lane
Charlotte, NC 28234 USA
Markets: Aerospace; Carpet; Electronic/Electrical;
   Furniture;

BARON CONSULTING CO.
273 Pepe's Farm Rd.
Milford, CT 06473 USA
Markets: Aerospace; Appliances; Automotive;
   Carpet;Chemical; Clothing; Construction;
   Defense/Military; Electrionic/Electrical; Footwear;
   Furniture; Industrial; Lumber; Marine; Medical;
   Recreational; Refrigeration; Wheels/Tires

BARRY INSTRUMENTS, INC.
1156 N. Batavia St.
Orange, CA 92667 USA

Finished Goods Manufacturer: Elastomer; TPU
Markets: Electronic/Electrical; Business Machines

BARTON SOLVENTS
P. 0. Box 221
Des Moines, IA 50301 USA
Systems Manufacturer: Rigid Foam Systems; Flexible
Foam Systems

B & B ROOFING & INSULATION
604 N. Tremont
Kewanee, IL 61443 USA
Markets: Construction

B. F. GOODRICH, CO.
Speciality Polymers & Chemicals Division.
9911 Brecksville Rd.
Brecksville, OH 44141 USA
Raw Material Supplier/Manufacturer: Resins;
Compounds Thermoplastic
Markets: Aerospace; Appliances; Automotive; Chemical;
Defense/Military; Electronic/Electrical; Footwear;
Industrial; Material Handling; Medical; Recreational;
Wheels/Tires

B. W. FREEMAN, INC.
P.O. Box 576
714 S. Franklin
Cuba, MO 65453 USA
Finished Goods Manufacturer: Elastomer; Cast; Flexible
Foam; Rigid Foam; Molded Two Component
Markets: Appliances; Automotive; Construction;
Footwear; Furniture; Industrial; Medical;
Pipe/Plumbing; Recreational

BACON INDUSTRIES, INC.
192 Pleasant St.
Watertown, MA 02172 USA

BACON INDUSTRIES OF CA, INC.
16731 Hale Ave.
Irvine, CA 92714 USA
Finished Goods Manufacturer:. Elastomer; Cast; Coating;
Adhesive; Sealant
Markets: Aerospace; Appliances; Defense/Military;
Electronic/Electrical; Medical

BAILEY-PARKS URETHANE INC.
184 Gilbert Ave.
Memphis, TN 38106 USA
Raw Material Supplier/Manufacturer: Resins
Systems Manufacturer: Elastomer Foam Systems; Rigid
Foam Systems
Equipment or Machinery Supplier/Manufacturer:
Dispensing Equipment; Tanks; Pumps; Automated
Controls
Finished Goods Manufacturer: Coating; Sealant; Rigid
Foam
Markets: Chemical; Construction; Industrial;
Recreational/Wheels/Tires

BASF CORPORATION
Automotive Urethane Division
1609 Biddle Ave.

Wyandotte, MI 48192 USA
Raw Material Supplier/Manufacturer: lsocyanate (TDI
and MDI); Polyol(Polyether and Graft)
Systems Manufacturer: Prepolymer Manufacturer;
Elastomer Systems; Rigid Foam Systems; Flexible
Foam Systems;
Markets: Automotive

BASF CORPORATION
Polymers Division
100 Cherry Hill Rd.
Parsippany, NJ 07045 USA
Raw Material Supplier/Manufacturer: Isocyanate; Polyol;
Prepolymer; Resins
Systems Manufacturer: Prepolymer Manufacturer:
Elastomer Systems; Rigid Foam Systems; Flexible
Foam Systems
Equipment or Machinery Supplier/Manufacturer:
Dispensing Equipment; Automated Controls

BASIC RUBBER & PLASTICS CO.
3295 Haggerty Rd.
Walled Lake, MI 48390 USA
Markets: Aerospace; Defense/Military;
Electronic/Electrical; Industrial; Transportation

BATA SHOE CO., INC.
4501 Pulaski Highway
Beicamp, MD 21017 USA
Finished Goods Manufacturer:. Elastomer; TPU

BAUMER OF AMERICA INC.
P.O. Box 235
Towaco, NJ 07082 USA
Equipment or Machinery Supplier/Manufacturer: Foam
Cutting Machinery
Markets: Aerospace; Automotive; Carpet; Clothing;
Construction; Defense/Military; Electronic/Electrical;
Footwear; Furniture; Industrial; Marine; Material
Handling; Medical; Packaging; Pipe/Plumbing;
Recreational; Refrigeration; Transportation

BAXTER RUBBER CO.
10 Spielman Rd.
Fairfield, NJ 07004 USA
Markets: Aerospace; Appliances; Chemical;
Defense/Military; Electronic/Electrical; Furniture;
Industrial; Marine; Material Handling; Medical;
Mining/Drilling; Packaging; Paper; Printing

BAY RUBBER CO.
404 Pendleton Way
Oakland, CA 94602 USA
Markets: Aerospace; Automotive; Chemical;
Construction; Defense/Military; Industrial; Lumber;
Marine; Material Handling; Packaging; Paper;
Pipe/Plumbing; Printing; Refrigeration;
Transportation

BEAN RUBBER MTG. CO.
1623 S. Tenth St.
San Jose, CA 95112 USA
Finished Goods Manufacturer: Elastomer

Markets: Aerospace; Chemical; Construction; Defense/
Military; Electronic/Electrical; Industrial; Marine;
Medical; Mining/Drilling; Packaging; Paper;
Pipe/Plumbing; Recreational; Refrigeration;
Transportation

BEAR ROOFING & INSULATTON, INC.
91 8 Harris St.
P.O. Box 1551
Charlottesville, VA 22902 USA
Markets: Construction

BEI PACKAGING
13540 Lake City Way NE
Seattle, WA 98125 USA
Markets: Aerospace; Appliances; Defense/Military;
Electronic/Electrical; Furniture; Industrial; Marine;
Medical; Packaging; Pipe/Plumbing

BELOIT MANHATTAN DIVISION
Ivy Industrial Park
P.O. Box 155
Clarks Summit, PA 18411 USA
Finished Goods Manufacturer: Elastomer; Cast
Markets: Industrial; Paper; Printing

BELT CORPORATION OF AMERICA
3455 Hutchinson Rd.
Cumming, GA 30130 USA
Finished Goods Manufacturer: Elastomer
Markets: Clothing; Defense/Military; Industrial; Material
Handling; Packaging; Paper; Printing; Office
Automation

BELTING IND. CO., INC.
20 Boright Ave.
Kenilworth, NJ 07033 USA
Finished Goods Manufacturer: Elastomer
Other: Laminator of Urethane to Subtrate
Markets: Carpet; Industrial; Material Handling; Medical;
Packaging; Printing; Banking; Currency; Business
Machines

BELTSERVICE CORPORATION
4143 Rider Trail N.
St. Louis, MO 63045 USA
Finished Goods Manufacturer: Elastomer; Cast; Coating
Markets: Chemical; Construction; Industrial; Lumber;
Material Handling; Mining/Drilling; Packaging

BELZONA MOLECULAR, INC.
2000 N.W. 88th Ct.
Miami, FL 33172 USA
Finished Goods Manufacturer: Elastomer
Markets: Industrial; Filtration

BEMEL INDUSTRIES, INC.
P.O. Box 224961
Dallas, TX 75222 USA
Markets: Industrial; Filtration

BERMAR ASSOCIATES, INC.
433 Minnesoto Ave.
Troy, MI 48083 USA

Finished Goods Manufacturer: Elastomer; TPU
Markets: Appliances; Automotive; Industrial

BERTEK, INC.
Mylan Pharmecuticals
110 Lake St.
St. Albans, VT 05478 USA
Finished Goods Manufacturer: Elastomer; TPU; Coated
Fabric
Markets: Chemical; Clothing; Construction;
Defense/Military; Footwear; Industrial; Medical;
Paper; Printing; Recreational

BEST FOAM FABRICATORS
9633 S. Cottage Grove
Chicago, IL 60628 USA
Finished Goods Manufacturer: RIM; Automotive; Non-
automotive; Flexible Foam
Other: Fabricator
Markets: Aerospace; Appliances; Automotive; Defense/
Military; Electronic/Electrical; Furniture; Industrial;
Medical; Packaging; Paper

BEST ROOFING SYSTEMS, INC.
P.O. Box 14574
Oklahoma City, OK 73113 USA
Markets: Construction

BF GOODRICH COMPANY
Walker and Moore Rd.
P.O. Box 122
Avon Lake, OH 44012 USA

BINGHAM
11101 W. Franklin Ave.
Franklin Park, IL 60131 USA
Finished Goods Manufacturer: Elastomer; Cast
Markets: Industrial; Paper; Printing; Metal Decorating

BINKS MANUFACTURING COMPANY
PRED
9201 W. Belmont Ave.
Franklin Park, IL 60131 USA
Equipment or Machinery Supplier/Manufacturer:
Laminators; Dispensing Equipment; Tanks; Pumps;
Automated Controls
Markets: Aerospace; Appliances; Automotive;
Construction; Defense/Military; Furniture; Industrial;
Marine; Material Handling; Mining/Drilling;
Packaging; Recreational; Refrigeration

BIXBY INTERNATIONAL CORP.
1 Preble Rd.
Newbury, MA 01 950 USA
Finished Goods Manufacturer: TPU; Coated Fabrics
Markets: Automotive; Construction; Defense/Military;
Footwear; Furniture; Industrial; Marine; Material
Handling; Medical: Recreational

BLACK BROTHERS SOUTHEAST, INC.
1315 Baker Rd.
High Point, NC 27263 USA
Finished Goods Manufacturer: Elastomer; Cast; Millable
Urethane Gums

Markets: Automotive; Furniture; Industrial; Lumber; Packaging; Printing

BLOCKSOM AND COMIPANY

2432 S. Indiana St.

Los Angeles, CA 90023 USA

Markets: Aerospace; Construction; Defense/Military; Furniture; Industrial; Marine; Packaging; Pipe/Plumbing; Refrigeration; Transportation

BOEDEKAR PLASTICS, INC.

Route 2, Box 5

Shiner, TX 77984 USA

Other: Distributor

Markets: Aerospace; Appliances; Automotive; Chemical; Clothing; Construction; Defense/Military; Electronic/Electrical; Footwear; Industrial; Lumber; Marine; Material Handling; Medical; Mining/Drilling; Packaging; Paper; Printing; Recreational; Refrigeration; Transportation

BONDAFOAM, INC.

Highway 32 E.

Water Valley, MS 38965 USA

Finished Goods Manufacturer: Bonded Carpet Cushion (Rebond)

Markets: Carpet

BONDED PRODUCTS, INC.

Spellite Corporation

439 S. Bolmar St.

West Chester, PA 19381 USA

Systems Manufacturer: Elastomer Systems; Rigid Foam Systems; Flexible Foam Systems FinishedGoods Manufacturer Flexible Foam; Integral Skin; coating; Adhesive; Sealant; Polyurea Spray Systems

Markets: Automotive; Construction; Defense/Military; Industrial; Packaging

BOSTIK, INC.

Boston St.

Middleton, MA 01949 USA

Finished Goods Manufacturer: Elastomer; TPU; Adhesive; Sealant

Markets: Aerospace; Appliances; Automotive; Carpet; Chemical; Construction; Defense/Military; Furniture; Industrial; Lumber; Marine; Medical; Pipe/Plumbing; Recreational; Refrigeration; Transportation

BOSTROM SEATING, INC.

P.O. Box 600

Cudahy, WI 53110 USA

Other: Fabricator

Markets: Automotive

BOTTOMS U.S.A., INC.

1891 Hotel Rd.

P.O. Box 1238

Auburn, ME 04211 USA

Finished Goods Manufacturer: Shoe Soles; Microcellular

Markets: Footwear

BOWEN RUBBER & URETHANE, INC.

4410 Maple Ave.

P.O. Box 993

Burlington, NC 27215 USA

Finished Goods Manufacturer:. Elastomer; Cast; Coating

Markets: Industrial; Material Handling; Wheels/Tires

BOYD CORPORATION

6630 Owens Dr.

Pleasanton, CA 94588 USA

Finished Goods Manufacturer: Elastomer; Flexible Foam; Integral Skin

Markets: Aerospace; Appliances; Automotive; Clothing; Construction; Defense/Military; Electronic/Electrical; Footwear; Industrial; Marine; Medical; Recreational; Refrigeration; Transportation; Telecommunications; Computer

BRADFORD INDUSTRIES

1857 Middlesex St.

Lowell, MA 01851 USA

Finished Goods Manufacturer: Elastomer; Cast

Markets: Aerospace; Appliances; Automotive; Chemical; Clothing; Construction; Defense/Military; Electronic/Electrical; Footwear; Industrial; Marine; Material Handling; Medical; Mining/Drilling; Pipe/Plumbing; Printing; Recreational; Refrigeration; Transportation; Noise Control; Static Control

BRADFORD-WHITE CORPORATION

200 Lafayette St.

Middleville, MI 49333 USA

Finished Goods Manufacturer: Rigid Foam

Markets: Appliances

BRADLEY CO., B.B.

7755 Crile Rd.

Painesville, OH 44077 USA

Other: Fabricator

Markets: Packaging

BRAINERD CHEMICAL

3510 S. Sheridan

P.O. Box 470010

Tulsa, OK 74147 USA

Raw Material Supplier/Manufacturer: Isocyanate; Polyol; Catalyst; Surfactant; Blowing Agent; Polymer

Markets: Chemical; Packaging

BRANDYWINE FIBRE PRODUCTS CO.

15th and Poplar St.s, Dept. PI

Wilmington, DE 19801 USA

Other: Fabricator

Markets: Aerospace; Appliances; Automotive; Chemical; Construction; Defense/Military; Electronic/Electrical; Furniture; Industrial; Marine; Material Handling; Medical; Packaging; Pipe/Plumbing; Printing; Recreational; Refrigeration; Transportation.

BRANSON ULTRASONICS CORP

41 Eagle Rd.

Danbury, CT 06813-1961 USA

Finished Goods Manufacturer: Elastomer; Cast

Markets: Industrial

BRC RUBBER GROUP, INC.
589 Highway 33S
Churubusco, IN 46723 USA
FinishedGoods Manufacturer: Elastomer; TPU
Markets: Aerospace; Appliances; Automotive;
    Construction; Defense/Military; Electronic/Electrical;
    Industrial; Marine

BREMEN CORPORATION
405 N. Industrial Dr.
Bremen, IN 46506 USA
Other: Fabricator
Markets: Appliances; Automotive; Construction;
    Defense/Military; Industrial; Medical; Recreational;
    Transportation; Sports Equipment

BRIN-MONT CHEMICALS, INC.
3921 Spring Garden St.
Greensboro, NC 27407 USA
Raw Material Supplier/Manufacturer: Polyol; Polymer
Systems Manufacturer: Rigid Foam Systems; Flexible
    Foam Systems
Markets: Appliances; Carpet; Construction; Furniture;
    Industrial; Marine; Material Handling; Packaging;
    Recreational; Refrigeration; Transportation

BROOKLYN PRODUCTS, INC.
9203 Wamplers Lake Rd.
Brooklyn, MI 49230 USA
Other: Fabricator

BRUCE PLASTICS, INC.
P.O. Box 4547
Pittsburgh, PA 15205 USA
Finished Goods Manufacturer: Elastomer; TPU
Markets: Appliances; Carpet; Furniture; Medical;
    Packaging; Recreational; Refrigeration

BRUCK PLASTICS CO.
990 E. 107th S.
Lemont, IL 60439 USA
Other: Distributor

BRUNSWICK SEAT CO.
120 Tices Lane
East Brunswick, NJ 08816 USA
Finished Goods Manufacturer: Flexible Foam; Rigid Foam
Other: Fabricator
Markets: Automotive

BRYANT RUBBER CORP.
1112 Lomita Blvd.
Harbor City, CA 90710 USA
FinishedGoods Manufacturer: Elastomer; Cast
Markets: Aerospace; Automotive; Defense/Military; Elec-
    tronic/Electrical; Medical; Packaging; Pipe/Plumbing;
    Transportation

BRYANT UNIVERSAL ROOFING, INC.
3311 N. 44th St., No. 100
Phoenix, AZ 85018 USA
Other: Roofing Contractor

BRYN HILL INDUSTRIES
Homes Corporate Center
Price/Pine Streets
Holmes, PA 19043 USA
Finished Goods Manufacturer: Flexible Foam
Markets: Automotive; Defense/Military;
    Electronic/Electrical; Furniture; Medical; Packaging;
    Recreational; Transportation

BUCKEYE RUBBER & PACKNG CO.
23940 Mercantile Rd.
Cleveland, OH 44122 USA
Finished Goods Manufacturer: Elastomer; Cast; Flexible
    Foam; Coating
Other: Distributor
Markets: Aerospace; Appliances; Automotive; Chemicals;
    Defense/Military; Electronic/Electrical; Industrial;
    Material Handling; Medical; Mining/Drilling;
    Pipe/Plumbing; Printing; Refrigeration;
    Transportation; Wheels/Tires; Hydraulics; Pneumatics

BUNKER CORPORATION
960 Calle Amanecer
San Clemente, CA 92672 USA
Finished Goods Manufacturer: Elastomer; Cast
Other: Fabricator
Markets: Automotive; Construction; Industrial; Marine;
    Material Handling; Pipe/Plumbing; Recreational;
    Transportation

BURKART FOAM, INC.
36th and Sycamore Streets
Cairo, IL 62914 USA
Finished Goods Manufacturer: Flexible Foam; Rigid Foam
Other: Consultant
Markets: Automotive; Carpet; Construction; Furniture;
    Industrial; Medical; Packaging; Recreational;
    Refrigeration; Transportation

BURLY SEAL PRODUCTS CO.
1026 S. Santa Fe Ave.
Los Angeles, CA 90021 USA
Finished Goods Manufacturer: Elastomer
Markets: Appliances; Automotive; Industrial; Hydraulic

BURNET ASSOCIATES LTD.
P.O. Box 331
Syracuse, NY 13206 USA
Other: Fabricator
Markets: Packaging

BURTIN CORPORATION
2550 Garnsey St.
Santa Ana, CA 92707 USA
Raw Material Supplier/Manufacturer: Polyol; Polymer;
    Prepolymer
Systems Manufacturer: Prepolymer Manufacturer;
    Elastomer Systems, Rigid Foam Systems; Flexible
    Foam Systems; Coatings Systems
Finished Goods Manufacturer: Elastomer; RIM; Non-
    automotive; TPU; Adhesive

Markets: Aerospace; Appliances; Automotive; Chemical; Construction; Defense/Military; Electronic/Electrical; Footwear; Furniture; Industrial; Marine; Medical; Mining/Drilling; Packaging; Pipe/Plumbing; Recreational; Refrigeration; Transportation; Wheels/Tires

BUSAK & SHAMBAN

Corporate Research and Development

2531 Bremer Dr.

Fort Wayne, IN 46803 USA

Finished Goods Manufacturer: Elastomer; TPU; Cast

Markets: Aerospace; Appliances; Automotive; Chemical; Construction; Defense/Military; Electronic/Electrical; Industrial; Marine; Medical; Mining/Drilling; Pipe/Plumbing; Refrigeration; Transportation

BUTLER MANUFACTURING CO.

P.O. Box 19917

Kansas City, MO 64141 USA

Finished Goods Manufacturer: Rigid Foam

Other: Fabricator

Markets: Construction; Refrigeration

BUTLER RUBBER PRODUCTS, INC.

P.O. Box 1019

Guntersville, AL 35976 USA

Finished Goods Manufacturer: Elastomer; Millable Polyurethane

Markets: Aerospace; Defense/Military; Electronic/Electrical; Industrial; Lumber; Material Handling; Pipe/Plumbing; Wheels/Tires

BVANS-ST CLAIR, INC.

2541 Mills St.

Marysville, MI 48040 USA

Finished Goods Manufacturer: Adhesive

C.A. LAWTON CO.

111 W. Walnut St.

Green Bay, WI 54303 USA

Markets: Aerospace; Appliances; Automotive; Defense/Military; Electronic, /Electrical; Furniture; Industrial; Marine; Material Handling; Medical; Mining/Drilling; Packaging; Paper; Recreational; Transportation

C.J. PRODUCTS INC.

100 Christmas Place

Weston, WV 26452 USA

Equipment or Machinery Supplier/Manufacturer: Molds; Dispensing Equipment; Conveyor; Turn Tables; Product Handling Equipment

Finished Goods Manufacturer: Elastomer; Flexible Foam; Rigid Foam

Markets: Aerospace; Appliances; Automotive; Construction; Defense/Military; Electronic/Electrical; Furniture; Industrial; Marine; Material Handling; Medical; Mining/Drilling; Packaging; Printing; Recreational; Refrigeration; Transportation

C.P. HALL CO., THE

7300 S. Central Ave.

Chicago, IL 60638 USA

Raw Material Supplier/Manufacturer: Polyol; Chain Extenders; Plasticizer

Markets: Aerospace; Appliances; Automotive; Chemical; Defense/Military; Electronic/Electrical; Footwear; Marine, Medical; Wheels/Tires

C.U.E., INC.

11 Leonberg Rd.

Mars, PA 16046 USA

Finished Goods Manufacturer: Elastomer; Cast

Markets: Aerospace; Defense/Military; Industrial; Material Handling; Mining/Drilling; Packaging; Paper; Transportation

C.W. FIFIELD COMPANY INC.

4 Keith Way

Hingham, MA 02043 USA

Finished Goods Manufacturer: Coating

Markets: Clothing; Packaging; Publishing; Personal Leather Goods; Fashion

CABLEC CONTINENTIAL CABLES COMPANY

2555 Kingston Rd.

York, PA 17402 USA

Finished Goods Manufacturer: Elastomer; TPU

Markets: Aerospace; Appliances; Automotive; Chemical; Construction; Defense/Military; Electronic/Electrical; Industrial; Mining/Drilling; Paper; Power Plants; Wire/Cable

CAL POLYMERS, INC.

2115 Gaylord St.

Long Beach, CA 90813, USA

Other: Fabricator; Recycler/Reprocessor

Markets: Carpet; Furniture

CAL-FIBER CO.

625 S. Anderson St.

Los Angeles, CA 90023 USA

CA GASKET & RUBBER CORP.

Rubber Division

1601 W. 134 St.

Gardena, CA 90249 USA

Finished Goods Manufacturer: Elastomer

Markets: Aerospace; Appliances; Automotive; Industrial

CA URETHANE

1821 Railroad St.

Corona, CA 91720 USA

Finished Goods Manufacturer: Flexible Foam; Integral Skin; Rigid Foam

CALLAHAN CHEMICAL CO.

P.O. Box 65

Broad St. and Filmore Ave.

Palmyra, NY 08065 USA

Raw Material Supplier/Manufacturer: Isocyanate; Polyol; Solvents

Other: Distributor

Markets: Chemical; Defense/Military; Industrial; Packaging; Paper; Transportation

CALUMET RUBBER CORP.
3545 S. Normal
Chicago, IL 60609 USA
Finished Goods Manufacturer: Elastomer
Markets: Appliances; Industrial; Printing

CANNON, USA
1235 Freedom Rd.
Mars, PA 16046 USA
Equipment or Machinery Supplier/Manufacturer:
   Laminators; Dispensing Equipment; Tanks; Pumps;
   Automated Controls
Markets: Aerospace; Appliances; Automotive; Chemical;
   Construction; Footwear; Furniture; Recreational;
   Refrigeration

CAPE INDUSTRIES
Subsidiary of Hoechst Celanese Corporation
Highway 421 N.
Wilmington, NC 28402-0327 USA
Raw Material Supplier/Manufacturer: Polyol
Markets: Aerospace; Appliances; Automotive;
   Construction; Defense/Military; Furniture; Industrial;
   Marine; Packaging; Pipe/Plumbing; Recreational;
   Refrigeration; Transportation; Agriculture

CAPITOL FOAM PRODUCTS, INC.
75 E. Union Ave.
East Rutherford, NJ 07073 USA
Other: Fabricator
Markets: Automotive; Clothing; Defense/Military;
   Electronic/Electrical; Footwear; Furniture; Industrial;
   Marine; Medical; Packaging; Recreational; Bedding;
   Display

CARDINAL INDUSTRIAL FINISHES
1329 Potrero Ave.
South El Monte, CA 91733 USA
Finished Goods Manufacturer: Coating
Markets: Aerospace; Appliances; Automotive;
   Defense/Military; Electronic/Electrical; Furniture;
   Industrial

CARDIOTECH INTERNATIONAL, INC.
11 State St.
Woburn, MA 01801

CARLISLE GEAUGA COMPANIES
100 Seventh Ave.
Chardon, OH 44024 USA
Finished Goods Manufacturer: Elastomer: TPU
Markets: Appliances; Automotive; Construction;
   Furniture; Industrial; Medical; Recreational;
   Refrigeration; Transportation; Wheels/Tires

CAROLINA RUBBER ROLLS
H.K. Porter Co., Inc.
Osage Dr.
Donaldson Center
Greenville, SC 29605 USA
Finished Good Manufacturer: Elastomer; Cast
Markets: Chemical; Furniture; Industrial; Paper;
   Printing; Textile

CARPENTER INSULATION & COATINGS, INC.
5016 Monument Ave.
P.O. Box 27205
Richmond, VA 23261 USA
System Manufacturer: Rigid Foam Systems
Finished Goods Manufacturer: Flexible Foam
Markets: Construction; Furniture; Industrial; Marine;
   Packaging; Refrigeration; Transportation

CARTEX CORP.
200 Rock Run Rd.
Fairless Hills, PA 19030 USA
Finished Goods Manufacturer: Elastomer; Flexible Foam
Markets: Automotive

CASCADE FOAM & COATINGS, INC.
Route 1, Box 281
Winthrop, WA 98862 USA
Other: Contractor
Markets: Construction

CASCHEM, INC.
40 Ave. A
Bayonne, NJ 07002 USA
Raw Material Supplier/Manufacturer: Polyol; Catalyst;
   Surfactant; Prepolymer; Plasticizer
Systems Manufacturer: Prepolymer Manufacturer;
   Elastomer Systems; Adhesives
Finished Goods Manufacturer: Adhesive; Sealant
Markets: Automotive; Carpet; Chemical; Clothing;
   Construction; Defense/Military; Electronic/Electrical;
   Industrial; Marine; Medical; Mining/Drilling; Pipe/
   Plumbing; Printing; Telecommunications Adhesives

CASTALL, INC.
P.O. Box 58
Weymouth Industrial Park
East Weymouth, MA 02189 USA
Raw Material Supplier/Manufacturer: Prepolymer
   Manufacturer; Elastomer Systems; Flexible Foam
   Systems
Markets: Aerospace; Automotive; Defense/Military;
   Electronic/Electrical

CASTLE RUBBER CO.
Park-Ohio Industries, Inc.
P.O. Box 589
Butler, PA 16001 USA
Finished Goods Manufacturer: Elastomer; Flexible Foam
Markets: Appliances; Automotive; Industrial

CBI (CHICAGO BRIDGE & IRON)
Project Services Insulation
1501 N. Division St.
Plainfield, IL 60544 USA
Other: Contractor; Applicator; Insulated Storage Tanks
   and Steel Plate Structures
Markets: Construction

CELLULAR TECHNOLOGY, INC.
6065 Roswell Rd. N.E.
Suite 1373
Atlanta, GA 30328-4019 USA

Raw Material Supplier/Manufacturer: Catalyst
Surfactant; Flame Retardants; Intestates; Fillers and
Extender Products
Markets: a Appliances; Automotive; Carpet; Chemical;
Clothing; Construction; Defense/Military;
Electronic/Electrical; Footwear; Furniture; Industrial;
Marine; Medical; Packaging; Paper

CELOTEX CORPORATION, THE
4010 Boy Scout Blvd.
Tampa, FL 33607-5736 USA
Finished Goods Manufacturer: Rigid Foam
Markets: Construction; Building Industry-Commercial
and Residential

CENTER FOR APPLIED ENGINEERING, INC.
Cellular Plastics and Material Testing Departments
10301 Ninth St. N
St. Petersburg, FL 33716 USA
Other: Consultant; Testing; Research Organization
Markets: Appliances; Automotive; Carpet; Chemical;
Construction; Furniture; Industrial; Refrigeration;
Transportation

CENTRAL RUBBER CO.
844 E. Jackson
Belvidere, IL 61008 USA
Finished Goods Manufacturer: Elastomer; TPU
Markets: Appliances; Construction; Electronic/Electrical;
Industrial; Mining/Drilling; Printing; Wheels/Tires

CENTURY RUBBER COMPANY INC.
21609 Parthenia St.
Canoga Park, CA 91304 USA
Finished Goods Manufacturer: Elastomer; Millables
Markets: Aerospace; Appliances; Defense/Military;
Electronic/Electrical; Industrial

CGR PRODUCTS
P.O. Box 2110
Greensboro, NC 27402 USA
Finished Goods Manufacturer: Elastomer; Cast
Other: Distributor
Markets: Appliances; Carpet; Chemical; Clothing/Textile;
Construction; Defense/Military; Electronic/Electrical;
Footwear; Furniture; Industrial; Lumber; Marine;
Material Handling; Medical; Mining/Drilling;
Packaging; Paper; Pipe/Plumbing; Printing;
Recreational; Refrigeration; Transportation;
Wheels/Tires

CHAMBERLAIN RUBBER COMPANY INC.
P.O. Box 22700
Rochester, NY 14692 USA
Other: Fabricator
Markets: Appliances; Chemical; Construction;
Electronic/Electrical; Industrial; Material Handling;
Medical; Packaging; Recreational; Transportation

CHAMBERS GASKET & MANUFACTURING COMPANY
4701 W. Rice St.
Chicago, IL 60651 USA

Other: Fabricator
Markets: Aerospace; Appliances; Automotive;
Construction; Electronic/Electrical; Industrial;
Marine; Material Handling; Medical; Mining/Drilling;
Packaging; Refrigeration; Transportation

CHARDON RUBBER CO.
373 Washington St.
Chardon, OH 44024 USA
Markets: Appliances; Automotive; Footwear; Industrial

CHARLES ENGINEERING & SERVICES, INC.
P.O. Box 428
Belcamp, MD 21017 USA
Finished Good Manufacturer: Flexible Foam; Rigid Foam
Markets: Chemical; Defense/Military; Medical;
Refrigeration; Art

CHEM-ELAST COATINGS, INC.
P.O. Box 19799
St. Louis, MO 63144 USA
Systems Manufacturer: Rigid Foam Systems
Other: Distributor
Markets: Construction

CHEMQUEST, INC.
6235 S. McIntosh Rd.
Sarasota, FL 34238 USA
Systems Manufacturer: Elastomer Systems; Rigid Foam
Systems; Flexible Foam Systems; Polyurea Spray
Finishes; Instant Set Adhesives
Markets: Aerospace; Carpet; Chemical; Construction;
Defense/Military; Furniture; Lumber; Marine;
Packaging; Pipe/Plumbing; Refrigeration;
Transportation; Taxidermy Forms; Chemical and
GEO-Containment

CHEMREX, INC.
Sonneborn
889 Valley Park Dr.
Shakopee, MN 55379 USA
Finished Good Manufacturer: Coating; Adhesive; Sealant
Markets: Construction

CHEMSTAR URETHANES
1148 California Ave.
Corona, CA 91719 USA
Systems Manufacturer: Elastomer Systems; Rigid Foam
Systems; Flexible Foam Systems
Other: Urethane Recycler
Markets: Carpet; Construction; Packaging

CHESTNUT RIDGE FOAM, INC.
P.O. Box 781
Latrobe, PA 15650 USA
Finished Goods Manufacturer: Flexible Foam
Markets: Aerospace; Defense/Military;
Electronic/Electrical; Furniture; Marine; Medical;
Packaging; Recreational; Transportation

CHICAGO-ALLIS MANUFACTURING CORPORATION
113 N. Green St.
Chicago, IL 60607 USA

Finished Goods Manufacturer: Elastomer; TPU; Millables
Markets: Hydraulics

CHILDERS PRODUCTS CO.
2061 Hartel St.
Levittown, PA 19057 USA
Systems Manufacturer: Rigid Foam Systems
Other: Distributor
Markets: Construction; Industrial

CHIVAS PRODUCTS LIMITED
HEADQUARTERS
Chivas Urethane Products
6200 19 Mile Rd.
Sterling Heights, MI 48314-3266 USA
Finished Goods Manufacturer: Flexible Foams; Integral
   Skin; Coating; Rigid Foam
Other: Fabricator
Markets: Automotive

CHROMA CORPORATION
3900 Dayton St.
McHenry, IL 60050 USA
Finished Goods Manufacturer: Elastomer

CIA COMMERICIAL & INDUSTRIAL
APPLICATION, INC.
P.O. Box 1250
Tomball, TX 77377-1250 USA
Other: Consultant; Contractor
Markets: Construction; Roofing

CIBA-GEIGY
Furane Aerospace Products Division
5121 San Fernando Rd. W.
Los Angeles, CA 90039 USA
Systems Manufacturer: Elastomer Systems
Finished Goods Manufacturer: Adhesive; Sealant;
   Syntactics *Markets:* Aerospace; Transportation

CIBA-GEIGY CORPORATION
Ciba Additives Division
7 Skyline Dr.
Hawthorne, NY 10532 USA
Raw Material Supplier/Manufacturer: Light Stabilizers;
   UV Absorbers; Antioxidants
Markets: Aerospace; Appliances; Automotive; Carpet;
   Chemical; Clothing; Construction; Defense/Military;
   Electronic/Electrical; Footwear; Furniture; Industrial;
   Lumber; Marine; Material Handling; Medical;
   Mining/Drilling; Packaging; Paper; Pipe/Plumbing;
   Printing; Recreational; Refrigeration; Transportation;
   Wheels/Tires; Light Stabilization; Heat Stabilization

CINCINNATI MILACRON
4165 Half Acre Rd.
Batavia, OH 45103 USA
Equipment or Machinery Supplier/Manufacturer:
   Dispensing Equipment; Clamps; Processing
   Equipment
Markets: Aerospace; Appliances; Automotive;
   Construction; Defense/Military; Furniture; Industrial;

Marine; Recreational; Refrigeration; Transportation;
Plastics

CLARK DOOR CO.
2366 Centerline Dr.
St. Louis, MO 63146 USA
Finished Goods Manufacturer: Rigid Foam
Markets: Construction; Refrigeration

CLARK FOAM PRODUCTS
4630 W. 53rd St.
Chicago, IL 60632 USA
Other: Fabricator
Markets: Aerospace; Appliances; Automotive;
   Construction; Defense/Military; Electronic/Electrical;
   Footwear; Furniture; Industrial; Marine; Material
   Handling; Medical; Packaging; Recreational;
   Refrigeration; Transportation

CLARK FOAM PRODUCTS
25887 Crown Valley Parkway
South Laguna, CA 92677 USA
Finished Goods Manufacturer: Rigid Foam
Markets: Industrial; Recreational

CLARK-SCHWEBEL DISTRIBUTION CORP.
698 Bryant Blvd.
Rock Hill, SC 29732 USA
Other: Distributor
Markets: Chemical; Construction; Electronic/Electrical;
   Industrial; Marine; Medical; Recreational;
   Refrigeration; Transportation; Cultured Marble

CLASSIC MODULAR SYSTEMS, INC.
1911 Columbus St.
Two Rivers, WI 54241 USA
Finished Goods Manufacturer: Molded Soft Polyurethane
   Edging
Other: Fabricator
Markets: Industrial; Medical; Universities

CLAYTON CORPORATION
866 Horan Dr.
Fenton, MO 63026 USA
Finished Goods Manufacturer: Valve Manufacturer

CLEVITE ELASTOMERS
33 Lockwood Rd.
Milan, OH 44846 USA
Finished Goods Manufacturer: Coating
Markets: Automotive

CO-PLAS, INC.
5112 Wheeler Ave.
Fort Smith, AR 72901 USA
Systems Manufacturer: Rigid Foam Systems; Flexible
   Foam Systems
Other: Distributor
Markets: Construction; Refrigeration; Transportation

COASTCRAFT RUBBER CO.
23340 S. Normandie Ave.
Torrance, CA 90502 USA
Finished Goods Manufacturer: Elastomer

Markets: Aerospace; Automotive; Defense/Military;
  Medical

COATED FABRICS GROUP, INC.
213 Central St.
Milford, MA 01757 USA
Finished Goods Manufacturer: Elastomer; Millable Gums
Markets: Defense/Military; Industrial

COATINGS SYSTEMS, INC.
55 Crown St.
Nashua, NH 03060 USA
Finished Goods Manufacturer: Coating
Markets: Appliances; Clothing; Footwear; Furniture;
  Industrial; Marine

COATINGS UNLIMITED, INC.
1150 Old US Highway One S.
Southern Pines, NC 28387 USA
Systems Manufacturer: Elastomer Systems

COLEMAN CO.
250 North St., Francis
P.O. Box 1762
Wichita, KS 67201 USA
Finished Goods Manufacturer: Flexible Foam; Rigid Foam
Markets: Recreational

COLOMBIA CEMENT CO., INC.
159 Hanse Ave.
Freeport, NY 11520 USA
Finished Goods Manufacturer: Adhesive
Markets: Appliances; Automotive; Carpet; Construction;
  Defense/Military; Furniture; Industrial; Marine;
  Recreational; Refrigeration; Transportation

COLORIM SYSTEMS, INC.
281 Enterprise Ct.
Suite 400
Bloomfield Hills, MI 48302 USA
Finished Goods Manufacturer: Elastomer; RIM
Markets: Automotive;

COLUMBIA RUBBER MILLS
P.O. Box 220
Clackamas, OR 97015 USA
Finished Goods Manufacturer: Elastomer; Cast
Markets: Industrial; Material Handling; Printing

COMBINED ENERGIES CORPORATION
Young Builders
22222 N. 22nd Ave.
Phoenix, AZ 85027 USA
Equipment or Machinery Supplier/Manufacturer:
  Dispensing Equipment
Finished Goods Manufacturer: Elastomer; Coating; Rigid
  Foam
Other: Fabricator; Contractor
Markets: Aerospace; Construction; Defense/Military;
  Industrial; Marine

**COMBINED SUPPLY & EQUIPMENT
COMPANY INC.**
21 5 Chandler St.

Buffalo, NY 14207 USA
Other: Distributor

COMCAST URETHANE COMEPANY
425 Leggitt Rd.
Marshall, MI 49068 USA
Finished Goods Manufacturer: Integral Skin;
  Microcellular
 Markets: Automotive; Industrial; Material Handling;
  Recreational

COMMERCIAL PLASTICS COMPANY
800 E. Allanson Rd.
Mundelein, IL 60060 USA
Finished Goods Manufacturer: Elastomer; TPU
Markets: Appliances; Construction; Electronic/Electrical;
  Industrial; Medical; Packaging; Pipe/Plumbing;
  Recreational; Refrigeration; Transportation; Tobacco

COMP TECH
3409 W. 14th St.
Erie, PA 16505 USA
Finished Goods Manufacturer: Elastomer; TPU
Markets: Electronic/Electrical; Medical; Business
  Equipment; Consumer Products

COMPONENT FINISHING CORP.
217 Edgerton St.
Bryan, OH 43506 USA
Finished Goods Manufacturer: Flexible Foam
Markets: Appliances; Automotive; Electronic/Electrical;
  Marine; Sports Equipment

COMPOSITE PRODUCTS, INC.
P.O. Box 241414
Charlotte, NC 28224 USA
Finished Goods Manufacturer: Elastomer; Cast
Other: Consultant
Markets: Footwear; Industrial; Medical; Recreational

COMSTOCK INDUSTRIES
185 Waukewan St.
Meredith, NH 03253 USA
Finished Goods Manufacturer: Elastomer; Cast
Markets: Construction; Industrial; Marine;
  Mining/Drilling; Packaging; Paper; Recreational;
  Wheels/Tires

CON-TEK MACHINE, INC.
3575 Hoffman Rd. E.
Saint Paul, MN 55110 USA
Equipment or Machinery Supplier/Manufacturer: Custom
  Foaming; Fixtures; RIM Clamps; Control Systems
Other: Fabricator
Markets: Aerospace; Appliances; Automotive; Carpet;
  Chemical; Defense/Military; Industrial; Refrigeration;
  Transportation

CONAP, INC.
1405 Buffalo St.
Olean, NY 14760 USA
Raw Material Supplier/Manufacturer: Isocyanate;
  Polymer; Prepolymer; Resins

Systems Manufacturer: Prepolymer Manufacturer;
  Elastomer Systems
Finished Goods Manufacturer: Elastomer; Coating;
  Adhesive; Sealant
Markets: Aerospace; Appliances; Automotive; Chemical;
  Defense/Military; Electronic/Electrical, - Industrial;
  Marine; Material Handling; Medical; Mining/Drilling;
  Packaging; Printing; Recreational; Wheels/Tires

CONCEPT COATING
Rural Route 1
Box 148
Beardstown, IL 62618 USA
Systems Manufacturer: Spray-in-Place Polyurethane
  Foam
Equipment or Machinery Supplier/Manufacturer: Pumps
Finished Goods Manufacturer: Coating
Other: Contractor
Markets: Construction; Industrial

CONE MILLS CORP.
1201 Maple St.
Greensboro, NC 27405 USA
Finished Goods Manufacturer: Flexible Foam
Markets: Automotive; Clothing; Furniture

CONITRON
P.O. Box 90
Trinity, NC 27370 USA
Equipment or Machinery Supplier/Manufacturer: Molds
Finished Goods Manufacturer: RIM; Automotive; Non-
  automotive; Flexible Foam; Integral Skin
Markets: Aerospace; Automotive; Defense/Military;
  Furniture; Industrial; Medical; Recreational;
  Transportation

COMKLIN COMPANY, INC.
Building Products
551 Valley Park Dr.
Shakopee, MN 55379 USA
Systems Manufacturer: Roof Coating Systems
Finished Goods Manufacturer: Coating
Markets: Construction

**CT RUBBERMOLDING
CORPORATION**
94A School St.
Danielson, CT 06239 USA
Finished Goods Manufacturer: Elastomer; Cast; Millable
  Gums
Markets: Defense/Military; Electronic/Electrical;
  Industrial; Marine; Material Handling; Medical;
  Pipe/Plumbing; Wheels/Tires

CONSULTMORT, INC.
6723 14th Ave. N.
St. Petersburg, FL 33710-5404 USA
Other: Consultant
Markets: Construction; Industrial- Recreational; New
  Product Development and Marketing

CONTACT RUBBER CORP.
8635 198th Ave.

Bristol, WI 53104 USA
Finished Goods Manufacturer: Elastomer; Cast
Markets: Appliances; Automotive; Industrial; Printing

CONTINENTAL INDUSTRIES GROUP, INC.
245 E. 58th St.
Suite 12A
New York, NY 10022 USA
Raw Material Supplier/Manufacturer: Isocyanate; Polyol;
  Blowing Agent
Other: Distributor
Markets: Chemical

CONTINENTAL SEAL
6723 S. Hanna St.
Fort Wayne, IN 46816 USA
Markets: Aerospace; Automotive; Defense/Military;
  Electronic/Electrical; Industrial; Refrigeration;
  Transportation; Wheels/Tires

CONTINENTAL WHITE CAP, INC.
2215 Sanders Rd.
Northbrook, IL 60062 USA
Finished Goods Manufacturer: Elastomer; Cast
Markets: Industrial

CONTOUR FABRICATORS, INC.
4100 E. Baldwin Rd.
P.O. Box 56
Grand Blanc, MI 48439 USA
Other: Fabricator
Markets: Furniture

CONTRAST EQUIPMENT
1449 N. Topping
Kansas City, MO 64120 USA
Other: Distributor
Markets: Industrial

CONTROLLED RUBBER PRODUCTS
1425 Kalamazoo St.
South Haven, MI 49090 USA
Finished Goods Manufacturer: Elastomer, RIM;
  Automotive
Markets: Automotive; Industrial; Mining/Drilling;
  Transportation

CONVENIENCE PRODUCTS
866 Horan Dr.
Fenton, MO 63026 USA
Systems Manufacturer: Rigid Foam Systems; Flexibie
  Foam Systems
Equipment or Machinery Supplier/Manufacturer:
  Dispensing Equipment; Portable Units
Markets: Construction; Industrial; Marine;
  Pipe/Plumbing; Refrigeration; Transportation

COWERSE, INC.
One Fordham Rd.
N. Reading, MA 01864 USA
Finished Goods Manufacturer: Elastomer; Cast
Markets: Footwear

CONVERTERS, INC.
1617 Republic Rd.
Huntingdon Valley, PA 19006 USA
Finished Goods Manufacturer: Flexible Foam
Other: Distributor; Fabricator
Markets: Aerospace; Appliances; Automotive;
Construction, Defense/Military, Electronic/Electrical;
Industrial; Material Handling; Medical; Packaging;
Paper; Printing; Recreational; Refrigeration;
Transportation

CONWAY INDUSTRMS, INC.
1365 Shuler Ave.
Hamilton, OH 45011 USA
Finished Goods Manufacturer: Flexible Foam
Markets: Furniture

COOK ASSOCIATES, INC.
Chemical
212 W. Kinzie St.
Chicago, IL 60610 USA
Other: Consultant
Markets: Chemical

COOK COMPOSITES AND POLYMERS
Urethane Products Division
217 Freeman Dr.
P.O. Box 996
Port Washington, WI 53074 USA
Raw Material Supplier/Manufacturer: Polyol; Polymer;
Prepolymer; Resins
Systems Manufacturer: Prepolymer Manufacturer;
Elastomer Systems; Rigid Foam Systems; Flexible
Foam Systems; Non-cellular Castables
Markets: Aerospace; Appliances; Automotive;
Construction; Defense/Military; Electronic/Electrical;
Footwear; Furniture; Industrial; Marine; Medical;
Mining/Drilling; Packaging; Recreational;
Refrigeration; Transportation; Wheels/Tires;
Taxidermy

COOLEY GROUP
50 Esten Ave.
Pawtucket, RI 02860 USA
Finished Goods Manufacturer: Coating
Markets: Aerospace; Automotive; Chemical; Clothing;
Construction; Defense/Military; Electronic/Electrical;
Industrial; Marine; Material Handling; Medical;
Packaging; Recreational; Refrigeration;
Transportation; Nuclear Industry; Oil Booms

COOLEY, INC.
50 Esten Ave.
Pawtucket, RI 02860 USA
Finished Goods Manufacturer: Coating
Markets: Coating; Aerospace; Automotive; Clothing;
Construction; Marine; Medical

CORAM INSTRUMENTS, INC.
37 Whinstone St.
Coram, NY 11727 USA

Equipment or Machinery Supplier/Manufacturer: Test
Equipment
Other: Testing
Markets: Aerospace; Automotive; Footwear; Furniture

CORBOND CORP
32404 Frontage Rd.
Bozeman, MT 59715 USA
Systems Manufacturer: Rigid Foam Systems; Flexible
Foam Systems
Other: Distributor

CORCORAN MANUFACTURING CO., INC.
1140 E. Howell St.
Anaheim, CA 92805 USA
Finished Goods Manufacturer: Flexible Foam
Other: Fabricator
Markets: Paint

CORETAPE, INC.
80 Marc Dr.
Cuyahoga Falls, OH 44223 USA
Finished Goods Manufacturer: Elastomer; Tapes
Other: Fabricator
Markets: Construction; Electronic/Electrical; Industrial;
Packaging; Refrigeration

CORK INSULATION SALES CO., INC.
1943 1st Ave. S.
Seattle, WA 98134 USA
Other: Fabricator

CORROSION CONTROL, INC.
1947 W. 28th St.
Cleveland, OH 44113 USA
Other: Contractor
Markets: Construction

CORSON RUBBER PRODUCTS
103-109 Smith St.
P.O. Box 154
Clover, SC 29710 USA
Finished Goods Manufacturer: Elastomer
Markets: Industrial

COSTAL FABRICATORS INC.
P.O. Box 906
Conway, SC 29526 USA
Other: Fabricator
Markets: Furniture; Marine; Bedding

COURTAULD AEROSPACE
5430 San Fernando Rd.
Glendale, CA 91203 USA
Finished Goods Manufacturer: Coating; Adhesive; Sealant
Markets: Aerospace; Defense/Military: Industrial;
Packaging

COURTIN GASKET & RUBBER CO., INC.
680 S. River St.
P.O. Box 1347
Aurora, IL 60507 USA
Finished Goods Manufacturer: Elastomer; Cast
Markets: Industrial

COWART FOAM CO., INC.
Trinity Investments, Inc.
405 Bell
Montgomery, AL 36104 USA
Finished Goods Manufacturer: Flexible Foam; Rigid Foam
Other: Distributor; Fabricator
Markets: Carpet; Defense/Military; Furniture; Industrial;
   Marine; Medical; Packaging; Recreational

CP INDUSTRIES
P.O. Box 524
12767 Industrial Dr.
Granger, IN 46530 USA
Equipment or Machinery Supplier/Manufacturer: Molds-
   Patterns; Models
Markets: Aerospace; Automotive; Furniture; Industrial;
   Material Handling; Medical; Recreational;
   Transportation

CPM INC.
10830 Sanden Dr.
Dallas, TX 75238-1337 USA
Finished Goods Manufacturer: Elastomer; TPU
Markets: Photographic; Graphic Arts

CRAFTED PLASTICS
P.O. Drawer 327
Sheboygan, WI 53082-0327 USA
Finished Goods Manufacturer: TPU; Extruded
Markets: Appliances; Automotive; Construction;
   Defense/Military; Furniture; Industrial; Marine;
   Mining/Drilling; Packaging; Pipe/Plumbing;
   Recreational; Refrigeration; Transportation

CRAFREX PROTECTIVE COATINGS
P.O. Box 020108
Tuscaloosa, AL 35402-0108 USA
Finished Goods Manufacturer: Coating
 Markets: Carpet; Construction; Furniture; Industrial;
   Lumber; Mining/Drilling; Paper; Wheels/Tires

CRAIN COMMUNICATIONS. INC.
Urethanes Technology
1725 Merriman Rd.
Suite 300
Akron, OH 44313-5251 USA
Other: Industry Publications

CRAIN INDUSTRIES
P.O. Box 6478
Fort Smith, AR 72903 USA
Finished Goods Manufacturer: Flexible Foam
Markets: Automotive; Carpet; Furniture

CRANZ RUBBER & GASKET, INC.
2671 Main St.
Buffalo, NY 14214 USA
Finished Goods Manufacturer: Elastomer
Markets: Appliances; Automotive; Industrial

CREATIVE FOAM CORP.
300 N. Alloy Dr.
Fenton, MO 48430 USA
Finished Goods Manufacturer: Flexible Foam; Rigid Foam

Other: Fabricator
Markets: Automotive; Medical; Floral

CREATIVE MACHINE CORPORATION
P.O. Box 1720
Auburn, ME 04211 USA
Equipment or Machinery Supplier/Manufacturer: Molds
Other: Mfg Rep; Mold Maker
Markets: Appliances; Automotive; Electronic/Electrical;
   Footwear; Industrial; Medical; Recreational

CREATIVE PRODUCTS, INC.
3917 Covington Highway
Decatur, GA 30032 USA
Other: Die Cutting
Markets: Packaging; Business Machinery

CREATIVE URETHANES, INC.
P.O. Box 919
310 N. 21st St.
Purcellville, VA 22132 USA
Finished Goods Manufacturer: Elastomer; RIM;
   Automotive; Non-automotive; Cast; Flexible Foam;
   Integral Skin; Rigid Foam
Markets: Aerospace, Appliances; Automotive;
   Construction; Defense/Military; Electronic/Electrical;
   Industrial; Marine; Material Handling; Medical;
   Mining/Drilling; Packaging; Paper; Pipe/Plumbing;
   Printing; Recreational; Transportation; Wheels/Tires;
   Custom Molding

CREME ART CORPORATION
3155 Dallavo Ct.
Walled Lake, MI 48390 USA
Markets: Automotive; Clothing; Footwear; Furniture;
   Luggage

CREST FOAM INDUSTRIES, INC.
100 Carol Place
Moonachie, NJ 07074 USA
Finished Goods Manufacturer: Flexible Foam; Reticulated
   Polyurethane Foams
Markets: Aerospace; Appliances; Automotive; Defense/
   Military; Electronic/Electrical; Footwear; Industrial;
   Marine; Medical; Printing; Recreational; Personal
   Care; Consumer Products; Filtration

CROWLEY CHEMICAL CO.
261 Madison Ave.
New York, NY 1001 6 USA
Raw Material Supplier/Manufacturer: Surfactant;
   Extender/Plasticizers
Markets: Automotive; Carpet; Chemical; Construction;
   Footwear; furniture; Industrial; Marine; Refrigeration;
   Transportation; Wheels/Tires

CROWN PRODUCTS CORP.
309 N. 17th St.
St. Louis, MO 63103 USA
Other: Fabricator
Markets: Packaging

CTL ENGINEERING, INC.
6301 A Augus

Raleigh, NC 27613 USA
Other: Consultant
Markets: Construction; Roofing

CUSTOM COATING, INC.
204 W. Industrial Blvd.
Dalton, GA 30720 USA
Finished Goods Manufacturer: Flexible Foam; Coating
Markets: Automotive; Carpet; Footwear; Recreational

CUSTOM DECORATIVE MOULDINGS
P.O. Box F
Greenwood, DE 19950 USA
Finished Goods Manufacturer: Rigid Foam; Architectural
   Features
Markets: Construction; Architectural Trim

CUSTOM MATERIALS, INC.
16865 Park Circle Dr.
Chagrin Falls, OH 44023 USA
Finished Goods Manufacturer: Microcellular
Other: Fabricator
Markets: Electronic/Electrical; Printing

CUSTOM PACK, INC.
3 Bacton Hill Rd.
Malvern, PA 19355 USA
Finished Goods Manufacturer: Flexible Foam
Other: Fabricator
Markets: Appliances; Electronic/Electrical; Furniture;
   Industrial; Medical; Packaging

CUSTOM RUBBER CORP.
1274 E. 55th St.
Cleveland, OH 44103-1029 USA
Finished Goods Manufacturer: Flexible Foam
Markets: Industrial

CUSTOM URETHANE APPLICATORS
6826 Hillsdale Ct.
Indianapolis, IN 46250 USA
Other: Distributor; Contractor
Markets: Chemical; Construction; Industrial

CYKO, L.C.
Box 926
Newton, IA 50208 USA
Finished Goods Manufacturer: Elastomer; Cast
Markets: Recreational

CYTEC INDUSTRIES
Five Garrett Mountain Plaza
West Paterson, NJ 07424 USA
Raw Material Supplier/Manufacturer: Isocyanate; Polyol;
   Catalyst; Surfactant; Polymer; Resins
Markets: Aerospace; Appliances; Automotive; Chemical;
Defense/Military; Electronic/Electrical; Furniture;
   Marine; Mining/Drilling; Paper; Printing;
   Wheels/Tires

CYTEC INDUSTRIES
Research Laboratory
1937 W. Main St.
Stamford, CT 06904 USA

Raw Material Supplier/Manufacturer: Aliphatic Diisocyanate; Unsaturated Isocyanate
Markets: Chemical

D&S DATA RESOURCES, INC.
P.O. Box H
Yardley, PA 19067 USA
Other: Consultant; Research Organization
Markets: Aerospace; Appliances; Automotive; Carpet;
   Chemical; Clothing; Construction- Defense/Military-
   Electronic/Electrical; Footwear; Furniture; Industrial;
   Lumber; Marine; Material Handling; Medical;
   Mining/Drilling; Packaging; Paper; Pipe/Plumbing;
   Printing; Recreational; Refrigeration; Transportation;
   Wheels/Tires

D.S. BROWN CO.
331 E. Cherry St.
North Baltimore, OH 45872 USA
Finished Goods Manufacturer: Elastomer; Cast
Markets: Industrial

D.S. BROWN CO.
Brown Building
P.O. Box 158
N.Baltimore, OH 45872 USA
Finished Goods Manufacturer: Elastomer
Markets: Construction; Bridges, Highway

DA PRO RUBBER, INC.
601 N. Poplar
Broken Arrow, OK 74012 USA
Finished Goods Manufacturer: Elastomer
Markets: Chemical- Industrial; Material Handling;
   Medical

DALCO INDUSTRIES
584 Chestnut Ridge Rd.
Chestnut Ridge, NY 10977 USA
Finished Goods Manufacturer: Coating
Markets: Construction; Industrial; Recreational

DALLAS FOAM, INC.
2171 E. Southlake Blvd.
Southlake, TX 79062 USA
Other: Fabricator
Markets: Marine; Packaging; Display

DALLAS URETHANE CONTRACTORS, INC.
406 Ranch Rd.
Forney, TX 75126 USA
Other: Contractor
Markets: Construction; Industrial

DANCO PLASTOCK, INC.
11 Danco Rd.
Putnam, CT 06260 USA
Finished Goods Manufacturer: Elastomer; RIM;
   Automotive; Non-automotive; Rigid Foam
Markets: Automotive; Electronic/Electrical; Furniture

DARCY RUBBER CO.
1915 Bronxdale
New York, NY 10462 USA

Finished Goods Manufacturer: Elastomer; TPU
Markets: Chemical; Material Handling; Mining/Drilling

DAVIDSON EXTERIOR TRIM TEXTRON
Route 6, Box 10
Americus, GA 31709 USA
Finished Goods Manufacturer: High Density Foam
Markets: Automotive

DAVIDSON INTERIOR TRIM TEXTRON
P. 0. Box 1502
Dover, NH 03820-1504 USA
Finished Goods Manufacturer: Flexible Foam
Markets: Automotive

DAVIDSON-TEXTRON CO.
Instrument Panels
Route 11
Farmington Manufacturing
Farmington, NH 03835 USA
Finished Goods Manufacturer: Elastomer; RIM;
    Automotive
Markets: Automotive

DAY INTERNATIONAL
333 W. First St.
Dayton, OH 45401-0338 USA
Finished Goods Manufacturer: Elastomer
Markets: Lumber; Packaging

DAYTON POLYMERIC PRODUCTS, INC.
Goshen Rubber Co.
85 Harrisburg Dr.
Englewood, OH 45322 USA
Finished Goods Manufacturer:. Flexible Foam; Integral
    Skin; Rigid Foam; Electronic Encapsulation
Markets: Appliances; Automotive; Construction;
    Electronic/Electrical; Furniture; Industrial; Marine;
    Material Handling; Medical; Packaging;
    Transportation

DECA RUBBER, INC.
1769 Bishop Rd.
P.O. Box 746
Chehalis, WA 98532 USA
Finished Good Manufacturer: Elastomer
Markets: Industrial; Transportation

DECKER INDUSTRIES, INC.
P.O. Drawer R
5051 S.E. Federal Highway
Port Salerno, FL 34992 USA
Equipment or Machinery Supplier/Manufacturer:
    Dispensing Equipment; Tanks; Pumps; Automated
    Controls
Other: Machine Manufacturer
Markets: Aerospace; Appliances; Automotive;
    Construction; Defense/Military; Electronic/Electrical;
    Footwear; Furniture; Industrial; Marine; Material
    Handling; Medical; Mining/Drilling; Packaging;
    Pipe/Plumbing; Printing; Recreational; Refrigeration;
    Transportation; Wheels/Tires

DEERFIELD URETHANE, INC.
Routes 5 and 10
Box 186
Deerfield, MA 01373 USA
Finished Goods Manufacturer: TPU
Markets: Industrial; Packaging; Paper

DE SEAT CO.
31 Blevins Dr.
New Castle, DE 19720 USA
Finished Goods Manufacturer: Flexible Foam
Markets: Automotive

DELCO PRODUCTS
General Motors Corporation
P.O. Box 1042
Dayton, OH 45401 USA
Finished Goods Manufacturer: Flexible Foam
Markets: Automotive

DELTA POLYMERS, INC.
281 E. Skip Lane
North Bay Shore, NY 11706 USA
Finished Goods Manufacturer: Coating
Markets: Construction; Electronic/Electrical

DENARCO, INC.
301 Industrial Dr.
Constantine, MI 49042 USA
Finished Goods Manufacturer: Sealant

DENCO INDUSTRIES, INC.
P.O. Box 73563
Houston, TX 77273 USA
Raw Material Supplier/Manufacturer: Catalyst;
    Surfactant; Inert Fillers
Other: Distributor
Markets: Carpet; Furniture; Recreational

DENNIS CHEMICAL COMPANY
2700 Papin St.
St. Louis, MO 63103 USA
Systems Manufacturer: Prepolymer Manufacturer;
    Elastomer Systems; Rigid Foam Systems; Flexible
    Foam Systems; Custom Manufacturer; Toll
    Manufacturer
Markets: Automotive; Construction;
    Electronic/Electrical; Furniture; Industrial; Medical;
    Mining/Drilling; Pipe/Plumbing; Recreational;
    Transportation

DETAIL MASTERS, INC.
330 North Park
San Antonio, TX 78216 USA
Other: Overspray Removal; Contractor
Markets: Construction

DETROIT RUBBER CO.
10401 Northlawn Ave.
Detroit, MI 48204 USA
Finished Goods Manufacturer: Elastomer
Markets: Aerospace; Automotive; Defense/Military; Elec-
    tronic/Electrical; Industrial; Printing

DIELECTRICS INDUSTRIES, INC.
300 Burnett Rd.
Chicopee, MA 01106 USA
Other: Fabricator; OEM/Contract Manufacturer
Markets: Aerospace; Appliances; Automotive; Defense/
Military; Electronic/Electrical; Footwear; Industrial;
Marine; Medical

DIKE-0-SEAL, INC.
3965 S. Keeler Ave.
Chicago, IL 60632 USA
Finished Goods Manufacturer:. Elastomer; Cast; Coating
Markets: Automotive; Defense/Military;
Electronic/Electrical; Packaging; Printing

DISCAS, INC.
567-1 S. Leonard St.
Waterbury, CT 06708 USA
Raw Material Supplier/Manufacturer: Resins
Systems Manufacturer: Elastomer Systems
Other: Research Organization
Markets: Footwear

DIVERSIFIED COMPOUNDERS, INC.
5701 E. Union Pacific Ave.
Los Angeles, CA 90022 USA
Systems Manufacturer: Elastomer Systems
Markets: Automotive; Construction;
Electronic/Electrical; Mining/Drilling

DIVERSIFIED FOAM PRODUCTS
134 Branch St.
St. Louis, MO 63147 USA
Other: Fabricator
Markets: Furniture; Industrial; Packaging;
Transportation

DIVERSIFIED FOAM, INC.
P.O. Box 1358
Yadkinville, NC 27055 USA
Finished Goods Manufacturer: Flexible Foam; Rigid Foam
Markets: Medical; Packaging

DIVISION SEVEN SYSTEMS
9843 Rist Canyon Rd.
P.O. Box 85
Bellvue, CO 80512 USA
Equipment or Machinery Supplier/Manufacturer:
Dispensing Equipment

DIXIE-NARCO
P.O. Drawer 719
Williston, SC 29853 USA

**DODGE-REGUPOL, INC. (FORMERLY
DODGE CORK CO., INC.)**
11 Laurel St.
Lancaster, PA 17603 USA
Systems Manufacturer: Elastomer Systems
Finished Goods Manufacturer: Elastomer; Cast
Markets: Aerospace; Automotive; Construction; Defense/
Military; Medical; Mining/Drilling

DOM DIMAGGIO, INC.
500 Broadway
P.O. Box 987
Lawrence, MA 01842 USA
Finished Goods Manufacturer: Molded and Flexible
Other: Distributor; Fabricator; Packaging; Retail Product
Design
Markets: Automotive; Defense/Military;
Electronic/Electrical; Furniture; Industrial; Marine;
Medical; Packaging; Recreational

DONALD W. BELLES & ASSOCIATES, INC.
1205 S. Graycroft Ave.
Madison, TN 37115 USA
Other: Consultant

DONNELLY CORPORATION
414 E. 40th St.
Holland, MI 49423 USA
Finished Goods Manufacturer: RIM; Automotive
Markets: Automotive

DONRAY CO.
500 S.O.M. Center Rd.
Cleveland, OH 44143 USA
Finished Goods Manufacturer: Flexible Foam

DOUGLAS AND LOMASON CO.
24600 Hallwood Ct.
Farmington Hills, MI 48331 USA
Finished Goods Manufacturer: Flexible Foam
Markets: Automotive

DOUGLASS INDUSTRIES
Urethane Foam Division
412 Boston Ave.
P.O. Box 701
Egg Harbor, NJ 08215 USA
Finished Goods Manufacturer: Flexible Foam
Other: Fabricator
Markets: Aerospace; Automotive; Carpet; Clothing;
Defense/Military; Electronic/Electrical; Furniture;
Industrial; Marine; Medical; Packaging; Recreational;
Transportation; Bedding

DOVE PRODUCTS, INC.
3341 State St.
Lockport, IL 60441 USA
Finished Goods Manufacturer: Elastomer; Cast; Flexible
Foam; Integral Skin
Other: Fabricator
Markets: Construction; Electronic/Electrical; Furniture;
Marine; Recreational

DOW CHEMICAL CO., THE
Polyurethanes Thermoset Applications
2040 Dow Center
Midland, MI 48674 USA
Raw Material Supplier/Manufacturer: lsocyanate; Polyol;
Catalyst; Polymer; Prepolymers Isocyanurate Foams;
Propylene Oxide; Ethylene Oxide; Butylene Oxide

Systems Manufacturer: Prepolymer Manufacturer; Elastomer Systems; Rigid Foam Systems; Flexible Foam Systems; RIM

Markets: Appliances; Automotive; Carpet; Construction; Footwear; Furniture; Industrial; Marine; Pipe/Plumbing; Recreational; Refrigeration; Wheels/Tires

DREW FOAM COMPANIES, INC.
P.O. Box 420
Highway 35 S.
Monticello, AR 71655 USA
Other: Fabricator
Markets: Construction; Packaging

DSM MELAMINE, US
Waterstone Suite 140
Atlanta, GA 30337 USA
Raw Material Supplier/Manufacturer: Resins; Melamine Capralactone-TPU
Finished Goods Manufacturer: Elastomer; TPU

DUKANE CORP.
2900 Dukane Dr.
St. Charles, IL 60174 USA
Finished Goods Manufacturer: Elastomer; Cast
Other: Fabricator

DUPONT CHEMICALS
Fluorochemicals Laboratory
Chestnut Run Plaza
Wilmington, DE 19898 USA
Raw Material Supplier/Manufacturer: Blowing Agent
Markets: Appliances; Automotive; Carpet; Chemical; Construction; Defense/Military; Electronic/Electrical; Footwear; Furniture; Industrial; Marine; Packaging; Refrigeration; Transportation

DUPRE ENTERPRISES
12623 New Brittany Blvd.
Fort Meyers, FL 33907 USA

DURA VENT CO.
1177 Markley Dr.
Plymouth, IN 46563 USA
Finished Goods Manufacturer: Elastomer; TPU
Markets: Industrial; Material Handling

DURAFOAM SEATING
Seating Division of Sears
1718 S. Palm Court St.
Davenport, IA 52802 USA
Finished Goods Manufacturer: Flexible Foam
Markets: Furniture

DUREX PRODUCTS, INC.
112 First Ave. W., Box 354
Luck, WI 54853 USA
Systems Manufacturer: Elastomer Systems
Finished Goods Manufacturer: Elastomer, Cast
Markets: Industrial; Material Handling; Mining/Drilling; Wheels/Tires

DURODYNE, INC.
850 E. Teton Hanger No.1
Tucson, AZ 85706 USA
Finished Goods Manufacturer: Elastomer; TPU
Markets: Industrial, Material Handling

DUROX COMPANY
12351 Prospect Rd.
Strongsville, OH 44136 USA
Finished Goods Manufacturer: Elastomer
Markets: Electronic/Electrical; Marine

DYNA SOURCE, INC.
P.O. Box 40337
Houston, TX 77240-0337 USA
Finished Goods Manufacturer: Elastomer
Markets: Automotive; Industrial

DYNAMIC FOAM PRODUCTS, INC.
P.O. Box 774861
Steamboat Springs, CO 80477 USA
Finished Goods Manufacturer: Flexible Foam
Markets: Footwear; Medical; Recreational

DYNAMIC PACKAGING, INC.
8601 73rd Ave. N.
Brooklyn, MN 55428 USA
Finished Goods Manufacturer: Flexible Foam
Markets: Packaging

DYNASAUER CORPORATION
3511 Tree Ct.
Industrial Blvd.
St. Louis, MO 63122 USA
Finished Goods Manufacturer: Elastomer; Cast
Markets: Aerospace; Automotive; Construction; Industrial; Marine; Material Handling; Mining/Drilling; Rollers

E-A-R SPECIALTY COMPOSITES
Cabot Safety Corporation
7911 Zionsville Rd.
Indianapolis, IN 46268 USA
Finished Goods Manufacturer: Elastomer; Automotive; Non-automotive; Cast; Flexible Foam
Markets: Aerospace; Appliances; Automotive; Defense/ Military; Electronic/Electrical; Footwear; Industrial; Marine; Material Handling; Medical; Mining/Drilling; Packaging; Recreational; Transportation;

E-K MANUFACTURING
1119 Mississippi Ave.
St. Louis, MO 63104 USA
Finished Goods Manufacturer: Flexible Foam
Other: Fabricator
Markets: Furniture; Packaging

E. D. BULLARD CO.
Route 7 White Oak Pike
Box 596
Cynthiana, KY 41031 USA
Other: Distributor

E. F. WHITMORE AND COMPANY
480 N. Indian Hill Blvd.
Suite 2A
Claremont, CA 91711 USA
Other: Mfg. Rep

E. G. SMITH CONSTRUCTION PRODUCTS, INC.
Smith-Steelight
100 Walls St.
Pittsburgh, PA 15202 USA
Finished Goods Manufacturer: Rigid Foam
Markets: Construction

E. I. DUPONT DE NEMOURS & CO.
1007 Market St.
C & P, BMP-1 321
Wilmington, DE 19898 USA
Raw Material Supplier/Manufacturer: Blowing Agent;
   Diisocyanates; Hydroflorocarbons
Markets: Aerospace; Appliances; Automotive; Carpet;
   Chemical; Clothing; Construction; Defense/Military;
   Electronic/Electrical; Footwear; Furniture; Industrial;
   Lumber; Marine; Material Handling; Medical;
   Mining/Drilling; Packaging; Paper; Pipe/Plumbing;
   Printing; Recreational; Refrigeration; Transportation;
   Wheels/Tires

E. J. DAVIS COMPANY, THE
IO Dodge Ave.
North Haven, CT 06473 USA
Other: Distributor; Fabricatoi
Markets: Aerospace; Appliances; Construction; Marine;
   Medical; Refrigeration

E. R. CARPENTER CO.
5016 Monument Ave.
Richmond, VA 23230 USA
Raw Material Supplier/Manufacturer: Polyol; Polymer;
   PrePolymer; Resins
Systems Manufacturer: Prepolymer Manufacturer;
   Elastomer Systems; Rigid Foam Systems; Flexible
   Foam Systems
Finished Goods Manufacturer: Elastomer; TPU; Cast;
   Flexible Foam; Integral Skin; Coating; Adhesive; Rigid
   Foam
Markets: Automotive; Carpet; Chemical; Clothing;
   Construction; Furniture; Marine; Material Handling;
   Medical; Packaging; Pipe/Plumbing; Recreational;
   Wheels/Tires

E. T. HORN CO.
16141 Heron Ave.
LaMirada, CA 92608 USA
Raw Material Supplier/Manufacturer: lsocyanate: Polyol;
   Catalyst; Prepolymer; Resins
Other: Distributor
Markets: Aerospace; Construction; Defense Military;
   Elec- tronic/Electrical: Furniture; Industrial;
   Wheels/Tires

E. T. OAKES CORPORATION
686 Old Willets Path

Hauppauge, NY 11788-4102 USA
Equipment or Machinery Supplier/Manufacturer: Mixing
Equipment

EAGLE PICHER AUTOMOTIVE GROUP
Orthane Division
1500 1-35 W.
Denton, TX 76207 USA
Finished Goods Manufacturer: Elastomer; Automotive;
   TPU; Flexible Foam
Other: Fabricator
Markets: Automotive; Industrial; Marine; Agriculture

EAGLE-PITCHER MAT
Eagle-Pitcher Industries
3911 Ben Hur Ave.
Willoughby, OH 44094 USA
Finished Goods Manufacturer: Elastomer; Cast
Markets: Appliances; Automotive; Construction;
   Defense/Military; Electronic/Electrical; Industrial;
   Mining/Drilling

EASTERN CHEMICAL CO.
5700 Tacony St.
Philadelphia, PA 19135 USA
Other: Distributor

EASTERN CONTAINERS
60 Maple St.
Mansfield, MA 02048 USA
Finished Goods Manufacturer: Flexible Foam; Rigid Foam
Markets: Aerospace; Defense/Military;
   Electronic/Electrical; Packaging

EASTERN FOAM CORP.
50 Hilton St.
Easton, PA 18042 USA
Finished Goods Manufacturer: Flexible Foam
Markets: Bedding

EASTERN MOLDING CO., INC.
597 Main St.
Belleville, NJ 07109 USA
Finished Goods Manufacturer: Elastomer
Markets: Appliances; Electronic/Electrical; Packaging;
   Paper; Printing

EASTMAN CHEMICAL CO.
Urethane Intermediates
P.O. Box 431
Kingsport, TN 37662 USA
Raw Material Supplier/Manufacturer: Polyol
Markets: Appliances; Automotive; Construction;
   Defense/Military; Electronic/Electrical; Footwear;
   Furniture; Marine; Packaging; Pipe/Plumbing;
   Recreational; Refrigeration; Transportation;
   Wheels/Tires

EBONITE INTERNATIONAL, INC.
1813 W. 7th St.
Hopkinsville, KY 42240 USA
Finished Goods Manufacturer: Elastomer; Cast
Markets: Recreational; Bowling Balls

EC/R INC.
University Tower, Suite 404
31 01 Petty Rd.
Durham, NC 27253 USA
Other: Consultant
Markets: State, Local, and Federal Environmental
    Agencies

ECKERT MFG., INC.
3820 Nicholson Rd.
Fowlerville, MI 48836 USA
Finished Goods Manufacturer: Elastomer; Cast
Markets: Automotive

EDCO, INC.
4261 Airlane Southeast
Kentwood, MI 49508 USA
Other: Fabricator
Markets: Furniture

EDGE-SWEETS COMPANY
2887 Three Mile Rd. NW
Grand Rapids, MI 49504 USA
Equipment or Machinery Supplier/Manufacturer:
    Laminators; Dispensing Equipment; Tanks; Pumps;
    Automated Controls; Molding Lines; Cutting and
    Fabricating; Saws; Bulk Storage; Slablines; Filler
    Metering Systems
Markets: Aerospace; Appliances; Automotive; Carpet;
    Chemical; Construction; Defense/Military; Electronic/
    Electrical; Footwear; Furniture; Industrial; Lumber;
    Marine; Material Handling; Medical; Packaging;
    Paper; Recreational; Refrigeration; Transportation;
    Wheels/Tires; Agricultural

EG GASKET & SUPPLY, INC.
1011 Centry Dr.
P.O. Box 31
Waukesha, WI 53186 USA
Finished Goods Manufacturer: Flexible Foam
Other: Fabricator
Markets: OEM

EJ THOMPSON, INC.
706 Linkfield Rd.
Watertown, CT 06795 USA
Other: Consultant

ELASCO, INC.
11377 Mardon Dr.
Garden Grove, CA 92641 USA
Finished Goods Manufacturer: Elastomer; Cast
Other: Consultant
Markets: Automotive; Industrial; Material Handling;
    Recreational; Transportation; Wheels/Tires

ELASTOGRAN MACHINERY (EMB)
1725 Biddle Ave.
Wyandotte, MI 48192 USA
Equipment or Machinery Supplier/Manufacturer:
    Dispensing Equipment; Conveyors; Processing
    Equipment
Markets: Plastic

ELASTOMER ENGINEERING, INC.
801 Steuben St.
Sioux City, IA 51101 USA
Finished Goods Manufacturer: Elastomer; RIM; Non-
    automotive; Cast
Markets: Industrial; Mining/Drilling; Paper;
    Recreational; Wheels/Tires; Textile Equipment

ELE CORPORATION
7847 W. 47th St.
Lyons, IL 60534 USA
Raw Material Supplier/Manufacturer: Polyol; Catalyst;
    Surfactant; Resins
Other: Distributor
Markets: Aerospace; Appliances; Automotive; Carpet;
    Chemical; Clothing; Construction; Defense/Military;
    Electronic/Electrical; Footwear; Furniture; Industrial;
    Lumber; Marine; Material Handling; Medical-
    Mining/Drilling; Packaging; Paper; Pipe/Plumbing;
    Printing; Recreational; Refrigeration; Transportation;
    Wheels/Tires

ELF ATOCHEM NORTH AMERICA
Fluorochemicals
2000 Market St.
Philadelphia, PA 19103-3222 USA
Raw Material Supplier/Manufacturer: Blowing Agent
Other: Raw Material Manufacturer
Markets: Appliances; Construction; Furniture; Industrial;
    Pipe/Plumbing; Refrigeration; Transportation

ELLIOTT COMPANY OF INDIANAPOLIS, INC.
9200 Zionsville Rd.
Indianapolis, IN 46268 USA
Finished Goods Manufacturer:. Rigid Foam
Markets: Appliances; Construction; Defense/Military;
    Industrial; Marine; Refrigeration; Transportation

ELLIOTT-WILLIAMS CO.
2900 N. Richardt Ave.
Indianapolis, IN 46219 USA
Finished Goods Manufacturer: Rigid Foam
Markets: Construction; Refrigeration

EMERSON & CUMING, INC.
77 Dragon Ct.
Woburn, MA 01888 USA
Raw Material Supplier/Manufacturer: Resins
Finished Goods Manufacturer: Elastomer; Cast; Coating;
    Encapsulants
Markets: Aerospace; Automotive; Electronic/Electrical;
    Marine

EMPIRE WEST FOAM CORP.
12104 1/2 E. Park St.
Cerritos, CA 90701 USA
Other: Distributor

EN-TECH INC.
2110 Reynolds Lane
Louisville, KY 40218 USA
Other: Contractor

Markets: Construction; General Mfg; Institutional
　　Roofing

ENDICOTT JOHNSON
1100 E. Main St.
Endicott, NY 13766 USA
Finished Goods Manufacturer: Microcellular
Markets: Footwear

ENERGY SUSPENSION
960 Calle Amanecer
San Clemente, CA 92672 USA
Finished Goods Manufacturer: Elastomer; Cast
Other: Consultant
Markets: Automotive; Transportation

ENGELHARD CORP.
Menlo Park, CN40
Edison, NJ 08818 USA
Raw Material Supplier/Manufacturer: Catalyst
Markets: Aerospace; Appliances; Automotive; Carpet;
　　Chemical; Clothing; Construction; Defense/Military;
　　Elec.tronic/Electrical; Footwear; Furniture; Industrial;
　　Lumber; Marine; Material Handling; Medical;
　　Mining/Drilling; Packaging; Paper; Pipe/Plumbing;
　　Printing; Recreational; Refrigeration; Transportation;
　　Wheels/Tires

ENGINEERED FABRICS CORP.
669 Goodyear St.
Rockmart, GA 30153 USA
Finished Goods Manufacturer: Elastomer; Cast; Adhesive
Markets: Aerospace

ENGINEERED POLYMERS CORP.
1020 E. Maple St.
Mora, MN 55051 USA
Finished Goods Manufacturer: Elastomer; RIM; Non-
　　automotive; Flexible Foam; Rigid Foam
Markets: Construction; Electronic/Electrical; Furniture;
　　Packaging; Agricultural Equipment

ENGINEERED RUBBER PRODUCTS
1745 Copley Rd.
Akron, OH 44320 USA
Finished Goods Manufacturer: Elastomer; Cast
Markets: Aerospace; Appliances; Automotive;
　　Construction; Electronic/Electrical; Marine

ENGINEERED SYSTEMS
United Technologies
1 641 Porter St.
Detroit, MI 48216 USA
Finished Goods Manufacturer: Flexible Foam; Rigid Foam
Markets: Automotive; Marine

ENGINEERING PLASTICS, INC.
P.O. Box 1440
Westboro, MA 01581-1440 USA
Finished Goods Manufacturer: Elastomer; Cast
Markets: Industrial

ENGINEERING POLYMIERS CORP.
1020 E. Maple St.

Mora, MN 55051 USA
Finished Goods Manufacturer: RIM; Non-automotive;
　　Flexible Foam; Rigid Foam
Markets: Construction; Electronic/Electrical; Packaging;
　　Agricultural Equipment

ENTERPRISE CHEMICAL CO.
4 Parkview Dr.
Dover, OH 44622-1168 USA
Raw Material Supplier/Manufacturer: Catalyst
Other: Consultant; Distributor
Markets: Aerospace; Appliances; Automotive; Carpet;
　　Chemical; Clothing; Construction; Defense/Military-
　　Electronic/Electrical; Footwear; Furniture; Industrial;
　　Lumber; Marine; Material Handling; Medical;
　　Mining/Drilling; Packaging; Paper; Pipe/Plumbing;
　　Printing; Recreational; Refrigeration; Transportation;
　　Wheels/Tires

ENTERPRISE RUBBER INC.
1070 Evans Ave.
Akron, OH 44305 USA
Finished Goods Manufacturer: Elastomer; Cast
Markets: Aerospace; Appliances; Automotive; Electronic/
　　Electrical

ENVIRONMENTAL ALTERNATIVES
2522 Blackwood Rd.
Wilmington, DE 19810 USA
Other: Consultant

ENVIROTECH RUBBER ENGINEERING, INC.
Baker Hughes Corporation
34595 700 W
Salt Lake City, UT 84119 USA
Finished Goods Manufacturer: Elastomer
Markets: Aerospace; Chemical; Construction; Defense/
　　Military; Industrial; Marine; Material Handling;
　　Mining/Drilling; Packaging; Paper; Pipe/Plumbing;
　　Wheels/Tires

EPHRATA PRECISION PARTS
Urethane Division
P.O. Box 323
Denver, PA 17517 USA
Finished Goods Manufacturer: Elastomer
Markets: Defense/Military; Industrial; Material
　　Handling; Mining/Drilling; Printing; Wheels/Tires

EPIC RESINS
1421 Ellis St.
Waukesha, WI 53186-5620 USA
Systems Manufacturer: Elastomer Systems
Finished Goods Manufacturer: Elastomer; RIM; Non-
　　automotive; Cast; Coating; Adhesive
Markets: Chemical; Defense/Military;
　　Electronic/Electrical; Furniture; Marine; Medical;
　　Recreational; Transportation

EPICHEM, INC.
447 Weybridge Dr.
Bloomfield Hills, MI 48304 USA
Systems Manufacturer: Elastomer Systems

Markets: Industrial; Mining/Drilling; Printing; Wheels/Tires

EPMAR CORP.
13210 E. Barton Circle
Santa Fe, CA 90670 USA
Finished Goods Manufacturer: Elastomer; Automotive; Non-automotive; Cast; Sealant
Markets: Automotive; Electronic/Electrical; Transportation

EPOLIN, INC.
358-364 Adams St.
Newark, NJ 07105 USA
Finished Goods Manufacturer: Coating
Markets: Defense/Military

ERA CORPORATION
2950 Niagara Lane N
Minneapolis, MN 554474835 USA
Finished Goods Manufacturer: Coating; Sealant
Other: Distributor
Markets: Construction

ERA INDUSTRIES, INC.
57 Folly Mill Rd.
P.O. Box 160
Seabrook, NH 03874 USA
Other: Fabricator
Markets: Construction; Industrial; Packaging

ERSKINE-JOHNS CO.
4677 Worth
Los Angeles, CA 90063 USA
Other: Distributor
Markets: Aerospace; Construction; Defense/Military; Industrial; Mining/Drilling; Transportation; Food

ESCO PLASTICS
P.O. Box 60721
16415 Waverley Rd.
Houston, TX 77203-0721USA
Finished Goods Manufacturer: Elastomer; Cast
Markets: Mining/Drilling; Packaging; Paper

ESSEX SPECIALTY PRODUCTS, INC.
Corporate Headquarters
No.1 Crossman Rd.
Sayreville, NJ USA
Finished Goods Manufacturer: Elastomer; Cast; Coating; Sealant
Markets: Automotive; Transportation

ETHYL CORPOPRATION
451 FL Blvd.
Baton Rouge, LA 70801 USA
Raw Material Supplier/Manufacturer: Catalyst; Curing Agents
Markets: Aerospace; Appliances; Automotive; Carpet; Chemical; Clothing; Construction; Defense/Military; Electronic/Electrical; Footwear; Furniture; Industrial; Lumber; Marine; Material Handling; Medical; Mining/Drilling, Packaging; Paper; Pipe/Plumbing;

Printing; Recreational; Refrigeration; Transportation; Wheels/Tires

EUTSLER TECHNICAL PRODUCTS, INC.
3718 Creekmont
P.O. Box 920818
Houston, TX 77292-0818 USA
Finished Goods Manufacturer: Elastomer; Cast
Markets: Construction; Mining/Drilling; Wheels/Tires

EVEREST COATING, INC.
P.O. Box 394
Spring, TX 77383 USA
Systems Manufacturer: Elastomeric Coatings
Markets: Construction

EVERGREEN MOLDING
Fiber Services, Inc.
340 Interstate Blvd.
Greenville, SC 29615 USA
Finished Goods Manufacturer: Elastomer; RIM; Automotive; Non-automotive; Cast; Flexible Foam; Integral Skin; Coating; Rigid Foam
Markets: Automotive; Defense/Military; Electronic/Electrical; Furniture; Industrial; Lumber; Marine; Medical; Mining/Drilling; Packaging; Printing; Recreational; Transportation; Wheels/Tires

EVODE-TANNER INDUSTRIES
Furman Hall Ct.
Greenville, SC 29600 USA
Finished Goods Manufacturer: Coating; Adhesive
Markets: Automotive; Transportation

EXCELSIOR, INC.
726 Chestnut St.
Rockford, IL 61105 USA
Other: Fabricator
Markets: Aerospace; Automotive; Chemical; Industrial

EXOTHERMIC MOLDING, INC.
50 Lafayette Place
Kennelworth, NJ 07033 USA
Finished Goods Manufacturer: RIM; Non-automotive;
Markets: Aerospace; Electronic/Electrical; Industrial; Medical

EXOTIC RUBBER & PLASTICS, INC.
P.O. Box 395
34700 Grand River Ave.
Farmington, MI 48332-0395 USA
Markets: Automotive; Defense/Military; Industrial; Printing; Wheels/Tires

EXPAN PAC CORP.
4816 Lakawana
Dallas, TX 75247 USA
Other: Fabricator
Markets: Packaging

EXPANDED RUBBER & PLASTICS
14000 S. Western Ave.
Gardena, CA 90249 USA
Systems Manufacturer: Elastomer Systems

Finished Goods Manufacturer: Elastomer; Cast; Flexible Foam; Coating; Rigid Foam
Other: Distributor
Markets: Aerospace; Automotive; Electronic/Electrical; Industrial; Recreational; Transportation

F.H. MAHONEY CO.
P.O. Box 287
Houston, TX 77001 USA
Finished Goods Manufacturer: RIM; Flexible Foam
Markets: Mining/Drilling; Petrochemical

F. P. WOLL & CO.
5216 E. Comly St.
Philadelphia, PA 19135 USA
Finished Goods Manufacturer: Flexible Foam
Other: Fabricator
Markets: Packaging

F/G PRODUCTS, INC.
P.O. Box 65
Rice Lake, WI 54868 USA
Markets: Refrigeration; Transportation

FABALL ENTERPRISES OF MARYLAND
2200-F Broening Highway
Baltimore, MD 21224 USA
Finished Goods Manufacturer: Elastomer; Cast
Other: Manufacturer
Markets: Recreational

FABREEKA INTERNATIONAL, INC.
1023 Turnpike St.
Stoughton, MA 02072 USA
Finished Goods Manufacturer: Elastomer; Cast
Markets: Aerospace; Automotive; Construction; Electronic/Electrical; Mining/Drilling; Paper; Printing

FABRICATED EXTRUSIONS COMPANY
2331 Hoover Dr.
Modesto, CA 95354 USA
Finished Goods Manufacturer: Elastomer; Extruded Elastomeric Tube; Profiled Elastomers
Markets: Medical

FACTORY MUTUAL RESEARCH CORP.
1151 Boston-Providence Turnpike
Norwood, MA 02062 USA
Other: Testing
Markets: Construction

FAIRLANE INDUSTRIES
3868 Washington
St. Louis, MO 63108 USA
Other: Fabricator
Markets: Chemical; Industrial

FALCON SHOE MFG COMPANY
2 Cedar St.
Lewiston, ME 04240 USA
Finished Goods Manufacturer: Elastomer; RIM; Non-automotive; Microcellular
Markets: Footwear

FALCON WHEEL DIVISION
407 Redondo Beach Blvd.
Gardena, CA 90247 USA
Finished Goods Manufacturer: Elastomer; Cast
Markets: Wheels/Tires

FALLS CITY PATTERN & MOLD CO.
1400 W. Madison
Louisville, KY 40203 USA
Equipment or Machinery Supplier/Manufacturer: Molds

FECKEN KIRFEL AMERICA, INC.
6 Leighton Place
Mahwah, NJ 07430 USA
Equipment or Machinery Supplier/Manufacturer: Cutting Equipment

FEDERAL HOSE MFG CORP.
25 Florence Ave., P.O. Box 480
Painesville, OH 44074 USA
Finished Goods Manufacturer: Elastomer
Markets: Chemical; Industrial

FEL-PRO CHEMICAL PRODUCTS
6120 E. 58th Ave.
Commerce City, CO 80022 USA
Systems Manufacturer: Elastomer Systems
Finished Goods Manufacturer: Sealant
Markets: Automotive; Chemical; Construction; Defense/Military; Electronic/Electrical; Industrial; Marine; Material Handling; Mining/Drilling; Transportation

FERRO CORPORATION
Plastic Color and Dispersion Division
54 Kellogg Ct.
Edison, NJ 18817-2509 USA
Raw Material Supplier/Manufacturer: Pigments; Dispersions
Markets: Automotive; Carpet; Footwear; Furniture; Industrial; Packaging; Coatings

FERRY/FEMCO
1050 W. Main St.
Kent, OH 44240 USA
Equipment or Machinery Supplier/Manufacturer: Splitter and Saws for Rigid and Flexible Foams; Casting Equipment

FET ENGINEERING, INC.
903 Nutter Dr.
Bardstown, KY 40004 USA
Equipment or Machinery Supplier/Manufacturer: Molds

FIB-CHEM INDUSTRIES, INC.
100 Third St.
Charleroi, PA 15022 USA
Systems Manufacturer: Rigid Foam Systems
Other: Distributor
Markets: Construction

FIBRE-GLAST DEVELOPMENTS CORP.
1944 Neva Dr.
Dayton, OH 45414 USA
Other: Distributor

FIBRE-RESIN CORPORATION
HB FULLER-MN
170 W. Providencia Ave.
Burbank, CA 91503 USA
Systems Manufacturer: Rigid Foam Systems
Equipment or Machinery Supplier/Manufacturer: Molds
Markets: Aerospace

FIELD RUBBER PRODUCTS, INC.
3211 E. Conner St.
Noblesville, IN 46060 USA
Finished Goods Manufacturer: Elastomer; Cast
Markets: Aerospace; Automotive; Defense/Military;
    Mining/Drilling; Packaging; Paper; Printing;
    Wheels/Tires

FILMASTER, INC.
561 US Highway 46
Fairfield, NJ 07006 USA
Finished Goods Manufacturer: Elastomer; Cast
Markets: Industrial

FIRESTONE BUILDING PRODUCTS CO.
525 Congressional Blvd.
Carmel, IN 46032 USA
Finished Goods Manufacturer: Rigid Foam
Markets: Construction

FIRST CHEMICAL CORPORATION
P.O. Box 1427
Pascagoula, MS 39568-1427 USA
Raw Material Supplier/Manufacturer: Supply Raw
    Materials For Manufacturing Isocyanate; Polyol;
    Catalyst; Surfactant; Blowing Agent; Polymer;
    Prepolymer; Resins
Other: Mfg. Rep.
Markets: Chemical

FLEX, INC.
7300 Industrial Park Blvd.
Mentor, OH 44060 USA
Finished Goods Manufacturer: Elastomer; TPU
Markets: Automotive; Construction; Defense/Military;
    Mining/Drilling

FLEXAN CORP.
6626 W. Dakin
Chicago, IL 60634 USA
Finished Goods Manufacturer: RIM; Automotive; Flexible
    Foam; Integral Skin
Markets: Automotive

FLEXAUST CO.
11 Chestnut St.
Amesbury, MA 01913 USA
Finished Goods Manufacturer: Elastomer
Markets: Chemical; Industrial

FLEXFIRM PRODUCTS, INC.
2300 N. Chico Ave.
South El Monte, CA 91733 USA
Finished Goods Manufacturer: Coating
Markets: Defense/Military

FLEXIBLE FLYER
P.O. Drawer 1296
100 Tubb Ave.
West Point, MS 39773 USA
Finished Goods Manufacturer: Elastomer; RIM; Cast;
    Rigid Foam
Markets: Recreational

FLEXIBLE FOAM PRODUCTS
Ohio Decorative Products, Inc.
P.O. Box 124
Spencerville, OH 45887 USA
Finished Goods Manufacturer: Flexible Foam
Other: Fabricator
Markets: Aerospace; Automotive; Construction;
    Furniture; Medical

FLEXIBLE INDUSTRIES
1105 Agency St.
Burlington, IA 52601 USA
Finished Goods Manufacturer: Elastomer; RIM; Cast;
    Flexible Foam; Integral Skin; Rigid Foam
Markets: Construction; Footwear; Industrial;
    Recreational; Refrigeration; Wheels/Tires;
    Agricultural; Sporting Goods

FLEXIBLE PRODUCTS CO.
Polyurethane Division
1007 Industrial Park Dr.
Marietta, GA 30061 USA
Systems Manufacturer: Elastomer Systems; Rigid Foam
    Systems; Flexible Foam Systems; Specialty Coatings
Equipment or Machinery Supplier/Manufacturer:
    Dispensing Equipment
Markets: Appliances; Automotive; Construction;
    Electronic/Electrical; Furniture; Marine; Medical;
    Mining/Drilling; Packaging; Recreational;
    Refrigeration; Transportation

FLEXIBLE PRODUCTS MFG CO., INC.
1613 Columbus St.
Manitowoc, WI 54241 USA
Finished Goods Manufacturer: Flexibled Foam
Markets: Furniture; Packaging; Transportation

FLEX STEEL INDUSTRIES
Brunswick Industrial Block
3400 Jackson St.
Dubuque, IA 52001 USA
Finished Goods Manufacturer: Flexible Foam
Markets: Furniture; Recreational, Transportation

FLEXTRON INDUSTRIES
720 Mount Rd.
Aston, PA 19014 USA
Finished Goods Manufacturer: Flexible Foam; Rigid Foam
Markets: Defense/Military; Industrial; Packaging

FLORIDA FOAM CO.
5523 Empire Dr.
Pensacola, FL 32505 USA
Finished Goods Manufacturer: Flexible Foam
Markets: Furniture; Bedding

FLORIDA INTERNATIONAL UNIVERSITY
Industrial Engineering Department
University Park, Ecs 444
Miami, FL 33199 USA
Other: Consultant; Testing; Research Organization
Markets: Aerospace; Defense/Military; Industrial;
　Marine; Transportation

FLORIFOAM
7485 N.W. 79th St.
Miami, FL 33166 USA
Finished Goods Manufacturer: Flexible Foam
Markets: Furniture; Bedding

FMC CORPORATION
1735 Market St.
Philadelphia, PA 19103 USA

FOAM MATERIALS & EQUIPMENT CO.
5216 N. Broadway
St. Louis, MO 63147 USA
Other: Distributor
Markets: Construction; Industrial; Marine; Packaging

FOAM MOLDERS & SPECIALTIES
20004 State Rd.
Cerritos, CA 90701 USA
Finished Goods Manufacturer: Flexible Foam
Other: Fabricator

FOAM PACKAGING LTD.
P.O. Box 5
Glenville, CT 06831 USA
Finished Goods Manufacturer:. Flexible Foam
Markets: Industrial; Packaging

FOAM PAINT & COATINGS, INC.
1944 W. North Lane
Suite 4
Phoenix, AZ 85021 USA
Other: Distributor

FOAM PRODUCTS
6530 Cambridge St.
Minneapolis, MN 55246 USA
Other: Distributor
Markets: Taxidermy

FOAM PRODUCTS CORP.
2525 Adie Rd.
P.O. Box 2217
Maryland Heights, MO 63043 USA
Finished Goods Manufacturer: Flexible Foam; Rigid Foam
Other: Fabricator
Markets: Industrial; Packaging

FOAM PRODUCTS CORPORATION
West Industrial Park
Calhoun, GA 30701 USA
Finished Goods Manufacturer: Flexible Foam
Other: Fabricator
Markets: Footwear; Furniture; Flooring

FOAM PRODUCTS OF TYLER
P.O. Box 4790

Tyler, TX 75712 USA
Other: Fabricator
Markets: Bedding

FOAM PRODUCTS, INC.
4749 Bronx Blvd.
Bronx, NY 10470 USA
Finished Goods Manufacturer: Flexible Foam
Other: Fabricator
Markets: Furniture; Packaging

FOAM RUBBER FABRICATORS, INC.
740 Washington Ave.
Belleville, NJ 07109 USA
Finished Goods Manufacturer: Flexible Foam
Other: Fabricator

FOAM RUBBER PRODUCTS
1423 S. Second
Milwaukee, WI 53204 USA
Finished Goods Manufacturer: Flexible foam; Rigid Foam

FOAM RUBBER PRODUCTS CO., INC.
2000 Troy Ave.
P.O. Box 525
New Castle, IN 47362 USA

FOAM SPECIALTIES
11110 Business Circle
Cerritos, CA 90701 USA
Finished Goods Manufacturer: Elastomer; RIM;
　Automotive; Non-automotive; Flexible Foam, Rigid
　Foam
Markets: Aerospace; Automotive; Electronic/Electrical;
　Medical

FOAM SUPPLIES, INC.
1510 Page Industrial Blvd.
St. Louis, MO 63132 USA
Systems Manufacturer: Elastomer Systems; Rigid Foam
　Systems
Equipment or Machinery Supplier/Manufacturer:
　Dispensing Equipment; Custom Blenders
Other: Testing; Research Organization
Markets: Appliances; Automotive; Defense/Military;
　Furniture; Marine; Mining/Drilling; Recreational;
　Refrigeration; Transportation

FOAM TFCHNOLOGIES, INC.
31 7 N.E. Harrison St.
Minneapolis, MN 55412 USA

FOAM TECHNOLOGY
20 Moshassuck Rd.
Lincoln, RI 02865 USA
Finished Goods Manufacturer: Rigid Foam
Markets: Construction- Theater Props

FOAM VISIONS, INC.
22 Ocean Ave.
Copiague, NY 11726 USA
Finished Goods Manufacturer: Rigid Foam; Foam
　Advertising; Displays and Signage

Markets: Packaging; Point-of-Purchase Advertising Displays and Signage (Indoor and Outdoor)

FOAM CONVERTERS T/A COMPLETE PKG.
111 Park Dr.
Box 337
Montgomeryville, PA 18936 USA
Finished Goods Manufacturer: Flexible Foam; Rigid Foam
Other: Distributor; Fabricator
Markets: Appliances; Automotive; Defense/Military; Electronic/Electrical; Industrial; Marine; Medical; Packaging; Paper

FOAM DESIGN, INC.
444 Transport Ct.
Lexington, KY 40581 USA
Finished Goods Manufacturer: Flexible Foam
Other: Fabricator
Markets: Packaging

FOAM ENTERPRLSES, INC.
13630 Watertower Circle
Minneapolis, MN 55441 USA
Raw Material Supplier/Manufacturer: Resins
Systems Manufacturer: Rigid Foam Systems; Flexible Foam Systems
Equipment or Machinery Supplier/Manufacturer: Dispensing Equipment; Tanks; Pumps; Automated Controls
Markets: Aerospace; Appliances; Chemical; Construction; Defense/Military; Electronic/Electrical; Industrial; Marine; Mining/Drilling; Packaging; Paper; Refrigeration; Transportation

FOAM EQUIPMENT & SUPPLY CO.
280 Arundel Rd.
Elkton, MD 21921 USA
Systems Manufacturer: Rigid Foam Systems; Flexible Foam Systems
Equipment or Machinery Supplier/Manufacturer: Dispensing Equipment; Cutting Equipment

FOAM EXPERTS ROOFING, INC.
Aztec Chemical Division
P.O. Box 1423
Mesa, AZ 85210 USA
Systems Manufacturer: Elastomer Systems; Rigid Foam Systems
Finished Goods Manufacturer: Elastomer; Rigid Foam
Markets: Construction

FOAM FABRICATORS OF MN
Route 2
Box 201 B
Maple Lake, MN 55358 USA
Other: Fabricator
Markets: Packaging

FOAM FABRICATORS, INC.
3850 E. 48th Ave.
Denver, CO 80216 USA
Equipment or Machinery Supplier/Manufacturer: Molds

Finished Goods Manufacturer: Flexible Foam; Adhesive; P.E.; Foam in Place
Other: Distributor; Fabricator
Markets: Aerospace; Automotive; Construction; Defense/Military; Electronic/Electrical; Footwear; Furniture; Industrial; Marine; Medical; Mining/Drilling; Packaging; Paper; Recreational; Transportation

FOAM FABRICATORS, INC.
6908 E. Thomas
Suite 302
Scottsdale, AZ 85251 USA
Finished Goods Manufacturer: RIM; Non-automotive; Rigid Foam
Other: Fabricator
Markets: Medical; Packaging

FOAM FACTORY, INC.
3510 N.W. 53rd St.
Fort Lauderdale, FL 33309 USA
Finished Goods Manufacturer: Flexible Foam; Rigid Foam
Other: Fabricator
Markets: Construction; Defense/Military; Electronic/Electrical; Industrial; Packaging

FOAM FAIR INDUSTRIES
P.O. Box 304
3 Merion Terrace
Aldan, PA 19018 USA
Finished Goods Manufacturer: Flexible Foam
Markets: Aerospace; Automotive; Construction; Defense/Military; Electronic/Electrical; Furniture; Industrial; Marine; Material Handling; Medical; Packaging; Printing; Recreational; Refrigeration; Transportation

FOAM TEK INC.
11650 Emerald Suite A
Dallas, TX 75229 USA
Finished Goods Manufacturer: Flexible Foam
Other: Distributor; Fabricator
Markets: Defense/Military; Furniture; Packaging

FOAM TO SIZE
302 S. Leadbetter Rd.
Ashland, VA 23005 USA
Finished Goods Manufacturer: Flexible Foam
Other: Fabricator

FOAM-FORM, INC.
TRU-PACK
410 Century Blvd.
Wilmington, DE 19809 USA
Finished Goods Manufacturer: Flexible Foam
Markets: Aerospace; Medical

FOAM-TEK IPI
1151 Atlantic Dr.
West Chicago, IL 60185 USA
Systems Manufacturer: Rigid Foam Systems; Flexible Foam Systems
Markets: Marine; Packaging; Refrigeration

FOAMADE INDUSTRIES
2550 Auburn Ct.

P.O. Box 215110
Auburn Heights, MI 48057 USA
Finished Goods Manufacturer: Flexible Foam; Rigid Foam
Other: Fabricator
Markets: Automotive, Medical; Packaging; Retail

FOAMCRAFT INC.
947 W. Van Buren
Chicago, IL 60607 USA
Finished Goods Manufacturer: Flexible Foam
Other: Fabricator
Markets: Automotive; Packaging;

FOAMCRAFT, INC.
5529-43 E. Bonna Ave.
Indianapolis, IN 46219 USA
Finished Goods Manufacturer: Flexible Foam
Other: Fabricator

FOAMCRAFTERS
1327 Levee St.
Dallas, TX 75207 USA
Finished Goods Manufacturer: Flexible Foam
Other: Fabricator
Markets: Furniture; Packaging; Displays

FOAMEDGE PRODUCTS
39 Alice Dr.
Akron, OH 44319 USA
Finished Goods Manufacturer: Flexible Foam
Markets: Chemical; Construction; Industrial

FOAMEX L.P.
823 Waterman Ave.
East Providence, RI 02914 USA
Finished Goods Manufacturer: Flexible Foam
Markets: Aerospace; Appliances; Automotive; Carpet;
   Defense/Military; Footwear; Furniture; Industrial;
   Marine; Medical; Packaging; Recreational;
   Transportation

FOAMFAB, INC.
409 Oakland St.
Mansfield, MA 02048 USA
Finished Goods Manufacturer: Flexible Foam
Other: Fabricator
Markets: Packaging

FOAMSEAL, INC.
2425 N. Lapeer Rd.
P.O. Box 455
Oxford, MI 48371 USA
Systems Manufacturer: Prepolymer Manufacturer;
   Elastomer Systems; Rigid Foam Systems; Flexible
   Foam Systems
Equipment or Machinery Supplier/Manufacturer:
   Dispensing Equipment; Pumps
Markets: Automotive; Construction; Marine; Medical;
   Packaging; Recreational; Transportation;
   Manufactured Housing

FOAMTEK-IPI
Division PML, Inc.
1151 Atlantic Dr., No. 5

West Chicago, IL 60185 USA
Systems Manufacturer: Prepolymer Manufacturer;
   Elastomer Systems; Rigid Foam Systems; Flexible
   Foam Systems
Equipment or Machinery Supplier/Manufacturer:
   Dispensing Equipment
Markets: Aerospace; Appliances; Automotive; Chemical;
   Construction; Defense/Military; Electronic/Electrical;
   Footwear; Furniture; Industrial; Marine; Material
   Handling; Medical; Mining/Drilling; Packaging;
   Pipe/Plumbing; Recreational; Refrigeration;
   Transportation; Wheels/Tires

FOCUS CHEMICAL CORPORATION
B9 Orchard Park
875 Greenland Rd.
Portsmouth, NH 03801 USA
Raw Material Supplier/Manufacturer: Polyol; Catalyst
Other: Distributor
Markets: Chemical; Footwear; Urethane; Adhesives;
   Sealants-Coatings; Reprographics

**FOGEL COMMERCIAL REFRIGERATOR
COMPANY**
5400 Eadom St.
Philadelphia, PA 19137-1399 USA
Other: Manufacturer of Commercial Refrigerators
Markets: Construction; Defense/Military; Industrial;
   Marine; Medical; Refrigeration

FOMO PRODUCTS
2775 Barber Rd.
P.O. Box 1078
Norton, OH 44203 USA
Systems Manufacturer: Flexible Foam Systems
Finished Goods Manufacturer: Sealant; Rigid Foam
Markets: Construction; Defense/Military; Industrial;
   Lumber; Marine; Mining/Drilling; Pipe/Plumbing;
   Recreational; Refrigeration; Transportation

FORD MOTOR, CO. UTICA TRIM OPERATIONS
50500 Mound Rd.
Utica, MI 48087 USA
Finished Goods Manufacturer: Elastomer; RIM;
   Automotive; Flexible Foam
Markets: Automotive

FOREST CITY TECHNOLOGIES
299 Clay St.
Wellington, OH 44090 USA
Finished Goods Manufacturer: Elastomer
Markets: Automotive

FOSS COMPANY
1751 Placentia Ave.
Costa Mesa, CA 92627 USA
Finished Goods Manufacturer: Elastomer; Cast; Rigid
   Foam
Markets: Marine

FOSS FOAM, INC.
4480 126th Ave. N.
Clearwater, FL 34622 USA

Finished Goods Manufacturer: Rigid Foam
Markets: Marine

FOULKE RUBBER PRODUCTS, INC.
600 Linnerud Dr.
Sun Prairie, WI 53590 USA
Finished Goods Manufacturer: Elastomer; Non-
    automotive; Millable Gums
Markets: Packaging; Paper; Pipe/Plumbing

FRANK LOWE RUBBER & GASKET CO, INC.
15149 7th Ave.
Whitestone, NY 11357-1236 USA
Finished Goods Manufacturer: Flexible Foam; Integral
    Skin
Other: Fabricator
Markets: Aerospace; Defense/Military;
    Electronic/Electrical; Furniture; Industrial; Packaging;
    Display; Glazing

FRANKLIN FIBRE-LAMITEX CORPORATION
901 E. 13th St.
Wilmington, DE 19899 USA
Equipment or Machinery Supplier/Manufacturer: Molds
Finished Goods Manufacturer: Elastomer; Cast; Rigid
    Foam
Markets: Electronic/Electrical

FREE ENERGY COMPANY
Horton International
544 Park Ave.
P.O. Box 828
Park City, UT 84060 USA
Finished Goods Manufacturer: Elastomer; RIM; TPU
Markets: Aerospace; Industrial; Recreational

FREMONT DIE CUT PRODUCTS
3177 E. U.S. Route 20
Fremont, OH 43420 USA
Systems Manufacturer: Coating Systems
Finished Goods Manufacturer: Coating
Markets: Automotive; Marine

FREUDENBERG-NOK
Plastic Products Division
Grenier Industrial Park
Manchester, NH 03103 USA
Systems Manufacturer: Flexible Foam Systems
Finished Goods Manufacturer: Elastomer; TPU; Cast;
    Integral Skin; Adhesive; Extruded
Markets: Aerospace; Automotive; Construction;
    Electronic/Electrical; Industrial; Material Handling;
    Medical; Mining/Drilling; Wheels/Tires; Food

FREUDENBERG-NOK
Headquarters
47690 E. Anchor Ct.
Plymouth, MI 48170 USA
Finished Goods Manufacturer: Elastomer; Automotive;
    Flexible Foam
Markets: Aerospace; Appliances; Automotive; Industrial

FRIGIDAIRE
701 33rd Ave. N

St. Cloud, MN 56301 USA
Finished Goods Manufacturer: Rigid Foam
Markets: Appliances; Refrigeration

FRONTIER RUBBER CO., INC.
2218 Rhode Island Ave.
Niagara Falls, NY 14305 USA
Finished Goods Manufacturer: Elastomer; Cast
Markets: Automotive

FROZEN NORTH CONSTRUCTION
HCO-1, Box 6825
Palmer, AK 99645 USA
Other: Contractor
Markets: Automotive; Construction; Refrigeration

FRUEHAUF TRAILER CORPORATION
26999 Central Park Blvd.
Southfield, MI 48076 USA
Equipment or Machinery Supplier/Manufacturer: Truck
    Trailers
Markets: Aerospace; Automotive; Chemical;
    Construction; Defense/Military; Electronic/Electrical;
    Furniture; Industrial; Lumber; Marine; Material
    Handling; Mining/Drilling; Paper; Pipe/Plumbing;
    Refrigeration; Transportation; Wheels/Tires

FSP RESEARCH, INC.
655 Plains Rd.
P.O. Box 28
Milford, CT 06460 USA
Other: Consultant; Testing; Research Organization
Markets: Aerospace; Automotive; Chemical;
    Construction; Defense/Military; Electronic/Electrical.
    Footwear; Industrial; Marine; Medical; Recreational;
    Refrigeration; Transportation

FUROM CO.
407 E. St.
P.O. Box 1911
New Haven, CT 06501 USA
Finished Goods Manufacturer: Flexible Foam

FURON CORP.
Dekoron Division
1199 S. Chillicothe Rd.
Aurora, OH 44202 USA
Systems Manufacturer: Elastomer Systems
Markets: Construction; Marine; Food

FUSA ENGINEERING
11301 S. Shore Rd.
Reston, VA 22090 USA
Equipment or Machinery Supplier/Manufacturer:
    Laminators; Dispensing Equipment; Multi-Opening
    Panel Presses
Other: Mfg. Rep.
Markets: Construction; Refrigeration

FUTURA COATINGS, INC.
9200 Latty Ave.
Hazelwood (St. Louis), MO 63042 USA
Systems Manufacturer: Prepolymer Manufacturer;
    Elastomer Systems; Flexible Foam Systems; Coatings

Equipment or Machinery Supplier/Manufacturer: Dispensing Equipment; Pumps

Finished Goods Manufacturer: Elastomer; RIM; Cast; Flexible Foam; Coating; Adhesive; Sealant

Markets: Appliances; Automotive; Chemical; Construction; Electronic/Electrical; Industrial; Marine; Mining/Drilling- Pipe/Plumbing; Recreational; Transportation

FUTURE FOAM, INC.
25 Main Place
Council Bluffs, IA 51503 USA
Finished Goods Manufacturer: Flexible Foam
Markets: Aerospace; Automotive; Carpet; Clothing; Furniture; Industrial; Material Handling; Medical; Packaging; Recreational; Toys; Ad Specialty

FYPON, INC.
22 W. PA Ave.
P.O. Box 365
Stewartstown, PA 17363 USA
Finished Goods Manufacturer: High Density Rigid Foam
Markets: Construction

G-FORCE AERODYNAMICS
4895 NW Salishan Dr.
Portland, OR 97229 USA
Finished Goods Manufacturer: Flexible Foam; Integral Skin
Markets: Automotive

G.S. MANUFACTURING
1760Monrovia C-1
CostaMesa, CA 92627 USA
Finished Goods Manufacturer: Elastomer; Automotive; Flexible Foam
Markets: Aerospace; Appliances; Automotive; Industrial

G.T. SALES & MANUFACTURING, INC.
Wichita KS-Kansas City MO
Omaha, NE
2202 S. West St.
Wichita, KS 67213 USA
Systems Manufacturer: Elastomer Systems; Rigid Foam Systems; Flexible Foam Systems
Equipment or Machinery Supplier/Manufacturer: Molds; Laminators
Finished Goods Manufacturer: Elastomer; Non-automotive, Flexible Foam; Integral Skin; Coating; Adhesive; Sealant; Rigid Foam
Other: Distributor; Mfg. Rep.; Fabricator
Markets: Aerospace; Automotive; Carpet; Chemical; Clothing; Construction; Defense/Military; Electronic/Electrical; Industrial; Lumber; Marine; Material Handling; Medical; Mining/Drilling; Packaging; Pipe/Plumbing; Recreational; Refrigeration; Transportation; Wheels/Tires

G.C.S. COATINGS, INC.
1999 Beaver Ave.
Monaca, PA 15061 USA
Systems Manufacturer: Elastomer Systems

Markets: Chemical; Construction; Defense/Military; Industrial; Lumber; Marine; Material Handling; Mining/Drilling; Pipe/Plumbing; Transportation

GACO WESTERN, INC.
P.O. Box 88698
Seattle, WA 98138 USA
Systems Manufacturer: Elastomer Systems; Rigid Foam Systems; Flexible Foam Systems

GAGNE, INC.
1080 Chenango St.
Binghamton, NY 13901 USA
Markets: Aerospace; Automotive; Construction; Defense/Military; Electronic/Electrical; Material Handling; Medical; Packaging; Paper; Recreational; Automation; Textile;

GALLAGHER CORPORATION
3966 Morrison Dr.
Gurnee, IL 60031-1284 USA
Finished Goods Manufacturer: Elastomer; TPU; Cast
Markets: Aerospace; Appliances; Automotive; Carpet; Construction; Defense/Military; Electronic/Electrical; Industrial; Lumber; Marine; Material Handling; Medical; Mining/Drilling; Paper; Printing; Recreational; Transportation; Food

GANT WESTERN, INC.
Injection Molding
P.O. Box 2796
Grass Valley, CA 95945 USA
Equipment or Machinery Supplier/Manufacturer: Molds
Markets: Electronic/Electrical; Medical; Recreational

GARDENA RUBBER CO., INC.
155 E. 157 St.
Gardena, CA 90248 USA
Finished Goods Manufacturer: Elastomer
Markets: Aerospace; Appliances; Automotive; Construction; Electronic/Electrical; Medical

GARLOCK MECHANICAL PACKING
1666 Division St.
Palmyra, NY 14522 USA
Finished Goods Manufacturer: Elastomer
Markets: Chemical; Construction; Industrial; Marine; Material Handling; Mining/Drilling

GASKA TAPE, INC.
181 0 W. Lusher Ave.
Elkhart, IN 46515 USA
Systems Manufacturer: Flexible Foam Systems
Finished Goods Manufacturer: Flexible Foam
Markets: Appliances; Automotive; Construction; Furniture; Industrial; Marine; Medical; Recreational; Refrigeration; Transportation

GASKET & MOLDED PRODUCTS
10302 S. Progress Way
Parker, CO 80134 USA
Finished Goods Manufacturer: Flexible Foam
Other: Distributor; Fabricator

Markets: Electronic/Electrical; Industrial; Medical;
  Packaging; Recreational

GATES MOLDED PRODUCTS CO.
Gates Rubber Co.
5229 Langfield
Houston, TX 77040 USA
Finished Goods Manufacturer: Elastomer; TPU; Cast
Markets: Appliances; Automotive; Construction; Mining/
  Drilling; Transportation; Food

GATES RUBBER COMPANY, THE
300 College St. Rd.
Elizabethtown, KY 42701 USA
Finished Goods Manufacturer: Elastomer; Cast
Markets: Aerospace; Appliances; Automotive; Mining/
  Drilling; Recreational; Business Machines

GEMINI LAMINATING CORP.
35 Balch St.
Beverly, MA 01915 USA
Markets: Footwear; Sports Equipment

GENCORP
5476 Port Chester Dr.
Hudson, OH 44236 USA

GENERAL ELECTRIC APPLIANCES
Appliance Park
AP-5 188-22
Louisville, KY 40225 USA
Finished Goods Manufacturer: Rigid Foam
Markets: Appliances

GENERAL ELECTRIC CORP.
1 Plastics Ave.
Pittsfield, MA 01201 USA
Finished Goods Manufacturer: Elastomer

GENERAL ELECTRIC SILICONES
260 Hudson River Rd.
Waterford, NY 12188 USA

GENERAL FIBERGLASS SUPPLY
Construction
23740 Cooper Dr.
Elkhart, IN 46514 USA
Equipment or Machinery Supplier/Manufacturer:
  Dispensing Equipment; Pumps;
Other: Distributor
Markets: Aerospace; Appliances; Automotive;
  Construction; Defense/Military; Electronic/Electrical;
  Furniture; Industrial; Marine; Medical; Packaging;
  Recreational; Refrigeration; Transportation

GENERAL FOAM
PMC Division
W. 100 Century Rd.
Paramus, NJ 07652 USA
Finished Goods Manufacturer: Flexible Foam; Vinyl
  Foam; PVC Expanded
Markets: Aerospace; Appliances; Automotive; Carpet;
  Clothing; Defense/Military; Electronic/Electrical;

Footwear; Furniture; Industrial; Marine; Medical;
  Packaging; Recreational

GENERAL FOAM
25 Jaycee Dr.
W. Hazelton, PA 18201-1193 USA
Finished Goods Manufacturer: Flexible Foam
Markets: Carpet; Furniture

GENERAL FOAM OF MINNESOTA
1800 Como Ave.
St. Paul, MN 55108 USA
Finished Goods Manufacturer: Flexible Foam
Markets: Aerospace; Automotive; Construction; Defense/
  Military; Furniture; Industrial; Marine; Packaging;
  Recreational

GENERAL GASKET CORP.
2322 S. Seventh St.
P.O. Box 12240
St. Louis, MO 63157-0240 USA
Finished Goods Manufacturer: Elastomer; Cast
Markets: Appliances; Automotive; Construction;
  Defense/Military; Electronic/Electrical; Industrial;
  Mining/Drilling; Food

GENERAL LATEX & CHEMICAL CORP.
2321 N. Davidson St.
Charlotte, NC 28205 USA
Systems Manufacturer: Rigid Foam Systems
Finished Goods Manufacturer: Adhesive
Markets: Appliances; Automotive; Carpet; Construction;
  Electronic/Electrical; Industrial; Marine; Packaging;
  Pipe/Plumbing; Recreational; Refrigeration;
  Transportation

GENERAL MOTORS CORP.
Rochester Production
500 Sibley Tower
25 North St.
Rochester, NY 14692 USA
Finished Goods Manufacturer: Elastomer; Cast
Markets: Automotive

GENERAL PLASTICS MFG. CO.
4910 Burlington Way
P.O. Box 9097
Tacoma, WA 98409 USA
Finished Goods Manufacturer: Flexible Foam; Rigid Foam
Markets: Construction; Defense/Military

GENERAL RUBBER & PLASTICS CO., INC.
1016 Majaun Rd.
Lexington, KY 40511 USA
Finished Goods Manufacturer: Elastomer; Cast; Flexible
  Foam
Other: Distributor; Fabricator
Markets: Appliances; Automotive; Construction;
  Electronic/Electrical; Industrial; Mining/Drilling;
  Packaging

GENERAL RUBBER CORP.
11 A Empire Blvd.
South Hackensack, NJ 07656 USA

Finished Goods Manufacturer: Elastomer; Cast;
Other: Fabricator
Markets: Defense/Military; Industrial; Marine;
Mining/Drilling; Paper; Pipe/Plumbing

GENEVA RUBBER CO.
5449 Bishop Rd.
P.O. Box 270
Geneva, OH 44041 USA
Finished Goods Manufacturer: Elastomer; Cast
Markets: Automotive

GEORGE MANN & CO., INC.
Formulated Products Division
P.O. Box 9066
Providence, RI 02940 USA
Raw Material Supplier/Manufacturer: Release Agents;
Coatings
Markets: Aerospace; Appliances; Automotive; Defense/
Military; Footwear; Furniture; Industrial; Marine;
Material Handling; Medical; Mining/Drilling;
Printing; Refrigeration; Transportation; Wheels/Tires

GA BONDED FIBERS, INC.
15 Nuttman St.
Newark, NJ 07103 USA
Finished Goods Manufacturer: Elastomer; TPU
Markets: Aeros

GIL-BAR RUBBER PRODUCTS CO., INC.
2525 S. 50 Ave.
Cicero, IL 60650 USA
Finished Goods Manufacturer: Elastomer; TPU
Markets: Aeorspace; Appliances; Automotive; Defense/
Military; Electronic/Electrical; Industrial; Medical;
Printing

GLACIER MACHINERY SALES CORP.
5211 N. 130th St.
St. Paul, MN 55110 USA
Equipment or Machinery Supplier/Manufacturer:
Laminators; Dispensing Equipment; Tanks; Pumps
Markets: Aerospace; Appliances; Automotive; Carpet;
Chemical; Construction; Defense/Military; Electronic/
Electrical; Footwear; Furniture; Industrial; Lumber;
Marine; Material Handling; Medical; Mining/Drilling;
Packaging; Pipe/ Plumbing; Printing; Recreational;
Refrigeration; Transportation

GLAS-CRAFT, INC.
5845 W. 82nd St.
Suite 102
Indianapolis, IN 46278 USA
Systems Manufacturer: Elastomer Systems

GLASCOAT, INC.
720 Collins Rd.
Elkhart, IN 46516 USA
Finished Goods Manufacturer: Rigid Foam

GLOBE MANUFACTURING CO.
456 Bedford St.
Fall River, MA 02720 USA
Finished Goods Manufacturer: Elastomer; Spandex Fiber

Markets: Clothing

GLOBE RUBBER WORKS, INC.
254 Beech St.
Rockland, MA 02370 USA
Finished Goods Manufacturer: Elastomer; TPU; Cast
Other: Fabricator
Markets: Aerospace; Appliances; Defense/Military;
Industrial; Marine; Medical; Mining/Drilling

GLOVERLATEX, INC.
118 W. Elm
Anaheim, CA 92805 USA
Finished Goods Manufacturer: Elastomer
Other: Manufacturer
Markets: Aerospace; Chemical; Electronic/Electrical;
Industrial; Lumber; Marine; Printing; Food

GOLDEN GRIP PRODUCTS, INC.
13800 Nelson Rd.
Detriot, MI 48227 USA
Finished Goods Manufacturer: Elastomer; Non-
automotive; Adhesive; Sealant
Markets: Automotive; Carpet; Construction

GOLDEMWEST MFG., INC.
P.O. Box 1148
Cedar Ridge, CA 95924 USA
Raw Material Supplier/Manufacturer: Resins; Fillers
Finished Goods Manufacturer: Rigid Foam
Other: Fabricator
Markets: Aerospace; Appliances; Automotive;
Construction; Defense/Military; Electronic/Electrical;
Furniture; Industrial; Marine; Medical;
Pipe/Plumbing; Recreational; Transportation; Pattern
Shops; Model Makers; Machine Shops

**GOLDSCHMIDT CHEMICAL
CORPORATION**
Polyurethane Additives
P.O. Box 1299
914 E. Randolph Rd.
Hopewell, VA 23860 USA
Raw Material Supplier/Manufacturer: Catalyst;
Surfactant; Specialty Additives
Markets: Aerospace; Appliances; Automotive; Carpet;
Chemical; Clothing; Construction; Defense/Military;
Footwear; Furniture; Industrial; Marine; Packaging;
Recreational; Refrigeration; Transportation

**GOODYEAR RUBBER CO. OF SOUTHERN
CALIFORNIA**
8833 Industrial Lane
Rancho Cucamonga, CA 91730 USA
Finished Goods Manufacturer: Elastomer; Non-
automotive
Other: Testing
Markets: Aerospace; Automotive; Chemical; Industrial;
Paper; Wheels/Tires

GOODYEAR TIRE & RUBBER CO.
Engineered Products
1689 E. Front St.

Logan, OH 43130 USA
Finished Goods Manufacturer: Elastomer; RIM;
  Automotive; TPU; Flexible Foam; Composites
Markets: Printing; Wheels/Tires; Waterproofing
  Membranes

GORDON RUBBER & PACKAGING CO., INC.
Cemetery Ave.
Derby, CT 06418 USA
Finished Goods Manufacturer: Elastomer; Cast
Markets: Aerospace; Construction

GOSHEN RUBBER CO., INC.
1525 South 10 St.
Goshen, IN 46526 USA
Finished Goods Manufacturer: Elastomer; TPU
Markets: Appliances; Automotive; Construction;
  Electronic/Electrical

GOULD MID-WEST CORP.
1302 Clarkson/Clayton Center
Suite 101
Ellisville, MO 63011 USA
Finished Goods Manufacturer: Flexible Foam
Markets: Packaging

GOULDSOUTHERN, INC.
3088 Mercer University Dr.
Suite 100
Atlanta, GA 30341 USA
Finished Goods Manufacturer: Flexible Foam
Markets: Packaging

GPI CORP.
101 Northern Rd.
Scofield, WI 54476 USA
Finished Goods Manufacturer: Elastomer; Cast; Integral
  Skin; Adhesive; Rigid Foam
Markets: Automotive; Defense/Military; Industrial;
  Marine; Paper; Recreational; Transportation

GRACO, INC.
Corporate Headquarters
P.O. Box 1441
Minneapolis, MN 55440 USA
Systems Manufacturer: Rigid Foam Systems
Equipment or Machinery Supplier/Manufacturer:
  Dispenisng Equipment; Pumps

GRACO, INC.
Industrial and Automotive Equipment
9451 W. Belmont Ave.
Franklin Park, IL 60131-2891 USA
Systems Manufacturer: Rigid Foam Systems
Equipment or Machinery Supplier/Manufacturer:
  Dispensing Equipment; Pumps; Automated Controls
Markets: Aerospace; Appliances; Automotive; Chemical;
  Construction; Defense/Military; Electronic/Electrical;
  Furniture; Industrial; Marine; Material Handling;
  Packaging; Printing; Recreational; Refrigeration;
  Transportation; Wheels/ Tires

GRAFT0N MACHINE & RUBBER CO., INC.
640 Cleveland Rd.

Ravenna, OH 44266 USA
Finished Goods Manufacturer: Elastomer; Cast
Markets: Aerospace; Construction

GRAIN PROCESSING CORPORATION
1600 Oregon St.
Muscatine, IA 52761 USA

GREAT LAKES CHEMICAL CORPORATION
P.O. Box 2200
West Lafayette, IN 47906 USA
Raw Material Supplier/Manufacturer: Polyol; Flame Retar
  dants; Anti-Oxidants; UV Absorbers
Markets: Aerospace; Automotive; Carpet; Chemical;
  Clothing; Construction; Defense/Military;
  Electronic/Electrical; Footwear; Furniture; Industrial;
  Material Handling; Medical; Mining/Drilling;
  Recreational; Refrigeration; Transportation;
  Wheels/Tires

GREAT LAKES TERMINAL & TRANSPORT CORP
1750 N. Kingsbury St.
Chicago, IL 60614 USA
Systems Manufacturer: Rigid Foam Systems
Other: Distributor

GREAT WESTERN FOAM PRODUCTS CO.
2060 N. Batavia St.
Orange, CA 92665 USA
Finished Goods Manufacturer: Flexible Foam
Markets: Furniture

GREEN MOUNTAIN FOAM PRODUCTS
P.O. Box 8000
Route 15
Underhill, VT 05489 USA
Finished Goods Manufacturer: Flexible Foam; Rigid Foam
Other: Fabricator
Markets: Furniture

GREENE, TWEED & CO.
Kulpsville, PA 19443 USA
Finished Goods Manufacturer: Elastomer; Cast
Markets: Aerospace; Automotive; Construction; Defense/
  Military; Electronic/Electrical; Paper; Food

GRIFFITH POLYMERS, INC.
1930 SE. Minter Bridge Rd.
Hillsboro, OR 97123 USA
Finished Goods Manufacturer: Elastomer; Cast
Markets: Aerospace; Automotive; Defense/Military; Elec-
  tronic/Electrical; Lumber; Marine; Material Handling;
  Medical; Mining/Drilling; Paper; Printing;
  Recreational; Wheels/Tires

GROENDYK MTG. CO., INC.
P.O. Box 278
Buchanan, VA 24066 USA
Markets: Aerospace; Appliances; Automotive; Defense/
  Military; Marine; Recreational; General Industry

GROTENRATH RUBBER PRODUCTS CO., INC.
9591 York Alpha Dr., Unit No.6
N. Royalton, OH 44133 USA

Other: Distributor
Markets: Aerospace; Appliances; Automotive; Defense/
Military; Electronic/Electrical; Industrial; Marine;
Medical; Pipe/Plumbing; Printing; Refrigeration;
Transportation

GUARDIAN PACKAGING, INC.
3615 Security St.
Garland, TX 75042 USA
Finished Goods Manufacturer: Flexible Foam
Markets: Aerospace; Automotive; Electronic/Electrical;
Marine; Packaging

GUF INDUSTRIES, INC.
347 National, NW.
Grand Rapids, MI 49504 USA
Equipment or Machinery Supplier/Manufacturer:
Reconditioned and Used Foam and Coating
Equipment; Foam Shaving Equipment

GUSMER, INC.
PMC Division
One Gusmer Dr.
Lakewood, NJ 08701 USA
Equipment or Machinery Supplier/Manufacturer:
Dispensing Equipment; Tanks; Pumps; Automated
Controls
Markets: Aerospace; Appliances; Automotive; Carpet;
Construction; Defense/Military; Furniture; Industrial;
Marine; Material Handling; Recreational;
Refrigeration; Transportation

H&W SYSTEMS, INC.
Main Office
419 Kimber Dr.
Holland, MI 49424 USA
Equipment or Machinery Supplier/Manufacturer:
Molding Machines and Equipment
Other: Mfg Rep
Markets: Automotive; Material Handling; Medical

H.B. FULLER COMPANY
ASC Division
3530 Lexington Ave., N
St. Paul, MN 55126-8076 USA
Finished Goods Manufacturer: Adhesive; Sealant; Hot
Melts; Epoxies; Water Based; Solvent Cements; Glue
Stick and Guns
Markets: Aerospace; Appliances; Automotive; Carpet;
Chemical; Clothing; Construction; Defense/Military;
Electronic/Electrical; Footwear; Furniture; Industrial;
Lumber; Marine; Medical; Packaging; Paper;
Pipe/Plumbing; Printing; Recreational; Refrigeration;
Transportation

H.H. ROBERTSON/CECO
Merchant St.
Ambridge, PA 15003 USA
Finished Goods Manufacturer: Rigid Foam
Markets: Construction

HACKNEY BROTHERS, INC.
P.O. Box 2728

Wilson, NC 27894 USA
Other: Fabricator
Markets: Refrigeration

HAMLIN INDUSTRIES
P.O. Box 1215
Fox Farm Rd.
Warsaw, IN 46580 USA

HAMPSHIRE CHEMICAL CORPORATION
Polymers Division
55 Hayden Ave.
Lexington, MA 02173 USA
Raw Material Supplier/Manufacturer: Prepolymer
Systems Manufacturer: Prepolymer Manufacturer;
Elastomer Systems; Flexible Foam Systems
Markets: Medical; Personal Care

HAMPTON INDUSTRIES
532 E. McNeil St.
Maguolia, AR 71753 USA
Systems Manufacturer: Rigid Foam Systems; Flexible
Foam Systems

HANSEN RUBBER CO., INC.
2215 E. Tioga St.
Philadelphia, PA 19134 USA
Finished Goods Manufacturer: Elastomer; Cast
Markets: Industrial; Waterproofing

HAPCO, INC.
353 Circuit St.
Hanover, MA 02339 USA
Raw Material Supplier/Manufacturer: Polymer;
Prepolymer; Resins
Systems Manufacturer: Prepolymer Manufacturer;
Elastomer Systems
Markets: Aerospace; Appliances; Automotive;
Construction; Electronic/Electrical; Industrial;
Marine; Medical; Mining/Drilling; Transportation;
Wheels/Tires

HARDMAN, INC.
600 Cortland St.
Belleville, NJ 07109 USA
Finished Goods Manufacturer: Elastomer

HARPER RUBBER PRODUCTS, CO., INC.
349 New Britain Ave.
Plainville, CT, 06062 USA
Finished Goods Manufacturer: Elastomer; Cast
Markets: Aerospace; Automotive; Electronic/Electrical;
Industrial; Medical; Paper; Printing; Wheels/Tires

HARWOOD RUBBER PRODUCTS, INC.
1365 Orien Ave.
Cuyahoga Falls, OH 44221 USA
Finished Goods Manufacturer: Elastomer; Cast

HASTINGS PLASTICS CO.
1704 Colorado Ave.
Santa Monica, CA 90404 USA
Raw Material Supplier/Manufacturer: Catalyst; Resins

Systems Manufacturer: Elastomer Systems; Rigid Foam
  Systems; Flexible Foam Systems
Markets: Aerospace; Chemical; Construction; Defense/
  Military; Electronic/Electrical; Furniture; Industrial;
  Marine; Recreational

HAWTHORNE RUBBER MFG CORP.
35 Fourth Ave.
P.O. Box 171
Hawthorne, NJ 07507 USA
Finished Goods Manufacturer: Elastomer; Cast
Markets: Wheels/Tires

HEDSTROM CO.
Plastics Division
710 Orange St.
Ashland, OH 44805 USA
Finished Goods Manufacturer: Flexible Foam; Integral
  Skin; Rigid Foam
Markets: Industrial; Marine; Material Handling; Medical;
  Recreational; Wheels/Tires

HELLER CO.
1001 E. Second St.
Dayton, OH 45402 USA
Finished Goods Manufacturer: Flexible Foam
Other: Fabricator
Markets: Furniture; Packaging

HENDERSON-JOHNSON, INC.
P.O. Box 6964
Syracuse, NY 13217 USA
Finished Goods Manufacturer: Rigid Foam
Other: Contractor
Markets: Construction

HENNECKE MACHINERY
Miles, Inc.
Mayview Rd. and Park Dr.
Lawrence, PA 15055 USA
Equipment or Machinery Supplier/Manufacturer: Molds;
  Laminators; Dispensing Equipment; Pumps;
  Automated Controls; Process Equipment

**HENRY COMPANY (DBA) RESIN
TECHNOLOGY CO.**
2270 Castle Harbor
Ontario, CA 91711 USA
Systems Manufacturer: Rigid Foam Systems
Finished Goods Manufacturer: Coating
Markets: Aerospace; Appliances; Construction; Defense/
  Military; Furniture; Marine; Packaging; Refrigeration;
  Transportation

HENTZEN COATINGS, INC.
6937 W. Mill Rd.
Milwaukee, WI 53218 USA
Finished Goods Manufacturer: Coating
Markets: Aerospace; Appliances; Construction; Defense/
  Military; Electronic/Electrical; Material Handling

HERCULES CHEMICAL CO., INC.
29 W. 38th St.
New York, NY 10018 USA

Finished Goods Manufacturer: Adhesive; Sealant
Markets: Industrial; Pipe/Plumbing

HERMAN A. GELMAN CO.
1251 DeKalb Ave.
Brooklyn, NY 11221 USA
Finished Goods Manufacturer: Flexible Foam; Rigid Foam

HERMAN MILLER
8500 Byron Rd.
Zeeland, MI 49464 USA
Finished Goods Manufacturer: Non-automotive; Flexible
  Foam; Integral Skin
Markets: Furniture

HEWITT-ROBINS CORP.
Conveyor Equipment Division
40 Fairfield Place
W. Caldwell, NJ 07006 USA
Finished Goods Manufacturer: Elastomer; Cast
Markets: Construction; Industrial; Material Handling;
  Mining/Drilling

HEXCEL CORPORATION
Resins Group
20701 Nordhoff St.
Chatsworth, CA 91311 USA
Raw Material Supplier/Manufacturer: Resins
Systems Manufacturer: Prepolymer Manufacturer;
  Elastomer Systems
Finished Goods Manufacturer: Coating; Adhesive-Sealant
Markets: Aerospace; Automotive; Chemical; Defense/
  Military; Electronic/Electrical; Footwear; Medical;
  Industrial; Marine; Telecommunications

HEXGON ENTERPRISES, INC.
Chemical Components Division
20 DeForest Ave.
E. Hanover, NJ 07936 USA
Systems Manufacturer: Prepolymer Manufacturer; Rigid
  Foam Systems
Finished Goods Manufacturer: Coating; Adhesive; Sealant.

HI LIFE PRODUCTS, INC.
13940 Magnolia Ave.
Chino, CA 91710 USA

HI-TECH ENGINEERING, INC.
4585 40th St. SE
Grand Rapids, MI 49512 USA
Equipment or Machinery Supplier/Manufacturer:
  Dispensing Equipment; Tanks; Pumps; Automated
  Controls; Fluidlear; Mixing Heads; Bulk Storage
  Systems; Mold Clamps/Carriers; Panel Lines
Markets: Aerospace; Appliances; Automotive;
  Construction; Defense/Military; Electronic/Electrical;
  Furniture; Industrial; Marine; Material Handling;
  Packaging; Pipe/Plumbing; Recreational;
  Refrigeration; Transportation; Research Laboratories;
  Specialty Molders

HI-TECH INC.
443 Shewville Rd.
Ledyard, CT 06339 USA

Finished Goods Manufacturer: Cast; Coating; Rigid Foam
Markets: Construction; Industrial; Marine; Refrigeration

HIWATHA RUBBER CO.
1700 67th Ave. N
Minneapolis, MN 55430 USA
Finished Goods Manufacturer: Elastomer; Cast
Markets: Aerospace; Defense/Military-
Electronic/Electrical; Industrial

HIBCO PLASTTCS, INC.
P.O. Box 157
Yadkinville, NC 27055 USA
Finished Goods Manufacturer: Flexible Foam; Rigid Foam

HICKORY SPRINGS MANUFACTURING CO.
P.O. Box 128
Hickory, NC 28603 USA
Finished Goods Manufacturer: Flexible Foam
Other: Fabricator
Markets: Carpet; Furniture; Packaging; Bedding

HIGH-TECH/SEA-LONG MEDICAL
1987 S. Park Rd.
Louisville, KY 40219 USA
Finished Goods Manufacturer: Elastomer; TPU; Injection/
Molding
Markets: Appliances; Automotive; Construction;
Electronic/Electrical; Furniture; Industrial; Material
Handling; Medical; Printing; Recreational

HK RESEARCH CORPORATION
P.O. Box 1809
Hickory, NC 28603 USA
Raw Material Supplier/Manufacturer: Polyol
Systems Manufacturer: Elastomer Systems
Other: Research Organization
Markets: Electronic/Electrical; Marine; Recreational

HOBART TAFA TECHNOLOGIES
146 Pembroke Rd.
Concord, NH 03301 USA
Equipment or Machinery Supplier/Manufacturer: Molds;
Metal Spray Equipment
Markets: Aerospace; Appliances; Automotive; Defense/
Military; Electronic/Electrical; Footwear; Furniture;
Industrial; Packaging; Pipe/Plumbing; Recreational;
Refrigeration; Wheels/Tires

HOBSON BROS. ALUMINIUM FOUNDARY & MOLD WORKS
Highway 3W
Shell Rock, IA 50670 USA
Equipment or Machinery Supplier/Manufacturer: Molds
Markets: Aerospace; Appliances; Automotive;
Construction; Defense/Military; Electronic/Electrical;
Furniture; Industrial; Lumber; Marine; Medical;
Packaging; Pipe/Plumbing; Recreational;
Refrigeration; Transportation; Wheels/Tires

HOECHST CELANESE CORPORATION
Engineering Plastics
26 Main St.

Chatham, NJ 07928 USA
Finished Goods Manufacturer: Elastomer

HOLMCO INDUSTRIES ROBIN INDUSTRIES, INC.
P.O. Box 188
Winesburg, OH 44690 USA
Finished Goods Manufacturer: Elastomer; TPU;
Composites
Markets: Aerospace; Appliances; Automotive; Defense/
Military; Electronic/Electrical; Medical

HOMASOTE CO.
P.O. Box 7240
West Trenton, NJ 08628 USA
Finished Goods Manufacturer: Rigid Foam

HORIZON PRODUCTS
Grain Processing Corporation
1600 Oregon St.
P.O. Box 349
Muscatine, IA 52761-0349 USA
Raw Material Supplier/Manufacturer: Polyol; Resins;
Adhesives
Markets: Appliances; Clothing; Lumber; Mining/Drilling;
Paper; Refrigeration

HOULDEN CONTRACTING, INC.
P.O. Box 306
Cambridge, NV 69022 USA
Equipment or Machinery Supplier/Manufacturer: Pumps
Finished Goods Manufacturer: Coating; Rigid Foam;
Other: Fabricator; Spray Foam Contractor Roofing and
Insulation
Markets: Construction; Industrial

HOUSE OF KOLOR, INC.
Kosmoski Specialty Co.
2521 27th Ave. S.
Minneapolis, MN 55406 USA
Finished Goods Manufacturer: Coating
Markets: Automotive

HOUSTON FOAM PLASTICS, INC.
2019 Brooks St.
Houston, TX 77026 USA
Finished Goods Manufacturer: Flexible Foam; Rigid Foam
Markets: Construction; Packaging

HOY SHOE CO.
4970 Kemper Ave.
St. Louis, MO 63139 USA
Finished Goods Manufacturer: Flexible Foam
Markets: Footwear

HUB FABRIC LEATHER CO., INC.
7 Chariton St.
Everett, MA 02149 USA
Finished Goods Manufacturer: Coating
Markets: Automotive; Clothing; Footwear; Marine;
Packaging

HUDSON INDUSTRIES, INC.
5250 Klockner Dr.
Richmond, VA 22231 USA

Raw Material Supplier/Manufacturer: Polymer
Systems Manufacturer: Flexible Foam Systems
Equipment or Machinery Supplier/Manufacturer: Molds
Finished Goods Manufacturer: Flexible Foam; Coating;
  Adhesive
Other: Fabricator
Markets: Furniture; Industrial; Marine; Medical;
  Packaging; Recreational; Sporting Goods; Pet Supplies

HUGH G. BROOKS IND.
Circle Dr.
P.O. Box 305
Sneedville, TN 37869 USA
Finished Goods Manufacturer: Spray Foam
Other: Contractor
Markets: Construction

HULL & CO.
P.O. Box 4250
Greenwich, CT 06830 USA
Other: Consultant; Research Organization
Markets: Aerospace; Appliances; Automotive; Carpet;
  Chemical; Clothing; Construction; Defense/Military;
  Electronic/Electrical; Footwear; Furniture; Industrial;
  Lumber; Marine; Material Handling; Medical;
  Mining/Drilling; Packaging; Paper; Pipe/Plumbing;
  Printing; Recreational; Refrigeration; Transportation;
  Wheels/Tires

HULS AMERICA, INC.
Turner Place
Piscataway, NJ 08855-0365 USA
Finished Goods Manufacturer: Coating
Markets: Aerospace; Appliances; Automotive

HUMISEAL
Chase Corp.
26-60 Brooklyn Queens Expressway
Woodside, NY 11377 USA
Finished Goods Manufacturer: Coating

HUNTER HYDRAULICS, INC.
P.O. Box 7117
Canton, OH 44705 USA
Finished Goods Manufacturer: Hydraulic Presses

HURON TECHNOLOGIES, INC.
3729 Trade Center Dr.
Ann Arbor, MI 48108 USA
Raw Material Supplier/Manufacturer: Mold Release
  Agents
Systems Manufacturer: Release Coating Systems
Markets: Automotive; Electronic/Electrical; Footwear;
  Industrial; Recreational; Wheels/Tires

HY-TECH INSULATORS, INC.
P.O. Box 232
Clarksburg, NJ 0851 0 USA
Other: Roof/Insulation; Contractor
Markets: Construction

HYDRA-MATIC PACKING CO.
2992 Frank Rd.

Bethayres, PA 19006 USA
Finished Goods Manufacturer: Flexible Foam; Rigid Foam

HYDRAJET SERVICES, INC.
141 S. 7th St.
Reading, PA 19602 USA
Other: Consultant; Fabricator
Markets: Aerospace; Automotive; Carpet; Clothing;
  Defense/Military; Electronic/Electrical; Furniture;
  Industrial; Medical

HYDRO SEAL COATINGS
1119 E. Mission Rd.
Fallbrook, CA 92028 USA
Systems Manufacturer: Elastomer Systems; Flexible Foam
  Systems;
Finished Goods Manufacturer: Coating
Markets: Aerospace; Construction

HYSOL COMMUNICATION PRODUCTS
P.O. Box 158
Exeter, NH 03833 USA
Systems Manufacturer: Elastomer Systems
Finished Goods Manufacturer: Elastomer; Cast
Markets: Defense/Military; Electronic/Electrical;
  Industrial; Material Handling; Communications

ICI KLEA
Tatnall 1
3411 Silverside Rd.
Wilmington, DE 19850 USA
Raw Material Supplier/Manufacturer: Blowing Agent
Markets: Appliances; Construction; Industrial;
  Packaging; Refrigeration

ICI POLYURETHANES
286 Mantua Grove Rd.
West Deptford, NJ 08066-1732 USA
Raw Material Supplier/Manufacturer: isocyanate; Polyol
Systems Manufacturer: Prepolymer Manufacturer;
  Elastomer Systems; Rigid Foam Systems; Flexible
  Foam Systems
Markets: Appliances; Automotive; Carpet; Construction;
  Footwear; Furniture; Industrial; Lumber; Material
  Handling; Packaging; Refrigeration; Transportation

ICIS-LOR
3730 Kirby Dr.
Suite 850
Houston, TX 77098 USA
Other: Reporting Agent
Markets: Aerospace; Appliances; Automotive; Carpet;
  Chemical; Clothing; Construction; Defense/Military;
  Electronic/Electrical; Footwear; Furniture; Industrial;
  Lumber; Marine; Material Handling; Medical;
  Mining/Drilling; Packaging; Paper; Pipe/Plumbing;
  Printing; Recreational; Refrigeration; Transportation;
  Wheels/Tires

ILLBRUCK INC.
3800 Washington Ave., N
Minneapolis, MN 55412 USA

Finished Goods Manufacturer: Elastomer; Flexible Foam;
Rigid Foam
Other: Fabricator
Markets: Automotive; Construction; Industrial

ILLIG INDUSTRIES, INC.
9750 Alden
Lenaka, KS 66215 USA
Finished Goods Manufacturer: Flexible Foam; Rigid Foam
Other: Fabricator

ILLINOIS FIBRE SPECIALITY CO., INC.
Foam Division
4301 S. Western Blvd.
Chicago, IL 60609 USA
Equipment or Machinery Supplier/Manufacturer:
Laminators
Other: Distributor; Fabricator
Markets: Appliances; Construction; Furniture; Industrial;
Marine; Packaging; Pipe/Plumbing; Recreational;
Transportation

ILLINOIS INSTITUTE OF TECHNOLOGY
Center of Excellence in Polymer Science and Engineering
10 W. 33rd St., I IT-Center
Chicago, IL 60616 USA
Systems Manufacturer: Rigid Foam Systems; Flexible
Foam Systems; Recycling of Foamed and Filled
Polymers
Finished Goods Manufacturer: Flexible Foam; Integral
Skin; Coating; Rigid Foam
Other: Consultant; Testing; Research Organization;
Training Courses; Workshops; Seminars on
Production and Recycling of PU Foams.
Markets: Aerospace; Automotive; Carpet; Construction;
Defense/Military; Electronic/Electrical; Furniture;
Lumber; Medical; Mining/Drilling; Packaging;
Pipe/Plumbing; Refrigeration; Transportation;
Wheels/Tires

IMCO, INC.
1819 W. Park Dr.
Huntington, IN 46750 USA
Finished Goods Manufacturer: Elastomer; Cast
Markets: Appliances; Automotive; Construction;
Electronic/Electrical; Industrial; Mining/Drilling

IMPERIAL ADHESIVES, INC.
6315 Wiehe Rd.
Cincinnati, OH 452374277 USA
FinishedGoods Manufacturer: Coating; Adhesive; Sealant
Markets: Appliances; Automotive; Carpet; Construction;
Defense/Military; Footwear; Furniture; Industrial;
Marine; Medical; Packaging; Paper; Pipe/Plumbing;
Recreational; Transportation

IMPERIAL RUBBER ENTERPRISES
15130 Illinois Ave.
Paramont, CA 90723 USA
Finished Goods Manufacturer: Elastomer; Millable
Urethane
Markets: Industrial

IMT CORP.
330 Greco Ave.
Coral Gables, FL 33146 USA
Equipment or Machinery Supplier/Manufacturer:
Dispensing Equipment; Tanks; RTM Manufacturing
Equipment
Markets: Aerospace; Appliances; Automotive; Carpet;
Chemical; Construction; Defense/Military; Footwear;
Furniture; Industrial; Material Handling; Packaging;
Recreational; Refrigeration; Transportation

INDIAN RUBBER CO., INC.
419 Dodson Lake Dr.
Arlington, TX 76012 USA
Finished Goods Manufacturer: Elastomer; Cast
Markets: Automotive; Defense/Military;
Electronic/Electrical; Food

INDUSOL, INC.
Depot St.
Sutton, MA 01590 USA
Systems Manufacturer: Elastomer Systems; Rigid Foam
Systems; Flexible Foam Systems
Markets: Automotive; Defense/Military;
Electronic/Electrical; Footwear; Furniture; Industrial;
Refrigeration; Wheels/Tires

INDUSTRIAL SCREEN AND MAINTENANCE, INC.
P.O. Box 373
750 E. F St.
Casper, WYOMING 82601 USA
Finished Goods Manufacturer: Elastomer; Cast
Markets: Chemical; Construction; Mining/Drilling

INDUSTRIAL & MILITARY TECHNOLOGY CORP.
P.O. Box 0346
Flourtown, PA 19031-0346 USA
Finished Goods Manufacturer: Elastomer; Cast; Flexible
Foam; Coating
Markets: Construction; Industrial; Mining/Drilling;
Wheels/Tires; Waterproofing

INDUSTRIAL ADHESIVES CO.
Bond-Plus Adhesives and Coatings
130 N. Campbell Ave.
Chicago, IL 60612 USA
Finished Goods Manufacturer: Coating; Adhesive

INDUSTRIAL COATINGS CO, INC.
P.O. Box 190
Rogers, MN 55374 USA
Finished Goods Manufacturer: Sprayed Foam for
Insulation and Roofing
Markets: Construction

INDUSTRIAL CUSTOM PRODUCTS
5200 Quincy St.
Minneapolis, MN 55112 USA
Finished Goods Manufacturer: Flexible Foam; Adhesive;
Sealant
Other: Fabricator
Markets: Aerospace; Appliances; Automotive; Defense/
Military; Electronic/Electrical; Industrial; Lumber;

Marine; Material Handling; Medical; Packaging;
Refrigeration

INDUSTRIAL DISTRIBUTORS

2 Townsend West
Nashua, NH 03063 USA
Finished Goods Manufacturer: Elastomer; Cast
 Markets: Aerospace; Automotive; Defense/Military;
 Industrial; Marine; Packaging; Paper; Printing;
 Wheels/Tires

INDUSTRIAL FILTER AND PUMP MFG.CO.

5900 W. Ogden Ave.
Cicero, IL 60650 USA
Equipment or Machinery Supplier/Manufacturer: Pressure
 Filters
Markets: Chemical

INDUSTRIAL POLYMERS, INC.

8980 Scranton St.
Suite B
Houston, TX 77075 USA
Raw Material Supplier/Manufacturer: Prepolymer
Systems Manufacturer: Prepolymer Manufacturer;
 Elastomer Systems
Finished Goods Manufacturer: Elastomer; Flexible Foam;
 Coating; Adhesive; Sealant
Markets: Aerospace; Construction; Defense/ Military;
 Electronic/Electrical; Industrial; Marine; Medical;
 Mining/Drilling; Packaging; Recreational

INDUSTRIAL ROLLER CO.

218 N. Main St.
Smithton, IL 62285 USA
Finished Goods Manufacturer: Elastomer; Cast
Markets: Packaging; Paper; Printing; Plastic

INDUSTRIAL ROOFING & INSULATION, INC.

8980 Scranton St.
Houston, TX 77075 USA
Finished Goods Manufacturer: Polyurethane Roofing
 Application and Franchisor of Roofing Application
 Companies
Markets: Aerospace; Chemical; Construction;
 Refrigeration

INDUSTRIAL RUBBER & PLASTICS CO., INC.

105 S. Prospect St.
P.O. Box 1478
Haverhill, MA 01831 USA
Finished Goods Manufacturer: Flexible Foam; Rigid Faom
Other: Fabricator

INDUSTRIAL RUBBER & SUPPLY, INC.

2307 E. D St.
Tacoma, WASHINGTON 98401 USA
Finished Goods Manufacturer: Flexible Foam
Other: Fabricator
Markets: Furniture; Packaging

INDUSTRIAL RUBBER WHEELS, INC.

29090 Anderson Rd.
Wickliffe, OH 44092 USA
Finished Goods Manufacturer: Elastomer; Cast

Markets: Aerospace; Appliances; Automotive; Carpet;
 Chemical; Construction; Electronic/Electrical;
 Industrial; Lumber; Marine; Material Handling;
 Medical; Mining/Drilling; Packaging; Paper; Printing;
 Recreational; Refrigeration; Wheels/Tires

INDUSTRIAL RUBBER WORKS, INC.

312 W. Chestnut St.
Chicago, IL 60610 USA
Finished Goods Manufacturer: Elastomer; Flexible Foam
Other: Distributor; Fabricator
Markets: Appliances; Automotive; Construction;
 Defense/Military; Electronic/Electrical; Industrial;
 Marine; Material Handling; Mining/Drilling;
 Packaging; Paper; Pipe/Plumbing; Recreational;
 Transportation

INDUSTRIAL THERMOSET PLASTICS

9355 Pineneedle Dr.
Mentor, OH 44060 USA
Finished Goods Manufacturer: Elastomer; Cast; Coating
Markets: Automotive; Construction;
 Electronic/Electrical; Industrial; Marine; Material
 Handling; Mining/Drilling; Transportation

INGLA RUBBER PRODUCTS, INC.

8800 Park St.
Bellflower, CA 90706 USA
Finished Goods Manufacturer: Elastomer; Cast
Markets: Automotive; Defense/Military;
 Electronic/Electrical; Industrial; Food

INJECTEC LTD.

2333 S. Cicero Ave.
Cicero, IL 60650 USA
Finished Goods Manufacturer: Elastomer; Integral Skin
Markets: Aerospace; Defense/Military

INJECTED RUBBER PRODUCTS CORP.

153 S. Main St.
Albion, NY 14411 USA
Finished Goods Manufacturer: Coating
Markets: Automotive

INLAND LEIDY

2225 Evergreen St.
Baltimore, MD 21216 USA
Raw Material Supplier/Manufacturer: Polyol
Other: Distributor
Markets: Chemical

INNOVATIVE ENGINEERING

941 W. Round Lake Rd.
Dewitt, MI 48820-0620 USA
Systems Manufacturer: Elastomer Systems; Flexible Foam
 Systems
Finished Goods Manufacturer:. Elastomer; Cast;
Markets: Aerospace; Appliances; Automotive;
 Construction; Defense/Military; Electronic/Electrical;
 Footwear; Furniture; Industrial; Lumber; Marine;
 Material Handling; Medical; Mining/Drilling;
 Packaging; Paper; Pipe/Plumbing; Printing;

Recreational; Refrigeration; Transportation; Wheels/Tires

INNOVATIVE PLASTICS CORP.
400 Route 303
Orangeburg, NY 10962 USA
Finished Goods Manufacturer: Elastomer; TPU
Markets: Appliances; Automotive; Chemical; Construction; Defense/Military; Electronic/Electrical; Industrial; Medical; Packaging; Recreational; Transportation

INOAC U.S.A., INC.
2380 Meijer Dr.
Troy, MI 48084 USA
Raw Material Supplier/Manufacturer: Isocyanate; Polyol; Catalyst; Polymer; Resins
Finished Goods Manufacturer: Elastomer; RIM; Automotive; Flexible Foam; Molded Semi-Rigid Foam
Other: Research Laboratory; Distributor; Mfg Rep; Fabricator; Exporter/Importer
Markets: Automotive; Construction; Electronic/Electrical; Footwear; Furniture; Industrial; Medical; Mining/Drilling; Packaging; Pipe/Plumbing; Recreational; Wheels/Tires; Cosmetics; Office Equipment; Baby and Nursery Products

INOLEX CHENUCAL CO.
Jackson and Swanson Streets
Philadelphia, PA 19148 USA
Raw Material Supplier/Manufacturer: Polyol
Markets: Appliances; Automotive; Carpet; Chemical; Clothing; Construction; Footwear; Furniture; Industrial; Recreational; Refrigeration; Transportation; Wheels/Tires

INOUE RUBBER INTERNATIONAL CO. LTD.
31 Franklin Ave.
P.O. Box 396
Hewlett, NY 11557 USA
Finished Goods Manufacturer: Elastomer; TPU; Composites
Markets: Aerospace; Automotive; Defense/Military; Footwear; Furniture; Mining/Drilling; Printing; Wheels/Tires

INPLEX, INC.
1663 S. Mount Prospect Rd.
Des Plaines, IL 60018 USA
Finished Goods Manufacturer: Elastomer; TPU
Markets: Aerospace; Appliances; Automotive; Chemical; Construction; Defense/Military; Electronic/Electrical; Industrial; Marine; Material Handling; Medical; Packaging

INSTA-FOAM PRODUCTS, INC.
1500 Cedarwood Dr.
Joliet, IL 60435 USA
Systems Manufacturer: ¬Rigid Foam Systems; Flexible Foam Systems; Froth Foam Systems; Packaging Foam Systems

Equipment or Machinery Supplier/Manufacturer: Dispensing Equipment; Tanks
Finished Goods Manufacturer: Flexible Foam; Adhesive; Sealant
Markets: Aerospace; Chemical; Construction; Defense/Military; Electronic/Electrical; Furniture; Industrial; Lumber; Marine; Mining/Drilling; Packaging; Pipe/Plumbing; Recreational; Refrigeration; Transportation; Utility; Pool/Spa

INSTITUTE FOR RESEARCH, INC.
8330 West Glen Dr.
Houston, TX 77063 USA
Other: Research Organization

INSUL FOAM, INC.
P.O. Box 1 1 0
Herington, KS 67449 USA
Finished Goods Manufacturer: Rigid Foam
Other: Contractor

INSULATED ROOFING CONTRACTORS
Urethane of Kentuckiana, Inc.
10801 Bluegrass Parkway
Louisville, KY 40299 USA
Other: Contractor
Markets: Construction

INSULDYNE CORP.
Box 52566
Houston, TX 77052 USA
Finished Goods Manufacturer: Rigid Foam

INTEGRAM ST. LOUIS FOAM OPERATIONS
110 Integram Dr.
P.O. Box 220
Pacific, MO 63069 USA
Finished Goods Manufacturer: Flexible Foam
Markets: Automotive

INTEK WEATHERSEAL PRODUCTS, INC.
800 E. 1Oth St.
Hastings, MN 55033 USA
Finished Goods Manufacturer: Flexible Foam
Markets: Construction

INTERNATIONAL ISOCYANATE INSTITUTE
119 Cherry Hill Rd.
Parsippany, NJ 07054 USA
Other: Trade Association

INTERNATIONAL PERMALITE, INC.
300 N. Haven Ave.
Ontario, CA 91761 USA
Finished Goods Manufacturer: Rigid Foam
Markets: Construction

INTERPLASTIC DISTRIBUTION GROUP
10940 Laureate Dr.
Suite 8360
San Antonio, TX 78249 USA
Other: Distributor
Markets: Construction; Marine; Mining/Drilling; Refrigeration; Transportation

INTREPID INDUSTRIES, INC.
P.O. Box 5460
Pasadena, TX 77508 USA
Finished Goods Manufacturer: RIM; Non-automotive
Markets: Chemical; Industrial; Marine

IPI, INC.
PMC Division
P.O. Box 70
Elton, MD 21922-0070 USA
Systems Manufacturer: Prepolymer Manufacturer;
Elastomer Systems; Rigid Foam Systems; Flexible
Foam Systems
Markets: Aerospace; Appliances; Automotive;
Construction; Defense/Military; Electronic/Electrical;
Furniture; Industrial; Marine; Material Handling;
Packaging; Recreational; Refrigeration; Transportation

IPN INDUSTRIES
151 Essex St.
Haverhill, MA 01831 USA
Systems Manufacturer: Encapsulants; Adhesives
Markets: Electronic/Electrical

IRATHANE SYSTEMS, INC.
4045 Sinton Rd.
Colorado Springs, CO 80907 USA
Systems Manufacturer: Rigid Foam Systems
Finished Goods Manufacturer: Elastomer; Cast
Markets: Construction; Electronic/Electrical;
Mining/Drilling; Paper; Food

ISELANN-MOSS INDUSTRIES, INC.
41 Slater Rd.
Cranston, RI 02920 USA
Finished Goods Manufacturer: Elastomer; Cast
Other: Distributor
Markets: Aerospace; Appliances; Industrial;
Mining/Drilling; Paper; Printing; Wheels/Tires;
Business Machines

ISENMANN AMERICA, INC.
642 S. Broadway
Lexington, KY 40508 USA
Finished Goods Manufacturer: Elastomer; Cast

ISOTEC
201 Longview St.
Canton, GA 30114 USA
Systems Manufacturer: Elastomer Systems; Rigid Foam
Systems; Flexible Foam Systems
Markets: Aerospace; Chemical; Construction; Electronic/
Electrical; Industrial; Marine; Packaging; Refrigeration

ISOTHERMAL PROTECTIVE COATINGS, INC.
13800 O'Day Rd.
Route 3, Box 3800
Pearland, TX 77581 USA
Finished Goods Manufacturer: Coating
Markets: Construction; Defense/Military; Industrial

**ITRAN CORP. NIAGARA RUBBER &
RUBBERCRAFTERS, INC.**
375 Metuchen Rd.

P.O. Box 98
South Plainfield, NJ 07080 USA
Finished Goods Manufacturer: Elastomer; Cast
Markets: Defense/Military; Marine; Medical; Agricultural

ITW DEVCON
30 Endicott St.
Danvers, MA 01923 USA
Finished Goods Manufacturer: Coating; Adhesive; Sealant
Other: Manufacturer
Markets: Aerospace; Appliances; Automotive;
Construction; Defense/Military; Electronic/Electrical;
Footwear; Furniture; Industrial; Marine;
Mining/Drilling; Paper; Pipe/Plumbing; Recreational;
Refrigeration; Transportation

ITW IRATHANE SYSTEMS
3516 East 13th Ave.
Hibbing, MN 55746 USA
Finished Goods Manufacturer: Elastomer; Cast
Markets: Chemical; Construction; Electronic/Electrical;
Industrial; Material Handling; Mining/Drilling;
Pipe/Plumbing; Coal; Coatings; Power

ITWC, INC.
S. Main St.
Box 247
Malcom, IA 50157 USA
Raw Material Supplier/Manufacturer: Prepolymer; Resins
Systems Manufacturer: Prepolymer Manufacturer
Equipment or Machinery Supplier/Manufacturer:
Dispensing Equipment
Markets: Chemical

J-VON
25 Litchfield St.
Leominster, MA 01453 USA
Raw Material Supplier/Manufacturer: Resins
Markets: Aerospace; Appliances; Automotive; Clothing;
Construction; Defense/Military; Electronic/Electrical;
Footwear; Furniture; Industrial; Marine; Material
Handling; Medical, Mining/Drilling; Packaging;
Pipe/Plumbing; Printing; Recreational; Refrigeration;
Transportation; Wheels/Tires

J. J. D. URETHANE CO.
855 Cherry St.
Norristown, PA 19460 USA
Other: Contractor
Markets: Construction

J. L. SCHROTH CO.
23500 Blackstone
Warren, MI 48089 USA
Finished Goods Manufacturer: Flexible Foam
Markets: Appliances; Automotive; Printing; Graphic Arts

J.R. FLUORO-PLASTICS, INC.
P.O. Box 244
Palmyra, NY 14522 USA
Finished Goods Manufacturer: Elastomer; Cast
Markets: Industrial

JADE ENGINEERED PLASTICS, INC.
Box 28
Bristol, RI 02809 USA
Finished Goods Manufacturer: Elastomer; Cast
Markets: Industrial

JADELIMITED
Route 3
Box 5
Kirksville, MO 63501 USA
Systems Manufacturer: Elastomer Systems; Rigid Foam
    System
Other: Distributor

JAMES H. RHODES & COMPANY
Route 12B
Franklin Springs, NY 13341 USA
Finished Goods Manufacturer: Elastomer; Cast;
    Microcellular

JAMES WALKER MFG COMPANY
511 W. 1 95th St.
P.O. Box 467
Glenwood, IL 60425 USA
Finished Goods Manufacturer: Elastomer; Cast
Markets: Aerospace; Appliances; Automotive;
    Construction; Mining/Drilling; Paper

JAMESON COMPANY LTD.
2200 Terminal Rd.
Niles, MI 49120-9398 USA
Systems Manufacturer: Rigid Foam Systems; Flexible
    Foam Systems
Equipment or Machinery Supplier/Manufacturer:
    Dispensing Equipment; Pumps; Hose and Fittings
Finished Goods Manufacturer: Coating
Other: Distributor
Markets: Construction; Industrial; Refrigeration

JASPER PLASTICS
Kimball Industrial Park
P.O. Box 520
Jasper, IN 47546 USA
Finished Goods Manufacturer: Rigid Foam

JASPER RUBBER PRODUCTS INC.
1010 First Ave.
Jasper, IN 47546 USA
Finished Goods Manufacturer: Elastomer; Cast
Markets: Aerospace; Appliances; Automotive;
    Construction; Defense/Military; Electronic/Electrical;
    Industrial; Medical; Wheels/Tires

JEDTCO CORP.
5899 Executive Dr. East
Westland, MI 48185 USA
Finished Goods Manufacturer: Elastomer; Cast; Adhesive
Markets: Aerospace; Defense/Military;
    Electronic/Electrical

JEFFCO UPHOLSTERY & FOAM PRODUCTS
50 Howe Ave.
Milbury, MA 01527 USA

Finished Goods Manufacturer: Flexible Foam
Markets: Carpet; Furniture; Bedding

JET RUBBER CO.
4457 Tallmadge Rd.
Rootstown, OH 44272 USA
Finished Goods Manufacturer: Elastomer
Markets: Appliances; Automotive; Construction;
    Defense/Military; Electronic/Electrical; Marine;
    Mining/Drilling; Agricultural

JM HUBER CORP.
Chemicals Division
907 Revel St.
P.O. Box 31 0
Havre De Grace, MD 21078 USA
Raw Material Supplier/Manufacturer: Blowing Agent;
    Polymer
Markets: Industrial

JOHN G. SHELLEY CO., INC.
16 Mica Lane
Wellesley Hills, MA 02181 USA
Finished Goods Manufacturer: Elastomer; Cast
Markets: Automotive; Clothing; Construction; Defense/
    Military; Electronic/Electrical; Furniture; Industrial;
    Paper-Printing

JOHSON BROTHERS RUBBER CO.
P.O. Box N
West Buckeye St.
West Salem, OH 44287 USA
Finished Goods Manufacturer: Elastomer; Flexible Foam
Markets: Appliances; Automotive; Construction;
    Agricultural

JOHNSON CONTROLS, INC.
135 E. Bennett St.
Saline, MI 48176 USA
Finished Goods Manufacturer: Elastomer; Flexible Foam
Markets: Automotive

JONES& VINING
P. O. Box 1903
Lewiston, ME 04240 USA
Finished Goods Manufacturer: Elastomer; RIM; Non-
    automotive; Microcellular
Markets: Appliances; Footwear; Industrial

JORDAN-WEST
4949 District Blvd.
Vernon, CA 90058 USA
Finished Goods Manufacturer: Rigid Foam
Markets: Appliances

JOWAT COPP
P.O. Box 1368
High Point, NC 27261 USA
Finished Goods Manufacturer: Adhesive
Markets: Automotive; Furniture; Industrial; Lumber;
    Packaging; Paper

JP HOGAN & CO
109 W. Fifth Ave

Knoxville, TN 37917 USA
Other: Research Organization; Marketing Organization
Markets: Automotive; Carpet; Chemical; Furniture;
Transportation; Bedding

JPS ELASTOMERICS CORP.
395 Pleasant St.
Northampton, MA 01060 USA
Systems Manufacturer: Elastomer Systems
Finished Goods Manufacturer: Industrial; Marine

JTL CONSULTING SERVICES
2070 Arbor Lane
Northfield, IL 60093-3349 USA
Other: Consultant

JUMPING JACKS SHOES, INC.
100 Fifth St.
Monett, MT 65708 USA
Finished Goods Manufacturer: Elastomer; Cast
Markets: Footwear

K. J. QUINN & CO., INC.
135 Folly Mill Rd.
Seabrook, NH 03874 USA
Raw Material Supplier/Manufacturer: Isocyanate; Pre-
Polymer; Waterbase Polymers
Systems Manufacturer: Elastomer Systems
Finished Goods Manufacturer: Coating; Adhesive
Markets: Aerospace; Chemical; Construction; Electronic/
Electrical; Footwear; Industrial; Marine; Medical;
Printing; Recreational; Transportation

K.P. ENTERPRISES
17912 Morrow Circle
Villa Park, CA 92667 USA

KALAMAZOO PLASTICS CO.
1811 Factory St.
Kalamazoo, MI 49001 USA
Finished Goods Manufacturer: Elastomer; Non-
automotive; Cast; Integral Skin; Rigid Foam;
Expanded Polystyrene
Markets: Aerospace; Appliances; Automotive;
Construction; Electronic/Electrical; Furniture;
Industrial; Material Handling; Medical; Packaging;
Refrigeration

KALPLAS, INC.
P.O. Box 21 0
Vicksburg, MI 49097 USA
Finished Goods Manufacturer: Elastomer
Markets: Industrial

KARMAN RUBBER CO.
2331 Copley Rd.
Akron, OH 44320 USA
Finished Goods Manufacturer: Elastomer; Cast
Markets: Aerospace; Appliances; Automotive;
Construction; Electronic/Electrical; Packaging;
Pipe/Plumbing; Printing; Agricultural

KASTALON, INC.
4100 W. 124th Place
Alsip, IL 60658 USA

Finished Goods Manufacturer:. Elastomer; Coating
Markets: Appliances; Automotive; Construction;
Defense/ Military; Footwear; Furniture; Industrial;
Lumber; Marine; Material Handling; Mining/Drilling;
Packaging; Paper; Printing; Wheels/Tires

KDI TECHNOLOGIES, INC.
5161 Woodfield Ct.
Grand Rapids, MI 49505 USA
Equipment or Machinery Supplier/Manufacturer: Steel
Rule Dies
Finished Goods Manufacturer: Elastomer; Automotive;
Non-automotive
Other: Fabricator
Markets: Aerospace; Appliances; Automotive; Carpet;
Chemical; Clothing; Defense/Military;
Electronic/Electrical; Footwear; Furniture; Industrial;
Lumber; Marine; Material Handling; Medical;
Packaging; Paper; Printing; Recreational;
Transportation; Wheels/Tires

KEENE CO.
Arlon Division
2811 S. Harbor Blvd.
Santa Ana, CA 92704 USA
Finished Goods Manufacturer: Flexible Foam
Other: Fabricator
Markets: Industrial; Medical; Decorative

KELLER & HECKMAN
1001 G St., NW
Suite 500 W.
Washington, DC 20001 USA
Other: Attorney

KEMCO PLASTTCS
724 S. Arlington
Akron, OH 44306 USA
Finished Goods Manufacturer: Flexible Foam; Integral
Skin; Rigid Foam
Markets: Aerospace; Appliances; Defense/Military;
Electronic/Electrical; Furniture; Industrial; Marine;
Material Handling; Medical; Mining/Drilling;
Packaging; Recreational

KENRICH PETROCHEMICALS, INC.
140 E. 22nd St.
P.O. Box 32
Bayonne, NJ 07002 USA
Raw Material Supplier/Manufacturer: Nucleating Agents;
Fillers

KENT LATEX PRODUCTS
1500 St. Clair Ave.
Kent, OH 44240 USA
Finished Goods Manufacturer: Elastomer
Markets: Industrial; Medical

KENT MFG CO.
1840 Oak Industrial Dr. NE
Grand Rapids, MI 49505 USA
Finished Goods Manufacturer: Flexible Foam
Markets: Automotive; Furniture

KERN FOAM PRODUCTS CORP.
1253 New Market Ave.
South Plainfield, NJ 07080 USA
Finished Goods Manufacturer: Flexible Foam; Integral
  Skin; Rigid Foam; Custom Molders HR Foams;
  Integral Skin Foams; Shock Absorbent Foams
Markets: Aerospace; Appliances; Automotive; Defense/
  Military; Electronic/Electrical; Furniture; Industrial;
  Marine; Material Handling; Medical; Packaging;
  Pipe/Plumbing; Printing; Recreational; Refrigeration;
  Transportation

KEYSTONE TOOL & DIE
P.O. Box 604
Weston Mills, NY 14788 USA
Equipment or Machinery Supplier/Manufacturer: Molds;
  Automated Controls
Other: Consultant
Markets: Aerospace; Appliances; Automotive; Chemical;
  Construction; Defense/Military; Electronic/Electrical;
  Industrial; Material Handling; Mining/Drilling;
  Pipe/Plumbing; Printing; Recreational;
  Transportation; Wheels/Tires

KEYSTONE URETHANE PRODUCTS
P.O. Box 604
Weston Mills, NY 14788 USA
Finished Goods Manufacturer: Flexible Foam
Markets: Aerospace; Appliances; Automotive; Chemical;
  Construction; Defense/Military; Electronic/Electrical;
  Industrial; Material/Handling; Mining/Drilling;
  Pipe/Plumbing; Printing; Recreational;
  Transportation; Wheels/Tires

KIEHL ENGINEERING CO.
212 Foster Ave.
Bensenville, IL 60106 USA
Finished Goods Manufacturer: Elastomer
Markets: Industrial

KIMBALL COMPANIES, THE
35 Industrial Dr.
East Longmeadow, MA 01028 USA
Other: Fabricator
Markets: Packaging

KING FIBERGLASS CORP
Box 25
Arlington, WASHINGTON 98223 USA
Systems Manufacturer: Elastomer Systems; Gel Coats
Markets: Industrial; Marine

KINGSLEY MFG COMPANY
1984 Placentia Ave.
Costa Mesa, CA 92627 USA
Finished Goods Manufacturer: Elastomer; Cast;
  Microcellular
Markets: Medical

KIRSCH CHEMICAL COMPANY, INC.
10411 Andiron Dr.
Matthews, NC 28105 USA

Raw Material Supplier/Manufacturer: Catalyst;
  Surfactant; Blowing Agent; Pigment Dispersions;
  Urethane Removers
Systems Manufacturer: Elastomer Systems; Rigid Foam
  Systems; Flexible Foam Systems
Finished Goods Manufacturer: Elastomer; RIM;
  Automotive; Flexible Foam
Other: Distributor; Mfg Rep
Markets: Appliances; Automotive; Carpet; Chemical;
  Construction; Furniture; Packaging; Refrigeration;
  Transportation

KLINE & COMPANY, INC.
165 Passaic Ave.
Fairfield, NJ 07004 USA
Other: Consultant

KLOCKNER FERROMATIK DESMA
KFD Sales and Service
283S Crescent Springs Rd.
Erlanger, KY 41018 USA
Equipment or Machinery Supplier/Manufacturer:
  Dispensing Equipment; Tanks; Pumps; Automated
  Controls
Markets: Automotive; Defense/Military; Furniture;
  Industrial

KNAPP SHOE
29 Lowell St.
Lewiston, ME 04240 USA
Finished Goods Manufacturer: Microcellular
Markets: Footwear

KORNYLAK CORPORATION
400 Heaton St.
Hamilton, OH 45011 USA
Equipment or Machinery Supplier/Manufacturer: Molds;
  Laminators; Dispensing Equipment; Tanks; Pumps;
  Automated Controls
Markets: Aerospace; Appliances; Automotive; Carpet;
  Construction; Defense/Military; Footwear; Furniture;
  Industrial; Marine; Material Handling; Packaging;
  Refrigeration; Transportation; Wheels/Tires

KRAUSS-MAFFEI CORPORATION
Rim Division
7095 Industrial Rd.
P.O. Box 6270
Florence, KY 41042 USA
Equipment or Machinery Supplier/Manufacturer:
  Dispensing Equipment; Tanks; Pumps; Automated
  Controls; Rim Units; Conveyors; Bulk and Blend
  Systems; Robots
Markets: Appliances; Automotive; Carpet; Chemical;
  Construction; Footwear; Furniture; Industrial;
  Recreational; Refrigeration; Transportation

KRC ROLLS
Salisbury Plant
P.O. Box 1618
1415 Jake Alex Blvd.
Salisbury, NC 28144 USA

Finished Goods Manufacturer: Cast; Roll Covers;
   Elastomer
Other: Roll Services
Markets: Paper; Industrial

KRYPTONICS, INC.
740 S. Pierce Ave.
Louisville, CO 80027 USA
Systems Manufacturer: Prepolymer Manufacturer
Finished Goods Manufacturer: Elastomer; Cast; Flexible
   Foam; Integral Skin; Rigid Foam
Markets: Industrial; Material Handling; Recreational;
   Transportation; Construction; Defense/Military;
   Electrronic/ Electrical; Lumber; Mining/Drilling;
   Wheels/Tires

KURABO INDUSTRIES LTD.
111 W. 40th St.
15th Floor
New York, NY 10018 USA

KURTH INTERNATIONAL
1333 10th Ave.
Columbus, GA 31901 USA
Finished Goods Manufacturer: Flexible Foam; Integral
   Skin

KUSTOM FOAM MFG., INC.
2301 A Ninth St.
Modesto, CA 95351 USA
Finished Goods Manufacturer: Flexible Foam; Integral
   Skin; Coating; Rigid Foam
Markets: Aeropsace; Automotive; Defense/ Military;
   Electronic/Electrical; Industrial; Medical; Packaging;
   Recreational; Transportation

KWS CO., INC.
6764 Preston Ave.
Suite D
Livermore, CA 94550 USA
Finished Goods Manufacturer: Elastomer

KYONAX CORPORATION
Kyonax S.A.
8020 N. E. 1 Oth Ave.
Miami, FL 33138 USA
Raw Material Supplier/Manufacturer: Polymer
Other: Distributor
Markets: Chemical; Clothing; Material Handling; Paper

L.P. YOUNG, INC.
9401 Roberts Dr.
Suite 38D
Dunwoody, GA 30350 USA
Other: Manufacturing Rep

L. W. REINHOLD PLASTICS, INC.
8763 Crocker St.
Los Angeles, CA 90003 USA
Finished Goods Manufacturer: Elastomer
Markets: Automotive

LA ROCHE CHEMICALS, INC.
P.O. Box 1031

Airline Highway
Baton Rouge, LA 70821 USA
Raw Material Supplier/Manufacturer: Blowing Agent
Markets: Aerospace; Appliances; Automotive; Carpet;
   Chemical; Construction; Defense/Military; Footwear;
   Furniture; Industrial; Marine; Medical;
   Mining/Drilling; Packaging; Pipe/Plumbing; Printing;
   Recreational; Refrigeration; Transportation

LACKS INDUSTRIES
1601 Galbrarth St., SE
Grand Rapids, MI 49506 USA
Finished Goods Manufacturer: Elastomer; RIM-
   Microcellular
Markets: Automotive; Industrial

LAKEVIEW INDUSTRIES, INC.
Formerly Alfa III Corp.
1225 Lakeview Dr.
Chaska, MN 55318-9506 USA
Other: Fabricator
Markets: Appliances; Defense/Military;
   Electronic/Electrical; Industrial; Recreational;
   Refrigeration; Transportation

LAMATEK, INC.
P.O. Box 236
511 N. Read St.
Cinnaminson, NJ 08077 USA
Finished Goods Manufacturer: Flexible Foam
Markets: Industrial

LANCER CORP.
6655 Lancer Blvd.
San Antonio, TX 78219 USA
Systems Manufacturer: Rigid Foam Systems
Markets: Food

LARSON PUBLISHING CO., INC.
56 Industrial Park Rd.
Suite 4
Saco, ME 04072 USA
Other: Database Publishing; Industry Publications; Mail
   Lists; Merge Mailings; Marketing
   Development/Services
Markets: Aerospace; Appliances; Automotive; Carpet;
   Chemical; Clothing; Construction; Defense/Military;
   Electronic/Electrical; Footwear; Furniture; Industrial;
   Lumber; Marine; Material; Medical; Mining/Drilling;
   Packaging; Paper; Pipe/Plumbing; Printing;
   Recreational; Refrigeration; Transportation;
   Wheels/Tires

LARSTAN, INC.
Route 3, Earley Dr.
Box 300
Hagerstown, MD 21740 USA
Finished Goods Manufacturer: Flexible Foam
Markets: Automotive

LATEX FOAM CUSHIONING CO.
1817 E. Venango
Philadelphia, PA 19134 USA

LATICRETE INTERNATIONAL, INC.
91 Amity Rd.
Bethany, CT 06524 USA
Finished Goods Manufacturer: Elastomer
Markets: Construction; Industrial

LAUREL RUBBER CO.
Star-Glo Industries
2 Carlton Ave.
East Rutherford, NJ 07470 USA
Finished Goods Manufacturer: Elastomer; TPU; Cast
Markets: Aerospace; Appliances; Automotive; Clothing; Construction; Electronic/Electrical; Mining/Drilling; Packaging; Paper; Printing; Agricultural; Food

LAVELLE IND., INC.
665 McHenry St.
Burlington, WI 53105 USA
Finished Goods Manufacturer: Elastomer; TPU
Markets: Aerospace; Appliances; Automotive; Defense/Military; Electronic/Electrical; Furniture; Industrial; Lumber; Material Handling; Medical; Packaging; Pipe/Plumbing; Refrigeration; Transportation; Wheels/Tires

LEEDOR
S6 Industrial Park Rd.
Suite #4
Saco, ME 04072 USA
Other: Executive Recruiters

LEGGETT & PLATT, INC.
Urethane Products Group
1301 Cold Springs Rd.
Fort Worth, TX 76113 USA
Finished Goods Manufacturer: Flexible Foam; Rigid Foam
Markets: Automotive; Carpet; Furniture; Industrial; Marine; Medical; Packaging; Recreational; Flooring

LEHIGH RUBBER WORKS, INC.
32 W. Bridge St.
Morrisville, PA 19067 USA
Finished Goods Manufacturer: Elastomer; Coating
Markets: Appliances; Automotive; Defense/Military; Electronic/Electrical; Industrial; Agricultural

LESCON, INC.
P.O. Box 687
Flemington, NJ 08822 USA
Raw Material Supplier/Manufacturer: Surfactant; Pre-Polymer; Color Dispersions; Curatives; Mold Releases
Systems Manufacturer: Elastomer Systems
Equipment or Machinery Supplier/Manufacturer: Dispensing Equipment; Automated Controls; Drum Warmers
Other: Mfg. Rep.
Markets: Health and Safety

LEWIS INDUSTRIES
10035 Geary Ave.
Santa Fe Springs, CA 90670 USA
Finished Goods Manufacturer: Flexible Foam
Other: Fabricator

Markets: Automotive; Furniture; Marine; Packaging; Recreational; Bedding

LEXINGTON COMPONENTS, INC.
250 Ridgewood Rd.
Jasper, GA 30143 USA
Finished Goods Manufacturer: Elastomer; Cast
Markets: Automotive; Industrial; Transportation

LIBERTY FOAM & PACKAGING
312 Luther St.
Liberty, NC 27298 USA
Finished Goods Manufacturer: Flexible Foam
Other: Fabricator
Markets: Aerospace; Appliances; Automotive; Construction; Electronic/Electrical; Industrial; Material Handling; Medical; Packaging; Recreational; Transportation

LINATEX CORPORATION OF AMERICA
Beckaey Operations
P.O. Box 32
Mount Hope, WV 25880 USA
Finished Goods Manufacturer: Elastomer; Cast; Coating
Other: Fabricator
Markets: Automotive; Construction; Defense/Military; Industrial; Material Handling; Mining/Drilling; Pipe/Plumbing; Recreational; Wheels/Tires

LINDEN INDUSTRIES, INC.
4020 Bellaire Lane
Peninsula, OH 44264 USA
Equipment or Machinery Supplier/Manufacturer: Laminators; Dispensing Equipment; Tanks; Pumps; Automated Controls-Presses
Other: Consultant
Markets: Aerospace; Appliances; Automotive; Carpet; Chemical; Construction; Defense/Military; Electronic/Electrical; Footwear; Furniture; Industrial; Lumber; Marine; Medical; Packaging; Pipe/Plumbing; Recreational; Refrigeration; Transportation; Wheels/Tires

LINING TECHNOLOGIES, INC.
4945 N. River Rd.
Port Allen, LA 70767 USA
Finished Goods Manufacturer: Elastomer; Coating
Markets: Construction; Paper; Waterproofing, Food

LION OIL COMPANY
1000 McHenry St.
P.O. Box 7005
El Dorado, AZ 71730 USA
Systems Manufacturer: Rigid Foam Systems

LION TRADING INTERNATIONAL
3619 Santa Maria Ave.
Laredo, TX 78041 USA
Raw Material Supplier/Manufacturer: Resins; T.P.R.
Equipment or Machinery Supplier/Manufacturer: Molds; Dispensing Equipment; Tanks; Pumps; Automated Controls
Finished Goods Manufacturer: Adhesive

Other: Distributor; Mfg. Rep.
Markets: Footwear; Industrial

LIQUID CONTROL CORP
7576 Freedom Ave. NW
P.O. Box 2747
North Canton, OH 44720 USA
Equipment or Machinery Supplier/Manufacturer:
    Dispensing Equipment; Mixing
Markets: Aerospace; Appliances; Automotive; Chemical;
    Defense/Military; Electronic./Electrical; Furniture;
    Industrial; Marine; Material Handling; Medical;
    Mining/Drilling; Package; Recreational; Refrigeration;
    Transportation; Wheels/Tires

LLOYD LABORATORIES, INC.
23 Caller St.
Peabody, MA 01960 USA
Systems Manufacturer: Elastomer Systems
Finished Goods Manufacturer: Coating
Markets: Footwear; Furniture

LONZA, INC.
Special Fine Chemicals
17-17 Route 208
Fair Lawn, NJ 0741 0 USA
Raw Material Supplier/Manufacturer: Curatives
Markets: Aerospace; Automotive; Chemical; Defense/
    Military; Electronic/Electrical; Printing

LORD CORPORATION
Chemical Products Division
Elastomer Products Unit
2000 W. Grandview Blvd.
Erie, PA 16514 USA
Finished Goods Manufacturer: Coating; Adhesive
Markets: Aerospace; Automotive; Chemical;
    Construction; Defense/Military; Electronic/Electrical;
    Footwear; Furniture; Industrial; Marine; Medical;
    Mining/Drilling; Printing; Recreational;
    Transportation; Wheels/Tires

LOYALTY FOAM CO.
P.O. Box 128
Okolona, MS 38860 USA
Finished Goods Manufacturer: Flexible Foam
Markets: Furniture; Packaging

LTK INDUSTRIES, INC.
807 Holland Rd.
Simpsonville, SC 29681 USA
Other: Fabricator
Markets: Aerospace; Clothing; Industrial; Medical;
    Packaging; Recreational

LTV ENERGY PRODUCTS
Oil States Industries
P.O. Box 670
Arlington, TX 76004 USA
Finished Goods Manufacturer: Elastomer; Cast
Markets: Aerospace; Automotive; Construction; Mining/
    Drilling; Paper

LUDLOW-SAYLOR
1402 E. Old Highway 40
Warrenton, MO 63383 USA
Finished Goods Manufacturer: Elastomer; Cast
Markets: Industrial; Mining/Drilling

LUDWIG, INC.
P.O. Box 450
Waldo, AR 71770 USA
Finished Goods Manufacturer: Flexible Foam; Integral
    Skin; Rigid Foam
Other: Consultant; Fabricator
Markets: Aerospace; Chemical; Defense/Military;
    Footwear; Marine; Medical

LUNDELL MFG CORP.
2702 Ranchview Lane
Minneapolis, MN 55447 USA
Finished Goods Manufacturer: Flexible Foam
Markets: Construction; Furniture; Industrial

M & C SPECIALTIES CO.
90 James Way
Southhampton, PA 18966 USA
Finished Goods Manufacturer: Flexible Foam
Markets: Industrial

M & K INTERNATIONAL, INC.
42 Montgomery Dr.
Griffin, GA 30223 USA
Systems Manufacturer: Elastomer Systems
Equipment or Machinery Supplier/Manufacturer:
    Dispensing Equipment
Finished Goods Manufacturer: Adhesive
Markets: Industrial

M & R FLEXIBLE PACKAGING, INC.
1881 Southtown Blvd.
Dayton, OH 45439 USA
Finished Goods Manufacturer: Flexible Foam
Markets: Aerospace; Appliances; Automotive; Chemical;
    Construction; Defense/Military; Electronic/Electrical;
    Industrial; Material Handling; Medical; Packaging;
    Wheels/Tires

M C & D CAPITOL CORP
9623 E. Imperial Highway
Downey, CA 90242 USA
Systems Manufacturer: Elastomer Systems; Rigid Foam
    Systems
Other: Distributor

M&H INDUSTRIES
32500 Capitol
Livonia, MI 48150 USA
Finished Goods Manufacturer: Elastomer- Flexible Foam
Markets: Automotive; Recreational

M&R RUBBER COMPANY
6583 Revlon Dr.
Belvidere, IL 61008 USA
Finished Goods Manufacturer: Elastomer
Markets: Aerospace; Appliances; Automotive; Defense/
    Military; Electronic/Electrical; Industrial; Printing

M. G. LUCKE RUBBER CO.
765 McGlincey Lane
Campbell, CA 95008 USA
Finished Goods Manufacturer: Elastomer; Cast
Markets: Construction; Industrial; Food

MAC SPECIALTIES LTD.
8 Maple Place
Freeport, NY 11520 USA
Finished Goods Manufacturer: RIM; Non-automotive;
Flexible Foam
Markets: Medical; Packaging; Recreational; Toys

MACKENZIE CHEMICAL WORKS, INC.
55 G Brook Ave.
Deer Park, NY 11729 USA
Raw Material Supplier/Manufacturer: Catalyst

MACKLANBURG DUNCAN
10950 S. Pipeline Rd.
Euless, TX 76040 USA
Finished Goods Manufacturer: Adhesive; Rigid Foam
Markets: Carpet; Construction; Industrial

MACROTECH FLUID SEALING, INC.
Polyseal
1750 W. 500 South
Salt Lake City, UT 84104 USA
Equipment or Machinery Supplier/Manufacturer: Sealing
Devices
Finished Goods Manufacturer: Elastomer; TPU
Other: Fabricator
Markets: Aerospace; Automotive; Chemical;
Construction; Industrial; Lumber; Marine; Material
Handling; Medical; Mining/Drilling; Paper;
Transportation

MADISON POLYMERIC ENGINEERING
495 Ward St. Extension
Wallingford, CT 06492 USA
Finished Goods Manufacturer: Flexible Foam
Markets: Aerospace; Appliances; Automotive;
Construction; Electronic/Electrical; Industrial;
Material Handling; Medical; Packaging; Recreational;
Transportation

MAINE RUBBER INTERNATIONAL
Permathane
21 Saco St.
Westbrook, ME 04092 USA
Finished Goods Manufacturer: Elastomer; Cast
Markets: Construction; Industrial; Material Handling;
Mining/ Drilling; Wheels/Tires

MAJORS INC.
10117 1 St.
Omaha, NE 68127 USA
Finished Goods Manufacturer: Elastomer
Markets: Automotive; Industrial

MANAGEMENT RECRUITERS
Polymer Specialties
105 E. Jefferson Blvd.
Suite 800

South Bend, IN 46601 USA
Other: Search Consultant for Technical and Sales Areas

MANCHESTER PLASTICS
845 Progress Ct.
Williamston, MI 48895 USA
Finished Goods Manufacturer: RIM; Automotive; Non-
automotive; Flexible Foam; Integral Skin; Coating;
Adhesive; Rigid Foam
Markets: Automotive; Medical; Recreational

MANDRELS INC.
621 North Chapel St.
Louisville, OH 44641 USA
Finished Goods Manufacturer: Elastomer; Integral Skin
Markets: Automotive; Construction; Industrial; Material
Handling; Transportation

MANVILLE RUBBER PRODUCTS, INC.
1009 Kennedy Blvd.
Manville, NJ 08835 USA
Finished Goods Manufacturer: Elastomer; TPU; Cast
Markets: Aerospace; Construction; Electronic/Electrical;
Industrial; Packaging; Food

MARCHEM CORPORATION
2500 Adie Rd.
Maryland Heights, MO 63043 USA
Raw Material Supplier/Manufacturer: Polymer;
Prepolymer
Systems Manufacturer: Prepolymer Manufacturer;
Elastomer Systems; Rigid Foam Systems; Flexible
Foam Systems; Froth Foam Systems
Equipment or Machinery Supplier/Manufacturer:
Dispensing Equipment
Finished Goods Manufacturer: Elastomer; Flexible Foam;
Integral Skin; Coating; Adhesive
Other: Consultant; Testing; Research Organization; Toll
Manufacturing
Markets: Appliances; Carpet; Chemical; Construction;
Defense/Military; Furniture; Industrial; Marine;
Packaging; Recreational; Refrigeration;
Transportation; Wheels/Tires

MARCHEM DUBLON, INC.
84 Waydell St.
Newark, NJ 07105 USA
Raw Material Supplier/Manufacturer: Polymer;
Prepolymer
Systems Manufacturer: Prepolymer Manufacturer;
Elastomer Systems; Rigid Foam Systems; Flexible
Foam Systems; Froth Foam Systems; One-Component
Adhesives
Equipment or Machinery Supplier/Manufacturer:
Dispensing Equipment
Finished Goods Manufacturer: Elastomer; Flexible Foam;
Integral Skin; Coating; Adhesives
Other: Distributor; Mfg Rep
Markets: Appliances; Carpet; Construction;
Defense/Military; Furniture; Industrial; Marine;
Packaging; Recreational; Refrigeration;
Transportation; Wheels/Tires

MARCHEM SOUTHEAST, INC.
400 N. Main St.
Adairsville, GA 30103 USA
Raw Material Supplier/Manufacturer: Polymer;
    Prepolymer
Systems Manufacturer: Prepolymer Manufacturer;
    Elastomer Systems; Rigid Foam Systems; Flexible
    Foam Systems; Froth Foams; One-Component
    Adhesives
Equipment or Machinery Supplier/Manufacturer:
    Dispensing Equipment
Finished Goods Manufacturer: Elastomer; Coating;
    Adhesive
Other: Consultant; Testing; Research Organization; Toll
    Manufacturer
Markets: Aerospace; Appliances; Carpet; Chemical;
    Construction; Defense/Military; Furniture; Industrial;
    Marine; Packaging; Recreational; Refrigeration;
    Transportation; Wheels/Tires

MARIAN RUBBER PRODUCTS CO., INC.
1212 E. Michigan St.
Indianapolis, IN 46202 USA
Finished Goods Manufacturer: Elastomer; Cast; Flexible
    Foam
Markets: Automotive; Construction; Defense/Military;
    Electronic/Electrical: Industrial; Paper; Printing

MARINE RUBBER, INC.
Marine Urethane, Inc.
P.O. Box 2438
Humble, TX 77347 USA
Finished Goods Manufacturer: Elastomer; Cast; Coating
Markets: Marine; Mining/Drilling; Printing

MARK TOOL CO.
P.O. Box 51422
Lafayette, LA 70505 USA
Finished Goods Manufacturer: Elastomer; Cast
Markets: Industrial; Wheels/Tires

MARKEL CORPORATION
P.O. Box 752
Norristown, PA 19404 USA
Finished Goods Manufacturer: Elastomer; TPU
Markets: Aerospace; Appliances; Automotive;
    Construction; Defense/Military; Electronic/Electrical

MARKO FOAM PRODUCTS, INC.
1411 S. Village Way
Santa Ana, CA 92705 USA
Finished Goods Manufacturer: Flexible Foam
Other: Fabricator

MARLOCK, INC.
200 Raccoon Valley Rd.
Maynardville, TN 37807-3143 USA
Finished Goods Manufacturer: High Density Rigid Foam
Markets: Furniture; Architectural Millwork; Advertising
    Specialty Signs

MARTEC PLASTICS
3201 W. Thompson Rd.

Fenton, MI 48430 USA
Finished Goods Manufacturer: Elastomer; RIM;
    Automotive; Non-automotive; Flexible Foam; Rigid
    Foam; Composites
Other: Consultant; Testing; Research Organization; Mfg
    Rep; Fabricator
Markets: Appliances; Automotive; Construction;
    Defense/ Military; Electronic/Electrical; Furniture;
    Marine; Recreational; Transportation; Wheels/Tires

MARTIN MARIETTA MANNED SPACE SYSTEMS
P.O. Box 29304
New Orleans, LA 70429 USA
Finished Goods Manufacturer: Rigid Foam
Other: Fabricator
Markets: Aerospace

MARVLEE
5 Sidney Ct.
N. Lindenhurst, NY 11 757 USA
Finished Goods Manufacturer: Flexible Foam
Markets: Recreational

MASON SHOE MANUFACTURING COMPANY
1251 First Ave.
Chippewa Falls, WI 54774 USA
Finished Goods Manufacturer: Elastomer; Cast
Markets: Footwear

MASTER-BILT PRODUCTS
Highway 15N
New Albany, MS 38652 USA
Finished Goods Manufacturer: Rigid Foam
Markets: Construction; Industrial; Refrigeration

MATERIALS TECHNOLOGY ASSOCIATES
2603 Artie St.
Suite 16-131
Huntsville, AL 35805 USA
Systems Manufacturer: Elastomer Systems; Rigid Foam
    Systems; Flexible Foam Systems
Equipment or Machinery Supplier/Manufacturer: Molds;
    Automated Controls
Finished Goods Manufacturer: Elastomer; Cast; Integral
    Skin; Coating; Adhesive; Sealant; Rigid Foam
Other: Consultant; Testing; Research Organization
Markets: Aerospace; Appliances; Construction; Defense/
    Military; Marine; Medical; Recreational;
    Transportation; Composite Materials;
    Prosthetics/Orthotics

MATHER SEAL CO.
52S Redman Rd.
Milan, MI 48160 USA
Finished Goods Manufacturer: Elastomer
Markets: Industrial

MAX MACHINERY, INC.
1420 Healdsburg Ave.
Healdsburg, CA 95448 USA
Equipment or Machinery Supplier/Manufacturer:
    Dispensing Equipment

Markets: Aerospace; Automotive; Clothing; Construction; Defense/Military; Electronic/Electrical; Footwear; Furniture; Industrial; Marine; Material Handling; Medical; Mining/ Drilling; Paper; Pipe/Plumbing; Printing; Recreational; Transportation; Wheels/Tires

MAYPAC INC.
582 Fairfield Rd.
P.O. Box 326
Wayne, NJ 07470 USA
Finished Goods Manufacturer: Flexible Foam
Markets: Packaging

MCGHEE INSULATION, INC.
719 S. Foster St.
Dothan, AL 36301 USA
Other: Contractor
Markets: Construction

MCNETT CORPORATION
P.O. Box 996
1405 Fraser St.
Bellingham, WASHINGTON 98227 USA
Finished Goods Manufacturer: Elastomer; Adhesive; Sealant
Markets: Aerospace; Automotive; Clothing; Construction; Footwear; Industrial; Marine; Recreational; Tent and Awning; Watersports; Fishing

MCP URETHANES
A Division of MCP industries, Inc.
P.O. Box 1839
Corona, CA 91718 USA
Systems Manufacturer: Elastomer Systems; Rigid Foam Systems
Finished Goods Manufacturer: Coating
Markets: Construction; Pipe/Plumbing

MEARTHANE PRODUCTS
16 Western Industrial Dr.
Cranston, RI 02921 USA
Finished Goods Manufacturer: Elastomer; Cast; Flexible Foam
Markets: Aerospace; Appliances; Automotive; Industrial; Mining/Drilling; Paper; Wheels/Tires; Business Equipment; Skate Wheels

MECHANICAL RUBBER PRODUCTS CORP.
77 Forester Ave.
Warwick, NY 10990 USA
Finished Goods Manufacturer: Elastomer; RIM; Non-automotive; Cast; Flexible Foam; Coating
Markets: Aerospace; Appliances; Automotive; Defense/Military; Electronic/Electrical; Marine

MEISNER ROLLER CO.
3103 Millers Lane
Louisville, KY 40216 USA
Finished Goods Manufacturer: Elastomer; Cast
Markets: Appliances; Clothing; Construction; Electronic/Electrical; Bedding; Food

MERAMEC GROUP, INC.
P.O. Box 279
Sullivan, MT 63080 USA
Finished Goods Manufacturer: Integral Skin
Markets: Footwear; Medical

MERECO PRODUCTS
1505 Main St.
W. Warwick, RI 02893 USA
Systems Manufacturer: Rigid Foam Systems
Finished Goods Manufacturer: Elastomer
Markets: Construction

MEREEN-JOHNSON MACHINE CO.
4401 Lyndale Ave. N
Minneapolis, MN 55412 USA
Equipment or Machinery Supplier/Manufacturer: Rigid Foam Processing Equipment
Other: Fabricator
Markets: Furniture; Industrial; Lumber; Material Handling

MERN MANUFACTURING CORPORATION
813 Jerusalem Rd.
Scotch Plains, NJ 07076 USA
Finished Goods Manufacturer: Elastomer; Cast
Markets: Aerospace; Appliances; Automotive; Carpet; Chemical; Construction; Defense/Military; Furniture; Industrial; Marine; Material Handling; Medical; Mining/Drilling; Packaging; Paper; Pipe/Plumbing; Printing; Recreational; Wheels/Tires

MERRY WEATHER FOAM, INC.
P.O. Drawer 752
Barberton, OH 44203 USA
Finished Goods Manufacturer: Flexible Foam; Rigid Foam
Markets: Construction; Industrial; Marine; Packaging

METAL RUBBER CORP.
1225 S. Shamrock Ave.
Monrovia, CA 91016 USA

METL-SPAN CORP.
1497 N. Kealy
Lewisville, TX 75057 USA
Finished Goods Manufacturer: Rigid Foam
Other: Fabricator
Markets: Construction; Defense/Miltary; Industrial; Mining/Drilling; Refrigeration

MHE, INC.
3866 N. Fratney St.
P.O. Box 12050
Milwaukee, WI 53212 USA
Finished Goods Manufacturer:. Elastomer; Cast

MI RUBBER PRODUCTS, INC.
1200 Eighth Ave.
Cadillac, MI 49601 USA
Finished Goods Manufacturer: Elastomer; TPU
Markets: Appliances; Automotive; Marine; Agricultural

MICON
25 Allegheny Square

Glassport, PA 15045-1649 USA
Equipment or Machinery Supplier/Manufacturer: Pumps
Other: Specialty Tunnel and Mining; Grouting and
Contracting Firm
Markets: Construction; Mining/Drilling

MICRODOT/POLYSEAL DIV.
1754 W. Fifth St. South
P.O. Box 26627
Salt Lake City, UT 84126 USA

MICROFOAM, INC.
Johnson Worldwide Association
31 Faass Ave.
Utica, NY 13502 USA
Systems Manufacturer: Flexible Foam Systems
Finished Goods Manufacturer: Flexible Foam
Other: Fabricator
Markets: Aerospace; Appliances; Automotive;
Defense/Military; Electronic/Electrical; Footwear;
Industrial; Lumber; Printing; Recreational;
Wheels/Tires

MICROMET INSTRUMENTS, INC.
A Geo-Centers Company
7 Wells Ave.
Newton Centre, MA 02159 USA
Equipment or Machinery Supplier/Manufacturer:
Automated Controls; Cure Monitoring
Instrumentation
Markets: Aerospace; Appliances; Automotive; Chemical;
Defense/Military; Electronic/ Electrical; Industrial;
Marine; Medical; Wheel/Tires

MID AMERICA ROLLER CO.
2035 Washington
Kansas City, MO 64108 USA
Finished Goods Manufacturer: Elastomer; RIM; TPU;
Cast; Coating; Adhesive; Sealant
Markets: Chemical; Industrial; Material Handling;
Packaging; Paper; Printing

MIDWEST INDUSTRIAL CHEMICAL CO.
1509 Sublette Ave.
St. Louis, MO 63110 USA
Finished Goods Manufacturer: Adhesive; Sealant
Markets: Chemical; Clothing; Construction; Footwear;
Furniture; Industrial; Lumber; Marine; Packaging;
Paper

MID-AMERICA DOOR CO.
P.O. Box 2423
Ponca City, OKLAHOMA 74602 USA
Finished Goods Manufacturer: Rigid Foam
Markets: Construction

MIDCO PACKAGING INDUSTRIES
6543 S. Laramie
Bedford Park, IL 60638 USA
Finished Goods Manufacturer: Flexible Foam
Markets: Packaging

MIDWEST CORTLAND, INC.
9535 W. River St.

Schiller Park, IL 60176 USA
Finished Goods Manufacturer: Elastomer; Cast; Flexible
Foam; Rigid Foam
Markets: Packaging; Recreational

MIDWEST URETHANE
P.O. Box 27
110 W. Main St.
Mulvane, KS 671 1 0 USA
Finished Goods Manufacturer: Flexible Foam
Markets: Chemical; Construction

MIKI SANGYO (USA), INC.
747 Third Ave.
New York, NY 10017 USA
Raw Material Supplier/Manufacturer: Activator-M;
BHEB; Polycaprolactone

MILCUT, INC.
4837 W. Woolworth Ave.
Milwaukee, WI 53218 USA
Finished Goods Manufacturer: Flexible Foam

MILES, INC. (now Bayer, Inc.)
Polymers Division
1 Mobay Rd.
Pittsburgh, PA 15205-9741 USA
Raw Material Supplier/Manufacturer: Isocyanate; Polyol;
Catalyst; Surfactant; Blowing Agent; Polymer;
Prepolymer; Resins
Systems Manufacturer: Prepolymers Manufacturer;
Elastomer Systems; Rigid Foam Systems; Flexible
Foam Systems; Equipment or Machinery
Supplier/Manufacturer Molds; Laminators;
Dispensing Equipment; Tanks; Pumps; Automated
Controls
Markets: Appliances; Automotive; Carpet; Construction;
Footwear; Furniture; Industrial; Lumber; Marine;
Mining/

MILFOAM CORPORATION
P.O. Box 4478
Hamden, CT 06514 USA
Finished Goods Manufacturer: Elastomer; RIM; Cast;
Flexible Foam; Integral Skin; Coating; Adhesive; Rigid
Foam
Markets: Aerospace; Appliances; Automotive; Chemical;
Defense/Military; Electronic/Electrical, Furniture;
Industrial; Marine; Medical; Pipe/Plumbing;
Recreational; Refrigeration; Transportation

MILLER URETHANE PRODUCTS, INC.
2523 Carl Jones Rd.
Moody, AL 35004 USA
Finished Goods Manufacturer: Elastomer
Markets: Industrial; Mining/Drilling

MILLIKEN CHEMICAL
Performance Chemicals
920 Milliken Rd.
P.O. Box 1927, M401
Spartanburg, SC 29304 USA
Raw Material Supplier/Manufacturer: Colorant

Markets: Automotive; Carpet; Chemical; Footwear; Medical; Packaging

MILSCO MANUFACTURING CO.
9009 N. 51st St.
Milwaukee, WI 53223 USA
Finished Goods Manufacturer: Non-automotive; Flexible Foam; Integral Skin
Markets: Construction; Defense/Military; Furniture; Industrial; Marine; Material Handling; Medical; Mining/Drilling; Recreational; Transportation

MN RUBBER
3630 Woodale Ave.
Minneapolis, MN 55416 USA
Finished Goods Manufacturer: Elastomer; Flexible Foam; Integral Skin; Rigid Foam
Markets: Industrial

MINOR RUBBER CO, INC.
49 Ackerman St.
Bloomfield, NJ 07003 USA
Other: Fabricator
Markets: Aerospace; Appliances; Construction; Defense/ Military; Electronic/Electrical; Industrial; Marine; Material Handling; Medical; Recreational; Transportation

MISSION RUBBER CO., INC.
1660 Leeson Lane
Corona, CA 91718 USA
Finished Goods Manufacturer: Elastomer; Coating
Markets: Construction; Recreational; Gymnasiums; Sports Tracks

MITSUBISHI HEAVY INDUSTRIES
3822 W. 13 Mile Rd.
Apt. #C-6
Royal Oak, MI 48073 USA

MODERN TOOLS
Libby-Owens/Ford
911 Matzinger Rd.
Toledo, OH 43695 USA
Equipment or Machinery Supplier/Manufacturer: Molds
Finished Goods Manufacturer: Elastomer; RIM; Automotive; Integral Skin
Other: Consultant
Markets: Automotive; Transportation

MOHICAN INDUSTRIES, INC.
306 E. Third St.
P.O. Box 592
Ashland, OH 44805 USA
Finished Goods Manufacturer: Elastomer; Cast
Markets: Industrial

MOLDED DIMENSIONS, INC.
701 Sunset Rd.
Port Washington, WI 53074 USA
Finished Goods Manufacturer: Elastomer; Cast
Other: Fabricator
Markets: Construction; Electronic/Electrical; Industrial; Material Handling; Medical; Mining/Drilling;

Packaging; Paper; Printing; Recreational; Transportation; Wheels/Tires

MOLDED MATERIALS, INC.
14555 Jib St.
Plymouth, MI 48170 USA
Finished Goods Manufacturer: Elastomer; TPU; Cast; Flexible Foam; Rigid Foam
Markets: Automotive; Chemical; Defense/Military; Industrial; Packaging; Recreational; Food; Pharmaceutical

MOLDED RUBBER & PLASTIC CORP.
13161 W. Glendale Ave.
Butler, WI 53007 USA
Finished Goods Manufacturer: Elastomer; TPU; Cast
Markets: Appliances; Automotive; Construction; Electronic/Electrical; Agricultural; Food

MOLDEX RUBBER CO., INC.
8052 Armstrong Rd.
Milton, FL 32583 USA
Finished Goods Manufacturer: Elastomer; RIM; Automotive; Cast
Markets: Automotive

MOLDING TECHNICAL SYSTEMS
Sub. of 7-Sigma, Inc.
2843 26th Ave. South
Minneapolis, MN 55406 USA
Finished Goods Manufacturer: Elastomer; Cast
Markets: Aerospace; Appliances; Automotive; Defense/ Military; Electronic/Electrical; Mining/Drilling; Paper; Printing; Food

MONARCH INDUSTRIES TIRE
Mono-Thane
61 State Route 43 North
Hartville, OH 44632 USA
Finished Goods Manufacturer: Elastomer; Cast
Markets: Material Handling; Wheels/Tires

MONO-THANE
Monarch Industrial Tire Corp.
1460 Industrial Parkway
Akron, OH 44310 USA
Finished Goods Manufacturer: Elastomer; Cast; Flexible Foam
Markets: Aerospace; Mining/Drilling; Wheels/Tires

MONSANTO CHEMICAL COMPANY
800 N. Lindbergh Blvd.
St. Louis, MO 63167 USA
Raw Material Supplier/Manufacturer: Resins
Finished Goods Manufacturer: Elastomer

MONTANA URETHANE SYSTEMS SUPPLY, INC.
2110 Lea Ave.
Bozeman, MT 59715 USA
Equipment or Machinery Supplier/Manufacturer: Dispensing Equipment; Pumps
Other: Distributor
Markets: Construction

MOONEYCHEMICALS, INC.
2301 Scranton Rd.
Cleveland, OH 44113 USA
Raw Material Supplier/Manufacturer: Catalyst
Markets: Aerospace; Appliances; Automotive; Carpet;
   Chemical; Clothing; Construction; Defense/Military;
   Electronic/Electrical; Footwear; Furniture; Industrial;
   Lumber; Marine; Material Handling; Medical;
   Mining/Drilling; Packaging; Paper; Pipe/Plumbing;
   Printing; Recreational; Refrigeration; Transportation;
   Wheels/Tires

MORGAN CHEMICALS
5502 Main St.
Buffalo, NY 14221 USA
Raw Material Supplier/Manufacturer: Isocyanate; Polyol;
   Surfactant
Other: Distributor
Markets: Construction; Packaging; Paper; Printing;
   Refrigeration; Transportation

MORTELL CO.
550 Hobbie Ave.
Kankakee, IL 60901 USA
Finished Goods Manufacturer: Adhesive; Sealant
Markets: Aerospace; Appliances; Automotive;
   Construction

MORTON INTERNATIONAL
Thermoplastic Polyurethane Division
137 Folly Mill Rd.
Seabrook, NH 03874 USA
Raw Material Supplier/Manufacturer: Resins
Markets: Automotive; Chemicals; Clothing; Electronic/
   Electrical; Footwear; Medical; Mining/Drilling;
   Recreational; Wheels/Tires

MORTON INTERNATIONAL
Industrial Adhesives
100 N. Riverside Plaza
Chicago, IL 60606 USA
Finished Goods Manufacturer: Adhesive
Other: Research Organization; Distributor; Fabricator
Markets: Construction; Lumber

MORTON INTERNATIONAL
10 S. Electric St.
West Alexandria, OH 45381 USA
Finished Goods Manufacturer: Elastomer; TPU; Cast;
   Adhesive
Markets: Aerospace; Appliances; Automotive; Defense/
   Military; Electronic/Electrical; Footwear; Industrial;
   Printing;

MOULDED CHEMICAL PRODUCTS. CO., INC.
10930 116th Ave. NE
Kirkland, WA 98033 USA
Finished Goods Manufacturer: Elastomer; Cast; Flexible
   Foam; Coating
Markets: Aerospace; Industrial; Printing

MUELLER BELTING AND SPECIAL
150 N. Midland Ave.

Saddle Brook, NJ 07662 USA
Other: Fabricator
Markets: Industrial

MUTH ASSOCIATES, INC.
53 Progress Ave.
Springfield, MA 01104 USA
Finished Goods Manufacturer: Flexible Foam; Rigid Foam
Markets: Defense/Military; Packaging

MVR CHEMICALS CORPORATION
152 Madison Ave.
New York, NY 10016 USA
Raw Material Supplier/Manufacturer: Curatives
Other: Distributor

MYERS INDUSTRIES
Patch Rubber Company Inc.
P.O. Box H
Roanoke Rapids, NC 27870 USA
Finished Goods Manufacturer: Elastomer
Markets: Wheels/Tires

NALGE COMPANY
75 Panorama Creek Dr.
Box 20365
Rochester, NY 14600 USA
Equipment or Machinery Supplier/Manufacturer: Molds
Finished Goods Manufacturer: Elastomer
Markets: Chemical; Electronic/Electrical

NASHVILLE RUBBER & GASKET
1900 Elm Tree Dr.
P.O. Box 110357
Nashville, TN 37211 USA
Finished Goods Manufacturer: Flexible Foam
Other: Fabricator
Markets: Industrial

NATION/RUSKIN, INC.
Commerce Dr. and Enterprise Rd.
Montgomeryville, PA 18936 USA
Finished Goods Manufacturer: Flexible Foam
Other: Fabricator

NATIONAL COATINGS CORPORATION
912 Pancho Rd.
Camarillo, CA 93012 USA
Systems Manufacturer: Elastomer Systems
Finished Goods Manufacturer: Elastomer; Coating;
   Sealant
Markets: Construction; Building Renovations; Building
   Maintenance

NATIONAL FIRE HOSE CORP.
516 E. Oakes St.
Compton, CA 90224-4969 USA
Equipment or Machinery Supplier/Manufacturer: Process
   Equipment
Finished Goods Manufacturer: Elastomer; TPU; Cast
Markets: Material Handling; Pipe/Plumbing

NATIONAL FOAM MFG. CO.
8216 North East Parkway

North Richland Hills, TX 76180 USA
Finished Goods Manufacturer: Flexible Foam
Other: Fabricator
Markets: Carpet; Furniture; Packaging

NATIONAL O-RING
11634 Patten Rd.
Downey, CA 90241 USA
Finished Goods Manufacturer: Elastomer; Millable Gums
Markets: Automotive; Chemical; Industrial

NATIONAL ROOFING CONTRACTORS ASSOCIATION
10255 W. Higgins Rd.
#600
Rosemont, IL 60018-5607 USA
Other: Contractor Association
Markets: Construction

NATIONAL STARCH & CHEM CO.
10 Finderne Ave.
Bridgewater, NJ 08807 USA
Finished Goods Manufacturer: Adhesive
Markets: Automotive; Packaging

NAZAR RUBBER CO.
2727 Avondale Ave.
P.O. Box 2848, Kenwood Station
Toledo, OH 43606 USA
Finished Goods Manufacturer: Elastomer; Cast
Markets: Industrial; Material Handling; Metal;
　Mining/Drilling; Printing

NEOGARD
6900 Maple Ave.
P.O. Box 35288
Dallas, TX 75235 USA
Systems Manufacturer: Elastomer Systems
Finished Goods Manufacturer: Elastomer
Markets: Construction

NETHERLAND RUBBER CO.
2931 Exon Ave.
Cincinnati, OH 45212 USA
Other: Distributor; Fabricator
Markets: Appliances; Clothing; Construction; Electronic/
　Electrical; Material Handling; Medical;
　Mining/Drilling; Pipe/Plumbing; Transportation

NEW ENGLAND FOAM PRODUCTS
Windsor St.
Hartford, CT 06095 USA
Finished Goods Manufacturer: Flexible Foam
Other: Fabricator
Markets: Appliances; Automotive; Furniture; Medical;
　Packaging

NEW ENGLAND ROLLER & SUPPLY
P.O. Box 348
Millbury, MA 01527 USA
Finished Goods Manufacturer: Elastomer; Cast
Other: Distributor
Markets: Printing

NEW ENGLAND URETHANE, INC.
105 Sackett Point Rd.
North Haven, CT 06473 USA
Raw Material Supplier/Manufacturer: Resins
Finished Goods Manufacturer: Elastomer; TPU
Other: Consultant; Testing; Custom Compounder
Markets: Appliances; Electronic/Electrical; Medical;
　Computer Related Components

NEW PROCESS FIBRE CO.
100 New Process Dr.
Greenwood, DE 19950 USA
Finished Goods Manufacturer: Elastomer; TPU;
Other: Fabricator
Markets: Aerospace; Appliances; Automotive; Defense/
　Military; Electronic/Electrical; Furniture; Industrial;
　Marine; Medical; Mining/Drilling; Pipe/Plumbing;
　Recreational; Refrigeration; Transportation

NIANTIC RUBBER CO.
International Supply Group
144 Kenwood St.
Cranston, RI 02907 USA
Finished Goods Manufacturer: Elastomer; TPU; Cast
Markets: Aerospace; Construction; Defense/Military;
　Electronic/Electrical; Industrial; Paper; Food

NO SAG PRODUCTS CORP./LEAR SIGLER
Foam Division
1750 W. Downs Dr.
West Chicago, IL 60185 USA
Finished Goods Manufacturer: Flexible Foam
Markets: Automotive; Furniture

NOK, INC.
47690 E. Anchor Ct.
Plymouth, MI 48170 USA
Systems Manufacturer: Elastomer Systems; Rigid Foam
　Systems
Finished Goods Manufacturer: Elastomer; RIM;
　Automotive
Other: Consultant; Testing; Research Organization; Mfg.
　Rep.
Markets: Aerospace; Appliances; Automotive; Electronic/
　Electrical; Industrial; Marine; Material Handling;
　Pipe/Plumbing; Recreational; Transportation

NOR-LAKE, INC.
P.O. Box 248
Hudson, WI 54016 USA
Raw Material Supplier/Manufacturer: Blowing Agent
Finished Goods Manufacturer: Rigid Foam
Markets: Refrigeration

NOREN PRODUCTS, INC.
1010 O'Brien Dr.
Menlo Park, CA 94025 USA
Equipment or Machinery Supplier/Manufacturer:
　Thermal Pin(Tm) Heat Conductors

NORTECH ENGINEERING ASSOCIATES., INC.
1701 Cannon Lane
Northfield, MN 55057 USA

Equipment or Machinery Supplier/Manufacturer:
Processing Machines

NORTH AMERICAN FOAM & PACKAGING, INC.
1859 Floradale Ave.
South El Monte, CA 91733 USA
Finished Goods Manufacturer: Flexible Foam
Markets: Furniture

NC FOAM INDUSTRIES, INC.
P.O. Box 1528
Mt. Airy, NC 27030 USA
Systems Manufacturer: Elastomer Systems; Rigid Foam
Systems, Flexible Foam Systems
Equipment or Machinery Supplier/Manufacturer:
Dispensing Equipment; Pumps; Automated Controls
Finished Goods Manufacturer: Elastomer; Non-
automotive; Flexible Foam; Integral Skin; Coating
Markets: Aerospace; Carpet; Construction;
Defense/Military; Electronic/Electrical; Furniture;
Industrial; Marine; Medical; Packaging; Recreational;
Refrigeration; Transportation

NORTH COAST COMPOUNDERS, INC.
4935 Mills Industrial Parkway
North Ridgeville, OH 44039 USA
Other: Compounder

NORTHERN INDUSTRIES
429 Tiogue Ave.
Coventry, RI 02816 USA
Systems Manufacturer: Moisture Cured Coatings
Finished Goods Manufacturer: Coating
Markets: Carpet; Construction; Coatings (hardwood
floors, cabinets, etc.)

NORTHERN INSULATION PRODUCTS
A Division of Lloyd Refrigeration, Inc.
414 E. 13th St.
Gibbon, MN 55335 USA
Other: Distributor; Contractor
Markets: Construction; Industrial; Lumber; Marine;
Recreational; Refrigeration

NORTHERN INSULATION, INC.
Box 496
202 N. Cedar
Tower, MN 55790 USA
Other: Consultant; Contractor
Markets: Construction; Industrial; Marine;
Mining/Drilling; Pipe/Plumbing; Refrigeration

NORTHROP CORPORATION
Aircraft Division
1 Northrop Ave.
Hawthorne, CA 90250 USA
Finished Goods Manufacturer: Elastomer; TPU
Markets: Defense/Military

NORTHWEST COATINGS CORPORATION
7221 South 10th St.
Oak Creek, WI 53154 USA
Finished Goods Manufacturer: Coating; Adhesive
Markets: Packaging; Printing

NORTON COMPANY
Coated Abrasives Division
P.O. Box 808
Troy, NY 12181 USA

NORTON PERFORMANCE PLASTICS CORP.
150 Dey Rd.
Wayne, NJ 07470 USA
Systems Manufacturer: Elastomer Systems; Flexible Foam
Systems; Gasket-In-Place
Finished Goods Manufacturer: Elastomer; Cast; Flexible
Foam; Adhesive; Sealant; Mounting Tapes
Markets: Aerospace; Appliances; Automotive;
Construction; Electronic/Electrical; Footwear;
Furniture; Industrial; Material Handling; Medical;
Refrigeration; Transportation

NORWALD MATTRESS CO.
145 Cedar St.
South Norwalk, CT 06854 USA
Other: Fabricator
Markets: Bedding

NORWOOD INDUSTRIES
57 Morehall Rd.
Frazer, PA 19355 USA
Finished Goods Manufacturer: Cast; Flexible Foam; Rigid
Foam
Markets: Aerospace; Appliances; Automotive;
Construction; Defense/Military; Electronic/Electrical;
Footwear; Furniture; Industrial; Marine; Medical;
Packaging; Pipe/Plumbing; Printing; Recreational;
Refrigeration; Transportation

NOSTER RUBBER COMPANY INC.
1481 Township Rd. 229
Van Buren, OH 45889 USA
Finished Goods Manufacturer: Elastomer
Markets: Automotive; Electronic/Electrical; Industrial;
Material Handling; Printing

NOVEX, INC.
258 S. Main St.
Wadsworth, OH 44281 USA
Finished Goods Manufacturer: Elastomers; Cast
Markets: Aerospace; Appliances; Automotive; Chemical;
Footwear; Furniture; Industrial; Lumber; Material
Handling; Paper

NRCA (NATIONAL ROOF CONTRACTORS ASSOC.)
10255 W. Higgins Rd.
Rosemont, IL 60018-5607 USA

NRG BARRIERS, INC.
15 Lund Rd.
Saco, ME 04072 USA
Finished Goods Manufacturer: Rigid Foam
Markets: Construction

NTD AMERICAN, INC.
764 Thomas Dirve
Bensonville, IL 60106 USA
Finished Goods Manufacturer: Flexible Foam
Markets: Packaging

NU-WOOD/GR PLASTICS, INC.
Goshen Rubber
P.O. Box 489
Goshen, IN 45627-0489 USA
Finished Goods Manufacturer: Rigid Foam
Markets: Construction; Lumber

NYCO(TM) MINERALS, INC.
124 Mountain View Dr.
Willsboro, NY 12996 USA
Raw Material Supplier/Manufacturer: Wollastonite and
   surface modified wollastonite
Markets: Aerospace; Appliances; Automotive;
   Construction; Defense/Military; Electronic/ Electrical

O'BRIEN ASSOCIATES, INC.
318 Commerce Dr., No.7
Easton, MD 21601-9106 USA
Systems Manufacturer: Elastomer Systems; Rigid Foam
   Systems; Flexible Foam Systems
Other: Distributor; Mfg Rep

OAK RIDGE NATIONAL LABORATORY
Metals and Ceramics Division
P.O. Box 2008
Bldg. 4508, MS 6092
Oak Ridge, TN 37831 USA
Other: Research Organization; Testing

OCCIDENTAL CHEMICAL CORPORATION
Basic Chemicals Group
Occidental Tower
5005 LBJ Freeway
Dallas, TX 75244 USA
Raw Material Supplier/Manufacturer: Blowing Agent;
   Resins; Clean-up Solvents, Adhesive Solvents, Flame
   Retardants
Markets: Aerospace; Appliances; Automotive; Carpet;
   Chemical; Clothing; Construction; Defense/Military;
   Electronic/Electrical; Footwear; Furniture; Industrial;
   Lumber; Marine; Material Handling; Medical;
   Mining/Drilling; Packaging; Paper; Pipe/Plumbing;
   Printing; Recreational; Refrigeration; Transportation;
   Wheels/Tires

OCEAN EDGE
7992 Miramar Rd.
San Diego, CA 92126 USA
Finished Goods Manufacturer: TPU
Markets: Marine; Recreational

OH DECORATIVE PRODUCTS, INC.
220 S. Elizabeth St.
Spencerville, OH 45887 USA

OH FOAM CORPORATION
1000 S. Kibler St.
New Washington, OH 44854 USA
Finished Goods Manufacturer: Flexible Foam
Other: Fabricator
Markets: Automotive; Furniture; Transportation;
   Bedding

OHNCO, INC.
200 Old Pond Rd.
Suite 105
Bridgeville, PA 15017 USA
Other: Consultant; Buyer/Supplier of Surplus

**OIL STATES INDUSTRIES LTV ENERGY
PRODUCTS CO.**
P.O. Box 670
Arlington, TX 76004-0670 USA
Finished Goods Manufacturer: Elastomer; Cast
Markets: Aerospace; Automotive; Construction; Defense/
   Military; Electronci/Electrical; Marine;
   Mining/Drilling

OLEA INTERNATIONAL
4020 W. Chandler Ave.
Santa Ana, CA 92704 USA
Finished Goods Manufacturer: Flexible Foam; Rigid Foam
Markets: Aerospace; Automotive; Marine; Medical;
   Packaging

OLIN CORPORATION
Performance Urethanes
120 Long Ridge Rd.
Stamford, CT 06904 USA
Raw Material Supplier/Manufacturer: Isocyanate; Polyol;
   Surfactant; Prepolymer
Markets: Automotive; Chemical; Construction; Coatings

OLYMPIC PRODUCTS CO.
4100 Pleasant Garden Rd.
Greensboro, NC 27406 USA
Systems Manufacturer: Flexible Foam Systems
Finished Goods Manufacturer: Flexible Foam
Markets: Automotive; Carpet; Furniture; Medical;
   Packaging; Transportation

OMEGA RUBBER PRODUCTS
P.O. Box 11555
Fort Wayne, IN 46859 USA
Finished Goods Manufacturer: Elastomer; Cast; Flexible
   Foam

OMNI PLASTICS, INC.
2961 Ariens Dr.
P.O. Box 330
Hebron, KY 41048 USA
Finished Goods Manufacturer: Elastomer; Cast; Flexible
   Foam; Coating
Markets: Aerospace; Automotive; Defense/Military;
   Industrial; Medical; Mining/Drilling; Paper; Printing;
   Food

OMNI TECHNOLOGIES, INC.
2961 Ariens Dr.
P.O. Box 330
Hebron, KY 41048 USA
Finished Goods Manufacturer: Elastomer; Cast
Markets: Aerospace; Appliances; Automotive; Industrial;
   Mining/Drilling; Packaging; Paper; Printing;
   Recreational; Business Machines; Skate Wheels; Food

OMNIFOAM INC.
3200 NW. 110th St.
Miami, FL 33167 USA

ONO INDUSTRIES
P.O. Box 150
Ono, PA 17077 USA
Finished Goods Manufacturer: Elastomer; TPU
Markets: Aerospace; Construction; Recreational;
Refrigeration

OPTIBELT CORPORATION
1120 W. National Ave.
Addison, IL 60101 USA
Finished Goods Manufacturer: Elastomer; TPU
Other: Distributor
Markets: Industrial

OSI SPECIALITIES, INC.
39 Old Ridgebury Rd.
Danbury, CT 06817 USA
Raw Material Supplier/Manufacturer: Catalyst;
Surfactant; Sert(TM) Mold Release System; Geolite
(TM) Modifiers
Markets: Chemical

OTTAWA RUBBER CO.
6553 Angola Rd.
Holland, OH 43528 USA
Finished Goods Manufacturer: Elastomer; Rigid Foam
Other: Fabricator
Markets: Industrial

OXID, INC.
101 Concrete St.
Houston, TX 77012 USA
Raw Material Supplier/Manufacturer: Polyol
Markets: Appliances; Chemical; Construction; Industrial;
Refrigeration

P.A.T. PRODUCTS, INC.
44 Central St.
Bangor, ME 04401 USA
Raw Material Supplier/Manufacturer: Isocyanate; Polyol;
Blowing Agent; Resins; Thermoplastic Urethanes
Systems Manufacturer: Elastomer Systems; Polyether
Systems
Other: Distributor; Mfg Rep
Markets: Automotive; Chemical; Footwear; Furniture;
Industrial; Mining/Drilling; Wheels/Tires

PAC FOAM PRODUCTS CORP.
1561 W. MacArthur Blvd.
Costa Mesa, CA 92626 USA

PACIFIC COAST BUILDING PRODUCTS
1735 24th St.
Box 24331
Oakland, CA 94623 USA
Systems Manufacturer: Elastomer Systems; Rigid Foam
Systems; Flexible Foam Systems

PACIFIC COMBINING CORPORATION
3055 E. Fruitland Ave.

Los Angeles, CA 90058 USA
Finished Goods Manufacturer: Elastomer; Automotive;
Non-automotive; TPU; Reinforced Polyurethane
Laminates
Markets: Aerospace; Clothing; Construction;
Defense/Military; Industrial; Marine; Recreational

PACIFIC MECHANICAL SUPPLY
Industrial Plastics of Pacific Mechanical Supply
13710 Milroy Place
Santa Fe Springs, CA 90670 USA
Other: Distributor
Markets: Chemical; Industrial; Waste

PACIFIC STATES FELT AND MANUFACTURING
23869 Clawiter Rd.
Hayward, CA 94544 USA
Finished Goods Manufacturer: Elastomer; TPU; Flexible
Foam; Rigid Foam
Markets: Aerospace; Electronic/Electrical; Industrial

PACKAGING FOAM FABRICATORS, INC.
1749 East Trafficway
Springfield, MO 65802 USA
Finished Goods Manufacturer: Flexible Foam
Other: Fabricator
Markets: Electronic/Electrical

PACKAGING TECHNOLOGY, INC.
183 Pickering Way
Lionville, PA 19353 USA
Finished Goods Manufacturer: Flexible Foam
Markets: Defense/Military; Electronic/Electrical;
Packaging

PACKATEERS, INC.
P.O. Box 204
Edgemont, PA 19028 USA
Finished Goods Manufacturer: Flexible Foam
Other: Fabricator

PAGE BELTING CO.
26 Commercial St.
P.O. Box 482
Concord, NH 03301 USA
Finished Goods Manufacturer: Elastomer; Cast; Flexible
Foam; Integral Skin
Markets: Automotive; Industrial; Lumber; Material
Handling; Paper; Printing; Recreational;
Transportation; Wheels/Tires; Communications;
Textile; Metal Working; Abrasives/Abrasive Forming

PALMER DAVIS SEIKA, INC.
20 W. Vanderventer
P.O. Box 222
Port Washington, NY 11050 USA
Raw Material Supplier/Manufacturer: Curatives;
Intermediates; Release Agents; Chemical Specialties
Other: Distributor

PANEK COATINGS
SPFCD
13998 County House Rd.
Albion, NY 14411 USA

Other: Fabricator; Contractor
Markets: Construction; Refrigeration

PAR-FOAM PRODUCTS
239 Van Rensselaer St.
Buffalo, NY 14210 USA
Finished Goods Manufacturer: Elastomer; Flexible Foam; Rigid Foam
Markets: Automotive; Industrial

PARKER HANNIFIN CORPORATION
17325 Euclid Ave.
Cleveland, OH 44112 USA
Equipment or Machinery Supplier/Manufacturer: Pumps
Markets: Aerospace; Automotive; Construction; Defense/Military; Industrial; Marine; Material Handling; Medical; Mining/Drilling; Refrigeration; Transportation

PARWAY PRODUCTS, INC.
10293 Burlington Rd.
Cincinnati, OH 45231 USA
Finished Goods Manufacturer: Elastomer; Cast; Microcellular
Markets: Aerospace; Appliances; Automotive; Industrial; Mining/Drilling; Paper; Recreational; Wheels/Tires

PASSAIC ADHESIVE & CHEMICAL CO., INC.
210 Delawanna Ave.
P.O. Box 982
Clifton, NJ 07014-0982 USA
Finished Goods Manufacturer: Coating; Adhesive
Markets: Automotive; Construction; Footwear; Furniture

PASSAIC RUBBER CO.
45 Demarest Dr.
Wayne, NJ 07474-0505 USA
Systems Manufacturer: Elastomer Systems
Markets: Industrial; Transportation

PATCO CORPORATION
51 Ballou Blvd.
Bristol, RI 02885 USA
Finished Goods Manufacturer: Elastomer
Markets: Automotive; Industrial

PATTERSON MACHINE COMPANY
2 Van Dyke St.
Brooklyn, NY 11231 USA
Finished Goods Manufacturer: Elastomer; TPU
Markets: Marine

PAWLING CORP.
157 Charles Colman Blvd.
Pawling, NY 12564 USA
Finished Goods Manufacturer: Elastomer; Cast
Markets: Aerospace; Appliances; Automotive; Construction; Electronic/Electrical

PAXAR CORPORATION
530 Route 3
Orangeburg, NY 10962 USA
Raw Material Supplier/Manufacturer: Polymer
Finished Goods Manufacturer: Coating

Markets: Clothing

PB & S CHEMICAL CO.
Highway 136
W. Henderson, KY 42420 USA
Other: Distributor

PCO
2514 W. Arnold St.
Marshfield, WI 54449 USA
Finished Goods Manufacturer: Elastomer; Cast
Markets: Appliances; Automotive; Construction; Defense/Military; Electronic/Electrical; Material Handling; Mining/Drilling; Paper; Printing; Recreational; Transportation

PEACHTREE DOORS, INC.
P.O. Box 5700
Norcross, GA 30091 USA
Finished Goods Manufacturer: Rigid Foam
Markets: Construction

PECTIN CHEMICALS, INC.
1 Shell Plaza
P.O. Box 4407
Houston, TX 77001 USA
Finished Goods Manufacturer: Elastomer

PELMOR LABORATORIES, INC.
401 Lafayette St.
Newtown, PA 18940 USA
Finished Goods Manufacturer: Elastomer; Cast
Markets: Aerospace; Electronic/Electrical; Industrial; Mining/Drilling

PELRON CORPORATION
7847 W. 47th St.
Box 6
Lyons, IL 60534 USA
Raw Material Supplier/Manufacturer: Polyol; Catalyst; Surfactant; Resins

PENN COLOR
400 Old Dublin Pike
Doylestown, PA 18901 USA
Raw Material Supplier/Manufacturer: Pigment Dispersions; Colorants
Markets: Automotive; Chemical; Construction; Furniture; Industrial; Printing; Transportation

PENN FOAM CORP.
2625 Mitchell Ave.
Allentown, PA 18103 USA
Finished Goods Manufacturer: Flexible Foam
Other: Fabricator
Markets: Aerospace; Appliances; Carpet; Clothing; Construction; Electronic/Electrical; Furniture; Industrial; Medical; Packaging; Recreational

PERIPHERAL PRODUCTS, INC.
338 W. Main St.
Tilton, NH 03276 USA
Finished Goods Manufacturer: Elastomer; RIM; Automotive; Non-automotive; Cast

Markets: Appliances; Automotive; Construction;
    Industrial;

PERIPHLEX USA LTD.
P.O. Box 568
Mooresville, NC 28115 USA
Equipment or Machinery Supplier/Manufacturer:
    Automated Controls
Markets: Automotive; Carpet; Clothing; Footwear;
    Furniture; Material Handling; Recreational

PERMA-FLEX MOLD CO., INC., THE
1919 E. Livingston Ave.
Columbus, OH 43209 USA
Systems Manufacturer: Elastomer Systems
Markets: Aerospace; Defense/Military;
    Electronic/Electrical; Footwear; Industrial; Flexible
    Elastomers; Tool and Pattern Shops

PERMA-FLEX ROLLERS, INC.
375 Bellevue Rd.
Newark, DE 19713 USA
Finished Goods Manufacturer: Elastomer; Cast
Markets: Clothing; Industrial; Mining/Drilling;
    Packaging; Paper; Printing; Wheels/Tires

PERMA-FOAM INC.
605 S. 21st St.
Irvington, NJ 07111 USA
Finished Goods Manufacturer: Elastomer; Flexible Foam;
Rigid Foam
Other: Distributor; Fabricator
Markets: Aerospace; Defense/Military;
    Electronic/Electrical; Medical; Packaging

PERMABOND
211 W. Sylvania Ave.
Neptune City, NJ 07753 USA
Finished Goods Manufacturer: Elastomer; Cast
Markets: Pet Supplies

PERMATHANE, INC.
Maine Rubber International
50 Eisenhower Dr.
Westbrook, ME 04092 USA
Finished Goods Manufacturer: Elastomer; Cast; Coating
Markets: Aerospace; Appliances; Automotive;
    Construction; Electronic/Electrical; Footwear;
    Furniture; Packaging; Paper; Printing; Wheels/Tires;
    Bedding

PERMUTHANE BUSINESS OF STAHL USA
Zeneca, Inc.
13 Corwin St.
Peabody, MA 01960 USA
Raw Material Supplier/Manufacturer: Polyol
Finished Goods Manufacturer: Coating
Markets: Automotive; Footwear; Furniture;
    Transportation

PERRY CHEMICAL & MANUFACTURING CO.
2335 S. 30th St.
P.O. Box 6419
Lafayette, IN 47903-6419 USA

Finished Goods Manufacturer: Elastomer; Cast; Flexible
    Foam; Rigid Foam
Markets: Aerospace; Automotive; Construction; Defense/
    Military; Industrial

PERSHING RUBBER ROLLER CORP.
135 Orange St.
Bloomfield, NJ 07003 USA
Finished Goods Manufacturer: Elastomer; Cast

PETCO, INC.
28041 N. Bradley Rd.
Lake Forest, IL 60045 USA
Finished Goods Manufacturer: Elastomer; Cast
Markets: Appliances; Automotive; Packaging; Paper;
    Printing; Wheels/Tires

PETRO EXTRUSION TECHNOLOGIES, INC.
490 South Ave.
Garwood, NJ 07027 USA
Finished Goods Manufacturer: Extruded Urethane Tubing
Markets: Aerospace; Appliances; Automotive; Chemical;
    Construction; Electronic/Electrical; Industrial;
    Marine; Medical; Packaging; Pipe/Plumbing;
    Transportation

PHILLIP TOWNSEND ASSOCIATES, INC.
P.O. Box 90327
Houston, TX 77290 USA
Other: Consultant
Markets: Research and consulting in all fields concerning
    PU and other chemicals and plastics

PHOENIX BUILDING PRODUCTS
529 E. Juanita 8
Mesa, AZ 85204 USA
Finished Goods Manufacturer: Cast; Rigid Foam
Markets: Construction; Furniture

PHOENIX FOAM & FIBERGLASS, INC.
111 E. 5th St.
North Bend, NE 68649 USA
Finished Goods Manufacturer: Elastomer; RIM; Flexible
    Foam
Other: Fabricator

PICKER PARTS, INC.
7575 E. Manning Ave.
P.O. Box 307
Fowler, CA 93625 USA
Finished Goods Manufacturer: Elastomer; Cast
Markets: Paper; Wheels/Tires; Agricultural

PIERCE & STEVENS CORP.
1710 Ohio St.
Buffalo, NY 14240 USA
Finished Goods Manufacturer: Coating; Adhesive
Markets: Industrial; Packaging

PILLOWTEX CORPORATION
4111 Mint Way
Dallas, TX 75237 USA
Other: Fabricator
Markets: Bedding

PIMA
1001 PA Ave. NW
5th Floor
Washington, DC 20005 USA

PINDER POLYURETHANE & PLASTICS
481 E. 151st St.
P.O. Box 433
East Chicago, IN 46312 USA
Finished Goods Manufacturer: Elastomer; Cast
Markets: Industrial; Mining/Drilling; Paper;
  Wheels/Tires

PIONEER PLASTICS CORPORATION
One Pionite Rd.
Auburn, ME 04211 USA
Raw Material Supplier/Manufacturer: Resins

PIONEER ROOFING, INC.
Polyurethane Division
151 Maple St.
Johnson Creek, WI 53038 USA
Other: Contractor
Markets: Construction

PIONEER WEST
1401 Walnut St.
Suite 104
Boulder, CO 80302 USA
Other: Business Services

PIPELINE PIGGING PRODUCTS
P.O. Box 692005-300
Houston, TX 77269 USA
Finished Goods Manufacturer: Flexible Foam
Markets: Chemical; Industrial; Material Handling;
  Mining/Drilling; Pipe/Plumbing; Transportation

PLABELL RUBBER PRODUCTS CORPORATION
P.O. Box 1008 M.S.
300 S. St. Clair St.
Toledo, OH 43697 USA
Finished Goods Manufacturer: Elastomer; Cast
Markets: Industrial; Material Handling; Paper;
  Wheels/Tires; Glass Industry

PLAN TECH, INC.
Shaker Rd., RD No. 8
Box 3945
Loudon, NH 03301 USA
Finished Goods Manufacturer: Elastomer; Cast
Other: Fabricator
Markets: Aerospace; Appliances; Automotive; Defense/
  Military; Electronic/Electrical; Industrial; Marine;
  Material Handling; Mining/Drilling; Packaging;
  Paper; Printing; Recreational; Transportation;
  Wheels/Tires; Food Processing

PLASTEK
3 Perkins Way
Newburyport, MA 01950 USA
Finished Goods Manufacturer: Elastomer; RIM

PLASIT-CHEM INC.
2980 Falco Dr.
Madera, CA 93637 USA
Finished Goods Manufacturer: Elastomer

PLASTIC COATINGS CORPORATION
P.O. Box 1068
St. Albans, WV 25177 USA
Raw Material Supplier/Manufacturer: Protective
  Coatings-Fibered Acrylic Latex Based
Systems Manufacturer: Elastomer Systems
Other: Manufacturer
Markets: Roofing; Insulation

PLASTIC ENGINEERING INC.
Plastic Engineered Components
1821 Vanderbilt Rd.
Kalamazoo, MI 49002 USA
Finished Goods Manufacturer: Elastomer; TPU
Markets: Medical

PLASTIC MANUFACTURING
Harrisburg, NC 28075-0579 USA
Finished Goods Manufacturer: Elastomer; TPU
Other: Fabricator
Markets: Aerospace-Construction-Electronic/Electrical;
  Industrial; Packaging; Recreational; Transportation

PLASTIC MATERIALS, INC.
1561 2nd East First St.
Irwindale, CA 91706 USA
Raw Material Supplier/Manufacturer: Isocyanate; Polyol
Systems Manufacturer: Elastomer Systems; Rigid Foam
  Systems
Equipment or Machinery Supplier/Manufacturer:
  Dispensing Equipment
Other: Distributor
Markets: Aerospace; Automotive; Marine; Recreational;
  Transportation

PLASTIC-CRAFT PRODUCTS
164 W. Nyack Rd.
West Nyack, NY 10994 USA
Other: Distributor; Fabricator
Markets: Industrial

PLASTICLAD CORPORATION
840 Mason St.
P.O. Box 775
Springfield, OH 45501 USA
Finished Goods Manufacturer: Elastomer; TPU
Markets: Appliances; Automotive; Defense/Military;
  Industrial; Marine; Material Handling; Medical;
  Mining/Drilling; Packaging; Paper; Pipe/Plumbing;
  Recreational; Refrigeration; Transportation

PLASTICOID CO.
249 W. High St.
Elkton, MD 21921 USA
Finished Goods Manufacturer: Elastomer; TPU
Markets: Defense/Military; Medical

PLASTICS AND CHEMICALS, INC.
Box 306

Cedar Grove, NJ 07009 USA
Raw Material Supplier/Manufacturer: Polyol; Surfactant;
Blowing Agent; Prepolymer

PLASTICS UNLIMITED
80 Winters St.
Worcester, MA 01613 USA
Other: Distributor
Markets: Aerospace; Automotive; Construction;
Defense/Military; Marine; Medical; Recreational;
Refrigeration; Transportation

PLASTOMER CORPORATION
37819 Schoolcraft Rd.
Livonia, MI 48150-1096 USA
Finished Goods Manufacturer: Flexible Foam; Die Cut
Gaskets and Sound Insulators (with or without
pressure sensitive adhesive backing)
Other: Fabricator
Markets: Appliances; Automotive; Carpet;
Electronic/Electrical; Industrial; Recreational;
Refrigeration

PLASTOMERIC INC.
21300 Doral Ave.
Waukesha, WI 53187 USA
Systems Manufacturer: Elastomer Systems; Rigid Foam
Systems; Felxible Foam Systems
Markets: Automotive; Electronic/Electrical; Footwear;
Furniture; Industrial; Marine; Packaging;
Recreational; Transportation; Wheels/Tires

PLEIGER PLASTICS COMPANY
Crile Rd.
Washington, PA 15301 USA
Finished Good Manufacturer: Elastomer; TPU; Flexible
Foam; Rigid Rd.
Markets: Automotive; Construction;
Electronic/Electrical; Industrial; Mining/Drilling;
Printing; Transportation; Wheels/Tires

PLYFOAM PRODUCTS
815 Eisenhower Dr. S.
Goshen, IN 46526 USA
Finished Goods Manufacturer: Flexible Foam
Markets: Automotive; Medical; Packaging; Recreational

PLYMOUTH FOAM PRODUCTS
1800 Sunset Dr.
Plymouth, WI 53073 USA
Other: Fabricator
Markets: Packaging

POLLY PIG BY KNAPP, INC.
1209 Hardy St.
Houston, TX 77020 USA
Finished Goods Manufacturer: Elastomer; Cast; Flexible
Foam; Coating
Markets: Mining/Drilling; Food

POLLY PRODUCTS
49 Uxbridge Rd.
Mendon, MA 01756 USA
Finished Good Manufacturer: Flexible Foam; Rigid Foam

Other: Fabricator
Markets: Appliances; Automotive; Furniture; Industrial

POLY FLEX, INC.
19660 Eight Mile Rd.
Southfield, MS 48075 USA
Finished Goods Manufacturer: Elastomer; TPU; Cast,
Coating; Microcellular
Other: Distributor; Fabricator
Markets: Aerospace, Appliances; Automotive; Industrial;
Material Handling; Mining/Drilling; Paper; Printing;
Packaging; Recreational; Food; Business Machines

POLY FOAM, INC.
114 Pine St.
Lester Prairie, MN 55354 USA
Finished Goods Manufacturer: Flexible Foam
Other: Distributor
Markets: Packaging

POLY-FOAM, INC.
AFM Affiliate 116 Pine St., South
Lester Prairie, MN 55354 USA
Finished Goods Manufacturer: Flexible Foam
Markets: Construction; Marine; Packaging

POLYAD COMPANY
13 Rose Terrace
Barrington, IL 60010-1320 USA
Other: Consultant; Mfg Rep

POLYCASTER TECHNOLOGIES
2 Brooks Ave.
P.O. Box 275
Willow St., PA 17584 USA
Finished Goods Manufacturer: Elastomer; Cast
Markets: Appliances; Industrial; Mining/Drilling; Paper;
Printing; Wheels/Tires; Business Machines

POLYCHEM CORP.
20 Fifth Ave.
Cranston, NJ 02910 USA
Finished Goods Manufacturer: Coating; UV Curable
Coatings
Markets: Automotive; Medical; Jewelry

POLYCRAFT PRODUCTS, INC.
551 1 State Route 128
Cleves, OH 45002 USA
Finished Goods Manufacturer: Elastomer; Cast

POLYDRIVE INDUSTRIES
1537-HE. McFadden Ave.
Santa Ana, CA 92705 USA
Finished Goods Manufacturer: Elastomer; Cast
Markets: Electronic/Electrical; Industrial; Medical;
Printing

POLYDYNE INTERNATIONAL
1606 Pearl St.
Waukesha, WI 53186 USA
Systems Manufacturer: Elastomer Systems; Rigid Foam
Systems

Markets: Construction; Marine; Mining/Drilling; Refrigeration

POLYFOAM PACKERS CORPORATION

2320 Foster Ave.

Wheeling, IL 60090 USA

Finished Goods Manufacturer: Flexible Foam

Markets: Aerospace; Appliances; Automotive; Construction; Defense/Military; Electronic/Electrical; Industrial; Material Handling; Medical; Packaging; Pipe/Plumbing; Refrigeration

POLY FOAM PRODUCTS

P.O. Box 1132

Spring, TX 77383-8888 USA

Systems Manufacturer: Rigid Foam Systems; Flexible Foam Systems

Markets: Aerospace; Appliances; Construction; Industrial; Pipe/Plumbing; Refrigeration; Transportation

POLYMER COMPOSITES, INC.

4610 Theurer Blvd.

Winona, MN 55987 USA

Finished Goods Manufacturer: Elastomer; TPU

Markets: Automotive; Industrial; Material Handling

POLYMER DESIGN CORPORATION

180 Pleasant St.

Rockland, MA 02370 USA

Finished Goods Manufacturer: Elastomer; Cast

Markets: Aerospace; Defense Military; Electronic/Electrical; Marine; Medical; Mining/Drilling; Packaging

POLYMER DEVELOPMENT LABORATORIES, INC.

212 W. Taft Ave.

Orange, CA 92665 USA

Systems Manufacturer: Elastomer Systems; Rigid Foam Systems; Flexible Foam Systems; Hybrids

Finished Goods Manufacturer: Elastomer; RIM; Non-automotive

Markets: Aerospace; Construction; Marine; Medical; Packaging; Pipe/Plumbing; Recreational; Refrigeration

POLYMER DYNAMICS, INC.

2200 South 12th St.

P.O. Box 4400

Allentown, PA 181054400 USA

POLYMER ENTERPRISES, INC.

1 Northgate Square

Greensburg, PA 15601 USA

Finished Goods Manufacturer: Elastomer; TPU; Cast; Coating

Markets: Automotive; Transportation

POLYMER INSTITUTE/POLYMER

University of Detroit/Mercy

4001 W. McNichols Rd.

P.O. Box 19900

Detroit, MI 48219-0900 USA

Other: Consultant; Testing; Research Organization; Training and Education

Markets: Aerospace; Appliances; Automotive; Carpet; Chemical; Clothing; Construction; Defense/Military; Electronic/Electrical; Footwear; Furniture; Industrial; Lumber; Marine; Medical; Packaging; Paper; Printing; Recreational; Refrigeration; Transportation; Wheels/Tires

POLYMER PLASTICS, INC.

Roofing Division

65 Davids Dr.

Hauppauge, NY 11788 USA

Systems Manufacturer: Prepolymers Manufacturer; Elastomer Systems

Finished Goods Manufacturer: Elastomer; Coating; Adhesive; Sealant

Other: Mfg Rep

Markets: Construction; Defense/Military; Recreational

POLYMER PRODUCTS, INC.

2613 Aviation Parkway

Grand Prairies, TX 75051 USA

Finished Goods Manufacturer: Elastomer; Cast

Markets: Mining/Drilling

POLYMER RESOURSES, U.S.A.

2620 S. Parker Rd.

Suite 155

Aurora, CO 80014 USA

Equipment or Machinery Supplier/Manufacturer: Molds; Dispensing Equipment; Tanks; Automated Controls

Finished Goods Manufacturer: Elastomer

Other: Consultant; Mfg. Rep.

Markets: Aerospace; Construction; Defense/Military; Electronic/Electrical; Furniture; Industrial; Material Handling; Mining/Drilling; Recreational; Wheels/Tires

POLYMER TECHNOLOGIES, INC.

3 Laurelwood Dr.

Milford, OH 45150 USA

POLYMER TECHNOLOGY GROUP, INC., THE

4561-A Horton St.

Emeryville, CA 94608 USA

Systems Manufacturer: Prepolymer Manufacturer; Elastomer Systems

Finished Goods Manufacturer: Elastomer; Coating; Adhesive; Specialty Polymers

Other: Consultant; Research Organization; Contract R/D

Markets: Aerospace; Chemical; Clothing; Construction; Defense/Military; Electronic/Electrical; Footwear; Furniture; Industrial; Medical; Packaging; Recreational; Biotechnology

POLYONICS RUBBER CO.

100 E. Park St.

P.O. Box 100

Poplar Grove, IL 61065 USA

Finished Goods Manufacturer: Elastomer; Cast

Markets: Aerospace; Appliances; Automotive;
Construction; Electronic/Electrical

POLYPLASTICS
Buckley Industries
10201 Metropolitan Dr.
Austin, TX 78758 USA
Other: Mfg. Rep.; Fabricator
Markets: Aerospace; Appliances; Automotive;
Construction; Defense/Military; Electronic/Electrical;
Footwear; Furniture; Industrial; Marine; Material
Handling; Medical; Packaging; Recreational

POLYSET CO., INC.
Upper N. Main St.
Mechanicville, NY 12118 USA
Systems Manufacturer: Prepolymer Manufacturer;
Elastomer Systems
Finished Goods Manufacturer: Coating; Adhesive; Sealant
Markets: Appliances; Automotive; Construction;
Defense/ Military; Electronic/Electrical; Industrial;
Packaging

POLYTEK DEVELOPMENT CORPORATION
P.O. Box 384
Lebanon, NJ 08833 USA
Systems Manufacturer: Prepolymer Manufacturer;
Elastomer Systems; Rigid Systems
Equipment or Machinery Supplier/Manufacturer:
Accessory Products
Finished Goods Manufacturer: Adhesive; Molding and
Casting Materials
Markets: Construction; Industrial; Art/Ceramics
Foundry

POLYTHANE SYSTEMS, INC.
P.O. Box 1452
Spring, TX 77383 USA
Systems Manufacturer: Rigid Foam Systems
Markets: Construction

POLYURETHANE CORP. OF AMERICA
624 Schuyler
Lyndhurst, NJ 07071 USA
Systems Manufacturer: Rigid Foam Systems; Flexible
Foam Systems
Markets: Automotive; Construction; Furniture; Marine

POLYURETHANE FOAM ASSOCIATION
P.O. Box 1459
Wayne, NJ 07474-1459 USA
Other: Trade Association
Markets: All Flexible Polyurethane Foam Markets

POLYURETHANE MANUFACTURERS
ASSOCIATION
800 Roosevelt Rd.
Building C #20
Glen Ellyn, IL 60137 USA
Other: Trade Association
Markets: Appliances; Automotive; Chemical;
Construction Industrial; lumber; Mining/Drilling;
Paper; Printing; Recreational; Wheels/Tires

POLYURETHANE SPECIALTIES, INC.
624 Schuyler Ave.
Lyndhurst, NJ 07071 USA
Raw Material Supplier/Manufacturer: Prepolymer
Systems Manufacturer: Rigid Foam Systems
Finished Goods Manufacturer: Coating; Adhesive

POLYURETHANES RECYCLE & RECOVERY
COUNCIL (PURRC)
c/o SPI/Polyurethane Division
355 Lexington Ave.
New York, NY 10017 USA
Other: Recycling Research

POLYWORKS, INC.
Affiliate of Atlantic Thermoplastics Co.
P.O. Box 457
Blackstone, MA 01504 USA
Equipment or Machinery Supplier/Manufacturer: Molds
Other: Consultant; Research Organization
Markets: Appliances; Automotive; Footwear; Furniture;
Industrial; Marine; Medical; Wheels/Tires

PONTIAC PLASTICS & SUPPLY
4260 Giddings Rd.
Auburn Hills, MI 48326 USA
Finished Goods Manufacturer: Elastomer, Cast
Other: Distributor
Markets: Automotive; Marine

PORTAGE CASTING & MOLD, INC.
2901 Portage Rd.
P.O. Box 53
Portage, WI 53901 USA
Equipment or Machinery Supplier/Manufacturer: Molds
Finished Goods Manufacturer: Elastomer
Markets: Automotive; Industrial

POTOMAC RUBBER CO.
9011 Hampton Overlook
Capitol Heights, MD 20743 USA
Finished Goods Manufacturer: Adhesive; Sealant
Other: Distributor
Markets: Aerospace; Construction; Footwear; Marine;
Printing

POW-R-TOW, INC
1207 Station Plaza
Hewlett, NY 11557 USA
Raw Material Supplier/Manufacturer: Polyol; Polymer;
PrePolymer; Resins
Systems Manufacturer: Prepolymer Manufacturer;
Elastomer Systems
Equipment or Machinery Supplier/Manufacturer: Molds;
Dispensing Equipment
Finished Goods Manufacturer: Elastomer; Automotive;
Non-automotive; TPU; Cast; Flexible Foam; Integral
Skin
Other: Consultant; Research Organization; Distributor;
Fabricator
Markets: Appliances; Automotive; Carpet; Clothing;
Construction; Electronic/Electrical; Footwear;

Furniture; Industrial; Medical; Mining/Drilling;
Packaging; Printing; Recreational; Wheels/Tires;
Screen Printing

PRECISE POLYMERS
3550 Silicia Rd.
Sylvania, OH 43560 USA
Finished Goods Manufacturer: Elastomer; TPU; Coating
Markets: Printing; Steel

PRECISION PLASTIC BALL
3002 N. Cicero Ave.
Chicago, IL 60641 USA
Finished Goods Manufacturer: Elastomer-Cast
Markets: Industrial; Material Handling

PRECISION URETHANE AND MACHINE
612 Third St.
Hempstead, TX 77445 USA
Finished Goods Manufacturer: Elastomer; Cast
Markets: Furniture; Mining/Drilling

PREFERRED FOAM PRODUCTS, INC.
P.O. Box 1005
27 Ciro Dr.
North Branford, CT 06471 USA
Systems Manufacturer: Rigid Foam Systems
Equipment or Machinery Supplier/Manufacturer:
Dispensing Equipment
Other: Distributor
Markets: Appliances; Construction; Furniture; Industrial;
Marine; Mining/Drilling; Packaging; Pipe/Plumbing;
Recreational; Refrigeration; Transportation

PREMIUM POLYMERS, INC.
9721 Highway 290 East
P.O. Box 141159
Austin, TX 78714-1159 USA
Systems Manufacturer: Elastomer Systems; Rigid Foam
Systems; Acrylic Coatings; Silicone Coatings; Polyurea
Coatings; Urethane Coatings
Equipment or Machinery Supplier/Manufacturer:
Dispensing Equipment
Markets: Construction; Refrigeration

PRINCE MASTERCRAFT, INC.
400 Brown Ave.
Syracuse, NY 13208 USA
Raw Material Supplier/Manufacturer: Resins
Equipment or Machinery Supplier/Manufacturer: Molds
Finished Goods Manufacturer: Elastomer; Non-
automotive; Flexible Foam; Integral Skin; Coating;
Rigid Foam
Other: Distributor; Fabricator
Markets: Clothing; Defense/Military;
Electronic/Electrical; Industrial; Marine; Medical;
Wheels/Tires

PRINCE RUBBER & PLASTICS CO., INC.
137 Arthur St.
Buffalo, NY 14207 USA
Finished Goods Manufacturer: Elastomer; TPU

Markets: Appliances; Automotive; Industrial; Paper;
Food

PRODUCTS RESEARCH & CHEMICAL CORP.
5430 San Fernando Rd.
Glendale, CA 91203 USA
Finished Goods Manufacturer: Elastomer; Flexible Foam;
Coating
Markets: Aerospace; Construction; Defense/Military;
Electronic/Electrical; Paper; Printing

PROFESSIONAL ROOFING MAGAZINE
Published by National Roofing Contractors Association
10255 W. Higgins Rd. #600
Rosemont, IL 60018-5607 USA
Other: Association/Trade Publication
Markets: Roofing

PROTECT MANUFACTURING, INC.
1251 Ferguson Ave.
St. Louis, MO 63133 USA
Finished Goods Manufacturer: Flexible Foam
Other: Fabricator

PROTECTIVE COATINGS OF LAKE COUNTY, INC.
P.O. Box 1806
Eustis, FL 32729-1806 USA
Other: Sub-Contractor
Markets: Construction

PSH INDUSTRIES
5346 East Ave.
La Grange, IL 60525 USA
Finished Goods Manufacturer: Elastomer; Cast
Markets: Aerospace; Appliances; Automotive;
Construction; Footwear; Packaging; Printing

PSH INDUSTRIES
5346 East Ave.
LaGrange, IL 60525 USA
Finished Goods Manufacturer: Elastomer; Cast
Markets: Aerospace; Appliances; Automotive;
Construction; Footwear; Packaging; Printing

PSI URETHANES, INC.
10701 Metric Blvd.
Austin, TX 78758 USA
Finished Goods Manufacturer: Elastomer; Cast
Markets: Aerospace; Appliances; Automotive; Chemical;
Defense/Military; Electronic/Electrical; Industrial;
Material Handling; Mining/Drilling; Packaging;
Paper; Printing; Recreational-, Refrigeration;
Transportation; Wheels/Tires

PTM AND WINDUSTRIES, INC.
10640 S. Painter Ave.
Santa Fe Springs, CA 90670 USA
Systems Manufacturer: Elastomer Systems
Finished Goods Manufacturer: Elastomer; Cast; Adhesive
Markets: Aerospace

PUFF, INC.
Route 1, Box 255
Charlottesville, VA 22903 USA

Other: Contractor
Markets: Construction; Industrial

PUMEX U.S.A., INC.
1907 Saddle Creek
Houston, TX 77090 USA
Raw Material Supplier/Manufacturer: Polyol; Resins
Other: Mfg Rep
Markets: Appliances; Automotive; Chemical;
Construction; Footwear; Furniture; Industrial;
Marine; Packaging; Pipe/ Plumbing; Refrigeration;
Transportation

PURE RUBBER PRODUCTS CO., INC.
7 Ray Place
Fairfield, NJ 07006 USA
Finished Goods Manufacturer: Elastomer; Cast
Markets: Appliances; Automotive; Defense/Military;
Electronic/Electrical; Marine; Medical; Paper;
Printing; Agricultural; Food

PURETHANE, INC.
2691 Coolidge Highway
Berkley, MI 48072 USA
Finished Goods Manufacturer: Elastomer; RIM;
Automotive; Non-automotive; Flexible Foam; Integral
Skin
Other: Fabricator
Markets: Appliances; Automotive; Furniture; Industrial;
Marine; Medical; Recreational; Refrigeration;
Transportation

PYRAMID, INC.
522 N. 9th Ave. East
Newton, IA 50208 USA
Finished Goods Manufacturer: Elastomer; TPU
Markets: Automotive; Marine; Transportation

Q.O. CHEMICAL CO.
Great Lakes Chemical, Inc.
2801 Kent Ave.
P.O. Box 2500
West LaFayette, IN 47906 USA
Raw Material Supplier/Manufacturer: Polyol
Markets: Aerospace; Automotive; Clothing;
Defense/Military; Footwear; Medical;
Mining/Drilling; Transportation; Wheels/ Tires;
Spandex; T.P.U.; Thermoset Elastomers

QUALITY RUBBER MANUFACTURING CO., INC.
P.O. Box 709
2510 Industrial Park Rd.
Hendersonville, NC 28793 USA
Finished Goods Manufacturer: Elastomer; Cast
Markets: Aerospace; Appliances; Automotive; Electronic/
Electrical; Marine; Agricultural

QUANTIC LTD.
211 W. 56th St.
Apt. 6-G
New York, NY 10019 USA

R & J COATINGS AND WATERPROOFING, INC.
1725 S. Nova Rd.

S. Daytona, FL 32119 USA
Other: Roofing Contractor/Waterproofing
Markets: Construction

R & R RUBBER MOLDING, INC.
2444 Loma Ave.
South El Monte, CA 91733 USA
Finished Goods Manufacturer: Elastomer; Millable PU
Gums
Markets: Aerospace; Appliances; Medical; Food

R. P. ASSOCIATES, INC.
Minturn Rd.
Bristol, RI 02809 USA
Systems Manufacturer: Elastomer Systems
Other: Mfg Rep

R.B.L. INDUSTRIES
4809 Benson Ave.
Baltimore, MD 21227 USA
Finished Goods Manufacturer: Flexible Foam; Rigid Foam
Other: Fabricator
Markets: Packaging; Displays

R.E. DARLING CO., INC.
3749 N. Romero Rd.
Tucson, AZ 85705 USA
Finished Goods Manufacturer: Elastomer; RIM; Non-
automotive; Cast
Markets: Defense/Military; Industrial

RE PURVIS & ASSOCIATES
7740 W. 78th St.
Bloomington, MN 55439 USA
Finished Goods Manufacturer: Elastomer; Cast
Other: Distributor; Fabricator
Markets: Industrial

R.K. HYDRO-VAC, INC.
1700 Mote Dr.
Covington, OH 45318 USA
Other: Service

RADCURE SPECIALTIES INC.
Parent Company Brussels, Belgium
2000 Lake Park Dr.
Smyrna, GA 30080 USA
Raw Material Supplier/Manufacturer: Catalyst; Resins

RADVA CORP.
301 1st St.
Radford, VA 24143 USA
Finished Goods Manufacturer: Flexible Foam; Rigid Foam
Other: Fabricator
Markets: Packaging

RAHCO RUBBER, INC.
2001 N. Parkside Ave.
Chicago, IL 60639 USA
Finished Goods Manufacturer: Elastomer
Markets: Electronic/Electrical; Industrial; Pipe/Plumbing

RAINFAIR INC.
3600 S. Memorial Dr.
P.O. Box 1647

Racine, WI 53401 USA
Other: Fabricator
Markets: Clothing

RANDALL TEXTRON
10179 Commerce Park
Cincinnati, OH 45246 USA
Finished Goods Manufacturer: Flexible Foam
Markets: Appliances; Automotive

RAPPA, INC.
104 Fieldstone Dr.
LaPorte, IN 46530 USA

RAYAN QUALITY RUBBER CO., INC.
243 Furnace St.
Akron, OH 44304 USA
Finished Goods Manufacturer: Elastomer; Cast
Markets: Appliances; Automotive; Defense/Military;
   Electronic/Electrical; Furniture; Food

RBK TOOL & DIE
1279 N. Emerald Ave.
Modesto, CA 95352 USA
Equipment or Machinery Supplier/Manufacturer: Molds
Markets: Automotive

RED ROCK RUBBER, INC.
102 Clark St.
Pella, IA 50219 USA
Finished Goods Manufacturer: Coating
Markets: Aerospace; Packaging; Printing; Agricultural;
   Food

REDCO
3000 Arrowhead Dr.
Carson City, NV 89706 USA
Finished Goods Manufacturer: Elastomer; Cast; Flexible
   Foam
Other: Manufacturer
Markets: Aerospace; Electronic/Electrical; Industrial;
   Material Handling; Medical; Printing

REED RUBBER PRODUCTS
5425 Manchester Ave.
St. Louis, MO 63110 USA
Finished Goods Manufacturer: TPU
Markets: Appliances; Construction; Industrial; Material
   Handling

REEDY INTERNATIONAL
42 First St.
Keyport, NJ 07735 USA
Raw Material Supplier/Manufacturer: Blowing Agent
Other: Consultant
Markets: Aerospace; Appliances; Automotive; Carpet;
   Chemical; Clothing; Construction; Defense/Military;
   Footwear; Furniture; Industrial; Lumber; Marine;
   Medical; Packaging; Pipe/Plumbing; Recreational;
   Refrigeration; Transportation

REEVES BROTHERS
Coatings Division
P.O. Box 1531

Spartansburg, SC 29304 USA
Finished Goods Manufacturer: Coating
Other: Applicator
Markets: Clothing; Textile

REICHHOLD CHEMICALS
Swift Adhesives
3100 Woodcreek Dr.
Downers Grove, IL 60515 USA
Finished Goods Manufacturer: Adhesive
Markets: Appliances; Automotive; Carpet; Construction;
   Defense/Military; Footwear; Furniture; Lumber;
   Packaging; Recreational; Transportation

REILLY FOAM CORP.
1101 Hector
Conshohocken, PA 19428 USA
Finished Goods Manufacturer: Flexible Foam; Rigid Foam
Other: Fabricator
Markets: Electronic/Electrical; Industrial; Packaging

REISS CORPORATION
Blackstone Group
1 Polymer Place
Box 72
Blackstone, VA 23824 USA
Finished Goods Manufacturer: Elastomer; Automotive;
   Non-automotive; TPU
Markets: Aerospace; Appliances; Automotive;
   Construction; Defense/Military; Electronic/Electrical;
   Industrial; Marine; Medical; Mining/Drilling;
   Refrigeration; Transportation

REISS INDUSTRIES, INC.
319 Hart St.
P.O. Box 524
Watertown, WI 55074 USA
Finished Goods Manufacturer: RIM; Non-automotive;
   Flexible Foam; Integral Skin
Markets: Automotive; Furniture; Transportation

RELIABLE PLASTICS, INC.
777 N. Ave. Extension
Dunelien, NJ 08812 USA
Finished Goods Manufacturer: Flexible Foam; Rigid Foam
Other: Fabricator
Markets: Construction; Packaging; Recreational

RELIANCE PATTERN WORKS CO.
4350 W. Chicago Ave.
Chicago, IL 60651 USA
Finished Goods Manufacturer: Flexible Foam
Other: Fabricator
Markets: Automotive; Industrial; Packaging

RELIANCE RUBBER IND., INC.
8130 Ronda Dr.
Canton, MI 48187 USA
Finished Goods Manufacturer: Elastomer; Cast
Markets: Automotive

RELIANCE UPHOLSTERY SUPPLY CO.
137 E. Alondra Blvd.
Gardena, CA 90248 USA

Finished Goods Manufacturer: Flexible Foam
Other: Fabricator
Markets: Carpet; Furniture

REMPAC FOAM CORP.
61 Kuller Rd.
Clifton, NJ 07015 USA
Finished Goods Manufacturer: Flexible Foam
Markets: Packaging

RENOSOL CORPORATION
P.O. Box 1424
Ann Arbor, MI 48106 USA
Systems Manufacturer: Flexible Foam Systems
Finished Goods Manufacturer: RIM; Automotive; Flexible
    Foam; Integral Skin; Adhesive
Markets: Automotive; Footwear; Material Handling

REPUBLIC PACKAGING CORP.
9160 S. Green St.
Chicago, IL 60620 USA
Finished Goods Manufacturer: Flexible Foam
Markets: Packaging

REPUBLIC ROLLER CORPORATION
1233 Millard St.
P.O. Box 330
Three Rivers, MI 49093 USA
Finished Goods Manufacturer: Elastomer
Markets: Appliances; Automotive; Carpet; Chemical;
    Clothing; Construction; Defense/Military; Furniture;
    Industrial; Lumber; Material Handling; Packaging;
    Paper; Printing; Recreational; Refrigeration;
    Transportation

RESIN DESIGN INTERNATIONAL CORP.
801 M. Blocklawn Rd.
Conyers, GA 30207 USA
Systems Manufacturer: Elastomer Systems; Rigid Foam
    Systems
Other: Distributor

RESIN TECHNOLOGY CORP.
2270 Castle Harbor Place
Ontario, CA 91761 USA
Systems Manufacturer: Elastomer Systems; Rigid Foam
    Systems
Other: Distributor

REUEL, INC.
P.O. Box 10561
Goldsboro, NC 27532 USA
Equipment or Machinery Supplier/Manufacturer: Molds;
    Dispensing Equipment; Tanks; Pumps; Automated
    Controls; Molding Presses; Autoclaves; Casting
    Chambers; Handling Systems; Ovens
Markets: Aerospace; Automotive; Chemical; Electronic/
    Electrical; Industrial; Transportation

REXHAM CUSTOM INDUSTRIES
Division of Rexham, Inc.
700 Crestdale Ave.
P.O. Box 368
Matthews, NC 28105 USA

Finished Goods Manufacturer: Coating
Markets: Automotive; Defense/Military;
    Electronic/Electrical; Medical

REXNORD CORPORATION
16350 W. Glendale Dr.
New Berlin, WI 53151 USA
Finished Goods Manufacturer: Elastomer; Cast
Markets: Industrial; Material Handling; Power
    Transmission; Environmental

RHEIN-CHEMIE CORPORATION
Polyurethane Products
1008 Whitehead Rd. Extension
Trenton, NJ 08638 USA
Raw Material Supplier/Manufacturer: Catalyst;
    Surfactant; Resins; Anti-Hydrolysis Agents;
    Dispersants; Combustion Modifiers; Cross-linkers;
    Emulsifiers; Stabilizers
Finished Goods Manufacturer: Millable Gums
Markets: Aerospace; Appliances; Automotive; Carpet;
    Construction; Footwear; Furniture; Industrial;
    Marine; Medical; Mining/Drilling; Packaging;
    Recreational; Refrigeration

RHH FOAM SYSTEMS, INC.
P.O. Box 752
6001 S. PA Ave.
Cudahy, WI 53110 USA
Systems Manufacturer: Rigid Foam Systems
Equipment or Machinery Supplier/Manufacturer:
    Dispensing Equipment
Markets: Aerospace; Appliances; Automotive;
    Construction; Defense/Military; Industrial; Lumber;
    Marine; Mining/Drilling; Packaging; Pipe/Plumbing;
    Refrigeration; Transportation

RHINO LININGS USA, INC.
9557 Candida St.
San Diego, CA 92126 USA
Raw Material Supplier/Manufacturer: Isocyanate; Resins
Systems Manufacturer: Elastomer Systems; Specialized
    Cast; Spray Systems
Equipment or Machinery Supplier/Manufacturer:
    Dispensing Equipment; Tanks; Spray Equipment
    Systems
Finished Goods Manufacturer: Elastomer; Automotive;
    Cast; Coating
Other: Distributor; Fabricator
Markets: Automotive; Chemical; Construction; Defense/
    Military; Industrial; Marine; Mining/Drilling;
    Recreational;
Transportation

RIBCO, INC.
1032 Niagara St.
Buffalo, NY 14213 USA
Other: Contractor
Markets: Construction

RICHARD O. SCHULTZ CO.
2425 N. 75th Ave.

Elmwood Park, IL 60635 USA
Equipment or Machinery Supplier/Manufacturer: Tool Makers

RICHARDS, PARENTS & MURRAY (RPM)
606 Franklin Ave.
Mount Vernon, NY 10550 USA
Finished Goods Manufacturer: Elastomer; Cast
Other: Fabricator

RICHARDS, PARENTS & MURRAY, INC.
606 Franklin Ave.
Mount Vernon, NY 10550 USA
Finished Goods Manufacturer: Automotive; Non-automotive; Flexible Foam; Coating; Adhesive
Markets: Appliances; Automotive; Construction; Defense/ Military; Electronic/Electrical; Industrial; Printing

RICKS ROOFING AND INSULATION
Rural Route 2
Box 23
Elkton, SD 57026 USA
Other: Contractor
Markets: Construction

RIMNETICS, INC.
433 Clyde Ave.
Mountain View, CA 94043 USA
Finished Goods Manufacturer: Elastomer; RIM; Non-automotive; Structural Foams
Markets: Industrial

RISER/ALMEIDA ASSOCIATES
250 Montgomery St.
San Francisco, CA 94104 USA

RIVERSIDE SEAT CO.
500 N.W. Platte Valley Dr.
Riverside, MO 64150 USA
Finished Goods Manufacturer: Flexible Foam
Markets: Automotive

RM ENGINEERED PRODUCTS, INC.
The lntertech Group, Inc.
Garco St. and O'Hear Ave.
P.O. Box 5205
North Charleston, SC 29406 USA
Finished Goods Manufacturer: Elastomer; TPU; Coating
Markets: Aerospace; Automotive; Clothing; Defense/Military; Mining/Drilling; Agricultural

RMAX, INC.
13524 Welch Rd.
Dallas, TX 75224 USA
Finished Goods Manufacturer: Rigid Foam
Markets: Construction

RO-LAB AMERICAN RUBBER CO., INC.
8830 W. Linne Rd.
P.O. Box 450
Tracy, CA 95376 USA
Finished Goods Manufacturer: Elastomer; TPU; Cast
Other: Fabricator

Markets: Aerospace; Automotive; Construction; Defense/ Military; Industrial; Material Handling; Mining/Drilling; Recreational; Transportation; Wheels/Tires

ROBBINS AND MEYERS
1895 W. Jefferson St.
Springfield, OH 45506 USA
Equipment or Machinery Supplier/Manufacturer: Pumps

ROBERTSON
Division of United Dominion Co.
400 Holiday Dr.
Pittsburgh, PA 15220 USA
Finished Goods Manufacturer: Rigid Foam
Markets: Construction

ROBINSON RUBBER PRODUCTS CO., INC.
4600 Quebec Ave. North
Minneapolis, MN 55428 USA
Finished Goods Manufacturer: Elastomer; Cast
Markets: Appliances; Automotive; Construction; Electronic/Electrical; Marine; Medical; Mining/Drilling; Packaging; Printing; Agricultural; Food

ROBOTICS, INC.
24-21 Route 9
Ballston Spa, NY 12020 USA
Equipment or Machinery Supplier/Manufacturer: Dispensing Equipment; Cutting Equipment
Markets: Aerospace; Appliances; Automotive; Defense/ Military; Electronic/Electrical; Footwear; Furniture; Industrial; Marine; Material Handling; Medical; Recreational; Refrigeration; Transportation; Wheels/Tires

ROCKET POLYMEERS
3000 Locust St.
St. Louis, MO 63103 USA
Finished Goods Manufacturer: Elastomer; Cast
Other: Fabricator
Markets: Aerospace; Appliances; Automotive; Carpet; Chemical; Clothing; Construction; Defense/Military; Electronic/Electrical; Footwear; Furniture; Industrial; Lumber; Marine; Material Handling; Medical; Mining/Drilling; Packaging; Paper; Pipe/Plumbing; Printing; Recreational; Refrigeration; Transportation; Wheels/Tires

ROCKMONT INDUSTRIES
10675 Empire Rd.
Lafayette, CO 80026 USA
Finished Goods Manufacturer: Flexible Foam
Markets: Aerospace; Recreational; Safety

ROCKY MOUNTAIN RUBBER MFG CO.
2375 S. Raritan St.
Englewood, CO 80110 USA
Finished Goods Manufacturer: Elastomer; Cast
Markets: Automotive; Electronic/Electrical; Food

RODAC RUBBER CO.
8693 Crosby Lake Rd.

Clarkston, MI 48346 USA
Finished Goods Manufacturer: Elastomer; Flexible Foam
Other: Fabricator
Markets: Appliances; Construction; Furniture; Industrial;
   Material Handling; Packaging; Paper; Printing

RODIC CHEMICAL & RUBBER CO.
The Belco Corp.
11931 Jericho Rd.
Kingsville, MD 21087 USA
Finished Goods Manufacturer: Elastomer; Cast
Markets: Appliance; Automotive; Electronic/Electrical;
   Industrial; Printing

RODMAR MFG & RESEARCH INC.
P.O. Box 68
Nelsonville, WI 54458 USA
Finished Goods Manufacturer: Flexible Foam; Rigid Foam
Markets: Industrial; Transportation

ROGERS CORP.
Engineering Products Group
P.O. Box 126
Willimantic, CT 06226 USA
Finished Goods Manufacturer: Elastomers
Markets: Medical; Packaging; Paper; Printing

ROGERS CORP. WILLIAMANTIC DIVISION
730 Windham Rd.
South Windham, CT 06266 USA
Finished Goods Manufacturer: Elastomer; TPU; Cast;
   Flexible Foam; Microcellular
Markets: Automotive; Electronic/Electrical; Printing

ROGERS CORPORATION
P.O. Box 158
East Woodstock, CT 06244 USA
Finished Goods Manufacturer: Flexible Foam
Markets: Furniture

ROGERS CORPORATION
One Technology Dr.
Rogers, CT 06263 USA
Finished Goods Manufacturer: Elastomer; Flexible Foam;
   Integral Skin
Markets: Automotive; Defense/Military; Industrial;
   Medical; Printing

ROGERS FOAM CORP.
20 Vernon St.
Somerville, MA 02145 USA
Finished Goods Manufacturer: Flexible Foam; Rigid Foam
Other: Fabricator
Markets: Aerospace; Appliances; Automotive; Carpet;
   Clothing; Construction; Defense/Military;
   Electronic/Electrical; Footwear; Furniture; Industrial;
   Lumber; Marine; Material Handling; Medical;
   Packaging; Paper; Printing; Recreational;
   Refrigeration; Transportation

ROHM TECH INC.
195 Canal St.
Malden, MA 02148 USA

Raw Material Supplier/Manufacturer: Isocyanate;
   Polymer; Prepolymer; Resins
Other: Distributor
Markets: Automotive; Carpet; Clothing;
   Electronic/Electrical; Footwear; Furniture; Packaging;
   Paper; Printing; Transportation

ROLL RITE CORP.
421 Pendleton Way
P.O. Box 2107
Oakland, CA 94621-0007 USA
Finished Goods Manufacturer: Elastomer; Cast
Markets: Appliances; Automotive; Clothing;
   Construction; Defense/Military; Electronic/Electrical;
   Furniture; Industrial; Mining/Drilling; Paper;
   Printing; Wheels/Tires

ROLLERCOAT INDUSTRIES
3416 E. Columbus Dr.
Tampa, FL 33605 USA
Finished Goods Manufacturer: Elastomer; Cast; Coating
Markets: Construction; Industrial; Mining/Drilling;
   Paper; Printing

ROMAC CORP.
P.O. Box 1350
Plumsteadville, PA 18949 USA
Finished Goods Manufacturer: Rigid Foam
Markets: Industrial

ROME RIM, INC.
74000 Van Dyke Ave.
Romeo, MI 48065 USA
Finished Goods Manufacturer: Elastomer; RIM;
   Automotive; Non-automotive; Flexible Foam; Integral
   Skin; Rigid Foam
Markets: Automotive; Material Handling; Recreational;
   Transportation

RON MANUFACTURING CO., INC.
960 Lively Blvd.
Wood Dale, IL 60191 USA
Equipment or Machinery Supplier/Manufacturer: Molds;
   Automated Controls
Markets: Medical

ROOFBOND SYSTEMS
620 W. Main
Herington, KS 67449 USA
Systems Manufacturer: Elastomer Systems; Rigid Foam
   Systems
Markets: Construction

ROOFING SYSTEMS, INC.
P.O. Box 9424
Birmingham, AL 35220 USA
Finished Goods Manufacturer: Coating; Rigid Foam;
   Urethane Foam Roofing Contractor
Markets: Construction; Retro Fit

ROSS & ROBERTS, INC.
1299 W. Broad St.
Stratford, CT 06497 USA
Finished Goods Manufacturer: Flexible Foam

ROYAL PRODUCTS CO.
P.O. Box 1827
Montgomery, AL 36102 USA
Finished Goods Manufacturer: Flexible Foam
Markets: Furniture; Packaging

ROYAL RUBBER CO.
500 Chippewa Ave.
P.O. Box 267
South Bend, IN 46624 USA
Finished Goods Manufacturer: Elastomer; Cast
Markets: Appliances; Automotive; Clothing;
　Construction; Electronic/Electrical; Industrial;
　Mining/Drilling; Paper; Printing

RPM, INC.
2628 Pearl Rd.
P.O. Box 777
Medina, OH 44258 USA
Finished Goods Manufacturer: Coating
Markets: Construction

RS RUBBER CORP.
218 River Dr.
Garfield, NJ 07026 USA
Finished Goods Manufacturer: Flexible Foam
Other: Fabricator
Markets: Industrial; Pipe/Plumbing; Wheels/Tires

RT VANDERBILT COMPANY, INC.
Plastics Department
30 Winfield St.
Norwalk, CT 06855 USA
Raw Material Supplier/Manufacturer: Antioxidants
Markets: Polyol Producers; Foam Manufacturers

RTP CO.
580 E. Front St.
Winona, MN 55987 USA
Finished Goods Manufacturer: Coating; Adhesive; Sealant
Other: Compounder
Markets: Aerospace; Appliances; Automotive; Defense/
　Military; Electronic/Electrical; Footwear; Marine;
　Medical; Packaging; Printing; Recreational;
　Transportation

RUBATEX CORP
1113 Monroe St.
P.O. Box 340
Bedford, VA 24523 USA
Finished Goods Manufacturer: Closed Cell Flexible Foam
Markets: Aerospace; Appliances; Automotive; Clothing;
　Construction; Defense/Military; Electronic/Electrical;
　Footwear; Industrial; Marine; Material Handling;
　Medical; Mining/ Drilling; Pipe/Plumbing;
　Recreational; Refrigeration; Transportation;
　Wheels/Tires

RUBBAIR DOOR CO.
100 Groton Shirley Rd.
Ayer, MA 01432 USA
Finished Goods Manufacturer: Rigid Foam
Other: Fabricator

Markets: Construction; Industrial

RUBBER & ACCESSORIES, INC.
2123 E. Edgewood Dr.
Lakeland, FL 33803 USA
Finished Goods Manufacturer: Rigid Foam
Other: Distributor
Markets: Chemical; Construction; Industrial; Lumber;
　Marine; Material Handling; Mining/Drilling;
　Pipe/Plumbing, Printing; Transportation

RUBBER & SAFETY SUPPLY CO.
195 W. 2950 South
Salt Lake City, UT 84115 USA
Finished Goods Manufacturer: Rigid Foam
Other: Distributor
Markets: Industrial; Mining/Drilling

RUBBER & SILICONE PRODUCTS CO., INC.
P.O. Box 1215
Caldwell, NJ 07007 USA
Finished Goods Manufacturer: Elastomer; Cast; Flexible
　Foam; Coating
Markets: Aerospace; Appliances; Automotive; Industrial;
　Mining/Drilling; Wheels/Tires; Business Machines

RUBBER ASSOCIATES, INC.
1522 W. Turkeyfoot Lake Rd.
Barberton, OH 44203 USA
Finished Goods Manufacturer: Elastomer; RIM;
　Automotive; Non-automotive
Markets: Aerospace; Appliances; Automotive; Defense/
　Military; Electronic/Electrical; Industrial;
　Mining/Drilling; Printing

RUBBER CORPORATION OF AMERICA
2545 N. Broad St.
Philadelphia, PA 19132 USA
Finished Goods Manufacturer: Elastomer; TPU
Markets: Construction; Defense/Military

RUBBER DEVELOPMENT, INC.
426 Perrymont Ave.
San Jose, CA 95125 USA
Finished Goods Manufacturer: Elastomer; Cast; Flexible
　Foam
Markets: Aerospace; Automotive; Defense/Military; Elec-
　tronic/Electrical; Medical; Printing; Agricultural

**RUBBER ENGINEERING GROUP BGA
INTERNATIONAL**
3459 S. 700 W.
P.O. Box 26188
Salt Lake City, UT 84126 USA
Finished Goods Manufacturer: Elastomer; Cast
Markets: Defense/Military; Industrial; Mining/Drilling;
　Paper; Food

RUBBER GROUP
30 Center Rd.
Unit 1
Somersworth, NH 03708 USA
Finished Goods Manufacturer: Elastomer; Cast; Rigid
　Foam

Markets: Aerospace; Defense/Military;
Electronic/Electrical; Industrial; Material Handling;
Paper; Printing; Transportation; Wheels/Tires

RUBBER INDUSTRIES, INC.
215 Cavanaugh Dr.
Shakopee, MN 55379 USA
Finished Goods Manufacturer: Elastomer; Cast
Markets: Automotive; Electronic/Electrical; Industrial;
Printing

RUBBER MILLERS, INC.
709 S. Caton Ave.
Baltimore, MD 21229 USA
Finished Goods Manufacturer: Elastomer; Cast; Tank
Linings
Markets: Automotive; Chemical; Defense/Military;
Industrial; Marine; Material Handling;
Mining/Drilling; Packaging; Printing; Waste Water
Treatment Plants

RUBBER SPECIALTIES, INC.
8117 Pleasant Ave. South
Minneapolis, MN 55420 USA
Finished Goods Manufacturer: Elastomer; TPU;
Composites
Markets: Wheels/Tires

RUBBER TECH
5208 Wadsworth Rd.
Dayton, OH 45414 USA
Finished Goods Manufacturer: Elastomer; Cast
Markets: Automotive; Defense/Military;
Electronic/Electrical; Industrial; Marine; Medical;
Recreational; Refrigeration; Transportation;
Wheels/Tires

RUBBER TECHNOLOGY, INC.
Polymer Enterprises, Inc.
Tri-County Industrial Park
P.O. Box 577
Piney Flats, TN 37686 USA
Finished Goods Manufacturer: Cast
Markets: Industrial; Material Handling; Mining/Drilling

RUBBER-RIGHT ROLLERS
51 Kelvin St.
Everett, MA 02149 USA
Finished Goods Manufacturer: Elastomer; Cast
Markets: Industrial

RUBBER-URETHANES, INC.
968 W. Foothill Blvd.
Azusa, CA 91702 USA
Finished Goods Manufacturer: Elastomer; Cast; Millable
Markets: Aerospace; Defense/Military;
Electronic/Electrical; Marine; Medical; Packaging;
Paper; Printing; Computers

RUBBER MILL
824 Winston St.
Greensboro, NC 27405 USA
Finished Goods Manufacturer: Elastomer; Cast

Markets: Appliances; Electronic/Electrical; Industrial;
Lumber; Material Handling; Mining/Drilling;
Packaging; Paper; Printing; Recreational;
Transportation; Wheels/Tires

RUCO POLYMER CORPORATION
New South Rd.
Hicksville, NY 11802 USA
Raw Material Supplier/Manufacturer: Polyol; Polymer;
Prepolymer
Markets: Aerospace; Appliances; Automotive; Chemical;
Clothing; Construction; Defense/Military; Footwear;
Furniture; Industrial; Marine; Packaging; Printing;
Recreational; Transportation, Wheels/Tires

RUDOLPH BROS. & COMPANY
P.O. Box 425
Canal Winchester, OH 43110-0425 USA
Equipment or Machinery Supplier/Manufacturer:
Dispensing Equipment
Other: Distributor
Markets: Aerospace; Appliances; Automotive; Defense/
Military; Electronic/Electrical; Furniture; Industrial;
Marine; Material Handling; Packaging

RYNEL LTD.
P.O. Box 298
Route 22, Boothbay Industrial Park
Boothbay, ME 04537 USA
Raw Material Supplier/Manufacturer: Prepolymer
Systems Manufacturer: Flexible Foam Systems
Equipment or Machinery Supplier/Manufacturer:
Dispensing Equipment
Finished Goods Manufacturer: Elastomer; Cast; Flexible
Foam
Markets: Footwear; Industrial; Medical; Cosmetic;
Household Cleaning

S F PRODUCTS, INC.
P.O. Box 18188
3860 Delp St.
Memphis, TN 38118 USA
Systems Manufacturer: Elastomer Systems
Other: Distributor

S&S PLASTICS
310 Sherman Ave.
Newark, NJ 07114 USA
Finished Goods Manufacturer: Flexible Foam; Sealant
Other: Fabricator
Markets: Aerospace; Appliances; Automotive;
Construction; Defense/Military; Electronic/Electrical;
Industrial; Marine; Material Handling; Medical;
Packaging; Paper; Pipe/Plumbing; Recreational;
Refrigeration; Transportation

SACKNER PRODUCTS
2573 Rochester Rd.
Suite 110
Rochester Hills, MI 48307 USA
Finished Goods Manufacturer: Flexible Foam
Markets: Automotive; Furniture; Industrial; Medical

SAF-T-GARD INTERNATIONAL, INC.
205 HuehlRoad
Northbrook, IL 60062 USA
Other: Safety Products
Markets: Aerospace; Appliances; Automotive; Carpet;
Chemical; Clothing; Construction; Defense/Military;
Electronic/Electrical; Footwear, - Furniture;
Industrial; Lumber; Marine; Material Handling;
Medical; Mining/Drilling; Packaging; Paper;
Pipe/Plumbing; Printing; Recreational; Refrigeration;
Transportation; Wheels/Tires

SAN ANTONIO FOAM FABRICATORS
13715 Topper Circle
San Antonio, TX 78233 USA
Finished Goods Manufacturer: Flexible Foam; Rigid Foam
Other: Fabricator
Markets: Aerospace; Appliances; Automotive; Chemical;
Clothing; Construction; Defense/Military;
Electronic/Electrical; Footwear; Industrial; Lumber;
Marine; Material Handling; Medical; Packaging;
Paper; Printing; Recreational; Refrigeration;
Transportation; Wheels/Tires

SANIGLASTIC MFG. CO.
P.O. Box 184
Hales Corner, WI 53130 USA
Finished Goods Manufacturer: Flexible Foam
Other: Fabricator
Markets: Medical

SANNCOR INDUSTRIES, INC.
300 Whitney St.
Leominster, MA 01453-3209 USA
Raw Material Supplier/Manufacturer: Surfactant;
Polymer; Resins; Cross-linkers
Markets: Aerospace; Automotive; Construction-
Footwear; Furniture; Industrial; Lumber; Marine;
Paper; Printing; Wood Coatings/Finishes;
Vinyl/Leather Fabrics

SANTA FE RUBBER PRODUCTS
12306 E. Washington Blvd.
Whittier, CA 90606 USA
Finished Goods Manufacturer: Elastomer; Cast
Markets: Aerospace; Appliances; Mining/Drilling;
Packaging

SAVON FOAM CORP.
Brooklyn Navy Yard Building No.3
Brooklyn, NY 11 205 USA
Finished Goods Manufacturer: Flexible Foam
Markets: Furniture; Packaging

SCANDURA, INC.
1801 N. Tyron St.
Charlotte, NC 28230 USA
Finished Goods Manufacturer: Coating; Adhesive
Markets: Industrial

SCCI CORP
P. 0. Box 2239
Danbury, CT 06810 USA

Finished Goods Manufacturer: Flexible Foam; Integral
Skin; Coating; Rigid Foam
Markets: Automotive; Carpet; Chemical; Construction;
Defense/Military; Industrial; Material Handling;
Medical; Packaging; Printing; Transportation;
Institutional

SCHERING-PLOUGH HEALTH CARE
3030 Jackson Ave.
Memphis, TN 38151 USA

SCHLEGEL CORPORATION
1555 Jefferson Rd.
Rochester, NY 14623 USA
Finished Goods Manufacturer: PU Foam, Flexible
Urethane: Molded
Markets: Aerospace; Automotive; Construction; Defense/
Military; Electronic/Electrical; Industrial; Medical

SCHNADIG CORP.
4820 Belmont Ave.
Chicago, IL 60641 USA
Finished Goods Manufacturer: Flexible Foam
Other: Fabricator
Markets: Furniture

SCHNEE MOREHEAD, INC.
Corp Head Office
111 N. Nursery Rd.
P.O. Box 171305
lrving, TX 75017-1305 USA
Finished Goods Manufacturer: Sealant
Markets: Aerospace; Construction

SCHULLERE INTERNATIONAL
P.O. Box 625005
Littleton, CO 80162 USA

SCOTT PORT-A-FOLD, INC.
100 Taylor Parkway
Archbold, OH 43502 USA
Finished Goods Manufacturer: Elastomer; RIM; Non-
automotive; Flexible Foam; Integral Skin
Markets: Appliances; Chemical; Footwear; Furniture;
Industrial; Marine; Material Handling; Recreational;
Transportation; Sporting Goods

SCOTTDEL INC.
400 Church St.
Swanton, OH 43558-1199 USA
Finished Goods Manufacturer: Flexible Foam
Markets: Carpet; Packaging

SCOUGAL RUBBER CORP.
6239 Corson Ave. 5th
P. 0. Box 80226
Seattle, WASHINGTON 98108 USA
Finished Goods Manufacturer: Elastomer; Cast
Markets: Automotive; Construction; Industrial; Lumber;
Material Handling; Packaging; Paper; Printing

SCULLY RUBBER MANUFACTURING INC.
4501 E. Lombard St.
Baltimore, MD 21224 USA

Finished Goods Manufacturer: Elastomer; Millable Gums
Markets: Electronic/Electrical; CV Seals for Cable Mfg.

SD POLY CORP.
24619 Broadway
Oakwood Villa, OH 44146 USA

SDM, INC. - ENVIRONMENTAL ALTERNATIVES
2522 Blackwood Rd.
Wilmington, DE 19810 USA
Other: Consultant

SEAL FLEX CO.
141 Pickering St.
Portland, CT 06480 USA
Finished Goods Manufacturer: Elastomer; Cast;
Markets: Industrial; Mining/Drilling; Packaging; Printing

SEAL REINFORCED FIBERGLASS
23 N. Bethpage Rd.
Copiague, NY 11726 USA
Finished Goods Manufacturer: Elastomer
Other: Fabricator
Markets: Aerospace; Automotive; Construction;
    Furniture; Industrial; Recreational

SEAL-FLEX CO.
141 Pickering St.
Portland, CT 06480 USA
Finished Goods Manufacturer: Elastomer; TPU
Markets: Appliances; Electronic/Electrical;
    Mining/Drilling; Printing; Wheels/Tires

SEALANT EQUIPMENT & ENGINEERING
45677 Helm St.
Plymouth Township, MI 48170-0965 USA
Equipment or Machinery Supplier/Manufacturer:
    Dispensing Equipment; Automated Controls
Finished Goods Manufacturer: Sealant

SEALED AIR CORPORATION
Engineered Products Division
10 Old Sherman Turnpike
Danbury, CT 06810 USA
Systems Manufacturer: Rigid Foam Systems; Packaging
Equipment or Machinery Supplier/Manufacturer:
    Dispensing Equipment; Pumps; Automated Controls
Markets: Aerospace; Construction; Defense/Military;
    Electronic/Electrical; Furniture; Industrial; Medical;
    Packaging; Refrigeration

SEARS MANUFACTURING CO.
1718 S. Concord Ave.
Davenport, IA 52802 USA
Finished Goods Manufacturer: Flexible Foam; Integral
    Skin; Flamebonding
Markets: Automotive; Construction; Industrial; Material
    Handling; Medical; Agricultural

SEAWAY PLASTICS CORP.
814 Degurse Ave.
Marine City, MI 48039 USA
Finished Goods Manufacturer: RIM; Automotive
Markets: Automotive

SEEGOTT INC. OF NEW JERSEY
140 Littleton Rd.
Parsippany, NJ 07054 USA
Finished Goods Manufacturer: Coating
Other: Distributor

SEMPERIT INDUSTRIAL PRODUCTS, INC.
103 Spring Valley Rd.
Montvale, NJ 07645 USA
Finished Goods Manufacturer: Elastomer; Cast
Markets: Automotive; Construction; Marine; Medical;
    Mining/Drilling; Food

SENCORP SYSTEMS, INC.
P. O. Box 6001
Hyannis, MA 02601 USA

SFT, INC.
8989 Yellow Brick Rd.
Baltimore, MD 21237 USA
Finished Goods Manufacturer: Flexible Foam
Markets: Furniture; Transportation

SHAMROCK PLASTICS & RUBBER
5330 E. 25th St.
Indianapolis, IN 46218 USA
Finished Goods Manufacturer: Elastomer
Other: Distributor

SHELDAHL INC.
801 North Highway
Northfield, MN 55057 USA
Other: Fabricator
Markets: Aerospace; Automotive; Construction; Defense/
    Military; Electronic/Electrical; Industrial

SHELL CHEMICAL CO.
One Shell Plaza
Houston, TX 77002 USA
Raw Material Supplier/Manufacturer: Surfactant

SHELL CHEMICALS (PUERTO RICO) LTD.
P. O. Box 2768
San Juan, PR 00936 USA
Raw Material Supplier/Manufacturer: Polyol; Resins
Markets: Aerospace; Appliances; Automotive; Carpet;
    Chemical; Clothing; Construction; Defense/Military;
    Electronic/Electrical; Footwear; Furniture; Industrial;
    Lumber, Marine; Material Handling; Medical;
    Mining/Drilling; Packaging; Paper; Pipe/Plumbing;
    Printing; Recreational; Refrigeration; Transportation;
    Wheels/Tires

SHELL CONTAINERS
3000 Marcus Ave.
Lake Success, NY 11042 USA
Markets: Defense/Military; Electronic/Electrical;
    Material Handling; Packaging; Paper

SHELLER-GLOBE CORP
1641 Porter St.
Detroit, MI 48216 USA
Finished Goods Manufacturer: Flexible Foam
Markets: Automotive

SHEPHERD CHEMICAL COMPANY, THE
4900 Beech St.
Cincinnati, OH 45212-2398 USA
Raw Material Supplier/Manufacturer: Catalyst
Markets: Automotive; Carpet; Industrial; Lumber

SHERWIN WILLIAMS CO.
601 Canal Rd.
Cleveland, OH 44113 USA
Finished Goods Manufacturer: Coating
Markets: Automotive; Chemical; Construction;
Furniture; Industrial; Packaging; Printing;
Transportation

SHORE INSTRUMENTS CO., INC.
80 Commercial St.
Freeport, NY 11520 USA
Equipment or Machinery Supplier/Manufacturer: Testing
Equipment

SHREINER SOLE CO., INC.
Taylor Dr.
P.O. Box 347
Killbuck, OH 44637 USA
Finished Goods Manufacturer: Elastomer; Cast
Markets: Footwear

SHURCLOSE SEAL RURBER & PLASTIC
P.O. Box 305
Lake Orion, MI 48361 USA
Finished Goods Manufacturer: Elastomer; Cast
Markets: Appliances; Automotive; Agricultural

SIGMA COATINGS
1401 Destrehan Ave.
P.O. Box 816
Harvey, LA 70059 USA
Finished Goods Manufacturer: Coating
Markets: Chemical; Defense/Military; Furniture;
Industrial; Marine; Packaging; Paper; Printing

SIGNAL INDUSTRIAL PRODUCTS CORPORATION
1601 Cowart St.
Chattanooga, TN 37408 USA
Finished Goods Manufacturer: Elastomer; Non-
automotive; Cast
Other: Distributor
Markets: Chemical; Industrial; Marine; Paper;
Refrigeration

SKG INDUSTRIES
Route 10
Morgantown Rd.
Reading, PA 19607 USA
Systems Manufacturer: Elastomer Systems; Rigid Foam
Systems; Flexible Foam Systems
Equipment or Machinery Supplier/Manufacturer:
Dispensing Equipment; Tanks; Pumps; Automated
Controls
Markets: Aerospace; Appliances; Automotive; Carpet;
Chemical; Clothing; Construction;
Electronic/Electrical; Footwear; Industrial; Lumber;
Medical; Packaging; Paper

SMITH GROUP, INC.
811 E. Cayuga St.
Philadelphia, PA 19104 USA
Finished Goods Manufacturer: Elastomer
Other: Distributor

SMOOTH-ON, INC.
Polyurethane Products Division
1000 Valley Rd.
Gillette, NY 07933 USA
Systems Manufacturer: Elastomer Systems; Rigid Foam
Systems; Flexible Foam Systems
Finished Goods Manufacturer: Elastomer
Other: Formulator
Markets: Construction; Furniture; Industrial; Marine;
Material Handling; Medical; Recreational;
Wheels/Tires; Art-Related/Ceramics; Flexible Mold
Compound

SMS INDUSTRIES
549 Route 11
Farmington, NH 03835 USA
Finished Goods Manufacturer: Elastomer; Cast
Markets: Clothing; Paper; Printing; Textiles

SNIDER MOLD CO., INC.
6303 W. Industrial Dr.
Mequon, WI 53092 USA
Equipment or Machinery Supplier/Manufacturer: Molds
Markets: Automotive

SNOW CRAFT CO., INC.
200 Fulton Ave.
Garden City Park, NY 11040 USA
Finished Goods Manufacturer: Flexible Foam; Rigid Foam
Markets: Furniture; Packaging

SNYDER PLASTICS, INC.
1707 E. Lewis St.
Bay City, MI 48706 USA
Equipment or Machinery Supplier/Manufacturer: Molds
Finished Goods Manufacturer: Elastomer; RIM;
Automotive; Non-automotive

SOCIETY OF THE PLASTICS INDUSTRY, INC., THE
SPI Polyurethane Division
355 Lexington Ave.
New York, NY 1001 7 USA
Other: Trade Association

SOLAR COMPOUNDS CORP.
1201 W. Blancke St.
Linden, NJ 07036 USA
Finished Goods Manufacturer: Elastomer; TPU; Flexible
Foam; Coating; Adhesive; Sealant
Markets: Aerospace; Appliances; Automotive; Carpet;
Clothing; Construction; Defense/Military;
Electronic/Electrical; Footwear; Furniture; Industrial;
Marine; Medical; Packaging; Paper; Pipe/Plumbing;
Printing

SOLAR CONTRACTING, INC.
P.O. Box 51044
St. Louis, MO 63129 USA

Other: Contractor
Markets: Construction

SOLVAY SPECIALTY CHEMICALS, INC.
Solvay Performance Chemicals, Inc.
41 W. Putnam Ave.
Greenwich, CT 06830 USA
Raw Material Supplier/Manufacturer: Blowing Agent;
 Fire Retardant Chemicals
Other: Distributor
Markets: Appliances; Chemical; Construction;
 Refrigeration

SONNEBORN
889 Valley Park Dr.
Shakopee, MN 55379 USA
Finished Goods Manufacturer: Coating; Adhesive; Sealant
Markets: Construction; Industrial

SORBOTHANE, INC.
2144 State Route 59
Kent, OH 44240 USA
Finished Goods Manufacturer: Elastomer; RIM;
 Automotive; Non-automotive; Cast; Flexible Foam
Markets: Automotive; Defense/Military; Footwear;
 Industrial; Medical; Recreational; Transportation

SOUNDCOAT CO., INC.
3002 Croddy Way
Santa Ana, CA 92799 USA
Finished Goods Manufacturer: Flexible Foam
Markets: Industrial

SOUTH ATLANTIC FOAM
P.O. Box 639
Apex, NC 27502 USA
Finished Goods Manufacturer: Flexible Foam
Other: Fabricator
Markets: Defense/Military; Electronic/Electrical;
 Industrial; Packaging

SOUTEASTERN FOAM RUBBER CO.
P.O. Box 7183
High Point, NC 27264 USA
Finished Goods Manufacturer: Flexible Foam
Other: Fabricator
Markets: Furniture; Textile

SOUTHWEST RUBBER INDUSTRIES
Dynamic Seal, Inc.
541 Industrial Blvd.
Grapevine, TX 76051 USA
Finished Goods Manufacturer: Elastomer; TPU; Cast
Markets: Aerospace; Automotive; Defense/Military; Elec-
 tronic/Electrical; Medical; Mining/Drilling

SPAN-AMERICA
P.O. Box 5231
Greenville, SC 29606 USA
Finished Goods Manufacturer: Flexible Foam
Markets: Furniture; Industrial; Medical

SPARKS BELTING CO.
JSJ Corp

3800 Stahl Dr.
Grand Rapids, MI 49546 USA
Finished Goods Manufacturer: Elastomer; Cast; Coating
Other: Fabricator
Markets: Automotive; Industrial; Material Handling;
 Packaging; Conveyor Belt Industry

SPECIAL DESIGN PRODUCTS
2699 Harrison Rd.
Columbus, OH 43204 USA
Finished Goods Manufacturer: Flexible Foam
Other: Fabricator
Markets: Appliances; Automotive; Defense/Military;
 Electronic/Electrical; Furniture; Industrial; Material
 Handling; Medical; Packaging; Recreational;
 Transportation

SPECIALTY CASTINGS, INC.
42 Curtis Ave.
Woodbury, NJ 08096 USA
Finished Goods Manufacturer: Elastomer; Cast; Rollers
Markets: Aerospace; Defense/Military; Industrial;
 Marine; Material Handling; Printing; Wheels/Tires;
 Mechanical Rubber Goods

SPECIALTY FOAM PRODUCTS
11561 Westminster Ave.
Garden Grove, CA 92643 USA
Finished Goods Manufacturer: Flexible Foam
Markets: Defense/Military; Electronic/Electrical;
 Medical; Packaging

**SPECIALTY INSULATION & WATERPROOFING CO.,
INC.**
4027 Wendy Dr.
Orlando, FL 32808 USA
Markets: Construction; Industrial; Refrigeration;
 Production Sets

SPECIALTY URETHANES, INC.
P.O. Box 598
21 OC N. 21st St.
Purcellville, VA 22132 USA
Finished Goods Manufacturer: Elastomer; Cast
Markets: Aerospace; Defense/Military;
 Electronic/Electrical; Industrial; Marine; Material
 Handling; Mining Drilling; Packaging- Printing;
 Transportation; Wheels/Tires; Custom Molder

SPECIFIED EQUIPMENT SYSTEMS
10017 Beckleyview
Dallas, TX 75232 USA
Systems Manufacturer: Elastomer Systems; Rigid Foam
 Systems
Other: Distributor

SPERRY RUBBER & PLASTICS CO., INC., THE
Division of Alco industries, Inc.
9146 U.S. Route 52
Brookeville, IN 47012 USA
Finished Goods Manufacturer: Custom Extrusions
Markets: Appliances; Automotive; Construction;
 Defense/Military; Furniture; Industrial; Marine;

Mining/Drilling; Pipe/Plumbing; Recreational; Transportation; Burial Caskets

SPM RESEARCH CO.
51 Seiter Hill Rd.
Wallingford, CT 06492 USA
Other: Consultant

SPRAY EQUIPMENT TECHNOLOGY
6826 Hillsdale Ct.
Indianapolis, IN 46250 USA
Systems Manufacturer: Rigid Foam Systems
Other: Distributor

SPRAY, INC.
P.O. Box 151
Teele Rd.
Bolton, MA 01740 USA

SPRAYFOAM SOUTHWEST, INC.
1510 W. Drake
Tempe, AZ 85283 USA
Other: Contractor
Markets: Construction

SPRAYING SYSTEMS CO.
North Ave.
P.O. Box 7900
Wheaton, IL 60189 USA
Equipment or Machinery Supplier/Manufacturer: Spray Nozzles; Liquid Strainers; Valves; Spray Guns
Markets: Aerospace; Appliances; Automotive; Carpet; Chemical; Construction; Defense/Military; Electronic/Electrical; Footwear; Furniture; Industrial; Lumber; Marine; Medical; Mining/Drilling; Packaging; Paper; Printing; Refrigeration; Transportation; Wheels/Tires; Food; Textile; Forestry; Leather Products; Utilities

SPRAYKOTE SYSTEMS
P.O. Box 638
Thousand Palms, CA 92276 USA
Other: Contractor
Markets: Construction

SPUHL ANDERSON MACHINE CO.
1610 Parallel St.
Chaska, MN 53318 USA
Equipment or Machinery Supplier/Manufacturer: Dispensing Equipment; Automated Controls
Markets: Appliances; Automotive; Electronic/Electrical; Footwear; Medical; Packaging

SRI INTERNATIONAL
333 Ravenswood Ave.
Menlo Park, CA 94025 USA

SSF MOLDERS, INC.
1613 Columbus St.
Two Rivers, WI 54241 USA
Equipment or Machinery Supplier/Manufacturer: Molds
Finished Goods Manufacturer: Elastomer; Flexible Foam; Integral Skin; Rigid Foam
Other: Fabricator

Markets: Construction; Defense/Military; Furniture; Industrial; Medical; Packaging; Recreational; Transportation; Wheels/Tires

STALLMAN CO. M.H.
292 Charles St.
Providence, RI 02904 USA
Finished Goods Manufacturer: Flexible Foam
Markets: Medical; Packaging

STANCHEM, INC.
401 Berlin St.
East Berlin, CT 06023 USA
Raw Material Supplier/Manufacturer: Blowing Agent; Polymer
Finished Goods Manufacturer: Coating; Adhesive; Sealant
Markets: Aerospace; Appliances; Chemical; Construction; Defense/Military; Electronic/Electrical; Furniture; Industrial; Lumber; Medical; Packaging; Paper

STAR BUILDING SYSTEMS
P.O. Box 9491 0
Oklahoma City, OK 73143 USA
Finished Goods Manufacturer: Rigid Foam
Markets: Construction

STAR-GLO INDUSTRIES, INC.
2 Carlton Ave.
East Rutherford, NJ 07073 USA
Finished Goods Manufacturer: Elastomer; TPU; Cast
Markets: Aerospace; Appliances; Automotive; Electronic/Electrical; Mining/Drilling; Packaging; Paper; Printing

STEINMETZ POLYMERS
P.O.Box 393
Spencer Rd.
Moscow, PA 18444 USA
Finished Goods Manufacturer: Elastomer; TPU, Cast
Markets: Aerospace; Appliances; Automotive; Chemical; Defense/Military; Electronic/Electrical; Industrial; Marine; Material Handling; Medical; Mining/Drilling; Packaging; Paper; Pipe/Plumbing; Printing; Recreational; Transportation; Wheels/Tires

STEPAN COMPANY
22 W. Frontage Rd.
Northfield, IL 60093 USA
Raw Material Supplier/Manufacturer: Polyester Polyols; Polyether Polyols
Systems Manufacturer: Prepolymer Manufacturer; Elastomer Systems; Rigid Foam Systems
Markets: Aerospace; Appliances; Chemical; Construction; Defense/Military; Electronic/Electrical; Furniture; Marine; Refrigeration; Transportation

STEPHEN GOULD CORP.
35 S. Jefferson Rd.
Whippany, NJ 07981 USA
Finished Goods Manufacturer: Flexible Foam
Markets: Electronic/Electrical; Packaging

STEPHENSON & LAWYER, INC.
3831 Patterson Ave. SE

Grand Rapids, MI 49518 USA
Finished Goods Manufacturer: Flexible Foam; Integral Skin; Adhesive
Other: Distributor; Fabricator
Markets: Aerospace; Appliances; Automotive; Chemical; Construction; Defense/Military; Electronic/Electrical; Industrial; Medical; Packaging; Refrigeration; Transportation

STERLING MANUFACTURING
1203 White St.
P.O. Box 7703
Houston, TX 77270 USA
Finished Goods Manufacturer: Flexible Foam
Markets: Furniture

STERLING TIRE & RUBBER CO.
Colbert Industrial Park
Highway 435
Muscle Shoals, AL 35662 USA
Finished Goods Manufacturer: Elastomer; Cast
Other: Testing
Markets: Automotive; Chemical; Construction; Defense Military; Furniture; Industrial; Lumber; Material Handling; Recreational; Wheels/Tires

STOCKDALE, INC.
P.O. Box 40428
Houston, TX 77240 USA
Systems Manufacturer: Rigid Foam Systems; Flexible Foam Systems
Other: Distributor

STONEHURST INDUSTRIES
250 Executive Dr.
Edgewood, NY 11717 USA
Finished Goods Manufacturer: Flexible Foam; Rebond
Other: Distributor
Markets: Packaging

STRUCKMEYER CORP.
4201 B Kellway Circle
Dallas, TX 75234 USA
Finished Goods Manufacturer: Flexible Foam
Markets: Medical; Packaging

STRUCTURAL FOAM PLASTICS
P.O. Box 5208
North Branch, NJ 08876 USA
Equipment or Machinery Supplier/Manufacturer: Molds
Finished Goods Manufacturer: Flexible Foam
Other: Consultant; Fabricator
Markets: Aerospace; Appliances; Automotive; Carpet; Chemical; Clothing; Construction; Defense/Military; Electronic/Electrical; Footwear; Furniture; Industrial; Lumber; Marine; Material Handling; Medical; Mining/Drilling; Packaging; Paper; Pipe/Plumbing; Printing; Recreational; Refrigeration; Transportation; Wheels/Tires

STRUX CORP.
100 E. Montauk Highway
Lindenhurst, NY 11757 USA

Finished Goods Manufacturer: Rigid Foam
Markets: Automotive; Marine; Recreational; Refrigeration; Transportation

STYLE-MARK, INC.
Sub. of Sauder Woodworking
P.O. Box 301
Archbold, OH 43502 USA
Finished Goods Manufacturer: RIM; Rigid Foam

STYLETEK, INC.
1857 Middlesex St.
Lowell, MA 01851-1198 USA
Equipment or Machinery Supplier/Manufacturer: Molds
Finished Goods Manufacturer: Elastomer; TPU
Markets: Automotive; Electronic/Electrical; Footwear; Recreational

STYRO-MOLDERS CORP.
P.O. Box 577
CO Springs, CO 80901 USA
Finished Goods Manufacturer: Flexible Foam; Rigid Foam
Markets: Packaging; Transportation

SUN ROLLER CORP.
1108 Enterprise Rd.
Arlington, TX 76017 USA
Finished Goods Manufacturer: Elastomer; TPU; Cast
Markets: Clothing; Furniture; Industrial; Lumber; Packaging; Paper; Printing

SUNRISE INDUSTRIES
P.O. Box 184
Dundee, NY 14837 USA
Systems Manufacturer: Rigid Foam Systems
Other: Distributor

SUPERIOR PRODUCTS CO
Glenfleld Rd.
New Albany, MS 38652 USA
Finished Goods Manufacturer: Flexible Foam
Markets: Furniture; Packaging

SUPERIOR TIRE & RUBBER CORP.
P.O. Box 308
Warren, PA 16365 USA
Systems Manufacturer: Elastomer; Cast
Markets: Construction; Defense/Military; Industrial; Material Handling; Mining/Drilling; Printing; Wheels/Tires

SUPRACOR
1135 E. Arques Ave.
Sunnyvale, CA 94088 USA
Systems Manufacturer: Elastomer Systems; Honeycomb
Finished Goods Manufacturer: Elastomer; Fusion Bonded Thermoplastic; Honeycomb Cores and Panels
Markets: Aerospace; Automotive; Carpet; Defense/Military; Footwear; Furniture- Industrial; Marine; Medical; Packaging; Recreational; Transportation

SUR-SEAL GASKET
1242 W. Mehring Way

Cincinnati, OH 45203 USA
Finished Goods Manufacturer: Elastomer; Cast
Markets: Industrial

SURCO, INC.
271 W. Broad St.
Hatfield, PA 19440 USA
Finished Goods Manufacturer: Elastomer; Cast
Markets: Appliances; Automotive; Construction;
Defense/Military; Electronic/Electrical;
Mining/Drilling; Packaging; Paper; Printing

SUTTER FOAM & COATING
710 Cooper Ave.
Yuba City, CA 95991 USA
Finished Goods Manufacturer: Coating; Sealant; Rigid
Foam
Other: Contractor
Markets: Construction

SWD URETHANE COMPANY
222 S. Date St.
P.O. BOX 1422
Mesa, AZ 85211 USA
Raw Material Supplier/Manufacturer: Polyol
Systems Manufacturer: Prepolymer Manufacturer;
Elastomer Systems; Rigid Foam Systems; Flexible
Foam Systems
Finished Goods Manufacturer: Coating
Markets: Aerospace; Automotive; Chemical;
Construction; Defense/Military; Electronic/Electrical;
Furniture; Industrial; Marine; Mining/Drilling;
Packaging; Agricultural Storage; Beverage Tanks

SWIFT ADHESIVES
Reichhold Chemicals, Inc.
31 00 Woodcreek Dr.
Downers Grove, IL 60515 USA
Finished Goods Manufacturer: Adhesive
Markets: Appliances; Automotive; Construction;
Footwear; Furniture; Marine; Packaging; Paper;
Recreational; Transportation

SWISS-TEX, INC.
P.O. Box 9258
Greenville, SC 29604 USA
Finished Goods Manufacturer: Flexible Foam
Other: Fabricator
Markets: Medical; Packaging; Recreational

SYNAIR CORPORATION
2003 Amnicola Highway
Chattanooga, TN 37406 USA
Raw Material Supplier/Manufacturer: Resins
Finished Goods Manufacturer: Elastomer; Cast; Flexible
Foams; Coating; Adhesive
Markets: Aerospace; Automotive; Construction; Defense/
Military; Footwear; Industrial; Lumber; Marine;
Mining/Drilling; Recreational; Transportation;
Wheels/Tires

SYNTHETIC SURFACES, INC.
P.O. Box 241

Scotch Plains, NJ 07076-0241 USA
Raw Material Supplier/Manufacturer: Prepolymer
Finished Goods Manufacturer: Coating- Adhesive; Sealant
Markets: Aerospace; Automotive; Carpet; Construction;
Defense/Military; Footwear; Furniture; Industrial;
Marine; Packaging; Recreational

SYSTEMS FOR THE BUILDING INDUSTRY, INC.
SPI/PFCD
3154 E. La Palma
Suite D
Anaheim, CA 92806 USA
Other: Contractor
Markets: Construction; Refrigeration

T&S DIE CUTTING
13301 Alondra Blvd.
Santa Fe Springs, CA 90670 USA
Finished Goods Manufacturer: Flexible Foam; Rigid Foam
Other: Fabricator; Die cutter

T. D. WILLIAMSON, INC.
10727 E. 55th Place
P.O. Box 1121
Tulsa, OKLAHOMA 74146 USA
Finished Goods Manufacturer: Elastomer; RIM; Non-
automotive; Cast
Markets: Industrial

T.A. DAVIES CO.
363 W. 133rd St.
Los Angeles, CA 90061 USA
Raw Material Supplier/Manufacturer: Polyol
Systems Manufacturer: Elastomer Systems; Casting
Systems
Markets: Aerospace; Automotive; Chemical; Electronic/
Electrical- Furniture; Mining/Drilling; Recreational

T.O. PLASTICS
2901 E. 78th St.
Minneapolis, MN 55425 USA
Finished Goods Manufacturer: Flexible Foam
Other: Fabricator
Markets: Marine; Packaging; Refrigeration

TAKEDA AMERICA, INC.
2550 Corporate Exchange Dr.
Columbus, OH 43231 USA

TANDEM PRODUCTS INC.
3444 Dight Ave. South
Minneapolis, MN 55406 USA
Finished Goods Manufacturer: Elastomer; Cast; Integral
Skin; Sealant
Markets: Automotive; Chemical; Construction; Defense/
Military; Electronic/Electrical; Industrial; Lumber;
Marine; Material Handling; Medical; Mining/Drilling;
Packaging; Paper; Recreational; Transportation;
Wheels/Tires

TECHNICAL COATING APPLICATORS, INC.
1419 W. 27th St.
Panama City, FL 32405 USA

Systems Manufacturer: Elastomer Systems; Rigid Foam Systems
Finished Goods Manufacturer: Elastomer; Coating; Adhesive; Sealant; Rigid Foam
Other: Consultant; Testing; Fabricator
Markets: Chemical; Construction; Defense/Military; Industrial; Marine; Recreational; Refrigeration

TECHNICAL INNOVATIONS, INC.
595 Bradford St.
Pontiac, MI 48053 USA
Finished Goods Manufacturer: Elastomer

TECHNICAL RUBBER & PLASTIC CORP.
180 Getty Ave.
Clifton, NJ 07011 USA
Finished Goods Manufacturer: Elastomer; Cast
Markets: Appliances; Automotive; Construction; Industrial; Mining/Drilling; Paper; Printing

TECHNICAL SPECIALTIES
8000 W. 47th St.
Lyons, IL 60534 USA
Equipment or Machinery Supplier/Manufacturer: Dispensing Equipment
Other: Distributor

TECHNICAL SPECIALTIES CO, INC.
2415 Destiny Way
Odessa, FL 33556 USA
Finished Goods Manufacturer: Elastomer; Cast; Flexible Foam
Other: Distrubutor; Fabricator
Markets: Aerospace; Appliances; Defense/Military; Electronic/Electrical; Material Handling; Medical; Packaging; Printing; Wheels/Tires

TECHNICON
4412 Republic Dr.
Concord, NC 28025 USA
Finished Goods Manufacturer: Flexible Foam
Other: Fabricator

TECHNOGRAPHICS FITCHBURG COATED PRODUCTS
P.O. Box 1106
Scranton, PA 18501 USA
Systems Manufacturer: Coatings
Finished Goods Manufacturer: Coating; Adhesive
Markets: Printing

TECHIFOAM
13800 24th Ave.
Plymouth, MN 55441 USA
Finished Goods Manufacturer: Elastomer; RIM; Flexible Foam; Integral Skin; Coating; Adhesive; Sealant
Markets: Aerospace; Appliances; Automotive; Carpet; Chemical; Construction; Defense/Military; Electronic/Electrical; Footwear; Industrial; Marine; Material Handling; Medical; Packaging; Paper; Pipe/Plumbing; Printing; Recreational; Refrigeration; Transportation; Wheels/Tires

TECTONIC SYSTEMS, INC.
3124 NW 16th Terrace
Pompano Beach, FL 33064 USA
Other: Contractor
Markets: Construction; Refrigeration

TELEDYNE MONO-THANE
1460 Industrial Parkway
Akron, OH 44310 USA
Finished Goods Manufacturer: Elastomer; RIM; Flexible Foam; Rigid Foam
Markets: Aerospace

TEMA ISENMANN INC.
7806 Redsky Dr.
Cincinnati, OH 45249 USA
Finished Goods Manufacturer: Elastomer; Cast; Adhesive
Markets: Construction; Industrial; Mining/Drilling

TEMPRESS, INC.
701 S. Orchard St.
Seattle, WA 98108 USA
Finished Goods Manufacturer: Elastomer; RIM; Non-automotive; Flexible Foam; Integral Skin; Rigid Foam

TETRA PLASTICS, INC.
Subsidiary of Nike, Inc.
620 Spirit of St. Louis Blvd.
Chesterfield, MO 63005 USA
Finished Goods Manufacturer: Elastomer; TPU
Markets: Automotive; Defense/Military; Footwear; Industrial; Medical; Packaging; Recreational; Refrigeration; Transportation

TEXACO CHEMICAL CO.
P.O. Box 27707
Houston, TX 77227-7707 USA
Raw Material Supplier/Manufacturer: Catalyst; Surfactant; Resins; Amine-Terminated Polyols
Markets: Appliances; Automotive; Chemical; Construction; Furniture; Industrial; Lumber; Packaging; Paper; Refrigeration; Transportation

TEXSTAR, INC.
802 Ave. J East
P.O. Box 534036
Grand Prairie, TX 75053 USA
Finished Goods Manufacturer: Flexible Foam
Markets: Aerospace; Furniture

TEXTILE RUBBER & CHEMICAL CO.
1400 Tiarco Dr.
Dalton, GA 30720 USA
Finished Goods Manufacturer: Flexible Foam; Coating
Markets: Clothing; Construction; Footwear; Furniture; Marine; Packaging

TEXTILE RUBBER CO.
14241 Alondra Blvd.
La Mirada, CA 90638 USA
Finished Goods Manufacturer: Flexible Foam; Rigid Foam
Other: Fabricator

THANEX CHEMIE
417 Upland Rd.
Havertown, PA 19083 USA
Finished Goods Manufacturer: Flexible Foam; Rigid Foam
Markets: Packaging

THARCO
222 Grant Ave.
San Lorenzo, CA 94580 USA
Finished Goods Manufacturer: Flexible Foam; Rigid Foam
Markets: Packaging

THERMA-TRU CORP.
108 Mutzfeld Rd.
Butler, IN 46721 USA
Systems Manufacturer: Rigid Foam Systems; Flexible
Foam Systems
Other: Distributor

THERMEDICS
470 Wildwood St.
Woburn, MA 01888 USA
Finished Goods Manufacturer: Elastomer; TPU
Markets: Automotive; Construction; Industrial; Medical

**THERMO-SHIELD URETHANE INSULATION
& ROOFING**
P.O. Box 1366
Burbank, CA 91507 USA
Other: Contractor
Markets: Construction

THERMOSET PLASTICS, INC.
5101 E. 65th St.
P.O. Box 20902
Indianapolis, IN 46220-0902 USA
Systems Manufacturer: Encapsulants; Adhesives
Finished Goods Manufacturer: Elastomer; RIM;
Automotive; Non-automotive; Adhesive
Markets: Automotive; Electronic/Electrical; Industrial

THOMBERT, INC.
Box 1123
31 6 E. 7th St. North
Newton, IA 50208 USA
Finished Goods Manufacturer: Elastomer; Cast
Markets: Appliances; Construction; Industrial; Material
Handling; Mining/Drilling; Packaging; Recreational;
Transportation; Wheels/Tires

THORODIN, INC.
5541 Central Ave.
Boulder, CO 80301 USA
Raw Material Supplier/Manufacturer: Prepolymer
Systems Manufacturer: Elastomer Systems
Equipment or Machinery Supplier/Manufacturer:
Dispensing Equipment; Tanks
Finished Goods Manufacturer: Flexible Foam
Other: Research Organization
Markets: Defense/Military; Footwear; Marine;
Recreational; Wheels/Tires; Acoustics

TI-KROMATIC PAINTS
2492 Doswell Ave.

St Paul, MN 55108 USA
Finished Goods Manufacturer: Coating; Sealant
Markets: Appliances; Automotive; Construction;
Defense/Military; Electronic/Electrical; Furniture;
Industrial; Lumber; Marine; Material Handling;
Medical; Pipe/Plumbing; Recreational; Refrigeration;
Transportation

TIDWELLS URETHANE FOAM SERVICES, INC.
P.O.Box 950
1818 W. Olive St.
Lakeland, FL 33802 USA
Finished Goods Manufacturer: Rigid Foam
Markets: Construction; Industrial; Refrigeration

TILTON ENGINEERED COMPONENTS
A Scapa Group Company
336 W. Main St.
Tilton, NH 03276 USA
Finished Goods Manufacturer: Elastomer; Cast; Flexible
Foam; Coating; Rigid Foam
Markets: Electronic/Electrical; Industrial; Medical; Paper;
Printing; Wheels/Tires

TIMBERLAND COMPANY
11 Merrill Dr.
Hampton, NH 03842-2050 USA
Finished Goods Manufacturer: Elastomer; Cast
Markets: Footwear

TIME TECH, INC.
409 Brandywine Blvd.
Wilmington, DE 19803 USA
Equipment or Machinery Supplier/Manufacturer:
Electronic Test Equipment/Ultrasonic Rate of Rise
Markets: Aerospace; Appliances; Automotive; Furniture;
Refrigeration-Transportation

TOOL CHEMICAL CO., INC.
2226 Burdette
P.O. Box 20040
Ferndale, MI 48220 USA
Finished Goods Manufacturer: Elastomer

TOOL MATE CORP.
P.O. Box 39141
Cincinnati, OH 45239 USA
Finished Goods Manufacturer: Elastomer; TPU; Cast;
Coating
Markets: Aerospace; Appliances; Automotive;
Wheels/Tires; Bedding

TOSHIBA/GE
GE AP5, 1 SC
Louisville, KY 40225 USA
Finished Goods Manufacturer: Rigid Foam
Markets: Appliances; Refrigeration

TOSOH USA
Suite 600
1100 Circle 75 Parkway
Atlanta, GA 30339 USA
Raw Material Supplier/Manufacturer: Catalyst; Additives

Markets: Aerospace; Appliances; Automotive; Carpet; Chemical; Clothing; Construction; Defense/Military; Electronic/Electrical; Footwear; Furniture; Industrial; Lumber; Marine; Material Handling; Medical; Mining/Drilling; Packaging; Paper; Pipe/Plumbing; Printing; Recreational; Refrigeration; Transportation; Wheels/Tires

TOTAL PLASTICS, INC.
851 47th St.
Grand Rapids, MI 49507 USA
Finished Goods Manufacturer: Elastomer; TPU: Cast; Coating; Adhesive; Sealant
Other: Distributor
Markets: Aerospace; Automotive; Construction; Defense/Military; Electronic/Electrical; Footwear; Furniture; Indusirial; Lumber; Marine; Material Handling; Medical; Mining/Drilling; Packaging; Paper; Pipe/Plumbing; Printing; Recreational; Refrigeration; Transportation; Wheels/Tires

TOWNLEY ENG & MFG CO., INC.
P.O. Box 221
Candler, FL 32111 USA
Finished Goods Manufacturer: Elastomer; Cast
Markets: Mining/Drilling

TRELLEX, INC.
Subsidiary of Svedala, Inc.
30700 Solon Ind. Parkway
Solon, OH 44139 USA
Finished Goods Manufacturer: Elastomer; TPU; Cast
Markets: Construction; Mining/Drilling; Paper

TREMCO
3777 Green Rd.
Beachwood, OH 44122 USA
Finished Goods Manufacturer: Sealant
Markets: Automotive; Construction

TREXLER RUBBER CO.
503 N. Diamond St.
P.O. Box 667
Ravenna, OH 44266-0667 USA
Finished Goods Manufacturer: Elastomer; Cast
Markets: Aerospace; Automotive; Defense/Military; Electronic/Electrical; Footwear; Medical; Mining/Drilling

TRI-SEAL INTERNATIONAL
217 Bradley Hill Rd.
Blauvelt, NY 10913 USA
Markets: Appliances; Automotive; Chemical; Construction; Defense/Military; Electronic/Electrical; Footwear; Industrial; Marine; Material Handling; Medical; Mining/Drilling; Packaging; Pipe/Plumbing; Recreational; Refrigeration; Transportation

TRIAD-FABCO, INC.
1325 Baker Rd.
High Poiunt, NC 27263 USA
Finished Goods Manufacturer: Elastomer; Cast

Markets: Automotive; Furniture; Industrial; Material Handling; Medical; Packaging; Recreational; Transportation

TRINITY FOAM OF CAROLINA, INC.
P.O. Box 4622
High Point, NC 27263 USA
Finished Goods Manufacturer: Flexible Foam
Markets: Furniture

TRIPLE "C" ROOFING
Box 15
Woodbine, IA 51 579 USA
Other: Contractor
Markets: Construction

TROSTEL PACKINGS LTD.
901 Maxwell St.
Lake Geneva, WI 53147 USA
Finished Goods Manufacturer: Elastomer; TPU; Cast
Markets: Appliances; Automotive; Clothing; Construction; Defense/Military; Furniture; Industrial; Lumber; Marine; Material Handling; Mining/Drilling; Packaging; Paper; Pipe/Plumbing; Printing; Recreational; Transportation; Wheels/Tires

TRUE MOLDED RUBBER, INC.
12908 S. Main St.
Los Angeles, CA 90061 USA
Finished Goods Manufacturer: Elastomer; Cast
Markets: Aerospace; Automotive; Food

TRUE PRECISION PLASTICS
129 Ashmore Dr.
Leola, PA 17540 USA
Finished Goods Manufacturer: Elastomer; TPU
Markets: Aerospace; Appliances; Automotive; Defense/Military; Electronic/Electrical; Industrial; Medical; Transportation

TRULY MAGIC PRODUCTS, INC.
1200 Northland Ave.
Buffalo, NY 14215 USA
Finished Goods Manufacturer: Flexible Foam
Other: Fabricator
Markets: Aerospace; Appliances; Automotive; Carpet; Electronic/Electrical; Footwear; Furniture; Industrial; Marine; Material Handling; Medical; Packaging; Printing; Recreational

TRUSTY-COOK, INC.
10530 E. 59th St.
Indianapolis, IN 46236 USA
Finished Goods Manufacturer: Elastomer; Cast; Flexible Foam; Coating
Markets: Automotive

TRW, OILWELL CABLE DIVISION
ENGINEERED ELASTOMERS
P.O. Box 945
Lawrence, KS 66044 USA
Finished Goods Manufacturer: Elastomer; Cast
Markets: Automotive; Electronic/Electrical; Industrial

TSE INDUSTRIES, INC.
5260 113th Ave. North
P.O. Box 17225
Clearwater, FL 34622 USA
Finished Goods Manufacturer: Millable Polyurethanes;
  Adhesives
Markets: Automotive; Footwear; Industrial; Material
  Handling; Packaging; Paper; Printing; Wheels/Tires

TUSCARORA PLASTICS, INC.
P.O. Box 448
Antioch, IL 60066 USA
Finished Goods Manufacturer: Flexible Foam
Other: Fabricator

U-FLOW ROOF DRAIN SYSTEMS, INC.
7 (Masterformat)
P.O. Box 6489
Buffalo, NY 14240-6489 USA
Finished Goods Manufacturer: Retrofit Roof Drain for
  Spray Foam Systems
Markets: Construction

U.S.A. DRIVES, INC.
281 Shore Dr.
Burr Ridge, IL 60521 USA
Finished Goods Manufacturer: Elastomer; Cast;
  Microcellular
Markets: Appliances; Automotive; Industrial;
  Mining/Drilling; Paper; Printing; Wheels/Tires;
  Business Machines; Skate Wheels

U.S.A. MACHINERY CO., INC.
P.O. Box 630
Rockville Centre, NY 11571 USA
Equipment or Machinery Supplier/Manufacturer:
  Injection; Extrusion; Blow Molding; Compression
  Molding Machines

UCSC
1208 N. Grand
Roswell, NM 88201 USA
Equipment or Machinery Supplier/Manufacturer:
  Dispensing Equipment; Pumps
Finished Goods Manufacturer: Coating
Other: Consultant; Distributor; Mfg Rep
Markets: Construction; Industrial; Packaging

UNICAST, INC.
17 McFadden Rd.
Easton, PA 18043-4627 USA
Finished Goods Manufacturer: Elastomer; Cast; Flexible
  Foam; Coating
Other: Fabricator
Markets: Aerospace; Appliances; Automotive; Chemical;
  Defense/Military; Electronic/Electrical; Industrial;
  Lumber; Marine; Material Handling; Paper; Printing;
  Transportation; Wheels/Tires

UNICHEMA NORTH AMERICA
4650 S. Racine Ave.
Chicago, IL 60609 USA

Raw Material Supplier/Manufacturer: Polyol
Markets: Aerospace; Automotive; Chemical; Printing

UNIFLEX, INC.
P.O. Box 406
Wixom, MI 48393-0406 USA
Finished Goods Manufacturer: Elastomer; TPU: Cast;
  Coating; Adhesive; Sealant
Other: Distributor; Fabricator
Markets: Aerospace; Appliances; Automotive; Chemical;
  Construction; Defense/Military; Electronic/Electrical;
  Industrial; Marine; Material Handling;
  Mining/Drilling; Packaging; Paper; Pipe/Plumbing;
  Printing; Recreational; Transportation; Wheels/Tires

UNION SPECIALTIES, INC.
3 Malcolm Hoyt Dr.
Newburyport, MA 01950 USA
Raw Material Supplier/Manufacturer: Resins
Finished Goods Manufacturer: Coating; Adhesive
Markets: Automotive; Chemical; Footwear; Industrial

UNIQUE FABRICATING, INC.
1601 W. Hamlin Rd.
Rochester Hills, MI 48309 USA
Finished Goods Manufacturer: Flexible Foam; Adhesive;
  Rigid Foam
Other: Fabricator
Markets: Automotive; Carpet; Electronic/Electrical;
  Industrial

UNIQUE URETHANES, INC.
3000 Old Alabama Rd.
Suite 119-326
Alpharetta, GA 30202-5820 USA
Raw Material Supplier/Manufacturer: Resins
Systems Manufacturer: Elastomer Systems; Rigid Foam
  Systems; Flexible Foam Systems; Quick Set Coatings;
  Custom Blended Systems
Equipment or Machinery Supplier/Manufacturer: Molds;
  Laminators; Dispensing Equipment; Tanks; Pumps
Finished Goods Manufacturer: Elastomer; RIM; Flexible
  Foam; Integral Skin; Coating; Adhesive; Sealant
Other: Consultant; Mfg Rep
Markets: Aerospace; Appliances; Automotive; Chemical;
  Construction; Defense/Military; Furniture; Industrial;
  Marine; Packaging; Refrigeration; Transportation

UNIROYAL ADHESIVES & SEALANTS
312 N. Hill St.
P.O. Box 2000
Mishawaka, IN 46544 USA
Finished Goods Manufacturer: Adhesive; Sealant
Markets: Appliances; Automotive

UNIROYAL CHEMICAL COMPANY, INC.
World Headquarters-Benson Rd.
Middlebury, CT 06749 USA
Systems Manufacturer: Prepolymer Manufacturer
Markets: Chemical; Construction; Defense/Military;
  Electronic/Electrical; Industrial; Mining/Drilling;
  Printing; Recreational; Wheels/Tires

UNITED COATINGS
E. 19011 Cataido
Greenacres, WA 99016 USA
Systems Manufacturer: Elastomer Systems
FinishedGoods Manufacturer. Elastomer; Coating;
Sealant
Markets: Aerospace; Chemical; Construction; Defense/
Military; Industrial; Marine; Mining Drilling;
Pipe/Plumbing; Recreational

UNITED FOAM PLASTICS CORPORATION
172 E. Main St.
Georgetown, MA 01833 USA
Systems Manufacturer: Rigid Foam Systems; Flexible
Foam Systems
Finished Goods Manufacturer: Flexible Foam; Rigid Foam
Markets: Aerospace; Appliances; Automotive; Defense/
Military; Electronic/Electrical; Footwear; Industrial;
Material Handling; Medical; Packaging; Recreational

UNIVERSAL ADHESIVES & CHEMICAL CO.
P.O. Box 1502
Dalton, GA 30720 USA
Finished Goods Manufacturer: Elastomer; TPU; Coating;
Adhesive
Markets: Automotive; Clothing; Footwear; Furniture;
Mining/Drilling; Paper; Printing

UNIVERSAL AIR FILTER
P. 0. Box 853
East St. Louis, IL 62203 USA
Finished Goods Manufacturer: Flexible Foam
Other: Fabricator
Markets: Aerospace; Appliances; Automotive; Defense/
Military; Electronic/Electrical; Industrial; Marine;
Medical; Refrigeration; Transportation

UNIVERSAL APPLICATORS, INC.
P.O. Box 310
Forest Lake, MN 55025 USA
Finished Goods Manufacturer: Rigid Foam Spray
Other: Design and Install Contractors
Markets: Construction

UNIVERSAL COATINGS
5449 E. Lamona
Fresno, CA 93727 USA
Systems Manufacturer: Elastomer Systems
Other: Distribution

UNIVERSAL PLASTICS
2587 S. Arlington Rd.
Akron, OH 44319 USA
Finished Goods Manufacturer: Elastomer; TPU: Cast;
Coating; Adhesive
Other: Consultant; Testing; Distributor; Fabricator
Markets: Aerospace; Appliances; Automotive; Defense/
Military; Electronic/Electrical; Industrial; Material
Handling; Medical; Mining/Drilling; Packaging;
Paper; Printing; Refrigeration; Transportation;
Wheels/Tires

UNIVERSAL PRODUCTS, INC.
224 N. Montello St.
Brockton, MA 02041 USA
Finished Goods Manufacturer: Flexible Foam
Other: Fabricator
Markets: Packaging

UNIVERSAL URETHANE PRODUCTS, INC.
410 First St.
P.O. Box 50617
Toledo, OH 43605 USA
Finished Goods Manufacturer: Elastomer; Cast

UNIVERSAL URETHANES
3919 Easttax Freeway
Houston, TX 77026-1125 USA
Finished Goods Manufacturer: Flexible Foam; Sealant
Other: Fabricator
Markets: Aerospace; Automotive; Carpet; Clothing;
Defense/ Military; Footwear; Furniture; Industrial;
Medical; Packaging; Pipe/Plumbing; Printing;
Recreational; Refrigeration; Transportation

**UNIVERSITY OF DETROIT/MERCY
(SEE POLYMER INSTITUTE) USA
UNIVERSITY OF MA/LOWELL**
Plastics Engineering Dept.
Lowell, MA 01854 USA
Other: Consultant; Testing; Research Organization;
Education

UNIVERSITY OF SOUTHERN CALIFORNIA
MC 1211
Los Angeles, CA 90089 USA
Other: Consultant; Testing; Research Organization
Markets: Aerospace; Automotive; Defense/Military;
Electronic/Electrical; Industrial; Marine; Packaging;
Printing; Transportation

UOP
25 E. Algonquin Rd.
Des Plaines, IL 60017 USA
Raw Material Supplier/Manufacturer: Unilink; Chain
Extenders; Cross Linkers; Curatives; Unisiv and Molsiv
Dissicants

UPACO ADHESIVES
Worthen Industries
3 E. Spit Brook Rd.
Nashua, NH 03060 USA
Finished Goods Manufacturer: Coating
Markets: Automotive; Construction; Paper; Graphic Arts

UPCOA
5 First Ave.
Box 3405
Peabody, MA 01960 USA
Finished Goods Manufacturer: Elastomer; TPU; Cast
Markets: Aerospace; Appliances; Automotive; Chemical;
Construction; Defense/Military; Electronic/Electrical;
Industrial; Marine; Material Handling; Medical;
Printing; Recreational; Refrigeration; Transportation

URASEAL CORPORATION
One Dexter Dr.
Seabrook, NH 03874 USA
Finished Goods Manufacturer: Elastomer; Cast
Markets: Electronic/Electrical; Telecommunications;
Power

URETECH CASTING, INC.
528 Hi-Tech Parkway
Oakdale, CA 95361 USA
Finished Goods Manufacturer: Elastomer; Non-
automotive—Cast
Markets: Construction; Defense/Military;
Electronic/Electrical; Industrial; Lumber; Material
Handling; Mining/Drilling; Packaging; Paper;
Pipe/Plumbing; Printing; Wheels/Tires

URETHANE APPLICATIONS, INC.
P.O. Box 12082
Birmingham, AL 35202 USA
Finished Goods Manufacturer: Elastomer; Cast
Other: Mfg. Rep.; Fabricator
Markets: Industrial; Marine; Mining/Drilling;
Pipe/Plumbing; Wheels/Tires

URETHANE CONSULTANTS INTERNATIONAL
Town Farm Rd.
West Springfield, NH 03284 USA
Equipment or Machinery Supplier/Manufacturer: Molds;
Dispensing Equipment; Tanks; Specialized Processing
Machines
Other: Consultant
Markets: Automotive; Chemical; Footwear; Lumber;
Medical; Mining/Drilling; Recreational; Wheels/Tires

**URETHANE CONTRACTORS SUPPLYING &
CONSULTING**
1208 N. Grand
Roswell, NM 88201 USA
Systems Manufacturer: Elastomer Systems; Rigid Foam
Systems
Equipment or Machinery Supplier/Manufacturer:
Dispensing Equipment; Pumps; Automated Controls
Other: Consultant; Distributor; Mfg. Rep.
Markets: Construction; Industrial; Packaging

URETHANE ENGINEERING CORP.
35438 Mound Rd.
Sterling Heights, MI 48310 USA
Finished Goods Manufacturer: Elastomer; Cast; Flexible
Foam; Integral Skin; Rigid Foam
Markets: Automotive; Chemical; Construction; Defense/
Military; Industrial; Material Handling;
Mining/Drilling; Recreational; Transportation

URETHANE PLASTICS, INC.
550 W. Crowther Ave.
Placentia, CA 92670 USA
Systems Manufacturer: Elastomer Systems; Rigid Foam
Systems; Flexible Foam Systems
Other: Distributor
Markets: Construction

URETHANE PROCESS TECHNOLOGY
14515 Wisteria Hollow
Houston, TX 77062 USA
Other: Consultant

URETHANE PRODUCTS CO., INC.
P.O. Box 308
1750 Plaza Ave.
New Hyde Park, NY 11040 USA
Finished Goods Manufacturer: Flexible Foam
Other: Distributor; Fabricator
Markets: Appliances; Automotive; Carpet; Clothing;
Construction; Electronic/Electrical; Furniture;
Industrial; Marine; Medical; Packaging;
Pipe/Plumbing; Recreational; Transportation

URETHANE SERVICE
2730 Monterey St., #103
Torrance, CA 90503 USA
Equipment or Machinery Supplier/Manufacturer:
Dispensing Equipment; Tanks; Pumps; Automated
Controls; RIM Equipment; Casting Equipment
Other: Consultant; Distributor; Mfg Rep
Markets: Aerospace; Appliances; Automotive; Carpet;
Chemical; Construction; Defense/Military; Electronic/
Electrical; Footwear; Furniture; Industrial; Marine;
Medical; Mining/Drilling; Packaging; Recreational;
Refrigeration; Transportation; Wheels/Tires

URETHANE TECHNOLOGIES, INC.
1202 E. Wakeham
Santa Ana, CA 92701 USA
Raw Material Supplier/Manufacturer: Polyurethane
Cleaning Solvent
Systems Manufacturer: Elastomer Systems; Flexible Foam
Systems; Microcellular and Integral Skin Systems;
Structural Foam; Rigid PU Plastics
Equipment or Machinery Supplier/Manufacturer: Molds;
Dispensing Equipment; Centrifugal Casting Units
Other: Research Organization; Distributor
Markets: Aerospace; Automotive; Chemical;
Construction; Defense/Military; Electronic/Electrical;
Footwear; Furniture; Industrial; Marine; Material
Handling; Medical; Mining/Drilling; Paper; Printing;
Recreational; Transportation; Wheels/Tires

URETHANE TECHNOLOGIES, INC.
P.O. Box 616
Port Allen, LA 70767 USA
Finished Goods Manufacturer: Elastomer; Cast; Coating
Other: Fabricator
Markets: Chemical; Defense/Military;
Electronic/Electrical; Marine; Mining/Drilling; Paper;
Recreational; Transportation

URETHANE TECHNOLOGY CO., INC.
59-77 Temple Ave.
Newburgh, NY 12550 USA
Raw Material Supplier/Manufacturer: Isocyanate; Resins
Systems Manufacturer: Elastomer Systems; Rigid Foam
Systems; Flexible Foam Systems

Markets: Aerospace; Appliances; Automotive; Carpet;
Chemical; Construction; Defense/Military; Electronic/
Electrical; Footwear; Furniture; Industrial; Marine;
Medical; Mining/Drilling; Packaging; Pipe/Plumbing;
Refrigeration; Transportation; Wheels/Tires

USA ENGINEERING
11301 S. Shore Rd.
Reston, VA 22090 USA
Equipment or Machinery Supplier/Manufacturer:
Laminators; Dispensing Equipment; Multi-Opening
Panel Presses
Other: Mfg Rep
Markets: Construction; Refrigeration

UTAH FOAM PRODUCTS, INC.
3609 S. 700 West
P.O. Box 70838
Salt Lake City, UT 84119 USA
Systems Manufacturer: Elastomer Systems; Rigid Foam
Systems; Flexible Foam Systems
Equipment or Machinery Supplier/Manufacturer:
Dispensing Equipment
Other: Distributor

UTEX INDUSTRIES
10810 Katy Rd.
P.O. Box 79227
Houston, TX 77279 USA
Finished Goods Manufacturer: Elastomer; Non-
automotive; TPU; Cast
Markets: Aerospace; Chemical; Construction; Defense/
Military; Industrial; Lumber; Marine;
Mining/Drilling; Paper; Recreational; Wheels/Tires

V L INDUSTRIES, INC.
13219 15-Mile Rd.
Marshall, MI 49068 USA
Finished Goods Manufacturer: Elastomer; TPU; Cast
Markets: Aerospace; Appliances; Automotive; Clothing;

VAIL RUBBER WORKS, INC.
521 Langley Ave.
P.O. Box 64
St. Joseph, MI 49085 USA
Systems Manufacturer: Elastomer Systems
FinishedGoods Manufacturer: Elastomer; Cast
Markets: Aerospace; Appliances; Automotive; Carpet;
Chemical; Clothing; Construction;
Electronic/Electrical; Footwear; Furniture; Industrial;
Lumber; Material Handling; Medical;
Mining/Drilling; Packaging; Paper; Pipe/Plumbing;
Printing; Recreational; Transportation; Wheels/Tires

VANGUARD FOAM & PACKAGING CO.
154 W. 131 St.
Los Angeles, CA 90061 USA
Finished Goods Manufacturer: Flexible Foam
Markets: Furniture

VELCO ENTERPRISES LTD.
101 Executive Blvd.
Elmsford, NY 10523 USA

Raw Material Supplier/Manufacturer: Isocyanate; Polyol;
Catalyst; Surfactant; Blowing Agent; Polymer;
Prepolymer
Systems Manufacturer: Rigid Foam Systems; Flexible
Foam Systems
Other: Consultant; Distributor
Markets: Appliances; Automotive; Carpet; Furniture;
Industrial; Packaging; Refrigeration; Transportation

VERNAY LABORATORIES, INC.
P.O. Box 310
Yellow Springs, OH 45387 USA
Finished Goods Manufacturer: Elastomer; Cast
Markets: Industrial

VERTEX
P.O. Box 2244
Fairbanks, AK 99701 USA
Systems Manufacturer: Rigid Foam Systems; Flexible
Foam Systems
Other: Distributor

VERTROD CORPORATION
2037 Utica Ave.
Brooklyn, NY 11234 USA
Equipment or Machinery Supplier/Manufacturer:
Thermal Impulse Heat Sealing Machinery
Markets: Aerospace; Appliances; Automotive; Carpet;
Chemical; Clothing; Construction; Defense/Military;
Electronic/Electrical; Footwear; Furniture; Industrial;
Lumber; Marine; Material Handling; Medical;
Mining/Drilling; Packaging; Paper; Pipe/Plumbing;
Printing; Recreational; Refrigeration; Transportation;
Wheels/Tires

VI-CAS MFG. CO., INC.
8407 Monroe Ave.
Cincinnati, OH 45236 USA
Finished Goods Manufacturer: Elastomer; Cast
Markets: Industrial; Material Handling; Packaging

VIP RUBBER CO., INC.
945 S. East St.
Anaheim, CA 92805 USA
Finished Goods Manufacturer: Elastomer; TPU
Markets: Aerospace; Appliances; Chemical;
Construction; Defense/Military; Industrial; Marine;
Material Handling; Mining/Drilling; Pipe/Plumbing;
Printing; Transportation

VIPIN N. TOLAT, PE, CONSULTING ENGINEER
5081 SW. 95th Ct.
Miami, FL 33165 USA
Other: Consultant
Markets: Construction

VISCOR, INC.
1266 Profit Dr.
Dallas, TX 75247 USA
FinishedGoods Manufacturer: Flexible Foam
Markets: Aerospace; Appliances; Automotive; Carpet;
Construction; Electronic/Electrical; Industrial;

Lumber; Marine; Medical; Paper; Printing; Recreational; Transportation

VITAFOAM INC.
POURING & FABRICATION
2222 Surrett Dr.
High Point, NC 27261 USA
Finished Goods Manufacturer: Flexible Foam
Markets: Furniture; Packaging; Bedding; Consumer Products

VIZIFLEX SEELS, INC.
16 E. Lafayette St.
Hackensack, NJ 07601 USA
Equipment or Machinery Supplier/Manufacturer: Molds
Other: Fabricator
Markets: Electronic/Electrical; Medical

W.H. PORTER, INC.
4240 N. 136 Ave.
Box 1138
Holland, MI 49423 USA
Finished Goods Manufacturer: Flexible Foam; Rigid Foam
Markets: Appliances; Automotive; Construction; Defense/Military; Marine; Medical; Packaging; Pipe/Plumbing; Refrigeration; Transportation

W.J. RUSCOE COMPANY
485 Kenmore Blvd.
Akron, OH 44301 USA
Systems Manufacturer: Elastomer Systems; Rigid Foam Systems
Other: Distributor

W.M. BARR AND COMPANY
Custom Manufacturing Division
P.O. Box 1879
Memphis, TN 38101 USA
Other: Custom Liquid Products Packaging
Markets: Appliances; Automotive; Chemical; Construction; Industrial; Marine; Transportation

W.S. SHAMBAN COMPANY SEALS DIVISION
2531 Bremer Dr.
P.O. Box 176
Fort Wayne, IN 46801USA
Finished Goods Manufacturer: Elastomer; TPU; Cast
Markets: Aerospace; Appliances; Automotive; Construction; Defense/Military; Industrial; Medical; Mining/Drilling; Paper; Printing

WABASH MPI
1569 Morris St.
P.O. Box 298
Wabash, IN 46992 USA
Equipment or Machinery Supplier/Manufacturer: Molds; Laminators; Compression molding presses
Markets: Aerospace; Automotive; Carpet; Chemical; Construction; Defense/Military; Electronic/Electrical; Footwear; Furniture; Industrial; Lumber; Marine; Medical; Packaging; Paper; Wheels/Tires

WALLACE COAST MACHINERY COMPANY
2151 Madison St.

Bellwood, IL 60614 USA
Equipment or Machinery Supplier/Manufacturer: Presser/ Mold Clamps RIM/RTM; Turnkey Systems
Markets: Aerospace; Appliances; Automotive; Construction; Defense/Military; Furniture; Industrial; Lumber; Marine; Material Handling; Packaging; Recreational; Transportation

WATTLE & DAUB CONTRACTORS
P.O. Box 427
Torrington, WYOMING 82240 USA
Systems Manufacturer: Rigid Foam Systems; Flexible Foam Systems
Other: Distributor

WEAVER INDUSTRIES, INC.
425 S. Fourth St.
Denver, PA 17517 USA
Equipment or Machinery Supplier/Manufacturer: Molds
Finished Goods Manufacturer: Elastomer; Cast; Flexible Foam; Integral Skin; Rigid Foam
Other: Fabricatoi
Markets: Automotive; Clothing; Construction; Defense/ Military; Electronic/Electrical; Furniture; Industrial; Lumber; Marine; Material Handling; Mining/Drilling; Packaging; Paper; Printing; Recreational; Wheels/Tires

WEINBRENNER SHOE COMPANY, INC.
108 S. Polk St.
Merrill, WI 54452 USA
Finished Goods Manufacturer: Flexible Foam; Integral Skin
Markets: Footwear

WEST COAST LAMINATING
1820 Embarcadero
Oakland, CA 94606 USA
Finished Goods Manufacturer: Flexible Foam; Rigid Foam; Laminating Foam Based Materials
Other: Laminator
Markets: Appliances; Automotive; Marine; Medical; Recreational; Transportation

WESTERN POLYFOAM CORP
3888 Commerce St.
Riverside, CA 92507 USA
Systems Manufacturer: Rigid Foam Systems
Other: Distributor

WESTERN TEXTILE PRODUCTS
594 Linden Ave.
P.O. Box 260
Memphis, TN 38101 USA
Finished Goods Manufacturer: Flexible Foam
Other: Fabricator
Markets: Automotive; Clothing; Furniture; Medical; Packaging; Recreational

WESTINGHOUSE ELECTRIC CORPORATION
401 E. Hendy Ave.
P.O. Box 4999 M/S EE-1
Sunnyvale, CA 94088-3499 USA

Finished Goods Manufacturer: Elastomer; Cast
Markets: Aerospace; Marine

WESTPHALIA ENGINEERING
P.O. Box 114
Westphalia, MO 65085 USA
Systems Manufacturer: Rigid Foam Systems
Other: Distributor

WHIRLPOOL CORPORATION
303 Upton Dr.
St. Joseph, MI 49085 USA
Markets: Appliances; Refrigeration

WHITEFIELD PLASTICS
3522 Pinemont
P.O. Box 920533
Houston, TX 77292-0533 USA
Finished Goods Manufacturer: Elastomer; Cast; Coating
Markets: Construction; Marine; Mining/Drilling

WILLIAMETTE VALLEY CO., THE
Specialty Urethane Products
P.O. Box 2280
Eugene, OR 97402 USA
Systems Manufacturer: Elastomer Systems; Rigid Foam
    Systems
Equipment or Machinery Supplier/Manufacturer: Molds;
    Dispensing Equipment; Pumps
Finished Goods Manufacturer: Elastomer; Cast; Coating;
    Adhesive
Other: Distributor; Fabricator
Markets: Automotive; Chemical; Construction;
    Furniture; Industrial; Lumber; Marine; Material
    Handling; Paper; Transportation; Wheels/Tires; Wood
    Products

WILLIAM C HART CO., INC.
236 25th St.
Brooklyn, NY 11232 USA
Finished Goods Manufacturer: Elastomer; Cast
Markets: Industrial; Paper; Printing; Graphic Arts

WILLIAM T. BURNETT & COMPANY
2112 Montevideo Rd.
Jessup, MD 20794 USA
Finished Goods Manufacturer: Elastomer; Cast; Flexible
    Foam
Markets: Carpet; Clothing; Furniture; Marine; Packaging

WILSHIRE ADVANCED MATERIAL
1240 E. 230th St.
Carson, CA 90745 USA
Equipment or Machinery Supplier/Manufacturer: Molds;
    Laminators
Finished Goods Manufacturer: Flexible Foam; Coating;
    Adhesive; Sealant
Markets: Aerospace; Automotive; Clothing;
    Defense/Military; Electronic/Electrical; Footwear;
    Industrial; Marine; Medical; Packaging; Printing;
    Recreational; Refrigeration; Transportation

WILSHIRE FOAM PRODUCTS, INC.
1240 E. 230th St.

Carson, CA 90745 USA
Finished Goods Manufacturer: Flexible Foam
Other: Manufacturer
Markets: Automotive; Carpet; Clothing;
    Defense/Military; Footwear; Furniture; Industrial,
    Medical

WOLF CREEK RUBBER CO.
206 Main St.
P.O. Box 200
Beaman, IA 50609 USA
Finished Goods Manufacturer: Elastomer; Cast
Markets: Appliances; Automotive; Electronic/Electrical;
    Industrial; Printing

WOLF ENGINEERING
8946 Loberg Rd.
Amherst Junction, WI 54407 USA
Equipment or Machinery Supplier/Manufacturer: Molds
Other: Consultant
Markets: Electronic/Electrical; Furniture; Industrial;
    Material Handling; Medical; Recreational;
    Refrigeration; Transportation

WOLVERINE COATINGS
9012 First St.
Baroda, MI 49101 USA
Systems Manufacturer: Elastomer Systems
Other: Distributor

WOLVERINE WORLD WIDE, INC.
9341 Courtland Dr.
Rockford, MI 49351 USA
Finished Goods Manufacturer: Elastomer; Molded
    Outsoles for Shoes
Markets: Construction; Defense/Military; Footwear;
    Marine; Medical

WOODBRIDGE FOAM FABRICATING, INC.
Commercial & Industrial Division
1120 Judd Rd.
P.O. Box 509
Chattanooga, TN 37401 USA
Finished Goods Manufacturer: Flexible Foam
Markets: Automotive; Carpet; Footwear; Furniture;
    Industrial; Medical

WOODBRIDGE GROUP, THE
Automotive, Commerical and Industrial
Automotive Headquarters
2500 Meijer Dr.
Troy, MI 48084 USA
Finished Goods Manufacturer: Elastomers; RIM;
    Automotive; Flexible Foam; Rigid Foam; Energy
    Absorbing Polyurethane Foams; Foam-in-Place Seat
    Assemblies
Markets: Automotive; Carpet; Furniture; Industrial;
    Recreational

WOODLAWN RUBBER CO.
11268 Williamson Rd.
Cincinnati, OH 45241 USA
Finished Goods Manufacturer: Elastomer; Cast

Markets: Appliances; Automotive; Construction; Electronic/Electrical; Mining/Drilling; Packaging

WORK DEVELOPMENTS, INC.
Forks of Coal Industrial Park
Alum Creek, WV 25003 USA
Finished Goods Manufacturer: Elastomer; Cast
Markets: Construction

ZED INDUSTRIES, INC.
3580 Lightner Blvd.
Box458
Vandalia, OH 45377 USA
Equipment or Machinery Supplier/Manufacturer: Molds; Thermoformer Heat Sealing Equipment; Skin Packaging Equipment
Markets: Aerospace; Appliances; Automotive; Defense/ Military; Electronic/Electrical; Industrial; Marine; Medical; Packaging; Pipe/Plumbing; Printing; Recreational

ZENITH PUMPS/PARKER HANNIFIN CORP.
5910 Elwin Buchanan Dr.
Sanford, NC 27330 USA
Equipment or Machinery Supplier/Manufacturer: Pumps

ZERILLO PRODUCTS, INC.
Frost Mill Rd.
P.O. Box 305
Mill Neck, NY 11765 USA
Finished Goods Manufacturer: Elastomer; Cast; Flexible Foam; Coating
Markets: Aerospace; Appliances; Automotive; Clothing; Construction; Electronic/Electrical; Mining/Drilling; Packaging; Paper; Printing; Agricultural

.

B

Urethane Processing Systems

Accumetric, Inc. (see Meter-Mix, Inc.)

Adapt-A-Pak Corp.

Portable air-operated, pour-in-place foaming system for polyurethane is designed for use with company's Adapt-a-pak foam, a 0.5-lb density, semirigid material. The system operates with compressed air and uses PTFE-coated gun with a fingertip-controlled air-purge mechanism.

Admiral Equipment Co., Sub. of Dow Chemical Co.

Designs, builds, and installs complete urethane processing lines. Areas included are:

- Bulk storage and blending, including fiber or filler blending systems
- Low- and high-pressure metering and mixing equipment, including RIM and RRIM units
- Mold carriers, including mold clamps, turntables and mold conveyor systems
- Auxiliary robotics, ovens, hydraulics and materials-handling equipment
- Structural RIM preform systems for producing multilayer glass-mat preforms

Advanced Process Technology (formerly Amplan, Inc.)

Manufactures automatic meter/mix/dispense equipment to process two-, three-, and four-component urethane elastomers and foams. The company's *Flying Wedge* features variable ratio, digital-readout temperature control to 300° F, vacuum degassing, positive-displacement metering, and output rates from 50 g to 50 lb/min. Unit handles laboratory, prototype, and production applications.

Automatic Process Control, Inc.

Offers metering/mixing/dispensing machines for casting urethanes, silicones, polyesters, and epoxies. Shot size is adjustable from very small to continuous at a rate of 2 to 3 gal/min. The company also supplies continuous on-the-fly resin degassers.

BASF Corporation

Polyurethane foam equipment line includes meter/mix/dispense units that apply rigid and semirigid foams in densities ranging from 0.35 to 10 lb/ft^3 at 3 to 60 lb/min. Systems include a dispensing gun and chemicals packaged in pressurized tanks and are said to be suitable when low capital cost, simplicity of operation, and low maintenance are desired. Two systems are offered: an Autofroth unit for producing rigid froth foams for building insulation, refrigeration and flotation applications, and an Autopak system for making foam-in-place, semirigid foams used in industrial packaging.

Binks Manufacturing Co., Plastics Resin Equipment Division

Manufactures systems and components for decorative and protective urethane, epoxy, or hybrid systems. A broad selection of fixed and variable-ratio proportioning systems deliver a wide range of volumes, from microshots to 60 lb/min, and are available in air-, hydraulic-, or electric-powered models. Application devices for manual or automatic spray, using high-volume, low pressure, airless, air-assisted airless, or conventional air atomization complement material delivery systems.

Urethane foam systems are available that utilize high-volume spray, pour, or frothing, as well as microshots to fill voids. Low-volume special applications are available in standard packages or can be designed for custom installations.

Binks Manufacturing Co., Poly-Craft System Division

Slingshot gun is available for either spraying or pouring polyurethane foams and other two-component reactive polymers. The gun adapts to virtually all existing plural-component equipment. Hand-held and machine-mounted models are offered with either manual (standard), semiautomatic, or fully automatic operation. Features include built-in lead/lag adjustment and choice of mechanical or static mixing. The gun is suited to applications requiring precise rates.

Cannon USA, Inc.

Offers a complete line of standard and custom-made polyurethane equipment and turnkey systems. Urethane equipment includes Models A, B, C, H, HE, and S metering units. A complete range of clamps 2 to 800 tons is available. Turnkey specifications include Capsotec (window) encapsulation, Insotec (sound absorption), Stratotec (headliners), Compotec (SRIM solutions), Rotoflex (flexible foam), Rimmodule (RRIM, polyurea), Zero-Time Change (appliance cabinet), laminators, drum units, and other systems.

Construction Technology Division, Investment Holdings Group

Represents Mondomix Holland multicomponent equipment for processing multicomponent reactive resin systems with various types of metering devices commensurate with viscosity characteristics, accuracy requirements, and matched to meet overall process control/demands of the production line. Metering/ratio accuracy is controlled by mass-flow meters with computer-monitored interlock responsive to the instant demand of the production line. Temperature control of all components is available, including the mixer rotor, and stator. A special application of the mixing head design permits the mechanical encapsulation of gases providing foam cell-size control down to the submicron range (in lieu of chemical blowing reagents). Upon dispensing this mechanically encapsulated glass blend, the blend emerges pre-expanded and may be titled frothing equipment.

With the required elimination of CFCs for urethane foaming operations and the potential substitution of HCFCs, the difference in the resultant thermal insulation properties can be eliminated by producing submicron-size cells using the special Mondomix mixing heads.

Materials in contact with the polymeric components are all stainless steel. Throughput capacities range from 10 lb/hr to more than 10,000 lb/hr. Laboratory-sized test/demonstration equipment is available for process evaluation.

Decker Industries, Inc.

Manufactures low-pressure and high-pressure RIM metering, mixing, and dispensing machines for processing polyurethane foams and elastomers. Low-pressure equipment features company's patented Pulse-Surge air/solvent purge timer that reduces purge solvent usage by up to 70%. Output ranges are from 50 g/min to 240 lb/min. Mix ratios are infinitely variable from 5:1 to 1:5. Process temperatures are controlled within ±1° F. Readings for pressure, temperature and throughput are available with statistical process control (SPC) communications as an option.

RIM equipment features impingement mixing with mechanical self-cleaning mix heads of either straight or angled design, and axial piston metering pumps and precise temperature control between 75 and 120° F. Outputs range from 10 to 220 lb/min; mix ratios are variable from 3:1 to 1:3.

Dow Chemical Company (see Admiral Equipment Co.)

EFY Corporation

The company's Plastic Processing Machinery Division designs, engineers, and installs a broad range of polyurethane processing machinery for the automotive, appliance and custom-molding industries. Equipment offered includes:

- High/low pressure metering machines from 50 g/min to 400 lb/min with fully adjustable ratios to 5:1 to 1:5
- Hydraulic and pneumatic presses, clamps, and mold carriers for all RIM, RRIM, SRIM, and RTM molded products
- Racetrack carousels, rotary tables and stationary press systems for rigid and flexible parts

The company also supplies low-density rigid urethane A and B components in returnable, reusable pressure vessels complete with a purge froth dispenser.

Edge Sweets Company

Manufactures a broad range of equipment for metering and dispensing plural component urethane systems. Low-pressure equipment covers a range from 0.3 lb/min up to 1000 lb/min and uses a patented valving system. The system reportedly can handle a wide range of materials in multistream inputs from 2 to 20 metered streams.

   The company builds RIM equipment, running at speeds of 50 to 800 lb/min. The mixing head incorporates a sleeved design said to be economical and is offered in a straight head as well as a free-pour design to allow impingement mixing into open molds without splash. Custom equipment is available.

Elastogran Machinery, Inc.

   The company supplies a range of high- and low-pressure urethane systems that include:

- Rotary table systems for molded parts
- Continuous double-belt conveyor systems for foam-core panel production
- Semicontinuous and discontinuous lines for foam-core panels
- Continuous production lines for pipe insulation
- Production lines for water heaters and refrigerator cabinets
- Single machines and complete production lines for automotive, furniture, technical parts (including computer cabinets), medical, recreational, and leisure applications

Various types of mold carriers and metal and epoxy molds are also available as well as bulk handling and blending systems for urethane systems.

Flexible Products Company, Polyurethane Division

The company's Delta-Therm rigid urethane foam dispensing equipment includes the following features:

- Urethane foam A and B components are shipped in returnable pressure vessels.
- Chemical dispensing is controlled by a timer and monitored by a module that sounds an alarm if the flow departs from the designed ratio.
- Fixed-rate flow controllers regulate the chemical system component flow as a timer activates on/off ball valves, which are the only moving parts on the gun.
- Chemical components meet in the gun body and are mixed in an attached static mixing chamber prior to dispensing. Gun is solventless and flushless.

Rigid, semirigid, and flexible urethane foam systems and elastomeric compounds are also offered.

Foam Enterprises

The company custom formulates rigid, flexible, semirigid, and elastomer urethane chemical systems and distributes a full line of low- and high-pressure urethane dispensing equipment for pour, spray, and RIM.

Complete turnkey systems are provided in which chemicals, equipment, and services are provided, along with a complete line of low or non-CFC foam systems for flotation, refrigeration, and roofing.

GS Manufacturing Co.

The company has added two new units to its line of FRP-related systems. Both units enable fabricators to expand application/lamination procedures.

1. A dual-component system is for use with two-component, 1:1 ratio materials such as two-part resins and urethane foams. The compact pumping unit can be used for spraying, pouring, or RTM processes. The company's Little Willie gun is used either as an external airless sprayer or with an internal-mix head attachment. Optional features of unit include in-line heaters, heated hoses, and digital timing systems.
2. A low-cost airless gelcoat, entry-level compact system for RTM is offered, and it can be wall-mounted or attached to a five-gallon pail. It is suitable for R&D or small production runs. The unit features a 5:1 ratio material pump, a slave catalyst system, and the Little Willie gun with an internal-mix head suitable for wet-out rollers, saturators, or RTM nozzles. A digital timer can be used for exact-time pours of material.

The company also provides pour equipment for foam packaging applications and spray equipment for insulation and waterproofing. The Pack-and-Foam unit features an air-purge gun, 20-ft heated hoses, and direct pumping from drums. The FSF-20-1-50 foam spray gun is complete with heated hose, external-mix spray gun, and drum pump.

Gateway Precision Technologies

Gateway manufactures machines for automated handling, slabbing, down cutting, shaping, and contour cutting of expanded polystyrene and flexible urethane foam. Equipment includes fully integrated, computer-controlled bun handling and cutting lines for the manufacture of building panels; automatic lines for producing a variety of EPS and foam components for seat cushions or packaging; and versatile manual cutting tables and hand-held foam cutting guns. CNC is standard on moving-gantry contour-cutting systems and is available as an option on most other machines. A computer retrofit program for existing foam processing equipment is also available. The company also offers two- and three-axis CNC foam milling machines and an automated embossing system designed for foam applications. With its in-house technology and computer integration expertise, the company can undertake custom engineering and prototype development projects.

Glas-Craft, Inc.

Glas-Craft manufactures urethane spray (round and flat pattern), pour, packaging, and froth dispensing equipment. The company's Probler gun has patented automatic air-purge, air-assisted trigger, and material shut-off valves, and it fits all company's proportioners. T-series units have outputs up to 30 lb/min and separate solid-core primary heaters with solid-state controllers. Optional automatic temperature-controlled hoses allow up to 400-ft extension. Timing and fluorocarbon injection are available on all models. Also offered are variable-ratio proportioners (1:1 to 10:1) with 4 to 30 lb/min output rates.

Graco, Inc.

Graco supplies polyurethane foam spray and froth systems. Spray systems with outputs up to 30 lb/min incorporate a Foam Cat proportioning pump, heaters, heated hose up to 208 ft, and a low-maintenance foam spray gun with mechanical or solvent purge. The company also offers a complete selection of feed pumps for handling raw materials directly from original containers.

Froth systems are available in standard models for outputs up to 60 lb/min, and custom systems can be built for up to 300 lb/min, such as for barge filling. These systems use larger pumping units and a static-mix dispense head.

The company installs complete turnkey systems and custom designs plural-component systems for special requirements.

Gusmer Corporation

Gusmer produces an integrated line of plural component meter, mix, and dispensing equipment for spray or pour applications and outputs ranging from 2 to 240 lb/min. The RIM division produces high-pressure metering units, mix heads, and optional equipment designed for RIM processing. The company can design and build turnkey operations, bulk storage systems for chemicals, RIM presses, clamps, turntables, and multistation conveyors to fit customers' plant production requirements. The company maintains a 5000 ft^2 facility, containing its full product line for customer demonstrations and trials in systems and process development. Products include

- *RimCell Series.* Three models are available with outputs of 50-, 100-, and 240-lb/min and operating pressures to 3000 psi. The RimCell proportioner utilizes super-charged, high-pressure axial piston pumps, and is designed for processing unfilled, nonabrasive, noncorrosive materials.
- *Delta RIM Series.* Four models are available with outputs of 40, 80, 120, and 240 lb/min and operating pressures to 3000 psi. They are specifically designed to dispense higher-viscosity filled and abrasive materials, polyurea, polyurethane, elastomers, composites, and other plural component RIM systems.
- *H-2000.* This model is designed to dispense materials at pressures up to 2000 psi with outputs up to 30 lb/min. It is used for spray or pour applications and available with variable ratio.
- *FF-1600.* This is an air-driven, compact, lightweight unit capable of outputs up to 16 lb/min and pressures up to 1600 psi.
- *H-3500.* The H-3500 is used for both pour and spray applications. It dispenses material at pressures up to 3500 psi with outputs up to 20 lb/min.

Hennecke GmbH (see Miles Inc.)

Hi-Tech Engineering, Inc.

The company designs and manufactures high-pressure, plural-component metering and mixing machines and related equipment. It manufactures machines for RIM, RRIM, open molding, and SRIM in the urethane, polyurea, and DCPD markets. The company markets high-pressure, lab-scale machines as well as machines for the RTM market. Bulk storage systems, as well as design and fabrication for complete molding systems, are also provided. Large RIM, SRIM and RTM presses are also available.

The company offers its Precision Flow pump-type continuous meter machine featuring high-pressure pumps mounted in the daytank with a built-in maintenance boom that allows leak-free operation with ease of maintenance. The unit has found applications where rapid shots or continuous metering are required. In addition, the HR series is offered for those who prefer Rexroth-type metering machines.

IFTA Canada, Inc.

IFTA Canada provides individual machines and turnkey systems from Impianti S.p.A. of Italy, including Series 2001-C low-pressure pouring systems for urethane foams, which have computer controls, recycling mix heads, rotary mechanical mixers (6000 rpm), and outputs from 3 g/s to 11 lb/s. Standard models can handle up to six components.

Series VHP high-pressure metering systems for foams, elastomers, and RIM have self-cleaning mix heads and axial piston pumps. A special version capable of rapid, automatic ratio changes is designed for producing dual-hardness auto seating. The 2LC models with plunger-type metering cylinders permit use of fillers and reinforcements on both polyol and isocyanate sides. The most sophisticated models are hydraulically driven. Special models have also been developed for nylon RIM.

Series EL elastomer pouring machines can work at temperatures to 265° F and provide high-vacuum degassing and outputs to 100 g/s at 1:1 ratio. Mechanically driven mix heads and self-balancing ratio and output control are additional features.

A variety of rotary mold carousels and mold-support presses, including tilting and other RIM-type presses, are available. Continuous-pouring lines for flexible and rigid buns and blocks can produce 220 to 1100 lb/min. The patented Top Foam lines produce blocks with square sections. Foam-core sandwich panel lines include special versions for phenolic foams. Turnkey refrigeration foaming systems are also offered.

Insta-Foam Products, Inc.

Insta-Foam manufactures self-contained, factory-pressurized, single- and twin-component disposable and refillable systems for pouring and spraying foams in a variety of densities. Applications include cryogenic, cold-storage, foam packaging, caulking and sealing, void-filling, flotation, and insulating irregular surfaces.

Italian Trade Commission

The commission office can provide information on a large range of equipment for flexible and rigid foams built by Italian companies. These include low-pressure machines, high-pressure pouring, RIM and RRIM systems, and complete turnkey plants.

IVEK Corporation

IVEK manufactures liquid dispensing, metering, and blending systems for urethanes, epoxies, additives, and other liquids. Complete systems with static mixers are available for low-flow applications. Features include solid-state electronics, stepping-motor drives, and maintenance-free operation.

Kornylak Corporation

Specializes in complete production line equipment and components for producing urethane, isocyanurate, phenolic, and inorganic foams. Equipment supplied includes

- Continuous foam slab, bun, and log production lines (from 2 to 96 × 30 to 60 in) that operate at 5 to 60 ft/min
- Continuous foam-core and honeycomb-core panel production lines (from 2 to 12 ft wide producing 1/8 to 6 in thick panels that run at 2 to 200 ft/min)
- Continuous scrap foam rebonding lines
- Rigid and flexible foam molding lines

Also available is the company's three- or four-component phenolic foam machine. The PhenoFlo meters, mixes, and pours or sprays phenolic formulations in a viscosity from 5000 to 100,000 cp. The company also offers a complete turnkey package, including operator training, equipment start-up assistance, and assistance in chemical formulation and selection.

The newest addition to the company's line of foam equipment is the Foamcutter, a computerized high-production profile foam-cutting machine.

Krass-Maffei Corporation

The company offers a complete range of RIM and RRIM machinery and plant equipment for molding flexible, semirigid, and rigid urethane foam parts, as well as elastomeric and solid parts made from reactive mixtures.

The company's Rim Star machine capacity range is approximately 10 to 1600 lb/min. Lance-type piston RIM and RRIM machines (type KK) allow shot weights up to 80 lb. Standard equipment includes color metering units and systems for blowing agent metering. Filler blending is available. RIM and RRIM machines have patented self-cleaning impingement mix heads, which recirculate through injection nozzles for proper fill in closed molds or splash-free dispensing into open molds. All machines can be operated with multiple mix heads.

In addition, company supplies mold carriers and clamps, racetrack conveyors, turntables, mix-head manipulators, and robots. Machines and equipment are controlled by programmable controllers and computerized systems.

Kymofoam Company

Kymofoam offers low-pressure dispensing equipment for processing rigid, semirigid, and flexible polyurethane foams (pour and forth systems); elastomers; epoxies; and silicones. Both standard and custombuilt machines are designed to handle up to eight components with varying viscosity ranges and dispensing capacities from 0.5 to 800 lb/min.

Metering systems with adjustable ratio range from 1:1 to 100:1 feature gear pumps with closed-loop recirculating systems, a lightweight mixing head with electrically or hydraulically driven mixer, and a unique calibration design. One-shot, nonrecirculating piston metering systems are also available. KF equipment is designed for both low- and high-volume production of molded parts, flexible/rigid foam slabs, and boardstock insulation.

LME-Beamech, Inc.

The company's dispensing systems for polyurethane foams have 300 to 800 lb/min capacity. The injector block has 16 ports, including a 2-in resin port. Systems are set up so that all functions are controlled and programmed from a single center. Systems are supplied for flexible and rigid foam slab, foam boardstock, and sandwich-panel laminators.

Foam slab storage and multiple-tier automatic handling systems cut foam to length using a belt and aluminum slat design. Other systems offered include automotive seat-pad routers, seat-pad dust-removal units, continuous-pattern routers for rolled material, and handling equipment. Complete polyurethane foam pour conveyor systems, automatic cut-off saws, and trimmer saws are also supplied.

Rigid foam laminator conveyors are available for continuous production of rigid urethane foam board to 48 in wide × 6 in high on a double-metal-slat conveyor with top and/or bottom paper or other substrates. Laminators are built from 30 to 100 ft long. Models are available for up to 20 psi operating pressure; special units are built to customer specifications.

Edge-trim saws and cut-off units are synchronized to conveyor speed, providing a finished product accurately cut to commercial length.

Linden Industries, Inc.

The company engineers, designs and manufactures plural-component systems and specializes in the following areas:

- Complete turnkey automotive and nonautomotive process lines.
- Bulk storage systems.
- Dust-free glass blending systems.
- Large-scale RIM, RRIM, SRIM, and DCPD metering machines based on high-pressure impingement. These can include up to 50% filler content, and special applications experience is available.
- Flexible foam and rigid foam metering machines.
- Laboratory-scale metering machinery and data acquisition software.
- RIM recycling equipment.
- Turntables, mold carriers, and special designs for production efficiency.
- Controls and SPC packages for automotive and nonautomotive requirements.
- Consulting and design for robotics and control systems and automation requirements.
- 3-D CAD design capabilities.
- RIM, RRIM, RTM clamps, C-frame, and four posters.
- Dual-stage nucleation and in-line density measurement.

Liquid Control Corporation

The company's Twinflo polyurethane dispensing machines for metering two- or three-components use patented Posiload volumetric piston pumps. The unit is said to repeat shots accurately from 5 cc and up at speeds ranging from less than 1 to 100 lb/min, and to dispense viscosities ranging from water-thin to thick, heavily filled pastes. Pump design offers ratios from 1:1 to 200:1. Standard and custom-designed units are offered. The equipment is suited to filling open molds or for injecting materials under constant pressure into closed cavities.

Machem Corporation

Machem offers two lines of equipment—the Mistafroth gun for froth foams and the Mistamixer for pour-in-place and packaging foams. The Mistamixer requires no purging solvent and can handle packaging foam, pour-in-place foam, or froth foam. It meters, mixes, and dispenses foam in 5 to 15 lb/min rates. Mistafroth gun is designed for use with pressurized tanks, does not require electricity, and has few moving parts to give simple, maintenance-free operation. The company also offers a complete line of rigid and flexible foam systems.

Max Machinery, Inc.

The company manufactures liquid mixing/metering systems for use with two or more components. Systems will heat materials up to the desired process temperature, degas, and meter with an accuracy of ±0.5% of the desired flow rate, mix, and dispense. Systems are suitable for dispensing of urethanes, silicones, epoxies, and other liquids, and they are designed around user requirements, using standard subassemblies.

Meter-Mix, Inc., sub. of Accumetric, Inc.

The company manufactures volumetric dispensing and mixing machines that reportedly employ unique cylinders designed so that seals are not subject to metering stroke pressure. Static or dynamic mixing heads are offered with valving systems that are said to eliminate lead/lag problems. Also offered is a range of equipment utilizing positive-displacement gear and piston pumps driven by dc motors with digital control for precise ratio control and metering accuracy. Such machines can process up to five chemical components. Temperature control of the chemical components is optional.

Miles, Inc., Hennecke Machinery Group

The company offers a broad line of standard and custom machinery for polyurethane processing. Equipment offered includes bulk handling facilities, blending systems, temperature-control and metering equipment, lay-down systems, and processing machinery. The company builds double-belt laminators for up to 150 ft/min production of rigid foam board, stationary fixtures for refrigerator applications, and turntable and conveyor lines for flexible foam molding and other applications.

The company also supplies RIM and RRIM metering units and mold clamps up to 250-ton capacity. Metering equipment covers a processing range of 1.5 to 600 lb/min or more and can include any of several proprietary impingement mix-head types such as the self-cleaning MQ mix head.

Hennecke Machinery is the manufacturer and representative of Hennecke GmbH polyurethane processing equipment for the North American market.

Mobay Corporation (see Miles Inc.)

Mondomix Holland (see Construction Technology)

Poly-Craft Systems Division (see Binks Mfg. Co.)

Pyles Business Unit, Graco, Inc.

Meter/mix/dispense equipment for pouring polyurethane foam at flow rates up to 30 lb/min can be arranged for shot-type applications or continuous flow setups. The line is available in fixed- or adjustable-ratio units with a positive-displacement metering system that is said to ensure accurate dispensing.

The Rotary Duo-Flo 9100 is currently used for RIM and with other rapid-cure urethane systems. Mixing/dispense rates are 5, 10, and 20 lb/min.

Quick-Foam, Inc.

Quick-Foam offers two lines of equipment: pouring systems for industrial foam-in-place packaging and spray systems for insulation applications. Two pieces of equipment are made for each line: a small portable unit that uses nitrogen propellant and has two 10-gal refillable tanks, and a larger unit that pumps directly from 55-gal drums. All equipment has separate preheat and hose heat controls, separate pressure controls for each side, and a patented self-cleaning gun. A full line of urethane foam chemical systems is offered, including semiflexible, flexible, and rigid foams.

Rynel (formerly Twin Rivers Engineering)

Rynel supplies meter/mix/dispense systems capable of variable-ratio processing of plural-component liquid systems such as urethanes, epoxies, polyesters, and silicones. The company offers components, groups of components incorporated into an existing system, or complete production systems created for a specific purpose. Indexing and fixtured equipment can be supplied as part of a system. Sizes include lab and pilot-plant to full-scale production capability. Equipment has been designed to incorporate up to 80% solid fillers in urethane foams.

Sealant Equipment & Engineering, Inc.

The company manufactures variable- and fixed-ratio meter, mix, and dispense equipment with volumetric or timed shots, or operator-controlled volume and flow rates to 20 lb/min. Equipment can be supplied with heated pressure tanks, gravity, pump, or positive-displacement feed. Applications include packaging and fireproofing.

Sealed Air Corporation

Sealed Air offers its Instapak line of foam-in-place dispensing systems. Several different models, including semiautomated systems, handle rigid, semirigid, and flexible foams. The company also offers a family of Instapak polyurethane packaging foams of varying density.

Stepan Co., Industrial Chemicals Department

This is the manufacturer of StepanPol polyester polyols and a complete line of Stepanfoam rigid polyurethane systems for insulation, molding, and aerospace applications. It also produces a line of low-pressure foam-dispensing equipment.

Urethane Service, Inc.

Urethane Service offers equipment for processing multicomponent thermosetting materials such as urethanes, polyesters, silicones, polysulfides, etc. Materials processed include elastomers in a wide range of hardness and also as syntactic or chemically blown foams over a wide range of densities. Filled and unfilled materials routinely process in viscosity ranges from 30 to 2000 cps. With special techniques, materials in excess of 1 million cps can be easily processed. Machines are offered that can deposit from less than 0.1 g of material to large void-filling equipment that dispenses mixed and metered materials at over 400 lb/min.

Types of machines supplied include equipment for pouring, casting, bead laying, frothing, potting, spraying, RIM, and resin-transfer molding (RTM). Self-cleaning impingement mix heads that use no solvent to purge can be used on most of the urethane materials. Disposable static mixers are used to process other materials to eliminate solvent usage. Dynamic mixing heads are furnished when required for the project.

C

Glossary

Abrasion cycle The number of repetitive abrasion motions that a specimen is subjected during an abrasion test.

Abrasion resistance The fundamental ability of a material to withstand surface rubbing, erosion, or scraping. Generally, polyurethanes display high abrasion resistance.

Abrasion Gradual erosion of surfaces by physical forces.

Abrasive A finely divided, hard substance capable of smoothing, cleaning, or polishing surfaces of other substrates. Abrasives may be used to polish the surface of polyurethane elastomers.

Accelerated test Any test procedure where specified conditions are intensified to theoretically reduce the time required to obtain a given deterioration.

Accelerator A substance used to increase the rate of a chemical reaction. Sometimes used in polymerization of thermoplastics, and rubber. Sometimes used as a synonym for catalysts.

Acid value A measure of the residual acidity of a substance, e.g., a polyester. Measured in mg KOH/g needed to neutralize the acidity.

Acid number A number expressing the amount of acidic residual material in the polyol. It is reported in terms of the number of milligrams of potassium hydroxide required to neutralize the acid present in one gram of a sample. The number is useful in indicating batch-to-batch uniformity; and as a correction factor in calculating hydroxyl numbers. For instance, the specifications for some polyols list the maximum acid number as 0.20 mg/KOH/g of sample.

Activator An agent added to enhance the action the accelerator in natural or synthetic resin to enhance the action of the accelerator in the vulcanization process.

Active hydrogen atoms Refers to the hydrogens that, because of their position in the molecule, display activity according to the Zerewitinoff test as described by Kohler in *J. Am. Chem. Soc.*, 49, 31–81 (1927). These active hydrogens readily react with isocyanates to produce urethane or urea linkages.

Addition polymers A large family of polymers formed by the simple combination of monomer units, without evolution of low-molecular-weight by-products, such as water or carbon dioxide. Polyethylene, polystyrene, vinyl, and acrylics are examples of addition polymers.

Additive Any substance added to polymers to alter certain properties. In the polyurethane field, these materials are added in small amounts to improve processing characteristics, enhance heat stability, provide antioxidant protection, stabilize polymers against gamma radiation, etc.

Adduct A reactive product or monomer formed by reacting two polyfunctional molecules. For example, TDI is reacted with polyols at low temperature to produce a reactive TDI adduct.

Adhere To cause two surfaces to be permanently held together by physical or chemical forces.

Adherend A substrate that is held to another body by an adhesive agent.

Adherometer An instrument designed to measure the strength of an adhesive bond.

Adhesion   The physicochemical state by which two surfaces are permanently held together by interfacial forces, which may consist of covalent forces, mechanical interlocking, or a combination of both.

Adhesion, mechanical   Adhesion between two surfaces in which the adhesive holds the parts together by an interlocking action.

Adhesion, specific   Adhesion between two surfaces in which the adhesive holds the parts together by covalent or ionic forces.

Adhesive, heat-activated   An adhesive, dry at room temperature, that is rendered tacky or fluid upon application of heat.

Adhesive, hot-melt   An adhesive that is applied in a molten state and is capable of forming a bond upon cooling to a solid state.

Adhesive, hot-setting   An adhesive that sets at temperatures at or above 100° C.

Adhesive, intermediate-temperature-setting   An adhesive that sets at temperatures of 31 to 99° C.

Adhesive, pressure-sensitive   A viscoelastic material that remains permanently tacky and is capable of adhering to most solid surfaces upon the application of minor pressure.

Adhesive, solvent   An adhesive containing a volatile organic vehicle or rheological agent.

Adhesive, solvent-activated   An adhesive that is dry at room temperature but rendered tacky upon contact with an organic solvent.

Adhesive, cold-setting   An adhesive capable of hardening (setting) at temperatures below 20° C.

Adhesive, contact   An adhesive that is apparently dry to the touch and capable of adhering to itself upon the application of minor mechanical pressure.

Adhesive, foamed   An adhesive whose apparent density has been decreased by the presence of uniformly distributed gaseous cells.

Adhesive, dispersion   A two-phase system in which the adhesive is suspended as small droplets in a liquid.

Adhesives   Materials capable of permanently joining one surface to another. Adhesives are widely used in the medical device industry to join a plastic device to another of the same polymer, a different polymer, or a nonplastic material. Adhesives used in these applications can be further classified into five types. (1) A *monomeric adhesive* contains a monomer and a catalyst so that a bond is produced by polymerization at the adhesive/substrate interface. (2) A *solvent adhesive* is one that mutually dissolves the polymers being joined, forming strong intermolecular bonds, prior to solvent vaporization. (3) *Bonded adhesives* are solvent solutions of polymers, sometimes containing tackifiers and plasticizers, that dry at room temperature. (4) *Elastomeric adhesives* contain elastomeric polymers dissolved in solvents or suspended in water (latexes) and are intended to bond at or near room temperature. (5) *Reactive adhesives* are those containing partially polymerized resins. They cure with the aid of catalysts or heat to form a permanent bond.

After cure   The period of time after all accelerated curing attempts have ceased and during which maximum final physical properties are attained.

Agglomerate   Large groups of pellets thermally bonded together. Usually, they are formed at the cutter hub under adverse conditions and break loose into the slurry line.

Anti-seize   A copper- or nickel-based compound designed to withstand high temperatures and prevent metallic parts from seizing together.

Aliphatic   (1) A term describing a chain-like molecule composed of carbon and hydrogen atoms, without the presence of benzene rings. (2) Organic compounds characterized by open-chain structures, whose molecules do not have their carbon atoms arranged in a ring structure. Contrasted with these are aromatic compounds, such as benzene, which have ring structures. Aliphatic compounds are named from the Greek *aleiphar*, which means fat or oil.

Alkali resistance   The important characteristic of many plastic materials to resist the corrosive effects of strong bases.

Alloy   Synergistic polymer combinations with property advantages derived from a high level of thermodynamic compatibility between components. Alloys exhibit strong intermolecular forces and form single-phase systems with unique glass transition temperatures. The most significant commercial alloys are the styrene-modified polyphenylene oxides. Polymer blends, by comparison, have less intense thermodynamic compatibility than alloys and thus result in less advantageous physical properties.

Alternating copolymer   A copolymer in which the two monomeric units that make up the chain alternate in a repeating, predictable fashion, such as -A-B-A-B-.

Ambient temperature   The temperature of the environment surrounding a specimen. Frequently used to denote normal, prevailing room temperature of 23° C.

Ambient   Completely surrounding a specimen. Indicative of the surrounding environmental conditions around a specimen, such as temperature, pressure, etc.

Amine equivalent   An analysis value determined for isocyanate materials to express their reactive assay, used in some methods for calculating *stoichiometric balance* of a formula. The value secured has an inverse relationship to the assay of the isocyanate.

Amine   A class of catalyst compound used in the urethane foam reaction; a compound derived by replacing one or more hydrogen atoms by organic radicals.

Amorphous   Devoid of molecular crystallinity, stratification or interchain orientation. Many polymers are amorphous at processing temperatures, with some retaining this state at room temperature, such as ABS, polycarbonate, polysulfone, and polyarylate. Amorphous polymers are generally clear, as opposed to crystalline polymers, which are generally opaque.

Aniline   A key raw material used to make MDI; made from benzene, a component of oil.

Anisotropic   Foam having different properties when tested along axes in different directions, e.g., parallel to foam rise as opposed to perpendicular to the foam rise.

Anisotropy   The tendency of a material to react differently to stresses applied in different directions. Medical devices are sometimes manufactured to match the natural anisotropy of skin, or blood vessels.

Annealing   The physical process of relieving internal stresses in a polymer by heating below softening temperatures and maintaining this temperature for a predetermined period of time. Annealing can be summarized as slow crystallization by heat treatment without large-scale melting. Annealing improves those properties of a given thermoplastic material that would normally be associated with high crystallinity.

Anti-foaming agent   (1) An additive that reduces the surface tension of a solution, thus inhibiting or modifying the formation of a urethane foam. (2) An additive that reduces the surface tension of an aqueous solution or emulsion, thereby inhibiting or preventing the formation of foam. Silicone antifoams are used during hemodialysis to reduce frothing and foaming of blood.

Antiblocking agents   Additives incorporated into polymers to reduce their tendency to stick during processing, storage, or use. Synthetic waxes and silicone oils are examples of antiblocking agents used in medical polymers.

Antimony trioxide   A white powder (Sb_2O_3) widely used as a flame retardant in urethane foams.

Antioxidant   Materials that, when added to a foam formulation, improve the resistance of the foam to oxidative type reactions.

Antioxidant   A substance capable of inhibiting the tendency of polymers to undergo oxidation when exposed to oxygen at normal or elevated temperatures. Antioxidants are most frequently used in conjunction with oxidation-susceptible polymers such as polyolefins, ABS, acetals, and polyurethanes.

Antiozonant   A substance added to polymers to retard or prevent deterioration caused by continuous exposure to an atmosphere containing ozone gas.

Apparent density   The weight per unit volume of a material, including all voids inherent in the material as tested. Used in connection with powders, granules, pellets, etc. See also *Bulk factor.*

Aromatic compound   An important class of organic compounds characterized by an unsaturated ring of carbon atoms. Most of the ring compounds are related to benzene (C_6H_6) or its derivatives. The name is associated with the penetrating odor of the first few compounds isolated; this odor was described as *aromatic.* This classification, however, is no longer limited to an odor or type of odor.

Aromatic hydrocarbon   A hydrocarbon containing one or more six-carbon rings and having properties similar to benzene. This large family of compounds includes many plastics solvents.

Aromatic   A term describing molecules that include at least one benzene ring in the structure.

Arrhenius principle   This principle is concerned with chemical reaction rates, stating that, in general, for every 10° C (18° F) rise in temperature, the first-order chemical reaction doubles, and vice versa. The Arrhenius equation has the form: $k = Ae^{-E/RT}$ where A = proportionality constant, e = base for natural logarithms, E = activation energy, and T= absolute temperature. Examples of procedures that utilize the Arrhenius aging techniques are ASTM D-3045 (heat aging of plastics without load) and ASTM D-2990 (tensile, compressive, flexural creep, and creep rupture of plastics).

Aryl   A compound whose molecules have the ring structure characteristic of benzene, naphthalene, phenanthrene, etc. An aryl group may be phenyl (C_6H5-), or naphthyl ($C_{10}H_9$-).

Assay   An analytical measurement of the reactive content of a chemical component.

ASTM   Abbreviation for American Society for Testing and Materials, a nonprofit corporation organized in 1898. It is a world leader in the development of voluntary standards for materials, products, systems, and services. ASTM publishes more than 8500 standards in diverse fields such as medical devices, plastics, biotechnology, etc. ASTM's membership of 32,000, of which 4,000 are international members, is organized into 134 different technical committees that do the actual work of writing standards. Committee members voluntarily contribute their time and effort.

Autoxidation   A spontaneous, self-catalyzed oxidation occurring in the presence of air. It usually involves a free-radical mechanism and is seen in the degradation of high polymers exposed to gamma radiation, UV exposure for prolonged periods, etc.

Auxiliary   A solvent with a low boiling point, added to assist foaming in a TDI formulation. Normally, 80% of the foaming would come from the water/TDI reaction and 20% from the auxiliary blowing agent.

Average molecular weight   The average molecular weight of the chains of a polymer, independent of the specific chain length. The value falls between weight average and number average molecular weight.

Ball viscometer   A viscometer that employs solid spheres of specified weight and diameter as the shearing mechanism.

Ball rebound test   A method for determining the elastic response of polymeric materials by measuring the energy absorbed when a steel ball impacts the material. The ball is dropped from a fixed height, and the rebound height is measured. The difference between the two heights indicates the energy absorbed. The more elastic the polymer, the greater the rebound height.

Banbury mixers   A mixing machine consisting of two counter-rotating spiral blades encased in segments of cylindrical housings, intersecting so as to leave a ridge between the blades. Banbury mixers are frequently used for compounding solid radiopacifiers into biomaterials.

Barrier Coat   A clear or pigmented coating sprayed into a mold, prior to foaming, that adheres to the foam and provides a base for further finishing steps.

Batch   A manufactured unit, or a blend of multiple units of the same formulation and processing.

Billet A cut-off segment of a continuously produced loaf of polyurethane foam. Also called *bun* or *block*.

Binder (1) An adhesive material used for holding particles of dry substances together. (2) The important component of an adhesive formulation that is responsible for the adhesive forces that hold two substrates together.

bis Prefix meaning t*wice* or *again*. Used in chemical nomenclature to indicate that a chemical grouping or radical occurs twice in a molecule (e.g., bisphenol A) where two phenols appear, or ethylene bis stearamide, where two stearic acids reacted with ethylene amine to form the amide linkage.

Blend A uniform combination of two or more materials, either of which could be used alone for the same purpose as the blend.

Block copolymer A copolymer with chains composed of shorter homopolymer chains that are covalently linked together. The blocks can be regular or randomly placed. Block copolymers usually display higher impact strengths than either homopolymer or plain physical mixtures of the two homopolymers.

Blocked curing agent A curing agent that can be activated by physical or chemical means. Activation usually results in molecular splitting in which one of the products is capable of curing a resin.

Blocking An undesirable adhesion between layers of plastic, which may develop during pressure of storage, heat, sterilization, etc. The tendency to block is reduced by chemicals added to the plastic called antiblocking agents.

Bloom An undesirable cloudy effect on the surface of a plastic article due to the incompatibility and migration of a compounding ingredient such as a lubricant, antioxidant, plasticizer, etc.

Blowing reaction One of the several chemical reactions occurring in the final mixture while it is foaming.

Blowing agent The chemical ingredient in the formulation that provides the gas that creates the expansion of the foam. Also called a *foaming agent*.

Boiling and flashing Boiling is said to occur in a urethane foaming reaction when the rate of formation and/or release of the blowing agent (CO_2, refrigerant 11) is much greater than the polymerization rate. The result will be a vigorous evolution of gas from the liquid material that may end in foam collapse.

Bond strength (l) The unit load applied in tension, compression, flexure, peel, impact, cleavage, or shear required to break an adhesive assembly, with failure occurring at the interfacial plane of the bond. (2) The degree of attraction existing between atoms within a molecule.

Bond, chemical An attractive force between atoms strong enough to permit the combined aggregate to function as a unit. Covalent bonding results most commonly when electrons are shared by two atomic nuclei.

Bottoming The characteristic of some flexible polyurethane foams, especially polyesters, to support an initial load with a small amount of deflection but to virtually collapse under any additional load. After this severe compression, any additional load will not cause much further deflection.

Branched polymer A polymer in which the molecules have side chains, or branching. The opposite of a linear polymer.

Branching The growth of a new polymer chain from an active site on an established chain, in a direction different from the original chain, similar to the branching of a tree. Branching occurs as a result of chain transfer processes or from the polymerization of difunctional monomers. It has important effects on the physical properties of the finished polymer.

Brittleness temperature The temperature at which polymers rupture by sudden impact under specified conditions. This temperature is related to the glass transition temperature, T_g.

Buffered catalyst Usually, a very alkaline amine that has been partially or wholly neutralized with acid so that the condensation reaction would not be affected by the extreme alkalinity of the catalyst.

Bulk factor   The ratio of volume of plastic particles to volume of the same mass after molding or compression.

Bulk polymerization (mass polymerization)   The polymerization of a monomer in the absence of any medium other than a catalyst or accelerator. Polystyrene, polymethyl methacrylate, polyhydroxy ethyl methacrylate, polyethylene, and styrene-acrylonitrile are prototype polymers produced by bulk polymerization. Most of the medical hydrogels are formed by this process.

Bulking agent   A material or chemical added to another chemical that increases the quantity of the mixture without changing the chemical reactivity of the total.

Bun   Synonym for billet, or block. A segment of foam cut off from continuously produced polyurethane foam slabs.

C stage   The final, cured state of a thermosetting resin.

Calender coating   The process of coating substrates with thermoplastic resins by forcing both the substrate and the thermoplastic resin through calender rolls.

Calender   A large machine performing the operation of calendering onto thermoplastic resins.

Carbon black   A generic name for the entire family of colloidal carbons, produced by the incomplete combustion of gas, oil, or another hydrocarbon. Carbon black is used to impart electrical conductivity to biomaterials.

Carbonyl group   The divalent group C=O, which occurs in a wide range of chemical compounds. It is present in aldehydes, ketones, organic acids, and sugars.

Carboxylic acid   Any organic acid composed chiefly of alkyl (hydrocarbon) groups, usually in a straight chain, terminating in a carboxylic acid group. Exceptions are formic and oxalic acids.

Carboxylic   Term for the COOH group, the radical found in organic acids.

Castable polyurethane   Castable polyurethanes are all liquid nonfoam urethane polymers, 100% (or nearly 100%) solids, reactive polymer systems.

Casting   (a) The process of forming solid or hollow articles from fluid plastic mixtures or resins by pouring or injecting the fluid into a mold or against a substrate with little or no pressure, followed by solidification and removal of the formed object. (b) The process of forming solid or hollow articles from fluid plastic mixtures in a mold with little or no pressure. Following solidification, the articles are removed by the process known as *demolding*.

Castor oil   A pale yellow oil derived from *Ricinus communis* beans. It is an important raw material for plasticizers, nylons, alkyd resins, and certain medical-grade polyurethane casting compounds.

Catalyst   A chemical that has the property of being able to change the speed of a chemical reaction without apparently taking part in the reaction. There are also negative catalysts or *inhibitors*.

Cell structure   Term often used to describe overall uniformity of foam cell diameter. *Open cells:* interconnecting cells in a foam. *Closed cells:* cells enclosed by continuous membranes and struts. *Cell opening:* breaking on intercellular membranes. Cell opening may occur during foam rise, by mechanical crushing, or by chemical treatment. *Cell count:* the number of cells per linear inch.

Cell size   The average diameter of the pores in the final product.

Cell collapse   A defect characterized by slumped and cratered surfaces, along with collapse of internal cells resembling a stack of leaflets when viewed in cross section under a microscope. It is caused by excessively rapid permeation of the blowing gas through the cell walls, or by weakening of the cell walls by plasticization.

Cell count   The number of cell or bubbles per linear inch or centimeter.

Cell membrane   Thin intact film that forms the bubble walls; also called *windows*.

Cell   Synonymous with bubble or pore. It refers to the cavities left in the foam structure after the bubble walls have completely polymerized and solidified or curled back and fused into the boundary joints to form a skeletal structure; a cavity formed by gaseous displacement in a plastic.

Cellular plastic   (1) A plastic composed of numerous microscopic cells evenly distributed throughout its mass. A cellular plastic may be created by (a) incorporation of a blowing agent that decomposes at elevated temperature to liberate a gas, (b) mechanical frothing of a gas, (c) incorporation of a soluble material that is subsequently leached, and (4) addition of a solvent-extractable material. (2) Plastic with numerous cells disposed throughout its mass. The terms *cellular*, *expanded*, and *foamed* plastic are used synonymously.

Cementing   The process of permanently joining plastics to themselves, or to other substances, by means of solvents, dopes, or chemical cements. Solvent cements constitute a mutual solvent for the cement and substrate; dope cements contain a solvent solution of a plastic similar to the plastic substrate; chemical cements are based on reactive species that polymerize to form strong bonds.

Centigrade (Celsius)   A measurement of heat based on a scale divided into 100 degrees, where water freezes at $0°$ and boils at $100°$. Normal body temperature on this scale is $37°$.

Centipoise   A unit of viscosity, composed of one hundredth of a *poise*. Water at room temperature has a viscosity of about 1 centipoise.

Centistoke   A unit of viscosity, composed of one hundredth of a *stoke*. A stoke is equal to the viscosity in poises of a liquid, times the density of the liquid in grams per cubic centimeter.

Chains   In thermoplastic extrusion, groups of pellets that are fused together in a single row.

Control panel   An enclosure (usually mounted near the pelletizer) that houses pushbuttons, selector switches, alarm devices, and a front door. Power is usually located in a separate enclosure.

Channeling   A small or narrow undercutting of the expanding or rising foam front by a stream of the mixed clear liquid.

Chemical nomenclature   The origin and use of the names of elements, compounds, and other chemical entities, individually and as a group, as well as the various accepted conventions for systematizing them. Present nomenclature follows the reforms adopted by the International Union of Pure and Applied Chemistry (IUPAC).

Chemical change   Rearrangement of atoms, ions, or radicals resulting in the formation of new substances, often displaying entirely different properties. Such a change is called a *chemical reaction*.

Chemical resistance   The fundamental ability of a polymer to withstand chronic exposure to acids, alkalis, organic solvents, and other corrosive chemical environments.

Chemical technology   A generic term covering the spectrum of physicochemical knowledge of materials, processes, and operations used in the chemical industry. It includes (1) basic phenomena such as catalysis and polymerization, (2) properties, behavior, and handling of materials, and (3) the formulation, compounding, fabrication, and testing of chemicals.

Chemical reaction   A change that may occur by combination, replacement, decomposition, or some suitable combination of these. Common reactions are oxidation, reduction, ionization, combustion, polymerization, hydrolysis, condensation, enolization, saponification, rearrangement, etc. Chemical reactions involve only rupture of the bonds that hold the molecules together and should not be confused with nuclear reactions, in which the atomic nucleus is involved.

Chemically foamed plastic   A cellular plastic in which the cells are formed by a chemical blowing agent or by the chemical reaction of constituents.

Chemisorption   The formation of bonds between the surface molecules of a material of high surface energy and a gas or liquid in contact with it.

CHEMTREC   Abbreviation for Chemical Transportation Emergency Center, established in Washington D.C., by the Chemical Manufacturers Association to provide emergency information on materials involved in transportation accidents.

Cis   A chemical prefix, from the Latin *on this side,* denoting an isomer in which certain atoms are on the same side of a plane. The opposite of *trans.*

Clean Air Act   This act addresses three broad types of air pollution problems. (1) the Ozone Protection Act forces manufacturing plants to attain norms deemed necessary to protect the Earth's ozone layer. (2) The acid rain provisions force plants to cut emissions of acid gases. (3) The toxic air section specifies that any pollution source, that imposes a cancer risk greater than a certain probability be eliminated.

Closed cells   A foam structure (as in rigid foams) wherein each individual cell has a complete cell membrane enclosing the cell and there are no open passageways between cells.

Closed molding   The practice of molding a foam object in a cavity in which the foaming liquid is 100% contained, with only gas being allowed to escape.

Closed cells   The property of a foam where each individual bubble is completely sealed from its neighbor so that no gas exchange can take place except by diffusion through the walls.

CO_2-blown foam   Foam in which all of the gas for expanding or blowing is generated by the chemical reaction between water and the isocyanate material. This type foam is called *water blown.*

Collapse   In foam terminology, the inadvertent and undesirable densification of a cellular plastic.

Colligative property   A property that is numerically equal for any group of substances, independent of their respective chemical structures.

Colloid chemistry   A subdivision of physical chemistry comprising the phenomenological study of matter when one or more of its dimensions lie in the range between 1 millimicron (nanometer) and 1 micron (micrometer). Natural colloidal systems include blood, milk, and rubber latex.

Colloid   Any substance capable of forming stable suspensions or emulsions with a liquid. Colloidal suspensions should not settle to any noticeable degree and should not diffuse readily through vegetable or animal membranes. High-molecular-weight substances capable of forming colloidal suspensions are usually in the range of 10^{-7} to 10^{-5} cm in diameter.

Colloidal mill   A piece of equipment capable of manufacturing emulsions and colloidal suspensions, comprising a high-speed rotor and a fixed or counter-rotating element in close proximity to the rotor.

Color stability   The maintenance of original color in a polymer. See also *Light resistance.*

Color migration   The movement (diffusion) of dyes and pigments through the matrix of a material.

Colorants   Dyes or pigments that impart color to polymers. Dyes are synthetic or natural compounds of microscopic size, soluble in organic solvents, yielding transparent colors. Pigments are organic or inorganic substances, with larger particle size, insoluble in organic solvents. Organic pigments produce nearly transparent colors, whereas inorganic pigments produce opaque colors (with some exceptions).

Color concentrate   A plastics compound containing a high percentage of pigments or dyes, used for blending with another resin so the correct and desired amount of color is easily achieved. The concentrate provides a dustless and convenient method of obtaining reproducible colors in plastics compounds. The term *master batch* is sometimes used for color concentrates as well as for concentrates of other important additives such as catalysts, radiopacifiers, and lubricants.

Colorfastness   See *Light resistance.*

Colorimeter   An instrument used for matching colors with results approximating those obtained by expert visual inspection. The sample is illuminated by light from three primary color filters and qualitatively scanned by an electronic detection system. Colorimeters are frequently used in conjunction with spectrophotometers for close color control in production operations.

Colorimetry   An analytical method based on the knowledge that polymers undergo characteristic color changes when exposed to certain chemicals.

Combustible liquid   An organic liquid that evaporates flammable vapors at 150° F or below. Many organic solvents used in the fabrication of medical devices are combustible according to this definition. Exceptions are the chlorinated and fluorinated solvents.

Combustion   An exothermic oxidation reaction that may occur with any organic compound as well as certain elements, e.g., hydrogen, sulfur, phosphorus, and magnesium. The heat evolved by combustion is due to rupture of chemical bonds and the formation of new compounds. Substances vary greatly in their combustibility, that is, in their ignition points (solids and gases) or their flash points (liquids).

Comminute   To pulverize or reduce solids to small sizes by mechanical methods such as grinding.

Compound   A mixture of polymer(s) and ingredient(s) necessary to modify the polymer to a form suitable for processing, fabrication, and stability in the intended implant site.

Compounding   The art of mixing polymers with additives such as lubricants, stabilizers, fillers, and pigments in a form suitable for production of finished products.

Compressibility   The change in volume per unit volume produced by a change in pressure. The reciprocal of bulk modulus.

Compression molding   A method of molding in which the preheated polymer is forced into a cavity. The material is subjected to pressure and (usually) heat until cure has been effected. The process most often employs thermosetting resins such as silicones for the production of medical devices.

Compression ratio   In extrusion terminology, the ratio of the volume of material held in the first flight at the feed zone to the volume held in the last flight in the metering section. This ratio provides an indication of the compaction performed on the material and the amount of mechanical work done on the melt by the screw.

Compression mold   A precision-machined mold used in the process of compression molding.

Compression zone   The portion of an extruder barrel in which melting of the extrudate is completed.

Compression set   A permanent deformation of a material resulting from the application of compressive stress.

Compressive modulus   The ratio of compressive stress to compressive strain below the proportional limit. Theoretically equal to Young's Modulus determined from stress-strain curves.

Compressive stress   The compressive load per unit area of original cross section carried by the specimen during a compression test.

Compressive strength   The maximal load sustained by a test specimen in a compressive test divided by the original area of the specimen.

Condensation agent   A chemical substance that acts as a catalyst during a polycondensation reaction. It may also be a substance that provides a complement reactant necessary for a polycondensation reaction to occur.

Condensation   (1) A chemical reaction in which two or more molecules combine, with the separation of water, alcohol, or other simple substance (condensate). If a polymer is formed by condensation, the process is referred as polycondensation. (2) The change of state of a substance from the vapor to the liquid state.

Conditioning   The act of subjecting a test specimen to standard environmental and/or stress history conditions prior to testing. Typical conditions for medical devices are 40–50% relative humidity at room temperature.

Conductance   The electrical conductivity of a solution, defined as the reciprocal of resistance. Most commonly used in connection with electrolytic solutions.

Conductivity, electrical   The reciprocal of volume resistivity; the conductance of a unit cube of material.

Configuration   In an organic molecule, the specific location or disposition of substituent atoms or groups around asymmetric carbon atoms.

Conformal coatings   Thin surface coatings of polymers designed to protect substrates from environmental degradation, abrasion, etc.

Conformation   The shapes or arrangements in three-dimensional space that an organic molecule assumes by rotating carbon atoms, or their substituents around single covalent bonds.

Conjugated double bonds   A chemical term signifying double bonds separated from each other by single bonds, such as 1,3 butadiene, CH_2=CH–CH=CH_2.

Conjugated   In chemistry, the regular alternation of single and double bonds between atoms and molecules, such as those seen in the benzene ring.

Contact adhesive   A liquid adhesive that dries to a film that is tack-free to other materials but not to itself. A typical medical-grade contact adhesive can be made of a silicone elastomer dissolved in chlorinated solvents. The adhesive is applied to both surfaces to be joined, then partially dried. When pressured together for a few seconds, a bond of high initial strength is obtained.

Contact pressure resins   Liquid resins that thicken on heating and, when used for bonding, require little or no pressure to effect bonding.

Continuous polymerization   An economically advantageous type of polymerization in which the monomers are continuously fed to a reactor, and the polymer produced is continuously removed. Medical-grade Pellethane polyurethane is produced by a continuous extruder-reactor polymerization process disclosed in U.S. Patent 3,642,964.

Continuous phase   In a suspension, the liquid medium in which the solid particles are dispersed. The solid particles are the discontinuous phase.

Control/power panel   A single, usually free-standing panel enclosure that houses pushbuttons, selector switches, alarm lights, instrumentation devices, etc. on the front door. Fuses, contactors, motor starters, transformers, etc. are mounted on the back plate.

Cooling water   The water supplied from a cooling tower and passed through the body of the heat exchanger portion of the tempered water system.

Converting   A term descriptive of processes by which a multitude of packaging products are produced. Converting involves coating, impregnating, laminating, embossing, printing operations, etc. that are necessary to produce finished packaged products.

Copolycondensation   The copolymerization of two or more monomers by the condensation polymerization process.

Copolymer   A high-molecular-weight substance containing several types of repeating structures. A styrene-methyl methacrylate copolymer is obtained by polymerizing styrene and methyl methacrylate together. It is sometimes used for terpolymers (acrylonitrile butadiene styrene), quadripolymers, etc. Three common types of copolymers are block copolymers, graft copolymers, and random polymers.

Corona discharge   The flow of electrical energy from a conductors to the surrounding air or gas. The discharge is produced by high voltage, >5000 V, resulting in a characteristic pale violet glow.

Corona discharge treatment   An important surface treatment that renders normally inert polymers, such as olefins and fluorocarbons, more receptive to coatings, adhesives, and inks. The corona discharge oxidizes the surface of the polymer by the formation of polar groups on reactive site.

Corrosion   (1) The degradation of polymers as a result of environmental forces. (2) The destruction of body tissues by strong acids and bases. (3) The electrochemical degradation of metals due to environmental factors.

CPR   Abbreviation for *controlled polymerization rate*. This concept is used as a means of indicating the reserve base of the polyol. It represents the residual weakly basic constituents not qualitatively defined but including weak acid salts of strong bases present in less than the 10 ppm range in the polyol. These residual basic materials have a catalytic effect on the polymerization rate, so it is important that the amount be controlled from batch to batch.

Crazing   Microscopic fine cracks that may extend in a network on or under the surface of a polymer. An undesirable effects in plastics articles characterized by a frosty appearance, due to (1) shrinkage, (2) flexing, (3) solvents, (4) temperature changes, and (5) environmental shocks.

Cream time   The time between the discharge of the foam ingredients from the foam head and the beginning of the foam rise. At this point, the surface of the liquid will change color (usually lighter) due to evolution of the blowing agent.

Creep rupture   A failure mechanism resulting in rupture of a polymer under continuously applied stress at a point below the normal tensile stress. This phenomenon is caused by the viscoelastic nature of many polymers.

Creep   The time-dependent dimensional change of a polymer under load. Creep at room temperature is also called *cold flow*.

Critical surface tension   The value of surface tension of a liquid below which a liquid will wet the surface of a polymer. The critical surface tension is measured by contact angle studies and is measured in dynes/cm. Values of some medically important polymers are acetal = 47, epoxy = 47, fluoroethylene propylene = 16, polyamide = 46, polycarbonate = 46, polyethylene = 46, polymethyl methacrylate = 49, polysulfone = 41, polytetrafluoroethylene = 18, and silicone = 39. Surface tension of common liquids are: glycerol = 63, petroleum oil = 29, epoxy adhesive = 47, and water = 73.

Cross-linking index   The average number of cross-linked units per primary polymer molecule in the system as a whole.

Cross-linking   As applied in polymer chemistry, the formation of covalent bonds between adjacent molecular chains. When extensive, as in most thermosetting resins, cross-linking makes one infusible supermolecule of all molecular chains, also rendering the material insoluble in most organic liquids. Cross-linking may be achieved by chemical agents that produce free radicals, by using multifunctional monomers, or by irradiation with high-energy electron beams.

Crude isocyanate   (1) An undistilled isocyanate mixture containing several different polymeric isocyanates. (2) Polymeric isocyanate derived from aniline and formaldehyde.

Crude TDI   Toluene diisocyanate containing various by-products designed to slow down its reactivity.

Cryogenic surgical device   A Class II device used to destroy nervous tissue by the application of extreme cold to the target site.

Cryogenic   Pertaining to very low temperatures, such as −150° F or below. Cryogenic temperatures are used to granulate elastomeric polymers.

Cryptometer   An instrument for measuring the opacifying power of surface coatings.

Crystal lattice   The spatial arrangement of atoms or radicals in a crystal.

Crystal   A homogeneous solid displaying an orderly and repetitive steric arrangement of its atoms.

Crystalline polymers   Polymers containing a portion of their atoms and molecules arranged in a perfect crystal lattice. These polymers exhibit excellent chemical resistance, sharp melting points, low melt viscosity, and significant tensile, flexural, and heat distortion improvements with reinforcement. Typical crystalline polymers are: polypropylene, polyacetal, nylon 6/6, and polybutylene terephthalate.

Crystallinity   A state of molecular structure in some polymers attributed to the existence of crystals with a definite geometric form.

Crystallization   The phenomenon of crystal formation by nucleation and accretion.

Curative   Any substance or agent that effects a fundamental and desirable change in a material to make it more suitable for practical purposes.

Cure time   The length of time required for sufficient reaction completion to develop desired polymer properties such as strength, dimensional stability, elongation, etc. The longer the pot life, the longer the cure time. Heat and catalysts facilitate the cure as they speed the chemical activity of the compound. Temperatures below 70° F lengthen pot life and lengthen cure time.

Cure   (1) To change the physicochemical properties of a material by polymerization, cross-linking, or vulcanization. Usually accomplished by the combined actions of heat, pressure, and a catalyst. The term is properly used when referring to thermosetting resins, although it is incorrectly used in conjunction with the polymerization of thermoplastic resins. (2) The completeness of the chemical reaction. At 100% completion, the polyurethane should have 100% of the maximum physical properties attainable with that particular formulation.

Curing agent   A chemical that is added to a polyurethane mixture to affect a cure (molecular extension) in a polymer.

Color, polyols   All polyols have color intensity analyses and are of two types: (1) APHA color, and (2) Gardner color scale.

Curing agents (hardeners, curatives)   Substances or mixtures of substances added to a compound to promote or control the curing reaction. Curing agents are reactive substances that become part of the molecular structure during cure.

Curing temperature   The temperature at which a thermosetting or elastomeric material is subjected to attain final cure. The term is used primarily in conjunction with thermosetting resins and rubbers.

Curing time   The time necessary to attain full cure in a thermosetting resin or rubber.

Cutter protector   The foam rubber cover used to protect the operator from the sharp cutter blades when the pelletizer is not in use.

Cutting chamber   The chamber in which the pellets are cut from the pelletizer's die plate.

Depth gauge   The *T* shaped dial calipers used to ensure that the front face of the water box is parallel with the die plate.

Die plate   The metal disc with through holes drilled in a circular pattern for directing molten polymer into strands.

Die pressure   The pressure drop across the die plate.

Die temperature   The steel temperature of the die plate.

Filter basket   The screening basket located in the water tank just below the dryer water outlet.

Cutter blades   The blades on the underwater pelletizer that cut plastic into pellets as it emerges from the die plate.

Cutter hub   The removable holder from which cutter blades are attached.

Damping   Hysteresis, or variations in properties resulting from dynamic loading conditions. Damping is a fundamental mechanism occurring in viscoelastic polymers, providing an important mechanism for dissipating energy, preventing brittle failure, and improving fatigue performance.

Deaerate   To remove air from a substance. Deaeration is an important step in the production of two-component castable polyurethanes to remove air that would cause objectionable bubbles.

Defoamer   An ingredient that, when added in small quantities to a fluid containing gas bubbles, causes the small bubbles to coalesce into larger bubbles that rise the surface and break.

Degree of polymerization   The average number of monomer units per polymer molecule, a measure of molecular weight. The degree of polymerization is given in percentage terms, i.e., 95% conversion.

Degree of cure   The extent to which cross-linking (curing) has progressed in a thermosetting resin or rubber.

Dehydration Removal of water from any substance, either by ordinary drying or heating, absorption, adsorption, chemical reaction, condensation of water vapor, centrifugal force, or hydraulic pressure.

Dehydrogenation The removal of hydrogen from a compound by chemical means.

Dehydrohalogenation The process of splitting hydrogen chloride from polymers containing chlorine atoms such as PVC, caused by excessive heat and/or light. This process may be slowed by addition of heat stabilizers and UV absorbers.

Delamination The undesirable separation of one or more layers in a laminate caused by failure at the adhesive interphase.

Delustrants Chemical agents used to produce dull surfaces on synthetic monofilaments to obtain a more natural silk-like appearance.

Denier The weight in grams of 9000 m of synthetic fiber in the form of a continuous monofilament.

Density Mass per unit volume of a substance at room temperature, expressed in metric units such as grams per cubic centimeter, English units such as pounds per cubic foot, or pounds per gallon.

Depolymerization The reversion of a polymer into its constituent monomers, or by chain scission into a polymer of lower molecular weight. Reversion may occur as a result of physical, chemical, and mechanical factors.

Demold time Time at which a molded object may be removed readily from the mold without altering the shape of the object and without post expansion due to incomplete cure.

Desiccant A substance having a great affinity for water and used as a drying agent.

Deterioration An undesirable permanent change in the physical or chemical properties of a polymer evidenced by an impairment of these properties.

Development Those technical activities of a nonroutine nature concerned with translating research findings or other scientific knowledge into commercial products or processes.

Diatomite (diatomaceous earth) Naturally occurring deposits of skeletal remains of aquatic plants called *diatoms*. Used as a reinforcing agent in some medical-grade silicone elastomers.

Di-tert-butyl peroxide A stable liquid used as a high temperature polymerization catalyst for a variety of olefin and vinyl monomers.

Dibasic Pertaining to acids or salts that have two active hydrogen atoms per molecule. Substances having one active hydrogen are called *monobasic*, and those with three are *tribasic*.

Dibutyltin di-2-ethylhexoate A waxy solid formed by reacting dibutyltin oxide with 2-ethylhexoic acid. Used as a catalyst for curing silicone resins and polyether-based urethane foams.

Dibutyltin dilaurate $(C_4H_9)_2Sn(OOCC_{11}H_{23})_2$ An important catalyst for the production of urethane elastomers. Also used as a stabilizer for vinyl resins when optical clarity is desired.

Dichlorodifluoromethane CCl_2F_2 (Freon 12) A nonhazardous blowing agent for foamed polymers.

Dichroism (1) A property of many refracting crystals exhibiting different colors when viewed from different directions. (2) The exhibition of different colors by some solutions with different degrees of concentration or dissolution.

Die lines (1) In blow molding, vertical defect lines caused by damaged die elements or compound contamination. (2) In extrusion, parallel defect lines caused by damaged dies or compound degradation.

Die block The part of an extrusion die that holds the forming bushing and core.

Die spider In extrusion terminology, the membranes or wires supporting a mandrel within the head and die assembly.

Die swell In extrusion terminology, the increase in diameter of the extrudate compared to the die opening through which it is extruded.

Die swell ratio   In extrusion terminology, and particularly in blow molding, the ratio of the outer parison to the outer diameter of the extrusion die.

Die adaptor   The portion of an extrusion die that holds the die block.

Die land   The final element of an extrusion die that imparts to the extrudate its final profile.

Die cutting   The process of cutting shapes from sheets of plastic by applying hydraulic or mechanical pressure to a knife edge.

Die cone (torpedo)   The tapered portion of an extrusion die that guides the extrudate to the webs of the spider.

Die plates   In injection molds, the members that are attached respectively to the fixed and moving heads of the compression press.

Die   A metal block containing a precision-machined orifice through which a molten plastic is extruded, shaping the extrudate to the desired profile.

Dielectric heating   Also known as *radio frequency heating* and *high frequency heating*. Heating a polymer by internal molecular friction and stress induced by an alternating current of sufficient frequency and intensity. Most polymers have dielectric loss characteristics sufficiently high to be heated in this highly efficient fashion.

Dielectric heat sealing   An important sealing method, widely used for sealing films made of thermoplastics with sufficient dielectric loss, in which the films are heated rapidly by dielectric heating, causing adhesion between the films. Medical packaging is frequently sealed by dielectric heat sealing.

Dielectric loss   A loss of energy resulting in the temperature rise of a dielectric material placed in an alternating electrical field of sufficient frequency and intensity.

Dielectric loss angle (dielectric phase difference)   The arithmetic difference between 90° and the dielectric phase angle.

Dielectric phase angle   The angular difference in phase between the sinusoidal voltage applied to the dielectric material and the resulting current.

Dielectric strength   A measure of the maximum voltage required to puncture a dielectric material, expressed in volts per mil of thickness. The voltage value is. the average root-mean-square voltage gradient between two electrodes under which the electrical breakdown occurs under standard test conditions.

Dielectric   A polymer with very weak electrical conductivity such that different parts of its surface display different electrical charge. These polymers typically exhibit conductivities in the order of one millionth of a reciprocal ohm per centimeter.

Diene polymers   A large family of polymers characterized by the presence of unsaturated hydrocarbons or diolefins having double bonds. In conjugated dienes the double bonds are separated by only a single bond; in unconjugated dienes, the double bonds are separated by at least two single bonds. Included in this family are copolymers of ethylene, propylene, isoprene, butadiene, and cyclopentadiene.

Differential thermal analysis (DTA)   An analytical method, similar in some respects to thermogravimetric analysis (TGA) except that the specimen is heated simultaneously with an inert control material, each having its own temperature sensing and recording apparatus. The curve that plots weight losses of both materials under the same heating rate is known as the DTA curve.

Diffusion couple   An assembly of two materials in such intimate contact that each may diffuse into the other.

Diffusion   The spontaneous mixing of one substance with another, resulting from the movement of molecules of each substance through the empty molecular spaces of the other substance. Diffusion occurs in and between gases, liquids, and solids. The new technology of controlled drug release is based in large part on the diffusion of drugs through nonporous polymer films.

Diisocyanate   One of the major chemicals used in the manufacture of polyurethane foams, elastomers, etc. The basis for all urethane synthetic reactions.

Dilatancy   A rheological phenomenon characterized by an increase in viscosity with increasing rates of shear. It is the opposite of pseudoplasticity.

Dilatant   A rheological term that describes a liquid that resists being moved but is quite fluid at rest; opposite of thixotropic.

Dilatometer   See *Pycnometer*.

Diluent   A substance that lowers the concentration, viscosity, or cost of another more expensive material. In adhesives, diluents are used for thinning out liquids, increasing or decreasing rates of evaporation, lowering costs, etc. If it is an inert powder added to an elastomer merely to increase its volume (and thereby reduce cost), it is called a *filler*.

Dilute solution viscosity   The viscosity of a dilute solution of a polymer, measured under standard conditions, as an indication of the molecular weight and degree of polymerization of a resin.

Dimensional stability   The innate ability of a plastic part to retain the precise dimensions in which it was extruded, molded, cast, or fabricated.

Dimer   (l) A molecule formed by the union of two identical simpler molecules. (2) A substance composed of dimers. For instance, propane C_4H_8 is a dimer of ethane C_2H_4.

Dimethylformamide (DMF) $(CH_3)NCOH$. A colorless, strong solvent for polyurethanes, vinyls, nylons and other resins. Its strong solvent action makes it suitable for adhesive and coating compositions.

Diol   A polyol or resinous material having two reactive hydroxyl (-OH) groups attached to each molecule. Diols have relatively low molecular weight (<600) as opposed to polyols, whose molecular weight ranges from 650 to 3500.

Dip coating   A coating process wherein the object to be coated is immersed in a vessel containing a solution, dispersion, or heated fluid coating material, then withdrawn and subjected to heat or drying to solidify the film deposit. Fluidized bed coating is a typical example of this technique.

Dip forming (dip coating)   A process similar to dip coating except that the cured or dried deposit is stripped from the dipping mandrel. Devices such as intra-aortic balloons are manufactured in this fashion. Also, many developmental prostheses are fabricated this way, the most notable being diaphragms for artificial heart devices.

Discoloration   (l) Any undesirable change from an initial coloration possessed by a polymer. (2) A lack of uniformity in color where color should be uniform over the whole surface of a plastic article. Also known as mottling, segregation, and two-toning.

Dispersant   Any liquid component capable of solvating a resin to aid in dispersing and suspending it.

Disperse phase   In a suspension, the disperse phase refers to the solid particles dispersed in the liquid medium, thus being the discontinuous phase. In contrast, the liquid is called the *continuous* phase.

Dispersing agents   Any substance added to a suspending medium to promote and maintain the separation of discrete fine particles of solids or liquids.

Dispersion resins   A special type of PVC resin with small spherical particles (one micron or less in diameter), suitable for compounding with liquid plasticizers by simple stirring techniques, forming plastisols and organosols.

Dispersion   A two-phase or multiphase system comprising a finely divided material (the discontinuous phase) uniformly distributed in another material (the continuous phase). In the medical industry, the term *dispersion* usually denotes a finely divided solid dispersed in a liquid. Dispersions are classified as (l) emulsions (liquids in liquids), (2) suspensions (solids in liquids), (3) foams (gases in liquids), and (4) aerosols (liquids in gases).

Dissipation factor (mechanical)   The ratio of the loss modulus to the modulus of elasticity.

Dissipation factor (electrical)   The ratio of conductance of a capacitor in which the material is the dielectric. Most polymers have a low dissipation factor, a desirable property because it minimizes the waste of electrical energy.

Dissolution   The process by which a chemical or drug becomes dissolved in a solvent. In biological systems, aqueous drug dissolution is a prerequisite of systemic absorption. The rate at which drugs with poor aqueous solubility dissolve from a solid dosage form often controls the rate of systemic absorption.

DMF   Abbreviation for dimethylformamide solvent.

Doctor bar (doctor blade, doctor knife)   A precision-machined flat bar used for regulating the amount of liquid material on the rollers of a coating machine, or to control the thickness of a coating after it has been applied to a substrate.

Doctor roll   A roll operating at a different speed or in the opposite direction, as compared to the primary roll of a coating machine, thus regulating and controlling the uniformity and thickness of coating material before it is applied to the substrate.

Doctor   Any device used to spread a coating onto a substrate in a layer of controlled uniformity and thickness.

Domain   A region in a polymer matrix which contains molecular segments of similar types. Most polyurethane elastomers contain domain structures.

Double bond   A type of intermolecular structure in which a pair of valence bonds joins a pair of carbon or other atoms, or a covalent linkage, in which atoms share two pairs of electrons. In general, double bonds are represented in chemical formulas by the symbols "=" or ":", as in ethylene, $H_2C=CH_2$ or $H_2C:CH_2$.

Drawdown ratio   In extrusion terminology, the ratio of thickness of die opening to the final thickness of the product.

Drawdown   In extrusion terminology, the process of stretching the hot extrudate away from the die at speeds higher than the merging melt, thus reducing the cross sectional dimensions of the extrudate.

Dry coloring   The compounding process of adding dyes and pigments to resins in particulate form. This process enables manufacturers to color match different products without carrying large inventories of colored resins.

Dry strength   The strength of an adhesive interface immediately after drying under specified conditions, as opposed to wet strength, where the strength of the interface is measured immediately after joining.

Dry blend   A dry, free-flowing powdery mixture of resin and plasticizer, prepared by blending at high shear and temperature, below the fluxing point. PVC dry blends are frequently used in the extrusion of vinyl tubing for extracorporeal applications.

Ductility   The maximum amount of strain that a material can withstand, before undergoing ductile fracture. Ductile materials usually exhibit a yield point in the stress-strain curve.

Durometer hardness   Measured by a hand-held instrument that determines penetration hardness. See also *Indentation load deflection*.

Dye   An intensely colored organic substance that imparts color to a plastic. Dyes used for coloring plastics usually dissolve in the plastic melt, resulting in transparent products, as opposed to pigments that remain as undissolved particles, resulting in opaque products.

Elasticity   The ability of a material to return to its original shape after removal of a load. Also referred as *memory* or *recovery*.

Elastomer   Generally speaking, any rubber-like substance. More specifically, the term refers to materials that can be stretched repeatedly to 150% or more of their initial length and that will return rapidly and with force to their approximate original length.

Emulsifier   Additives that aid in stabilizing a mixture between the time it is mixed and poured and the time it starts to foam.

Equivalent weight (EW)   Mass of polyol per reactive OH. EW = molecular weight/functionality.

Exotherm   The heat liberated by some of the chemical reactions occurring during urethane synthesis. This heat accelerates the synthetic processes.

Exothermic reaction   A chemical reaction that produces heat.

Fascia   Elastomeric covering for the energy-absorbing system of an automobile, such as the bumper or even the entire front and rear ends. Microcellular urethane foam is a leading contender for this application.

Filled foams   A name applied mostly to flexible foams containing various levels of a fairly inert powdery material. Also can refer to foams containing high levels of plasticizers or inert oils.

Filler   An inert material added to a formulation to reduce cost, and change the final physical properties. It may improve hardness, stiffness, and impact strength. If it enhances these properties, it is called a *reinforcing filler.*

Fines   Plastic particles that have broken away from pellets during the cutting or drying process. Excessive fines usually result from poor alignment, but a certain amount of fines are always generated.

Fingernail test   A test to determine the recovery of the flexible foam when indented with a fingernail or sharp object. A test of high-density rigid foam to determine the hardness or wood-like quality of the skin or surface.

Finishing   (1) The secondary operation involving the removal of flash, gate marks, and surface defects from plastic articles. (2) The development of desired surface textures, ranging from smooth to embossed.

Fish eye   A fault in transparent plastics, appearing as a globular mass of polymer caused by incomplete melting or blending with the bulk. In polyurethanes, fish eyes may also occur as a result of incomplete drying of the resin prior to extrusion.

Flame retardants   (1) A substance having the ability to resist combustion. A flame retardant is considered to be one that will not continue to bum or glow after the source of ignition has been removed. (2) Substances usually incorporated as additives into polymers to reduce the tendency of plastics to burn. Nonreactive flame retardants include0 antimony trioxide, phosphate esters, and chlorinated paraffins. Reactive flame retardants include bromine- or phosphorous-containing polyols and a variety of chlorinated compounds.

Flammability   The relative burnability of a material in a specified situation. Meanings vary according to the test method used.

Flash gate   A wide gate extending from a runner that runs parallel to an edge of a molded part along the parting line of a mold.

Flatting agent   Minute particles of irregular size incorporated into polymers, used to disperse incident light rays so that a *flat* or dull surface effect is achieved. Flatting agents most frequently used are silica, diatomaceous earths, and heavy metal salts.

Flexible foams   Open-celled materials that allow free movement of air throughout the materials when flexed. Typical densities range from 0.93 to 2.8 lb/ft^3.

Flexible molds   Molds made of elastomeric materials (such as silicones, rubbers, or soft thermoplastics), capable of stretching to permit easy demolding of hard cured pieces with undercuts.

Flight   In extrusion, the outer surface of the helical ridge of metal left after machining the screw channels. The clearance between the screw flights and the inner diameter of the extruder barrel provide the desired compression to the molten polymer.

Flowable polymerizable resins   Flowable polymerizable resins refer to compositions that, prior to undergoing a significant degree of polymerization, are liquid, whether at ambient or elevated temperature, and that, following polymerization, provide a shape-retaining resin material that may be of the thermoplastic or thermosetting category. Examples of flowable polymerizable resin-forming compositions include polyurethanes (both thermoplastic and thermosetting), polyesters and copolyesters, alkyds, polyamides, epoxies, silicones, allyls, amino resins, vinyl resins, interpenetrating polymer networks (IPNs), etc.

Flow line   A mark on a molded piece made by the meeting of two flow fronts. Also called *weld line.*

Fluorocarbons   The general family of fluorinated hydrocarbons to which belong some of the most popular low-boiling-point chemicals used as urethane blowing agents. The most common fluorocarbons used in polyurethane foams are volatile liquids such as CFC-11 (CCl_3F) and CFC-12 (CCl_2F_2). CFC is the abbreviation for chlorofluorocarbon.

Foam casting   The process of foaming a fluid resin or prepolymer/catalyst system before or during molding by mechanical frothing, or by gas dissolved in the mixture or released from a low-boiling-point liquid.

Foam   A product, either flexible or rigid, that has been produced by the internal generation or liberation of a gas in a fluid medium that is simultaneously polymerizing while expanding in volume. The final product is either an open- or closed-cell foam.

Foamed plastics   Plastics with internal void or cells. The foam may be flexible, semirigid, or rigid, the cells closed or connected, and the density anything below that of the solid parent resin down to 2 lb/ft² or less.

Foamed-in-place   The technology for permitting the foaming reaction to take place in a mold or form that itself will be the final product or a part of the final product.

Foaming agents   Chemicals or gases added to plastics that cause the resin to assume a cellular structure. Also called *blowing agents.*

Foam-in-place   Refers to the deposition of foams at the pour site. It requires that the foaming machine be brought to the work site, which is *in place,* as opposed to bringing the work to a forming machine located in the factory.

Free rise   The unhampered expansion of a foam sample or product in a container with no top, and a height of side wall not greater than twice the diameter.

Freeze-off   Polymer solidification in the die holes, thereby restricting or blocking polymer flow.

Friability   A term used in rigid urethane foam to indicate a crumbling, flaking, or powdering effect of the foamed surface upon rubbing.

Froth   A manufacturing process incorporating an unusually low-boiling material into the final foam mixture. When this liquid is discharged from the pressurized chamber, it expands instantly into a semiliquid foam. Fluorocarbon 12 is the most common low-boiling liquid used for frothing.

Frothing   Technique of applying foam by introducing blowing agents or shallow air bubbles, under pressure, into the liquid mixture of foam ingredients.

Functionality   (1) As applied to urethane chemistry, functionality means the number of hydroxyl groups (reactive sites) per molecule of the polyol which are available as reaction sites (i.e., diol, triol). It is an indication of the degree or order of reactivity. (2) The number of reactive groups in a chemical molecule. Polyurethane reactants commonly have functionalities of 2, 3, 4, and 5.

Heat exchanger   The device on the tempered water system used to remove heat from the process water. The process water absorbs excess heat from the molten polymer.

Gel point   The point in a multifunctional reaction when an essentially infinite polymer network is formed.

Gel time   The point at which, when the surface foam is touched with a stick and pulled upward, the string that forms from this action breaks.

Gel count   A thin coating of high-quality polyester plastic applied to the surface of a mold prior to filling the mold with foam. The foam adheres to the gel coat so that the part, when removed from the mold, has the desired surface finish.

Graft   A polymeric arrangement in which side chains of one composition are attached to a polymeric backbone of a different composition.

Green strength   The strength of a part at the time of demolding. A subjective measure of the ease of demolding and handling of an injection-molded article.

Hand   A subjective description of the feel of the foam as the hand is rubbed lightly over the surface. If the foam is harsh or rough to the touch, it is described as having poor hand. Good hand has a velvet feel.

Hardness   The resistance of a material to penetration, usually expressed in *Shore A* or *Shore D* units.

HDI   An abbreviation for hexamethylene diisocyanate and 1,6-diisocyanato-hexane.

High pelletizer load switch   An auxiliary switch that is provided with the pelletizer load indicator to shut the extruder down in the event of pelletizer motor overload.

HMDI   An abbreviation for hydrogenated MDI and 4,4'-diisocyanato-dicyclohexylmethane.

Homopolymer   A polymer consisting substantially of a single type or repeating unit, resulting from the polymerization of a single monomer.

Hydrogen bonding   Points of intermolecular attraction caused by polarized links involving hydrogen atoms (interhydrogen) or hydrogen and other nitrogen or oxygen atoms.

Hydrolyzable chloride   Hydrolyzable chloride refers to chlorinated compounds that will react with water- or OH-containing materials. In TDI, it is present as either carbonyl chloride or phosgene. These products will react with water in the formulated system and will not liberate CO_2.

Hydrophilic   The propensity of substances (urethane foams) to have an affinity for water. In general, polyether foams are not as hydrophilic as polyester foams.

Hydroxyl number   This number indicates, to some extent, the reactive amount of hydroxyl groups available for reaction in the polyol. It is expressed in terms of the number of milligrams of potassium hydroxide that is equivalent to the hydroxyl content of one gram of the sample. For instance, Voranol CP 3000 has an average hydroxyl number of 56.1.

Hydroxyl (OH)   (1) A chemical functional group composed of a hydrogen atom bonded to an oxygen atom. Hydroxyls react with isocyanates to form urethane linkages. (2) A group containing an (OH) moiety. In polyurethanes, a primary hydroxyl group has the structure -CH_2-OH and is more reactive than a secondary hydroxyl group.

Hydroxyl equivalent weight   This is the number of grams of sample required so that one equivalent weight (17.008 g) of hydroxyl will be present in the sample.

In situ foaming   The technology wherein the liquid polymer mixture is poured into the place where it is intended to have the foaming reaction occur.

Indentation load deflection (ILD)   A measure of the load-bearing ability of the foam. Standard test is to depress a 50 in^2 indenter foot into the foam and read the number of pounds required to achieve a desired deflection. This is described in ASTM Test Method D-1564.

Inhibitor   A negative catalyst, i.e., a substance that slows down a chemical reaction.

Initiation time   Synonymous with cream time. In frothing, it usually refers to the delay time between the beginning of the initial expansion and the beginning of the secondary or final expansion.

Initiator   A substance capable of reacting with ethylene or propylene oxide to form a polyether polyol.

Integral skin foam   A molded urethane foam product having a dense, tough, outer foam skin and a lower-density core.

Interpenetrating polymer network (IPN)   The term IPN denotes a class of materials that contain at least two polymers, each in network form. The two polymers must have been synthesized or cross-linked in the presence of each other, and they cannot be physically separated. IPNs may be thought of as two polymer networks that have been interwoven or have become inextricably tangled together in three dimensions. IPNs have a unique set of physical properties, often different from the properties of either constituent polymer itself.

Isocyanate equivalent weight   The number of grams of sample required so that one gram equivalent weight (42.02) will be present in the sample.

Isocyanate index   A measure of the relationship between the equivalent weight of the isocyanate on the one side and the polyol and water equivalent weight on the other side. An index of 100 indicates that both equivalents are equal. An index of 105 indicates a 5% surplus of isocyanate equivalents.

Isocyanate neutralizer   A combination of materials used to cover and neutralize isocyanate spills. An effective method for treating small spills is to absorb the isocyanate with sawdust or other absorbent, shovel it into suitable unsealed containers, transport it to a well ventilated outside area, and treat it with neutralizing solution consisting of a mixture of water and 3 to 8% concentrated ammonium hydroxide (or 5 to 10% sodium carbonate). To neutralize, about 10 parts of neutralizer are needed per part of isocyanate with mixing. The mixture is allowed to stand for 48 hr while letting CO_2 escape.

Isocyanate   (1) Those compounds having one or more reactive (-NCO) radicals or groups attached to the main molecule. (2) A substance containing the highly reactive -NCO- group. Isocyanates are the starting point for all commercial reactions resulting in polyurethane products.

Isocyanurate   A trimeric reaction product of an isocyanate having a ring structure.

Isomer ratio   The ratio between the 2,4 isomer and the 2,6 isomer in commercial TDI.

Isotropic foam   Foam characterized by having the same strength properties in all directions.

k factor   The *k* factor is a term used for thermal conductivity and represents a measure of the heat transmission characteristics (i.e., insulating properties) of a material. These *k* values are used for comparison of the conductivity of different kinds of material. The value of the factor depends on the material through which the heat is flowing and the mean temperature, and it is a constant for materials of identical physical and chemical properties.

L/D   The ratio of the die hole land length to the die hole diameter.

Land length   The length of the final die hole diameter of the die plate, viewed from the raised face side of the die.

Lacquer   A solution of a film-forming polymer in a volatile organic solvent that, when applied to a surface, forms an adherent film that solidifies solely by solvent evaporation. The dried film should have the same physical properties as the polymer used in making the lacquer.

Laminar flow   Laminar flow of thermoplastic resins in a mold is accompanied by solidification of the layer in contact with the mold surface that acts as an insulating shell through which molten material flows to fill the remainder of the cavity.

Light resistance   The ability of a plastic to resist fading, darkening, and other degradation on exposure to sunlight or UV light. Also known as light fastness and color fastness.

Liquid injection molding (LIM)   A process that involves an integrated system for proportioning, mixing, and dispensing two-component liquid resin formulations and directly injecting the resultant mix into a mold, which is clamped under pressure.

Master batch   A premixed system wherein the catalysts, blowing agents, fire retardants, and perhaps the silicone surfactant are added to the polyol (according to a predetermined formula). Also called (B side) cross-linker, catalyst side. This side does not contain isocyanate.

MDI   In the U.S., abbreviation for 4,4'-diphenyl-methane diisocyanate. Crude MDI refers to polymeric isocyanates derived from MDI.

Mechanically foamed plastic   A cellular plastic whose structure is produced by physically incorporating gases into the polymer mass.

Melt temperature   The temperature of the molten polymer.

Motor adapter flange   The flange bolted onto the face of the pelletizer motor that couples up to the water box flange.

Modulus   Usually used in urethane technology to represent the result of dividing two points on the load-deflection curve for flexible urethane foam. An indication of the slope of the load-deflection curve. For flexible foams, the modulus is expressed as the ratio of the indentation load deflection in pounds at 65% deflection to the indentation load deflection in pounds at 25% deflection.

Mold release agent   One of many compounds that, when applied to the mold surface, serve to prevent the cured foam form sticking to the mold. Useful release agents are telomers, silicones, synthetic and natural waxes, and various sorts of soaps.

Mold holder   The removable sleeve on the front of the pelletizer shaft threaded for cutter hub installation.

Molecular weight   In polyols, the measure of the average length of a polyol. The molecular weight of polyols is only a weight average molecular weight, because a polymer will contain various chain lengths of molecules rather than all being the same size.

NCO/OH ratio   The ratio between the available isocyanate groups and the available hydroxyl groups in a polyurethane formulation. A related term, *index,* is more often used in connection with a complete polyurethane formula.

Nose cone   A cone-shaped piece of steel that bolts to the die plate. Its function is to direct molten polymer toward the die holes.

Nucleation   Physical assistance used in the generation of many small, uniform bubbles in a polyurethane foam, as opposed to a few large bubbles in the same volume of space. This assistance may involve the addition of special molecules, small rough particles, dissolving high-vapor pressure gas in the liquid, or a shortening of the time interval during which bubble formation occurs to minimize migration of gas molecules toward existing bubbles.

OH number   The number of milligrams of KOH that is chemically equivalent to the activity of a specified weight (in grams) of the polyol.

Olamine   Olamine is an organic compound having one or more hydroxyl groups and also one or more amine groups. Examples of olamines (alkanolamines) include monoethanolamine, diethanolamine, dimethylethanolamine, triethanolanine, N-methylethanolamine, N-ethylethanolamine, N-butylethanolamine, N-methyldiethanolamine, N-ethyldiethanolamine, N-butyldiethanolamine, monoisopropanolamine, diisopropanolamine, triisopropanolamine, N-methylisopropanolamine, N-ethylisopropanolamine, and N-propylisopropanolamine. Olamines are used in the synthesis of PIPA polyols.

One-shot system   A term describing the technique for the simultaneous mixing at the foam head of the polyol, TDI, catalyst, blowing agent, and silicone to make urethane foam. No prior reaction product between the TDI and polyol is used.

Open cell structure   A permeable structure (as in flexible foam) wherein there is no barrier between cells, and gases or liquids can pass freely through the substance. All cell walls have been broken.

Organotin catalysts   Catalysts noted for their specific activity toward the NCO/OH reaction. This influence is observed in the *tack-free time* of elastomers, *gel time* of foams, etc.

Pelletizer   The machine that converts molten polymer forced through a die plate into a usable form (pellets). The pelletizer must be used in conjunction with a tempered water system.

Pelletizer load indicator   An ammeter that monitors the pelletizer motor load. This indicator is usually mounted on the electrical panel.

PIPA polyols   PIPA polyol is a dispersion in which the polyol acts as a substantially inert carrier for PIPA particles. The PIPA particles are formed by the reaction of an isocyanate and a trifunctional alkanolamine in the presence of an organotin catalyst. The reaction product is an array of alkanolamine and isocyanate groups having pendant hydroxyl groups for further reaction. These polyols are used in the production of flexible foams.

Pipe spools   Pipe fittings that have been welded to a piece of pipe or to other fittings.

pH   A measure of the apparent acidity or basicity. This is a simple and fast analysis but probably leads to more confusion than any other analysis. There are three common standards for pH used in the urethane chemical industry, as follows: (1) methyl alcohol, H_2O mixture in a 10:1 ratio, (2) isopropyl alcohol, H_2O mixture in a 10:6 ratio, and (3) 5% polyol in water. Method (1) is the most common in the urethane foam industry.

PHD polyols   Poly Harnstoff dispersion (PHD) polyols are dispersions of polyurea particles in conventional polyols. These polyols are prepared by the reaction of diamine (hydrazine) with a diisocyanate (toluene diisocyanate) in the presence of a polyether polyol to produce a polyurea dispersion.

Plasticizer   Chemicals that generally serve to increase the flexibility of polyurethane formulations by lowering the T_g of the final polymer.

Plugs   Small, bullet-shaped pieces of brass used to temporarily block off unneeded die holes. These plugs are inserted from the nose cone side of the die.

PMDI   An abbreviation for polymeric methylene diphenyl diisocyanate. A brown liquid composed of approximately 50% methylene diisocyanate, with the remainder composed of MDI oligomers.

Pockets   The undesirable formation of large cavities or pockets in the foam structure. Pocketing is usually caused by rapid formation and/or release of the blowing agent before the polymer structure has gained sufficient strength to contain the gas.

Polar   A molecule having positive or negative electrical charges that are permanently separated.

Polyester foams   Polyurethane foams that have been made by reacting the isocyanate with a polyester polyol rather than a polyether or other resin component.

Polyether foams   Polyurethane foams that have been made by reacting the isocyanate with a polyether polyol rather than a polyester or other resin component.

Polyethers   Compounds containing terminal alcoholic groups used as reactants in the production of polyurethane elastomers, foams, etc.

Polyahl   A polyahl is a polyol, polyamine, polyamide, polymercaptan, polyacid, or a compound containing a mixture of active hydrogen groups.

Polymer polyol   Polymer polyols are ethylene oxide propylene oxide adducts of propylene glycol, containing ethylene oxide as a cap, and between 15 to 45 percents of styrene, acrylonitrile, or a combination of both as a dispersion. Polymer polyols are used in the manufacture of flexible foams.

Polyol mix   The product resulting from premixing many of the compatible minor ingredients into the polyol component. Also called *master batch* and *premix*.

Polyol   A substance, usually a liquid, containing at least two hydroxyl (-OH) groups attached to a single molecule. The two most common types of polyols used in the manufacture of polyurethanes are the polyesters and polyethers.

Polyurethane foams   A term gradually replaced by *urethane foams* in the U.S.

Polyurethanes   A large family of polymers based on the reaction products of an organic isocyanate with compounds containing a hydroxyl group. Polyurethanes are also called *urethanes,* a name that sometimes is confused with similar-sounding but totally different chemicals, including urea-formaldehyde and urethane (chemical name for ethyl carbamate). The general term *polyurethane* is not limited to those polymers containing only urethane linkages but includes polymers containing allophanate, biuret, carbodiimide, oxazolinyl, isocyanurate, uretidinedione, and urea linkages in addition to urethane.

Pore diameter   Synonymous with cell size. The term is primarily used in Europe.

Postcure   Time and temperature history of a molded article after removal from a mold.

Pot life   Total time at which a mixed elastomer system is usable for its intended purpose from the standpoint of pourability (50,000 cps).

Pour in place   The practice of pouring a liquid into a cavity and having it foam, fill the cavity, harden, and cure without having to remove it from the cavity and without having to shape the product by cutting or sawing. Sometimes called *potting*, but this term is more properly used in reference to the solid or nonfoamed pour in place.

Power panel   An enclosure, usually wall mounted, that houses fuses, contactors, motor starters, transformers, etc. on the back plate. Control circuitry is located in a separate panel.

PPDI   An abbreviation for p-phenylenediisocyanate and 1,4 diisocyanato-benzene.

Precure   Time and temperature history mold.

Premix   The mixture resulting from blending many of the minor ingredients with polyol in an effort to reduce the final number of components or to allow more time for mixing or blending chemicals that may not be readily miscible in the short period of exposure to the final mixing. Also called *master batch* or *polyol mix*.

Prepolymer   (1) A prepolymer is the reaction product of a polyol or blend of polyols with *excess* isocyanate. A prepolymer is used to describe a polyol-isocyanate adduct with a free isocyanate content of 1 to 15 percent by weight. (2) A cooked or nonpolymerized reactive product formed by reacting polyols or blends or polyol with excess isocyanate. This product is further reacted with polyols and/or water to produce a polymer.

Primer   A coating applied to a substrate to improve the adhesion, gloss, or durability of a subsequently applied coating.

Purge   Extrusion of molten polymer through the die plate while the pelletizer is disengaged from the water box.

Purging sleeve   A sheet metal tube that inserts into the water box to prevent molten polymer from sticking to the interior of the water box during purging.

Pycnometer   A standard vessel for measuring and comparing the densities of solids or liquids.

Quick-disconnect coupling   The coupling that is used to connect the water box to the motor adapter flange.

Quasi-prepolymers   A quasi-prepolymer is the reaction product of a polyol or blend of polyols with a large *excess* of isocyanate. The term quasi-prepolymer is used to describe a polyol-isocyanate adduct with free isocyanate contents between 16 and 32 percent by weight. Quasi-prepolymers are extensively utilized in the production of foams.

Raised face   The raised surface or that part of the die plate where the polymer exits and the cutter blades track.

Reaction   A chemical change in which two or more atoms or molecules produce a new substance.

Reaction injection molding (RIM)   A process that involves the high-pressure impingement mixing of two or more reactive liquid components and injecting into a closed mold at low pressure. The RIM process includes four interdependent elements: the chemical system, the RIM machine, a mold support, and mold temperature-control system. *Reaction injection molding* refers to any molding system that involves filling a mold with a flowable polymerizable resin-forming composition, which may contain a reinforcement component, the resin-forming composition then undergoing polymerization to provide an article of desired molded configuration. The expression *reaction injection molding* as used is synonymous with those operations referred to as *liquid reaction molding* (LRM), *reinforced reaction injection molding* (RRIM), *liquid injection molding* (LIM), and *liquid resin molding*.

Reactivity   Comparison of total time required for an elastomer system to cure to a finished product. Determined by viscosity buildup, gel time, hardness, and demold time.

Regrind   Waste material, from injection molding, blow molding, and extrusion operations, which has been reclaimed by shredding or granulating.

Relaxation (sighing)   When foam rises to a maximum height then settles. This is usually caused by poor surfactant activity or an incompatible mixture.

Reticulate   The process of removing residual membrane or cell windows from the foam structure so that only a skeletal web-like network remains. Several techniques can be employed to obtain reticulation.

Ribs   Thread-like structures formed at the joint between adjacent bubbles in a polyurethane foam that becomes open-celled. The ribs are usually reinforced by the remains of the cell membranes in good quality foam. Also called *struts*.

RIM   Abbreviation for reaction injection molding. The process involves the rapid metering, mixing, and reaction of polyurethane ingredients, followed by their injection into a mold.

Rise time   The time between the beginning of the foam rise (cream time) and the point at which the foam rise is complete.

Sandwich panel   The terminology for a panel that has been formed by permanent, indwelling, foamed urethane between sheets of a material such as wood, aluminum, etc.

Scorch   A yellow or brown discoloration of the foam, particularly in the center area. Scorching is caused by excessive heat during the exothermic reaction.

Screen changer   A machine that allows the change polymer of screen packs quickly, thus avoiding interruption of the extrusion line.

Screen changer hydraulic unit   The hydraulic unit that supplies the fluid energy to the screen changer cylinder.

Sealant   Low-modulus polyurethane elastomers that utilize high molecular weight polyols, low levels of cross-linking, and very little chain extension.

Segment   A short part of a polymer chain composed of identical units. In polyurethanes, a hard segment has a high softening or melting point with a T_g lower than room temperature. A soft segment is the reverse.

Self-skinning   The characteristic of a foam mixture that forms a skinned surface upon molding at a specified temperature and pressure.

Sight glass   A cylindrical assembly with flanges and a borosilicate glass element. When installed in the slurry pipeline, it permits visual slurry inspection.

Self-extinguishing   The ability of a material to cease burning once the source of flame has been removed.

Shelf life   The length of time that a product will be fit for use during storage under specific conditions. Also called *storage life* and *working life*.

Shiners   In flexible foams, intact cell walls as evidenced by reflected light of finished foam.

Shore hardness   The measure of firmness (resistance to indentation) of a compound determined by means of a durometer hardness gauge, measured on a Shore A or D scale.

Short-shot method   The process of partially filling a cavity with foamable resin and allowing the foaming reaction to complete the filling of the cavity.

Shrinkage   When foam rises to maximum height and then forms wrinkles and indentations.

Skin   The higher-density outer membrane of a foam article, usually the result of the outer surface cooling more rapidly than the core.

Slab foam   Foam made by the continuous pouring of mixed liquids on a conveyor and generating a continuous loaf of foam for as long as the machine is operating. It would generally be classed as free rise or unconfined, although fixed sides guide give the loaf a generally rectangular cross section with a slightly rounded top.

Slurry   The mixture of pellets and cycle water traveling from the pelletizer to the tempered water system. This is done during underwater pelletizing runs.

Smoking   Vapors rising from the surface of foam bun.

Sparklers   A term used to describe the undesirable bursting of the rising foam mass. This is usually caused an incompatible mixture.

Splits   The undesirable formation of fissures or cracks by too rapid an evolution of the blowing agent for the strength of the polymerized structure.

Sprue gate   The passage through which molten resin flows from the nozzle to the mold cavity.

Sprue   (1) The main feed channel that connects the mold-filling orifice with the runners leading to each cavity gate. (2) The piece of solidified plastic material formed in that channel.

Stabilizer   An additive that protects a polymer from changing its form or chemical structure. Stabilizers are used to protect polymers from degradation reactions such as hydrolysis, thermolysis, oxidation, etc.

Stabilizers   Additives that assist in maintaining the quality of polyurethane products in use, such as antioxidants, ultraviolet absorbers, acid absorbers, etc.

String time   A measure of gel time. The time between pouring the liquid foaming ingredient and the time when long *strings* of tacky material can be pulled away from the surface of the polyurethane foam.

Stringers   Pellets that are fused in a single row. May occur during underwater pelletization runs.

Structural foam molding   The process of molding thermoplastic articles with a cellular core and integral solid skins in a single operation.

Surfactant   An additive to the formulation that either helps or hinders the formation of fine, uniform cell structure in the resulting polyurethane foam.

Surfactants   A term to describe substances that markedly lower the liquids, thus permitting greater penetration or dispersion.

Syntactic foams   Composites of tiny, hollow spheres of glass or phenolic resins. The spheres form a fluid mass that can be cast into molds, trawled onto a surface, or incorporated into laminates. After forming, the mass is cured by heating.

System   The specific mixture of chemical components needed to produce a polyurethane product.

Tack-free time   The time between the beginning of the foam pour and the outer skin of the foam mass loses its stickiness or adhesive quality.

TDI   (1) An abbreviation for toluene diisocyanate or tolylene diisocyanate. (2) The abbreviation for toluene diisocyanate, particularly the 80:20 isomer blend, although it can be used for other isomer blends as well.

TDI index   This figure indicates the amount of TDI (Toluene Diisocyanate) available for reaction with the poly and water. An index of 105 indicates that there are 5% excess equivalents of TDI available in the system. In addition to the stoichiometric amount required by the polyol and water.

Tear strength   The load required to tear apart an elastomer or foam specimen, expressed in pounds per linear inch (pli).

Temperature regulating valve   The valve that regulates the flow of cooling water to the heat exchanger to maintain the desired temperature of the cycle water flowing to the underwater pelletizer.

Tempered water system   The system that removes the cycle water from the pellets, then filters, pressurizes, and cools it before returning it to the pelletizer.

Telomer   An addition polymer, usually of low molecular weight, that has had its chain length terminated to limit the molecular weight. Also called an *oligomer*.

Tensile strength   Resistance of a material to a stretching force. The cohesive strength of a material expressed in pounds per square inch (psi).

Theoretical molecular weight of a polyol   The theoretical molecular weight of a polyol denotes a molecular weight calculated using a mathematical equation based on the functionality of the starter used to produce the polyol and the experimentally determined hydroxyl number of the polyol.

Thixotropic   Describes a liquid that has much lower viscosity when stirred than when at rest.

Tin catalyst   A catalyst whose main function in polyurethanes is to control the rate of gelation. It is used in most flexible foam formulations, the most popular being stannous octoate. Only occasionally is tin used for rigid foam production.

TMXDI   An abbreviation for m- and p-tetramethylxylene diisocyanate, and 1,4- and 1,3-diisocyanato-dimethyl-methyl-benzene.

Total prepolymer   A polyurethane foam intermediate in which all the isocyanate material is premixed with all the polyol material and reacted under controlled conditions.

Unicellular plastic   Synonymous with closed-cell plastic.

Unsaturation determination   A test to measure the amount of carbon-to-carbon unsaturated compounds in a polyol sample. The results are reported in terms of the number of milliequivalent of potassium hydroxide per gram of polyol sample (meq/gm).

Viscosity   A very important physical property essential in maintaining batch-to-batch consistency in manufacture of polyols. This is a fast and very accurate analysis and is determined at 100° F. For more viscous polyols, the viscosity is usually determined at 210° F. There are a few special customers who require 77° F viscosities. Viscosities for polyols are usually expressed in centistokes. If the results are needed in centipoises, the two are related by the following equation:

$$\text{centistokes} \times \text{polymer density} = \text{centipoise}$$

Void   An unfilled space in a cellular plastic that is substantially larger than the individual cells. Can also be an empty space in any material or medium.

Voids   The undesirable formation of large cavities or pockets in the foam structure. Voids are usually caused by poor moldability or incorrect mold filling technique.

Voranol   Registered trade mark of The Dow Chemical Company for polyether polyols.

Water   Water in poly is free in the poly and not in a chemically combined state. It is expressed as weight percent water. The amount of water is taken from manufacture calculations for setting up a foam run.

Water-blown foam   Foam in which the gas for expansion was generated by the reaction between water and an isocyanate-bearing material.

Weigh percent isocyanate (% NCO)   The weight percentage of NCO units in the total weight (isocyanate equivalent weight, *IEW*) of an isocyanate. Percent NCO = 42.02/IEW.

Weight percent hydroxyl (% OH)   This is one of the most important analyses for polyols. It is simply a weight percentage of OH units in the total weight of the polymer.

D

Conversions and Formulas

D.1   Conversions

Metric Conversions			
Measurement	From	To	Multiply by
Mass	pound	kilogram (kg)	0.454
Length	inch	meter (m)	0.0254
Temperature	°F	°C	see below
Density	g/cc	g/cm³	1
Tensile strength	psi	megapascals (MPa)	0.006895
	psi	kilopascals (kPa)	6.895
Stiffness	psi	megapascals (MPa)	0.006895
	psi	kilopascals (kPa)	6.895
	inch-pounds	newton-meters (N-m)	0.1129848
Tear	pounds/linear inch (pli)	kilonewtons/meter (kN/m)	0.175117
Impact	foot-pounds (ft-lb)	Joules (J)	1.3558
	foot pounds (ft-lb)	newton-meters/meter	53.379

Temperature Conversions		
From	To	Formula
°C	°F	(°C + 40) × (9/5) − 40 or (°C × 1.8) + 32
°F	°C	(°F + 40) × (5/9) − 40 or (°F − 32)/1.8

D.2   Formulas for Polyurethane Calculations

Note that the following calculations are suitable for any polyol or isocyanate. However, all example calculations are based on a 3000 MW triol and TDI (toluene diisocyanate).

D.2.1   Equivalent Weight of Any Product

$$\frac{\text{molecular weight}}{\text{number of reactive sites}} = \frac{\text{MW}}{\text{f}} \tag{1}$$

For example, $\text{equivalent weight of triol} = \dfrac{3100}{3} = 1033$

D.2.2   Equivalent Weight of Any Polyol

$$\frac{\text{atomic weight of OH}}{\% \text{ OH in polyol}} = \frac{17}{\% \text{ OH (as a decimal fraction)}} \tag{2}$$

For example,   Equivalent weight of a triol with 1.65% OH $= \dfrac{17}{1.65/100} = 1033$

D.2.3   Equivalent Weight of Any Polyol

$$\frac{(\text{equivalent weight of KOH(mg/1.0g)})}{\text{OH number}} = \frac{(56.1)(1000)}{\text{OH number}} \tag{3}$$

For example, equivalent weight of a triol with 54.45 OH number* =

$$\frac{(56.1)(1000)}{54.45} = 1033$$

D.2.4   Percent Hydroxyl (% OH)

$$\% \text{ OH} = \frac{(17)(100)}{\text{eq. wt. polyol}}^{1} = \frac{(17)(100)(\text{OH number})}{(56.1)(1000)}^{2} = \frac{\text{OH number}}{33}^{3} \tag{4}$$

^{1}By rearrangement of Eq. (2)
^{2}By substituting from Eq. (3)
^{3}Reduction of constants

For example, % OH for a triol with 54.56 OH number =

$$\frac{54.45}{33} = 1.65$$

D.2.5   Hydroxyl Number (OH Number)

$$(33)(\% \text{ OH})   \text{[by rearrangement of Eq. (4)]} \tag{5}$$

For example, OH number for a triol with 54.45 OH number =

$$(33)(1.65) = 54.45$$

D.2.6   Molecular Weight of Any Polyol

$$\frac{(\text{functionality of polyol}(56.1)(1000))}{\text{OH number}}   \text{[from Eqs. (1) and (3)]} \tag{6}$$

Therefore, diol MW = $\dfrac{(2)(56.1)(1000)}{\text{OH number}} = \dfrac{112200}{\text{OH number}}$

Triol MW = $\dfrac{(3)(56.1)(1000)}{\text{OH number}} = \dfrac{168300}{\text{OH number}}$

For example, molecular weight of a triol with 54.45 OH number =

$$\frac{168300}{54.45} = 3090$$

---

*Hydroxyl number (OH number) = the number of milligrams of potassium hydroxide (KOH) equivalent to the hydroxyl content 1.0 g of polyol.

D.2.7 Stoichiometric Weight of TDI* Required for 100 Parts Polyol

$$\text{weight required} = (\text{eq. wt. TDI})\left(\frac{100}{\text{eq. wt. polyol}}\right) \tag{7}$$

D.2.8 Stoichiometric Weight of TDI Required for X Parts Water†

$$(\text{eq. wt. TDI})\left(\frac{X}{\text{eq. wt. water}}\right) \tag{8}$$

D.2.9 Total Amount of TDI

The total amount of TDI used in the formulation is the product of the TDI index and the sum of the TDI necessary for the reaction with the polyol and water components. TDI index indicates the amount of TDI available for reaction with polyol and water and any other reactive ingredients. An index of 105 means that there are 5% excess equivalents of TDI available in the system (in excess of the stoichiometric amount required by the polyol and water and any other reactive ingredients).

$$\text{Total wt. of TDI} = \left(\frac{\text{TDI index}}{100}\right)(\text{eq. wt. TDI})\left(\frac{100}{\text{eq. wt. polyol}} + \frac{\text{parts } H_2O}{\text{eq. wt. } H_2O}\right) \tag{9}$$

The following example illustrates how this equation may be used to calculate the weight of TDI to be used in the formulation shown. In this example, triol (OH number = 64.45) = 100.0 parts, water added to formulation = 3.6 parts, and TDI index (105) = X parts.

$$X = \text{total weight TDI} = \left(\frac{105}{100}\right)(87)\left(\frac{100}{\frac{(56.1)(1000)}{54.56}} + \frac{3.6}{9}\right) = 45.4 \text{ parts by weight}$$

D.2.10 Equivalent Weight of Adduct, Prepolymer, or Polymeric MDI

$$\frac{\text{formula wt. NCO}}{\% \text{ excess NCO}} = \frac{42.00 \times 100}{\% \text{ excess NCO}} \tag{10}$$

For example, for an adduct or prepolymer containing 30% NCO, the equivalent weight,

$$\frac{4200}{30} = 140$$

D.2.11 Parts Adduct Required for 100 Parts Polyol (eq. wt. adduct/eq. wt. polyol)

$$\frac{\left(\dfrac{42}{\% \text{ excess NCO}}\right)}{\left(\dfrac{17}{\% \text{ OH}}\right)}(100) = (247)\left(\frac{\% \text{ OH}}{\% \text{ NCO}}\right) \tag{11}$$

For example, for a triol and an adduct or prepolymer containing 30% NCO, parts adduct required for 100 parts polyol =

---

*Equivalent weight of TDI = 87.

†Molecular weight of H_2O = 18, but each molecule of water results in the rapid consumption of two NCO groups. Therefore, the equivalent weight of water is 9.

$$247\left(\frac{1.65}{30}\right) = 13.6 \text{ parts by weight}$$

D.2.12   TDI Index

$$\text{TDI index} = \frac{\text{no. equivalents TDI used}}{\text{no. equivalents polyol + water used}} \tag{12}$$

In this example, triol (OH no. 54.45) (equivalent weight = 1030.3) = 100.0 parts, water added to formulation = 3.6 parts, and TDI (eq. wt. = 87), 45.4 parts.

$$\text{TDI index (as a decimal fraction)} = \frac{(45.4)}{\left(\frac{100}{1033}\right)\left(\frac{3.6}{9}\right)} = 1.05$$

D.2.13   Calculations for Formulating a Prepolymer to a Specified Percentage NCO[*]

In this example, eq. wt. triol (1.65% OH) = 1030, eq. wt. TDI = 87, mixture (one equivalent of each) at 0% NCO = 1120. Let X = amount of TDI needed to bring contents up from 0 to 31.3% NCO.

$$\frac{48.27\%}{1120 + X} = 31.3\% \ (\% \text{ NCO in TDI} = 48.27) \tag{13}$$

$$X = 2066$$

2066 + 87 = 2153 g of TDI per 1033 g of VORANOL 3137 polyol.

D.2.14   Calculations for Physical Property Tests

Formula		English Units	Metric Units
$\text{Density} = \dfrac{\text{weight of sample}}{\text{length} \times \text{width} \times \text{height}}$		$1.0\dfrac{\text{lb}}{\text{ft}^3}$	$\dfrac{1}{16}\dfrac{\text{kg}}{\text{m}^3}$
$\text{Tensile} = \dfrac{\text{pulling force}}{\text{cross-sectional area of sample}}$		$1.0\dfrac{\text{lb}}{\text{in}^2}$	$\dfrac{1}{14.2}\dfrac{\text{kg}}{\text{cm}^2}$
$\text{Tear resistance} = \dfrac{\text{pulling force}}{\text{thickness of sample}}$		$1.0\dfrac{\text{lb}}{\text{in}}$	$\dfrac{1}{5.6}\dfrac{\text{kg}}{\text{cm}}$
$\text{Elongation} = \dfrac{\text{final length} - \text{original length}}{\text{original length}}(100)$		%	%
$\text{Compression set} = \dfrac{\text{original thickness} - \text{final thickness}}{\text{original thickness}}(100)$		%	%
$\text{Hysteresis return} = \dfrac{25\% \text{ return}}{25\% \text{ original}}(100)$		%	%
$\text{Modulus} = \dfrac{65\% \text{ IFD}}{25\% \text{ IFD}}$		unitless number	unitless number

References

Aggarwal, S.L. *Block Copolymers.* New York: Plenum, 1970.

Allport, D.C., and Janes, W.J. *Block Copolymers.* London: Applied Science, 1973.

Becker, W.E. *Reaction Injection Moulding.* New York: Van Nostrand Reinhold, 1972.

Bruins, P.F., ed. *Polyurethane Technology.* New York: Interscience, 1969.

---

[*]This is a calculation only, not a procedure for preparing a prepolymer.

Buist, J.M., and Gudgeon, H. *Advances in Polyurethane Technology.* London: Maclaren & Sons, 1968.

Burke, J.J., and Weiss, V., eds. *Block and Graft Copolymers.* Syracuse, New York: Syracuse University, 1973.

Ceresa, R.J. *Block and Graft Copolymers.* London: Butterworths, 1962.

David, D.J., and Staley, H.B. *Analytical Chemistry of Polyurethanes.* New York: Wiley Interscience, 1969.

Dombrow, B. *Polyurethanes.* New York: Van Nostrand Reinhold, 1995.

Encyclopaedia of Polymer Science and Technology, Polyurethanes, Vol. H. New York: Wiley-Interscience, 1969.

Evans, R. *Polyurethane Sealants.* Lancaster, Pennsylvania: Technomic, 1993.

Journal of Cellular Plastics. Lancaster, Pennsylvania: Technomic.

Ledwith, A., and North, A.M., eds. *Molecular Behavior and the Development of Polymer Materials.* London: Chapman & Hall, 1975.

Moacanin, J., Holden, G., and Tschoegl, N.W., eds. *Block Copolymers.* New York: Interscience, 1969.

Modern Plastics Encyclopedia. New York: McGraw-Hill.

Molau, G.E., ed. *Colloidal and Morphological Behavior of Block and Graft Copolymers.* New York: Plenum, 1971.

Noshay, A., and McGrath, J.E. *Block Copolymers.* New York: Academic Press, 1977.

Oertel, G., ed. *Polyurethane Handbook.* Munich: Carl Hanser Verlag, 1994.

Saunders, J.H., and Frisch, K.C. *Polyurethanes—Chemistry and Technology,* Vols. I and II. New York: Interscience, 1962.

Saunders, K.J. *Organic Polymer Chemistry.* London: Chapman & Hall, 1973.

Sweeney, F.M. *Introduction to Reaction Injection Molding.* Westport, Connecticut: Technomic, 1979.

Szycher, M., ed. *High Performance Elastomers.* Lancaster, Pennsylvania: Technomic, 1991.

Urethane Abstracts. Lancaster, Pennsylvania: Technomic.

Walker, B.M., ed. *Handbook of Thermoplastic Elastomers.* New York: Van Nostrand Reinhold, 1979.

Woods, G. *The ICI Polyurethanes Book.* New York: John Wiley & Sons, 1990.

Wright, P., and Cumming, A.P.C. *Solid Polyurethane Elastomers.* London: Maclaren and Sons, 1969.

Index